Smart Innovation, Systems and Technologies

Volume 27

Series editors

Robert J. Howlett, KES International, Shoreham-by-Sea, UK
e-mail: rjhowlett@kesinternational.org

Lakhmi C. Jain, University of Canberra, Canberra, Australia
e-mail: Lakhmi.jain@unisa.edu.au

For further volumes:
http://www.springer.com/series/8767

About this Series

The Smart Innovation, Systems and Technologies book series encompasses the topics of knowledge, intelligence, innovation and sustainability. The aim of the series is to make available a platform for the publication of books on all aspects of single and multi-disciplinary research on these themes in order to make the latest results available in a readily-accessible form. Volumes on interdisciplinary research combining two or more of these areas is particularly sought.

The series covers systems and paradigms that employ knowledge and intelligence in a broad sense. Its scope is systems having embedded knowledge and intelligence, which may be applied to the solution of world problems in industry, the environment and the community. It also focusses on the knowledge-transfer methodologies and innovation strategies employed to make this happen effectively. The combination of intelligent systems tools and a broad range of applications introduces a need for a synergy of disciplines from science, technology, business and the humanities. The series will include conference proceedings, edited collections, monographs, handbooks, reference books, and other relevant types of book in areas of science and technology where smart systems and technologies can offer innovative solutions.

High quality content is an essential feature for all book proposals accepted for the series. It is expected that editors of all accepted volumes will ensure that contributions are subjected to an appropriate level of reviewing process and adhere to KES quality principles.

Malay Kumar Kundu · Durga Prasad Mohapatra
Amit Konar · Aruna Chakraborty
Editors

Advanced Computing, Networking and Informatics – Volume 1

Advanced Computing and Informatics Proceedings of the Second International Conference on Advanced Computing, Networking and Informatics (ICACNI-2014)

Springer

Editors
Malay Kumar Kundu
Machine Intelligence Unit
Indian Statistical Institute
Kolkata
India

Amit Konar
Dept. of Electronics and
Tele-Communication Engineering
Artificial Intelligence Laboratory
Jadavpur University
Kolkata
India

Durga Prasad Mohapatra
Dept. of Computer Science and Engineering
National Institute of Technology Rourkela
Rourkela
India

Aruna Chakraborty
Dept. of Computer Science and Engineering
St. Thomas' College of Engineering
& Technology
Kidderpore
India

ISSN 2190-3018 ISSN 2190-3026 (electronic)
ISBN 978-3-319-38227-2 ISBN 978-3-319-07353-8 (eBook)
DOI 10.1007/978-3-319-07353-8
Springer Cham Heidelberg New York Dordrecht London

Softcover reprint of the hardcover 1st edition 2014

Printed on acid-free paper

Springer is part of Springer Science+Business Media (www.springer.com)

Foreword

The present volume is an outcome, in the form of proceedings, of the 2nd International Conference on Advanced Computing, Networking and Informatics, St. Thomas' College of Engineering and Technology, Kolkata, India, June 24–26, 2014. As the name of the conference implies, the articles included herein cover a wide span of disciplines ranging, say, from pattern recognition, machine learning, image processing, data mining and knowledge discovery, soft computing, distributed computing, cloud computing, parallel and distributed networks, optical communication, wireless sensor networks, routing protocol and architecture to data privacy preserving, cryptology and data security, and internet computing. Each discipline, itself, has its own challenging problems and issues. Some of them are relatively more matured and advanced in theories with several proven application domains, while others fall in recent thrust areas. Interestingly, there are several articles, as expected, on symbiotic integration of more than one discipline, e.g., in designing intelligent networking and computing systems such as forest fire detection using wireless sensor network, minimizing call routing cost with assigned cell in wireless network, network intrusion detection system, determining load balancing strategy in cloud computing, and side lobe reduction and beam-width control, where the significance of pattern recognition, evolutionary strategy and soft computing has been demonstrated. This kind of interdisciplinary research is likely to grow significantly, and has strong promise in solving real life challenging problems.

The proceedings are logically split in two homogeneous volumes, namely, Advanced Computing and Informatics (vol. 1) and Wireless Networks and Security (vol. 2) with 81 and 67 articles respectively. The volumes fairly represent a state-of-the art of the research mostly being carried out in India in these domains, and are valued-additions to the current era of computing and knowledge mining.

The conference committee, editors, and the publisher deserve congratulations for organizing the event (ICACNI-2014) which is very timely, and bringing out the archival volumes nicely as its output.

Kolkata, April 2014

Sankar K. Pal
Distinguished Scientist and former Director
Indian Statistical Institute

Message from the Honorary General Chair

It gives me great pleasure to introduce the *International Conference on Advanced Computing, Networking and Informatics (ICACNI 2014)* which will be held at St. Thomas' College of Engineering and Technology, Kolkata during June 24–26, 2014. ICACNI is just going to cross its second year, and during this small interval of time it has attracted a large audience. The conference received over 650 submissions of which only 148 papers have been accepted for presentation. I am glad to note that ICACNI involved top researchers from 26 different countries as advisory board members, program committee members and reviewers. It also received papers from 10 different countries.

ICACNI offers an interesting forum for researchers of three apparently diverse disciplines: Advanced Computing, Networking and Informatics, and attempts to focus on engineering applications, covering security, cognitive radio, human-computer interfacing among many others that greatly rely on these cross-disciplinary research outcomes. The accepted papers are categorized into two volumes, of which volume 1 includes all papers on advanced computing and informatics, while volume 2 includes accepted papers on wireless network and security. The volumes will be published by Springer-Verlag.

The conference includes plenary lecture, key-note address and four invited sessions by eminent scientists from top Indian and foreign research/academic institutes. The lectures by these eminent scientists will provide an ideal platform for dissemination of knowledge among researchers, students and practitioners. I take this opportunity to thank all the participants, including the keynote, plenary and invited speakers, reviewers, and the members of different committees in making the event a grand success.

Thanks are also due to the various Universities/Institutes for their active support towards this endeavor, and lastly Springer-Verlag for publishing the proceedings under their prestigious *Smart Innovation, Systems and Technologies (SIST) series.*

Wish the participants an enjoyable and productive stay in Kolkata.

Kolkata, April 2014

Dwijesh Dutta Majumder
Honorary General Chair
ICACNI -2014

Preface

The twenty first century has witnessed a paradigm shift in three major disciplines of knowledge: 1) Advanced/Innovative computing ii) Networking and wireless Communications and iii) informatics. While the first two are complete in themselves by their titles, the last one covers several sub-disciplines involving geo-, bio-, medical and cognitive informatics among many others. Apparently, the above three disciplines of knowledge are complementary and mutually exclusive but their convergence is observed in many real world applications, encompassing cyber-security, internet banking, healthcare, sensor networks, cognitive radio, pervasive computing and many others.

The International Conference on *Advanced Computing, Networking and Informatics* (ICACNI) is aimed at examining the convergence of the above three modern disciplines through interactions among three groups of people. The first group comprises leading international researchers, who have established themselves in one of the above three thrust areas. The plenary, the keynote lecture and the invited talks are organized to disseminate the knowledge of these academic experts among young researchers/practitioners of the respective domain. The invited talks are also expected to inspire young researchers to initiate/orient their research in respective fields. The second group of people comprises Ph.D./research students, working in the cross-disciplinary areas, who might be benefited from the first group and at the same time may help creating interest in the cross-disciplinary research areas among the academic community, including young teachers and practitioners. Lastly, the group comprising undergraduate and master students would be able to test the feasibility of their research through feedback of their oral presentations.

ICACNI is just passing its second birthday. Since its inception, it has attracted a wide audience. This year, for example, the program committee of ICACNI received as many as 646 papers. The acceptance rate is intentionally kept very low to ensure a quality publication by Springer. This year, the program committee accepted only 148 papers from these 646 submitted papers. An accepted paper has essentially received very good recommendation by at least two experts in the respective field.

To maintain a high standard of ICACNI, researchers from top international research laboratories/universities have been included in both the advisory committee and the program committee. The presence of these great personalities has helped the conference to develop its publicity during its infancy and promote it quality through an academic exchange among top researchers and scientific communities.

The conference includes one plenary session, one keynote address and four invited speech sessions. It also includes 3 special sessions and 21 general sessions (altogether 24 sessions) with a structure of 4 parallel sessions over 3 days. To maintain good question-answering sessions and highlight new research results arriving from the sessions, we selected subject experts from specialized domains as session chairs for the conference. ICACNI also involved several persons to nicely organize registration, take care of finance, hospitality of the authors/audience and other supports. To have a closer interaction among the people of the organizing committee, all members of the organizing committee have been selected from St. Thomas' College of Engineering and Technology.

The papers that passed the screening process by at least two reviewers, well-formatted and nicely organized have been considered for publication in the Smart Innovations Systems Technology (SIST) series of Springer. The hard copy proceedings include two volumes, where the first volume is named as *Advanced Computing and Informatics* and the second volume is named as *Wireless Networks and Security*. The two volumes together contain 148 papers of around eight pages each (in Springer LNCS format) and thus the proceedings is expected to have an approximate length of 1184 pages.

The editors gratefully acknowledge the contribution of the authors and the entire program committee without whose active support the proceedings could hardly attain the present standards. They would like to thank the keynote speaker, the plenary speaker, the invited speakers and also the invited session chairs, the organizing chair along with the organizing committee and other delegates for extending their support in various forms to ICACNI-2014. The editors express their deep gratitude to the Honorary General Chair, the General Chair, the Advisory Chair and the Advisory board members for their help and support to ICACNI-2014. The editors are obliged to Prof. Lakhmi C. Jain, the academic series editor of the SIST series, Springer and Dr. Thomas Ditzinger, Senior Editor, Springer, Heidelberg for extending their co-operation in publishing the proceeding in the prestigious SIST series of Springer. They also like to mention the hard efforts of Mr. Indranil Dutta of the Machine Intelligence Unit of ISI Kolkata for the editorial support. The editors also acknowledge the technical support they received from the students of ISI, Kolkata and Jadavpur University and also the faculty of NIT Rourkela and St. Thomas' College of Engineering and Technology without which the work could not be completed in right time. Lastly, the editors thank Dr. Sailesh Mukhopadhyay, Prof. Gautam Banerjee and Dr. Subir Chowdhury of St. Thomas' College of Engineering and Technology for their support all the way long to make this conference a success.

Kolkata
April 14, 2014

Malay Kumar Kundu
Durga Prasad Mohapatra
Amit Konar
Aruna Chakraborty

Organization

Advisory Chair

Sankar K. Pal	Distinguished Scientist, Indian Statistical Institute, India

Advisory Board Members

Alex P. James	Nazarbayev University, Republic of Kazakhstan
Anil K. Kaushik	Additional Director, DeitY, Govt. of India, India
Atilla Elci	Aksaray University, Turkey
Atulya K. Nagar	Liverpool Hope University, United Kingdom
Brijesh Verma	Central Queensland University, Australia
Debajyoti Mukhopadhyay	Maharashtra Institute of Technology, India
Debatosh Guha	University of Calcutta, India
George A. Tsihrintzis	University of Piraeus, Greece
Hugo Proenca	University of Beira Interior, Portugal
Jocelyn Chanussot	Grenoble Institute of Technology, France
Kenji Suzuki	The University of Chicago, Chicago
Khalid Saeed	AGH University of Science and Technology, Poland
Klaus David	University of Kassel, Germany
Maode Ma	Nanyang Technological University, Singapore
Massimiliano Rak	Second University of Naples, Italy
Massimo Tistarelli	University of Sassari, Italy
Nishchal K. Verma	Indian Institute of Technology, Kanpur, India
Pascal Lorenz	University of Haute Alsace, France
Phalguni Gupta	Indian Institute of Technology Kanpur, India
Prasant Mohapatra	University of California, USA
Prasenjit Mitra	Pennsylvania State University, USA
Raj Jain	Washington University in St. Louis, USA
Rajesh Siddavatam	Kalinga Institute of Industrial Technology, India

Rajkumar Buyya	The University of Melbourne, Australia
Raouf Boutaba	University of Waterloo, Canada
Sagar Naik	University of Waterloo, Canada
Salvatore Vitabile	University of Palermo, Italy
Sansanee Auephanwiriyakul	Chiang Mai University, Thailand
Subhash Saini	The National Aeronautics and Space Administration (NASA), USA

ICACNI Conference Committee

Chief Patron

Sailesh Mukhopadhyay	St. Thomas' College of Engineering and Technology, Kolkata, India

Patron

Gautam Banerjea	St. Thomas' College of Engineering and Technology, Kolkata, India

Honorary General Chair

Dwijesh Dutta Majumder	Professor Emeritus, Indian Statistical Institute, Kolkata, India Institute of Cybernetics Systems and Information Technology, India Director, Governing Board, World Organization of Systems and Cybernetics (WOSC), Paris

General Chairs

Rajib Sarkar	Central Institute of Technology Raipur, India
Mrithunjoy Bhattacharyya	St. Thomas' College of Engineering and Technology, Kolkata, India

Programme Chairs

Malay Kumar Kundu	Indian Statistical Institute, Kolkata, India
Amit Konar	Jadavpur University, Kolkata, India
Aruna Chakraborty	St. Thomas' College of Engineering and Technology, Kolkata, India

Programme Co-chairs

Asit Kumar Das	Bengal Engineering and Science University, Kolkata, India
Ramjeevan Singh Thakur	Maulana Azad National Institute of Technology, India
Umesh A. Deshpande	Sardar Vallabhbhai National Institute of Technology, India

Organizing Chairs

Ashok K. Turuk	National Institute of Technology Rourkela, Rourkela, India
Rabindranath Ghosh	St. Thomas' College of Engineering and Technology, Kolkata, India

Technical Track Chairs

Joydeb Roychowdhury	Central Mechanical Engineering Research Institute, India
Korra Sathyababu	National Institute of Technology Rourkela, India
Manmath Narayan Sahoo	National Institute of Technology Rourkela, India

Honorary Industrial Chair

G.C. Deka	Ministry of Labour & Employment, Government of India

Industrial Track Chairs

Umesh Chandra Pati	National Institute of Technology Rourkela, India
Bibhudutta Sahoo	National Institute of Technology Rourkela, India

Special Session Chairs

Ashish Agarwal	Boston University, USA
Asutosh Kar	International Institute of Information Technology Bhubaneswar, India
Daya K. Lobiyal	Jawaharlal Nehru University, India
Mahesh Chandra	Birla Institute of Technology, Mesra, India
Mita Nasipuri	Jadavpur University, Kolkata, India
Nandini Mukherjee	Chairman, IEEE Computer Society, Kolkata Chapter Jadavpur University, Kolkata, India
Ram Shringar Rao	Ambedkar Institute of Advanced Communication Technologies & Research, India

Web Chair

Indranil Dutta	Indian Statistical Institute, Kolkata, India

Publication Chair

Sambit Bakshi	National Institute of Technology Rourkela, India

Publicity Chair

Mohammad Ayoub Khan	Center for Development of Advanced Computing, India

Organizing Committee

Amit Kr. Siromoni	St. Thomas' College of Engineering and Technology, Kolkata, India
Anindita Ganguly	St. Thomas' College of Engineering and Technology, Kolkata, India
Arindam Chakravorty	St. Thomas' College of Engineering and Technology, Kolkata, India
Dipak Kumar Kole	St. Thomas' College of Engineering and Technology, Kolkata, India
Prasanta Kumar Sen	St. Thomas' College of Engineering and Technology, Kolkata, India
Ramanath Datta	St. Thomas' College of Engineering and Technology, Kolkata, India
Subarna Bhattacharya	St. Thomas' College of Engineering and Technology, Kolkata, India
Supriya Sengupta	St. Thomas' College of Engineering and Technology, Kolkata, India

Program Committee

Vinay. A	Peoples Education Society Institute of Technology, Bangalore, India
Chunyu Ai	University of South Carolina Upstate, Spartanburg, USA
Rashid Ali	Aligarh Muslim University, Aligarh, India
C.M. Ananda	National Aerospace Laboratories, Bangalore, India
Soumen Bag	International Institution of Information Technology, Bhubaneswar, India
Sanghamitra Bandyopadhyay	Indian Statistical Institute, Kolkata, India
Punam Bedi	University of Delhi, Delhi, India
Dinabandhu Bhandari	Heritage Institute of Technology, Kolkata, India
Paritosh Bhattacharya	National Institute of Technology, Agartala, India
Malay Bhattacharyya	University of Kalyani, Kolkata, India
Sambhunath Biswas	Indian Statistical Institute, Kolkata, India
Darko Brodic	University of Belgrade, Bor, Serbia
Sasthi C. Ghosh	Indian Statistical Institute, Kolkata, India
R.C. Hansdah	Indian Institute of Science, Bangalore, India
Nabendu Chaki	University of Calcutta, Kolkata, India
Goutam Chakraborty	Iwate Prefectural University, Takizawa, Japan

Mihir Chakraborty	Indian Statistical Institute, Kolkata, India
Amita Chatterjee	Jadavpur University, Kolkata, India
Debasis Chaudhuri	Indian Institute of Management, Kolkata, India
Lopamudra Chowdhury	Jadavpur University, Kolkata, India
Nabanita Das	Indian Statistical Institute, Kolkata, India
Soumya Kanti Datta	Institute Eurecom, Sophia Antipolis, France
Rajat De	Indian Statistical Institute, Kolkata, India
Utpal Garain	Indian Statistical Institute, Kolkata, India
Ashis Ghosh	Indian Statistical Institute, Kolkata, India
Mukesh Goswami	Dharmsinh Desai University, Gugrat, India
Yogesh H. Dandawate	Vishwakarma Institute of Information Technology, Pune, India
Biju Issac	Teesside University, Middlesbrough, UK
Lakhmi C. Jain	University of South Australia, Adelaide, Australia
R. Janarthanan	T. J. S. Engineering College, Chennai, India
Boleslaw K. Szymanski	Rensselaer Polytechnic Institute, New York, USA
Tienfuan Kerh	National Pingtung University of Science and Technology, Pingtung, Taiwan
Zahoor Khan	Dalhouise University, Halifax, Canada
Dakshina Ranjan Kisku	Asansole Engineering College, Asansole, India
Sotiris Kotsiantis	University of Patras, Hellas, Greece
Dipak Kr. Kole	St. Thomas' College of Engineering and Technology, Kolkata, India
Aswani Kumar Cherukuri	VIT University, Vellore, India
Swapan Kumar Parui	Indian Statistical Institute, Kolkata, India
Prasanta Kumar Pradhan	St. Thomas' College fo Engineering and Technology, Kolkata, India
B. Narendra Kumar Rao	Sree Vidyanikethan Engineering College, Tirupati, India
Flavio Lombardi	Roma Tre University of Rome, Rome, Italy
Pradipta Maji	Indian Statistical Institute, Kolkata, India
Raghvendra Mall	University of Leuven, Leuven, Belgium
Amiya Kr. Mallick	St. Thomas' College fo Engineering and Technology, Kolkata, India
Debaprasad Mandal	Indian Statistical Institute, Kolkata, India
Ujjwal Maulik	Jadavpur University, Kolkata, India
Pabitra Mitra	Indian Institute of Technology, Kharagpur, India
Imon Mukherjee	St. Thomas' College fo Engineering and Technology, Kolkata, India
Nandini Mukherjee	Jadavpur University, Kolkata, India
Jayanta Mukhopadhyay	Indian Institute of Technology, Kharagpur, India
C.A. Murthy	Indian Statistical Institute, Kolkata, India
M. Murugappan	University of Malayesia, Malayesia
Mita Nasipuri	Jadavpur University, Kolkata, India
Rajdeep Niyogi	Indian Institute of Technology, Roorkee, India

Steven Noel	George Manson University, Fairfax, USA
M.C. Padma	PES College of Engineering, Karnataka, India
Rajarshi Pal	Institute for Development and Research in Banking Technology, Hyderabad, India
Umapada Pal	Indian Statistical Institute, Kolkata, India
Anika Pflug	Hochschule Darmstadt - CASED, Darmstadt, Germany
Surya Prakash	Indian Institute of Technology, Indore, India
Ganapatsingh G. Rajput	Rani Channamma University, Karnataka, India
Anca Ralescu	University of Cincinnati, Ohio, USA
Umesh Hodeghatta Rao	Xavier Institute of Management, Bhubaneswar, India
Ajay K. Ray	Bengal Engineering and Science University, Shibpur, India
Tuhina Samanta	Bengal Engineering and Science University, Shibpur, India
Andrey V. Savchenko	National Research University Higher School of Economics, Molscow, Russia
Bimal Bhusan Sen	St. Thomas' College fo Engineering and Technology, Kolkata, India
Indranil Sengupta	Indian Institute of Technology, Kharagpur, India
Patrick Siarry	Universite de Paris, Paris
Nanhay Singh	Ambedkar Institute of Advanced Communication Technologies & Research, Delhi, India
Pradeep Singh	National Institute of Technology, Raipur, India
Vivek Singh	South Asian University, New Delhi, India
Bhabani P. Sinha	Indian Statistical Institute, Kolkata, India
Sundaram Suresh	Nanyang Technological University, Singapore
Jorge Sá Silva	University of Coimbra, Portugal
Vasile Teodor Dadarlat	Technical University of Cluj Napoca, Cluj Napoca, Romania
B. Uma Shankar	Indian Statistical Institute, Kolkata, India
M. Umaparvathi	RVS College of Engineering, Coimbatore, India
Palaniandavar Venkateswaran	Jadavpur University, Kolkata, India
Stefan Weber	Trinity College, Dublin, Ireland
Azadeh Ghandehari	Islamic Azad University, Tehran, Iran
Ch. Aswani Kumar	Vellore Institute of Technology, India
Cristinel Ababei	University at Buffalo, USA
Dilip Singh Sisodia	National Institute of Technology Raipur, India
Jamuna Kanta Sing	Jadavpur University, Kolkata, India
Krishnan Nallaperumal	Sundaranar University, India
Manu Pratap Singh	Dr. B.R. Ambedkar University, Agra, India
Narayan C. Debnath	Winona State University, USA
Naveen Kumar	Indira Gandhi National Open University, India

Nidul Sinha	National Institute of Technology Silchar, India
Sanjay Kumar Soni	Delhi Technological University, India
Sanjoy Das	Galgotias University, India
Subir Chowdhury	St. Thomas' College of Engineering and Technology, Kolkata, India
Syed Rizvi	The Pennsylvania State University, USA
Sushil Kumar	Jawaharlal Nehru University, India
Anupam Sukhla	Indian Institute of Information Technology, Gwalior, India

Additional Reviewers

A.M., Chandrashekhar
Acharya, Anal
Agarwal, Shalabh
B.S., Mahanand
Bandyopadhyay, Oishila
Barpanda, Soubhagya Sankar
Basu, Srinka
Battula, Ramesh Babu
Bhattacharjee, Sourodeep
Bhattacharjee, Subarna
Bhattacharya, Indrajit
Bhattacharya, Nilanjana
Bhattacharyya, Saugat
Bhowmik, Deepayan
Biswal, Pradyut
Biswas, Rajib
Bose, Subrata
Chakrabarti, Prasun
Chakraborty, Debashis
Chakraborty, Jayasree
Chandra, Helen
Chandra, Mahesh
Chatterjee, Aditi
Chatterjee, Sujoy
Chowdhury, Archana
Chowdhury, Manish
Dalai, Asish
Darbari, Manuj
Das, Asit Kumar
Das, Debaprasad
Das, Nachiketa
Das, Sudeb
Datta, Biswajita
Datta, Shreyasi
De, Debashis
Dhabal, Supriya
Dhara, Bibhas Chandra
Duvvuru, Rajesh
Gaidhane, Vilas
Ganguly, Anindita
Garg, Akhil
Ghosh Dastidar, Jayati
Ghosh, Arka
Ghosh, Lidia
Ghosh, Madhumala
Ghosh, Partha
Ghosh, Rabindranath
Ghosh, Soumyadeep
Ghoshal, Ranjit
Goyal, Lalit
Gupta, Partha Sarathi
Gupta, Savita
Halder, Amiya
Halder, Santanu
Herrera Lara, Roberto
Jaganathan, Ramkumar
Kakarla, Jagadeesh
Kar, Mahesh
Kar, Reshma
Khasnobish, Anwesha
Kole, Dipak Kumar
Kule, Malay
Kumar, Raghvendra
Lanka, Swathi
Maruthi, Padmaja
Mishra, Dheerendra
Mishra, Manu
Misra, Anuranjan
Mohanty, Ram
Maitra, Subhamoy
Mondal, Jaydeb
Mondal, Tapabrata
Mukherjee, Nabanita
Mukhopadhyay, Debajyoti
Mukhopadhyay, Debapriyay
Munir, Kashif
Nasim Hazarika, Saharriyar Zia
Nasipuri, Mita
Neogy, Sarmistha
Pal, Monalisa
Pal, Tamaltaru
Palodhi, Kanik
Panigrahi, Ranjit
Pati, Soumen Kumar
Patil, Hemprasad
Patra, Braja Gopal
Pattanayak, Sandhya
Paul, Amit
Paul, Partha Sarathi
Phadikar, Amit
Phadikar, Santanu
Poddar, Soumyajit
Prakash, Neeraj
Rakshit, Pratyusha
Raman, Rahul
Ray, Sumanta
Roy, Pranab
Roy, Souvik
Roy, Swapnoneel

Rup, Suvendu
Sadhu, Arup Kumar
Saha Ray, Sanchita
Saha, Anuradha
Saha, Anushri
Saha, Dibakar
Saha, Indrajit
Saha, Sriparna
Sahoo, Manmath N.
Sahu, Beeren
Sanyal, Atri
Sardar, Abdur
Sarkar, Apurba
Sarkar, Dhrubasish
Sarkar, Ushnish
Sen, Sonali
Sen, Soumya
Sethi, Geetika
Sharma, Anuj
Tomar, Namrata
Umapathy, Latha
Upadhyay, Anjana
Wankhade, Kapil

Contents

Signal and Speech Analysis

Machine Learning

Pattern Analysis and Recognition

Document Analysis

Biometric and Biological Data Analysis

Application of Bilinear Recursive Least Square Algorithm for Initial Alignment of Strapdown Inertial Navigation System

Bidhan Malakar and B.K. Roy

Department of Electrical Engineering
National Institute of Technology Silchar, Assam, India
bidhan.nits@gmail.com

Abstract. To improve the alignment accuracy and convergence speed of Strapdown inertial navigation system, an initial alignment which is based on Bilinear Recursive Least Square adaptive filter is proposed. The error model for the Strapdown Inertial Navigation System (SINS) is derived from the dynamic model by considering a small misalignment angle. In the literature many algorithms are proposed for the proper estimation of alignment accuracy for INS and it is still a challenge. In this paper, two algorithms which are mainly based on nonlinear adaptive filter viz. Volterra Recursive Least Square (VRLS) and Bilinear Recursive Least Square (BRLS) are proposed and compared for proper estimation of accurate azimuth alignment error. The comparative performances of the aforesaid algorithms are studied and the performance of proposed BRLS algorithm is found to be effective which is obtained in existence of two different white Gaussian noises. The simulation work is done in MATLAB simulating environment. For the realization and validation of proposed BRLS algorithm, the comparative analysis is also precisely presented.

1 Introduction

The Strapdown Inertial Navigation System (SINS) [1] has special advantages and is widely being adopted for the accurate positioning and navigation of missiles, aeroplanes, ships, and railway vehicles etc. INS is advantageous when compared with (Global Positioning System) GPS, as it is unaffected by the external sources. However the INS output is subjective to the errors in the data which is supplied to the system during long durations and also can be due to inaccurate design and construction of the system components. Three types of errors are mainly responsible for the rate at which navigation error grow over long distances of time and these are the initial alignment errors, sensor errors and computational errors. The initial alignment [2-5] is the main key technology in SINS and depending upon the sensors configuration it must provide accurate result. The main purpose of initial alignment is to get the initial coordinate transformation matrix from the body frame to computer coordinates frame and the misalignment angle is considered to be zero during the mathematical modeling. The performance of inertial navigation system is affected by

M.K. Kundu et al. (eds.), *Advanced Computing, Networking and Informatics - Volume 1*,
Smart Innovation, Systems and Technologies 27,
DOI: 10.1007/978-3-319-07353-8_1, © Springer International Publishing Switzerland 2014

the alignment accuracy directly as well as initial alignment time which is mainly responsible for the rapid response capability. Therefore there is a requirement of shorter alignment time with a high precision in initial alignment.

At present, Kalman filtering [2] techniques are basically used in order to achieve the initial alignment of inertial navigation system due to its simplicity and also considered as an effective method. But in conventional Kalman filtering technique one must have the future knowledge of the mathematical models [6] and the noise statistics must also be studied and considered. In case of conventional Kalman filter, it is unable to provide a better and efficient result.

After introduction in section 1, section 2 briefly describes initial SINS alignment. The description of estimation algorithms are given in section 3. Section 4 deals with the dynamic modeling used for the simulation. Section 5 deals with the dynamic simulation of fine alignment. Section 6 presents the results, discussions and the comparison between the proposed algorithms. Finally, we conclude the paper in section 7.

2 SINS Initial Alignments

2.1 Coarse Alignment

An INS determines the position of the body frame by the integration of measured acceleration and rotation rate. Since the position is always relative to the starting position and for this reason the INS must know the position, attitude and heading before the navigation begins. The position is assumed to be known but the attitude and heading needs to be determined and is the process of alignment [7], [8].

In the coarse alignment stage the measured acceleration and rotation rate in body frame is compared with the gravity vector G along with the Earth's rotation rate and it directly estimates transformation matrix of carrier coordinates to geographical coordinates. It estimates the attitude and heading accurately enough to justify the small angle approximations made in the error model, and hence allow fine alignment to use an adaptive filter using the error model to obtain a more precise alignment which helps to determine the direction cosine matrix or attitude matrix C_n^b relating the navigation frame (n) and body frame (b). For determining the orientation of the body frame INS makes the use of the accelerometers and gyroscopes for measurement with respect to a reference frame and it is required for the estimation of the measured value of C_n^b.

The basic of alignment in SINS is discussed which plays an important role to improve the initial alignment. Many algorithms are being proposed recently in the literature for estimation and optimization of errors for SINS [9-11]. A theoretical background of the proposed algorithms is discussed in the next section.

3 Theoretical Background

The general set up of an adaptive-filtering [12] environment is illustrated in Fig. 1.

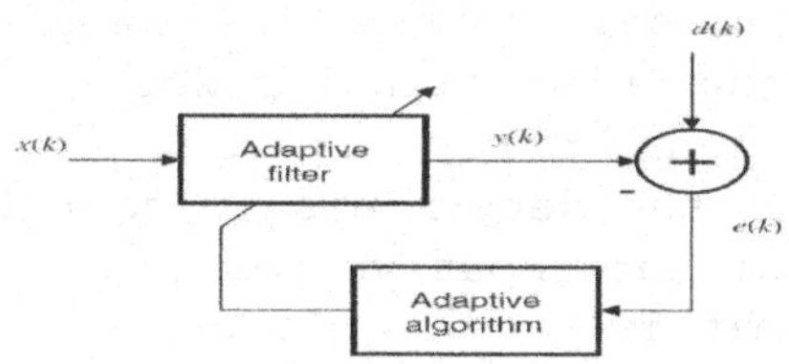

Fig. 1. Adaptive filtering structure

Here k is the iteration number, $x(k)$ denotes the input signal, $y(k)$ is the adaptive-filter output signal, and $d(k)$ defines the desired signal [12]. The error signal $e(k)$ is calculated as $d(k)$-$y(k)$. In order to determine the proper updating of the filter coefficients, the error signal is then used to form a performance (or objective) function that is required by the adaptation algorithm. The minimization of the objective function implies that the adaptive-filter output signal is matching the desired signal in some sense.

3.1 Bilinear RLS Algorithm (BRLS)

The most widely accepted nonlinear difference equation model used for adaptive filtering is the so-called bilinear equation given by (1).

$$y(k) = \sum_{m=0}^{M} b_m(k)x(k-m) - \sum_{j=1}^{N} a_j(k)y(k-j) + \sum_{i=0}^{I}\sum_{l=1}^{L} C_i, lx(k-i)y(k-l) \quad (1)$$

Here $y(k)$ is the adaptive-filter output and for this case, the signal information vector is defined by (2) and (3).

$$\phi(k)=\left[x(k)\ x(k-1)...\ x(k-M)\ y(k-1)\ y(k-2)\ y(k-N)\ x(k)y(k-1)\ x(k-I)y(k-L+1)\ x(k-I)y(k-L)\right]^T \quad (2)$$

$$\theta(k)=\left[b_0(k)\ b_1(k)...\ b_M(k)\ -a_1(k)\ -a_2(k)...\ -a_N(k)\ C_{0,1}(k)\ CI,L-1(k)\ CI,L(k)\right]^T \quad (3)$$

Here N, M, I and L are the orders of the adaptive-filter difference equations.

After the explanation of the two nonlinear adaptive algorithms that are used in the estimation of initial alignment angles error in SINS. The next section presents the dynamic model of SINS.

```
1. Initialization (set parameters for input vector
                  initialization)
   ai(k)= bi(k)= ci,1(k)= e(k)=0.
   y(k)= x(k)= 0, k<0
   Sd(0)=δ^-1 I where δcan be the inverse of an estimate of
                  the input signal power
   e(k) is the signal error
   Sd= inverse of the deterministic correlation matrix
                  of the input vector
   I= identity matrix
2. For each x(k), d(k), k≥0, do
    y(k)= θ^T(k)θ(k)
    e(k)= d(k) - y(k)
    e(k)= signal error, y(k)=adaptive filter output,
    θ(k)= coefficient vector
3. Calculate e(k) using the relation
```

$$S_D(k+1) = \frac{1}{\lambda}[S_D(k) - \frac{S_D(k)\varphi(k)\varphi^T(k)S_D(k)}{\lambda + \varphi^T(k)S_D(k)\varphi(k)}]$$

$$\theta(k+1) = \theta(k) - S_D(k+1)\varphi(k)e(k)$$

4 Dynamic Modeling

The model of a local level NED (North-East-Down) [2] frame is used in this paper as the navigation frame. From the alignment point of view the East and North axes are referred to as leveling axes and "Down" axis are called the azimuth axis. The position and velocity errors are ignored. The state equations [2] can be represented as given.

$$\dot{X} = AX + FW = \begin{pmatrix} P_1 & P_2 \\ 0_{5X5} & 0_{5X5} \end{pmatrix} X + \begin{pmatrix} P_2 & \\ 0_{5X5} & \end{pmatrix} W \tag{4}$$

where $X = [\delta V_N \ \delta V_E \phi_N \phi_E \phi_D \nabla_x \nabla_y \varepsilon_x \varepsilon_y \varepsilon_z]^T$

$\delta V_N, \delta V_E$ are the east and north velocity error respectively and ∇_x, ∇_y for accelerometer error.

ϕ_N, ϕ_E are the two level milianment angles and ϕ_U for the misalignment angle.

$\varepsilon_x, \varepsilon_y, \varepsilon_z$ are the gyro error and ω_{ie} for the Earth rotation rate.

The observation equation for Initial alignment of SINS is given by,

$$Y = TX + V \tag{5}$$

where, $Y = \begin{pmatrix} \delta V_E \\ \delta V_N \end{pmatrix}$, $T = \begin{pmatrix} I_{2X2} \\ 0_{2X8} \end{pmatrix}^T$ and V is the assumed to be the white noise.

The system noise is as follows,

$$W = [w_{ax} w_{ay} w_{gx} w_{gy} w_{gz}]^T$$

where *x*, *y*, *z* correspond to the Northeast coordinates (NED).

$$P_1 = \begin{pmatrix} 0 & 2\omega_{ie} Sin(L_0) & 0 & -g & 0 \\ -2\omega_{ie} Sin(L_0) & 0 & g & 0 & 0 \\ 0 & -\frac{1}{R} & 0 & \omega_{ie} Sin(L_0) & -\omega_{ie} Cos(L_0) \\ \frac{1}{R} & 0 & -\omega_{ie} Sin(L_0) & 0 & 0 \\ \frac{\tan L_0}{R} & 0 & \omega_{ie} Cos(L_0) & 0 & 0 \end{pmatrix}$$

and,

$$P_2 = \begin{pmatrix} C_{11} & C_{12} & 0 & 0 & 0 \\ C_{21} & C_{22} & 0 & 0 & 0 \\ 0 & 0 & C_{11} & C_{12} & C_{13} \\ 0 & 0 & C_{21} & C_{22} & C_{23} \\ 0 & 0 & C_{31} & C_{32} & C_{33} \end{pmatrix}$$

With the help of transformation matrix i.e. posture matrix P_2 the NED coordinate system can be transformed into body coordinate system and is given by,

$$P_2 = \begin{pmatrix} C_{11} & C_{12} & C_{13} \\ C_{21} & C_{22} & C_{23} \\ C_{31} & C_{32} & C_{33} \end{pmatrix}$$

$C_{11} = \cos(\phi_N)\cos(\phi_U) - \sin(\phi_E)\sin(\phi_N)\sin(\phi_U)$, $C_{12} = \sin(\phi_E)\sin(\phi_N)\cos(\phi_U) + \cos(\phi_N)\sin(\phi_U)$, $C_{13} = -\sin(\phi_N)\cos(\phi_E)$,
$C_{21} = -\cos(\phi_E)\sin(\phi_U)$, $C_{22} = \cos(\phi_E)\cos(\phi_U)$, $C_{23} = \sin(\phi_E)$,
$C_{31} = \sin(\phi_N)\cos(\phi_U) + \cos(\phi_N)\sin(\phi_E)\sin(\phi_U)$, $C_{32} = \sin(\phi_N)\sin(\phi_U) - \cos(\phi_N)\sin(\phi_E)\cos(\phi_U)$, $C_{33} = \cos(\phi_N)\cos(\phi_E)$

The systems observation equation by considering level velocity error as an external observation value is

$$Y = \begin{pmatrix} 1\,0\,0\,0\,0\,0\,0\,0\,0\,0 \\ 0\,1\,0\,0\,0\,0\,0\,0\,0\,0 \end{pmatrix} X + V \tag{6}$$

The continuous system in equation (4) and (5) must be converted into discrete system before the MATLAB simulation.

After modeling of the SINS the simulation of the model is given in the next section.

5 Dynamic Simulation of Fine Alignment

In order to evaluate the performance of the proposed Volterra RLS and Bilinear RLS algorithm, an example of a static self aligned stationary-base filtering program is considered, which use the NED coordinate frame [2]. The initial parameters used for the simulation are as follows.

Inertial navigation system location of longitude 125°, north latitude 45°, the initial value X_0 of the state variable X are assigned to zero; P_0, Q and R are taken as the corresponding value of the middle-precision gyroscopes and accelerometers; and the initial misalignment $\delta V_N\, \delta V_E\, \phi_N\, \phi_E\, \phi_U\, \nabla_x\, \nabla_y\, \varepsilon_x\, \varepsilon_y\, \varepsilon_z$ angles ϕ_N ϕ_E ϕ_U are taken as $1°$; gyro drift are often taken as 0.02°/h, random drift taken as 0.01°/h; and accelerometers taken as 100μg, velocity errors taken as $0.1m/s$, then

X_0 = diag{0,0,0,0,0,0,0,0,0,0}; P_0 = diag [(0.1 m/s)^2, (0.1 m/s)^2, (1°)^2, (1°)^2, (1°)^2, (100μg)^2, (100μg)^2, (0.02°/h)^2, (0.02°/h)^2, (0.02°/h)^2]; Q = diag [(50μg)^2 ,(50μg)^2, (0.01°/h)^2, (0.01°/h)^2, (0.01°/h)^2, 0, 0, 0, 0, 0, 0]; R = diag [([(0.1 m/s)^2, (0.1 m/s)^2)]

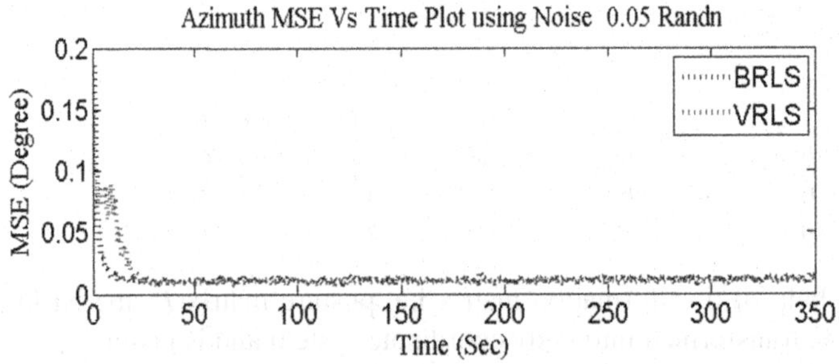

Fig. 2. Azimuth MSE plot vs time samples using BRLS and VRLS

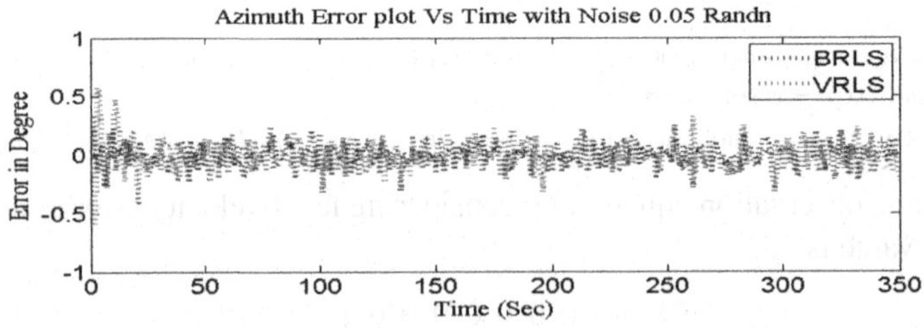

Fig. 3. Azimuth error plot vs time samples using BRLS and VRLS

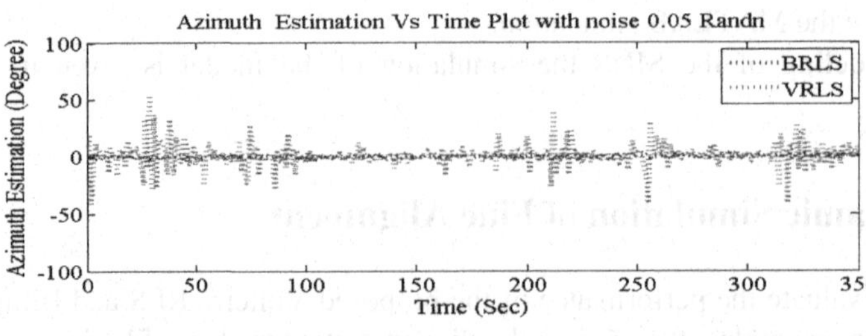

Fig. 4. Azimuth estimation plot vs time samples using BRLS and VRLS

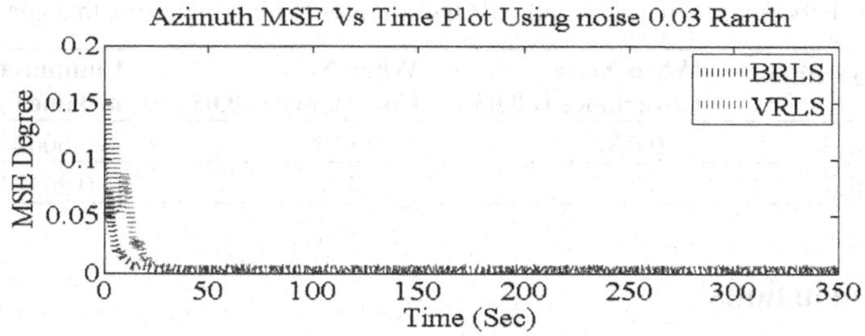

Fig. 5. Azimuth MSE plot vs time samples using BRLS and VRLS

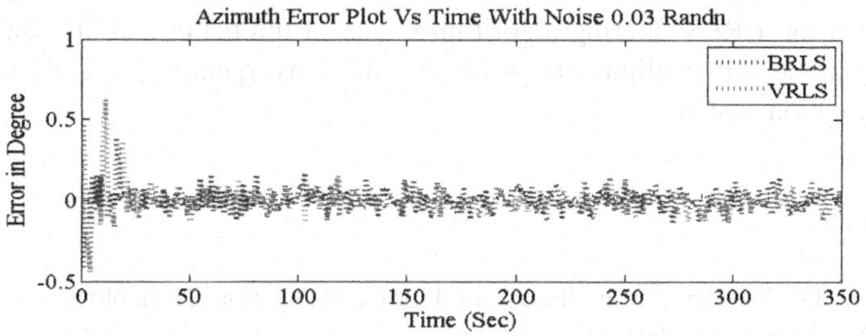

Fig. 6. Azimuth error plot vs time samples using BRLS and VRLS

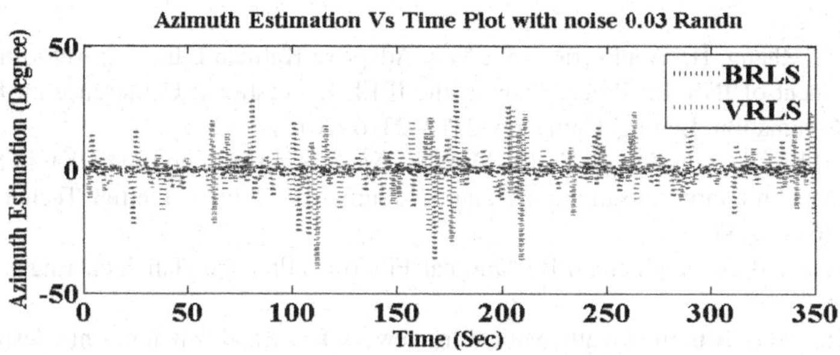

Fig. 7. Azimuth estimation plot vs time samples using BRLS and VRLS

In this section the various outputs obtained from the MATLAB on a laptop with 1GB RAM and 1.50 GHz processor, are shown and their performances characteristic are discussed in the next section.

6 Results and Discussions

The two algorithms are compared to each other by considering two different values of noises i.e. for 0.05 *rand*(*n*) and 0.03 *rand*(*n*). The performance of two algorithms for estimating Azimuth angle error is shown in Fig. 3, Fig. 6. Azimuth error estimation is shown in Fig. 4 and Fig. 7 and the MSE Azimuth angle error is shown in Fig. 2 and Fig. 5. The comparison performance is given in Table 1.

Table 1. Performance of The Proposed Algorithm for Estimating Azimuth angle Error

Algorithms	When Noise Covariance is 0.03	When Noise Covariance is 0.05	Computation al Time (Sec)
VRLS	0.630	0.810	0.60624
BRLS	0.372	0.451	0.26374

7 Conclusions

This paper reports a research on the adaptive filtering technique applied to improve the estimation accuracy of SIN'S azimuth angle error. From the simulation result it is observed that the BRLS filtering algorithm is one of the better and also an effective algorithm for the initial alignment accuracy and convergence speed of strap down inertial navigation system.

References

1. Titterton, D., Weston, J. In: Strapdown Inertial Navigation Technology, Institution of Electrical Engineers (2004)
2. Wang, X., Shen, G.: A Fast and Accurate Initial Alignment Method for Strapdown Inertial Navigation System on Stationary Base. Journal of Control Theory and Applications, 145–149 (2005)
3. Sun, F., Zhang, H.: Application of a New Adaptive Kalman Filitering Algorithm in Initial Alignment of INS. In: Proceedings of the IEEE International Conference on Information and Automation, Beijing, China, pp. 2312–2316 (2011)
4. Gong-Min, V., Wei-Sheng, Y., De-Min, X.: Application of simplified UKF in SINS initial alignment for large misalignment angles. Journal of Chinese Inertial Technology 16(3), 253–264 (2008)
5. Anderson, B.D.O., Moore, J.B.: Optimal Filtering. Prentice-Hall Inc., Englewood Cliffs (1979)
6. Savage, P.G.: A unified mathematical framework for strapdown algorithm design. J. Guid. Contr. Dyn. 29, 237–249 (2006)
7. Silson, P.M.: Coarse alignment of a ship's strapdown inertial attitude reference system using velocity loci. IEEE Transcript Instrumentation Measurement (2011)
8. Li, Q., Ben, Y., Zhu, Z., Yang, J.: A Ground Fine Alignment of Strapdown INS under a Vibrating Base. J. Navigation 1, 1–15 (2013)
9. Wu, M., Wu, Y., Hu, X., Hu, D.: Optimization-based alignment for inertial navigation systems: Theory and algorithm. Aeros. Science & Technology, 1–17 (2011)
10. Salychev, O.S.: Applied Estimation Theory in Geodetic and Navigation Applications. Lecture Notes for ENGO 699.52, Department of Geomatics Engineering, University of Calgary (2000)
11. Julier, S.J., Uhlmann, J.K.: Unscented filtering and nonlinear estimation. In: Proc. of the IEEE Aerospace and Electronic Systems, pp. 401–422 (2004)
12. Paulo, S.R.: Adaptive Filtering Algorithms and Practical Implementation, 3rd edn. Springer

Time-Domain Solution of Transmission through Multi-modeled Obstacles for UWB Signals

Sanjay Soni[1], Bajrang Bansal[1], and Ram Shringar Rao[2]

[1] Delhi Technological University, Delhi, India
{sanjoo.ksoni,bajrangbnsl}@gmail.com
[2] Ambedkar Institute of Advanced Communication Technologies and Research, Delhi, India
rsrao08@yahoo.in

Abstract. In this work, the time-domain solution for transmission through multi-modeled obstacles has been presented. The transmission through dielectric wedge followed by a dielectric slab has been analyzed. The analytical time-domain transmission and reflection coefficients for transmission through the conductor-dielectric interface, considering oblique incidence, are given for both soft and hard polarizations. The exact frequency-domain formulation for transmitted field at the receiver has been simplified under the condition of low-loss assumption and converted to time-domain formulation using inverse Laplace transform. The time-domain results have been validated with the inverse fast Fourier transform (IFFT) of the corresponding exact frequency-domain results. Further the computational efficiency of both the methods is compared.

Keywords: Ultra wideband, Propagation model, Transmission, Frequency-domain, Time-domain.

1 Introduction

In recent years, research in ultra wideband (UWB) propagation through indoor scenario where non line of sight (NLOS) communication is more dominant, has received great attention because of unique properties of UWB communication like resilient to multipath phenomena, good resolution and low power density. In radio propagation of UWB signals, especially in NLOS communication in deep shadow regions, transmitted field component proves to be very significant [1, 2]. Considering the huge bandwidth (3.1-10.6 GHz) of UWB signals, it is more efficient to study UWB propagation directly in time-domain (TD) where all the frequencies are treated simultaneously [3, 4]. The TD solution of transmitted field through a dielectric slab was presented in [5]. A simplified TD model for UWB signals transmitting through a dielectric slab was presented in [6, 7]. The TD solutions for the reflection and transmission through a dielectric slab were presented in [8].

In this paper, we present an approximate TD solution for the field transmitted through multi-modeled obstacles under the low-loss assumption, for UWB signals. In other words, an accurate TD solution for modeling the transmission of UWB signals through multi-modeled obstacles is presented. Transmission model is called multi-

M.K. Kundu et al. (eds.), *Advanced Computing, Networking and Informatics - Volume 1*,
Smart Innovation, Systems and Technologies 27,
DOI: 10.1007/978-3-319-07353-8_2, © Springer International Publishing Switzerland 2014

modeled because the transmission is considered through multiple shaped obstacles like a wedge followed by a slab. The analytical TD transmission and reflection coefficients, for transmission through the conductor-dielectric interface, considering oblique incidence, are presented for both soft and hard polarizations. Through these TD formulations, the expression of the TD transmitted field is obtained. The TD results are validated against the IFFT [9, 10] of the corresponding exact frequency-domain (FD) results. At last, the computational efficiency of the two approaches is compared to emphasize the significance of the TD solutions presented.

The paper is organized as follows. In section 2, the propagation environment is presented. Section 3 presents the TD transmission and reflection coefficients and the TD formulations for the computation of the transmitted field component. In section 4, the aforementioned numerical TD results are calculated and validated against the numerical IFFT of the corresponding exact FD results and finally section 5 concludes the paper.

2 Propagation Environment

The propagation environment is shown in Fig. 1, where a single dielectric wedge is followed by a dielectric slab. The parameters $r_i, i = 1,2...,5$ are the distances traversed by the transmitted field through the structure from the transmitter (Tx) up to the receiver (Rx).

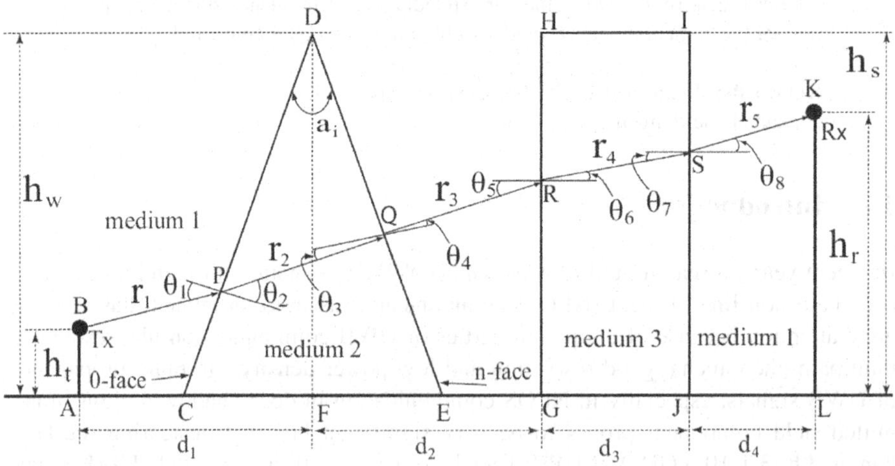

Fig. 1. Transmission through a dielectric wedge followed by a dielectric slab

Angles $\theta_1, \theta_3, \theta_5, \theta_7$ are the incidence angles with $\theta_2, \theta_4, \theta_6, \theta_8$ as the angles of refraction at points 'P', 'Q', 'R' and 'S' respectively and a_i is the internal wedge angle. The parameters h_t and h_r are the heights of the transmitter and the receiver, h_w is the height of the wedge and h_s is the height of the slab. Tx is at a distance d_1 from the wedge, d_2 is distance between wedge and slab, d_3 is width of slab and d_4 is distance between slab and Rx.

3 Proposed Transmission Model

3.1 TD Transmission Coefficient

The actual refracted angle across conductor-dielectric interface is given by [11]

$$\psi_t(\omega) = \tan^{-1}\{t(\omega)/q(\omega)\} \tag{1}$$

with $t(\omega) = \beta_1(\omega)\sin\theta_1$ and $q(\omega) = s(\omega)\{\alpha_2(\omega)\sin\zeta(\omega) + \beta_2(\omega)\cos\zeta(\omega)\}$.

where α_i and β_i are the attenuation constant and phase shift constant of ith medium. θ_1 is the incident angle and $\cos\{\theta_2(\omega)\} = s(\omega)\exp[j\zeta(\omega)]$ with $\theta_2(\omega)$ as the complex refracted angle.

From (1) it is clear that the true refraction angle is also frequency dependent in nature, which means that for different frequency components of the UWB signal, the true real angles of refraction are different. However for low-loss dielectric obstacles (i.e. $\sigma/\omega\varepsilon << 1$), the different true refracted angles reduce to an effective, constant real angle [12]. Thus the real refracted angle ψ_t can be treated as constant and frequency independent, to obtain the approximate analytical TD transmission coefficients. Now for hard polarization case, transmission coefficient while propagating from conductor to dielectric medium, is given by [6, 13, 14]

$$\gamma_h(t) \approx \delta(t) - r_h(t) \tag{2}$$

$$r_h(t) = [K_h\delta(t)] + \frac{4k_h}{(1-k_h^2)}e^{-pt}\left[\frac{K_h}{2}X_h + \frac{(1-X_h)}{2K_h} - \frac{(pt)X_h}{4}\right] \tag{3}$$

with $X_h = e^{-\left(\frac{K_h pt}{2}\right)}, K_h = \left(\frac{1-k_h}{1+k_h}\right)$, $k_h = (\cos\theta_i/\cos\psi_t)(1/\sqrt{\varepsilon_{r2}})$, and $p = \tau/2$ with $\tau = \sigma/\varepsilon_2$.

where, ε_2 and εr_2 are dielectric permittivity and relative dielectric permittivity of the dielectric medium respectively. The TD transmission coefficient for a soft polarized wave propagating from conductor to dielectric medium is given as

$$\gamma_s(t) \approx [\delta(t) - r_s(t)]\left(\frac{\cos\theta_i}{\cos\psi_t}\right) \tag{4}$$

where, $r_s(t) = K_s\delta(t) + \frac{4k_s}{(1-k_s^2)}e^{-pt}\left[\frac{K_s}{2}X_s + \frac{(1-X_s)}{2K_s} - \frac{(pt)X_s}{4}\right]$ with

$X_s = e^{-\left(\frac{K_s pt}{2}\right)}$, $K_s = \left(\frac{1-k_s}{1+k_s}\right)$, $k_s = (\cos\psi_t/\cos\theta_i)(1/\sqrt{\varepsilon_{r2}})$.

3.2 Transmitted Field through the Propagation Environment

The FD expression for the transmitted field at Rx through the structure shown in Fig. 1 is given by [2, 12]

$$E_{RX}(\omega)=\left(E_i(\omega)/r_{total}(\omega)\right)\left(\prod_{i=1}^{4}T_{i,s,h}(\omega)\right)\left(\exp(-jk_0r_1)\prod_{j=2}^{5}\exp\left\{-\left(\alpha_{ej}(\omega)+j\beta_{ej}(\omega)\right)\right\}\right) \tag{5}$$

with $r_{total}(\omega)=\sum_{i=1}^{5}r_i$ and $T_{total,s,h}(\omega)=\prod_{i=1}^{4}T_{i,s,h}(\omega)$ where $T_{i,s,h}(\omega), i=1,2...,4$ are the FD transmission coefficients with respect to points 'P', 'Q', 'R' and 'S' (See Fig. 1). $\alpha_{ej}(\omega), \beta_{ej}(\omega)$ are total effective attenuation constants and phase shift constants for different j^{th} regions [12]. Actual FD path-loss expression from (5) is given by

$$L_{total,s,h}(\omega)=\left(\exp(-jk_0r_1)\prod_{j=2}^{5}\exp\left\{-\left(\alpha_{ej}(\omega)+j\beta_{ej}(\omega)\right)\right\}\right) \tag{6}$$

Now the corresponding TD expression for the received field at Rx based on the FD transmission model of [2] is as follows

$$e_{RX}(t)\approx\left(\frac{e_i(t)}{r_{total}}\right)*\Gamma_{1,s,h}(t)*\Gamma_{2,s,h}(t)*\Gamma_{3,s,h}(t)*\Gamma_{4,s,h}(t)*l_{total,s,h}(t) \tag{7}$$

with '$*$' representing the convolution operator, $\Gamma_{1,s,h}(t)*\Gamma_{2,s,h}(t)*\Gamma_{3,s,h}(t)*\Gamma_{4,s,h}(t)=\Gamma_{total,s,h}(t)$ is the TD counterpart of $T_{total,s,h}(\omega)$, $l_{total,s,h}(t)$ [6, 7] is the TD counterpart of $L_{total,s,h}(\omega)$. The TD expressions for $\Gamma_{i,s,h}, i=1,2...,4$ can be obtained using (2) and (4) for different polarizations.

For loss tangent much less than unity $(\sigma/\omega\varepsilon<<1)$, the FD path-loss expression (6) reduces to the following approximate form with constant values of angles of refraction and along a single effective path for transmission:

$$L_{total,s,h}(\omega)\approx\exp\left(-jk_0(r_1+r_3+r_5)\right)\exp\left[-j\omega\sqrt{\mu\varepsilon}\left(1+\frac{\sigma}{2j\omega\varepsilon}\right)\left\{r_2+\left(d3\sqrt{\frac{\varepsilon_r}{\varepsilon_r-\sin^2\theta_5}}\right)\right\}\right] \tag{8}$$

Here $r_1+r_3+r_5$ is the total distance travelled by the field in free space. ε and σ are the parameters of the dielectric mediums in Fig. 1 (assuming same for wedge and slab). The term $l_{total,s,h}(t)$ in (7) is then given by

$$
\begin{aligned}
l_{total,s,h}(t) &\approx \exp\left[-\sqrt{\mu/\varepsilon}\left(\sigma/2\right)\left\{r_2+\left(d3\sqrt{\frac{\varepsilon_r}{\varepsilon_r-\sin^2\theta_5}}\right)\right\}\right] \\
&\delta\left[t-\sqrt{\mu\varepsilon}\left\{r_2+\left(d3\sqrt{\frac{\varepsilon_r}{\varepsilon_r-\sin^2\theta_5}}\right)\right\}\right]*\delta\left(t-\frac{r_1+r_3+r_5}{c}\right)
\end{aligned}
\tag{9}
$$

This approximated TD path-loss expression will be used in (7) to compute the TD transmitted field and the accuracy will be proved by the comparison of TD transmitted field with the IFFT of the exact FD results, as shown in next section.

4 Results and Discussions

In this section, our goal is to compare the proposed TD solution with conventional IFFT-FD method. Table 1 shows the electromagnetic properties of the considered materials.

Table 1. Electromagnetic properties of different dielectric materials

Material	Relative Permittivity	Conductivity (S/ m)
Wood[6]	2	0.01
Drywall[15]	2.4	0.004
Glass[15, 16]	6.7	0.001

Fig. 2 shows the transmitted field through the propagation environment discussed in section 2, for both hard and soft polarizations. The transmitted field at Rx suffers no distortion in shape in comparison to the shape of the excited UWB pulse. This is because of small magnitude of the loss tangent with respect to unity. However, the amplitude of the transmitted field is attenuated because of the transmission loss through the dielectric mediums.

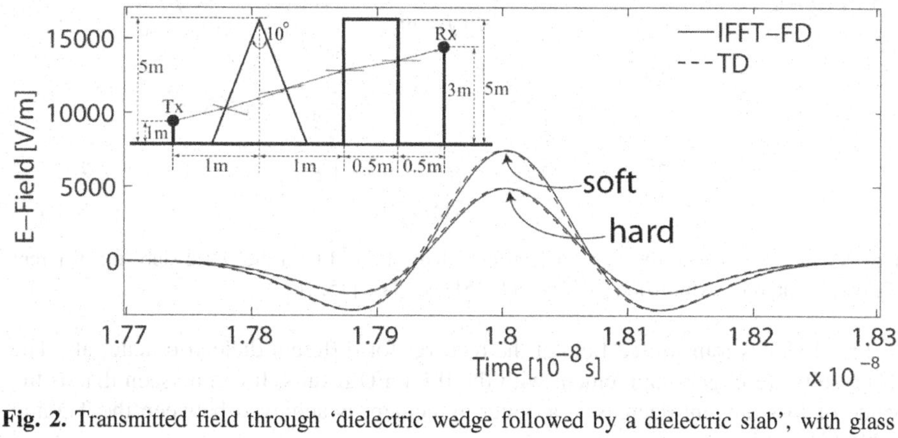

Fig. 2. Transmitted field through 'dielectric wedge followed by a dielectric slab', with glass [15, 16]

The TD results for the transmitted field for both the polarizations are in excellent agreement with corresponding IFFT of exact FD results, thus providing validation to the proposed TD solution.

Fig. 3 shows the effect of varying Rx position (changing distance d_4 in Fig. 1) on transmitted field at the receiver. Transmitted field gets more attenuated as Rx moves away from the obstacles. The results for soft and hard polarized fields come closer to each other as the distance d_4 increases. Also the TD results match closely with the IFFT-FD results.

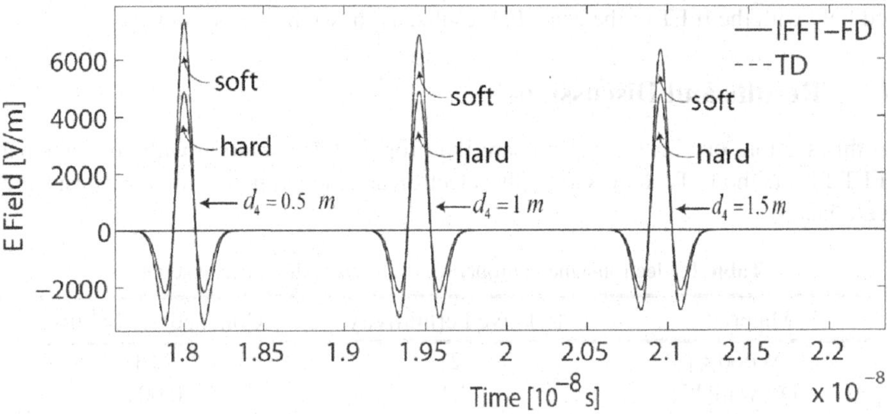

Fig. 3. Transmitted field through 'dielectric wedge followed by a dielectric slab' for different receiver positions, with glass [15, 16]

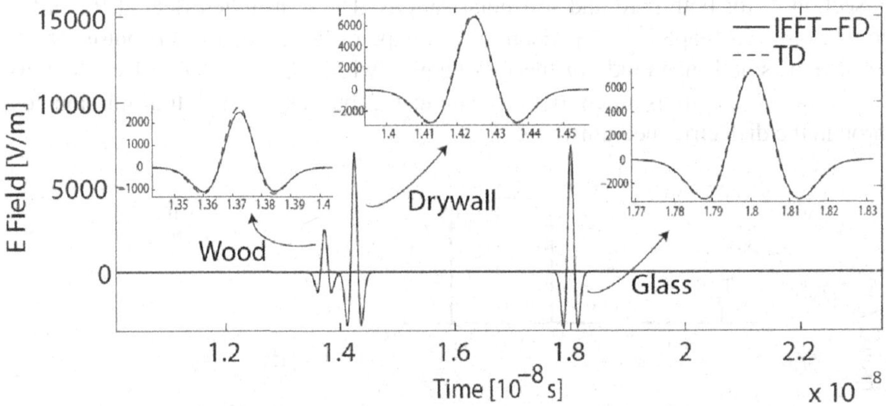

Fig. 4. Transmitted field through 'dielectric wedge followed by a dielectric slab' for different dielectric materials, with wood [6], drywall [15] and glass [15, 16]

Fig. 4 shows transmitted field at the receiver for different dielectric materials. The TD results are in good agreement with the IFFT-FD results. It can be seen that as the value of loss tangent decreases, a better agreement is achieved between the TD and IFFT-FD results.

A comparison between the computation times of the IFFT-FD method and the proposed TD solution for propagation profile considered in Fig. 1 is presented in Table 2. The presented results in Table 2 establish that the proposed TD analysis is computationally very efficient in comparison to the IFFT-FD solution.

Table 2. Efficiency comparison of two methods

Propagation profile	$T_{IFFT\text{-}FD}/T_{TD}$
For soft polarization	~198
For hard polarization	~191

The two main reasons for such a significant reduction in the computational time in TD are: (i) the efficient convolution technique [9] due to which few number of time samples suffice to provide accurate results. (ii) Approximation of the multiple transmission paths in FD by a single effective path for low-loss dielectric case.

Given the excellent agreement between proposed TD solution and IFFT of FD solution, it can be concluded that the proposed method is accurate for low loss tangent values in the UWB bandwidth. The presented work also establishes that the proposed TD solution is computationally more efficient than the conventional IFFT-FD method.

5 Conclusion

An analytical TD solution has been presented for the transmitted field through multi-modeled obstacles made up of low-loss dielectric materials. Analytical TD transmission and reflection coefficients for transmission through an interface between conductor and dielectric mediums are presented for soft and hard polarizations. The results of the proposed TD solution are validated against the corresponding IFFT-FD results and the computational efficiency of two methods is compared. The TD solution outperforms the IFFT-FD analysis in terms of the computational efficiency. The TD solution is vital in the analysis of UWB communication as it can provide a fast and accurate prediction of the total transmitted field in microcellular and indoor propagation scenarios.

References

1. de Jong, Y.L.C., Koelen, M.H.J.L., Herben, M.H.A.J.: A building-transmission model for improved propagation prediction in urban microcells. IEEE Trans. Veh. Technol. 53(2), 490–502 (2004)
2. Soni, S., Bhattacharya, A.: An analytical characterization of transmission through a building for deterministic propagation modeling. Microw. Opt. Techn. Lett. 53(8), 1875–1879 (2011)
3. Karousos, A., Tzaras, C.: Multiple time-domain diffraction for UWB signals. IEEE Trans. Antennas Propag. 56(5), 1420–1427 (2008)

4. Qiu, R.C., Zhou, C., Liu, Q.: Physics-based pulse distortion for ultra-wideband signals. IEEE Trans. Veh. Technol. 54(5), 1546–1555 (2005)
5. Chen, Z., Yao, R., Guo, Z.: The characteristics of UWB signal transmitting through a lossy dielectric slab. In: Proc. IEEE 60th Veh. Technol. Conf., VTC 2004-Fall, Los Angeles, CA, USA., vol. 1, pp. 134–138 (2004)
6. Yang, W., Qinyu, Z., Naitong, Z., Peipei, C.: Transmission characteristics of ultra-wide band impulse signals. In: Proc. IEEE Int. Conf. Wireless Communications, Networking and Mobile Computing, Shanghai, pp. 550–553 (2007)
7. Yang, W., Naitong, Z., Qinyu, Z., Zhongzhao, Z.: Simplified calculation of UWB signal transmitting through a finitely conducting slab. J. Syst. Eng. Electron. 19(6), 1070–1075 (2008)
8. Karousos, A., Koutitas, G., Tzaras, C.: Transmission and reflection coefficients in time-domain for a dielectric slab for UWB signals. In: Proc. IEEE Veh. Technol. Conf., Singapore, pp. 455–458 (2008)
9. Brigham, E.O.: The Fast Fourier transform and Its Applications. Prentice-Hall, Englewood Cliffs (1988)
10. Sevgi, L.: Numerical Fourier transforms: DFT and FFT. IEEE Antennas Propag. Mag. 49(3), 238–243 (2007)
11. Balanis, C.A.: Advanced engineering electromagnetic. Wiley, New York (1989)
12. Tewari, P., Soni, S.: Time-domain solution for transmitted field through low-loss dielectric obstacles in a microcellular and indoor scenario for UWB signals. IEEE Trans. Veh. Technol. (2013) (under review)
13. Barnes, P.R., Tesche, F.M.: On the direct calculation of a transient plane wave reflected from a finitely conducting half space. IEEE Trans. Electromagn. Compat. 33(2), 90–96 (1991)
14. Tewari, P., Soni, S.: A comparison between transmitted and diffracted field in a microcellular scenario for UWB signals. In: Proc. IEEE Asia-Pacific Conf. Antennas Propag., Singapore, pp. 221–222 (2012)
15. Muqaibel, A., Safaai-Jazi, A., Bayram, A., Attiya, A.M., Riad, S.M.: Ultrawideband through-the-wall propagation. IEE Proc.-Microw. Antennas Propag. 152(6), 581–588 (2005)
16. Jing, M., Qin-Yu, Z., Nai-Tong, Z.: Impact of IR-UWB waveform distortion on NLOS localization system. In: ICUWB 2009, pp. 123–128 (2009)

Indexing and Retrieval of Speech Documents

Piyush Kumar P. Singh, K.E. Manjunath, R. Ravi Kiran,
Jainath Yadav, and K. Sreenivasa Rao

Indian Institute of Technology Kharagpur,
West Bengal - 721302, India
piyushks@live.com, {ke.manjunath,r.ravi.kiran.88,jaibhu38}@gmail.com,
ksrao@iitkgp.ac.in

Abstract. In this paper, a speech document indexing system and similarity-based document retrieval method has been proposed. K-d tree is used as the index structure and codebooks derived from speech documents present in the database, are used during retrieval of desired document. Each document is represented as a sequence of codebook indices. The longest common subsequence based approach is proposed for retrieving the documents. Proposed retrieval method is evaluated using a speech database of 3 hours recorded by a male speaker and speech queries from 5 male and 5 female speakers. The accuracy of retrieval is found to be about 88% for the queries given by male speakers.

Keywords: Indexing and Retrieval, codebook, MFCC, k-d tree, retrieval, longest common subsequence.

1 Introduction

Due to rapid advancement in technology, there has been explosive growth in the generation and use of multimedia data, such as video, audio, and images. A lot of audio and video data is generated by internet, mobile devices, TV and radio broadcast channels. Indexing systems are desirable for managing and supporting usage of large databases of multimedia. Audio indexing finds applications in digital libraries, entertainment industry, forensic laboratories and virtual reality.

Many kinds of indexing schemes have been developed, and studied by researchers working in this field [13]. An approach for indexing audio data is the use of text itself. Transcripts from the audio data are generated, which are used for indexing. This approach is effective for indexing broadcast news, video lectures, spoken documents, etc., where the clean speech data is available. The retrieval from such indexing system is performed using keywords as a query. For effective use of multimedia data, the users should be able to make content based queries or queries-by-example, which are unrestricted and unanticipated. Content based retrieval systems accept data type queries i.e., hummed, sung or original clip of a song for a song retrieval. The methods used for audio indexing can be broadly classified as under:

- Signal parameter based systems [1,3,9,14] - In this scheme, the signal statistics such as mean, variance, zero crossing rate, autocorrelation, histograms

M.K. Kundu et al. (eds.), *Advanced Computing, Networking and Informatics - Volume 1*,
Smart Innovation, Systems and Technologies 27,
DOI: 10.1007/978-3-319-07353-8_3, © Springer International Publishing Switzerland 2014

of samples/difference of samples, energy contour, loudness contour, etc. either on the whole audio signal data or blocks of data are used. This type of indexing supports query by example. Audio Fingerprinting [1] is an example of this type of system.

– Musical parameter based systems [2,7,8,10] - In this scheme, the parameters of signal along with acoustical attributes such as melody contour, rhythm, tempo, etc. extracted from the audio signal are used for indexing. This type of indexing scheme supports both query by example and query by humming. This method requires extensive calculations.

In the design of an audio indexing system following two aspects need to be addressed. (1). Derivation of good features from the audio data to be used as indices during search. (2). Organization of these indices in a suitable multidimensional data structure with efficient search. Selection of a good measure of similarity (distance measure). The objective of this work is to design a content-based speech indexing system which supports query by example. In this work, the features used for indexing purpose is a codebook derived from MFCC feature vectors for every 10 seconds of speech data present in the database. The k-d tree is used for providing the indexing structure [4] and longest common subsequence (LCS) [11], [5] has been used as a measurement for similarity for retrieval purpose.

Rest of the paper is organized as follows - Section II describes the framework of the proposed indexing and retrieval system. In Section III, experimental results and their analysis are discussed. The summary of the contents presented in the paper, and conclusions drawn from the observations are presented in Section IV.

2 Proposed Speech Indexing and Retrieval System

A speech file is a concatenated sequence of sounds. So finding a matching speech clip in the database to a query clip provided, can be thought of as finding the clip having the similar sounds in the same sequence. But since the query can be provided by any person, the two sequences will not be the exact duplicate of each other. This problem of matching the sequence of sounds can be mapped to the well known problem of determining the longest common subsequence from two sequences of characters. The only difference is that in place of characters there are sounds. By mapping the sound sequence matching problem to the longest common subsequence problem, a different issue arises. How do we map sounds to characters? Determining the longest common subsequence directly from the sounds of the speech files is very difficult because a speaker can produce many variants of a single sound. These variations increase tremendously with change in speaker. So we build codebooks to bring uniformity in these sound sequences. The process to derive a codebook is explained in Section 2.1.

The framework of the proposed speech indexing and retrieval system is shown in the Fig. 1. The tasks of the system and the challenges they pose, can be explained in three phases.

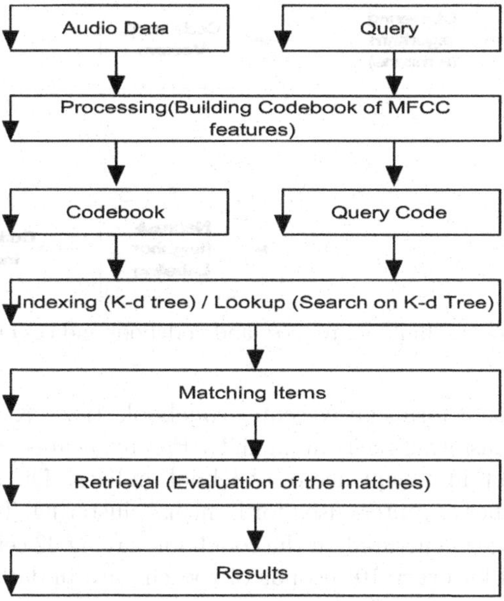

Fig. 1. Proposed speech indexing and retrieval system

Building the Codebook. Phase 1 (Preprocessing phase) converts all raw speech data into indexable items, typically, points in a high-dimensional vector space with an appropriate distance measure. In other words, in this phase we extract features from the speech files and the query to be processed. The features considered here are Mel-Frequency Cepstral Coefficients (MFCCs). MFCCs are determined from speech signal using the following steps:

1. Pre-emphasize the speech signal.
2. Divide the speech signal into sequence of frames with a frame size of 20 ms and a shift of 10 ms. Apply the Hamming window over each of the frames.
3. Compute magnitude spectrum for each windowed frame by applying DFT.
4. Mel spectrum is computed by passing the DFT signal through mel filter bank.
5. DCT is applied to the log mel frequency coefficients (log mel spectrum) to derive the desired MFCCs.

From each frame, 13 MFFCCs are extracted using 24 filter banks. This results in 1000 MFCC vectors for 10 seconds of speech. These 1000 vectors are then converted to a vector codebook using k-means clustering [12].

Fig. 2 shows the basic steps for building a vector quantization codebook and extraction of code book indices. Training data consists of 1000 MFCC vectors. The training data set is used to create an optimal set of codebook vectors for representing the spectral variability observed in the training data set. A centroid computation procedure is used as the basis of partitioning the training data set

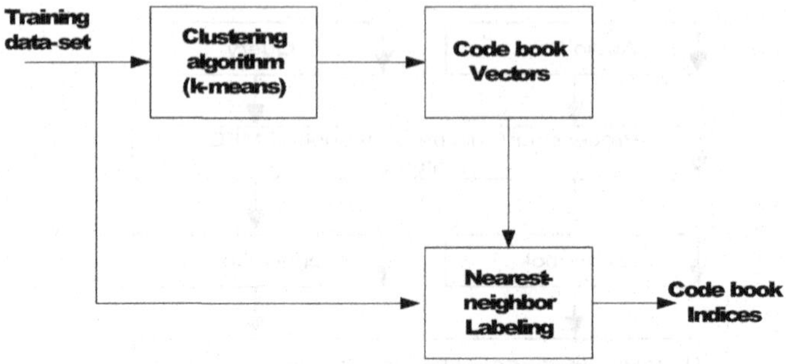

Fig. 2. Codebook building process and codebook indices extraction

into N clusters. Thus it forms an N-center codebook. Here, K-means clustering method is used for clustering the training data. Finally, nearest-neighbor labeling has been used to obtain a sequence of codebook indices. The size of codebook varies with the number of centers used for k-means clustering. The results shown in this paper have been generated on the codebook having 32 centers. A separate codebook is created for every 10 seconds of speech data in database.

Building the Index. Phase 2 (Indexing / Lookup phase) does the task of indexing the features extracted from the speech database. This phase also performs a lookup in the index to retrieve the best matching items for a query vector. The matching can be done in various ways, say, exact matching, nearest neighbour matching, etc. which depends on the application and the efficiency required.

In this work, MFCC feature vectors derived from speech consists of 13 dimensions. Various data structures can be used for multi-dimensional indexing, for example, quad tree, k-d tree, optimized k-d tree. In this work, the index is created from the codebook in the form of a k-d tree. The discriminating key for each level of the k-d tree is selected according to Bentleys definition of the k-d tree i.e., D = L mod k + 1, where D is the discriminating key number for level L and the root node is defined to be at level zero. However, the selection of the partition value is done in the way suggested by Friedman [4]. The partition value is selected as the median value in the dimension of the discriminator(D).

The codebooks created in this work have 32 codebook vectors. Each of these vectors is assigned an index number (1-32). This numbering of the codebook entries is utilized in Phase 3. We are required to preserve this numbering even when the codebook is converted to a k-d tree index. This is achieved by appending the index number as an additional dimension to the codebook vectors while creating the k-d tree. This 14th dimension is never used as a discriminating key while building the k-d tree index. The Algorithm 1 is used for k-d tree creation in this work.

The procedure *median(j,subfile)* returns median of the jth key values. The procedures *make_terminal* and *make_non_terminal* store their parameters as values of the node in the k-d tree and return a pointer to that node. The $leftsubfile$ and $rightsubfile$ procedures, partition the file along the dth key with respect

Algorithm 1. Algorithm used for k-d tree creation

```
Result: root ← build_tree(entirefile, 1)
procedure build_tree(file, dim)
    local p
    if size(file) ≤ b then
        return (make_terminal(file))
    end
    p ← median(dim, file)
    dim ← dim Mod k + 1
    left ← build_tree(left_subfile(d, p, file))
    right ← build_tree(right_subfile(d, p, file))
    return make_non_terminal(d, p, left, right)
end procedure
```

to the partition value p and return left and right subfiles, respectively. Heapsort algorithm having complexity of *O(NlogN)*, is used to compute the median as well as the left and right sub file at each level. Thus the time complexity of building a k-d tree from N vectors is $O(Nlog^2N)$. The searching algorithm proposed by Friedman [4] is used in this work for searching the k-d tree. This search algorithm is also a part of Phase 2 of the system.

Evaluation of Matches. Phase 3 (Retrieval phase) evaluates all the matching items and decides which files among the database are the best candidates, to be returned as the query response. As mentioned earlier, in this work we use longest common subsequence determination technique for finding the matching speech fragments.

The sequence of codebook indices is generated by using the MFCC feature vector sequence of the speech clip. For each MFCC vector, the index number (1-32) of nearest codebook vector is determined by searching the k-d tree. These

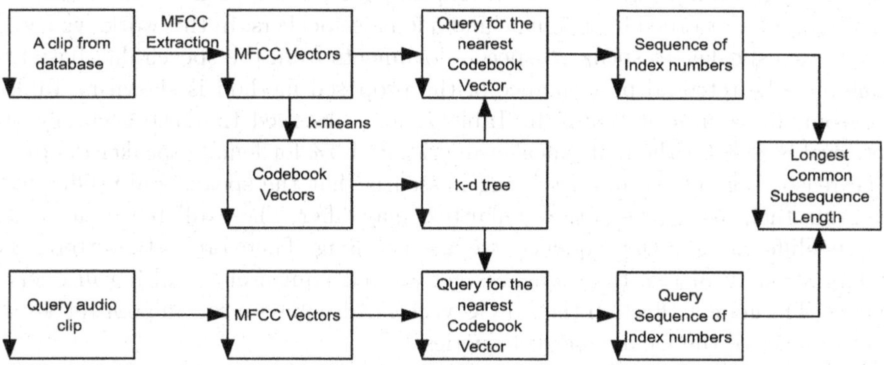

Fig. 3. Retrieval using longest common subsequence approach

index numbers are then concatenated to form a sequence of index numbers. The retrieval method finds out that clip in the database, whose MFCC vectors follow the similar codebook vector sequence, as that of the query clip. Steps in retrieval method are as follows:

1. After building the codebook for a speech document in the database, the MFCC vectors are scanned, in order of their occurrence, to determine the index number (1-32) of the nearest codebook vector.
2. By concatenating these index numbers obtained in the above step, we obtain an approximate sequence of sounds present in the clip.
3. MFCC vectors are extracted from the query speech clip provided.
4. By using the same technique of nearest neighbour search among the codebook vectors, we can obtain another sequence of index numbers. The length of the longest common subsequence between the sequences obtained in steps 2 and 4 is determined.
5. The length of the longest common subsequence obtained in step 4, can act as the similarity measure while comparison with other clips in the database, since the process of sequence determination will be repeated for all the clips in the database.

Longest Common subsequence is a classic problem in computer science. Many solutions are available in literature. In this work, longest common subsequence is determined by using dynamic programming technique (DP). DP is well known for its space efficient implementation and lower complexity. The entire process is summarized in Fig. 3.

3 Performance Evaluation

The database used for evaluation, consists of 3 hours of speech recorded by a single male speaker. The speech consists of articles about History of India available on Wikipedia. This recorded speech data was divided into 1080 speech documents, each of duration 10 seconds. Proposed retrieval approach was evaluated by 200 queries recorded from 5 male and 5 female speakers. In this work, we have asked each speaker to speak 20 speech documents from the speech database at random. The retrieval performance of the proposed method is shown in Table 1. From the results presented in Table 1, it is observed that the accuracy of retrieval is 88% for the male speaker queries and 3% for female speaker queries. The reason behind the queries being missed is that the speakers are different, and thus their utterances of the similar text may differ. These differences account for the differences in the sequences we are matching. Therefore, if the utterance differs at many places, the sequences get altered immensely resulting in query misses. The gender effect on the MFCC vectors, also plays a role in poor retrieval performance of the female speaker queries.

Table 1. Retrieval performance with longest common subsequence matching

Speaker Gender			
Male		Female	
Speaker Id	Queries matched	Speaker Id	Queries matched
1	18	6	0
2	20	7	0
3	15	8	2
4	20	9	0
5	15	10	1

The poor retrieval performance for the queries of female speakers may be due to the influence of gender characteristics. As the database contains only male speaker's speech utterances, therefore in case of female speakers queries, the indices generated by the queries is entirely different due to differences in shape and size of the vocal tract between the genders.

4 Summary and Conclusions

In this paper, a k-d tree based speech document indexing system has been proposed. For retrieving the desired speech document for a given query, the sequence of codebook indices, generated by the speech document and the query are compared using LCS approach. By computing the LCS scores between a query and all the speech documents, the retrieval system retrieves the desired document based on the highest LCS score. For evaluating the proposed retrieval approach, 3 hours of speech database recorded by a male speaker was used. The performance accuracy in the retrieval process is found to be better for the queries spoken by male speakers. In the case of female speakers, the performance is observed to be very poor.

In this work, the codebooks are generated from each 10 second speech document. The codebook captures the local characteristics of speech present in that 10 second segment. Therefore when we generate the indices from this codebook, the sequence of indices may be differ for different speakers even if the spoken message content is same. This problem may be addressed by building a single codebook from large amount of speech data instead of smaller codebooks for each 10 second segment. From the generalized codebook, if we derive the sequence of indices for the speech document and the query, the retrieval accuracy may be improved. It may also resolve the problem risen due to gender dependent query. Query gender dependency may also be resolved by adapting appropriate vocal tract length normalization(VTLN) feature transformation techniques in the codebook building procedure. Also, in the current experimental setup the queries given were similar to clips present in the speech database. In future, work can be done to perform retrieval based on queries having only few keywords and other connecting words.

References

1. Cha, G.-H.: An Effective and Efficient Indexing Scheme for Audio Fingerprinting. In: Proceedings of the 2011 Fifth FTRA International Conference on Multimedia and Ubiquitous Engineering, Washington, DC, USA, pp. 48–52 (2011)
2. Chen, A.L.P., Chang, M., Chen, J., Hsu, J.-L., Hsu, C.-H., Hua, S.Y.S.: Query by music segments: an efficient approach for song retrieval. In: 2000 IEEE International Conference on Multimedia and Expo (2000)
3. Foote, J.T.: Content-Based Retrieval of Music and Audio. In: Proceedings of SPIE, Multimedia Storage and Archiving Systems II, pp. 138–147 (1997)
4. Friedman, J.H., Bentley, J.L., Finkel, R.A.: An Algorithm for Finding Best Matches in Logarithmic Expected Time. ACM Transactions on Mathematical Software 3(3), 209–226 (1977)
5. Hirschberg, D.S.: A linear space algorithm for computing maximal common subsequences. ACM Communucations, 341–343 (1975)
6. Deller Jr., J.R., Hansen, J.H.L., Proakis, J.G.: Discrete-Time Processing of Speech Signal. IEEE Press (2000)
7. Kosugi, N., Nishihara, Y., Sakata, T., Yamamuro, M., Kushima, K.: A practical query-by-humming system for a large music database. In: Proceedings of the Eighth ACM International Conference on Multimedia, pp. 333–342 (2000)
8. Lemström, K., Laine, P.: Musical information retrieval using musical parameters. In: Proceedings of the 1998 International Computer Music Conference (1998)
9. Li, G., Khokhar, A.A.: Content-based indexing and retrieval of audio data using wavelets. In: 2000 IEEE International Conference on Multimedia & Expo, pp. 885–888 (2000)
10. Lu, L., You, H., Zhang, H.-J.: A new approach to query by humming in music retrieval. In: ICME 2001, pp. 595–598 (2001)
11. Maier, D.: The Complexity of Some Problems on Subsequences and Supersequences. J. ACM, 322–336 (1978)
12. Rabiner, L., Juang, B.-H.: Fundamentals of speech recognition. Prentice-Hall, Inc. (1993)
13. Rao, K.S., Pachpande, K., Vempada, R.R., Maity, S.: Segmentation of TV broadcast news using speaker specific information. In: NCC 2012, pp. 1–5 (2012)
14. Subramanya, S.R., Youssef, A.: Wavelet-based Indexing of Audio Data in Audio/Multimedia Databases. In: Proceedings of MultiMedia Database Management Systems (1998)

An Improved Filtered-x Least Mean Square Algorithm for Acoustic Noise Suppression

Asutosh Kar[1], Ambika Prasad Chanda[1], Sarthak Mohapatra[1], and Mahesh Chandra[2]

[1] Department. of Electronics and Telecommunication Engineering
Indian Institute of Information Technology, Bhubaneswar, India
[2] Department of Electronics and Communication Engineering
Birla Institute of Technology, Mesra, India
{asutosh,b210038}@iiit-bh.ac.in, chanda.iiit@gmail.com,
shrotriya@bitmesra.ac.in

Abstract. In the modern age scenario noise reduction is a major issue, as noise is responsible for creating disturbances in day-to-day communication. In order to cancel the noise present in the original signal numerous methods have been proposed over the period of time. To name a few of these methods we have noise barriers and noise absorbers. Noise can also be suppressed by continuous adaptation of the weights of the adaptive filter. The change of weight vector in adaptive filters is done with the help of various adaptive algorithms. Few of the basic noise reduction algorithms include Least Mean Square algorithm, Recursive Least Square algorithm etc. Further we work to modify these basic algorithms so as to obtain Normalized Least Mean Square algorithm, Fractional Least Mean Square algorithm, Differential Normalized Least Mean Square algorithm, Filtered-x Least Mean Square algorithm etc. In this paper we work to provide an improved approach for acoustic noise cancellation in Active Noise Control environment using Filtered-x LMS (FXLMS) algorithm. A detailed analysis of the algorithm has been carried out. Further the FXLMS algorithm has been also implemented for noise cancellation purpose and the results of the entire process are produced to make a comparison.

Keywords: adaptive filter, active noise control, Least Mean Square, Mean Square Error, FXLMS.

1 Introduction

Active noise control (ANC) is based on the principle of destructive superposition of acoustic waves [1, 2]. Using ANC concept in noise cancellation, a noise signal is generated which is correlated to the noise signal but opposite in phase. By adding both the original signal can be made noise free. The generation of the signal is controlled by an adaptive algorithm which adaptively changes the weight of the filter used. Active noise control is mainly classified into two categories i.e. "feed-forward ANC" and "feedback ANC". In feedback ANC method a controller is used to modify the response of the system. But in feed-forward ANC method a controller is used to

M.K. Kundu et al. (eds.), *Advanced Computing, Networking and Informatics - Volume 1*,
Smart Innovation, Systems and Technologies 27,
DOI: 10.1007/978-3-319-07353-8_4,

adaptively calculate the signal that cancels the noise. In this paper "feed-forward ANC" approach is implemented to cancel the noise with the help of FxLMS algorithm. This is because the amount of noise reduction achieved by feed-forward ANC system is more than that of feedback ANC system. A single channel feed-forward ANC system comprises of a reference sensor, a control system, a cancelling loudspeaker and an error sensor. The primary noise present is measured by the reference sensor and is cancelled out around the location of the error microphone by generating and combining an anti-phase cancelling noise that is correlated to the spectral content of the unwanted noise [3]. The reference sensor produces a reference signal that is "feed-forward" to the control system to generate "control signals" in order to drive the cancelling loudspeaker to generate the cancelling noise. The error microphone measures the residual noise after the unwanted noise and controls signal. It combines and sends an error signal back to the controller to adapt the control system in an attempt to further reduction of error. The control system adaptively modifies the control signal to minimize the residual error. The most famous adaptation algorithm for ANC systems is the FXLMS algorithm [4, 5] as mentioned earlier, which is a modified and improved version of the LMS algorithm [6-10].

Although FXLMS algorithm is widely used due to its stability but it has slower convergence. Various techniques have been proposed to improve the convergence of FXLMS algorithm with some parameter trade-offs. Other versions of the FXLMS include Filtered-x Normalized LMS (FXNLMS) [11], Leaky FXLMS [12], Modified FXLMS (MFXLMS) [13-15] etc. However, the common problem with all of these algorithms is the slow convergence rate, especially when there is large number of weights [16]. To overcome this problem, more complex algorithms such as Filtered-x Recursive Least Square (FXRLS) [17] can be used. These algorithms have faster convergence rate compared to the FXLMS; however they involve matrix computations and their real-time realizations might not be cost effective.

2 Problem Formulation

Acoustic noise problems are not only based on high frequency noise but many are also dominated by low frequency noise. Various passive noise control techniques such as noise barriers and absorbers do not work efficiently in low frequency environments. ANC works very efficiently in case of low frequency noise & it is also cost effective than bulky, heavy barriers and absorbers.

The basic LMS algorithm fails to perform well in the ANC framework. This is due to the assumption made that the output of the filter is the signal perceived at the error microphone, which is not the case in practice. The presence of the A/D, D/A converters, actuators and anti aliasing filter in the path from the output of the filter to the signal received at the error microphone cause significant change in the output signal. This demands the need to incorporate the effects of this secondary path function in the algorithm. But the convergence of the LMS algorithm depends on the phase response of the secondary path, exhibiting ever-increasing oscillatory behavior as the phase increases and finally going unstable at 90° [11]. The solution to this problem was either to employ an inverse filter in the cancellation path or to introduce a filter in the reference path, which is ideally equal to the secondary path impulse response. The former technique is referred to as the "filtered-error" approach, while the later is now known as the "filtered-reference" method, more popularly known as the FXLMS algorithm. The FXLMS solution is by far the most widely used due to its stable as well as predictable operation [3-5].

3 Simulation Setup

1) Feed-forward ANC system:

The basic idea of feed-forward ANC is to generate a signal (secondary noise), that is equal to a disturbance signal (primary noise) in amplitude and frequency, but has opposite phase. Combination of these signals results in cancellation of the primary (unwanted) noise. This ANC technique is well-known for its use in cancelling unwanted sound as shown in [2].

2) ANC system description using adaptive filter:

In this paper instead of the adaptive algorithm block an improved FXLMS algorithm approach is employed to suppress noise. The most popular feed-forward control algorithm for acoustic noise cancellation is FXLMS algorithm [7] which is the main area of focus in this paper. A control signal is created by the FXLMS algorithm by filtering the reference signal $x(n)$ with an adaptive control filter. The control filter is updated via a gradient descent search process until an ideal filter that minimizes the residual noise found. This is because the existence of a filter in the auxiliary and the error- path is shown which generally degrade the performance of the LMS algorithm. Thus, the convergence rate is lowered, the residual power is increased and the algorithm can even become unstable. In order to stabilize LMS algorithm, the reference signal $x(n)$ is filtered by an estimate of the secondary path transfer function $S(n)$ which is the propagation path from the controller to the error sensor giving the filtered-x signal. Hence the algorithm is known as FXLMS algorithm.

The noise measured at the reference sensor propagates through the primary path represented by $P(n)$ and arrives at the error sensor or receiver as the signal, $d(n)$. This is the unwanted noise to be cancelled which is sometimes referred to as the "desired" signal, meaning the signal that the controller is trying to duplicate with opposite phase. The desired signal will be correlated with the noise characterized by the reference. The LMS algorithm then generates control signal at the cancelling loudspeaker by filtering the reference signal $x(n)$ with an adaptive FIR control filter, $W(n)$. The control signal, $y(n)$, is the convolution of $x(n)$ with $W(n)$. The control signal is filtered by the secondary path transfer function, $S(n)$, and arrives at the error sensor as the output signal, $y'(n)$. The output signal combines with the unwanted noise to give the residual error signal, e(n), measured by the error sensor. An adaptive process searches for the optimal coefficients for the control filter which minimizes the error. This optimal filter is designated W_{opt} [7-10].

Step:1

For ANC system containing a Secondary Path transfer function $S(n)$ the residual error can be expressed as:

$$e(n) = d(n) - y'(n) \tag{1}$$

where, $y'(n)$ is the output of the secondary path $S(n)$.

Step:2

If $S(n)$ be an IIR filter with denominator coefficients $[a_0......a_n]$ and numerator coefficient $[b_0.........b_{M-1}]$, then the filter output $y'(n)$ can be written as the sum of the filter input $y(n)$ and the past filter output:

$$y'(n) = \sum_{i=1}^{N} a_i y'(n-i) + \sum_{j=0}^{M-1} b_j y(n-j) \tag{2}$$

Step:3

Hence the gradient estimate becomes:

$$\nabla \hat{\xi}(n) = -2x'(n)e(n), \tag{3}$$

where,
$$x'(n) = \sum_{i=1}^{N} a_i x'(n-i) + \sum_{j=0}^{M-1} b_j x(n-j) \tag{4}$$

In practical applications, $S(n)$ is not exactly known, therefore the parameters a_i and b_j are the parameters of the Secondary Path Estimate $\hat{S}(n)$.

Step:4

The weight update equation of the FXLMS algorithm is:

$$w(n+1) = w(n) + \mu x'(n)e(n) \tag{5}$$

Equation (5) is the filter weight update equation of the FXLMS algorithm. As it can be seen that the equation is identical to the LMS algorithm but here instead of using x(n) i.e. the reference signal, $x'(n)$ is used which is the filtered output of the original signal called as "filtered-x signal" or "filtered reference signal".

The FXLMS algorithm is very tolerant to modelling errors in the secondary path estimate. The algorithm will converge when the phase error between $S(n)$ and $H(n)$ is smaller than +/- 90 degrees. Computer simulations show that phase errors less than 45° do not significantly affect performance of the algorithm [8]. The gain applied to the reference signal by filtering it with $S(n)$ does not affect the stability of the algorithm and is usually compensated for by modifying the convergence parameter, μ. Convergence will be slowed down though, when the phase error increases. $H(n)$ is estimated through a process called system identification. Band-limited white noise is played through the control speaker(s) and the output is measured at the error sensor.

The measured impulse response is obtained as a FIR filter $S(n)$ in the time domain. The coefficients of $S(n)$ are stored and used to pre-filter the reference signal and give the input signal to the LMS update.

The performance of the FXLMS adaptation process is dependent on a number of factors. Those factors include the characteristics of the physical plant to be controlled, the secondary path impulse response and the acoustic noise band-width. Hence the FXLMS will not function properly if the secondary path has a long impulse response and/or the acoustic noise has a wide band-width. Among the design parameters, steady state performance increases, increasing the filter length but convergence of FXLMS algorithm is degraded by this. So the filter length should be chosen carefully. After the filter length is set, the maximum achievable performance is limited by a scalar parameter called the adaptation step-size. The speed of convergence will generally be higher when choosing a higher sample frequency and higher value of μ. [14-16]

It is advantageous to choose a large value of μ because the convergence speed will increase but too large a value of μ will cause instability. It is experimentally determined that that the maximum value of the step size that can be used in the FXLMS algorithm is approximately:

$$\mu_{\max} = \frac{1}{p_{x'}(L+\Delta)} \tag{6}$$

where $p_{x'} = E[x'^2(n)]$ is the mean-square value or power of the filtered reference signal $x'(n)$, L is the number of weights and Δ is the number of samples corresponding to the overall delay in the Secondary Path.

After convergence, the filter weights $W(n)$ vary randomly around the optimal solution W_{opt}. This is caused by broadband disturbances, like measurement noise and impulse noise on the error signal. These disturbances cause a change of the estimated gradient $\nabla\hat{\xi}(n)$, because it is based only on the instantaneous error. This results in an average increase of the MSE, that is called the excess MSE, defined as, [9]

$$\xi_{excess} = E[\hat{\xi}(n)] - \xi_{\min} \tag{7}$$

This excess MSE is directly proportional to step size. It can be concluded that there is a design trade-off between the convergence performance and the steady-state performance. A larger value of μ gives faster convergence but gives bigger excess MSE and vice-versa. Another factor that influences the excess MSE is the number of weights.

4 Simulation and Results

MATLAB platform is chosen for simulation purpose. The room impulse response is measured for an enclosure of dimension 12×12×8 ft^3 using an ordinary loudspeaker and an uni-dynamic microphone kept at a distance of 1 ft which has been sampled at 8 KHZ. The channel response is obtained by applying a stationary Gaussian stochastic signal with zero mean and unit variance as input on the measured impulse response. Fig. 1 shows the MATLAB simulation of noise cancellation.

Fig. 1 (a) is the original noise, (b) is the noise observed at the headset. Fig. 1(c) is the inverse noise generated by an adaptive algorithm using the DSK, and Fig. 1(d) is the error signal between the noise observed at the headset and the inverse noise generated by the DSK. Fig. 1(d) shows that the error signal is initially very high, but as the algorithm converges, this error tends to zero thereby effectively cancelling any noise generated by the noise source.

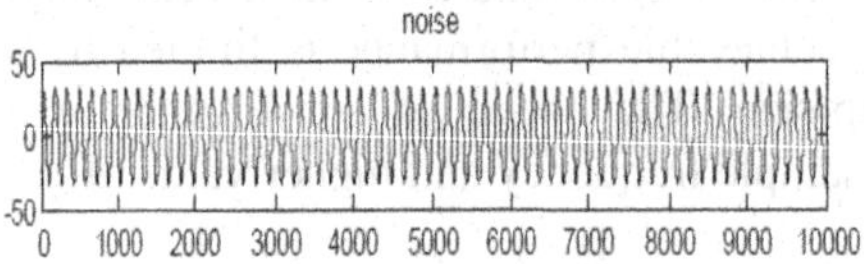

Fig. 1(a). Original noise

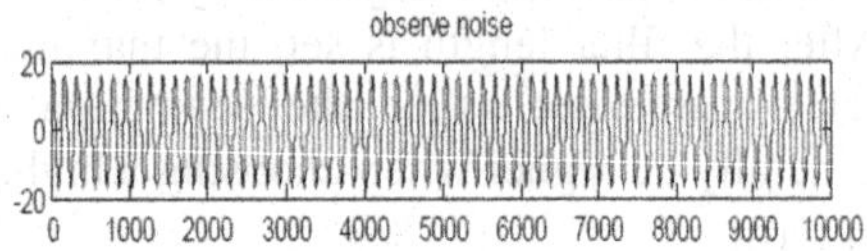

Fig. 1(b). Observed noise

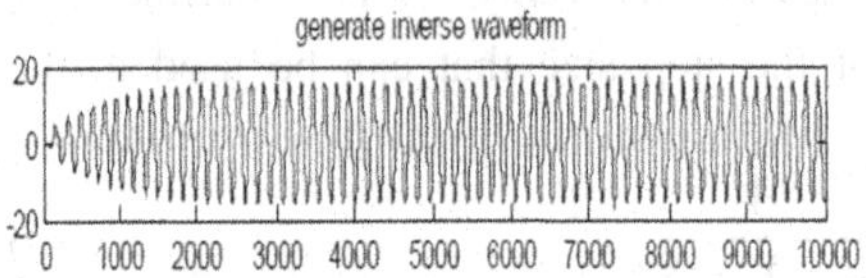

Fig. 1(c). Inverse noise generated by adaptive algorithm

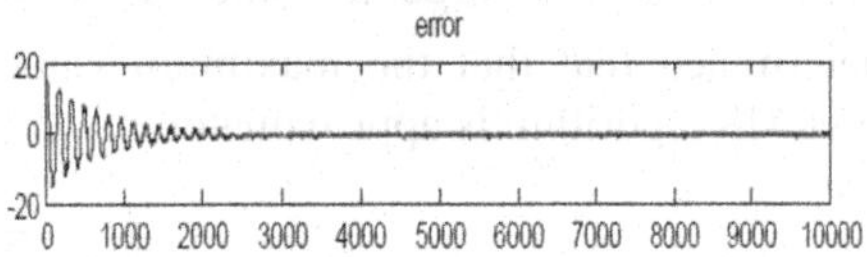

Fig. 1(d). Error signal

The primary and secondary path impulse response with respect to time is presented in Fig. 2 and Fig. 3 respectively. The first task in active noise control is to estimate the impulse response of the secondary propagation path. This is usually performed prior to noise control using a random signal played through the output loudspeaker while the unwanted noise is not present. Fig.4 shows the true estimation of secondary path impulse response with the variation in coefficient value for the true, estimated and error signal with respect to time. The performance of FXLMS algorithm for ANC can be judged from the discussed results pertaining to an adaptive framework.

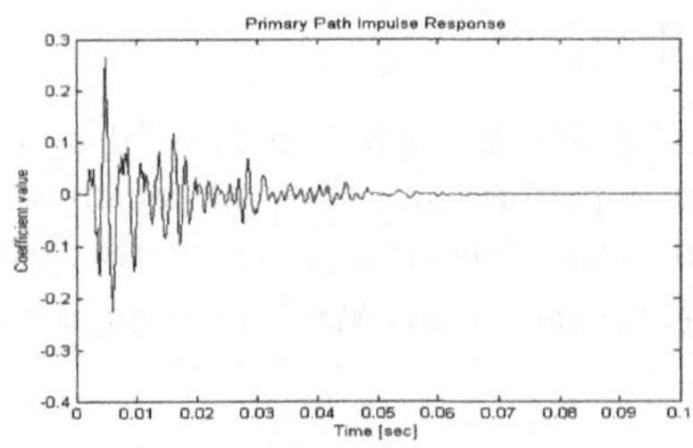

Fig. 2. Primary path impulse response

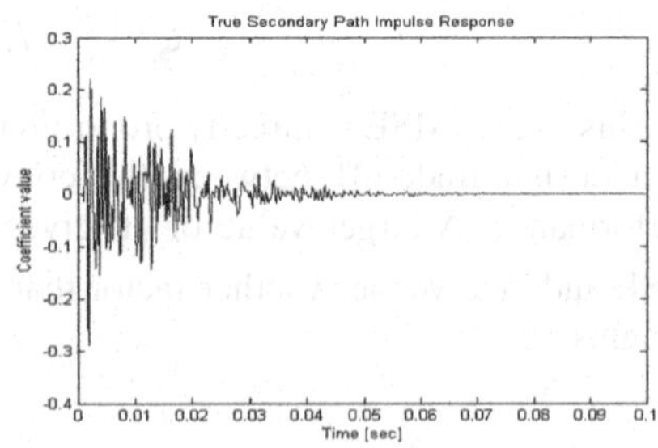

Fig. 3. True secondary path impulse response

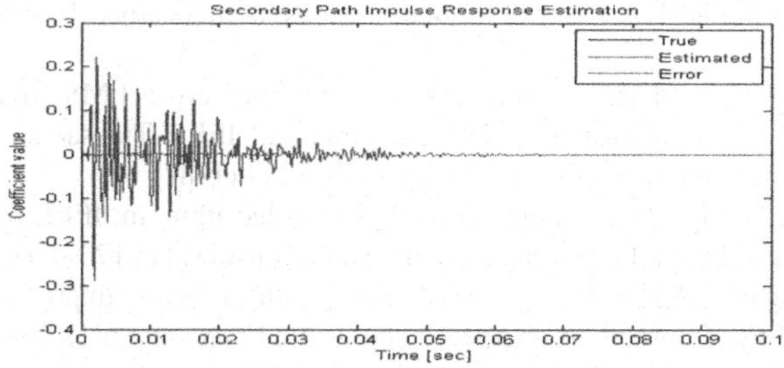

Fig. 4. True estimation of Secondary path impulse response

5 Conclusion

FXLMS Algorithm is widely used in acoustic noise cancellation environment due to its simple real time implementation & low computational complexity. Although the algorithm has slow convergence but various modifications in the existing circuit can be done to improve the convergence. FXLMS algorithm has greater stability than other algorithms used in case of acoustic noise cancellation as shown in the paper. There is a wide range of modification scopes available in case of FXLMS algorithm which can improve the convergence. FXLMS algorithm can also be used in very high noise environment.

References

1. Elliot, S.J.: Signal Processing for Active Control. Academic Press, London (2001)
2. Elliot, S.J., Nelson, P.A.: Active noise control. IEEE Signal Processing Magazine 10, 12–35 (1993)
3. Lueg, P.: Process of silencing sound oscillations. U.S. Patent, 2043416 (1936)
4. Kuo, S.M.: Morgan, D.R.: Active Noise Control Systems-Algorithms and DSP Implementations. Wiley (1996)
5. Kuo, S.M., Morgan, D.R.: Active Noise Control: A tutorial review. Proceedings of IEEE 87, 943–973 (1999)
6. Widrow, B., Stearns, S.D.: Adaptive Signal Processing. Prentice Hall, New Jersey (1985)
7. Morgan, D.R.: An analysis of multiple correlation cancellation loops with a filter in the auxiliary path. IEEE Transactions in Acoustics, Speech, and Signal Processing 28, 454–467 (1980)
8. Boucher, C.C., Elliot, S.J., Nelson, P.A.: Effects of errors in the plant model on the performance of algorithms for adaptive feedforward control. IEE Proceedings F138(4), 313–319 (1991)
9. Widrow, B., McCool, J.M., Larimore, M., Johnson, C.R.: Stationary and non-stationary learning characteristics of the LMS adaptive filter. IEEE Proceedings 64(8), 1151–1162 (1976)
10. Butterweck, H.: A wave theory of long adaptive filters. IEEE Transactions on Circuits and Systems I:Fundamental Theory and Applications 48(6), 739–747 (2001)

11. Warnaka, G.E., Poole, L.A., Tichy, J.: Active acoustic attenuators. U.S. Patent 4473906 (1984)
12. Elliott, S.J., Stothers, I.M., Nelson, P.A.: A multiple error LMS algorithm and its applications to active control of sound and vibration. IEEE Transactions on Acoustic, Speech and Signal Processing Processing 35, 1423–1434 (1987)
13. Rupp, M., Sayed, A.H.: Two variants of the FxLMS algorithm. In: IEEE ASSP Workshop on Applications of Signal Processing to Audio and Acoustics, pp. 123–126 (1995)
14. Rupp, M., Sayed, A.H.: Robust FxLMS algorithms with improved convergence performance. IEEE Transactions on Speech and Audio Processing 6(1), 78–85 (1998)
15. Davari, P., Hassanpour, H.: A variable step-size FxLMS algorithm for feedforward active noise control systems based on a new online secondary path modelling technique. In: IEEE/ACS International Conference on Computer Systems and Applications, pp. 74–81 (2008)
16. Kunchakoori, N., Routray, A., Das, D.: An energy function based fuzzy variable step size fxlms algorithm for active noise control. In: IEEE Region 10 and the Third International Conference on Industrial and Information Systems, pp. 1–7 (2008)
17. Eriksson, L., Allie, M., Melton, D., Popovich, S., Laak, T.: Fully adaptive generalized recursive control system for active acoustic attenuation. In: IEEE International Conference on Acoustics, Speech, and Signal Processing, vol. 2, pp. II/253–II/256 (1994)

A Unique Low Complexity Parameter Independent Adaptive Design for Echo Reduction

Pranab Das[1], Abhishek Deb[1], Asutosh Kar[1], and Mahesh Chandra[2]

[1] Department. of Electronics and Telecommunication Engineering
Indian Institute of Information Technology, Bhubaneswar, India
[2] Department of Electronics and Communication Engineering
Birla Institute of Technology, Mesra, India
{b210017,b210001,asutosh}@iiit-bh.ac.in,
shrotriya@bitmesra.ac.in

Abstract. Acoustic echo is one of the most important issues in full duplex communication. The original speech signal is distorted due to echo. For this adaptive filtering is used for echo suppression. In this paper our objective is to cancel out the acoustic echo in a sparse transmission channel. For this purpose many algorithms have been developed over the period of time, such as Least Mean Square (LMS), Normalized LMS (NLMS), Proportionate NLMS (PNLMS) and Improved PNLMS (IPNLMS) algorithm. Of all these algorithms we carry out a comparative analysis based on various performance parameters such as Echo Return Loss Enhancement, Mean Square Error and Normalized Projection Misalignment and find that for the sparse transmission channel all these algorithm are inefficient. Hence we propose a new algorithm modified - μ - PNLMS, which has the fastest steady state convergence and is the most stable among all the existing algorithms, this we show based on the simulation results obtained.

Keywords: Acoustic Echo, Adaptive Filter, Echo Return Loss Enhancement, LMS, Mean Square Error, Sparse Transmission Channel.

1 Introduction

Echo is a delayed and distorted version of the original signal. Acoustic echo is mainly present in mobile phones, Hands free phones, Teleconference or Hearing aid systems which is caused due to the coupling of microphone & loudspeaker. Echo largely depends on two parameters: amplitude & time delay of reflected waves. Usually we consider echoes with appreciable amplitude & larger delays of above 1 ms, but in certain cases if the generated echo is above 20 ms then it becomes a major issue and needs to be cancelled. Thus, developers are using the concepts of Digital Signal Processing for echo cancellation to stop the undesired feedback and allow successful full duplex communication. [1]

To achieve echo free systems numerous methods have been proposed like echo absorbers, echo barriers, echo cancellers to name a few. But considering the

M.K. Kundu et al. (eds.), *Advanced Computing, Networking and Informatics - Volume 1*,
Smart Innovation, Systems and Technologies 27,
DOI: 10.1007/978-3-319-07353-8_5, © Springer International Publishing Switzerland 2014

advancement of signal processing echo cancellation can be best done by the help of adaptive filtering. Adaptive filters are widely used due to its stability and wide scope for improvements. Adaptive filters can also be applied to low frequency noise. Echo can be suppressed in the most effective manner by continuous adaptation of the weights of the adaptive filter till the echo is completely cancelled out. System identification is generally used to generate a replica of the echo that is subtracted from the received signal. The echo canceller should have a fast convergence speed so that it can identify and track rapidly the changes in the unknown echo path. The convergence rate depends on the adaptive algorithm as well as the structure of adaptive filter used in AEC. In AEC, a signal is generated which is correlated to the echo signal but opposite in phase. By adding both the signals that is the signal corrupted by noise and the generated signal which is correlated and opposite phase to the actual echo signal, original signal can be made echo free. The generation of the signal is controlled by an adaptive algorithm which adaptively changes the weight of the filter used. AEC is classified into two categories i.e. "feed-forward AEC" and "feedback AEC". In feedback AEC method we use a controller to modify the response of the system. The controller may be like an addition of artificial damping. But in feed-forward AEC method a controller is used to adaptively calculate the signal that cancels the noise. In this paper "feed-forward AEC" approach is implemented to cancel the echo. This is because the amount of echo reduction achieved by feed-forward AEC system is more than that of feedback AEC system [2].

Further the nature of the transmission channel in AEC being sparse i.e. only few weight coefficients are active and all others are zero or tending to zero. The basic fundamental algorithms like LMS and NLMS suffer from slow convergence speed in these sparse channels. Hence a new modified algorithm was developed by Duttweiler named Proportionate NLMS for the sparse transmission channel systems [3]. The concept behind this is to assign each coefficient a step size proportionate to its estimated magnitude and to update each coefficient of the adaptive filter independently. But the main issue with the algorithm is that the convergence speed is reduced excessively after the initial fast period. A new kind of adaptive filtering process μ PNLMS was proposed to encounter the aforesaid problem. Here, the logarithm of the magnitude is used as the step gain of each coefficient instead of magnitude only. So this algorithm can consistently converge over a long period of time.

In this paper, we propose an algorithm to improve the performance of the MPNLMS algorithm to a greater extent in the sparse channel. The proposed algorithm adaptively detects the channel's sparseness and changes the step size parameter to improve the ERLE, Mean Square Error (MSE) and Normalized Projection Misalignment (NPM).

2 Existing Algorithms for AEC

A. LMS & NLMS:

It is one of the most widely used algorithms which has a weak convergence but is easy to use and is stable. It has two inputs one of which is the reference noise that is related with the noise that exits in the distorted input signal. The weight update equation for the LMS algorithm is given by:

$$\hat{h}(n+1) = \hat{h}(n) + 2x(n+1)e(\mathrm{n}+1) \tag{1}$$

In case of LMS algorithm if the step size is too small then the adaptive filter will take too much time to converge, and if it is too large the adaptive filter becomes unstable and its output diverges .So it fails to perform well in echo cancellation. The recursive formula for Normalized Least mean Square (NLMS) algorithm is

$$\hat{h}(n+1) = \hat{h}(n) + 2\mu \frac{x(n+1)e(\mathrm{n}+1)}{\| x(n+1) \|_2^2 + \delta_{NLMS}} \tag{2}$$

Here δ_{NLMS} is the variance of the input signal $x(n)$ which prevents division by zero during initialization stage when $x(n) = 0$ .Further to maintain stability the step size should be in the range $0 < \mu < 2\frac{E\{| \mathrm{x}(\mathrm{n}+1) |^2\}D(n+1)\}}{E\{| e(n+1) |^2\}}$.

Here $E\{| x(n) |^2\}$ is the power of input signal, $E\{| e(n) |^2\}$ is the power of error signal and $D(n)$ is the mean square deviation . The NLMS algorithm though being efficient does not take into consideration sparse impulse response caused to bulk delays in the path and hence needs to adapt a relatively long filter. Also unavoidable noise adaptation occurs at the inactive region of the filter. To avoid this we need to use sparse algorithms, where adaptive step size are calculated from the last estimate of the filter coefficients in such a way that the step size is proportional to step size of filter coefficients. So active coefficients converge faster than non-active ones and overall convergence time gets reduced. [4]

B. PNLMS

In order to track the sparseness measure faster PNLMS algorithm was developed from NLMS equation. Here the filter coefficient update equation is different from the NLMS in having a step size update matrix Q, with rest of the terms remaining same and is given below as:

$$\hat{h}(n+1) = \hat{h}(n) + \mu \frac{Q(n)x(n+1)e(\mathrm{n}+1)}{\| x(n+1) \|_2^2 + \delta_{PNLMS}} \tag{3}$$

Here, $\delta_{PNLMS} = \delta_{NLMS}/L$, and the diagonal matrix Q (n) is given by,

$$Q(n) = diag\{q_0(n), q_1(n)............q_{l-1}(n)\}$$

Here $q_l(n)$ is the control matrix and is given by ,

$$q_l(n) = \frac{k_l(n)}{\frac{1}{L}\sum_{i=0}^{L-1} k_i(n)} \tag{4}$$

Also $k_l(n) = \max\{\rho \times \max\{\gamma, |\hat{h}_0(n)|, \ldots |\hat{h}_{l-1}(n)|\}, |\hat{h}_l(n)|\}$

Here ρ and λ have values 5/L and 0.01 respectively. ρ prevents coefficients from stalling when they are smaller than the largest coefficient and also prevents $\hat{h}_l(n)$ from stalling during the initialization stage. [5]

C. IPNLMS:

It is an improvement over the PNLMS algorithm and employs a combination of proportionate (PNLMS) and non-proportionate (NLMS) updating technique, with the relative significance of each controlled by a factor α . Thus we have value of $k_l(n)$ as

$$k_l(n) = \frac{1-\alpha}{2L} + (1+\alpha)\frac{|h_l(n)|}{2\|h(n)\|_1 + \varepsilon} \tag{5}$$

here ε is a small positive constant .Also results so that good choice for α are 0.-0.5,-0.75. The regularization parameter should be taken such so that same steady state misalignment is achieved compared to that of NLMS using same step size. [6] We have

$$\delta_{IPNLMS} = \frac{1-\alpha}{2L}\delta_{NLMS} \tag{6}$$

D. MPNLMS:

This algorithm is an efficient one for a sparse transmission channel and has a steady convergence over a period of time. Unlike the previously discussed proportionate algorithms which have a faster convergence during the initial period but slows down over a period of time. In this algorithm we calculate the step size proportionate to the logarithmic magnitude of the filter coefficients. [7]

3 Proposed Algorithm

Among all the algorithms discussed above for a sparse channel, which is our transmission channel of interest, the μ-PNLMS algorithm has the fastest convergence. The existing classical NLMS algorithm's filter weight coefficients converge slowly in such a channel. Hence PNLMS and IPNLMS algorithms were designed specifically for echo cancellation in sparse transmission channels, but after initial fast convergence of filter weight vectors, these algorithms too fail.[6]-[7] So we propose μ-law PNLMS algorithm for a sparse channel. Here instead of using filter weight magnitudes directly, its logarithm is used as step gain of each coefficient. Hence μ-PNLMS algorithm can converge to a steady state effectively for a sparse channel. This algorithm calculates the optimal proportionate weight size in order to achieve fastest convergence during the whole adaptation process until the adaptive filter reaches its steady state.

Implementation:

Step 1: The weight update equation for the proposed algorithm is given as:

$$\hat{h}(n+1) = \hat{h}(n) + \mu \frac{Q(n)x(n+1)e(n+1)}{\| x(n+1) \|_2^2 + \delta_{MPNLMS}} \tag{7}$$

where;

$$\delta_{MPNLMS} = \delta_{NLMS} / L \tag{8}$$

Step 2: The control matrix $q_l(n)$ which is the matrix containing the diagonal elements of the $Q(n)$ is given as,

$$q_l(n) = \frac{k_l(n)}{\frac{1}{L}\sum_{i=0}^{L-1} k_i(n)} \tag{9}$$

Step 3: Finally the value of $k_l(n)$, which is the main differentiating factor in case of the proposed algorithm is given as:

$$\mathrm{k}_l(\mathrm{n}) = \max\{\rho \times \max\{\gamma, \mathrm{F}(|\hat{\mathrm{h}}_0(\mathrm{n})|) \ldots\ldots\ldots \mathrm{F}(|\hat{\mathrm{h}}_{l-1}(\mathrm{n})|)\}, F(|\hat{\mathrm{h}}_l(n)|)\} \tag{10}$$

Here,

$$F(|\hat{\mathrm{h}}_l(n)|) = \frac{\ln(1+\mu|\hat{h}_l(n)|)}{\ln(1+\mu)}, \tag{11}$$

$|\hat{\mathrm{h}}_L(n)| \leq 1$ and $\mu = 1/\varepsilon$. Constant 1 is taken inside the algorithm to avoid infinity value when the value of $|\hat{\mathrm{h}}_l(n)| = 0$ initially. The denominator $\ln(1+\mu)$ normalizes $F(|\hat{\mathrm{h}}_L(n)|)$ in the range [0, 1].The value of ε should be chosen based on the noise level and usually it is taken as 0.001.

4 Results and Discussion

MATLAB platform is chosen for simulation purpose. The room impulse response is measured for an enclosure of dimension 12×12×8 ft^3 using an ordinary loudspeaker and an uni-dynamic microphone kept at a distance of 1 ft. which has been sampled at 8 KHZ. The channel response is obtained by applying a stationary Gaussian stochastic signal with zero mean and unit variance as input on the measured impulse response.

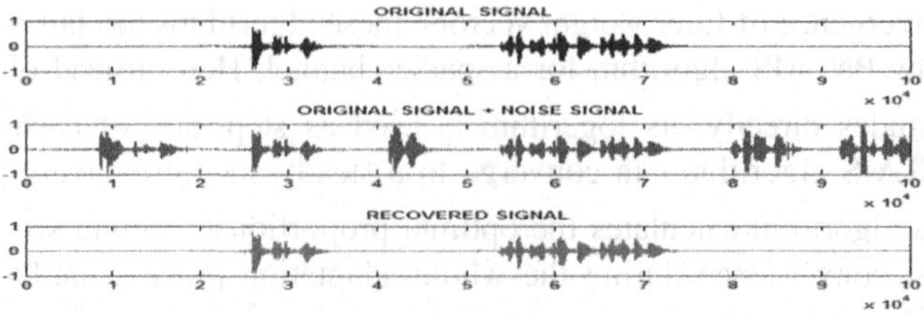

Fig. 1. AEC using proposed Algorithm

Fig. 1 shows the MATLAB simulation of acoustic echo cancellation using the proposed algorithm. Fig.1 has three separate graphs where the first is the original noise present in the system ,the second one is the graph of original signal along with the noise signal and the third graph is the original signal recovered using the proposed algorithm.

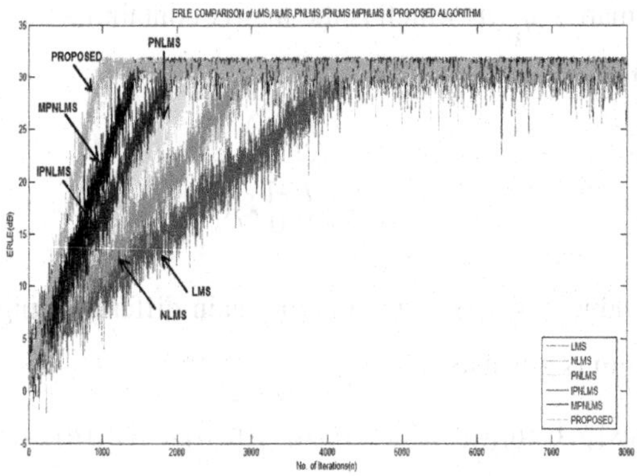

Fig. 2. ERLE Comparison of existing and proposed algorithms

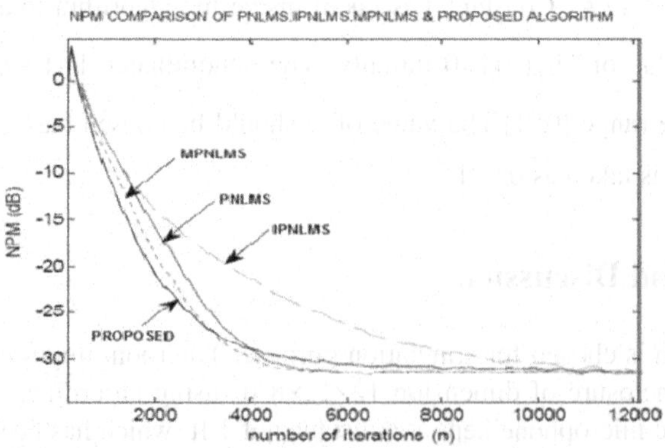

Fig. 3. NPM comparison of existing and proposed algorithms

In Fig. 2 we show the ERLE comparison of PNLMS, IPNLMS, μ – PNLMS and the proposed algorithm, the result clearly shows that the ERLE for the proposed algorithm is maximum. Further in Fig. 3 we show the comparative analysis of NPM of PNLMS, IPNLMS, μ – PNLMS and the proposed algorithm which clearly indicates that has the least value for our proposed algorithm.

The analysis is carried out by taking a signal $x(n)$ as an input to the various adaptive algorithms and the desired signal is taken as $d(n)$ .The error signal thus generated is minimized continuously in an adaptive manner by setting the step size to an optimum value. Our aim is to obtain the quickest steady state convergence for minimizing the error or the acoustic echo, which is obtained using our proposed algorithm. This can be shown through the comparison of ERLE and NPM for all the algorithms including our proposed algorithm. And based on our simulation results we obtain that our proposed algorithm has the maximum ERLE and the least NPM value among all the algorithms such as PNLMS, μ – PNLMS and IPNLMS.

5 Conclusion

In this paper we do a detailed study of the existing algorithms for acoustic echo cancellation and find out that only proportionate algorithms are capable of acoustic echo cancellation in sparse channel. But among the existing proportionate algorithms PNLMS, IPNLMS though have initial fast convergence they fail to adapt as time progresses [7]. And only μ – PNLMS algorithm has a good steady state convergence. Now to further improve the performance of μ – PNLMS algorithm we propose our own algorithm which we have shown to have the most efficient steady state convergence for transmission channels having sparse impulse response. This we ascertain based on our simulation results, where we take into consideration various performance measurement parameters like ERLE and NPM.We have obtained that our algorithm gives the maximum ERLE amongst all the existing algorithm and also the least NPM value. So we can safely conclude that our proposed algorithm is the most efficient one for acoustic echo cancellation in a sparse medium.

References

1. Verhoeckx, N.A.M.: Digital echo cancellation for base-band data transmission. IEEE Trans. Acoustic, Speech, Signal Processing 27(6), 768–781 (1979)
2. Haykin, S.: Adaptive Filter Theory, 2nd edn. Prentice-Hall Inc., New Jersey (1991)
3. Khong, A.W.H., Naylor, P.A., Benesty, J.: A low delay and fast converging improved proportionate algorithm for sparse system identification. EURASIP Journal of Audio Speech Music Processing 2007(1) (2007)
4. Diniz, P.S.R.: Adaptive Filtering, Algorithms and Practical Implementation. Kluwer Academic Publishers, Boston (1997)

5. Wang, X., Shen, T., Wang, W.: An Approach for Echo Cancellation System Based on Improved NLMS Algorithm. School of Information Science and Technology, pp. 2853–2856. Beijing Institute of Technology, China (2007)
6. Paleologu, C., Benesty, J., Ciochin, S.: An Improved Proportionate NLMS Algorithm based on the Norm. In: International Conference on Acoustics, Speech and Signal Processing, pp. 309–312 (2010)
7. Deng, H., Doroslovachi, M.: Proportionate adaptive algorithms for network echo cancellation. IEEE Trans. Signal Processing 54(5), 1794–1803 (2006)

On the Dissimilarity of Orthogonal Least Squares and Orthogonal Matching Pursuit Compressive Sensing Reconstruction

Arvinder Kaur and Sumit Budhiraja

Department of Electronics and Communications Engineering,
UIET, Panjab University, Chandigarh, India
sharma.arri@gmail.com, sumit@pu.ac.in

Abstract. Compressive sensing is a recent technique in the field of signal processing that aims to recover signals or images from half samples that were used by Shannon Nyquist theorem of reconstruction. For recovery using compressed sensing, two well known greedy algorithms are used- Orthogonal matching pursuit and orthogonal least squares. Generally these two algorithms are taken as same by the researchers which is not true. There is a remarkable difference between the two algorithms that is pointed out in this paper with the simulation results. The previous article clarifying the difference between these two algorithms are emphasized on theoretical difference and does not show any reconstruction simulation difference with these two algorithms and reason to preference over basis pursuit method . The key aim of this paper is to remove the confusion between the two algorithms on the basis of theory and reconstruction time taken with the output PSNR.

Keywords: Compressive sensing, Reconstruction, Orthogonal least squares, orthogonal matching pursuit.

1 Introduction

Compressed sensing is an innovative theory in the field of signal processing that aims to reconstruct signals and images from what was previously treated as incomplete information. It uses relatively few measurements than those were needed in the Shannon Nyquist theory of signal reconstruction. This theory comes from the fact that most of the signal information is carried by only few of coefficients and all other are get wasted while reconstructing that signal [1]. Then it gives a way to collect only those components in the first stage that adds up in signal recovery. The practical need of this study comes with the shortage of storage space for increasing data transferred from here to there.

Compressed sensing has been used in many fields like medical imaging, Radar, Astronomy, speech processing where the task is to reconstruct a signal. For example, in medical imaging CT (computerized Tomography), by exposing x-rays one can generate image of inside of human body. For the complete scanning process, patient is

M.K. Kundu et al. (eds.), *Advanced Computing, Networking and Informatics - Volume 1*,
Smart Innovation, Systems and Technologies 27,
DOI: 10.1007/978-3-319-07353-8_6,

exposed to radiations for a large span of time, which are harmful for him. By using compressive sensing, only the desired samples are to be taken, so scanning time reduces to a great extent by decreasing the number of samples to be taken. The mechanism of reconstruction using sparse matrix is shown in Fig. 1.

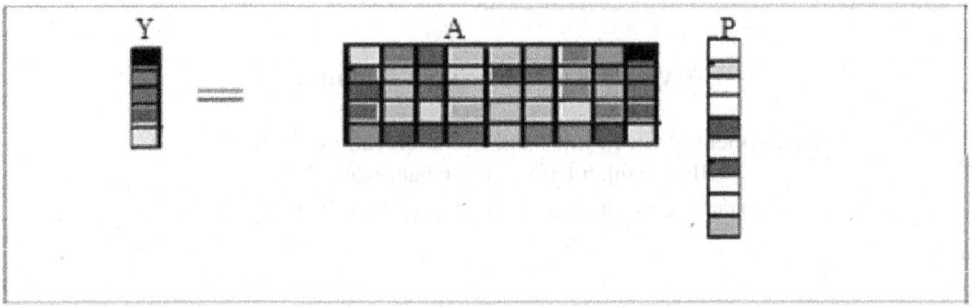

Fig. 1. Compressed sensing framework

Where Y is the reconstructed signal and A' is the random measurement matrix. Gaussian and Bernoulli matrices are mostly used as random measurement matrices. For Gaussian matrices the elements are chosen randomly as independent and identically distributed. These elements can have the variance as from [3]

$$\text{Variance}=1/K$$

For the Bernoulli matrices [3] with equal probability each element takes value of $1/\sqrt{K}$ or $+1/\sqrt{K}$

For reconstruction using compressed sensing two type of algorithm are used basis pursuit and greedy algorithms. Basis pursuit algorithms are quite simpler to implement as compared to greedy pursuit. The greedy algorithms generally used are orthogonal matching pursuit and orthogonal least squares give faster recovery. Generally these two algorithms are taken as same but there is certain difference in them [8]. For the greedy algorithms, the projection on selected signal elements is orthogonal. OLS selects the entity that minimizes the residual norm after signal's projection onto the selected entities [4]. It has been proved by Thomas Blumensath and Mike E. Davies in [4]. This particular entry is the one which best approximates the current residual. This marks the difference of OLS from OMP. Generally OLS is little bit more complex than OMP, but due to the same output results given by two algorithms, both are taken as same by the researchers. There had been previous work on clarification between two algorithms, but most of it is based on theoretical concepts. In this paper the confusion on the similarity part of both algorithms is removed by showing the simulation results of both the algorithms for same results. The reconstruction time taken by both is also considered.

2 Orthogonal Matching Pursuit

Tropp and Gilbert had given an idea for orthogonal matching pursuit algorithm for reconstruction with less number of measurements [5-6]. It has been proved in the literature by Huang and Rebollo that reconstruction time can be reduced to a great extent by using this particular algorithm [13], [15]. Suppose that the column vector of A are normalized such that

$$\| A_i \|_2=1, \text{ for } i=1,2,\ldots k$$

A(x) be a subset of A for x C {1,2,..,k}. To start the algorithm, signal support is needed to be calculate from pseudo inverse A\` of the measurement matrices A, as

$$P= (A)^{`}Y$$

Where A' is the complex conjugate of measurement matrices A and we know that

$$A^{`}= (A^{*}A)^{-1}A^{*}$$

During the implementation a matching operation is performed between the matrix A entries and the calculated residuals of P [2, 14]. At the end of all iterations, the complete signal is generated. The implemented algorithm includes following steps

1. Initialize the residual of Y as $s_0 = y$
2. Initialize the set of entries that are to be selected as A(c)= null
3. Start an iteration counter. Let i=1
4. Generate an estimate of given signal per iteration.
5. Solve the maximization problem

   $$\max| A_j ,\ s_{i-1} |$$

   In which A_j is any variable.
6. To A(c), the set of selected variables, add A_j, and update c after every result.
7. It is to be noted that the algorithm does not minimize the residual error after its orthogonal projection to the values.
8. Update result after the projection. The entry selected have a minimum error as

   $$r^n=S- S^n$$
9. One coordinate for signal support for P is calculated at the end after computing the residual.
10. Set i=i+1, to end the iterations of algorithm.

3 Orthogonal Least Squares

On the contrary to matching pursuit algorithm, there is another method in the literature based on least squares [7], [9]. In OLS, after the projection into the orthogonal subspace, OLS selects the element with smallest angle [4]. This element best approximates the current residual. For the least squares method, a vector is needed to be calculated of few non-zero elements such that the squared error is small [4]. That's why the name least squares is there. The vector calculated must be with dimensions

$$P \in D_K \text{ and}$$

$$A \in D^{Np * Ny}$$

An approximation to the signal P is calculated at the end of an iteration and result is updated after every sequence. The complete procedure goes same for least squares algorithm [12], [14]. Initially a counter is started and a residual is assumed and residual is updated after every projection result .This algorithm selects the entry having a minimum error from the sub matrix [4]. Just after the orthogonal projection step, the algorithm minimizes the residual error. The approximation error is given by

$$e^n = P - P^n$$

After the projection onto the signal entries, calculate the maximization problem as

$$j_{max} = \|A - (A^*_i A_i)^{-1} A\|_2$$

Solving A_i^{-1} is an inverse problem here. The pseudo inverse A_i^{-1} of the random measurement matrices is taken from

$$P_n = A_i^{-1} A$$

Thus the signal is reconstructed from the inverse problem solution [10].

4 Simulation Results and Analysis

The focus of this paper is to highlight difference between the two mentioned greedy algorithms. An estimate for matrix A is calculated and its columns are used to calculate the inner products required to select elements from signal. When the number of elements to be selected is quite smaller than the number of column vectors of matrix A; OLS is complex because all elements need to be orthogonalized [4].

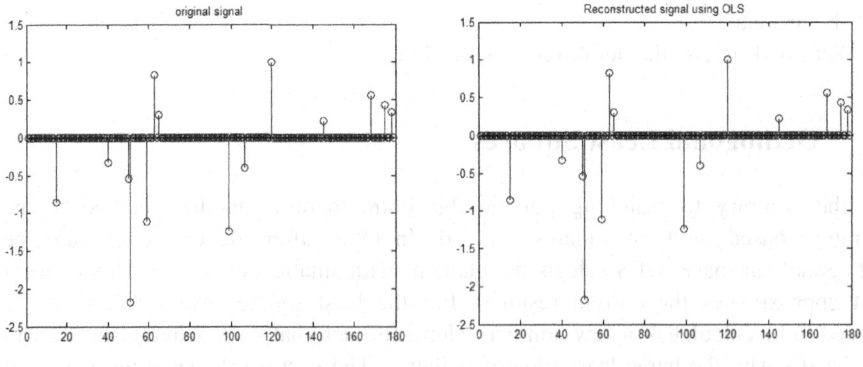

Fig. 2. (a) Original randomly generated signal (b) original signal reconstructed using Orthogonal Least Squares

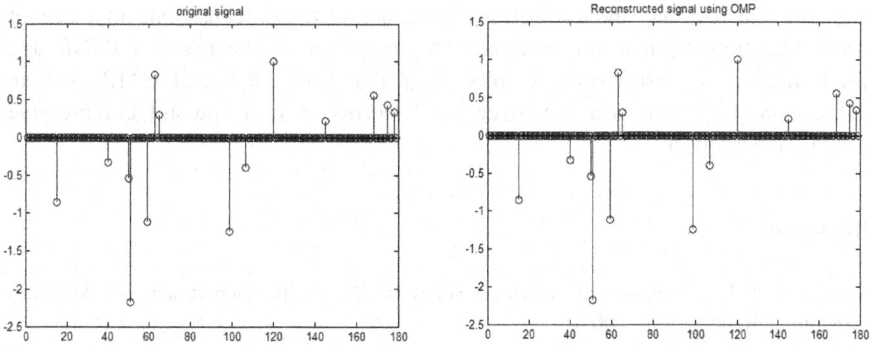

Fig. 3. (a) Original randomly generated signal (b) original signal reconstructed using Orthogonal Matching Pursuit

It has been proved previously that the greedy algorithms are faster than the basis pursuit compressive sensing reconstruction. This is shown in Fig.2(a),(b) and Fig.3(a),(b) A comparison of elapsed time for reconstruction and for PSNR for the output signal is shown in Table 1. A comparison of both the algorithms is done with Justin Romberg's Basis pursuit's algorithm [11].

Table 1. Comparison of elapsed time and PSNR for both the algorithms

Algorithm	Elapsed Time(s)	PSNR(db)
OLS	0.037765	83.5091
OMP	0.052453	83.5217
BP	0.731953	81.7943

In Table 1, the reconstruction time is given in seconds and is calculated from the beginning of the generation of sparse matrix up to the generation of output signal. A slight difference in PSNR is observed for the same signal recovery by using the two algorithms, which marks the difference between the two algorithms, i.e. proves that both are not same. The reconstruction results are obtained by implementing the greedy algorithms in a multi stage manner, with less number of dimensions. The PSNR obtained is higher than the L-norms or the basis pursuit method of compressive sensing reconstruction. It is clear from the table that greedy algorithms are faster than the basis pursuit algorithm.

5 Conclusions

The results from this paper demonstrate that both OLS and OMP algorithms gives same recovery results but there is a significant difference between the two in terms of minimizing the residual and projections. It is not justified to use one in place of another. The simulation results obtained are compared on the basis of PSNR and elapsed time of reconstruction. Results show that both OLS and OMP give an impressive way of reconstruction over the L-norms or basis pursuit compressive sensing reconstruction.

References

1. Donoho, D.L.: Compressed sensing. Stanford University Department of Statistics Technical Report (2004-2005)
2. Fornasier, M., Rauhut, H.: Compressive sensing. IEEE Transactions on Information Theory (2010)
3. Donoho, D.: Compressed sensing. IEEE Transactions on Information Theory 52(4), 1289–1306 (2006)
4. Blumensath, T., Davies, M.E.: On the difference between orthogonal matching pursuit and orthogonal least squares (2007)
5. Gillbert, T.: Signal recovery from random measurements via orthogonal matching pursuit. IEEE Transactions on Information Theory 53(12) (2007)
6. Beck, T.M.: Fast gradient based algorithms for constrained total variation image denoising and deblurring problems. IEEE Transactions on Image Processing 18, 2419–2434
7. Ehler, M., Fornasier, M., Sigl, J.: Quasi-Linear Compressed Sensing
8. Soussen, C., Gribnovel, R.: Joint k-step analysis of orthogonal matching pursuit and orthogonal least squares. IEEE Transactions on Information Theory 59(5)
9. Vaswani, N.: LS-CS-Residual (LS-CS): Compressive Sensing on Least Squares Residual. IEEE Transactions on Signal Processing 58(8) (2010)
10. Vehkapera, M., Kabashima, Y., Chatterjee, S.: Analysis of Regularized LS Reconstruction and Random Matrix Ensembles in Compressed Sensing
11. Candes, E., Romberg, J.: Sparsity and incoherence in compressive sampling. Inverse Problems 23(3), 969–985 (2007)
12. Gribonval, S., Herzet, I.: Sparse recovery conditions for orthogonal least squares. In: HAL 2 (2011)
13. Gharavi, H.T.S.: A fast orthogonal matching pursuit algorithm. In: IEEE International Conference on Acoustics, Speech and Signal Processing, vol. 3 (1998)
14. Kaur, A., Budhiraja, S.: In: Sparse signal reconstruction via orthogonal least squares. In: IEEE International Conference on ACCT (2014)
15. Rebollo-Neira, L., Lowe, D.: Optimized orthogonal matching pursuit approach. IEEE Signal Processing Letters 9(4) (2002)

Machine Learning Based Shape Classification Using Tactile Sensor Array

Dennis Babu, Sourodeep Bhattacharjee,
Irin Bandyopadhyaya, and Joydeb Roychowdhury

CSIR-CMERI, Mahatma Gandhi Avenue, Durgapur-713209, India
{denniskanjirappally,sourodeepbhattacharjee,
mom.tanuja}@gmail.com, jrc@cmeri.res.in

Abstract. Contact shape recognition is an important functionality of any tactile sensory system as it can be used to classify the object in contact with the tactile sensor. In this work we propose and implement an affine transformation invariant 2D contact shape classification system. A tactile sensor array gives the contact pressure data which is fed to an image enhancement system which sends the binary contact image to the global, region and contour based feature selection block. The five element feature vectors is used as input to a voting based classifier which implements a C4.5 algorithm and a naïve Bayes classifier and combine the results to get an improved classifier. The system is designed and tested offline in a Pressure Profile system based tactile array using both cross validation and separate dataset. Results indicate that combining simple classifiers increases the accuracy of the system while being computationally efficient.

Keywords: Tactile, Machine learning, contact-shape, classification, voting classifier.

1 Introduction

Humans interact with and explore the environment through five main senses viz. touch, vision, hearing, olfaction and taste. Exploiting one or a combination of these senses, humans discover new and unstructured environments. Touch or tactile sensors are an efficient way to replicate the human touch sensation which includes different parameters such as texture, shape, softness, shear and normal forces etc. Development of the artificial touch sensation system can broadly be categorized into four parts - Development of contact shape and size detection system, Development of hardness/rigidity analysis system, Development of texture analysis system and bio-mimicking of human exploration. This paper deals with the first part of tactile 3D data mapping i.e. shape classification using tactile sensor array by machine learning algorithms.

2 Related Work

A comprehensive review of the tactile sensing and its potential applications was done by Tiwana Mohsin *et al.* [1] in 2012 which covered the significant works in Tactile

M.K. Kundu et al. (eds.), *Advanced Computing, Networking and Informatics - Volume 1*,
Smart Innovation, Systems and Technologies 27,
DOI: 10.1007/978-3-319-07353-8_7, © Springer International Publishing Switzerland 2014

sensing based system design. Several results have been reported in the literature for contact shape classification. Primitive shape classification using tactile sensor array has been implemented with naïve Bayes [2] in the domain of Robotics in which the authors extracted structural nature of the object from its pressure map using covariance between pressure values and the coordinates of the map in which the values occur. In [2], Liu *et al.* used neural networks to classify shapes based on 512 feature vector derived from the pressure map [3]. Further works reported by Tapamayukh *et al.* [4] and Pezzementi *et al.* [5] discusses a robotic gripper which used tactile contact shape as a feature for object classification whereas in [6] recognition and localization of 2D shapes with minimal tactile data is analysed in an analytical perspective. Seiji Aoyoga [7] used a neural network model for shape classification into 8 shape categories with the data from arrayed capacitive tactile sensor. The issues of tactile array development and shape estimation methodologies are detailed in the survey conducted by Dahiya *et al.* [8].

Shape classification using machine learning techniques requires a set of geometric features to be extracted and transformed (into Fourier descriptors, Autoregressive models, Features) before feeding the training data to the machine learning algorithm [9], to learn relevant distinguishing features of a shape rather than taking the entire set of pixel intensities. The techniques employed to represent and classify shapes can be broadly divided into contour-based or region-based [10], depending on whether the attributes related to shape was derived from its boundary or the entire area covered by the shape. Contour based and region based techniques are further sub-divided into global and structural techniques, based on whether the entire contour or region was taken into account or subsections ("segments") of the same were used.

In this work we focus on the challenges posed when two dimensional pressure profile data needs to processed as a digital image (including removal of outliers and applying appropriate thresholding to convert the pressure values to a black and white binary image) to identify shapes that lie under the sensor area. The processed digital image is further classified using a learning system in order to cater for irregularities associated the translation and rotation. The rest of the paper is organised as follows Section 3 elaborates the methodology of approach, Section 4 details the experimental setup. Results and interpretations are analysed in Section 5 and finally the paper concludes in Section 6.

3 Methodology

The architecture of the system uses a 2 D capacitive tactile array with 256 elements arranged in a square shape. The signal excitation from the sensors are fed to a signal processing block which simultaneously processes the output signals and converts it to analog signal of 0-5V signal level with a subsequent analog to digital converter which converts the analog signal to 8 bit digital data for further data processing. The tactile array used for the proposed system was developed by Pressure profile systems with pressure range of 0-60kPa having minimum sensitivity of 0.8 kPa. The 16×16 tactile array data is further used for contact shape recognition and classification. The block schematic of the methodology is shown in Fig. 1.

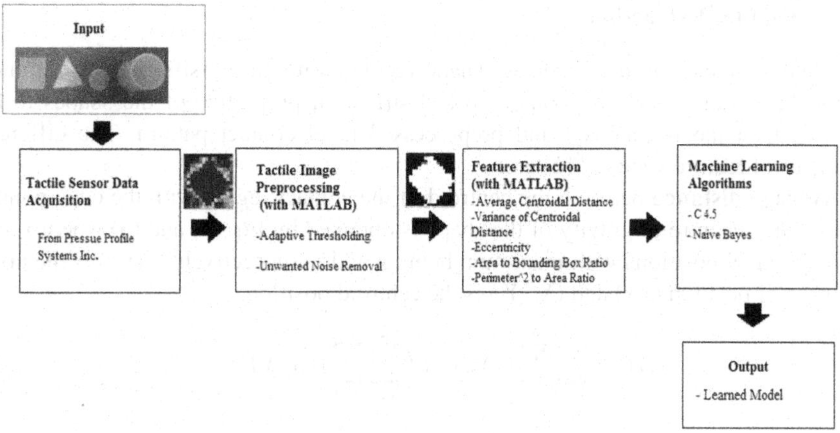

Fig. 1. Block diagram of the methodology

3.1 Tactile Image Enhancement

The tactile image processing block samples the signals at 8 frames per second and averages the sampled signals to get the averaged signal. This averaging is essential as tactile data unlike the vision data is prone to fluctuations as the tiny release or grabbing can cause pressure variations which alter the actual contact shape. The intermediate tactile image processing block performs the following operations before feeding the data to a voting machine learning unit for shape classification into basic structural shapes such as circular, square, triangular and ring.

The uneven tactile sensor performance causes a touch sensation even outside the area of contact. Thus a 3*3 matrix is used to convolve the 2D -256 point tactile data to avoid uneven signal excitation [1]. The convolving matrix used for enhancement is given below:

$$\begin{pmatrix} 0 & -1 & 0 \\ -1 & 4 & -1 \\ 0 & -1 & 0 \end{pmatrix} \tag{1}$$

The average taxel value at each point of the array with and without contact object is calculated and the difference in pressure variations between touched and untouched parts are used for obtaining a thresholding limit for shape classification. The low repeatability and response of the tactile sensor for slight touch causes uneven patches of high pressure regions in the taxel data outside the area of interest which is addressed in the signal processing block using morphological operations to remove the noise data. The image erosion followed by dilation using a disc structural element with radius one unit removes noise occurred due to non-uniform pressure data [11]. The areas of the high pressure regions are found and those thresholded taxels with maximum area in a taxel array is only accepted for feature extraction in each frame for shape recognition.

3.2 Feature Extraction

The feature vector used for contact shape recognition and classification is obtained using the region based and contour based structural properties of the shapes to be classified on the normalized and preprocessed taxel contact pattern. The different elements of feature vectors [12] are:

Average distance between the centroid of the contact region with the edge points. The centroid/centre of gravity of the shape is obtained by Eq. (2) and (3) where x and y are the taxel positions with maximum being n and m respectively , $I(x,y)$ is the taxel intensity at position (x,y) and (X_c ,Y_c) is the centroid position.

$$Xc = \sum_{x=1}^{n}\sum_{y=1}^{m} I(x,y)x \Big/ \sum_{x=1}^{n}\sum_{y=1}^{m} I(x,y) \tag{2}$$

$$Yc = \sum_{x=1}^{n}\sum_{y=1}^{m} I(x,y)y \Big/ \sum_{x=1}^{n}\sum_{y=1}^{m} I(x,y) \tag{3}$$

and the average centroidal distance is obtained by equation (4) where (X_i,Y_i) is the location of the pixels across the shape contour.

$$Avgdist = \sum_{i=1}^{c} \sqrt{((Xc - Xi)^2 + (Yc - Yi)^2)} \tag{4}$$

For different shapes the variations of the contour edges with the average radius gives us the second moment of the contact shape. The **average and maximum variance** of the contour points with the average radial distance can be used as a structural feature for object classification which is used for analytical distinguishing given by the following equations.

$$P\,\mathrm{var} = \sum_{i=1}^{c} (Avgdist - \sqrt{(Xc - Xi)^2 + (Yc - Yi)^2)}) \tag{5}$$

$$Pf\,\mathrm{max}\,fluct = MAX\,(Avgdist - \sqrt{(Xc - Xi)^2 + (Yc - Yi)^2)}) \tag{6}$$

The **eccentricity** of the foreground image-the ratio between the length of major and minor axes invariant of the rotation and translation of the contact shape.

Ratio of actual area of the contact taxels to the bounding box area is another parameter of the feature vector. The bounding box is the minimum rectangular region that covers the entire contact region thus has max value for square and the least being for ring.

The ratio of **perimeter^2 to that of area** gives the information on the solidity of the contact surface and thereby gives a region specific parameter rather than a contour based approach [13].

3.3 Machine Learning System

In this work we implement and compare the following machine learning models for accurately classifying between the primitive models.For our experiments we have used decision tree (J48), naïve Bayes and combined voting of J48 and naive Bayes strategies to classify the shapes [14]. The reason for choosing these algorithms is that they are relatively easy to analyse and should be sufficient to solve the problem of classifying regular shapes, given the fact that we have already identified and extracted the translation, rotation and scale invariant features of the shape.

Decision trees are logic based classifiers and work by sorting features values to classify instances. Nodes in trees represent various features of the instance and the edges of the tree denote the classes of values the nodes can assume. The root for every sub-tree is chosen such that features divide the data in best possible way. On the other hand Naïve Bayes networks is a statistical learning algorithm which generates directed acyclic graphs such that there is only one parent (the feature to be classified) and several children consisting of observed features. It is assumed here that the child nodes are not dependent on one another [15].

There are various strategies to combine classifiers to achieve better results; Voting being one of them [16]. In this paper we have used a very simple voting mechanism that combines the predictions of two or more classifiers using average of probabilities. The final prediction of the voting classifier is calculated at run time after the models of candidate algorithms have been generated.

4 Experimental Set Up

The basic shapes into which the objects are classified are square, triangle, circle and ring. The objects are carved out of wood. During data collection, the objects are placed in different orientations and positions on the tactile sensor and 200 gm weight is placed on them. Since we are interested in classifying two dimensional shapes only, we have placed the sensor array on a flat wooden table (as opposed to placing the sensor array on a sponge to get surface contour information). Appropriate class labels are attached to each record in the processed data derived after image processing. We have collected completely separate records with same objects (varying orientation and position) for training and testing. In this paper we have reported the percentage of correctly classified instances, model building time and the confusion matrices for the separate test cases. All experiments were performed on a Pentium 4 processor 2.7GHz with 1 GB RAM. Fig. 2 (a) shows the conformable tactile array sensor from Pressure Profile system while Fig. 2 (b) gives the wooden pieces of sample shapes with different sizes used for the experiment. WEKA learning system [17] is used for the initial training and testing before real time implementation.

Fig. 2.(a) Conformable tactile sensor array;

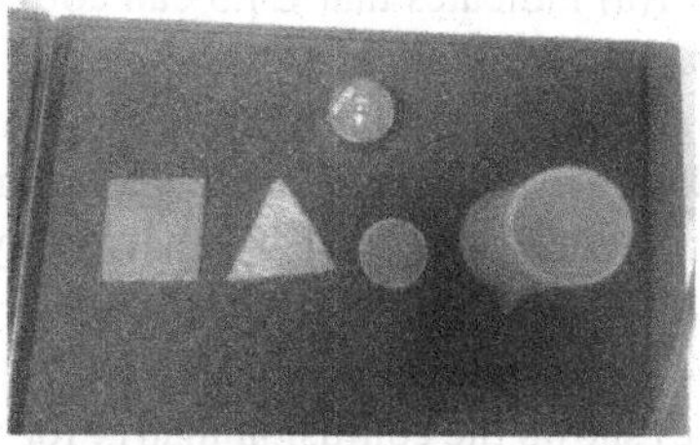

Fig. 2.(b) Sample shapes and weight used for experimentation

5 Result and Discussion

The raw tactile data obtained from tactile array is fed to the Mathworks environment for image enhancement and feature extraction which is further fed to the WEKA with corresponding class of shape. Fig. 3 shows the different contact patterns for different objects at different stages of image enhancement before feature extraction.

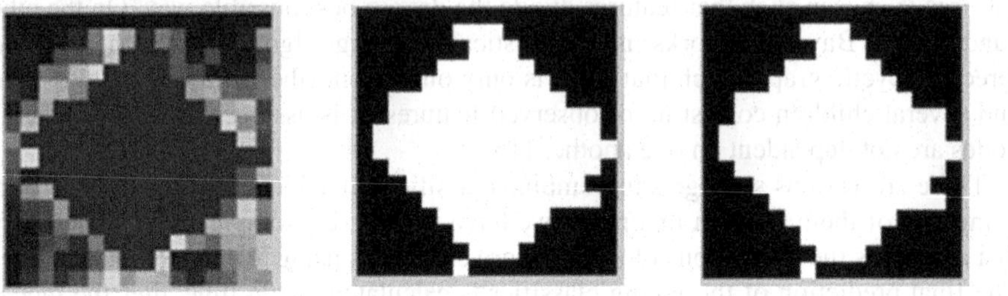

Fig. 3. (a) 8 bit tactile image of square object; (b) Tactile image after convolution and thresholding ; (c) Final image after morphological operations

The enhanced image gives an eroded and dilated image with the single largest connected element in tactile image In this section we have reported the percentage of correctly classified instances, relative absolute error and confusion matrix for C4.5, naive Bayes and Voting for 66% split (of entire data set for training and rest for testing) as well as with 10 fold cross-validation. The Table 1 shows the confusion matrix for C4.5 algorithm for classifying square (a), triangle (b), circle (c) and ring (d).

Table 1. Confusion Matrix for separate test dataset for (a) C4.5 classifier (b) Naive Bayes classifier (c) voting classifier

a	b	c	d	
8	0	0	0	a
1	8	0	1	b
0	0	5	0	c
0	0	0	4	d

(a)

a	b	c	d	
1	7	0	0	a
0	10	0	0	b
0	0	5	0	c
0	0	0	4	d

(b)

a	b	c	d	
8	0	0	0	a
0	9	0	1	b
0	0	5	0	c
0	0	0	4	d

(c)

Table 1(a) indicates that C4.5 can correctly classify square, circle and ring, but not triangles. C4.5 has an accuracy of 92 % for correctly classifying instances and a relative absolute error of 12 %. On the other hand the confusion matrix for naïve Bayes classifier indicates that the classifier correctly classifies triangle but not square as shown in the Table 1 (b). The naïve Bayes classifier has an accuracy of 74 % with relative absolute error of 34 %. This led us to apply the voting strategy using C4.5 and the naïve Bayes classifier and applying average probabilities for class predictions. Table 1(c) shows the confusion matrix for voting classifier.

The voting classifier has an accuracy of 96 % with relative absolute error of 23 %. While this may not be the best strategy to classify the shapes, it provides better results with minor code additions while implementing the system for real time applications. We have also recorded results for 10 fold cross validation. For C4.5 algorithm, the percent of correctly classified instances is 84 % with relative absolute error of 19 %. Table 2(a) summarizes the confusion matrix for C4.5 algorithm. Naive Bayes classifier correctly classifies 72 % of the instances with relative absolute error of 36 %. The confusion shown in Table 2(b) is consistent with the results received with 66 % sample split for testing The Voting classifier again out performs the individual algorithms with 87% correctly classified instances as shown in Table 2(c).

Table 2. Confusion Matrix for for 10 fold Cross validation for (a) C4.5 classifier (b) Naive bayes classifier (c) voting classifier

A	b	c	d	
17	4	0	0	a
7	20	0	1	b
0	0	18	0	c
0	0	0	12	d

(a)

a	b	c	d	
7	14	0	0	a
6	22	0	0	b
2	0	16	0	c
0	0	0	12	b

(b)

a	b	c	d	
17	4	0	0	a
5	22	0	1	b
0	0	18	0	c
0	0	0	12	d

(c)

The results for percent split and cross validation being consistent shows that our methodology is sound and the data we have gathered is reliable. The results with simple algorithms provide an initial idea regarding the proper approach to address the problem of shape classification using tactile sensing. Simple algorithms such as C4.5 and naive Bayes are easy to implement, analyze and deploy for real time systems as the model building time is negligible. Hence, we have used the voting strategy to increase the accuracy while keeping the system computationally efficient and analyzable.

6 Conclusion and Future Work

In this paper we have attempted to solve the problem of shape classification (circle, square, triangle and ring) using tactile sensors, with the help of feature extraction and machine learning algorithms. The methodology comprises of capturing pressure profile of an object and storing it into a 16 by 16 array corresponding to the 256 tactile elements. This array is then processed as an image and converted into black-and-white monochromatic image using thresholding and subsequently, unwanted noise is removed from the image using erosion and dilation. Distinguishing features are then extracted from the image using geometric operations on the data. We have chosen features which are independent of scaling, translation and rotation. Moreover, the feature extraction step has successfully reduced the size of the dataset from 256 attributes for 79 instances to 5 attributes for the same.

The dataset consisting of these features have been used as input for C4.5 and naive Bayes algorithms for shape classification. A voting strategy whereby a combination of predictions of C4.5 and naive Bayes using average of probabilities is implemented. For validation and testing we have used 66% split and 10 fold cross-validation on the entire dataset. Percentage of correctly classified instances for percent split for C4.5, naive Bayes and voting are 92%, 74% and 96% respectively. While for 10 fold cross validation, we have observed 84%, 72% and 87% for C4.5, naive Bayes and voting respectively.

Initial results are promising given the fact that we have used elementary machine learning algorithms and hence practical implementations of the system will return results in real time. However, the results are prone to error due to un-removable noise and other factors such as sensitivity of the sensor and uneven contour of the shape. Our main claim in this paper is that in order to get real time performance with acceptable accuracy, we have to use two or more elementary algorithms in tandem using voting strategy or other means such as stacking. Moreover, proper features needs to be extracted in a pre-processing step to reduce errors. Future work will consist of three dimensional object recognition with tactile sensors using minimal pressure profile samples of the object in question. More sophisticated machine learning techniques, such as Support Vector Machines, can also be incorporated in the system provided real time performance is not degraded.

References

1. Tiwana, M., Redmond, S., Lovell, N.: A review of tactile sensing technologies with applications in biomedical engineering. Sensors and Actuators A: Physical 179, 17–31 (2012)
2. Liu, H., Song, X., Nanayakkara, T., Seneviratne, L., Althoefer, K.: A computationally fast algorithm for local contact shape and pose classification using a tactile array sensor. In: Proceedings of International Conference on Robotics and Automation (ICRA), pp. 1410–1415 (2012)
3. Liu, H., Greco, J., Song, X., Bimbo, J., Seneviratne, L., Althoefer, K.: Tactile image based contact shape recognition using neural network. In: Multisensor Fusion and Integration for Intelligent Systems (MFI), pp. 138–143 (2012)
4. Bhattacharjee, T., Rehg, J., Kemp, C.: Haptic classification and recognition of objects using a tactile sensing forearm. In: Intelligent Robots and Systems (IROS), pp. 4090–4097 (2012)
5. Pezzementi, Z., Plaku, E., Reyda, C., Hager, G.: Tactile-object recognition from appearance information. Robotics 27, 473–487 (2011)
6. Ibrayev, R., Jia, Y.: Tactile recognition of algebraic shapes using differential invariants. Robotics and Automation 2, 1548–1553 (2004)
7. Aoyagi, S., Tanaka, T., Minami, M.: Recognition of Contact State of Four Layers Arrayed Type Tactile Sensor by using Neural Network. In: Information Acquisition, pp. 393–397 (2006)
8. Dahiya, R., Metta, G., Valle, M., Sandini, G.: Tactile sensing—from humans to humanoids. Robotics 26, 1–20 (2010)
9. Kauppinen, H., Seppanen, T., Pietikainen, M.: An experimental comparison of autoregressive and Fourier-based descriptors in 2D shape classification. IEEE Transactions in Pattern Analysis and Machine Intelligence 17, 201–207 (1995)

10. Zhang, D., Lu, G.: Review of shape representation and description techniques. Pattern Recognition 37, 1–19 (2004)
11. Gonzalez, R., Woods, R., Eddins, S.: Digital image processing using MATLAB, 2nd edn. Gatesmark Publishing, Knoxville (2009)
12. Peura, M., Iivarinen, J.: Efficiency of simple shape descriptors. In: Aspects of Visual Form, pp. 443–451 (1997)
13. Lu, G., Sajjanhar, A.: Region-based shape representation and similarity measure suitable for content-based image retrieval. Multimedia Systems 7, 165–174 (1999)
14. Quinlan, J.: C4. 5: Programs for Machine Learning 1 (1993)
15. Good, I.: Probability and the Weighing of Evidence, London (1950)
16. Lim, T., Loh, W., Shih, Y.: A comparison of prediction accuracy, complexity, and training time of thirty-three old and new classification algorithms. Machine Learning 40, 203–228 (2000)
17. Hall, M., Frank, E., Holmes, G., Pfahringer, B., Reutemann, P., Witten, I.: The WEKA Data Mining Software: An Update. ACM SIGKDD Explorations Newsletter 11, 10–18 (2009)

Multi-view Ensemble Learning for Poem Data Classification Using SentiWordNet

Vipin Kumar and Sonajharia Minz

School of Computer and Systems Sciences
Jawaharlal Nehru University, New Delhi, INDIA
{rt.vipink,sona.minz}@gmail.com

Abstract. Poem is a piece of writing in which the expression of feeling and ideas is given intensity by particular attention to diction, rhythm and imagery [1]. In this modern age, the poem collection is ever increasing on the internet. Therefore, to classify poem correctly is an important task. Sentiment information of the poem is useful to enhance the classification task. *SentiWordNet* is an opinion lexicon. To each term are assigned two numeric scores indicating positive and negative sentiment information. Multiple views of the poem data may be utilized for learning to enhance the classification task. In this research, the effect of sentiment information has been explored for poem data classification using Multi-view ensemble learning. The experiments include the use of Support Vector Machine (SVM) for learning classifier corresponding to each view of the data.

Keywords: Classification, Multi-view Ensemble Learning, Poem, SentiWordNet.

1 Introduction

Literature is a representation of high level of intellectual activity for human culture. The poem is one of the ways to read, write and enjoy various aspects of life. Poems are meant to reflect emotions such as angry, love, sad, happiness. Corresponding to the emotion (sentiment), many poem have been categorized in the literature, namely love, happy, sad, childhood, god, etc. In the present time, an increase in the poem collection can be observed on the internet. Increase in volume is turned as curse of high dimensionality for the classification task. Therefore, it is difficult task to classify poem data accurately in their labels. High dimensional data can handle by using dimension reduction data mining techniques. The problem with high dimensional data is that not all features are relevant for the classification task. High dimensional data can be viewed in multi-view perspective for the purpose of modeling a classifier. Multi-view learning is learning approach which utilizes redundant features (unimportant features) of the dataset to enhance the classification task. Manufactured splits of the data for multi-view learning may improve the performance of classification or clustering task. Therefore, multi-view learning is a very promising topic with widespread applicability [25]. Many techniques have been proposed for multi-view

M.K. Kundu et al. (eds.), *Advanced Computing, Networking and Informatics - Volume 1*,
Smart Innovation, Systems and Technologies 27,
DOI: 10.1007/978-3-319-07353-8_8, © Springer International Publishing Switzerland 2014

learning, where Canonical correlation analysis (CCA) [26] and Co-training [27] are two common representative techniques. Multi-view learning is strongly connected to other machine learning topics such as active learning, ensemble learning, and domain adoption [24]. It has been applied in supervised learning [33], semi-supervised learning [34], ensemble learning [35], active learning [36], transfer learning [37] and clustering [38] and dimensionality reduction [39]. Many applications of Multi-view learning for text document classification are mentioned in literature [6- 8].

Poems have short textual paragraphs to express the emotion. The feelings of the poet are expressed by word sentiments. Lexicons have been used to extract the sentiment information of the individual words. Sentiment information of the document can consider as document features to enhance the classification task. SentiWordNet [2] has sentiment information of the English words. Sentiment information comprises of numeric scores (positive and negative) of corresponds the words. Therefore, it has been utilized to extract the document sentiment information. Many applications of the SentiWordNet has been done in literature to extract the sentiment information for the textual documents [3-5]. The work in this paper has two objectives, the use of document sentiment information as document features, which is extracted from SentiWordNet and application of multi-view ensemble learning for the poem data to enhance the classification task.

This paper is organized as follows: Section 2 briefs outline of related work. In Section-3, preprocessing has been presented. Classification using Multi-view Ensemble Learning has been described in Section 4. Experimental Setup and their results are described in Section 5. Analysis of the results is described in Section 6. Finally, the conclusion of this research has mentioned in the Section 7.

2 Related Work

Many text classification algorithms have been proposed in the literature to extract the knowledge from unstructured data. One of the applications of the text classification algorithms is used to categorize emails [9], patent [10], magazine articles [11], news [12] etc. To classify text document, many machine learning algorithms have been applied such as k-nearest neighbor classifiers [13], Bayesian classifiers [14], Regression models [15], and Neural networks [16]. Malay poetry has been classified pantun by theme and poetry & none-poetry as well [17] using support vector machine (Radial Basic Function and linear kernel function). In different case studies, the various measures of poetry are introduced using fuzzy logic, Bayesian approach and decision [18]. It is important to identify which machine learning techniques are appropriate for the poem data. SVM classifier has performed best among the K-nearest neighbor (KNN) and Naïve Bayesian (NB) classifiers [19].

Wei and Pal [28] and Wan et al. [29] have proposed domain adaptation techniques. To solve the cross-language text classification problem, the target language has original documents which are included by the target domain and the source languages have the translated documents that are included by source domain. The different views of the original document can be seen in different language. Multi-view majority voting

[30], co-training [31], and multi-view co-classification [32] have been designed and successfully applied. Multiple views of the text data have been proposed for learning purpose using multi-view learning [8], [20-21].

The mood of the lyrics has been classified using SentiWordNet by adding supplementary sentiment features of the lyrics documents [22]. SentiWordNet has been applied to sentiment classification of reviews [3], financial news [4], Multi-domain sentiment analysis [23], and multilingual analysis. A domain independent sentiment analysis method using SentiWordNet has been proposed. A method is also proposed to classify subjective sentence and objective sentence from reviews and blog comments [5]. According to the current research and information related to the poem, analysis of poem data with multiple views having sentiment information of the document has not done for classification purpose. Therefore, this paper investigates sentiment information of the document as a document features and multi-view learning for classification task of the poem data.

3 Preprocessing

Data preprocessing includes transformation of the unstructured text document to structured dataset, feature extraction and other such processes to prepare the data for the aimed analyses. In this work three tasks have been included in the preprocessing stage; transformation of the text data into term-document vector, sentiment extraction of the document feature using SentiWordNet, and the creation of multiple views respectively. The transformation tasks included in this paper include, changing the upper case words to lower case, words tokenization, stop-word removal, and stemming using WordNet [40], [41]. The features extracted from the poem dataset include the term-document vector $(tf - \boldsymbol{idf})$, that represents the weight of the term in the documents [19] and the sentiment feature.

The sentiment measures of the document have been extracted using the SentiWordNet. Using the positive and negative scores of all the words of the document, the document sentiment feature has also been extracted. The following three sentiment information of the documents have been used as document sentiment features in the document matrix,

1) PosNegis the pair of normalized positive and negative scores of the document.
2) PosNeg_Ratiois the normalized ratio of the positive and negative score of the document.

Normalization has been done using Min-Max normalization. To rank the features, Gain Ratio (GR) has been applied.

The multidimensional data have been defined as information system $I = (U, A, V, F)$ where U is set of objects, A is the feature set which describe each object, V is the domain of the attributes and $F: U \times A \rightarrow V$, the function map a pair $(x, a) \in U \times A$ to $v_a \in V$. Therefore, for the data object $x \in U$and the value of a feature $a \in A$ describing x is $F\,(x, a) = v_a$. A view of a data set is the information system representing the set of data objects with respect to a subset of features.

Therefore, for a subset of attributes say $A' \subseteq A$, U the set of objects may be represented with respect to A', by the view $I' = (U, A', V, F)$. Let π_A be a partition of set of attributes A, with $A_1, A_2, A_3, \ldots, A_k$ the blocks of the partition. Then for $\pi_A = \{A_1, A_2, A_3, \ldots, A_k\}$, $A_i \subseteq A$, $1 \le i \le k$, and $\bigcup_{i=1}^{k} A_i = A$ where, $A_i \cap A_j = \phi$ for, $1 \le i, j \le k$ and $i \ne j$. Let I_i be a view corresponding the block $A_i \epsilon\, \pi_A$ such that $I_i = (U, A_i, V, F)$.

Given a partition π of the set of attributes A, there exists a view i.e. $I_l = (U, A_l, V, F)$. Then a multi-views representation of an information system with respect to π may be given as, $I = I_\pi = I_1 \circledcirc I_2 \circledcirc I_3 \circledcirc \ldots \circledcirc I_l$, where $\pi = \{A_1, A_2, A_3, \ldots, A_l\}$ consider the case that A is a large set i.e. the cardinality of A is large.

4 Classification Using Multi-view Ensemble Learning

For a given decision system $D = [X; Y]$, where X is the conditional part of the dataset and Y the corresponding class labels, in the proposed learning method the vectors X have been used for multi-view ensemble learning. The multiple views of a dataset have been proposed to be created for view-wise classifier modeling. Further, the classifier ensemble methods have been considered to ensemble the view-wise classifiers. A partition of the attributes defining the samples from the dataset has been considered to extract the corresponding views of the dataset. The construction of the multiple views and their evaluation is significant for efficient classification. A general framework for multi-view ensemble learning is presented in this section. Therefore, corresponding to a partition of the attributes of the given dataset, the multi-view ensemble learning is diagrammatically presented in Fig.1. Each $X_i, 1 \le i \le k$ represents the ith view of the dataset. Let $\circledcirc$ denote view-concatenation operator then the multiple views of a given dataset may be presented as,

$$X = X_1 \circledcirc X_2 \circledcirc X_3 \circledcirc \ldots \circledcirc X_k \quad (1)$$

Let the classification model be a function $f{:}\,X \rightarrow Y$. Then the multi-view ensemble classifier may be represented as in eq. 2:

$$f(X_1) \oplus f(X_2) \oplus f(X_3) \oplus \ldots \oplus f(X_k) = \bigoplus_{i=1}^{k} f(X_i) \quad (2)$$

where, $\oplus$ may represent an ensemble function to the combine the predictive models $f(X_i)$ using each of the views, say X_i for $i = 1,2,3, \ldots, k$.

In the literature, ensemble by vote or ensemble by weight has been commonly cited. Therefore, in this paper ensemble by weight has been considered. It can be mathematically represented as in eq. 3.

$$C(x) = argmax_{c_i \in dom(Y)}\left(\sum_k g(f_k(x), c_i)\right) \quad (3)$$

where, $f_k(x)$ is the prediction of the kth classifier of $x \in U$, as c_k and $g(c_k, c)$ is the indicator function defined as in eq. 4:

$$g(c_k, c) = \begin{cases} w_k, & c_k = c \\ 0, & c_k \neq c \end{cases} \tag{4}$$

A loss function is needed to access the classification accuracy. Zero-one Loss is commonly used loss function that is defined as eq. 5:

$$\mathcal{L}(y, C(x)) = \begin{cases} 0, & C(x) = y \\ 1, & C(x) \neq y \end{cases} \tag{5}$$

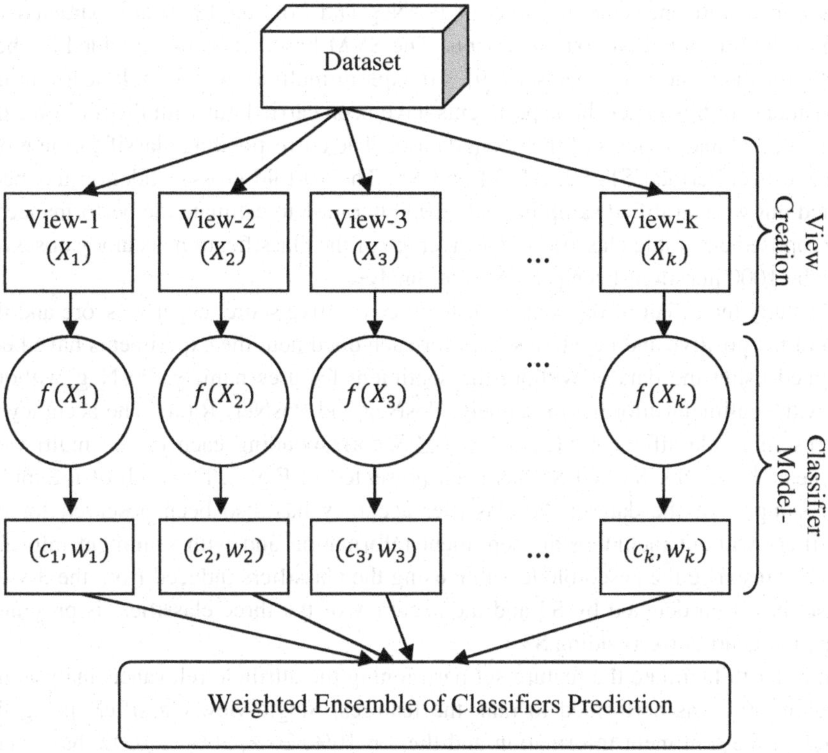

Fig. 1. A framework of multi-view ensemble learning

A single unit is assigned by the loss function for each misclassification. Therefore, accuracy of the classification can be defined as in eq. 6:

$$Acc(X) = 1 - \min_{D}\left(\frac{1}{|D|}\sum_{i=1}^{|D|}\mathcal{L}(y_i, C(x_i))\right) \tag{6}$$

Eq. (6) shows that the minimization of loss leads maximum classification accuracy. Generally, in multi-view ensemble learning, optimal classification accuracy depends upon three factors: 1) Feature set partitioning criteria, 2) Ensemble method, and 3) Classifier. Therefore, for a specific problem, the empirical study may help in selection of best feature set partitioning criteria, ensemble method and the classification algorithms.

5 Experimental Setup and Results

In [19], a dataset consisting of four hundred poems collected from various websites over the internet. The poems belong to the one of the eight classes namely alone, childhood, God, hope, love, success, valentine and world. Each class is represented by fifty poem documents. The preprocessing of the test result is 3707 features. Further sentiment feature measures in terms of PosNeg and PosNeg_Ratio are extracted for each of the four hundred text documents. The SVM-based classifiers yielded the best results for the same poem dataset [19]. To explore multi-view ensemble learning the experiments in this paper the experiments have been carried out with the SVM for the single, 2, 3, 4 and 5-views of the poem dataset. The corresponding classification models have been termed S1, S2, S3, S4 and S5. The 10-fold cross validation has been carried out with stratified sampling over 1000 iteration to estimate the performance of the proposed ensemble classifiers. The average of the classification accuracy observed over the 1000 iterations been presented for analysis.

To study the effect of the sentiment features positive score, negative score and the ratio of the positive and negative scores for each document the experiments have been designed using the dataset without the sentiment features namely, PosNeg_without, and with sentiment information namely, PosNeg and PosNeg_Ratio. The accuracy of the ensemble classifiers $Acc(Sj)$, $1 \leq j \leq 5$, corresponding each of the multi-view samples S1, S2, S3, S4 and S5 has been presented in Fig. 2. For each of the multi-view samples of the dataset the classifier accuracy has also been observed for the classifiers without including the sentiment information and with sentiment information. For instance the ensemble learning using the classifiers induced from the 3-view dataset has been denoted by S3 and the accuracy of the three classifiers is presented by the bar chart corresponding S3.

In order to facilitate the feature-set partitioning the attribute relevance analysis using gain-ratio was performed to rank the features. Single-view classifiers using the including the sentiment information and the top 10%, 20%, 30%… 100% have been induced using the SVM. The performance of all the ten classifiers has been shown in the Fig. 3. Since the performance of the classifier using the sentiment information and the top 20%attributes is marginally less than the classifier using the sentiment information and the top 30%attributes, therefore the experiments using the five multi-views ensemble classifiers has been performed on the dataset with respect to the 20% of the top ranked attributes. The performance comparison of these classifiers has been exhibited in the Fig. 4.

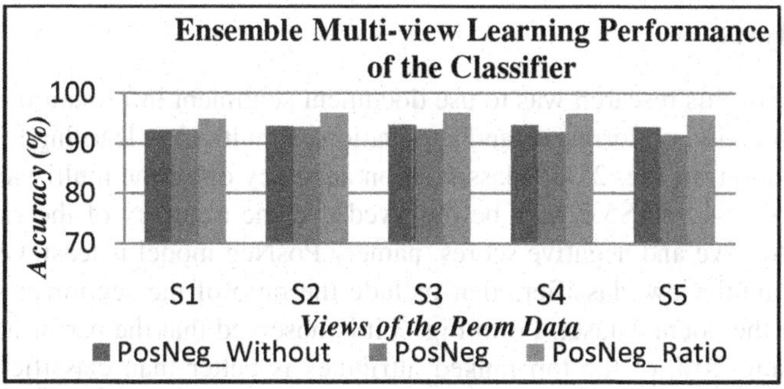

Fig. 2. Classifier performances using percentage of top ranked attribute

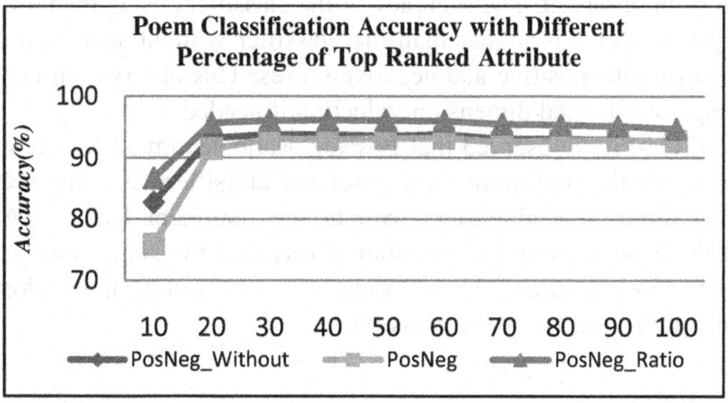

Fig. 3. Performance of the classifier for percentage top ranked attribute

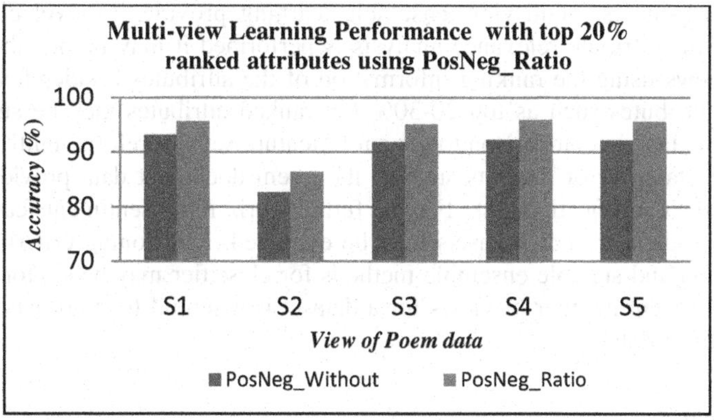

Fig. 4. Ensemble multi-view learning performance of the classifier using 20% top ranked attribute

6 Analysis

The objective of this research was to use document sentiment information for enhancing the classification performance and application of multi-view learning for the poem data classification. In Fig. 2, the classification accuracy of all the multi-view classifiers, S1, S2, S3, S4 and S5 it may be observed that the accuracy of the classifier including the positive and negative scores, namely PosNeg model is least. On the other hand all the multi-view classifiers that include the ratio of the sentiment scores perform best for the poem dataset. From Fig. 3 it is observed that the performance of the classifiers using 20% of the top ranked attributes is better than classifier using top 10%attributes. Secondly, the performance of the classifier based on top 20% is marginally less than the accuracy of the classifier using the top 30% attributes. Further no improvement in the accuracies of the classifier using top 30%, 40% up to 100% attribute has been observed. The accuracy of the classifiers using the ratio of the positive and negative scores is higher than the classifier without sentiment features and the classifier using the positive and negative scores. This observation may be utilized as partitioning criterion and dimension reduction if needed.

From the Fig. 4, it is observed that PosNeg_Ratio sentiment feature enhances the performance of all the five multi-view ensemble classifiers as compared to the five corresponding multi-view classifiers without the sentiment features. Although the accuracy of the 2-view ensemble classifier is less than the single-view classifier and the 3, 4 and 5-view classifiers, the performance in terms of accuracy alone is not impacted by the multi-view ensemble classifier.

7 Conclusion

The ratio of the positive and negative scores of a document is useful sentiment features for classification. For the datasets with large number of attributes in comparison to the data samples, multi-view ensemble learning provides a novel classification method. If the attribute relevance analysis is performed it may be possible to create multiple views using the ranking information of the attributes besides feature reduction. Few attributes such as top 20-30% top ranked attributes may be sufficient for classification task in comparison to the entire feature set. Therefore, multi-view learning for the classifier of datasets such as the poem document data provides a useful option for the classification task. For the future work, more sentiment features of the poem data and other such datasets may be extracted. Partitioning criteria for multi-view creation and suitable ensemble methods for classifier may be explored. Empirical study for right number of views for a dataset is essential for improved classification for a given dataset.

References

1. http://oxforddictionaries.com/definition/english/poem
2. Andrea, E., Sebastiani, F.: SentiWordNet: A Publicly Available Lexical Resource for Opinion Mining. In: Proceedings of the 5th Conference on Language Resources and Evaluation, pp. 417–422 (2006)

3. Hamouda, A., Rohaim, M.: Reviews Classification Using SentiWordNet Lexicon. In: World Congress on Computer Science and Information Technology (2011)
4. Devitt, A., Ahmad, K.: Sentiment Polarity Identification in Financial News: A Cohesion-based Approach. In: Proceedings of the 45th Annual Meeting of the Association of Computational Linguistics, Prague, pp. 984–991 (2007)
5. Aurangzeb, K., Baharum, B., Khairullah, K.: Sentence Based Sentiment Classification from Online Customer Reviews. In: FIT (2010)
6. Amini, M., Usunier, N., Goutte, C.: Learning from Multiple Partially Observed Views - An Application to Multilingual Text Categorization. In: Advances in Neural Information Processing Systems (2009)
7. Yuhong, G., Min, X.: Cross Language Text Classification via Subspace Co-Regularized Multi-View Learning. In: Proceedings of the 29th International Conference on Machine Learning, Edinburgh, Scotland, UK (2012)
8. Ping, G., QingSheng, Z., Cheng, Z.: A Multi-view Approach to Semi-supervised Document Classification with Incremental Naive Bayes. Computers & Mathematics with Applications 57(6), 1030–1036 (2009)
9. Androutsopoulos, I., Koutsias, J., Chandrinos, K.V., Spyropoulos, C.D.: An Experimental Comparison of Naive Bayesian and Keyword-Based Anti-Spam Filtering with Personal E-mail Messages. In: Proceedings of the 23rd Annual Int. ACM SIGIR Conference on Research and Development in Information Retrieval, pp. 160–167 (2000)
10. Richter, G., MacFarlane, A.: The Impact of Metadata on the Accuracy of Automated Patent Classification. World Patent Information 37(3), 13–26 (2005)
11. Moens, M.F., Dumortier, J.: Text Categorization: the Assignment of Subject Descriptors to Magazine Articles. Information Processing and Management 36(6), 841–861 (2000)
12. Shih, L.K., Karger, D.R.: Learning Classifiers: Using URLs and Table Layout for Web Classification Tasks. In: Proceedings of the 13th International Conference on World Wide Web, pp. 193–202 (2004)
13. Yang, Y.: An Evaluation of Statistical Approaches to Text Categorization. Information Retrieval 1(1-2), 67–88 (1999)
14. Scheffer, T.: Email Answering Assistance by Semi- Supervised Text Classification. Intelligent Data Analysis 8(5), 481–493 (2004)
15. Zhang, T., Oles, F.: Text Categorization Based on Regularized Linear Classifiers. Information Retrieval 4, 5–31 (2001)
16. Lee, P.Y., Hui, S.C., Fong, A.C.: Neural networks for web content filtering. IEEE Intelligent System 17(5), 48–57 (2002)
17. Noraini, J., Masnizah, M., Shahrul, A.N.: Poetry Classification Using Support Vector Machines. Journal of Computer Science 8(9), 1441–1446 (2012)
18. Tizhoosh, H.R., Dara, R.A.: On Poem Recognition. Pattern Analysis Application 9(4), 325–338 (2006)
19. Vipin, K., Sonajharia, M.: Poem Classification using Machine Learning Approach. In: Babu, B.V., Nagar, A., Deep, K., Bansal, M.P.J.C., Ray, K., Gupta, U. (eds.) SocPro 2012. AISC, vol. 236, pp. 675–682. Springer, Heidelberg (2014)
20. Yuhong, G., Min, X.: Cross Language Text Classification via Subspace Co-regularized Multi-view Learning. In: ICML (2012)
21. Massih-Reza, A., Nicolas, U., Cyril, G.: Learning from Multiple Partially Observed Views - an Application to Multilingual Text Categorization. In: NIPS, pp. 28–36 (2009)
22. Vipin, K., Sonajharia, M.: Mood Classification of Lyrics using SentiWordNet. In: ICCCI 2012. IEEE Xplore (2013)

23. Kerstin, D.: Are SentiWordNet Scores Suited for Multi-Domain Sentiment Classification? In: ICDIM, pp. 33–38 (2009)
24. Chang, X., Dacheng, T., Chao, X.: A Survey on Multi-view Learning. CoRR abs/1304.5634 (2013)
25. Shiliang, S.: A Survey of Multi-view Machine Learning. Neural Computing & Application. Springer, London (2013)
26. Hotelling, H.: Relations between Two Sets of Variates. Biometrika 28, 321–377 (1936)
27. Blum, A., Mitchell, T.: Combining Labeled and Unlabeled Data with Co-training. In: Proceedings of the 11th Annual Conference on Computational Learning Theory, pp. 92–100 (1998)
28. Wei, B., Pal, C.: Cross Lingual Adaptation: An Experiment on Sentiment Classification. In: Proceedings of the ACL 2010 Conference Short Papers, pp. 258–262. Association for Computational Linguistics (2010)
29. Wan, C., Pan, R., Li, J.: Bi-weighting Domain Adaptation for Cross-language Text Classification. In: Twenty-Second International Joint Conference on Artificial Intelligence (2011)
30. Massih-Reza, A., Nicolas, U., Cyril, G.: Learning from Multiple Partially Observed Views - an Application to Multilingual Text Categorization. In: NIPS 2009, pp. 28–36 (2009)
31. Wan, X.: Co-training for Cross-lingual Sentiment Classification. In: Proceedings of the Joint Conference of the 47th Annual Meeting of the ACL and the 4th International Joint Conference on Natural Language Processing of the AFNLP, vol. 1, pp. 235–243. Association for Computational Linguistics (2009)
32. Amini, M.R., Goutte, C.: A Co-classification Approach to Learning from Multilingual Corpora. Machine Learning 79(1), 105–121 (2010)
33. Chen, Q., Sun, S.: Hierarchical Multi-view Fisher Discriminant Analysis. In: Leung, C.S., Lee, M., Chan, J.H. (eds.) ICONIP 2009, Part II. LNCS, vol. 5864, pp. 289–298. Springer, Heidelberg (2009)
34. Sun, S.: Multi-view Laplacian Support Vector Machines. In: Tang, J., King, I., Chen, L., Wang, J. (eds.) ADMA 2011, Part II. LNCS, vol. 7121, pp. 209–222. Springer, Heidelberg (2011)
35. Xu, Z., Sun, S.: An Algorithm on Multi-view Adaboost. In: Wong, K.W., Mendis, B.S.U., Bouzerdoum, A. (eds.) ICONIP 2010, Part I. LNCS, vol. 6443, pp. 355–362. Springer, Heidelberg (2010)
36. Muslea, I., Minton, S., Knoblock, C.: Active Learning with Multiple Views. Journal of Artificial Intelligence Ressearch 27, 203–233 (2006)
37. Xu, Z., Sun, S.: Multi-view Transfer Learning with Adaboost. In: Proceedings of the 23rd IEEE International Conference on Tools with Artificial Intelligence, pp. 399–340 (2011)
38. De Sa, V., Gallagher, P., Lewis, J., Malave, V.: Multi-view Kernel Construction. Machine Learning 79, 47–71 (2010)
39. Chen, X., Liu, H., Carbonell, J.: Structured Sparse Canonical Correlation Analysis. In: Proceedings of the 15th International Conference on Artificial Intelligence and Statistics, pp. 199–207 (2012)
40. `http://wordnet.princeton.edu/`
41. Jiawei, H., Micheline, K.: Data Mining Concepts and Techniques, 2nd edn. Elsevier (2006)

A Prototype of an Intelligent Search Engine Using Machine Learning Based Training for Learning to Rank

Piyush Rai, Shrimai Prabhumoye, Pranay Khattri,
Love Rose Singh Sandhu, and S. Sowmya Kamath

Department of Information Technology,
National Institute of Technology Karnataka, Surathkal, India
{piyushrocks.rai,shrimai19,prakhattri92,lovesingh25}@gmail.com,
sowmyakamath@ieee.org

Abstract. Learning to Rank is a concept that focuses on the application of supervised or semi-supervised machine learning techniques to develop a ranking model based on training data. In this paper, we present a learning based search engine that uses supervised machine learning techniques like selection based and review based algorithms to construct a ranking model. Information retrieval techniques are used to retrieve the relevant URLs by crawling the Web in a Breadth-First manner, which are then used as training data for the supervised and review based machine learning techniques to train the crawler. We used the Gradient Descent Algorithm to compare the two techniques and for result analysis.

1 Introduction

Ranking is a crucial functionality that is inherent to any application offering search features to users. Hence, a lot of research in has been carried out the area of ranking. However, it is also a known fact that it is difficult to design effective ranking functions for free text retrieval. Often, a ranking function that works very well with one application will require major modifications in order to achieve the same level of quality with another application. Given a query, documents relevant to the query have to be ranked according to their degree of relevance to the query. The data that is retrieved usingan algorithm should be ranked appropriately irrespective of the data set it has been extracted from. Hence, the need to teach the system to rank instead of doing the same intuitively makes it unsuitable for most of database systems.

Learning to rank is a relatively new field in which machine learning algorithms are used to generate an effective ranking function. Learning to rank can be employed in a wide variety of applications in Information Retrieval (IR) and Natural Language Processing (NLP).A Learning to Rank algorithm should incorporate various tags/criteria to rank the links in order of their appropriateness, like user ratings, time stamp, associated frequency/weights of the information to be retrieved etc. Since same criteria cannot be used for all kinds of data-set like political/sports news or

M.K. Kundu et al. (eds.), *Advanced Computing, Networking and Informatics - Volume 1*,
Smart Innovation, Systems and Technologies 27,
DOI: 10.1007/978-3-319-07353-8_9,

various advancements in IT technology or historical events etc. more than one algorithm should be taken into account and results of most appropriate one should be referred to for making a raw data set.

In this paper, we present the prototype of a Learning to Rank system where in machine learning is incorporated for not only teaching the system to rank different data sets but also to teach it to choose the most appropriate algorithm. The paper is organized as follows: Section 2 presents a discussion on relevant literature to this field of work, Section 3 discusses the proposed system in detail and in Section 4, we present some experimental results. Section 5 presents a comparative case study with conventional search engine results, conclusion and future work in Section 6, followed by references.

2 Related Work

In any ranking model, the ranking task is performed by using a ranking model*f (q, d)* to sort the documents, where *q* denotes a query and *d* denotes a document. Traditionally, the ranking model *f (q, d)* is created without training. In the well known Okapi BM25 model [3], for example, it is assumed that *f (q, d)* is represented by a conditional probability distribution *P (r / q, d)* where *r* takes on 1 or 0 as value and denotes being relevant or irreverent, and *q* and *d* denote a query and a document respectively. In Language Model for IR (LMIR),*f (q, d)* is represented as a conditional probability distribution *P (q / d)*. The probability models can be calculated with the words appearing in the query and document, and thus no training is needed (only tuning of a small number of parameters is necessary) [1].

It is fact that most users are uninterested in more than the first few results in a search engine results page (SERP). Taking this into consideration, the Learning to Rank from user feedback model [2] considers each query independently. Radlinski et al [2] proposed the model where the log files only provide implicit feedback on a few results at the top of the result set for each query. They referred to this as "a sequence of reformulated queries or query chains". They state that these query chains that are available in search engine log files can be used to learn better retrieval functions.

Several other works present various techniques for collecting implicit feedback from click-through logs [4, 8,11]. All are based on the concept of the Click through rate (CTR) considers the fact that documents clicked on in search results are highly likely to be very relevant to the query. This can then be considered as a form of implicit feedback from users and can be used for improving the ranking function. Kemp et al. [4] present a learning search engine that is based on actually transforming the documents. They too use the fact that results clicked on are relevant to the query and append the query to these documents. However, other works [9, 10] showed that implicit click-through data is sometimes biased as it is relative to the retrieval function quality and ordering. Hence this cannot be considered to be absolute feedback. Some studies [13, 14] have attempted to account the position-bias of click. Carterette and Jones [12] proposed to model the relationship between clicks and relevance so that clicks can be used to unbiasedly evaluate search engine performance when there is a lack of editorial relevance judgment. Other research [14, 5, 6] attempted to model user click behavior during search so that future clicks may be accurately predicted based on observations of past clicks.

RankProp [7] is a neural net based ranking model. It basically uses two processes - a MSE regression on the current target values, and an adjustment of the target values themselves to reflect the current ranking given by the net. The end result is a mapping of the data to a large number of targets which reflect the desired ranking. RankProp has the advantage that it is trained on individual patterns rather than pairs; however the authors do not discuss the conditions at which it converges, and also it does not provide a probabilistic model.

3 Proposed System

Fig. 1 depicts the architecture of the proposed system. The Swappers search engine API is built using HTML 5 and AJAX. The database is created using MySQL and PHP is used to connect API to the database.

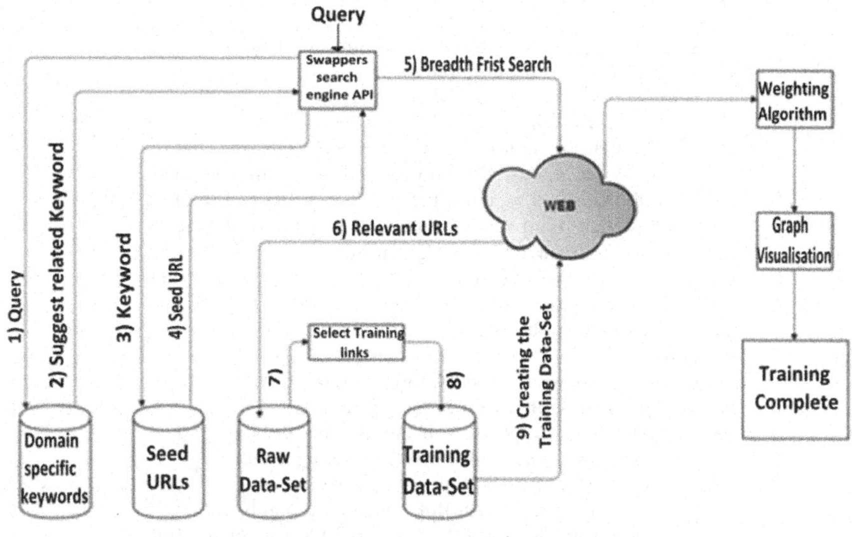

Fig. 1. Proposed Training based Search Engine Prototype System

The search engine is trained using two supervised machine learning algorithms namely selection based and review based. The tags/weights are calculated to rank the links in the training data-set. Both the algorithms follow the inclusion of different heuristics for the same. The weight of the link is determined by the frequency of the keyword in content of the link and the position where it occurs. Also, heuristics like whether the keyword is written in bold or italics; position where it occurs, for e,g. in page title, headings, metadata etc; and the number of outgoing links having the keyword in the URL are considered while calculating the weight of the link.

In the review based module, the user selects the links he wants to train the search engine with and also rates those selected links. The review based module then normalizes the two weights, one given by the user and the other calculated from the keyword density. The weighting algorithm module finds a best fit line using gradient descent technique and assigns weights to links which are relevant to the query.

Algorithm 1 Training Algorithm

```
Choose the option for training the search engine
Choose type of training supervised or review based
Initialize a variable with the choice
Initialize a string with keyword
Append a string with entered keyword
Suggest words completing the appended keyword
Save the keyword in cache
Apply pattern matching in all Seed URLs
Find for keyword as substring in the seed URL
Crawl using BFS technique in the Seed URL
Search for additional URLs in crawled pages
for each URL found in crawled page do
    Add the link in the database
    Continue with next link in Seed Page
end for
Take input from user for the links in raw-dataset
Append links in training database
if choice := selection based training then
    Crawl through training dataset links
    Calculate weight of each link against the keyword
    Associate the weight with the link in training database
    Initialize final weight of the link as weight of the link
    Associate final weight with link in training database
    Use Gradient Descent algorithm
    Generate line equation with weight on X-axis and rank
      on Y-axis
    Plot the graph
end if
if choice := review based training then
    Crawl through training dataset links
    Calculate weight of each link against the keyword
    Associate weight with link in training database
    Input rating for the link from user
    Associate rating with link in training database
    Normalize rating and weight final weight for the link
    Associate final weight with link in training database
    Use Gradient Descent algorithm
    Generate line equation with weight on X-axis and rank
      on Y-axis
    Plot the graph
end if
if either selection or review based training complete then
    Display message to acknowledge training completion
end if
```

Fig. 2. The System's Training Algorithm

In the stochastic gradient descent algorithm we are trying to find the solution to the minimum of some function $f(\mathbf{x})$. Given some initial value x0 for x, we can change its value in many directions (proportional to the dimension of **x**: with only one dimension, we can make it higher or lower). To figure out what is the best direction to minimize f, we take the gradient ∇f of it (the derivative along every dimension of **x**). Intuitively, the gradient will give the slope of the curve at that **x** and its direction will point to an increase in the function. So we change **x** in the opposite direction to lower the function value[20].

$$x_k+1 = x_k - \lambda\, \nabla f(x_k) \tag{1}$$

The $\lambda > 0$ is a small number that forces the algorithm to make small jumps. That keeps the algorithm stable and its optimal value depends on the function. Given stable conditions (a certain choice of λ), it is guaranteed that $f(\mathbf{x}_{k+1}) \leq f(\mathbf{x}_k)$.

The algorithm shown in Fig. 2 is used to train the system to rank the relevant documents. In this method, the training set is given as an input and a best fit line is found passing through maximum of these input training set points by minimizing the distance between the ordinates of each of the training set points and a given line.After getting the equation of the line, next time when the ranking of the data set is required, the testing value should be inputted and the system will return the rank of the testing set. The method to calculate the weight of the link is explained above. This weight will be the y co-ordinate of the line. This weight will be substituted in the equation of the line and the x- coordinate will be calculated and the x co-ordinate will be the rank of the document which will be returned. The link with highest weight is labeled as rank 1 and so on.

The equation of the line to be plotted is given by formula (2), where m is the slope of the line and c is a constant. Both the values are calculated by the gradient descent technique using the inputs of the training set.

$$Rank = m * weight + c \tag{2}$$

Once the line and its slope is computed, the Graph Visualization module takes input of 'm' and 'c' calculated by the gradient module and draws the graph. The rank of the link is plotted on Y-axis and the weight of the link is plotted on X-axis. The basic equation of the line plotted is using the equation (2). The graph module uses canvas of the HTML to draw the graph and also makes use of JavaScript to draw the graph. Graph visualization helps in understanding the training process better and in comparing the two training techniques.

4 Experimental Results

The training of the selection based and the review based systems use the same set of links as selected by the user to get trained with. Our database consists of around a few thousand seed URLs from which more links are discovered by crawling these web pages. Since the crawler uses BFS technique andhence will fetch the links ofrelevant web pages till a particular average depth, hence providing with an average of 2^(h+1) links where 'h' depends on the total number of links extracted from each seed URL to fetch around 20-25 links, which makes 'h' to vary from 3-4, to be added depending on the keyword. A thousand of these relevant set of URLs were chosen to be appended into the training dataset.

Adatabase query is used to fetch keyword related links from Seed URL. The system extracts links from the table named "training_data" in the database which gets generated based on users choice during training. Each of this link is then crawled upon by the various systems connected via wireless network. The keyword is then searched on the content of the page given by the link. Processing is done based on the places where it is present and other parameters explained later. The result is updated back into the table to be looked up later during further processing.

Links	Rank	Weight
http://www.babelgraph.org/links.html	1	6.6
http://www.utm.eu/departments/math/graph	2	3.4
http://deisel-graph-theory.com/index.html	3	1.2

Fig. 3. Rank and Weight of Links given by Selection based training

Feedback Form		
Help us to rank better. Rate the links based on relevance with the KEYWORD you entered		
Links	Rating	Weight
http://www.utm.edu/departments/math/graph/	8	5
http://www.babelgraph.org/links.html	7	3.9
http://diesel-graph-theory.com.index.html	1	1.1
		Submit

Fig. 4. Feedback form of the review based technique

Links	Rank	Weight
http://www.utm.eu/departments/math/graph	1	6.5
http://www.babelgraph.org/links.html	2	5.45
http://deisel-graph-theory.com/index.html	3	4.55

Fig. 5. User rating and weight of Links given by Review based technique

In the selection based training, the ranking of the links were derived by taking only the weights (includes the frequency of the keywords in the pages and the tags in the HTML source code of the page) of the links as found by the weighing algorithm. In the review based training, the ranking of the links were derived from the weights of the links as in the selection based and the user assigned weights (user rating) to the links. The user rating is collected through the feedback form provided to the user.

Finally, the values from these criteria are normalized to get the final weight for each link in the training set which is further used to plot the graph using the gradient descent algorithm. It plots all the points (weights) and finds the best fit line for both the algorithms. The characteristics of the lines (slope and constant) formed and used after training is done.

Fig. 6 shows the best-fit line made by the selection based technique. The line corresponds to the data of ranks and weights in Fig. 3. The line covers a short range of ranks.Fig. 7 shows the best fit line made by the review based technique. The line corresponds to the data of ranks and weights in Fig. 5. The line covers a wide range of ranks. This makes the review based technique of training more useful than selection based technique in training to rank wide range of sets.

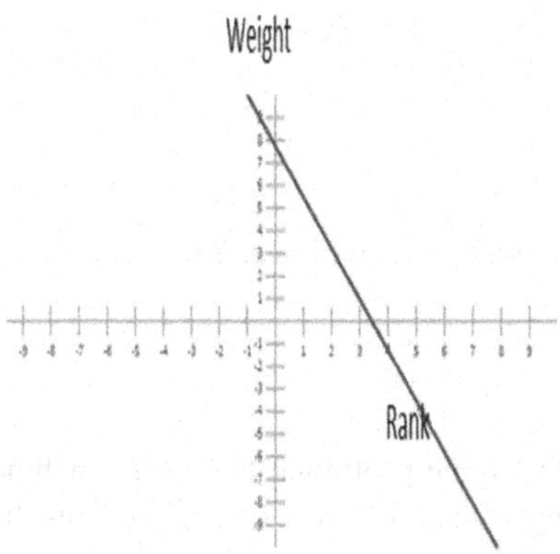

Fig. 6. Best-fit line from the selection based technique

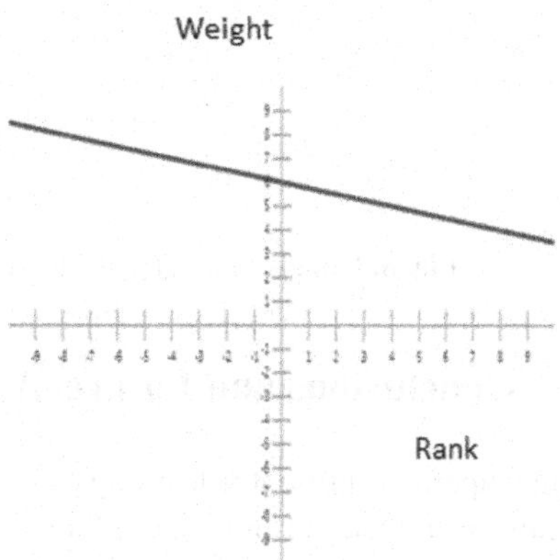

Fig. 7. Best-fit line from the review based technique

5 Case Study

As a part of this case study, we compared the relative ranking of the links given by selection based technique and review based technique with the search results of Google. Google gives its search results based on many parameters and hence we have not compared the absolute ranks of the links. By comparing the relative ranks of the links, we are comparing the rank of the link given by our system to its rank given by Google search results relative to the ranks of the other links.

The links and ranks given in Fig. 3and Fig. 4 are compared with the links and their relative ranks in Fig. 5. The relative ranking of the review based technique is matching with the relative ranking of the Google search as seen in Fig. 8.

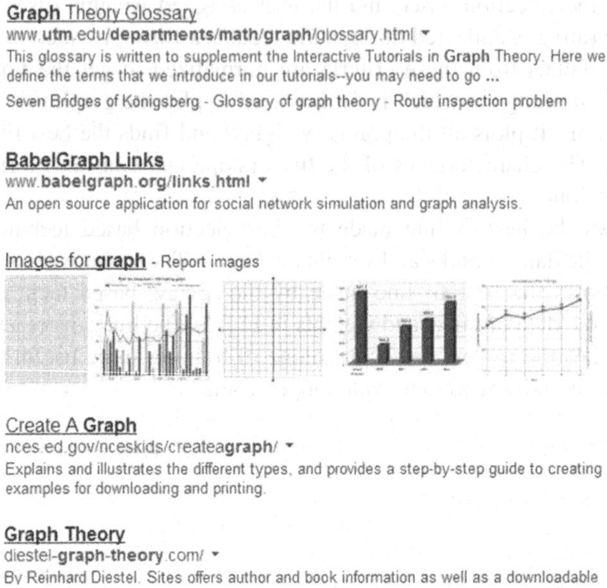

Graph Theory Glossary
www.utm.edu/departments/math/graph/glossary.html
This glossary is written to supplement the Interactive Tutorials in Graph Theory. Here we define the terms that we introduce in our tutorials--you may need to go ...
Seven Bridges of Königsberg - Glossary of graph theory - Route inspection problem
BabelGraph Links
www.babelgraph.org/links.html
An open source application for social network simulation and graph analysis.
Images for graph - Report images
Create A Graph
nces.ed.gov/nceskids/createagraph/
Explains and illustrates the different types, and provides a step-by-step guide to creating examples for downloading and printing.
Graph Theory
diestel-graph-theory.com/
By Reinhard Diestel. Sites offers author and book information as well as a downloadable

Fig. 8. Google search results for keyword "graph" (as on 01˙Sept-2013)

6 Conclusions and Future Work

In this paper, we present a learning based search engine prototype that uses machine learning techniques to infer the effective ranking function. The experimental results show that the review based technique of training performs better and is more accurate than the selection based learning technique. The review based technique is more accurate because the two weights - the weight assigned by the user and the weight of the keyword density are considered and normalized.

We also tried to match the ranking given by selection based and review based learning system with Google ranking. Google had many more results to offer based on different parameters but we tried to compare the ranking of the links given by our system to the ranking of the same links on Google. The review based learning technique had ranking matching with the ranking of Google.Supervised training is used in the system, and this could be made into a semi-supervised or unsupervised training. More parameters for considering the weight of the document can be used which would give a better result.

As future work, we are focusing on using these trained systems to determine the rank of a link. The values of slope 'm' and the constant 'c' mentioned in equation (2) will be stored in the database for each query. When the rank of the new link is to be determined, the weight of the link will be calculated by the weighing algorithm. The rank of the link will be calculated by substituting the weight of the link in equation (2) and retrieving the values of 'm' and 'c' from the database.

References

1. Croft, W.B., Metzler, D., Strohman, T.: Search Engines -Information Retrieval in Practice. Pearson Education (2009)
2. Radlinski, F., Joachims, T.: Query Chains: Learning to Rank from Implicit Feedback. In: Proceedings of the Eleventh ACM SIGKDD International Conference on Knowledge Discovery in Data Mining, pp. 239–248 (2005)
3. Baeza-Yates, R., Ribeiro-Neto, B.: Modern Information Retrieval. Addison Wesley (1999)
4. Kemp, C., Ramamohanarao, K.: Long-term learning for web search engines. In: Elomaa, T., Mannila, H., Toivonen, H. (eds.) PKDD 2002. LNCS (LNAI), vol. 2431, pp. 263–274. Springer, Heidelberg (2002)
5. Caruana, R., Baluja, S., Mitchell, T.: Using the future to "sort out" the present: Rankprop and multitask learning for medical risk evaluation. In: Advances in Neural Information Processing System, pp. 959–965 (1996)
6. Gradient Descent Methods, `http://webdocs.cs.ualberta.ca/~sutton/book/ebook/node87.html`
7. Joachims, T., Granka, L., Pan, B., Hembrooke, H., Gay, G.: Accurately interpreting clickthrough data asimplicit feedback. In: Annual ACM Conference on Research and Development in Information Retrieval, pp. 154–161 (2005)
8. Tan, Q., Chai, X., Ng, W., Lee, D.-L.: Applying co-training to clickthrough data for search engine adaptation. In: Lee, Y., Li, J., Whang, K.-Y., Lee, D. (eds.) DASFAA 2004. LNCS, vol. 2973, pp. 519–532. Springer, Heidelberg (2004)
9. Carterette, B., Jones, R.: Evaluating search enginesby modeling the relationship between relevance andclicks. In: Advances in Neural Information ProcessingSystems, vol. 20, pp. 217–224 (2008)
10. Craswell, N., Zoeter, O., Taylor, M., Ramsey, B.: Anexperimental comparison of click position-bias models. In: Proceedings of the International Conference on Web Search and Web DataMining, pp. 87–94 (2008)
11. Dupret, G., Piwowarski, B.: User browsing model to predict search engine click data from past observations. In: Proceedings of the 31st Annual International Conference on Research and Development in Information Retrieval (2008)
12. Richardson, M., Dominowska, E., Ragno, R.: Predicting clicks: estimating the click-through rate for new ads. In: Proceedings of the 16th International Conference on World Wide Web, pp. 521–530 (2007)
13. Zhou, D., Bolelli, L., Li, J., Giles, C.L., Zha, H.: Learning user clicks in web search. In: International Joint Conference on Artificial Intelligence (2007)
14. Ponte, J.M., Croft, W.B.: A language modeling approach to information retrieval. In: Proceedings of the 21st Annual International ACM SIGIR Conference on Research and Development in Information Retrieval, pp. 275–281 (1998)

Vegetable Grading Using Tactile Sensing and Machine Learning

Irin Bandyopadhyaya, Dennis Babu,
Sourodeep Bhattacharjee, and Joydeb Roychowdhury

CSIR-CMERI, Durgapur-713209, West Bengal, India
{mom.tanuja,denniskanjirappally,
sourodeepbhattacharjee}@gmail.com,
jrc@cmeri.res.in

Abstract. With the advent of e-Commerce, online automated fruit and vegetable grading has acquired significant research interests recently. To get an idea of freshness of the fruits and vegetables the softness or firmness estimation is one of the basic steps of the grading process which has its roots grounded in bio-mimicking. This paper proposes and implements a basic touch sensitive robotic system for ripeness classification of two vegetables using machine learning approaches. Two piezoresistive force sensors mounted on a robotic gripper, controlled by a PIC32 microcontroller, are used to receive tactile feedback on predefined palpation sequence of an object. Eight statistical parameters are generated from the force values from each instance of the dataset during tactile palpation. These parameters serve as the input for Support Vector Machine (SVM) and K-Nearest Neighbor (KNN) approaches. A study and analysis of these Machine Learning methodologies has been conducted in this paper.

1 Introduction

The surge in online marketing, coupled with state of the art of food storage facilities has sparked the online trade of fruits and vegetables in India ("Now, fruits and vegetables are also just a click away", The Hindu, COIMBATORE, May 12, 2012) and elsewhere very recently. Reliable and automatic grading and sorting of objects is an essential requirement for the realtime deployment of such economic systems. The current trend in fruit and vegetable grading systems focuses mostly on imaging systems rather than a fusion of tactile and visual inspection as done by human beings [1], [2]. With the advent of low cost tactile pressure/force arrays and single point sensors, the pattern of human palpation and imaging for sorting can be performed. The aim of this work is to analyze the role of touch sensation for the above mentioned goal, thereby realizing a mechanized palpation system using machine learning techniques. In [3] we studied the response of deterministic and probabilistic learning methodologies for robot assisted fruit grading using Decision Tree and Naïve Bayes Classifiers. In this paper advanced learning techniques such as Support Vector

M.K. Kundu et al. (eds.), *Advanced Computing, Networking and Informatics - Volume 1*,
Smart Innovation, Systems and Technologies 27,
DOI: 10.1007/978-3-319-07353-8_10, © Springer International Publishing Switzerland 2014

Machine (SVM) and K-Nearest Neighbor (KNN) methods have been used for the same along with a comparative performance analysis. The rest of the paper is organized as follows. Section II presents the Background. Section III elaborates the proposed methodology of vegetable grading. The extracted parameters, the applied machine learning methodologies and experimental setup are detailed in section IV , V and VI, while results of the classification is discussed in section VII, followed by conclusion and acknowledgement in section VIII and IX respectively.

2 Background

The researchers and scientists have been inventing different procedures of automatic grading with increasing efficiency and decreasing computational cost [4],[5]. Most of the investigations in this field are done by image processing. In [6] the authors described a method to investigate the images of different phases of apples' blemishes based on neural network methods which also acted as classification method for segmenting oranges [7], where more relevant features like Red/Green color component ratio, Feret's diameter ratio and textural components were used. Spectral imaging technique for fruit grading is another effective method used [8], [9]. From these spectral images fruits are classified using soft computing techniques [8], whereas Alexios *et al.* [9] made it a realtime approach. Also Blasco *et al.* elaborated a method of automatic fruit and vegetable grading through image based features associated with a machine learning approach-Bayesian discriminant method whose accuracy is similar at par with human being [10]. Other new features such as gabor features using PCA kernel, color quantization, texture, shape and size of objects; extracted through image processing for sorting of fruits and vegetables are described in [11-13].

With the advent of tactile sensing technology, it is possible to observe the local or internal quality of an object [14-18] because vision gives only the global information of any object. The vision and tactile information were merged by Edan *et al.* to detect the freshness of the fruits [15] whereas Steinmetz *et al.* [14] proposed a sensor fusion method to detect the firmness of the peaches. Kleynen *et al.* in [16] demonstrated a compression test of apples to determine their quality. In recent years India also had taken a step forward in the automatic fruit and vegetable grading industries as seen from the works of Arivazhagan *et al.* [17] and Suresha *et al.* [18].

3 Proposed Methodology

The proposed methodology for the automated grading of fruit and vegetable consists of two parts: Hardware unit for robotic palpation and force data acquisition and Software unit for data analysis and classification.

A 1 Degree of Freedom (DOF) robotic gripper equipped with force sensors is used to palpate the objects in the predefined sequence, during which force data is acquired online by the PIC microcontroller. The predefined sequence of grasping is generated using pulse width modulated (PWM) signal by programming the same microcontroller along with the driver circuit. The PWM is generated in such a way as to initially grasp the object (shown in Fig.1) and further increase the pulse width thereby incrementing the grasp force until the force value saturates (for hard and moderately soft objects) or decreases (for soft objects) and the online force data is sent serially into PC for analysis. The acquired force data is used to estimate eight statistical parameters which can be used to distinguish between the high and low quality fruits and vegetables. Further these parameters are fed to SVM and KNN modules to generate models for real-time implementation.

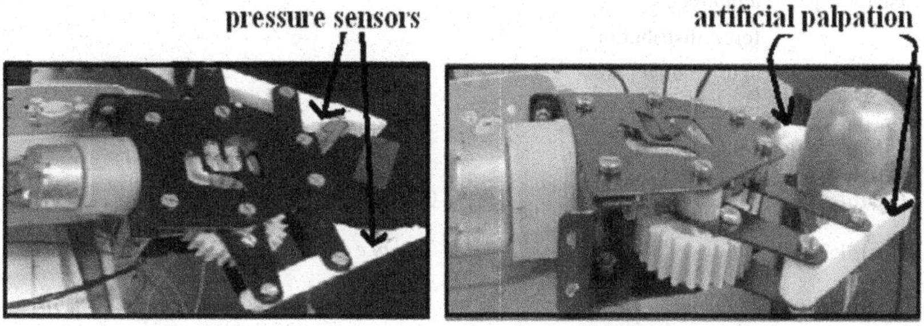

Fig. 1. Artificial palpation created by the robotic gripper

4 Parameter Selection

The calibration of the sensors is done in MATLAB to constitute a relationship between the applied force on the sensor and the relative voltage data from the OPAMP circuit. Eight statistical parameters are chosen by analyzing the force values of each instance, for the classification algorithms, briefly described in Table1, elaborated in [3]. The parameters are selected with the assumption that the force distribution for each instance of the objects is Gaussian in nature [19]. The classification between the force data of the different instances is done with the help of data mining software, WEKA and MATLAB.

5 Background of the Classifier Used

Machine learning is the way of mechanizing the process of day to day experience of human being on different objects and their way of recognizing them. Researchers have used different machine learning approaches to model human sensation applied in discriminating ripe or rotten vegetable/fruits from the rest. In this experiment two machine learning methods: SVM and KNN are used for vegetable classification. They are briefly discussed below.

Table 1. Parameters used for classification

Sl. No.	Parameter	Description	Equation
1	Mean	Average of force data	$\mu = \sum (f1 + f2)/n$
2	Variance	Dispersion in force distribution	$\sigma = \sum ((f_i - \mu)^2)/n$
3	Maximum Force	Maximum value of the force data	$Maxm = \max(f_i)$
4	Quality Factor [3]	Joint effect of the amplitude of force distribution and range of interest	$QF = Maximum(absolute(gaussiandistribution)) * range$
5	Quartile2	Outliers in force distribution	$Q2 = median(f_i);$
6	Quartile1		$Q1 = median(f_i(1:Q2));$
7	Quartile3		$Q3 = median(f_i(Q2:end))$
8	Quartile Factor [3]	Joint effect of IQR and Median of the force distribution	$Qrtlfac = IQR * Q2$

5.1 Support Vector Machine

SVM is a supervised, non probabilistic, parametric, binary learning method; which can be combined with other associate machine learning algorithms [18]. The principle is to separate a dataset into two classes through a hyperplane which maximize the distance between the support vectors (nearest data point from the boundary of a class) of each class. There are two types of SVM classification namely: linear and non linear. The nonlinear approach is required when the dataset is critical; which can be conducted through a method called kernel trick, which creates another feature dimension where the classification is linear. In this work, we used polynomial kernel trick (shown in Eqn.1) for classifying the vegetables into three different classes: Unripe, Moderately Ripe and Ripe.

$$K(x_i, x_j) = (x_i^T x_j + c)^d \quad (1)$$

Where, x_i and x_j are input data in the form of feature vector and c=constant term [20].

5.2 K-Nearest Neighbor

KNN is an instance based non-parametric approach in machine learning. The principle is: the training data is separately set aside in the memory. When a test data comes, it examines the relationship of the test data to the training data sets using distance functions. This distance function (Euclidean distance function is selected for this work, given in Eqn.2) gives the idea of the neighbors of the test data. K denotes the number of neighbors to be taken into consideration. According to the selected neighbors the algorithm assigns the class of those neighbors to the test data.

$$d(x_i, x_j) = \sqrt{\sum_{l=1}^{n} (a_l(x_i) - a_l(x_j))^2} \tag{2}$$

Where a_l(x) denotes the value of the lth attribute of instance x and $d(x_i, x_j)$ denotes the Euclidean distance between test and training data [20].

6 Experimental Setup

The visual concept of experimental setup is depicted in Fig.2. 2 types of vegetables, tomatoes and ladies-fingers were procured from the market in 8 to 10 numbers of samples of two different maturity levels, green/unripe and moderately ripe. After obtaining force data of grasping each of those samples, they were kept to become riper which resulted in the class of instances: ripe or moderately ripe tomatoes and ladies fingers. The force readings were taken in a similar way for these samples which later on helped to make the dataset of 44 tomatoes and 30 ladies-fingers. Based on eight parameters selected from the force pattern of grasping each instance the classification algorithms (both SVM and KNN) sorted out the Ripe, Moderately Ripe and Unripe tomatoes and ladies-fingers. To apply the classification algorithms, first the model of the classifier was generated through training dataset and validation was done through test dataset applying percentage split of 66%; further proceeding towards 10 fold cross validation on the whole dataset. During generation and validation of the classifier model five performance factors namely Accuracy (ACC), Relative Absolute Error (RAE), True Positive Rate (TPR), False Positive Rate (FPR), Precision were formulated which further helped to compare and analyze the efficiency of those two machine learning based classifiers. The experiments are performed on Pentium(R) 4 CPU (2.66GHz) and Intel(R) Core(TM) 2 CPU (1.73GHz), 1GB RAM associated with Windows XP Operating System.

7 Results and Discussions

According to the percentage split of the dataset, taken for classification of the two vegetables into Unripe/Green (denotes the harder objects), Moderately Ripe/Moderate (denotes the objects of medium softness) and Ripe (denotes the softer objects), 30 and 20 datasets were taken for training and 14 and 10 datasets are taken for testing respectively for tomatoes and ladies fingers. Among the datasets, three types of each of the vegetables were chosen to plot a graph of displacement of the gripper versus

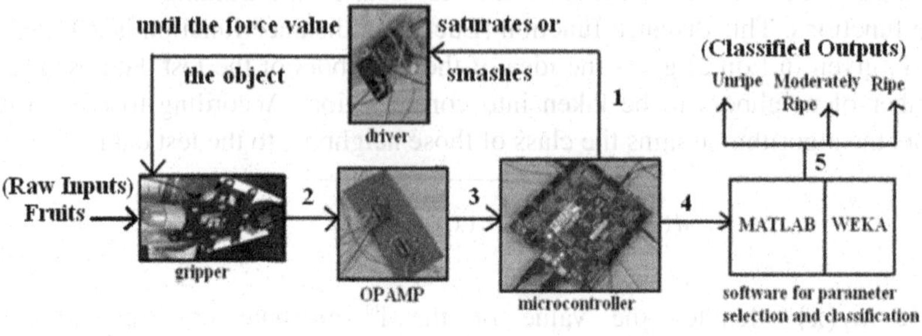

Fig. 2. Block diagram of experimental set up

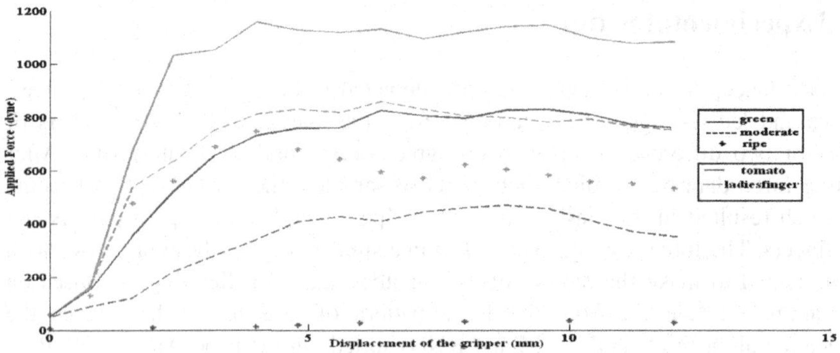

Fig. 3. Plot of displacement of the gripper versus applied force for tomatoes and ladies fingers

The applied force value on the object, delineated in Fig.3. Training means to fit a model or a hypothesis to a dataset and testing on the other hand refers to the application of the model or hypothesis to another set of data. Further cross validation refers to validate the generated model by randomly partitioning the whole dataset.

Based on the training model and the test datasets the way of classification have been discussed in the previous section. The comparison of the machine learning techniques based on the performance parameters are tabulated in Table.2.

Considering Fig.3, the green and red lines are chosen in the graph for ladies-fingers and tomatoes respectively. It is evident from the graph (Fig.3) itself that these two vegetables can be classified from their force distribution plot because the tomatoes are softer and required lesser force range to be operated on than the ladies fingers. For individual case study, it can be contrived that the Unripe samples of vegetables require the largest force then the Moderate instances (required medium force) and the Ripe samples (requires least force).

The effectiveness of SVM and KNN for vegetable grading based on their softness is analyzed below from the result rendered in Table 2. The accuracy of SVM during testing of the vegetables are 64.23% and 60% respectively for tomato and ladies-finger whereas those of KNN are 92.86% and 80%. Also accuracy during cross validation is about 86% for SVM and 86.64% and 83.33% for tomato and ladies-finger respectively. Besides, the relative absolute error is 60%-75% in the case of SVM and 19%-43% for KNN; much lesser than those obtained in SVM. Other factors like TPR, FPR and Precision also support the results obtained for accuracy and relative absolute error. Thus for this scope KNN shows better performance than SVM used with polynomial kernel.

Table 2. Comparative study of SVM during Testing and Cross validation

Type	**Performance Factors**	**Method: SVM**						**Method: KNN**					
		Tomato			Ladies-Finger			Tomato			Ladies-Finger		
		G	M	R	G	M	R	G	M	R	G	M	R
Testing	ACC		64.23			60			92.86			80	
	RAE		69.9			75.59			19.4			42.25	
Cross-validation	ACC		86.36			86.67			86.64			83.33	
	RAE		63.53			60.93			25.01			34.95	
	TPR	.91	.86	.71	.88	1	.71	.96	.86	.71	.94	.71	.71
	FPR	.1	.13	0	0	.17	0	.05	1	.03	.71	.13	.04
	Precision	.91	.75	1	1	.64	1	.96	0.8	.83	.94	.63	.83

8 Conclusion

The paper presents a mechanized way of sorting two vegetables, tomatoes and ladies-fingers of different softness depending on two machine learning based classification approaches, SVM and KNN and their performance study. The artificial grasping of objects is created through a robotic gripper, controlled by a PIC32 microcontroller and the acquired force data is treated on to generate 8 statistical parameters which help to build a model of each classifier using the training set and additionally tested and cross-validated through separate test set and whole dataset respectively.

The vegetables are classified into three categories namely Unripe, Moderately Ripe and Ripe according to their maturity stages using both SVM and KNN. The performance of SVM and KNN were considered in constrained number of dataset as the widespread deployment of such a system will be possible if system performs with limited training data, provided the data is uniform. The outcome of this classification in the form of 5 performance factors which leads to the conclusion that the classification through the instance based approach- KNN is better than the model based approach- SVM, in the aspect of mainly Accuracy, Relative Absolute Error and other factors. The polynomial SVM is used for classification which did not perform well during classification. KNN outperforms the polynomial SVM in this experiment of vegetable grading.

The paper explored a set up for mechanized vegetable grading using a robotic gripper, PIC32 microcontroller as hardware for data generation and acquisition and WEKA and MATLAB as software support for analyzing the data. The secured data is assumed to follow a Gaussian distribution to extract the parameters for classification through two dominantly uncorrelated methods, SVM (function or model based approach) and KNN (lazy or instance based approach). Future work will involve online fruit and vegetable grading technique using tactile sensing technology assisted with vision. Thus the whole system will become more feasible and reliable to overcome the real world problems.

Acknowledgements. The authors deem it a pleasure to acknowledge their sense of gratitude to Mr. Anirudh Kumar (Scientist, CSIR-CMERI, Embedded Systems Lab.), Mr. Biswarup Naiya (M.Tech. student, BESU) who helped us in every step of the work with their willingness. We are also grateful to Mr. Mukesh Prabhakar; Mr. Rajarshi Raj Ratnam Singh; Mr. Swarup Karmakar; Ms. Tanushree Banerjee (B.Tech 4th year students, NIT Durgapur) for their incisive support and association in this job.

References

1. Garcia-Ramos, F.J., Valero, C., Homer, I.: Margarita: Non-destructive fruit firmness sensors: A review. Spanish Journal of Agricultural Research 3, 61–73 (2005)
2. Khalifa, S., Komarizadeh, M.H., Tousi, B.: Usage of fruit response to both force and forced vibration applied to assess fruit firmness: A review. Australian Journal of Crop Science 5 (2011)
3. Bandyopadhyaya, I., Babu, D., Kumar, A., Roychowdhury, J.: Tactile sensing based softness classification using machine learning. In: 4th IEEE International Advance Computing Conference (2014)
4. Mahendran, R., Jayashree, G.C., Alagusundaram, K.: Application of Computer Vision Technique on Sorting and Grading of Fruits and Vegetables. Journal of Food Processing & Technology (2012)
5. Wills, R.B.H.: An introduction to the physiology and handling of fruit and vegetables. Van Nostrand Reinhold (1989)
6. Yang, Q.: Classification of apple surface features using machine vision and neural networks. Computers and Electronics in Agriculture 9, 1–12 (1993)
7. Kondo, N., Ahmad, U., Monta, M., Murase, H.: Machine vision based quality evaluation of Iyokan orange fruit using neural networks. Computers and Electronics in Agriculture 29, 135–147 (2000)
8. Guyer, D., Yang, X.: Use of genetic artificial neural networks and spectral imaging for defect detection on cherries. Computers and Electronics in Agriculture 29, 179–194 (2000)
9. Aleixos, N., Blasco, J., Navarron, F., Molto, E.: Multispectral inspection of citrus in real-time using machine vision and digital signal processors. Computers and Electronics in Agriculture 33, 121–137 (2002)
10. Blasco, J., Aleixos, N., Molto, E.: Machine vision system for automatic quality grading of fruit. Biosystems Engineering 85, 415–423 (2003)
11. Zhu, B., Jiang, L., Luo, Y., Tao, Y.: Gabor feature-based apple quality inspection using kernel principal component analysis. Journal of Food Engineering 81, 741–749 (2007)
12. Lee, D.J., Chang, Y., Archibald, J.K., Greco, C.G.: Color quantization and image analysis for automated fruit quality evaluation. In: IEEE International Conference on Automation Science and Engineering, pp. 194–199 (2008)
13. Mustafa, N., Fuad, N., Ahmed, S., Abidin, A., Ali, Z., Yit, W., Sharrif, Z.: Image processing of an agriculture produce: Determination of size and ripeness of a banana. In: International Symposium on Information Technology (2008)

14. Steinmetz, V., Crochon, M., Maurel, V., Fernandez, J., Elorza, P.: Sensors for fruit firmness assessment: comparison and fusion. Journal of Agricultural Engineering Research 64, 15–27 (1996)
15. Edan, Y., Pasternak, H., Shmulevich, I., Rachmani, D., Guedalia, D., Grinberg, S., Fallik, E.: Color and firmness classification of fresh market tomatoes. Journal of Food Science 62, 793–796 (1997)
16. Kleynen, O., Cierva, S., Destain, M.: Evolution of Pressure Distribution during Apple Compression Tests Measured with Tactile Sensors. In: International Conference on Quality in Chains. An Integrated View on Fruit and Vegetable Quality, pp. 591–596 (2003)
17. Arivazhagan, S., Shebiah, R., Nidhyanandhan, S., Ganesan, L.: Fruit recognition using color and texture features. Journal of Emerging Trends in Computing and Information Sciences, 90–94 (2010)
18. Ravikumar, M., Suresha, M.: Dimensionality Reduction and Classification of Color Features data using SVM and kNN. International Journal of Image Processing (2013)
19. Mitchell, T.: Machine learning. McGraw Hill, Burr Ridge (1997)
20. Alpaydin, E.: Introduction to Machine Learning. MIT Press (2004)

Neural Networks with Online Sequential Learning Ability for a Reinforcement Learning Algorithm

Hitesh Shah[1] and Madan Gopal[2]

[1] Department of Electronicsand Communication Engineering, G H Patel College of Engineering and Technology, Gujarat, India
[2] School of Engineering, Shiv Nadar University, Uttar Pradesh, India
iitd.hitesh@ gmail.com, mgopal@snu.edu.in

Abstract. Reinforcement learning (RL) algorithms that employ neural networks as function approximators have proven to be powerful tools for solving optimal control problems. However, neural network function approximators suffer from a number of problems like learning becomes difficult when the training data are given sequentially, difficult to determine structural parameters, and usually result in local minima or overfitting. In this paper, a novel on-line sequential learning evolving neural network model design for RL is proposed. We explore the use of minimal resource allocation neural network (mRAN), and develop a mRAN function approximation approach to RL systems. Potential of this approach is demonstrated through a case study. The mean square error accuracy, computational cost, and robustness properties of this scheme are compared with static structure neural networks.

Keywords: Reinforcement learning, online sequential learning, resource allocation neural network.

1 Introduction

Reinforcement learning (RL) paradigm is a computationally simple and direct approach to the adaptive optimal control of nonlinear systems [1]. In RL, the learning agent (controller) interacts with an initially unknown environment (system) by measuring states and applying actions according to its policy to maximize its cumulative rewards. Thus, RL provides a general methodology to solve complex uncertain sequential decision problems, which are very challenging in many real-world applications.

Often the environment of RL is typically formulated as a Markov Decision Process (MDP), consisting of a set of all states $\mathcal{S}$, a set of all possible actions $\mathcal{A}$, a state transition probability distribution $P:\mathcal{S}\times\mathcal{A}\times\mathcal{S}\rightarrow[0,1]$, and a reward function $R:\mathcal{S}\times\mathcal{A}\rightarrow\mathbb{R}$. When all components of the MDP are known, an optimal policy can be determined, e.g., using dynamic programming.

There has been a great deal of progress in the machine learning community on value-function based reinforcement learning methods [2]. In value-function based reinforcement learning, rather than learning a direct mapping from states to actions,

Smart Innovation, Systems and Technologies 27,
DOI: 10.1007/978-3-319-07353-8_11, © Springer International Publishing Switzerland 2014

the agent learns an intermediate data-structure known as a value function that maps states (or state-action pairs) to the expected long-term reward. Value-function based learning methods are appealing because the value function has well-defined semantics that enable a straight forward representation of the optimal policy, and theoretical results guaranteeing the convergence of certain methods [3], [4].

Q-learning is a common model-free value function strategy for RL [2]. Q-learning system maps every state-action pair to a real number, the Q-value, which tells how optimal that action is in that state. For small domains, this mapping can be represented explicitly by a table of Q-values. For large domains, this approach is simply infeasible. If, one deals with large discrete or continuous state and action spaces, it is inevitable to resort to function approximation, for two reasons: first to overcome the storage problem (*curse of dimensionality*), second to achieve data efficiency (i.e., requiring only a few observations to derive a near-optimal policy) by generalizing to unobserved states-action pairs. There is a large literature on RL algorithms using various value-function estimation techniques. Neural network function approximation (NN Q-learning) is one of the competent RL frameworks to deal with continuous space problem. NN generalizes among states and actions, and reduces the number of Q-values stored in lookup table to a set of NN weights. The back propagation NN with sigmoidal activation function can be used to learn the value function [5]. However, neural network function approximators suffer from a number of problems like learning becomes difficult when the training data are given sequentially, difficult to determine structural parameters, and usually result in local minima or over fitting. Consequently, NN function approximation can fail at finding the correct mapping from input state-action pairs to output Q-values.

In NN, the methods used to update the network parameters can broadly be divided into batch learning and sequential learning. In batch learning, it is assumed that the complete training data are available before training commences. The training usually involves cycling the data over a number of epochs. In sequential learning, the data arrive one by one or chunk by chunk, and the data will be discarded after the learning of the data is completed. In practical applications, new training data may arrive sequentially. In order to handle this using batch learning algorithms, one has to retrain the network all over again, resulting in a large training time.

Neural networks with on-line sequential learning ability employ a procedure that involves growing and/or pruning networks iteratively as the training data are presented. Learning is achieved through a combination of new neuron allocation and parameter adjustment of existing neurons. New neurons are added if presented training patterns fall outside the range of existing network neurons. Otherwise, the network parameters are adapted to better fit the patterns.

A significant contribution was made by Platt [6] through the development of an algorithm called resource-allocating network (RAN) that adds hidden units to the network based on the novelty of the new data in the process of sequential learning. An improved approach called RAN via extended Kalman filter (RANEKF) [7] was provided to enhance the performance of RAN by adopting an extended Kalman filter (EKF) instead of the least means square (LMS) method for adjusting network parameters. They all start with no hidden neurons and allocate the new hidden units when some criteria are satisfied. However, once the hidden neuron is generated, it will never be removed. The minimal resource allocating network (mRAN) [8, 9] is an improved version of RAN, as growing and pruning neurons are achieved with a certain criteria based on sliding windows. All the weights and the center of each hidden neuron are tuned until certain error condition is satisfied. So, a compact network can be implemented. Other improvements of RAN developed in [10] and [11], take into the consideration the pseudo-Gaussian (PG) function and orthogonal techniques including QR factorization and singular value decomposition (SVD), and have been applied to the problem of time series analysis. In [12, 13], a growing and pruning RBF (GPA-RBF) neural network, and generalized growing and pruning RBF(GGPA-RBF) neural network approaches have been presented. The idea is to only adjust the weights of the neuron that is nearest to the most recently received input, instead of the weights of all the neurons. Significance of a neuron is measured by the average information contained in that neuron. It requires estimating the distributions and the range of input samples, as well as choosing some of the parameters appropriately before training. An online sequential extreme learning machine (OS-ELM) [14] has been proposed for learning data one-by-one and/or chunk-by-chunk with fixed or varying chunk size. However, the parameters of hidden nodes are randomly selected and only the output weights are analytically updated based on the sequentially arriving data. It should be noted that the structure in OS-ELM is fixed in advance by user.

Vamplew and Ollington in [15] compare the global (static structure such as multi-layer perceptron) versus local constructive (dynamic structure such as resource allocation network) function approximation for on-line reinforcement learning. It has been shown that the globally-constructive algorithms are less stable, but that on some tasks they can achieve similar performance to a locally-constructive algorithm, whilst producing far more compact solutions. In contrast, the RAN performed well on three benchmark problems—Acrobot, Mountain-Car, and Puddleworld, for both the on-line and off-line measures. However, this performance was only achieved by choosing parameters that allowed the RAN to create a very large number of hidden neurons.

Shiraga *et al.*. [16] proposed a neural network model with incremental learning ability for reinforcement learning task, with resource allocating network with long-term memory (RAN-LTM). From the simulation results, they showed that their proposed model could acquire more accurate action-values as compared with the following three approaches to the approximation of action–value functions: tile coding, a conventional NN model, and a version of RAN-LTM [17].

In this paper, a novel online sequential learning neural network model design for RL is proposed. We explore the use of constructive approximators (such as mRAN,an improved version of RAN) which build their structure on-line during learning/training, starting from minimal architecture. Use of mRAN We develop a mRAN function approximation approach to RL system and demonstrate its potential through a case study – two-link robot manipulator. The mean square error accuracy, computational cost and robustness properties of this scheme are compared with the scheme based on global function approximation with static structure (such as NN).

The remaining part of the paper is organized as follows: Section 2 presents architecture of mRAN and value function approximation for RL systems. Section 3 gives details of on-line sequential learning of mRAN algorithm. Section 4 compares the empirical performance on the basis of simulation results. Finally, in Section V, the conclusions are presented.

2 Approximation of Value Function Using mRAN

A minimal resource allocation network (mRAN) proposed by Yingwei *et al.* [8] is a sequential learning RBF network, which combines the growth criteria of the resource allocation network (RAN) of Platt [6] with a pruning strategy to realize a minimal network structure. It is an improvement on RAN and the RANEKF algorithm of [7] that has the ability to grow and prune the hidden neurons to ensure a parsimonious structure.

A novel value function approximator for online sequential learning on continuous state domain based on mRAN is proposed in this paper. Fig. 1(a) shows architectural view of the mRAN function approximation approach to RL (commonly known as, *Q*-learning) system.

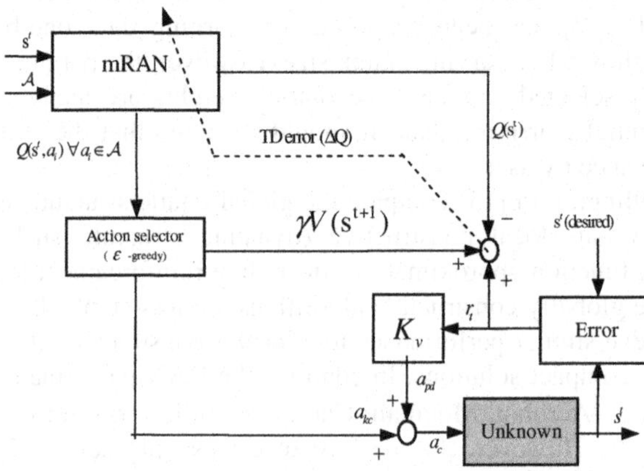

Fig. 1(a). mRAN controller architecture

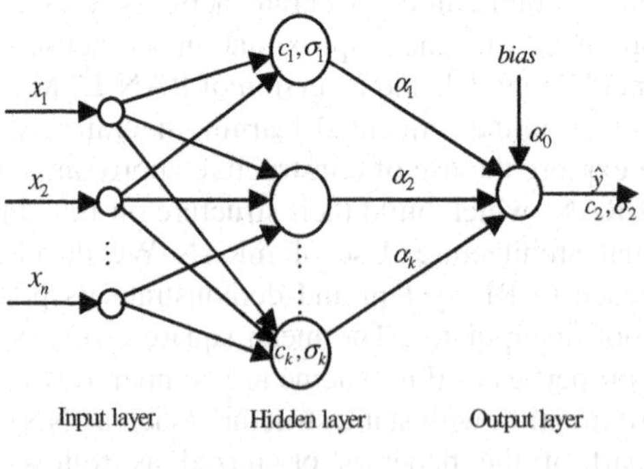

Fig. 1(b). The structure of mRAN

The state-action pair (s^t, a_t); where $\mathrm{s}^t = \{s_1^t, s_2^t, \ldots, s_n^t\} \in \mathcal{S}$ is the current system state and a_t is each possible discrete control action in action set $\mathcal{A} = \{a_i\};\ i = 1, \ldots, m$ is the input of mRAN. The output of the network (Fig. 1(b)), the estimated Q-value corresponding to (s^t, a_t) with K hidden neurons, has the following form:

$$Q(\mathrm{s}^t, a_t) = \hat{y}^t = f(\mathrm{x}^t) = \alpha_0 + \sum_{k=1}^{K} \alpha_k \phi_k(\mathrm{x}^t) \tag{1}$$

where x is the input vector ($\mathrm{x}^t = [x_1, x_2, \ldots, x_n]^T = (\mathrm{s}^t, a_t)$) of the network, α_k is the weight connecting the kth hidden neuron to the output neuron, α_0 is the bias term, and $\phi_k(\mathrm{x})$ is the response of the kth hidden neuron to the input vector x, defined by the following Gaussian function:

$$\phi_k(\mathrm{x}^t) = \exp\left(-\frac{\left\| \mathrm{x}^t - \mathrm{c}_k \right\|^2}{\sigma_k^2} \right) \tag{2}$$

c_k and σ_k refer to the center and width of the kth hidden neuron respectively, and $\|.\|$ indicates the Euclidean norm.

Training samples of mRAN should be obtained sequentially as the interaction between the learning agent (controller) and its environment (plant). The learning process of mRAN involves the allocation of new hidden neurons as well as adaptation of network parameters. Its learning begins with no hidden neurons. As the training samples are sequentially received, the network built based on certain growing and pruning criteria. The agent's action is selected based on the outputs of mRAN. The *temporal difference* (TD) error e^t for the output at time t is defined as follows:

$$e^t = \Delta Q = \eta[r_t + \gamma \max_{a \in \mathcal{A}} Q(\mathrm{s}^{t+1}, a) - Q(\mathrm{s}^t, a_t)] \tag{3}$$

It is an online sequential learning algorithm that learns an approximate state-action value function $Q(\mathrm{s}^t, a_t)$ that converges to the optimal function Q^* (commonly called Q-value). Online version is given by

$$Q(\mathrm{s}^t, a_t) \leftarrow Q(\mathrm{s}^t, a_t) + \Delta Q \tag{4}$$

where $\mathrm{s}^t \xrightarrow{r_t} \mathrm{s}^{t+1}$ is the state transition under the control action $a_t \in \mathcal{A}$ at instant t. In specific, control actions are selected using an exploration/exploitation policy in order to explore the set of possible actions and acquire experience through the online RL signals [2]. We use a pseudo-stochastic (ε-greedy)exploration as in [18].In ε-greedy exploration, we gradually reduce the exploration (determined by the ε parameter) according to some schedule; we have reduced ε to its 90 percent value after every 50 iterations. The lower limit of parameter ε has been fixed at 0.002 (to maintain exploration). Learning rate parameter, $\eta \in (0,1]$, can be used to optimize the speed of learning, and $\gamma \in (0,1]$ is the discount factor which is used to control the trade-off between immediate and future reward.

3 mRAN Learning Algorithm

mRAN learning algorithm conducts the following two basic growing and pruning steps [8,9]:

Growing step: The following three conditions are checked to decide on allocation of a new hidden neuron:

$$\left\|e^t\right\| = \left\|y^t - f(\mathrm{X}^t)\right\| > e_{\min} \tag{5}$$

$$\left\|\mathrm{X}^t - \mathrm{c}^*\right\| > \varepsilon_t \tag{6}$$

$$e_{rms}^t = \sqrt{\frac{\sum_{i=n-(M-1)}^{n} \left\|e_i\right\|^2}{M}} > e'_{\min} \tag{7}$$

where c^* is the center of the hidden neuron which is closest to the current input X^t; $e_{\min}$, ε_t, and $e'_{\min}$ are thresholds to be selected appropriately. $e_{\min}$ is an instantaneous error check and is used to determine if the existing nodes are insufficient to obtain a network output. ε_t ensures that the new input is sufficiently far from the existing centers and is given by $\varepsilon_t = \max[\varepsilon_{\max}\lambda^t, \varepsilon_{\min}]$ where λ is a decay constant ($0 < \lambda < 1$). $e'_{\min}$ is used to check the root mean square error value (e_{rms}^t) of the output error over a sliding window size (M) before adding a hidden neuron.

If the three error criteria in the growing step are satisfied, a new hidden neuron is added to the network. The parameters associated with the new hidden neuron are assigned as follows:

$$\begin{aligned} \alpha_{k+1} &= e^t \\ c_{k+1} &= \mathrm{X}^t \\ \sigma_{k+1} &= \kappa\left\|\mathrm{X}^t - \mathrm{c}^*\right\| \end{aligned} \tag{8}$$

where κ is an overlap factor (kappa), which determines the overlap of the hidden neurons responses with the hidden units in the input space.

If the observation does not meet the criteria for adding a new hidden neuron, extended Kalman filter (EKF) learning algorithm is used to adjust the network parameters w, given as:

$$\mathrm{w} = [\alpha_0, \alpha_1, \mathrm{c}_1^T, \sigma_1, \ldots, \alpha_K, \mathrm{c}_K^T, \sigma_K]^T \tag{9}$$

The update equations are given by

$$\mathrm{w}^t = \mathrm{w}^{t-1} + \mathrm{k}^t e^t$$

where k^t is the Kalman gain vector given by: $\mathrm{k}^t = P^{t-1}B^t[R^t + B^{t^T} P^{t-1}B^t]^{-1}$ and B^t is the gradient vector and has the following form:

$$B^t = \nabla_w \hat{y}^t = [I, \phi_1(\mathrm{X}^t)I, \phi_1(\mathrm{X}^t)\frac{2\alpha_1}{\sigma_1^2}(\mathrm{X}^t - c_1)^T, \phi_1(\mathrm{X}^t)\frac{2\alpha_1}{\sigma_1^2}\left\|\mathrm{X}^t - c_1\right\|^2, \ldots$$
$$\phi_K(\mathrm{X}^t)I, \phi_K(\mathrm{X}^t)\frac{2\alpha_K}{\sigma_K^2}(\mathrm{X}^t - c_K)^T, \phi_K(\mathrm{X}^t)\frac{2\alpha_K}{\sigma_K^2}\left\|\mathrm{X}^t - c_K\right\|^2]^T \tag{10}$$

R^t is the variance of the measurement noise, P^t is the error covariance matrix which is updated by $P^t = [I - k^t B^{t^T}]P^{t-1} + QI$; where Q is a scalar that determines the allowed random step in the direction of gradient vector, $K(p+q+1)+p$ is size of the positive definite symmetric matrix P^t. Here $q, K,$ and p denote the number of input, hidden, and output neurons, respectively. When a new hidden neuron is added, the dimensionality of P^t is increased by

$$P^t = \begin{pmatrix} P^{t-1} & 0 \\ 0 & P^0 I \end{pmatrix} \tag{11}$$

The new rows and columns are initialized by P^0, where P^0 is an estimate of the uncertainty in the initial values assigned to the parameters. The dimension of identity matrix I is equal to the number of new parameters introduced by adding new hidden neuron.

Pruning step: To keep mRAN network to the minimal size and make sure that there are no superfluous hidden neurons existing in the network, a pruning strategy is employed.

For each observation, the normalized hidden neuron output value u_k^t should be examined to decide whether it should be removed or not. The normalized output u_k^t is expressed by the following equation:

$$u_k^t = \left\|\frac{O_k^t}{O_{\max}^t}\right\|; \; k = 1, 2, \ldots, K \tag{12}$$

$$O_k^t = \alpha_k \exp\left(-\frac{\left\|\mathrm{x}^t - \mathrm{c}_k\right\|^2}{\sigma_k^2}\right) \tag{13}$$

where O_k^t is the output for the kth hidden neuron at time instance *t*, and $\left\|O_{\max}^t\right\|$ is the largest absolute hidden neuron output value at *t*.These normalized values are then compared with a threshold δ . If any of them falls below this threshold for *M* consecutive observations, then this particular hidden neuron is removed from the network; and the dimensionality of the covariance matrix (P^t) in the EKF learning algorithm is adjusted by removing rows and columns which are related to the pruned unit.

4 Simulation Experiments

To demonstrate the usefulness of mRAN function approximator in reinforcement learning framework, we conducted experiments using the well-known two-link robot manipulator tracking control problem.

4.1 Two-Link Robot Manipulator

We consider a two-link robot manipulator as the plant. There are two links (rigid) of lengths l_1 and l_2, and masses m_1 and m_2, respectively. The joint angles are θ_1 and θ_2, and g is the acceleration due to gravity. $\tau = [\tau_1 \quad \tau_2]^T$ is the input torque vector, applied on the joints of the robot. Dynamic equations for a two-link robot manipulator can be represented as [19]

$$\begin{bmatrix} \alpha+\beta+2\eta\cos\theta_2 & \beta+\eta\cos\theta_2 \\ \beta+\eta\cos\theta_2 & \beta \end{bmatrix}\begin{bmatrix} \ddot{\theta}_1 \\ \ddot{\theta}_2 \end{bmatrix} + \begin{bmatrix} -\eta(2\dot{\theta}_1\dot{\theta}_2+\dot{\theta}_2^2)\sin\theta_2 \\ \eta\dot{\theta}_1^2\sin\theta_2 \end{bmatrix} + \begin{bmatrix} \alpha e_1\cos\theta_1+\eta e_1\cos(\theta_1+\theta_2) \\ \eta e_1\cos(\theta_1+\theta_2) \end{bmatrix} + \begin{bmatrix} \tau_{1dis} \\ \tau_{2dis} \end{bmatrix} = \begin{bmatrix} \tau_1 \\ \tau_2 \end{bmatrix} \quad (14)$$

where $\tau_{dis} = [\tau_{1dis} \quad \tau_{2dis}]^T$ is the disturbance torque vector, and $\alpha = (m_1 + m_2)l_1^2$; $\beta = m_2 l_2^2$; $e_1 = g / l_1$ .Manipulator parameters are: $l_1 = l_2 = 1$ m ; $m_1 = m_2 = 1$ kg. Desired trajectory is: $\theta_{1d} = \sin(\pi t)$, and $\theta_{2d} = \cos(\pi t)$

4.2 Controller Setup

In order to setup mRAN function approximator in RL framework (mRAN controller), we used simulation parameters and learning details as follows:

System state space (continuous) has four variables, i.e., $s^t = [\theta_1 \quad \theta_2 \quad \dot{\theta}_1 \quad \dot{\theta}_2]^T = [s_1 \quad s_2 \quad \dot{s}_1 \quad \dot{s}_2]^T$. Each link state has two input variables: $(\theta_i^k, \dot{\theta}_i^k)$; $i = 1,2$. We define tracking error vector as: $e_r^t = \theta_d^t - \theta^t$ and cost function $r_t = \dot{e}_r^t + \Lambda e_r^t, \Lambda = \Lambda^T > 0$ with $\Lambda = \mathrm{diag}\{30,20\}$. $\theta_d^t = [\theta_{d1}^t \quad \theta_{d2}^t]$ represents robot manipulator's desired trajectory vector. Controller action sets for link1 and link2 are $|U|(1) = [-20 \quad 0 \quad 20]$ Nm, and $|U|(2) = [-2 \quad 0 \quad 2]$ Nm, respectively. Exploration level ε decays from $0.5 \to 0.002$ over the iterations. The discount factor γ is set to 0.8; learning-rate parameter η is set to 0.2, and PD gain matrix $K_v = \mathrm{diag}\{20,20\}$. We deliberately introduce deterministic noise of $\pm 1\%$ of control effort with a probability of (1/3), for stochastic simulation.

4.3 Network Topologies and Learning Details

mRAN Q-learning Controller **(mRANQC):** In implementation, the mRAN has as input the state-action pair and as output, the Q-value corresponding to the state-action pair. In particular, the mRAN network begins with no hidden neurons. As the data are sequentially received, the network is built-up based on certain growing and pruning criteria. The input and output layers use the identity as transfer function, the hidden layer uses the Gaussian RBF as the activation function. For simplicity, the controller uses two function approximators; one each for the two links. The initial free parameters are selected for constructing mRAN network as follows: The size of sliding data window, $M = 25$, the thresholds, $\varepsilon_{max} = 0.8$, $\varepsilon_{min} = 0.5$, $e_{min} = 0.02$, $e'_{min} = 0.002$, $\delta = 0.05$, overlap factor, $\mathcal{K} = 0.87$, the decay constant, $\lambda = 0.97$, and the estimate of the uncertainty in the initial values assigned to the parameters, $P^0 = 1.01$.

***Neural Network Q-learning Controller* (NNQC):** Structurally, NNQC remains same as mRANQC; the major difference is that the NN configuration comprises of one or two hidden layers containing a set of neurons with tan-sigmoidal activation function. The number of layers and neurons depends on the complexity and the dimensions of the value-function to be approximated. We consider 18 hidden neurons. The initialization of the NN weights is done randomly, and length of training samples (l) for batch mode processing is chosen as 100.

In mRANQC (or NNQC) controller implementations, we have used controller structure with an inner PD loop. Control action to the robot manipulator is a combination of an action generated by an adaptive learning RL signal through mRAN (or NN) and a fixed gain PD controller signal. The PD loop will maintain stability until mRAN (or NN) controller learns, starting with zero initialized Q-values. The controller, thus, requires no offline learning.

5 Results and Discussion

Simulations were carried out to study the learning performance, and robustness against uncertainties, for mRAN learning approach on two-link robot manipulator control problem. MATLAB 7.10 (R2010a) has been used as simulation tool. To analyze the mRAN algorithm for computational cost, accuracy, and robustness, we compare the proposed approach with NN reinforcement learning approach.

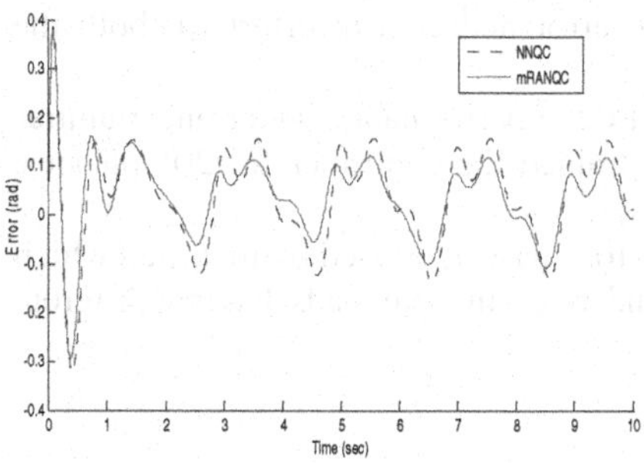

Fig. 2(a). Output tracking error (link1)

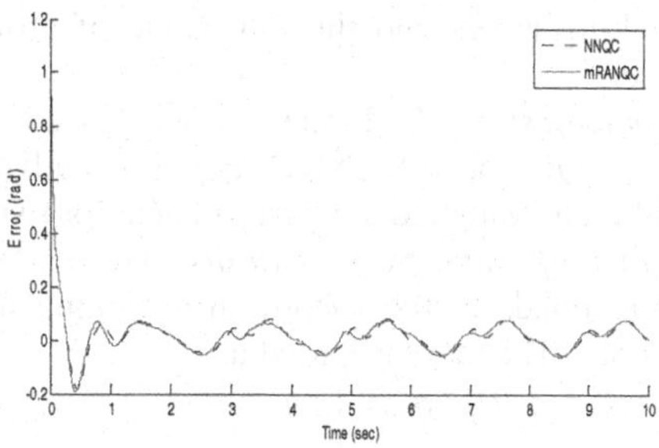

Fig. 2(b). Output tracking error (link2)

Learning performance study: The physical system has been simulated for a single run of 10 sec using fourth-order Runge-Kutta method, with fixed time step of 10 msec. Fig. 2 and Fig. 3 show the output tracking error (both the links) and control torque (both the links) for both the controllers • NNQC and mRANQC, respectively.

Table I tabulates the mean square error, absolute maximum error (max $|e(t)|$), and absolute maximum control effort (max $|\tau|$) under nominal operating conditions.

Table 1. Comparison of controllers

Controller	MSE (rad)		max $\|e(t)\|$ (rad)		max $\|\tau\|$ (Nm)		Training Time (sec)
	Link 1	Link 2	Link 1	Link 2	Link 1	Link 2	------
NNQC	0.0120	0.0065	0.1568	0.0779	83.5281	33.6758	376.78
mRANQC	0.0084	0.0065	0.1216	0.0772	81.3286	32.5861	31.55

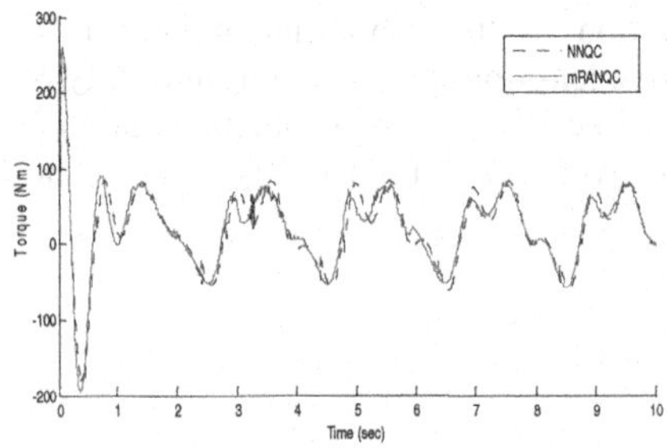

Fig. 3(a). Control torque (link1)

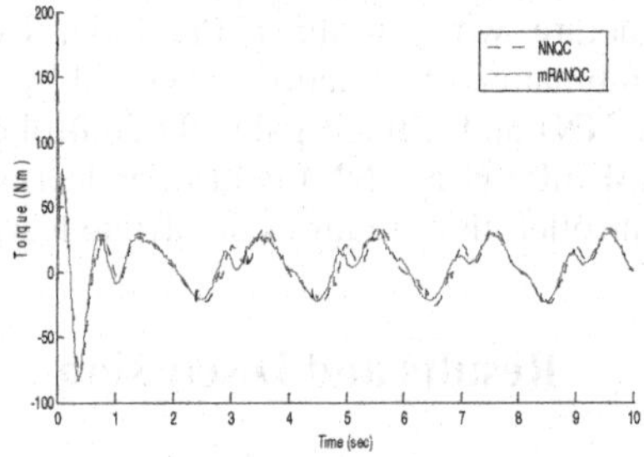

Fig. 3(b). Control torque (link2)

From the results (Fig. 2 and Fig. 3, Table 1), we observe that training time for mRANQC is lesser than NNQC. mRANQC outperforms NNQC, in terms of lower tracking errors and the low value of absolute error and control effort for both the links.

Robustness Study: In the following, we compare the performance under uncertainties of NNQC and mRANQC. For this study, we trained the controller for 20 episodes, and then evaluated the performance for two cases.

Effect of Payload Variations: The end-effector mass is varied with time, which corresponds to the robotic arm picking up and releasing payloads having different masses. The mass is varied as:

(a) $t < 2$ s ; $m_2 = 1$ kg

(b) $2 \le t < 3.5$ s ; $m_2 = 2.5$ kg

(c) $3.5 \le t < 4.5$ s ; $m_2 = 1$ kg

(d) $4.5 \le t < 6$ s ; $m_2 = 4$ kg

(e) $6 \le t < 7.5$ s ; $m_2 = 1$ kg

(f) $7.5 \le t < 9$ s ; $m_2 = 2$ kg

(g) $9 \le t < 10$ s ; $m_2 = 1$ kg .

Figs. 4(a) and (b) show the output tracking errors (both the links) and Table II tabulates the mean square error, absolute maximum error (max $|e(t)|$), and absolute maximum control effort (max $|\tau|$) at payload variations with time.

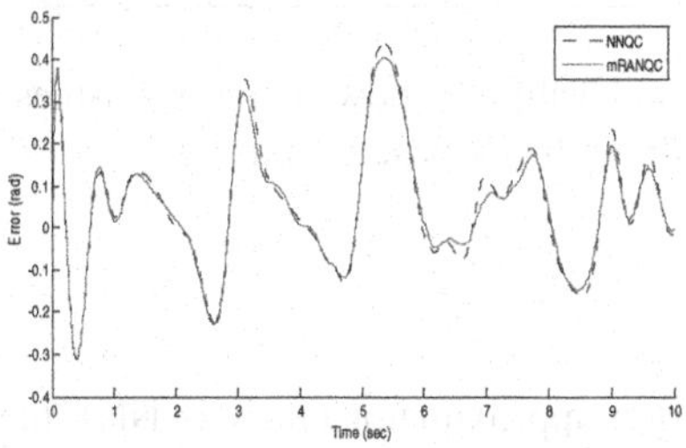

Fig. 4(a). Output tracking error (link1)

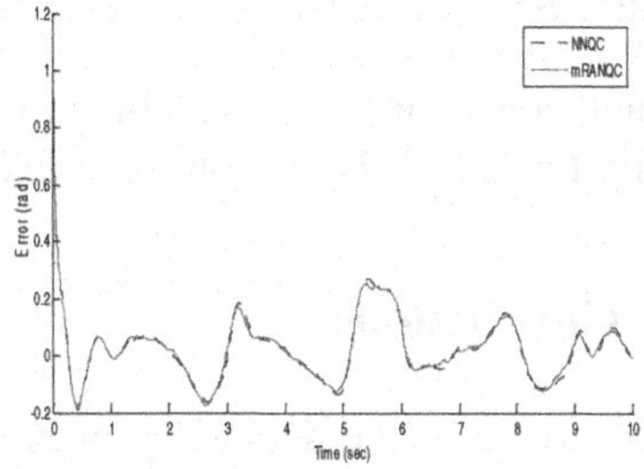

Fig. 4(b). Output tracking error (link2)

Table 2. Comparison of controllers

Controller	MSE (rad)		max $|e(t)|$ (rad)		max $|\tau|$ (Nm)	
	Link 1	Link 2	Link 1	Link 2	Link 1	Link 2
NNQC	0.0261	0.0141	0.4379	0.9034	276.5605	400.3960
mRANQC	0.0227	0.0134	0.4034	0.9060	281.1594	400.4040

***Effects of External Disturbances*:** A torque disturbance τ_{dis} with a sinusoidal variation of frequency 2π rad/sec, was added with time to the model. The magnitude of torque disturbance is expressed as a percentage of control effort. The magnitude is varied as: (a) $t < 2$ s ; 0% (b) $2 \leq t < 3.5$ s ; 0.2% (c) $3.5 \leq t < 4.5$ s ; 0% (d) $4.5 \leq t < 6$ s ; 0.8% (e) $6 \leq t < 7.5$ s ; 0% (f) $7.5 \leq t < 9$ s ; −0.2% (g) $9 \leq t < 10$ s ; 0%. Figs. 5(a) and (b) show the output tracking errors (both the links) and Table III tabulates the mean square error, absolute maximum error (max $|e(t)|$), and absolute maximum control effort (max $|\tau|$) for torque disturbances added with time to the model variation.

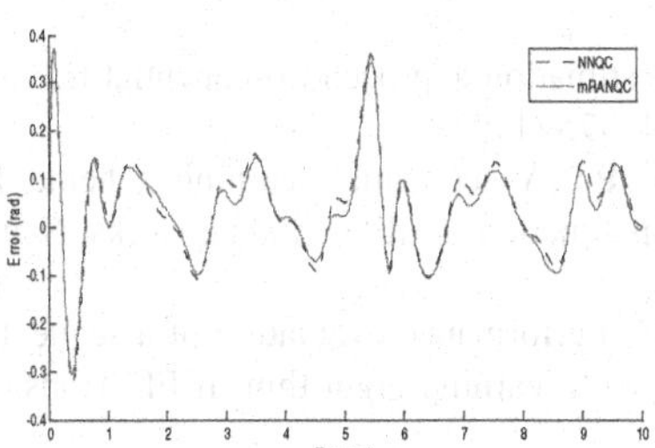

Fig. 5(a). Output tracking error (link1)

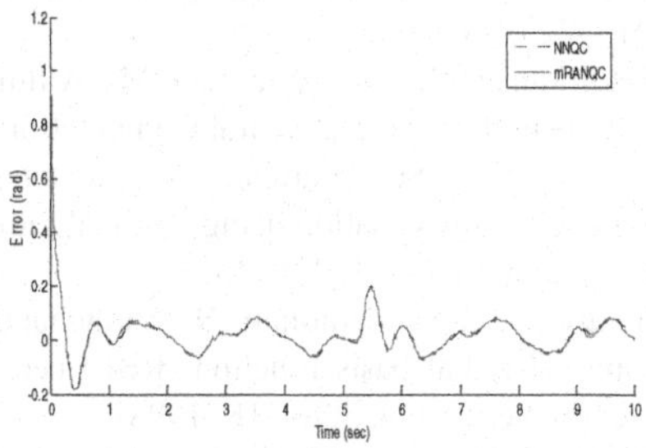

Fig. 5(b). Output tracking error (link2)

Table 3. Comparison of controllers

Controller	MSE (rad)		max $\|e(t)\|$ (rad)		max $\|\tau\|$ (Nm)	
	Link 1	Link 2	Link 1	Link 2	Link 1	Link 2
NNQC	0.0127	0.0075	0.3736	0.9078	275.4872	399.7907
mRANQC	0.0110	0.0073	0.3715	0.9062	270.0240	400.3965

Simulation results (Fig. 4 and Fig. 5, Table 2 and Table 3) shows better robustness property for mRAN based controller in comparison with NN based controller.

6 Conclusions

In order to tackle the deficiency of global function approximator (such as NN), the minimal resource allocation network (mRAN) is introduced in RL control system and a novel online sequential learning algorithm based on mRAN presented. mRAN is a sequential learning RBF network and has ability to grow and prune the hidden neurons to ensure a parsimonious structure that is well suited for real-time control applications.

From the simulation results, it is obvious that training time in mRAN based RL system is much shorter compared to the NN based RL system. This is an important feature for RL control systems from stability considerations. This feature is achieved without any loss of performance.

References

1. Sutton, R.S., Barto, A.G., Williams, R.J.: Reinforcement learning is direct adaptive optimal control. IEEE Control Syst. Mag. 12(2), 19–22
2. Sutton, R.S., Barto, A.G.: Reinforcement learning: An introduction. MIT Press, Cambridge
3. Watkins CJCHLearning with delayed rewards. Ph. D. Thesis, University of Cambridge (1989)
4. Singh, S., Jaakkola, T., Littman, M., Szpesvari, C.: Convergence results for single step on-policy reinforcement learning algorithms. Machine Learning 38, 287–308 (2000)
5. Hagen, S.T., Kröse, B.: Neural Q-learning. Neural Comput. & Applic. 12, 81–88 (2003)
6. Platt, J.: A resource-allocating network for function interpolation. Neural Computation 3, 213–225 (1991)
7. Kadirkamanathan, V., Niranjan, M.: A function estimation approach to sequential learning with neural networks. Neural Computation 5, 954–975 (1993)
8. Yingwei, L., Sundararajan, N., Saratchandran, P.: A sequential learning scheme for function approximation using minimal radial basis function (RBF) neural networks. Neural Computation 9, 461–478 (1997)
9. Yingwei, L., Sundararajan, N., Saratchandran, P.: Performance evaluation of a sequential minimal radial basis function (RBF) neural network learning algorithm. IEEE Trans. on Neural Network 9, 308–318 (1998)
10. Rojas, I., Pomares, H., Bernier, J.L., Ortega, J., Pino, B., Pelayo, F.J., Prieto, A.: Time series analysis using normalized PG-RBF network with regression weights. Neurocomputing 42, 267–285 (2002)

11. Salmeron, M., Ortega, J., Puntonet, C.G., Prieto, A., Improved, R.A.N.: sequential prediction using orthogonal techniques. Neurocomputing 41, 153–172 (2001)
12. Huang, G.B., Saratchandran, P., Sundararajan, N.: An efficient sequential learning algorithm for growing and pruning RBF (GAPRBF) networks. IEEE Transcript on System Man and Cybern. B 34, 2284–2292 (2004)
13. Huang, G.B., Saratchandran, P., Sundararajan, N.: A generalized growing and pruning RBF (GGAP-RBF) neural network for function approximation. IEEE Transcript on Neural Network 16, 57–67 (2005)
14. Liang, N.Y., Huang, G.B., Saratchandran, P., Sundararajan, N.: A fast and accurate online sequential learning algorithm for feed forward networks. IEEE Trans. on Neural Network 17, 1411–1423 (2006)
15. Vamplew, P., Ollington, R.: Global versus local constructive function approximation for on-line reinforcement learning. In: Zhang, S., Jarvis, R.A. (eds.) AI 2005. LNCS (LNAI), vol. 3809, pp. 113–122. Springer, Heidelberg (2005)
16. Shiraga, N., Ozawa, S., Abe, S.: A reinforcement learning algorithm for neural networks with incremental learning ability. In: Proceeding of the 9th International Conference on Neural Information Processing, vol. 5, pp. 2566–2570 (2002)
17. Kobayashi, M., Zamani, A., Ozawa, S., Abe, S.: Reducing computations in incremental learning for feed-forward neural network with long-term memory. In: Proc. International Joint Conference on Neural Networks, pp. 1989–1994 (2001)
18. Shah, H., Gopal, M.: A fuzzy decision tree based robust Markov game controller for robot manipulators. International Journal of Automatic and Control 4(4), 417–439 (2010)
19. Green, S.J.Z.: Dynamics and trajectory tracking control of a two-link robot manipulator. J. Vibration Control 10(10), 1415–1440 (2004)

Scatter Matrix versus the Proposed Distance Matrix on Linear Discriminant Analysis for Image Pattern Recognition

E.S. Gopi and P. Palanisamy

Department of Electronics and Communication Engineering
National Institute of Technology Trichy
Trichy-620015, India
esgopi@nitt.edu

Abstract. In this paper, we explore the performance of Linear Discriminant Analysis (LDA) by replacing the scatter matrix with the distance matrix for image classification. First we present the intuitive arguments for using the distance matrix in LDA. Based on the experiments on face image database, it is observed that the performance in terms of prediction accuracy is better when the distance matrix is used instead of scatter matrix in Linear Discriminant Analysis (LDA) under certain circumstances. Above all, it is observed consistently that the variation of percentage of success with the selection of training set is less when distance matrix is used when compared with the case when scatter matrix is used. The results obtained from the experiments recommend the usage of distance matrix in place of scatter matrix in LDA. The relationship between the scatter matrix and the proposed distance matrix is also deduced.

Keywords: Principal component analysis (PCA), Linear discriminant analysis (LDA), Inner product, Similarity measurements.

1 Introduction

In image recognition, the techniques like Principal Component Analysis [PCA] and Independent Component Analysis [ICA] which uses second order statistics [1] and higher order statistics [2] respectively are used for dimensionality reduction to reduce the computational complexity of the data. These techniques do not use the knowledge of the data labels. But, Linear Discriminant Analysis [LDA] bases are used to map the original higher dimensional feature vectors to the lower dimensional space for better class separability with the preservation of the class differences [3], [4]. Fisher's Linear Discriminant was first developed by Robert Fisher in 1936 for taxonomic classification [5]. It has been further used in many applications which includes, speech recognition [6], document classification [7], and face recognition [8]. The LDA basis are obtained by maximizing the objective function $J_S(W) = \frac{trace(W^T S_{BW} W)}{trace(W^T S_W W)}$ [3](refer Section 2 for notations). Maximizing $J_S(W)$ is equivalent to maximizing the sum of the variances of the

M.K. Kundu et al. (eds.), *Advanced Computing, Networking and Informatics - Volume 1*, Smart Innovation, Systems and Technologies 27,
DOI: 10.1007/978-3-319-07353-8_12,

elements of the vectors belonging to the identical class in the lower dimensional space. However, it minimizes the sum of the variances of the elements of the centroid vectors of all the classes in the lower dimensional space. This helps in bringing the vectors closer to each other if it belongs to the identical class and simultaneously separating the classes in the lower dimensional space.

The optimal solution (LDA bases) to the objective function is the set of Eigenvectors corresponding to the significant Eigenvalues of the matrix $S_{BW}^{-1}S_W$ [11]. The columns of the matrix S_W are in the column space spanned by the feature vectors used to obtain the scatter matrix. The rank of the scatter matrix S_W should be at the most equal to the number of vectors used in the training set. Hence, if the number of elements in the feature vector is larger than the number of vectors used to compute the scatter matrix S_W, the matrix S_W becomes singular and hence S_W^{-1} cannot be computed. This problem is known as small sample size problem. There have been many methods attempted to solve this problem [3], [8], [9].

In this paper, we explore the performances of the LDA using the proposed distance matrix as the replacement of scatter matrix in LDA. Inner-product LDA (special case of kernel LDA [11]) is used to reduce computation time. The problem of "small sample size problem" was taken care during realization by using the method proposed in [9].

2 Idea Behind the Proposed Distance Matrix

The traditional LDA [3] uses the within-class scatter matrix ($S_W = \sum_{i=1}^{r} n_i S_i$) and the between-class scatter matrix with scaling ($S_{BW} = \sum_{i=1}^{r} n_i(c_i - c)(c_i - c)^T$), where S_i is the scatter matrix of the ith class, n_i is the number of vectors in the ith class, c_i is the centroid vector of the ith class, c is the mean vector of all the centroids, x_{ij} is the jth column vector of the ith class and r is the total number of classes.

Slightly modified scatter matrix without scaling [10] can also be used in place of S_{BW} as mentioned in (1).

$$S_B = \sum_{i=1}^{r}(c_i - c)(c_i - c)^T \tag{1}$$

It is noted that the trace of the S_W matrix gives the measurement about how the vectors are closer to each other within the class in the higher dimensional space and similarly, the trace of the S_{BW} and S_B matrices measure the separation of various classes in the higher dimensional space. This is because the trace of the scatter matrix S_W used in classical LDA measures the summation of the squared distances between the vectors and the centroid of the corresponding classes. In classical LDA, this is the parameter to be minimized in the lower dimensional space. Similarly the trace of the scatter matrices S_{BW} and S_B measure the summation of the squared distances between the centroid vectors and the mean of the centroid vectors .

Consider three vectors in the particular class as shown in the Fig. 1(a). Let the centroid of the class be represented as C in Fig. 1(a). Using the conventional method (scatter matrix) ,the sum of the distances between the vectors with its class is used as the measure of closeness. Instead,the proposed technique (distance matrix) involves in computing the sum of the Euclidean distances between the vectors 1 with 2, 1 with 3 and 2 with 3.

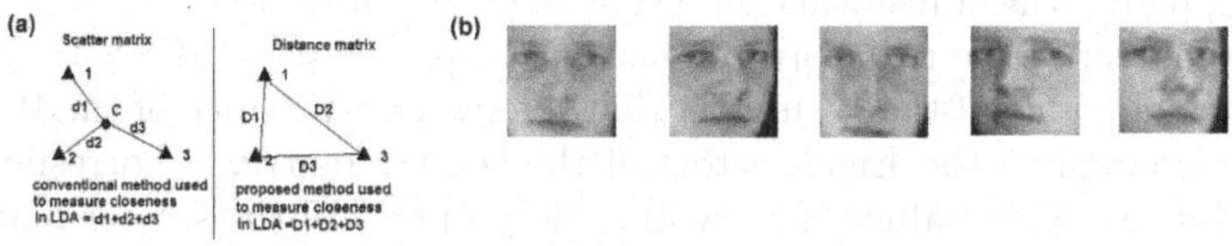

Fig. 1. (a)Illustration of the proposed technique, (b)Sample Images of the ORL face database corresponding to the identical person with different poses

Let the summation of squared distances of all the vectors from the centroid (conventional technique) and the actual squared Euclidean distances between the vectors within the class (proposed technique) be represented as $D1$ and $D2$ respectively. By direct expansion of the equations, it can be easily shown that, for the arbitrary single class with m vectors, $D2 = mD1$. Thus by intuition ,we understand that for the case of multiple classes with different size (number of vectors in the class), Euclidean distance metric gives more weightage to the class with more number of vectors and comparatively less importance to the classes that has less number of vectors.

Hence Euclidean distance based measurement can be used to measure the closeness of the vectors within the class and can also be used to measure the separation of classes effectively. To make use of this Euclidean distance based measurement in LDA, we are in need of symmetric matrix (whose trace is the Euclidean distance between the vectors) as the replacement of the traditional scatter matrices that are used in LDA. By intuition, we formulated the distance matrices whose traces measure the corresponding parameters. It can be easily shown that the proposed distance matrix (symmetric matrix) satisfies the properties that are exploited in LDA as described below.

Hence Euclidean distance based measurement can be used to measure the closeness of the vectors within the class and can also be used to measure the separation of classes effectively. To make use of this Euclidean distance based measurment in LDA, we are in need of symmetric matrix (whose trace is the Euclidean distance between the vectors) as the replacement of the traditional scatter matrices that are used in LDA. We formulated the distance matrices whose traces measure the corresponding parameters. It can be easily shown that the proposed distance matrix (symmetric matrix) satisfies the properties that are exploited in LDA as described below.

1. The trace of the arbitrary scatter matrix S is the sum of variances of the individual elements of the training set vectors. Therefore, the trace of the

within-class scatter matrix S_W gives the information about the proximity of the vectors to each other within the class. The trace of the between-class scatter matrix S_B gives the information about how the centroids of various classes are far away from each other.
2. The column vectors of the scatter matrix lie in the column space of the training set vectors. Hence, the column space of the matrix is the same as the column space of the training set vectors [12].
3. After applying the transformation using the transformation matrix W^T, the scatter matrix of the transformed vectors is represented as $W^T SW$.
4. The optimal value of the W matrix that maximizes the trace of the $W^T SW$ matrix is equal to the Eigenvectors of the scatter matrix S corresponding to the highest Eigenvalues. The symmetric property of the scatter matrix is used to obtain the above-mentioned optimal solution. Note that the solution to the classical LDA exploits that symmetric property of the scatter matrix [11]. Instead of using the scatter matrix, symmetric matrices based on the distance measure, that satisfies the above listed three properties are proposed as described in the next section.

3 Formulation of LDA Using Distance Based Symmetric Matrix

The square of the Euclidean distance between the column vectors X and Y is given as the trace of the matrix $(X-Y)(X-Y)^T$. If the vectors X and Y belong to the same class, the trace of this matrix gives the information about how the vectors are closer within the class. Similarly, if the vectors X and Y are the centroids of the two different classes, the trace of this matrix gives the information about how the vectors are distributed between the classes. The symmetric matrix thus obtained is named as distance matrix and is represented as D.

The LDA problem is redefined using the Euclidian distance-based similarity measure and the transformation matrix W^T is obtained by minimizing the trace of the matrix $W^T D_W W$ and maximizing the trace of the matrix $W^T D_B W$,where
$D_W = \sum_{i=1}^{r} \sum_{j=1}^{n_i-1} \sum_{k=j+1}^{n_i} (X_{ij} - X_{ik})(X_{ij} - X_{ik})^T$
and $D_B = \sum_{i=1}^{r-1} \sum_{j=i+1}^{r} (c_i - c_j)(c_i - c_j)^T$
are the within-class distance matrix and between-class distance matrix respectively.

$$J_D(W) = \frac{trace(W^T D_B W)}{trace(W^T D_W W)} \tag{2}$$

Note that the sum of the diagonal elements of the matrix $(X_{ij} - X_{ik})(X_{ij} - X_{ik})^T$ is the distance between the vectors X_{ij} and X_{ik}. Note that the above matrices are symmetric. It is also not difficult to show that the proposed distance-within class matrix and distance-betweeen class matrix satisfies the properties listed in Section 2. Thus the proposed distance matrix based LDA involves in estimating the optimal value of W such that $J_D(W)$ mentioned in (2) is maximized.

4 Experiments

The experiments have been performed with "ORL face database" [13] (refer Fig. 1(b)) to compare the performance in terms of the prediction accuracy of the proposed technique using distance matrix with the scatter matrix. The inner-product based LDA are used in both the cases to reduce computation time. "Small sample size problem" is also taken care using the technique proposed in [9] in both the cases. Thus the overall steps to map the arbitrary feature vector to the lower dimensional space is as described below.

Steps to map the arbitrary feature vector to the lower dimensional space:

1. Compute all the inner-product training feature vectors corresponding to all the training feature vectors in the feature space. Compute S_B and S_W matrices using the inner-product training sets.
2. Project all the inner-product training feature vectors to the null space of S_W as described as follows. Compute the Eigenvectors corresponding to the significant Eigenvalues using Eigen decomposition. The Eigenvectors thus obtained are arranged colum wise to obtain the transformation matrix T. The arbitrary inner-product training feature vector v is projected to the null space of S_W using the projection matrix TT^T as $v - TT^Tv$. This is repeated for all the inner-product training feature vectors.
3. Compute the between-class scatter symmetric matrix $S_B^{'}$ (say) using the projected vectors.Compute the transformation matrix that maximizes the trace of the matrix $U^TS_B^{'}U$, i.e., the Eigenvectors of the matrix $S_B^{'}$ are arranged rowwise to form the transformation matrix U^T. Compute the inner-product vector corresponding to the arbitrary feature vector in the higher dimensional space y and is represented as z.
4. Once the matrix U is obtained, transformation of the arbitrary vector y in the higher dimensional space to the lower dimensional space is obtained as U^Tz.

In a similar fashion, transformation of the arbitrary vector to the lower dimensional space is obtained using the proposed distance matrix just by replacing the scatter matrices S_B and S_W by distance matrices D_B and D_W. Also note that the scatter matrices and distance matrices mentioned in the above steps are normalized such that the maximum value in the matrices is 1 to easify the selection of the significant Eigen vectors. It is also noted that the nearest neigbour (based on the Euclidean distance) classifier is used to compute the percentage of success(POS) of the training set.

5 Results and Conclusion

5.1 Experiment 1: Case Study on the Comparison of the Performances of Distance Matrix with Scatter Using "ORL Face Database"

Let n_i be the number of classes from which, i number of training vectors are collected. Traing database consists of 5 vectors per class. There is the need to collect atleast two vectors from each classes to compute the matrix. so i varies from 2 to 5. There are 40 classes in the database. so $\sum_{i=2}^{5} n_i = 40$. The vector $[n_2\ n_3\ n_4\ n_5]$ of the particular trial is defined as the training vector distribution for that trial. Experiments are performed for all the valid vector distributions (ie 12431 trials). For the arbitrary trial, classes are randomly chosen from which the particular number of training vectors are collected. Also, the selection of vectors collected from the particular class are chosen randomly.

Table 1. Comparison of performance of scatter matrix with distance matrix under various significant numbers (n). $*$ Distance matrix performs better. $**$ Scatter matrix performs better

2*n	Iteration 1* (trials)	Iteration 1** (trials)	Iteration 2* (trials)	Iteration 2** (trials)
≥9	880	866	873	880
≥10	617	654	662	639
≥11	456	457	489	454
≥12	332	339	334	339
≥13	239	234	240	238
≥14	**169**	**156**	**171**	**112**
≥15	**110**	**107**	**117**	**112**
≥16	**76**	**65**	**76**	**75**

The absolute difference between the POS obtained using distance matrix and the scatter matrix for the particular trial is defined as the significant number. Number of trials when greater POS is obtained using distance matrix under various significant numbers is listed along with the cases when the scatter matrix performs better (greater POS) in the Table 1. The number of trials is greater in case of when the distance matrix is used when compared with the case of scatter matrix with high significant numbers (highlighted as bold in Table 1). This suggests the importance of using the distance matrix in place of scatter matrix in LDA.

5.2 Experiment 2: Study on the Effect of Selection of Images on the Performance of Distance Matrix on Keeping the Vector Distribution Fixed Using "ORL Face Database"

The set of trials during which distance matrix performs better with significant values greater than 7.5 are selected and the corresponding vector distributions are collected. Number of trials during which the distance matrix performs better with the significant number greater than 7.5 is 76 and the for the case of scatter matrix ,it is 65.

Thus the total number of trials belonging to the cases when the significant values greater than 7.5 is 65+75=141. For every such collected vector distribution,

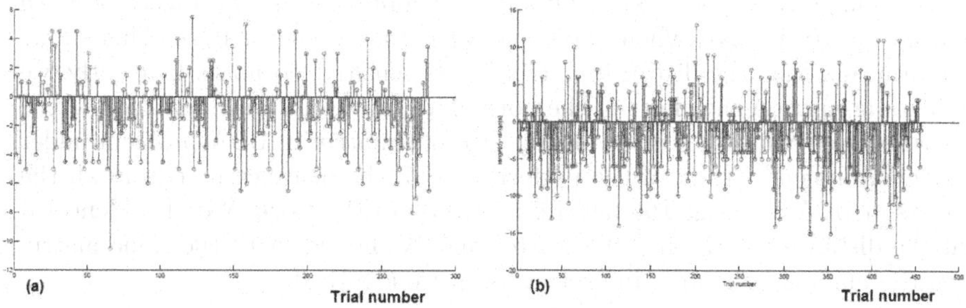

Fig. 2. (a) Illustration that the variation of POS (75 positives, 190 negatives, 17 zeros) is less in case of when distance matrix is used when compared with the case of scatter matrix. Experiments are performed using the trials belonging to the significant number $n > 7.5$. More number of negative values with higher magnitude values indicate that the distance matrix performs better, (b)Illustration (131 positive values,288 negative values and 37 zeros values) for the case with significant number $n > 5$

number of vectors from the individual classes are chosen randomly and is kept as fixed distribution. Actual vectors collected from the individual classes satisfying the fixed distribution are chosen randomly and are subjected to LDA problem using both scatter and distance matrices. This is repeated for 40 different combinations of selection of actual vectors. POS is noted for 40 different combinations using both scatter matrix and the distance matrix. This is repeated for all the collected vector distributions. The range of POS when scatter matrix is used is calculated as the difference between minimum POS and the maximum POS obtained for the particular vector distribution when scatter matrix is used.Let it be range(s). Similarly the range of POS when distance matrix is used is calculated as range(d) for every 40 different combinations.

The difference between the ranges i.e. range(d)-range(s) gives the information about the variation of POS when scatter matrix is used and when the distance matrix is used. If the value is positive,variation is more in case of when distance matrix is used. Similarly if the value is negative, variation is more in case of when scatter matrix is used. The complete experiment (mentioned above) is repeated again as the second iteration. The difference between the range of POS obtained using distance matrix and the scatter matrix obtained in both the iterations is plotted in the Fig. 2(a) for 2*141=282 trials (2 indicate two iterations,141 indicate number of trails in one iteration) . From the graph (75 positives, 190 negatives, 17 zeros) we understand that the variation is more, mostly when the scatter matrix is used and the variation is less when the distance matrix is used. The experiment mentioned above is repeated for the trials corresponding to the significant value of greater than 5 when distance matrix performs better. The result is plotted in Fig. 2(b). In this case,only one iteration is performed as the number of trials for one iteration are large i.e. 456 values. From the graph (131 positive values,288 negative values and 37 zeros values) justifies that the variation of POS is less when distance matrix is used.

To conclude, the variation of POS on the training set is less in case of when distance matrix is used when compared with the scatter matrix. This shows from intuition that the distance matrix is less sensitive to noise. Currently, we are investigating on the effect of the noisy data on the distance matrix and the importance of replacing scatter matrix distance matrix in LDA problem under noisy environment. Study of the performance of the distance matrix in all the variants of LDA is left as the natural extension of the work. We also identified that the distance matrix and the scatter matrix ends up with the same matrix when the number of vectors in every classes are identical.

Acknowledgment. Authors would like to thank the reviewers for their constructive comments. First author would like to thank Prof. K.M.M. Prabhu, Department of Electrical Engineering, Indian Institute of Technology Madras, India for his support.

References

1. Jolliffe, T.: Principal Component Analysis. Springer Verlag (1986)
2. Hyvarinen, A.: Survey on Independent Component Analysis. Neural Computing Surveys 2, 94–128 (1999)
3. Park, C., Park, H.: A Comparison of generalized linear discriminant analysis algorithms. Pattern Recognition 41, 1083–1097 (2008)
4. Nenadic, Z.: Information discriminant analysis:Feature Extraction with an Information-Theoritic objective. IEEE Transactions on Pattern Analysis and Machine Intelligence 29(8), 1394–1407 (2007)
5. Fisher, R.A.: The use of Multiple Measures in Taxonomic Problems. Ann. Eugenics 7, 179–188 (1936)
6. Kumar, N., Andreou, A.G.: Heteroscedastic Discriminant Analysis and Reduced Rank HMMs for Improved Speech Recognition. Speech Comm. 26, 283–297 (1998)
7. Torkkola, K.: Discriminative Features for Document Classification. In: Proceedings of 16th International Conference on Pattern Recognition, pp. 472–475 (2002)
8. Chen, L.-F., Liao, H.-Y.M., Ko, M.-T., Lin, J.-C., Yu, G.-J.: A New LDA-Based Face Recognition System which Can Solve the Small Sample Size Problem. Pattern Recognition 33(10), 1713–1726 (2000)
9. Cevikalp, H., Wilkes, M.: Discriminative Common vectors for Face Recognition. IEEE Transactions in Pattern Analysis and Machine Intelligence 27(1), 4–13 (2005)
10. Martinez, A.M., Kak, A.C.: PCA versus LDA. IEEE Transactions in Pattern Analysis and Machine Intelligence 23(2), 228–233 (2001)
11. Bishop, C.M.: Pattern recognition and Machine intelligece (2006)
12. Strang, G.: Linear algebra and its applications, pp. 102–107. Thomson, Brooks/Cole (2006)
13. Source: `http://www.orl.co.uk/facedatabase.html`

Gender Recognition Using Fusion of Spatial and Temporal Features

Suparna Biswas[1] and Jaya Sil[2]

[1] Department of Electronics & communication Engineering,
Gurunanak Institute of Technology, Sodepur, Kolkata, India
[2] Department of Computer Science and Technology,
Bengal Engineering & Science University, Shibpur, West Bengal, India
suparna_b80@yahoo.co.in, js@cs.becs.ac.in

Abstract. In the paper, a gender recognition scheme has been proposed based on fusion of spatial and temporal features. As a first step, face from the image is detected using Viola Jones method and then spatial and temporal features are extracted from the detected face images. Spatial features are obtained using Principal Component Analysis (PCA) while Discrete Wavelet Transform (DWT) has been applied to extract temporal features. In this paper we investigate the fusion of both spatial and temporal features for gender classification. The feature vectors of test images are obtained and classified as male or female by Weka tool using 10 fold cross validation technique. To evaluate the proposed scheme FERET database has been used providing accuracy better than the individual features. Experimental result shows 9.77% accuracy improvement with respect to spatial domain recognition system.

Keywords: Feature extraction, Gender classification, Discrete wavelet transform, Principal component analysis.

1 Introduction

Gender recognition is widely used in human-computer interaction, facial expression recognition, emotion repression and age estimation [1]. Currently several kinds of gender classification methods are available in the literature such as face based [2-7], gait based [8-11], hand based [12] and many more. However, facial images are widely used for gender classification because facial images probably contain the most common biometric characteristic used by humans to make a personal recognition [13].

Among different face based gender recognition methods, Mousavi *et al.* [18] proposed a novel method for gender detection using Fuzzy Inference System (FIS). Here Sugeno type FIS is used, which has 2-inputs and 1-output. It is compared with Neural Network, Threshold Adaboost, SVM, SVM+ Local Binary Patterns (LBP) and show better result compared to other classifiers. Sun *et al.*[14] showed that feature selection is an important issue for gender classification and genetic algorithm (GA) works well for selecting optimum feature set. In their work, first feature vectors are

M.K. Kundu et al. (eds.), *Advanced Computing, Networking and Informatics - Volume 1*,
Smart Innovation, Systems and Technologies 27,
DOI: 10.1007/978-3-319-07353-8_13,

obtained using principal component analysis (PCA) algorithm and then a subset of the features is selected using genetic algorithm. Performance of the method is compared using classifiers namely, Bayesian, Neural Network, Support Vector Machine (SVM) and Linear Discriminant analysis (LDA). Among the four classifiers SVM achieved the best result.

Wavelet transform [19] is an ideal tool to analyze images of different gender. It discriminates among several spatial orientations and decomposes images into different scale orientations, providing a method for in space scale representation. Continuous Wavelet Transform (CWT) and Support Vector Machine (SVM) are used for classifying the gender [20] from the facial images and compared with DWT, RADON along with SVM.

However, the existing gender recognition methods do not consider most important multi view problem and focus on frontal view only. There are some few works on aligned faces [15-17]. Another important problem of gender detection is illumination changes. To overcome the limitations of the existing methods, fusion technology is used to combine different types of features obtained from different face images. The proposed scheme motives to develop a face based gender recognition algorithm to deal with the existing problems.

The rest of the paper is organized as follows. In Section 2, the proposed method is described while feature extraction algorithms are illustrated in Section 3. Experimental results are discussed in Section 4 and finally concluding remarks are summarized in Section 5.

2 Proposed Method

The block diagram of the proposed method for gender recognition is shown in Fig. 1 consisting of five main modules: face detection, pre-processing, feature extraction, fusion and classification.

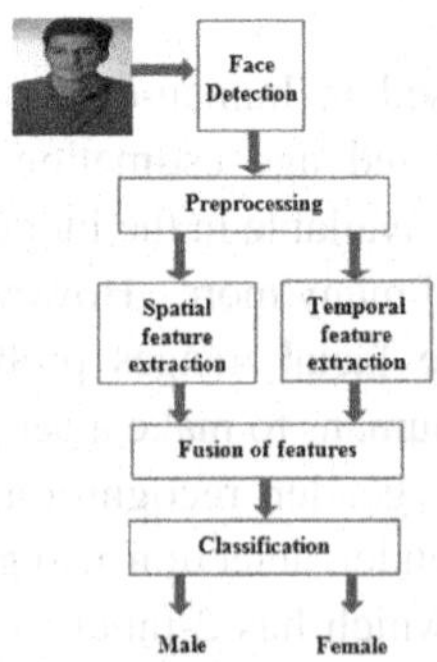

Fig. 1. Flow diagram of Gender Detection

2.1 Face Detection

The first step of gender recognition is face detection. In the paper face detection has been carried out using Viola Jones method, a real time face detection method implemented in openCV. This face detection system is based on Haar-like features. A Detail of this face detection system is discussed in [21]. This detection technique achieved very satisfactory results with a correct detection rate of 99%, tested on Caltech Image Database (CID) [22]. Fig. 2(a) represents the detected face from its input image.

2.2 Preprocessing

After face detection, the feature area is extracted by cropping the face image 15% from right and left and 20% from top of the image for removing ears and hairs.

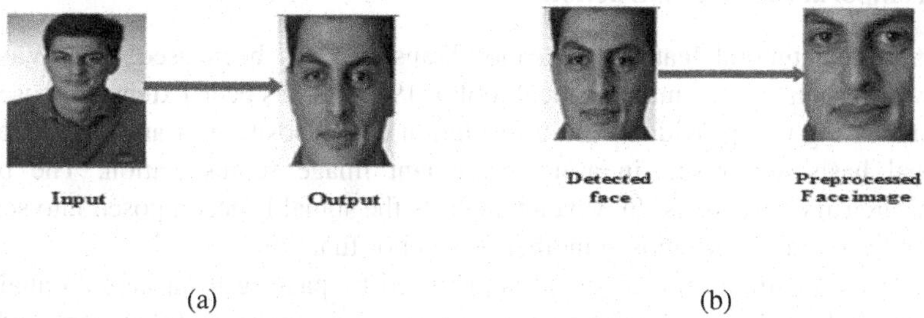

Fig. 2. Results of (a) face detection by Viola and Jones technique (b) preprocessing

Then in the next step Gaussian smoothing filter is applied and image is resized. At the last step of preprocessing, histogram equalization is performed to overcome the illumination differences. Fig. 2(b) represents the preprocessed face image from its detected face image.

3 Feature Extraction Techniques

Both spatial and temporal features are obtained from the preprocessed face images for gender recognition. The features are fused to form a compact feature representation of the facial images.

3.1 Spatial Feature Extraction

Following steps are used to extract the spatial feature:

Step1: Resize M number face images ($I_1, I_2, \ldots I_M$) to 128×128.
Step2: Convert each input image into column vector of size $N^2 \times 1$ and placed into the set :$\{\Gamma_1, \Gamma_2, \Gamma_M\}$.
Step3: Find the average face vector using following equation.

$$\Psi = \frac{1}{M}\sum_{i=1}^{M}\Gamma_i \quad (1)$$

and obtain $$\Phi_i = \Gamma_i - \Psi \quad (2)$$

Step4: Find the covariance matrix C using eq(3).

$$\mathrm{C} = \mathrm{AA}^{\mathrm{T}} \quad (3)$$

where $$\mathrm{A} = \{\Phi_1, \Phi_2, \ldots, \Phi_M\} \quad (4)$$

Step5: Compute Eigenvectors and Eigen values. Sort Eigen values by decreasing order.

Step6: Choose the Eigen vectors with first P no. largest Eigen values.

Step7: Project all training images into the Eigen Space, associated with P largest Eigen values. The projected Eigenvectors are treated as feature vectors.

3.2 Temporal Feature Extraction

To extract the temporal features, Wavelet Transform has been used. The Wavelet theory was introduced as a mathematical tool in 1980s. It has been extensively used in image processing that provides a multi resolution decomposition of an image in a bi-orthogonal basis and result in a non-redundant image representation. The basis functions are called wavelets. In wavelet analysis the signal is decomposed into scaled and shifted versions of the chosen mother wavelet or function.

In DWT domain, an image signal is analyzed by passing it through an analysis filter bank followed by a decimation operation. This analysis filter bank, which consists of a low pass and high pass filter at each decomposition stage, is commonly used in image compression. When a signal passes through these filters, it is split into two bands. The low pass filter, which corresponds to an averaging operation, extracts the coarse information (CA) or low frequency information of the signal. The high pass filter, which corresponds to a differencing operation, extracts the detail information of the signal.

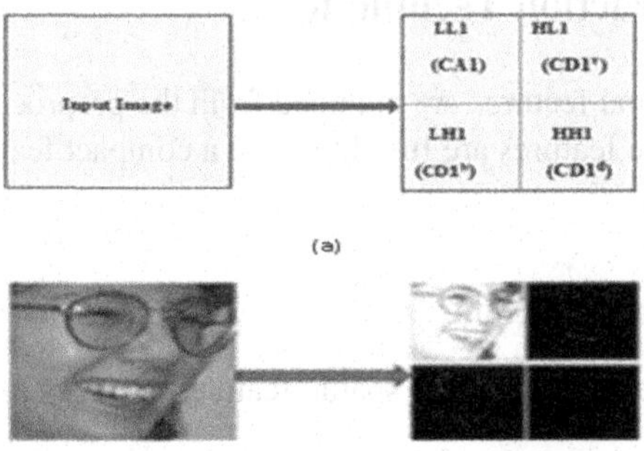

Fig. 3. Example of image from FERET database (sideview) into sub-bands

In two dimensional signal we get three sets of detail information, i.e. horizontal(CD^h), vertical(CD^v) and diagonal(CD^d). In the first level of decomposition, the image is split into four sub-bands, namely HH1 (CD^d), HL1 (CD^v), LH1 (CD^h), and LL1(CA1), as shown in Fig. 3(a). Fig. 3(b) shows an image from FERET database (aligned face) and after wavelet decomposition with level1 by db1 filter.

Following steps are used to extract temporal features

Step1: Apply 2D-DWT on each of the resized face images using db1 mother wavelet.
Step2: After 1st level decomposition, extract only the approximation coefficients (CA1) for each test images
Step3: Perform the operations from Step2 to Step7 of spatial feature extraction techniques, on the CA1 of test images to extract the temporal features.

3.3 Fusion of Features

Image fusion is a process of combining two or more images into a single image. In other words, image fusion is a technique to integrate information from multiple images with an expectation that it provides more information compare to a single one.

In this paper a feature level image fusion technique has been proposed. The information flow diagram of PCA based fusion algorithm is shown in Fig. 4.

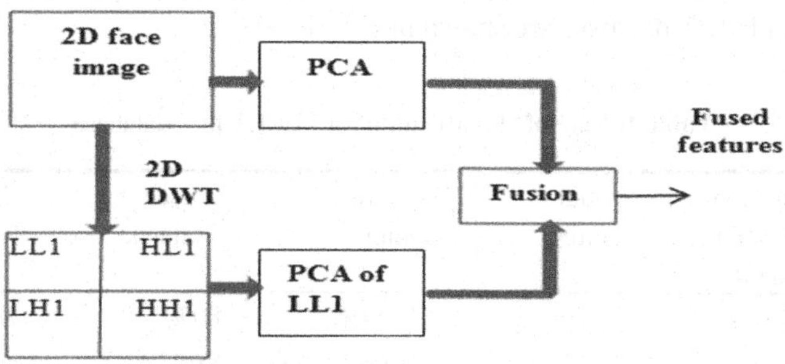

Fig. 4. Information flow diagram for the fusion scheme employing PCA

Following fusion strategies are invoked to obtain the fused feature:
i) Fused feature = {Spatial feature, Temporal feature}
ii) Fused feature = μ_1×Spatial feature + μ_2×Temporal feature, where μ_1=0.7 and μ_2 = 0.3 is used.

4 Experimental Studies

The first step of any image classification technique is to representing the face images in terms of input-output feature vectors. In the present work, Neural Network(NN),

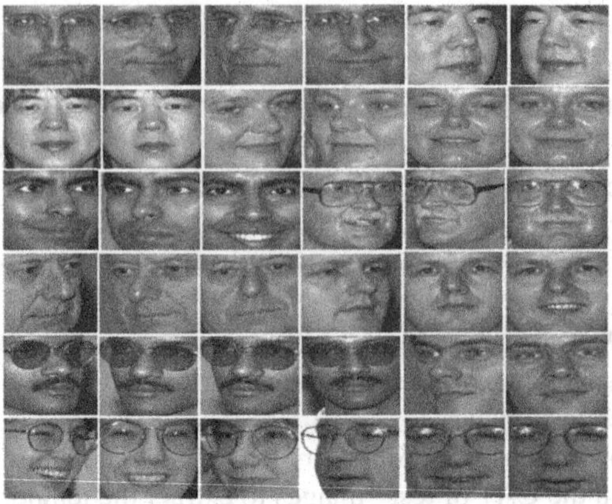

Fig. 5. Test images from FERET Database

Stochastic gradient descent(SGD), Naive Bayes, Simple logistic and SVM method are used for classification of images as male or female.

To demonstrate the effectiveness of the proposed algorithm, it is applied on FERET face image databases.

The FERET database contains frontal, left or right profile images and could have some variations in pose, expression and lightning. In the experiments, we have used frontal, aligned, pose variant and different expression face images. Some of the test images from FERET database are shown in Fig. 5.

Table 1. Rate of classification for FERET face database

Classification for FERET face database	Spatial features	Temporal features	Fused Rule(i)	Fused Rule(ii)
Naive Bayes	72.77%	74.44%	80%	78.88%
SVM	83.33%	81.66%	88.22%	86.66%
Neural Network	80.55%	86.55%	93%	90.55%
SGD	80%	78.88%	87.66%	83.66%
Simple logistic	82.77%	81.11%	85%	88.33%

Classification accuracy using the fused feature vectors are obtained from 200 test images. By applying ten-fold cross validation technique and using Neural network, SGD, Naive Bayes and SVM classifier, classification accuracy is computed and presented in Table 1 for 45 Eigen vectors (P). We have investigated the accuracy of gender classification by increasing the numbers of Eigenvectors up to 60 and observed that 45 Eigenvectors gives the best result.

Table 2. Comparison with other techniques

Method	Maximum Classification Accuracy	Database	Ref.
Neural Network with Gabor wavelet	91.5%	FERET	[23]
Neural Network +SVM	83.38%	FERET	[24]
PCA+DCT+Fuzzy	82%	FERET	[25]
SVM with Gabor wavelet	96.5%	FERET	[23]
Haar wavelet with AAM	92.53%	FERET	[26]
Gait based	96.77%	CASIA	[9]
Gait based	95.97%	CASIA	[8]
Gait based	87%	Soton & CASIA	[9]
Proposed method	93%	FERET	

5 Conclusions

In this paper, we have presented a fusion based gender detection scheme. Classification accuracy shows effectiveness of the proposed feature extraction methods. Experimental result shows that fused features improved the classification accuracy. This technique also can be used for the gender detection of aligned faces. Also it has been concluded that fusion using wavelets showed better performance.

References

1. Vending machines recommend based on face recognition. Biometric Technology Today 2011(1) (2011)
2. Ylioinas, J., Hadid, A., Pietikäinen, M.: Combining contrast information and local binary patterns for gender classification. Image Analysis, 676–686 (2011)
3. Shan, C.: Learning local binary patterns for gender classification on real-world face images. Pattern Recognition Letters 33(4), 431–437 (2012)
4. Wu, T.-X., Lian, X.-C., Lu, B.-L.: Multi-view gender classification using symmetry of facial images. Neural Computing and Applications, 1–9 (2011)
5. Wang, J.G., Li, J., Lee, C.Y., Yau, W.Y.: Dense SIFT and Gabor descriptors-based face representation with applications to gender recognition. In: 2010 11th International Conference on Control Automation Robotics & Vision (ICARCV), pp. 1860–1864 (2010)
6. Lee, P.H., Hung, J.Y., Hung, Y.P.: Automatic Gender Recognition Using Fusion of Facial Strips. In: 20th International Conference on Pattern Recognition (ICPR), pp. 1140–1143 (2010)

7. Rai, P., Khanna, P.: Gender classification using Radon and Wavelet Transforms. In: International Conference on Industrial and Information Systems, pp. 448–451 (2010)
8. Shiqi, Y., Tieniu, T., Kaiqi, H., Kui, J., Xinyu, W.: A Study on Gait- Based Gender Classification. IEEE Transactions Image Processing 18, 1905–1910 (2009)
9. Hu, M., Wang, Y., Zhang, Z., Wang, Y.: Combining Spatial and Temporal Information for Gait Based Gender Classification. In: 20th International Conference on Pattern Recognition, pp. 3679–3682 (2010)
10. Chang, P.-C., Tien, M.-C., Wu, J.-L., Hu, C.-S.: Real-time Gender Classification from Human Gait for Arbitrary View Angles. In: 2009 11th IEEE International Symposium on Multimedia, pp. 88–95 (2009)
11. Chang, C.Y., Wu, T.H.: Using gait information for gender recognition. In: 10th International Conference on Intelligent Systems Design and Applications, pp. 1388–1393 (2010)
12. Amayeh, G., Bebis, G., Nicolescu, M.: Gender classification from hand shape. In: Proc. IEEE Conference on Computer Vision and Pattern
13. Jain, A.K., Ross, A., Prabhakar, S.: An Introduction to Biometric Recognition. IEEE Transactions on Circuits and Systems 14(1), 4–20 (2004)
14. Sun, Z., Bebis, G., Yuan, X., Louis, S.J.: Genetic feature subset selection for gender classification: A comparison study. In: Proceedings of IEEE Workshop on Applications of Computer Vision, pp. 165–170 (2002)
15. Wu, T.-X., Lian, X.-C., Lu, B.-L.: Multi View gender classification using symmetry of facial image. In: ICONIP (2010)
16. Huang, J., Shao, X., Wechsler, H.: Face pose discrimination using support vector machines (svm). In: Proceedings of 14th International Conference on Pattern Recognition, ICPR 1998 (1998)
17. Raisamo, M.E.: Evaluation of gender classification methods with automatically detected and aligned faces. IEEE Trans. Pattern Analysis and Machine Intelligence 30(3), 541–547 (2008)
18. Somayeh, B., Mousavi, H.A.: Gender classification using neuro fuzzy system. Indian Journal of Science and Technology 4(10) (2011)
19. Basha, A.F., Shaira, G., Jahangeer, B.: Face gender image classification using various wavelet transform and support vector machine with various Kernels. IJCSI International Journal of Computer Science Issues 9(6), 2 (2012)
20. Ullah, I., Hussain, M., Aboalsamh, H., Muhammad, G., Mirza, A.M., Bebis, G.: Gender Recognition from Face Images with Dyadic Wavelet Transform and Local Binary Pattern. In: Bebis, G., Boyle, R., Parvin, B., Koracin, D., Fowlkes, C., Wang, S., Choi, M.-H., Mantler, S., Schulze, J., Acevedo, D., Mueller, K., Papka, M. (eds.) ISVC 2012, Part II. LNCS, vol. 7432, pp. 409–419. Springer, Heidelberg (2012)
21. Viola, P., Jones, M.: Rapid object detection using a boosted cascade of simple features. In: CVPR (2001)
22. Caltech Image Database, `http://www.vision.caltech.edu/htmlfiles/archive.html`
23. Leng, X.M., Wang, Y.: Improving Generalization For Gender Classification. In: ICIP (2008)
24. Makinen, E., Raisamo, R.: Evaluation of Gender Classification Methods with Automatically Detected and Aligned Faces. IEEE Transactions on PAMI 30(3) (2008)
25. Akbari, R., Mozaffari, S.: Performance Enhancement of PCA-based Face Recognition System via Gender Classification Method. In: MVIP (2010)
26. Xu, Z., Lu, L., Shi, P.F.: A hybrid approach to gender classification from face images. In: ICPR, pp. 1–4 (2008)

The Use of Artificial Intelligence Tools in the Detection of Cancer Cervix

Lamia Guesmi[1], Omelkir Boughzala[1], Lotfi Nabli[2], and Mohamed Hedi Bedoui[1]

[1] Laboratoire Technology and Medical Imaging (TIM), Faculty of Medicine, Monastir, Ecole National Engineering of Monastir Box 5000
Monastir, Tunisia
[2] Laboratoire of Automation and Computer Engineering (Lille LAIL), (CNRS UPRESA is 8021),
Ecole National Engineering of Monastir Box 5000,
Monastir, Tunisia
lamia_guesmi0107@yahoo.com, omelkhiboughzala@gmail.com, lotfinabli@yahoo.fr, medHedi.Bedoui@fmm.rnu.tn

Abstract. Recently, research into pathological cytology were intended to put in places of artificial intelligence systems based on the development of new diagnostic technologies and the cell image segmentation. These technologies are not intended to substitute the human expert but to facilitate his task. The objective of this work is to develop a method for diagnosing cancer cervical smears using cervical - vaginal segmented to build our database and a human supervisor and as an automatic tool manage and monitor the execution of the operation of diagnostic and proposing corrective actions if necessary. The Supervisor Smart is manufactured by the technique of neural networks with a success rate of 43.3% followed by the technique of fuzzy logic with a success rate equal to 56.7% and finally to improve this rate we used neuro-fuzzy approach which has a rate which reaches 94%.

1 Introduction

In Pathological Anatomy and Cytology, we distinguish two types of tests: The histology is the observation of the cutting of tissue and cytology is the examination of a spreading cell. We are particularly interested in cytology. The samples are spread on a slide and then fixed and stained to recognize the different cells present. The smears are then examined under a microscope by a cyto-technician to identify cells of interest. This step of reading the blade consists of a visual assessment of these cells onto a cytological blade. The purpose of this step is either the detection of abnormal or suspicious cells, namely the quantification of cells. This is of vital interest to the pathologist who must establish a reliable and valid diagnosis especially in the case of classification CSVs for the diagnosis of cancer cervix. That is why we introduced artificial intelligence tools to facilitate this task with a very high success rate based on the technical supervisor of human and automatic after illustrating a priori information used to recognize cells are size, shape, texture but also and mainly the color.

M.K. Kundu et al. (eds.), *Advanced Computing, Networking and Informatics - Volume 1*, Smart Innovation, Systems and Technologies 27,
DOI: 10.1007/978-3-319-07353-8_14, © Springer International Publishing Switzerland 2014

2 Segmentation of CSV

2.1 Elimination of Inflammatory Cells

In order to clean CSVs of inflammatory cells that invaded the image to the Treaty (Fig. 1), in which these cells recognize the same tent staining than nuclei of cells tested is why we spent at gray image to properly locate and illustrate the inflammatory cells using applications programming software MATLAB.

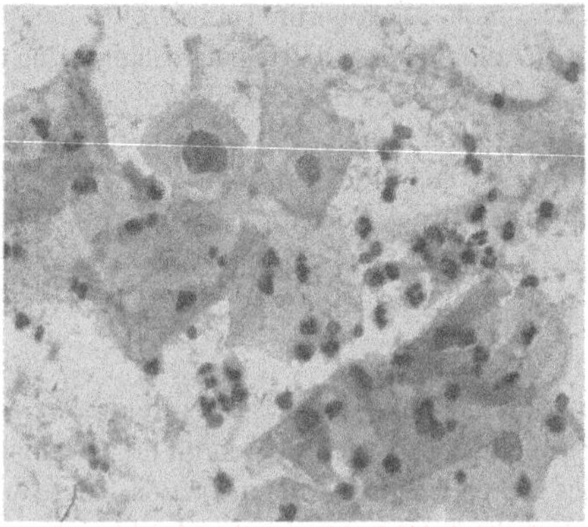

Fig. 1. Vaginal Smear: the background is inflammatory and contains blue parasites. The squamous cells show clear perinuclear halo (Papanicolaou staining)

We turn to the gray level of the smear to the elimination of certain inflammatory cells that exhibit a level of RGB (red - green - blue) is not perfectly zero as the case of carcinogenic cell nucleus. We note that the cleaning of the smear is not clear that essentially means the occupation of certain sensitive places of inflammatory cells into the cell nucleus as a carcinogen or the borders of the cytoplasm of the cells in question, all this may give false decisions. And to avoid confusion, it is recommended to make a levy as sharp as possible by the cyto patologist, so do the cleaning cell by cell.

We turn to the gray level of the smear to the elimination of certain inflammatory cells that exhibit a level of RGB (red - green - blue) is not perfectly zero as the case of carcinogenic cell nucleus. We note that the cleaning of the smear is not clear that essentially means the occupation of certain sensitive places of inflammatory cells into the cell nucleus as a carcinogen or the borders of the cytoplasm of the cells in question, all this may give false decisions. And to avoid confusion, it is recommended to make a levy as sharp as possible by the cyto patologist, so do the cleaning cell by cell.

2.2 Image Segmentation in Experimental Phase

Segmentation plays a very important role in pattern recognition. It was after this stage that we can extract the parameters of quality will be defined with the continuation. Sometimes we have more than one cell in the image examined. Then we have to separate the whole cell from other cells that may appear on the edges of the image, that is to say, incomplete cells. Here are enumerated the steps required to segment such an image: Playing the picture; Detection internal and external contours of a cell linearization outlines of objects; Dilatation contours of objects filling internal objects; Removing objects exceeding the edge of the image, and the softening of the object and segmentation (outer contour only). We ended up following a CSV segmented (Fig. 2):

Fig. 2. Vaginal smear segmented: we can determine the overall shape of the cell carcinogenic and size of the nucleus and the cytoplasm (*X, Y, index : punctual coordinates*)

3 Shape Parameters

After segmentation of the image we had in carcinogenic cell sorting parameters to locate the most appropriate forms, each grade of cancer (class) is described by the parameters of quality based on the shape of the nucleus and cytoplasm more significant. They represent the best indicators used for the diagnosis and to achieve the classification of cancer. Their values are measured directly and manually to get them out. We have specified eight parameters of the form using the physician are:

$$E1 = (DFMax)\ kernel\ /\ (DFMax)\ cytoplasm. \quad (1)$$

$$E2 = (DFMin)_{kernel} / (DFMin)_{cytoplasm.} \quad (2)$$

$$DF1 = (DFMin / DFMax)_{cytoplasm}. \quad (3)$$

$$DF2 = (DFMin / DFMax)_{kernel}. \quad (4)$$

$$U1 = SN1/SC1. \quad (5)$$

$$U2 = SN2/SC2. \quad (6)$$

$$SC1/SC2. \quad (7)$$

$$SN1/SN2. \quad (8)$$

With: DFMax: Maximum Feret diameter; DFMin: Minimal Feret diameter; SN1: Size of the smallest circle included the nucleus; SN2: Size of the largest circle circumscribing the nucleus SC1: Size of the smallest circle including the cytoplasm and SC2: Size of the largest circle circumscribing the cytoplasm.

4 Methodologies of Classification per the Tools of Artificial Intelligence of the Vaginal Smears Cervico

In our case, we are interested diagnosis-based models and database of a hybrid system because our support, equal to 120 CSVs distributed in equal shares to four classes of cancer (Cancer (C), High Grade Dysplasia (HGD), Low Grade Dysplasia (LGD) and Normal (N)). These smears will be processed and exploited, in order to extract qualitative and quantitative information. These will form the basis of data for the development and testing of the tool to develop diagnostic techniques based on Artificial Intelligence (AI).

4.1 The Technique of Neural Networks (Multilayer Perceptron (MP))

Neural networks have several types; the most used is the MP which is characterized by its learning algorithm and back propagation of errors [1]. The MP is an artificial neural network-oriented organized in layers, where information flows in one direction, from input layer to output layer. The input layer is always a virtual layer containing the entries of the system [2]. The following layers are the hidden layers that admit "m" number of layers according to the need for resolution of the system. It ends with the output layer representing the results of classification system [3]. In the case suggested, the MP consists of an input layer which consists of eight shape parameters that are specified above, a hidden layer which consists of 26 neurons and an output layer is composed of four neurons formant the four main grades of cancer cervix. This type of perceptron is chosen after several tests, in which by varying the number of neurons in the hidden layer, the learning period and the vector normalization. We used the activity as a function of "Log-Sigmoid" with the normalization vector is V = [0.01, 0.99], because the database is formed by values that are between 0 and 1.

4.2 The Technique of Fuzzy Logic

The construction of a fuzzy model for the diagnosis and classification is to move mainly by three main stages: the fuzzification, the inference and deffuzzification. To achieve these three steps have to be set membership functions of input variables and those of output.

Definition of membership functions of input and output variables:The membership functions are most commonly used form: Singleton, Triangular, Trapezoidal, Gaussian. We retained the trapezoid for their simplicity of coding and manipulation, and it includes the range of variation of shape parameters. The number of trapezoidal shape in the representation of the membership function will be limited as much as possible for four to describe the scope of a variable. In this case they will be associated with terms: low - medium - large-very important.

We associate a variable (Z) the following values of Table 1: minimum (Zmin), optimistic (ZTopt), optimistic (Zopt) average optimism (ZmoyO) average pessimistic (ZmoyP), pessimistic (Zpes), very pessimistic (Ztpes) and maximum (Zmax).

Table 1. Membership functions of input and output

	Variable (Z)							
	Z_{min}	Z_{Topt}	Z_{opt}	Z_{moyO}	Z_{moyP}	Z_{pes}	Z_{tpes}	Z_{max}
	Membership functions of input parameters							
E1	0.000	0.134	0.265	0.407	0,517	0,668	0,881	1.000
E2	0.000	0.186	0,339	0,429	0,609	0,611	0,998	1.000
DF1	0.000	0.383	0.391	0.484	0.833	0.936	0.960	1.000
DF2	0.000	0.403	0.493	0.519	0.903	0.977	0.983	1.000
U1	0.000	0.019	0.081	0.174	0.176	0.622	0.963	1.000
U2	0.000	0.015	0.081	0.165	0.290	0.477	0.810	1.000
SC1/SC2	0.000	0.154	0.174	0.210	0.754	0.802	0.913	1.000
SN1/SN2	0.000	0.208	0.210	0.242	0.883	0.914	0.991	1.000
	Membership function of output variable "S"							
S	0.000	0.150	0.170	0.250	0.270	0.400	0.500	0.700

Fuzzification: After defining the membership functions according to the expert, we had to convert our input parameters for numerical values of physical quantities in translating linguistic variables and fuzzy since the algorithm needs exact numerical variables (1, 2, 3, 4, 5, ...) to be usable by the following algorithm while the eight input parameters will admit four lexical classes: LOW, MEDIUM, "" IMPORTANT "and" VERY IMPORTANT "where we have assigned their respective values '1 ', '2', '3 'and '4'. For the output vector allows four lexical classes " HGD ", " LGD ", "C" and "N" where we have respectively their associated values '1 ', '2', '3 'and '4' [4].

Rules of Inference or Inference: This block is composed by all the fuzzy rules that exist between input variables and output variables expressed both as linguistic. We have chosen as a method of inference method "MAX-MIN". It is based on the use of two logical operations: the "AND" logic associated with the minimum and the "OR" logic associated to the maximum. These rules are of the form: IF (AND ... .) THEN (decision) OR .... [5-7]. We have 185 possible rules of inference in the form of the maximum number of rules.

Defuzzification: The inference methods provide a membership function resulting "μ" for the output variable "x". It is therefore fuzzy information that must be transformed into physical quantity. This conversion to reverse the phase of "fuzzification" [5,6]

4.3 The Hybrid Approach: Neuro-Fuzzy

Such a hybrid system is characterized by the following features: learning is performed by an algorithm derived from a neural network, the general architecture of this system is represented by a recurrent network, this system is interpreted in terms of rules the form "if ... then ..." Learning is done from the semantics of the underlying fuzzy model, thus preserving the linguistic interpretability of the model, this model performs a function approximation. Therefore, the use of such an approach neuro-fuzzy overcomes the disadvantages inherent in each of the previous approaches (neural networks and fuzzy logic), while retaining their benefits [7,8].

Table 2. The success rate of intelligenttools

		Intelligent tools for Artificial Intelligence		
		Neural network (%)	**Fuzzy (%)**	**Neuro – Fuzzy (%)**
Success rate	**HGD**	56,7	43,3	100
	LGD	40	66,7	96,7
	C	33,3	56,7	86,7
	N	43,3	60	93
Total Success Rate (%)		**43,3**	**56,7**	**94**

5 Simulation Results

The results of the use of an intelligent screening for cancer cervix in the three recurrent artificial intelligence techniques are recapitalized in Table 2 after which it appears the success rate each technique for the same database (the 120 cases).

6 Discussion

Furthermore, some data parameters are not clearly defined which are important for the evaluation of the general behavior of each method.

Table 3. Comparison of the proposed method and other methods appeared in the literature

Author	Year	Method	Advantages	Disadvantages
T.Chankong *et al.*[9]	2013	- Fuzzy c-means clustering	-Can find the threshold values automatically -Can handle uncertainty in data values(color, circularity, and object dimension) well	-Cannot separate overlapped cells -Fuzzy logic rules are fixed and cannot be adapted to changing conditions.
S.Issac Niwas *et al.* [10]	2012	- Complex Daubechies wavelets(CDW)	- Good selection -CDW configuration easy and inexpensive -Very successful and well used especially in face recognition.	-Storage of information Extracted during learning.
Marina E. Plissiti *et al.* [11]	2011	-K-means, spectral clustering - Support vector machines SVM	-Can be applied directly in Pap smear images obtained by an optical microscope, without any observer interference, for the accurate automated identification of the cell nuclei boundaries.	- Lack of precision Difficulty when considering several cell measures.
Lezoray *et al.* [12]	2002	- Watershed	-Incorporates color information on the watershed segmentation -High rate of accurate segmentation	-A training set is needed for the achievement of best result

The success rate obtained by the three approaches of artificial intelligence is very important for all cases of neuro-fuzzy can be improved if we increase our database used. The method of current areas of neuro-fuzzy confusion between two classes that are close to the formal point of view of the cell is carcinogenic for couples (HGD, C) and (LGD, N). To remedy this, we can add another shape parameter that expresses aspecific quality indicator or as previously mentioned we develop our database is the goal our next study we will increase our database up to 180 CSVs (minimum) and we will set a new quality setting for better qualify our classification. Despite these false positives, this method has the refuge of screening cervical cancer at the moment.

Beyond the comparison of our method with pixel classification schemes, Table 3 shows a comparison of our method and other methods appeared in the literature. In general, it is difficult to compare the methods directly since many of them do not include quantitative results and the performance criteria extensively vary.

7 Conclusion

The supervisor directed by the hybrid approach neuro - fuzzy won the record classification CSVs a success rate of around 94% which we can improve if we increase our data down by other cases of injuries precancerous and cancerous encountered.

Acknowledgments. I especially thank my supervisor Mr. Mohamed Hedi BEDOUI and my co- encadror Mr. Lotfi Nabli for their help in setting up my database and the moral support throughout my project.

References

1. Buniet, L.: Traitement automatique de la parole en milieu bruité: étude de modèles connexionnistes statiques et dynamiques. Thèse doctorat. Université Henri Poincaré - Nancy 1. Spécialité informatique, 40–39 (1997)
2. El Zoghbi, M.: Analyse électromagnétique et outils de modélisation couplés. Application à la conception hybride de composants et modules hyperfréquences. Thèse doctorat. Université de Limoges. Discipline: Electronique des Hautes fréquences et Optoélectronique. Spécialité: Communications Optiques et Microondes, 37 (2008)
3. Lamine, H.M.: Conception d'un capteur de pression intelligent. Mastère en micro électrique. Option IC Design. Université de Batna: faculté des sciences de l'ingénieur, 18 (2005)
4. Millot, P.: Systèmes Homme – Machine et Automatique. Université de Valenciennes et du Hainaut – Cambrésis. Laboratoire : LAMIH CNRS. In: Journées Doctorales d'Automatique JDA 1999, Conférence Plénière, Nancy, pp. 23–21 (1999)
5. Nabli, L.: Surveillance Préventive Conditionnelle Prévisionnelle Indirecte d'une Unité de Filature Textile : Approche par la Qualité. Thèse Doctorat. Discipline: Productique, Automatique et Informatique Industrielle Université des Sciences et Technologies de Lille, 34 – 20 (2000)
6. Cimuca, G.-O.: Système inertiel de stockage d'énergie associe à des générateurs éoliens. Thèse doctorat. Spécialité: Electrique. Ecole nationale supérieure d'arts et métiers centre de Lille, 162– 163 (2005)
7. Baghli, L.: Contribution à la commande de la machine asynchrone, utilisation de la logique floue, des réseaux de neurones et des algorithmes génétiques. Thèse doctorat. UFR Sciences et Techniques: STMIA. Université Henri Poincaré, Nancy-I, 16 (1999)
8. Främling, K.: Les réseaux de neurones comme outils d'aide à la décision floue. Rapport de D.E.A. Spécialité : informatique. Equipe Ingénierie de l'Environnement.Ecole Nationale Supérieure des Mines de Saint-Etienne. Juillet, 12-11 (1992)
9. Chankonga, T., Heera-Umpon, N., Auephanwiriyakul, S.: Automatic cervical cell segmentation andclassification in Pap smears (2013)
10. IssacNiwas, S., Palanisamy, P., Sujathan, K., Bengtsson, E.: Analysis of nuclei textures of fine needle aspirated cytology images for breast cancer diagnosis using Complex Daubechies wavelets13 (2012)
11. Plissiti, M.E., Nikou, C., Charchanti, A.: Combining shape, texture and intensity features for cell nuclei extraction in Pap smear images
12. Lezoray, O., Cardot, H.: Cooperation of Color Pixel ClassificationSchemes and Color Watershed: A Study for Microscopic Images

A Scalable Feature Selection Algorithm for Large Datasets – Quick Branch & Bound Iterative (QBB-I)

Prema Nedungadi[1] and M.S. Remya[2]

[1] Amrita Create
[2] Department of Computer Science,
Amrita Vishwa Vidyapeetham
prema@amrita.edu

Abstract. Feature selection algorithms look to effectively and efficiently find an optimal subset of relevant features in the data. As the number of features and the data size increases, new methods of reducing the complexity while maintaining the goodness of the features selected are needed. We review popular feature selection algorithms such as the probabilistic search algorithm based Las Vegas Filter (LVF) and the complete search based Automatic Branch and Bound (ABB) that use the consistency measure. The hybrid Quick Branch and Bound (QBB) algorithm first runs LVF to find a smaller subset of valid features and then performs ABB with the reduced feature set. QBB is reasonably fast, robust and handles features which are interdependent, but does not work well with large data. In this paper, we propose an enhanced QBB algorithm called QBB Iterative (QBB-I).QBB-I partitions the dataset into two, and performs QBB on the first partition to find a possible feature subset. This feature subset is tested with the second partition using the consistency measure, and the inconsistent rows, if any, are added to the first partition and the process is repeated until we find the optimal feature set. Our tests with ASSISTments intelligent tutoring dataset using over 150,000 log data and other standard datasets show that QBB-I is significantly more efficient than QBB while selecting the same subset of features.

Keywords: Feature Selection, Search Organization, Evaluation Measure, Quick Branch & Bound, QBB-Iterative.

1 Introduction

In today's information age, it is easy to accumulate data and inexpensive to store it. The ability to understand and analyze large data requires us to more efficiently find the optimal subset of features. Feature selection is the task of searching for an optimal subset of features from all available features [15]. Feature selection aims to find the optimal subset of features by removing redundant and irrelevant features that contribute noise and to reduce the computational complexity.

Existing feature selection algorithms (FSA) for machine learning typically fall into two broad categories: wrappers and filters [13-15]. Filter methods choose the best

M.K. Kundu et al. (eds.), *Advanced Computing, Networking and Informatics - Volume 1*, Smart Innovation, Systems and Technologies 27,
DOI: 10.1007/978-3-319-07353-8_15,

individual features, by first ranking the features by some informativeness criterion [21], such as their Pearson Correlation with the target and then selecting the top features. Finally, this subset of features is presented as input to the classification algorithm. Wrapper methods [21] use a search procedure in the space of possible feature subsets using some search strategy such as Sequential Forward Selection (SFS) or Sequential Backward Selection (SBS), and various subsets of features are generated and evaluated. There is also a third category named embedded methods [21], where the search for an optimal subset of features is built into the classifier construction. Features are selected as a part of the building the particular classifier, in contrast to the wrapper approach, where a classification model is used to evaluate a feature subset that is selected without using the classifier.

Within these categories, algorithms can be further differentiated by the precise nature of their evaluation function, and by how the space of feature subsets is explored. According to Dash *et al.*, evaluation measures can be divided into 5 categories, such as: distance (e.g. Euclidean distance), information (e.g. information gain), dependency (e.g. correlation coefficient), classifier error rate and consistency (e.g. inconsistency rate) [17]. By training and testing a definite classification model, the evaluation of a specific subset of features is performed. The search for the desired feature subset is wrapped around a specific classifier and training algorithm.

In this paper, we review a hybrid algorithm called Quick Branch and Bound (QBB) [16], which is a combination of two popular algorithms, the probabilistic search algorithm based Las Vegas Filter (LVF) and the complete search based Automatic Branch and Bound (ABB). QBB is reasonably fast, robust and handles features which are inter-dependent, but does not work well with large data [17]. Las Vegas Iterative (LVI) is a scalable version of LVF that first partitions the dataset into two partitions and uses the first partition to learn the features and verifies the goodness of the selected features with the second partition using the inconsistency measure [7]. We propose an enhanced version of QBB, called QBB Iterative (QBB-I) based on some of the concepts used in LVI and show that QBB-I is more efficient than QBB for larger datasets without sacrificing the quality of the features selected.

2 Feature Selection Algorithms

2.1 Feature Selection

A feature selection algorithm (FSA) is a computational solution that is motivated by a certain definition of relevance [19]. The FSAs can be classified according to the kind of output they yield; the algorithms that give a weighed linear order of features and the algorithms that find a subset of the original features [18]. Both types can be seen in a unified way by noting that in the latter case the weighting is binary. In this paper, we consider algorithms that find a subset of the original features. Fig. 1 describes the characteristics of FSA [19].

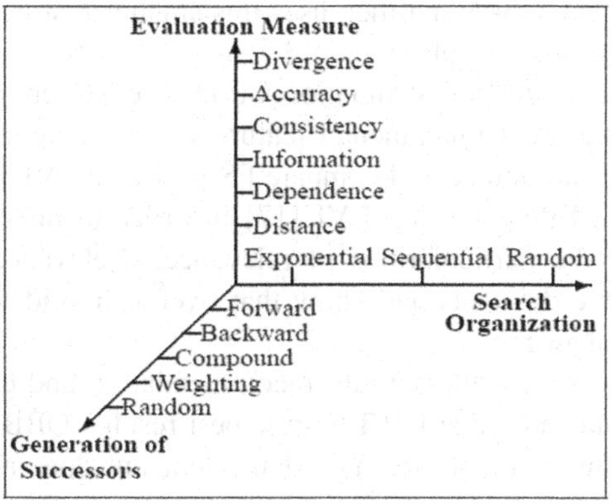

Fig. 1. Characterization of a FSA [19]

The two important aspects of a FSA are the search strategies to find the subset of features and the evaluation measure used to find the goodness of feature subsets. The search strategies employed by FSA include the exhaustive, complete, heuristic and random search. The search process may start with the empty set and increase by one feature at a time (forward search) or start with the full set and drop one feature at a time (backward search). The search may also start with a randomly generated feature set and change either probabilistically or deterministically. The evaluation measures include distance measures, information gain, correlation measures, consistency, and classifier error rate.

The algorithms used in this paper use the consistency measure. The inconsistency rate of a data is defined as follows [6]: (a) if the two patterns match all but their class labels, then they are considered inconsistent. (b) The inconsistency count is the total matching patterns minus largest number of patterns of different class label. For example, if there are n matching patterns, among them, c1 patterns belongs to label1, c2 to label2, c3 to label3 where $n = c1 + c2 + c3$. If c3 is largest among three, then inconsistency count = n - c3. (c) Inconsistency rate is equal to sum of all inconsistency counts divided by total number of patterns.

The QBB Algorithm [16] is a hybrid of two popular algorithms, the LVF [6] and the ABB [5] algorithms and exploits the advantage and disadvantage of both algorithms. LVF is probabilistic search algorithm which in the initial phase quickly reduces the number of features to a subset as initially many subsets can satisfy the consistency measure. However, after a certain point, fewer subsets can satisfy this and thus increase the computational complexity as it spends resource on randomly generating subsets that are obviously not good.

ABB starts with the full feature set and uses the inconsistency measure as the bound. It reduces the feature set by one feature at a time to generate subsets and is a complete search that guarantees optimal subset. However, its performance deteriorates when the difference between the total number of features and the optimal subset of features is large. QBB was proposed to exploit the advantages and overcome the disadvantages of both LVF and ABB. LVF quickly reduces the size of the initial feature set to a consistent subset and then uses this smaller feature set as the input to ABB to find the optimal feature subset.

In a survey and experimental evaluation of feature selection algorithms [19], Molina *et al.* review different fundamental feature selection algorithms and assesses their performances. They evaluate and compare FSA such as LVF [6], ABB [5], QBB [16], FOCUS [1], RELIEF [11], and LVI [17] in order to understand the general behavior of FSAs on the particularities of relevance, irrelevance, redundancy and sample size of synthetic data sets and show that overall hybrid algorithms perform better than individual algorithms.

In [23], the authors present the activity recommendation and compare QBB with different algorithms and show that QBB has the best results. QBB is reasonably fast, robust and handle features which are inter-dependent but does not work well with large datasets [17].

2.2 Quick Branch and Bound Iterative (QBB-I) Algorithm

We propose the Quick Branch and Bound Iterative (QBB-I) algorithm so as to retain the advantages of QBB while reducing the time complexity. QBB-I divides the data to two parts, D0 (p %) and D1 (1-p %) and finds a subset of features for D1 using the consistency measure. It is designed such that the features selected from the reduced data will not generate more inconsistencies with the whole data.

Algorithm 1. Algorithm for QBB-iterative

```
Procedure QBB-I (D, δ, S, p)
        δ = inConCal(S, D); T = S
        D0 = p% of Dataset; D1 = D - D0
        loop
                S' = LV F(S, MaxTries, D0);
                /*All legitimate subsets from LVF are in S'*/
                MinSize=card(S);
                loop
                        T' = ABB (Sj, D0)
                        if (MinSize > card (T') then
                                MinSize = card (T');
                                T = Sj;
                        end if
                end loop
                if (checkIncon (subset, D1, inconData) ≤ δ) then
                        Return T;
                else
                        append (inconData, D0);
                        remove (inconData, D1);
                end if
        end loop
end procedure
```

QBB-I finds a subset of features for D0 and checks this with D1 using the consistency measure. If the inconsistency rate exceeds the threshold, it appends the patterns from D1 which cause the inconsistency to D0 and deletes them from D1. This selection processes repeats until a solution is found. If no subset is found, the complete set of attributes is returned as a solution.

Table 1 describes the type of Search Organization, Generation of Successors and Evaluation Measure used by the four algorithms. A search algorithm takes responsibility for driving the feature selection process.

Table 1. Comparison of feature selection algorithms used

Characteristics	LVF	LVI	QBB	QBB-I
Search Organization	Random	Random	Random\ Exponential	Random\ Exponential
Generation of Successor	Random	Random	Random	Random
Evaluation Measures	Consistency	Consistency	Consistency	Consistency
Learning Speed	High	Medium	Medium	Medium
Data Size	Small	Large	Small	Large
Execution Time	High	Medium	High	Low

The time complexity of the feature selection algorithms is based on two parameters, the number of instances (N) and number of attributes (S). As N is much larger than S, we can significantly improve the performance with QBB-I over QBB by partitioning the data and starting with a smaller N′ << N.

3 Experimental Results of QBB-I

We want to verify a) QBB-I run faster than QBB on large datasets, b) the feature set returned by QBB-I is identical to QBB, c) QBB-I is faster than LVF, ABB, and QBB in general and d) QBB-I is particularly suitable for huge datasets.

Our experiments compare QBB-I with other baseline existing algorithms such as LVF, LVI and QBB in terms of execution time and selected features. All algorithms use the consistency measure with the inconsistency rate set to 0 as there is no prior knowledge about the data. Using eight different datasets, we find that QBB-I takes less execution time compared to other algorithms. Our tests with ASSISTments intelligent tutoring dataset using 150,000 log data and other standard datasets show that QBB-I is significantly more efficient than QBB while selecting the same set of optimal features.

3.1 Dataset Used for Feature Selection

For testing the proposed approached, we use the Cognitive Tutor dataset, from the Knowledge Discovery and Data Mining Challenge 2010 KDD Cup 2010 [12] from two different tutoring systems from multiple schools over multiple school years. The dataset contains 19 attributes. We used the Challenge dataset for this work. These datasets are quite large which contains 3,310 students with 9,426,966 steps [12]. There are many technical challenges such as the data matrix is sparse, there is a strong temporal dimension to the data and the problem a given student sees is determined in part by student choices or past success history.

A total of seven datasets, both artificial and real, are chosen for the experiments from the UC Irvine data repository [4]. The Lymphography dataset [3] contains 148 instances and the test set is selected randomly with 50 instances. It contains 19 attributes. The lung-cancer dataset [8] describes two types of pathological lung cancers and contains 32 instances and 57 attributes. Mushroom dataset [22] contains 8124 instances and the dataset has 22 discrete attributes. Parity5+5 dataset [10], the concept is the parity of five bits. The dataset contains 10 features, of which five are uniformly random ones.

Splice junctions [20] are points on a DNA sequence at which superfluous DNA is removed during the process of protein creation in higher organisms. Led24 dataset [2] has a total of 3200 instances, of which 3000 are selected for testing. It contains 24 attributes. All attribute values are either 0 or 1, according to whether the corresponding light is on or not for the decimal digit.

3.2 Experimental Setup

QBB and QBB-I were implemented using the C language. In all algorithms, the inconsistency rate is taken as zero. And, the partition percentage used in QBB-I is 27% and that of LVI is 24%. The experiment is done with over 150,000 log data from the ASSISTments dataset.

3.3 Results

A total of seven datasets, both artificial and real, are chosen for the experiments from the UC Irvine data repository to check the effectiveness of QBB-I. These datasets are either commonly used for testing feature selection algorithms or artificially designed so that relevant attributes are known. Fig. 2 shows the graphical representation for the comparison of LVF, QBB and QBB-I for different datasets. For the first three dataset, there is no much reduction in the execution time due to small set of data. But for parity and KDD datasets, it is much clearer. So from this analysis, we can conclude that QBB-I algorithm is faster than the other algorithms for larger datasets and less beneficial for smaller datasets.

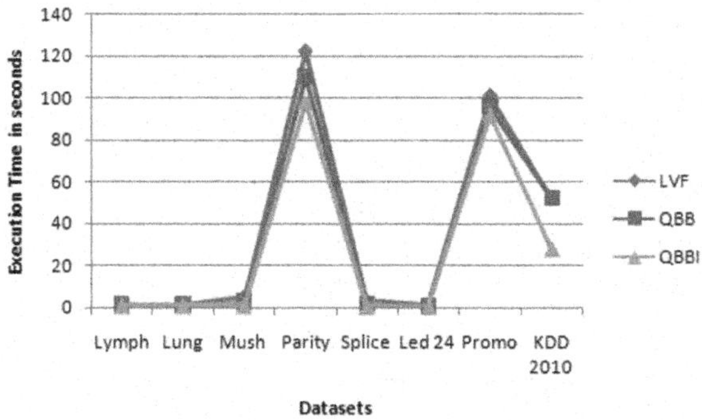

Fig. 2. Comparison of LVF, QBB and QBB-I in different datasets

We then compare the features selected by LVF, QBB and QBB-I. This experiment is mainly focused on checking the accuracy of our Enhanced algorithm. Table 2 shows the features selected by three algorithms. Table 2 shows that the features selected by QBB-I are identical to that of QBB while the features selected by LVI is identical to that of LVF.

Table 2. Features Selected By Three Algorithms

LVF	LVI	QBB	QBB-I
Row	Row		
StudentId	StudentId	StudentId	StudentId
Problem Hierarchy	Problem Hierarchy	Problem Hierarchy	Problem Hierarchy
Problem Name	Problem Name	Problem Name	Problem Name
Problem View	Problem View	Problem View	Problem View
Step Name	Step Name	Step Name	Step Name
CFA	CFA	CFA	CFA
InCorrects	InCorrects		
Hints	Hints		
Corrects	Corrects		
Opportunity	Opportunity	Opportunity	Opportunity
KC	KC	KC	KC

We also compare the execution time of the four algorithms (Fig. 3).The result shows that the QBB-I takes less execution time compared to all other algorithms.

Among the three existing algorithms, LVI was the fastest while QBB gave the better feature set. Our QBB-I algorithm, which uses the scalability concepts from LVI to scale the QBB algorithm, returns the same feature set as QBB and is faster than LVI.

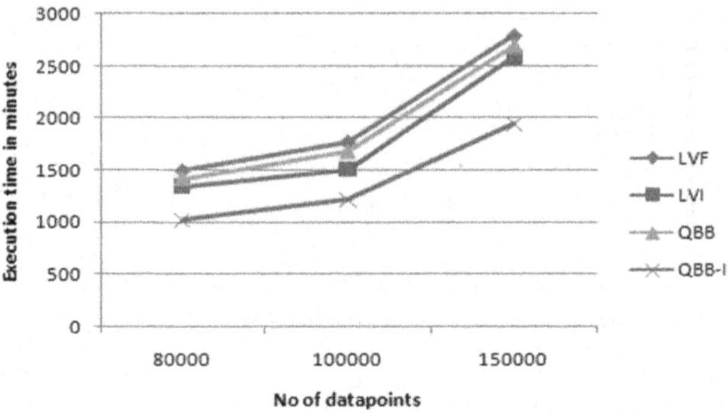

Fig. 3. Comparison of Execution time of different algorithms

For the ASSISTments dataset, Yu. *Et al*[9] manually identify useful combinations of features and then experimentally shows that some of the feature combinations effectively improve the Root Mean Squared Error (RMSE). While the previous experiments were with individual features, we also compare the features selected by QBB and QBB-I with the following combinations of features suggested by this paper. (Student Name, Unit Name), (Unit Name, Section Name), (Section Name, Problem Name), (problem Name, step Name), (Student Name, Unit Name, Section Name), (Section Name, Problem Name, Step Name), (Student Name, Unit Name, Section Name, Problem Name) and (Unit Name, Section Name, Problem Name, Step Name).

Both QBB and QBB-I selected the last four combinations with a subset of the ASSISTments dataset consisting of 150,000 log records, which have previously been shown to be the best combinations [9]. Fig. 4 shows the comparison of the time taken by QBB and QBB-I for selecting the combinations for 80,000, 100,000 and 150,000 log records. The result shows that QBB-I takes less time for execution compared to QBB.

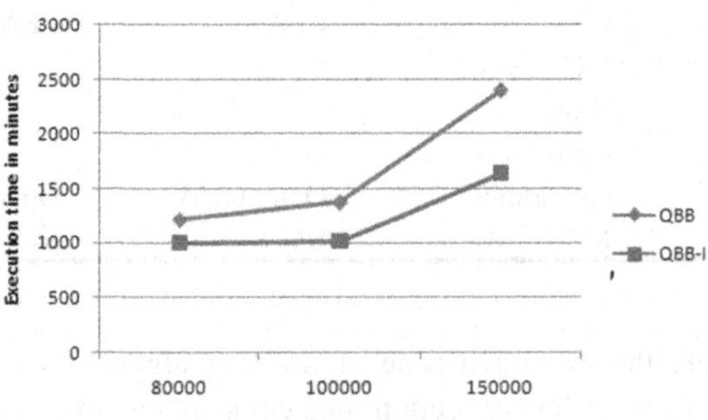

Fig. 4. Comparison of time for selecting combinations

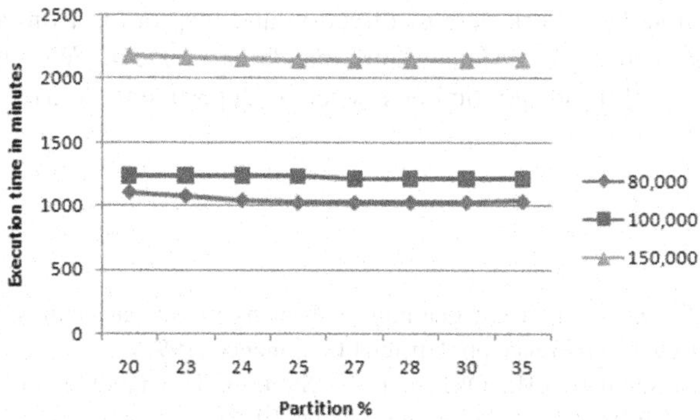

Fig. 5. Experiment to find out the optimum partition size (%)

Finally we vary the partition size (Refer Fig. 5) as we expect that as the partition size increases; the execution time reduces and then becomes a constant. We find that for 80000 log records of the ASSISTments dataset, this is reached with a partition size of around 25% while with 100,000 and 150,000 log records of the same dataset, this is reached with a partition size of 27%.

4 Conclusion

We proposed an algorithm (QBB-I) to feature selection which is a scaled version of QBB and show that it is more efficient than QBB while selecting the same feature set. We increase the efficiency of QBB-I by partitioning the dataset such that the proportion of the first partition is small with respect to the entire dataset. As inconsistent entries are moved to the first partition, QBB-I learns and avoids checking inconsistent entries in future iterations.

QBB-I improves the execution time of QBB algorithm when the dataset is large. When the dataset is small the performance improvement is negligible as the inconsistency checking is of order (N) and the time saving is not much if N is not large. Similar to LVI, the second issue is the size of the partition to start QBB-I. If D0 is too small, QBB-I may select features where majority of the data in D1 are inconsistent. In the worst case, D0 may become the entire dataset in the next iteration. If D0 is too large, the overheads of the additional loops may be larger and it may be better to just use QBB. The optimum partition size needs to be determined by testing with the data.

Using eight different datasets, we find that QBB-I takes less execution time compared to other algorithms. Our tests with ASSISTments intelligent tutoring dataset using over 150,000 log data and other standard datasets shows that QBB-I is significantly more efficient than QBB while selecting the same subset of optimal features. Future work involves testing this with larger datasets and the use of parallel feature selection algorithms.

Acknowledgments. This work derives direction and inspiration from the Chancellor of Amrita University, Sri Mata Amritanandamayi Devi. We thank Dr. M Ramachandra Kaimal, head of Computer Science Department, Amrita University for his valuable feedback.

References

1. Almuallim, H., Dietterich, T.G.: Learning with many irrelevant features. In: Proceedings of the 9th National Conference on Artificial Intelligence (1991)
2. Breiman, L., Friedman, J.H., Olshen, R.A., Stone, C.J.: Classification and Regression Trees. Wadsworth International Group, Belmont (1984)
3. Cestnik, G., Konenenko, I., Bratko, I.: Assistant-86: A Knowledge- Elicitation Tool for Sophisticated Users. In: Progress in Machine Learning, pp. 31–45. Sigma Press (1987)
4. Merz, C.J., Murphy, P.M.: UCI Repository of Machine Learning Batabases. University of California, Department of Information and Computer Science, Irvine (1996), `http://www.ics.uci.edu/mlearn/MLRepository.html`
5. Liu, H., Motoda, H., Dash, M.: A monotonic measure for Optimal Feature Selection. In: Proceedings of the European Conference on Machine Learning, pp. 101–106 (1998)
6. Liu, H., Setiono, R.: A probabilistic approach to feature selection: a Filter solution. In: Proceedings of the 13th International Conference on Machine Learning, pp. 319–327 (1996)
7. Liu, H., Setiono, R.: Scalable feature selection for large sized databases. In: Proceedings of the 4th World Congress on Expert System, p. 6875 (1998)
8. Hong, Z.Q., Yang, J.Y.: Optimal Discriminant Plane for a Small Number of Samples and Design Method of Classifier on the Plane. Pattern Recognition 24, 317–324 (1991)
9. Yu, H.-F., Lo, H.-Y., Hsieh, H.-P.: Feature Engineering and Classifier Ensemble for KDD Cup 2010. In: JMLR: Workshop and Conference Proceedings, vol. 1, pp. 1–16 (2010)
10. John, G.H., Kohavi, R., Pfleger, K.: Irrelevant features and the subset selection problem. In: Proceedings of the Eleventh International Conference in Machine Learning (1994)
11. Kira, K., Rendell, L.: A practical approach to feature selection. In: Proceedings of the 9th International Conference on Machine Learning, pp. 249–256 (1992)
12. Koedinger, K., Baker, R., Cunningham, K., Skogsholm, A., Leber, B., Stamper, J.: A data repository for the EDM community: the pslc datashop (2010)
13. Kohavi, K.: Wrappers for performance enhancement and oblivious decision graphs. Phd thesis, Stanford university (1995)
14. Kohavi, R., John, G.H.: Wrappers for feature subset selection. Artificial Intelligence 97(12), 273–324 (1996)
15. Dash, M., Liu, H.: Feature selection for classification. Intelligent Data Analysis 1(1-4), 131–156 (1997)
16. Dash, M., Liu, H.: Hybrid search of feature subsets. In: Lee, H.-Y., Motoda, H. (eds.) PRICAI 1998. LNCS, vol. 1531, pp. 238–249. Springer, Heidelberg (1998)
17. Dash, M., Liu, H., Motoda, H.: Feature Selection Using Consistency Measure. In: Arikawa, S., Nakata, I. (eds.) DS 1999. LNCS (LNAI), vol. 1721, pp. 319–320. Springer, Heidelberg (1999)
18. Kudo, M., Sklansky, J.: A Comparative Evaluation of medium and largescale Feature Selectors for Pattern Classifiers. In: Proceedings of the 1st International Workshop on Statistical Techniques in Pattern Recognition, pp. 91–96 (1997)

19. Molina, L.P., Belanche, L., Nebot, A.: Feature selection algorithms: a survey and experimental evaluation. Universitat Politcnica de catalunya. departament de llenguatges i sistemes informtics (2002)
20. Noordewier, M.O., Towell, G.G., Shavlik, J.W.: Training Knowledge-Based Neural Networks to Recognize Genes in DNA Sequences. In: Advances in Neural Information Processing Systems, vol. 3 (1991)
21. Saeys, Y., Inza, I., Larranaga, P.: A review of feature selection techniques in bioinformatics. Bioinformatics 23(19), 2507–2517 (2007)
22. Schlimmer, J.S.: Concept Acquisition Through Representational Adjustment (Technical Report 87-19). Doctoral disseration, Department of Information and Computer Science, University of California, Irvine (1987)
23. Nguyen, T.: A Group Activity Recommendation Scheme Based on Situation Distance and User Similarity Clustering. M. Thesis, Department of Computer Science KAIST (2012)

Towards a Scalable Approach for Mining Frequent Patterns from the Linked Open Data Cloud

Rajesh Mahule and O.P. Vyas

Department of Information Technology
Indian Institute of Information Technology, Allahabad, India
rmahule@rediffmail.com,
dropvyas@gmail.com

Abstract. In recent years, the linked data principles have become one of the prominent ways to interlink and publish datasets on the web creating the web space a big data store. With the data published in RDF form and available as open data on the web opens up a new dimension to discover knowledge from the heterogeneous sources. The major problem with the linked open data is the heterogeneity and the massive volume along with the preprocessing requirements for its consumption. The massive volume also constraint the high memory dependencies of the data structures required for methods in the mining process in addition to the mining process overheads. This paper proposes to extract and store the RDF dumps available for the source data from the linked open data cloud which can be further retrieved and put in a format for mining and then suggests the applicability of an efficient method to generate frequent patterns from these huge volumes of data without any constraint of the memory requirement.

Keywords: Linked Data Mining, Data Mining, Semantic Web data Mining, RDF data mining.

1 Introduction

In recent years, the linked data principles [10] have become one of the prominent ways to interlink and publish datasets on the web creating the web space a big data store. With the data published in RDF form and available as open data on the web opens up a new dimension to produce knowledge from these data, called semantic web data. Linked Open data cloud [10] is a heterogeneous source of semantic web data, contains information from diverse fields like music, medicine & drugs, publications, people, geography etc. and most often the large government data represented in the RDF triple structure consisting of Subject, Predicate and Object (SPO). Each statement represents a fact and expresses a relation (represented by the predicate resource) between the subject and the object. Formally, the Subject and the Predicate resource are represented by a URI and the object by a URI or a literal such as a number or string; the Subject and Object being interconnected by predicates. The interconnection between the statements harbor many hidden relationships and can lead to

M.K. Kundu et al. (eds.), *Advanced Computing, Networking and Informatics - Volume 1*, Smart Innovation, Systems and Technologies 27,
DOI: 10.1007/978-3-319-07353-8_16, © Springer International Publishing Switzerland 2014

insights about the data and domain. Resources in the semantic web datasets are connected by multiple predicates, co-occurring in multiple implicit relations. The co-occurrence and the frequencies of triples in the datasets are a matter of investigation and give a pursuit for mining such SPO data such as Association rule mining [1]. Association rule mining has been widely studied in the context of basket analysis and sale recommendations [2]. In fact, association rules can be discovered in any domain, with many items or events among which interesting relationships can be discovered from co-occurrence of those item or events in the existing subsets (transactions). The discovered association rules may have many applications such as the enhancement of information source recommendation in the semantic web, association rule based classification and clustering of semantic web resources etc.

In this paper an approach on mining association rules based on the SPO view of RDF data from the linked open data cloud has been proposed using the Co-Occurrence Frequent Item Tree (COFI-tree) algorithm [3]. The advantages of using COFI tree for mining semantic web data over other methods (including FP-tree) is that "the algorithm builds a small tree called COFI-tree for each frequent 1-itemset and mine the trees with simple non-recursive traversals and only one COFI-tree resides in memory at one-time and it is discarded as soon as it is mined to make room for the next COFI-tree and hence dramatically reduces the requirement of the memory space so as to handle millions of transactions with hundreds of thousands of dimension" [3].

To illustrate the process of mining association rules from semantic web data using COFI-Tree mining algorithm, we integrate the publication data collected from different sources like DBLP [17], ACM Digital library [16] and Cite Seer [18] available in the publication domain of the linked open data cloud as downloadable repositories of RDF dump. We confine our study to identify association between the authors of the academic network for the identification of frequently collaborating co-authors in a particular domain so as to get an idea of the collaborative efforts made by the authors; as the present domain is confined only to the publication count of individual members.

The rest of the paper is structured as follows. Section 2 discusses related work and briefly describes the basics of association rule generation. Section 3 details the mining process using COFI-Tree algorithm with an example for illustration. Section 4 depicts the experimental work and evaluation and finally section 5 concludes the paper with a look on the work for the future.

2 Related Work

In the past couple of years, data mining has been an area of active researches in both the traditional datasets and the semantic datasets. In the semantic web datasets, there are approaches to mine association rules, logical rules, generating schema or taxonomies for the knowledge base etc. for application purposes. However, the majority of work in the arena of semantic web focuses on clustering and classification [4-5], there is also some work on Inductive Logic Programming (ILP) based on logics from ontology [6].

Association Rule mining (ARM) [8], mines frequent patterns that appear together on a list of transactions, where a transaction is set of items. For example in the context of market basket analysis or sales analysis, "a transaction is the set of items purchased together by a customer in a specific event". The mined rules represented in the form {bread, butter} → milk, meaning that people who bought bread and butter also bought milk. Many ARM algorithms have been proposed so far which works well in traditional datasets and can be classified into two types: Apriori based, and FP-tree based.

The apriori algorithm which serves as the base for most of the algorithms uses an anti-monotone property stating that "for a K-itemset to be frequent, all its (*k*-1) item-sets should also have to be frequent; thereby reducing the computational cost of candidate frequent itemset generation". But, in case of very large datasets having big frequent 1-itemset, this algorithm suffers from two main bottlenecks viz. "repeated I/O scanning and high computational cost". Another problem related to this algorithm is the huge size of candidate frequent 2-itemset and 3-itemset as observed for most real datasets.

Another approach for discovering frequent patterns in transaction datasets was FP-Growth [9]. The algorithm solves the multi-scan problem by creating a compact tree structure, FP-Tree, representing frequent patterns thereby improving the candidate itemset generation. This algorithm requires two full I/O scans of the dataset and buildsa prefix tree in main memory, which is then mined. The algorithm performs faster than the Apriori algorithm. The process of mining in the generated FP-Tree is performed recursively by creating conditional trees having the same magnitude in order of number, as the frequent patterns. Due to the huge mass of the created conditional trees, the algorithm is not scalable and hence suitable to mine large datasets.

An improvement to this above is the Co-Occurrence Frequent Item tree (COFI-tree) Mining algorithm [3] in which the author divides the algorithm in two phases. Phase I, in which the FP-Tree structure is built with the two full I/O scans of the transactional database. Phase II, small Co-occurrence Frequent trees for each frequent item is built which are first pruned, eliminating the non-frequent items in regard to the COFI-tree based frequent item, followed by the mining process. The most important advantage here is the pruning technique, which truncates the large memory space requirement of the COFI-trees.

The above association rule mining algorithms are proposed and well suited for mining with traditional dataset, i.e. those datasets which are based on transactions and cannot be applied directly to the semantic web data as the semantic web dataset are of the form of SPO, they need to be shaped into a form that can be suitable for mining, i.e. the transactions from the semantic web data has to be generated and the mining can be performed on those data.

Also, there are approaches for generating association rules from the semantic web datasets. In [7], an algorithm is proposed that mines association rules from the semantic web dataset by mining patterns which are provided by the end users. The mining patterns are based on the result of a query language SPARQL, the query language for the semantic web data. The algorithm mines the association rules in a semi-supervised manner as per the user provided input patterns.

In one of the recent algorithms [11], proposed by R. Ramezani et al., the author proposed to mine association rules from a single and centralized semantic web dataset, in an unsupervised environment and with the approach similar to the apriori method, as proposed for traditional datasets.

Also, there are similar works of performing datamining on the linked data available on the LOD cloud. One approach LiDDM [12], the authors applied techniques of data mining (Clustering, Classification and association rules) on linked data [10] by first acquiring the data from the LOD datasets using user defined queries, converting the result into traditional data, applying preprocessing to format the data for mining and then performing the data mining process. Similar to LIDDM is the Rapid Miner Semantic web plug-in [14], in which the end-user has to provide a suitable SPARQL query for retrieving desired data from the linked open data cloud. The software then converts data in a tabular feature format and then the data mining process is carried out.

3 Mining Data from LOD Cloud Using COFI Tree Algorithm

The approach to construct COFI tree involves the following steps –"the construction of the Frequent Pattern Tree, the building of small Co-Occurent Frequent trees for each frequent items which are then pruned for the elimination of any non-frequent items in regard to COFI-tree based frequent item, and finally, the mining work is carried out".

The work start with the collection of publication data from three sources DBLP [17], ACM Digital library [16] and CiteSeer [18] available as downloadable RDF dumps available at datahub.org [15], the authors names and the paper title is tokenized and retrieved for each publication available in the dump after parsing the RDF files individually. The parsing is done using Jean [13] which is available as a Java API. Each publication title and individual authors retrieved from the RDF/XML files are assigned a unique ID and are stored in a output text file. A Horizontal data format as a vector representation with the publication-id (Pi) and author-id (Ai) for some publications is formed and is as shown in Table -1 for illustration purpose. The output text file (enumerated file), which is in a table format is then used for the construction of the frequent pattern tree. The process comprising two modules, each module requiring a complete I/O scan of the table. In module 1, the support for all items in the transaction table is accumulated (Refer to Fig. 1: Step-1) followed by the removal of the infrequent authors i.e. authors with a support less than the support threshold(say 4 in this example) (Refer to Fig. 1: Step-2), followed by sorting of the remaining frequent authors according to their frequency to get frequent 1-itemset (Refer to Fig. 1: Step-3).The list containing this is arranged in a header table, having the authors, their respective support along with a pointer to the author's first occurrence in the frequent pattern tree. The second module constructs a frequent pattern tree.

In module 2, the first transaction (A1, A7, A4, A3, A2) is read and scanned for the frequent items present in the header table (i.e. A1, A4, A3and A2), which again is sorted by their authors' support (A1, A2, A3 and A4) and first path of the FP-Tree is generated using this ordered transaction with an initial item-node support value of 1. Between each item-node in the tree, a link is established and in the header table, corresponding item entry is done. Similar process is performed for the next transaction (A2, A3, A8, A5 and A4), yielding sorted frequent item list (A2, A3, A4, A5), which builds the second path for the FP-Tree. Next transaction i.e. Transaction 3 (A2, A4, A5, A1 and A13) yields the sorted frequent item list (A1, A2, A4,A5) that shares the same prefix (A1, A2) with the existing path in tree. The support for Item-nodes (A1 and A2) is incremented, setting the support of A1 and A2 with a value of 2 and creating a new sub-path with the left items on the list (A4, A5) with support value of 1. The process is repeated for the transactions in Sample Table for Publications and Authors as shown in Table 1. The final FP-Tree constructed after processing all the transactions of the table is shown in Fig. 2.

Table 1. Table for Publications and Authors

Pub-id	Auth-1	Auth-2	Auth-3	Auth-4	Auth-5
P1	A1	A7	A4	A3	A2
P2	A2	A3	A8	A5	A4
P3	A2	A4	A5	A1	A13
P4	A3	A5	A6	A1	A14
P5	A1	A2	A14	A15	A16
P6	A1	A3	A17	A18	A7
P7	A1	A3	A8	A9	A7
P8	A12	A5	A6	A11	A2
P9	A1	A6	A13	A14	A15
P10	A3	A6	A16	A7	A18
P11	A1	A4	A2	A8	A9
P12	A4	A5	A2	A11	A12
P13	A13	A4	A3	A7	A15
P14	A3	A6	A16	A17	A10
P15	A2	A4	A5	A6	A9
P16	A10	A5	A2	A1	A4
P17	A1	A11	A5	A6	A3
P18	A3	A4	A12	A2	A1

Auth	Count	Auth	Count
A1	11	A14	3
A2	10	A15	3
A3	10	A16	3
A4	9	A17	2
A7	4	A18	2
A5	8	A9	3
A8	3	A11	3
A6	7	A12	3
A13	3	A10	3

Step 1

Auth	Count
A1	11
A2	10
A3	10
A4	9
A5	8
A6	7

Step 2

Auth	Count
A6	7
A5	8
A4	9
A3	10
A2	10
A1	11

Step 3

Fig. 1. Steps of phase 1

The FP-Tree (Fig. 2) for the example table illustrated above is now used for the construction of independent small trees called COFI-trees for every frequent header table item of the FP-Tree. These trees are built independently for each individual item in the header table in the least frequent order of their occurrence i.e. COFI-tree for author A6 is created first followed by the building of COFI-trees for A5, A4, A3, A2, A1.. The COFI-trees so built are mined and are discarded separately as soon as they are built, releasing the memory for the construction of the next COFI-tree. The process for the construction of the COFI tree starts first for the author A6 which being least frequent author of the header table. In the tree for A6, all the authors in the header table which are more frequent than A6 and share transactions with A6 takes part in building of the tree. The construction starts with the root node containing the author for which the tree is created i.e. A6 and for each edge in the FP-tree containing author A6 with other frequent items in the header list which are parent nodes of A6 and are more frequent than A6, a edge is formed with origin as the root i.e. A6.

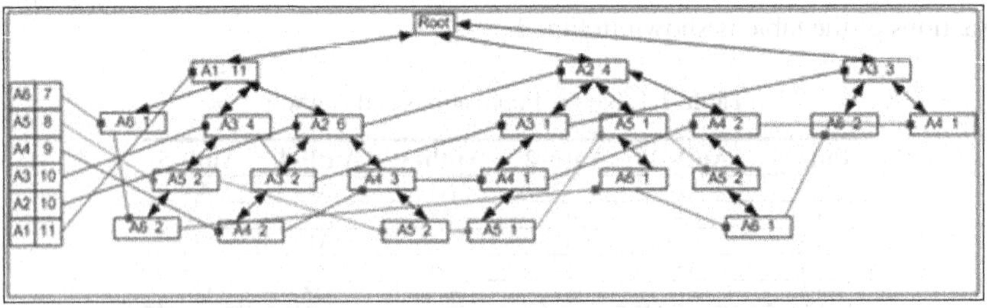

Fig. 2. Frequent pattern Tree

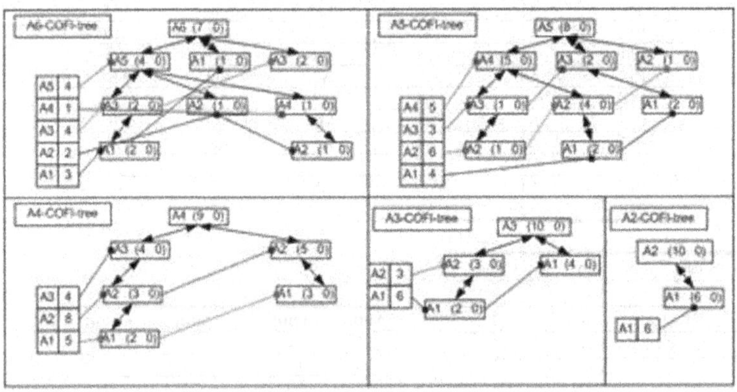

Fig. 3. COFI-trees

The support-count of the edge being equal to the support-count of the node for author A6 in the corresponding edge in FP-Tree. For multiple frequent authors that share the same node-item, they are put into same branch and a counter for each node-item of the COFI-tree is incremented. Below Fig. 3 depicts the COFI tree for all the frequent items of the FP-tree in Fig. 2.

In the above figure (Fig. 3), rectangular nodes represent the nodes in the tree with author label along with two numeric values representing counters for support-count and participation-count respectively. The first counter is the counter for the particular node and the second counter is the counter, initially initialized to zero, which is used during the mining process. Also in the figure above, there are horizontal and vertical bidirectional links used for pointing next node of same author and for establishing child-parent-child node-links respectively. The squares in the figure represent the cells of the header-table, which is sorted listing of the frequent items, used to build the tree and each list has author-id, frequency and a link pointing to the first node of tree with similar id as the author-id.

The mining process is performed independently as soon as the COFI-tree for individual authors are constructed and are discarded before the construction of the next COFI tree with the target of generating all frequent K-item set for the root-author of the tree. Prior to the mining process, with the support-count and the participation-count, the candidate-frequent patterns are extracted and put in a list for each branch of the tree followed by the removal of all non-frequent patterns from each branch. In the example above using the COFI-Tree algorithm[3], the frequent patterns generation is performed first from A5-COFI-Tree resulting, the frequent patterns as A5A4:5, A5A2:6 and A5A4A2:5.The A5-COFI-tree is then removed from the memory and the other COFI-trees are generated and mined to extract the frequent patterns for the root node-item. After performing mining process for A5 COFI-tree, A4-COFI tree is built and mined followed by the creation and mining A3-COFI-tree and A2-COFI-tree generating patterns A5A4:5, A5A2:6, A5A4A2:5, A4A2:8, A4,A1:5, A4A2A1:5 and A2A1:6.

4 Experimental Evaluation and Results

For our experiment we evaluated the applicability of the COFI-Tree algorithm on the datasets from the LOD Cloud, which we retrieved in RDF format. The transactions were generated by parsing the data from the DBLP [17], ACM Digital library [16] and CiteSeer [18] available as downloadable RDF dumps available at datahub.org [15] and publication title along with authors name were extracted and stored in a single row. The authors and publication data were converted into the form suitable for the mining process and a transaction dataset was created with a mix of data from the above three datasets with a total of around 10,000 record titles of publication and about 6800 distinct authors for this experiment purpose. The experiment was performed with very low support value (0.0025%), (0.005%) and (0.0075%) and the frequent patterns generated were found to be in a descending progression. The time needed for the execution of the experiment was also very less. The experiments were run on a 2.5 GHz core 2 machines with a RAM of 2GB.

5 Conclusion and Future Work

Finding scalable algorithm for extracting association rules from the heterogeneous dataset available as linked open data is the main goal of our research. As a lot data has been generated as linked open data content and there are very less application to consume the same. Consumption of linked data, like the work above requires a lot of efforts. The main effort required for this type of work is in the refinement of the available heterogeneous dataset and converting them in a common platform ready for the mining task or the application. The experiments were conducted on a small sample mix of the heterogeneous data as the refined data with a very-very large number of transactions (millions) could not be refined for the experiment. But looking into the advantages of the method used and its performance it is concluded that the method will prove to be efficient for scalable amount of data which we propose to experiment with in the near future. The domain selected by us for performing the experiment is just a single dimension of the mining activities that can be performed with the data available from the linked open data cloud. With the heterogeneity and the right selection of inter related domains varied dimensions of knowledge can be discovered with this type of approach.

References

1. Abedjan, Z., Naumann, F.: Context and Target Configurations for Mining RDF data. In: International Workshop on Search and Mining Entity-Relationship Data (2011)
2. Agrawal, R., Srikant, R.: Fast Algorithms for mining association rules in large databases. In: International Conference on Very Large Databases (1994)
3. El-Hajj, M., Zaiane, O.R.: COFI-tree Mining: A New Approach to Pattern Growth with Reduced Candidacy Generation. In: Workshop on Frequent Itemset Mining Implementations (FIMI 2003) in conjunction with IEEE-International Conference on Data Mining (2003)
4. Bloehdorn, S., Sure, Y.: Kernel methods for mining instance data in ontologies. In: Aberer, K., Choi, K.-S., Noy, N., Allemang, D., Lee, K.-I., Nixon, L.J.B., Golbeck, J., Mika, P., Maynard, D., Mizoguchi, R., Schreiber, G., Cudré-Mauroux, P. (eds.) ASWC 2007 and ISWC 2007. LNCS, vol. 4825, pp. 58–71. Springer, Heidelberg (2007)
5. Fanizzi, N., Amato, C., Esposito, F.: Metric-based stochastic conceptual clustering for ontologies. Information System 34(8), 792–806 (2009)
6. Amato, C., Bryl, V., Serafini, L.: Data-Driven logical reasoning. In: 8th International Workshop on Uncertainty Reasoning for the Semantic Web (2012)
7. Nebot, R.B.V.: Finding association rules in semantic web data. Knowledge-based System 25(1), 51–62 (2012)
8. Agrawal, R., Swami, A.N.: Mining association rules between sets of items in large databases. In: ACM SIGMOD International Conference on Management of Data (1993)
9. Han, J., Pei, J., Yin, Y.: Mining frequent patterns without candidate generation. In: ACM SIGMOD International Conference on Management of Data (2000)
10. Bizer, T.H.C., Berners-Lee, T.: Linked Data - The Story so Far. International Journal on Semantic Web and Information Systems (2009)
11. Ramezani, R., Saraee, M., Nematbakhsh, M.A.: Finding Association Rules in Linked Data a centralized approach. In: 21st Iranian Conference on Electrical Engineering (ICEE) (2013)
12. Narasimha, R.V., Vyas, O.P.: LiDDM: A Data Mining System for Linked Data. In: Workshop on Linked Data on the Web. CEUR Workshop Proceedings, vol. 813. Sun SITE Central Europe (2011)
13. The Jena API, `http://jena.apache.org/index`
14. Potoniec, J., Ławrynowicz, A.: RMonto: Ontological extension to RapidMiner. In: Poster and Demo Session of the ISWC 2011 - 10th International Semantic Web Conference, Bonn, Germany (2011)
15. The Data Hub, `http://thedatahub.org`
16. The Association for Computing Machinery (ACM) Portal, `http://portal.acm.org/portal.cfm`
17. The DBLP Computer Science Bibliography, `http://dblp.uni-trier.de/`
18. The Scientific Literature Digital Library and Search Engine, `http://citeseer.ist.psu.edu/`

Automatic Synthesis of Notes Based on Carnatic Music Raga Characteristics

Janani Varadharajan, Guruprasad Sridharan,
Vignesh Natarajan, and Rajeswari Sridhar

Department of Computer Science and Engineering
College of Engineering, Guindy
Anna University – 600 025, India
{kvv.emails,guruprasad.sridharan,
vignesh.natarajan,rajisridhar}@gmail.com

Abstract. In this paper, we propose two methods to automatically generate notes (Swaras) conforming to the rules of Carnatic Music, for a Raga of the user's choice as the input. The proposed methods are purely statistical in nature. The system requires training examples for learning the probability model of the chosen Raga and no hand-coded rules are required. Hence, it is easy to extend this method to work with a large number of Ragas. Each proposed method involves a Learning Phase and Synthesis Phase. In the Learning Phase, an already existing composition of the desired Raga is used to learn the transition probabilities between swara sequences based on the Raga lakshana characteristics. In the Synthesis Phase, using the previously constructed transition table, swaras are generated for the desired Raga. We describe two methods - one based on First Order Markov Models and the other based on Hidden Markov Models. We also provide comparison of the performance of both the approaches based on feedback from Carnatic experts.

1 Introduction

1.1 Basic Terms and Definitions

Swara - A Swara is the fundamental unit of Classical Music [1]. A Swara can be analogously understood as a musical note. The basic Swaras in Carnatic music are seven in number and they are as follows: Sa, Ri, Ga, Ma, Pa, Da, Ni.

Raga - A Raga is an arrangement of notes in a predefined manner where the position and neighbourhood of each Swara(note) depends on the Raga lakshana of a Raga. The Raga lakshana includes Arohana (ascending arrangement of notes), Avarohana (descending sequence of notes), the Amsa (characteristic phrase), the frequently used starting note, most frequently used note, ending note, note not used, etc.

Arohana - Arohana, Arohanam or Aroha, in the context of Indian classical music, is the ascending scale of notes in a Raga. The notes ascend in pitch from the lower tonic towards the upper tonic.

M.K. Kundu et al. (eds.), *Advanced Computing, Networking and Informatics - Volume 1*, Smart Innovation, Systems and Technologies 27,
DOI: 10.1007/978-3-319-07353-8_17, © Springer International Publishing Switzerland 2014

Avarohana - Avarohana, Avarohanam or Avaroha, in the context of Indian classical music, is the descending scale of any Raga. The notes descend in pitch from the upper tonic down to the lower tonic, possibly in a crooked manner.
In this work, we use the Raga lakshana like Arohana, Avarohana and Amsa (Characteristic phrase) of each Raga as features to construct a probabilistic transition model which is used for synthesis.

1.2 Fundamental Idea

The Ragas are defined not only by the different Swaras that are used to compose it, but also the order of occurrence of these Swaras and one or more characteristic Swara phrases occurring frequently in the composition, thereby giving the Raga an unique signature or an identity. Thus, it is not possible to simply use permutations and combinations of different Swaras to compose a melody in a particular Raga. For instance, the Ragas Shankarabharanam and Kalyani are both Melakartha Ragas, meaning they have all seven basic Swaras in their Arohana and Avarohana.

Shankarabharanam :
Arohana : S R_2 G_3 M_1 P D_2 N_3 S
Avarohana : S N_3 D_2 P M_1 G_3 R_2 S

Kalyani :
Arohana : S R_2 G_3 M_2 P D_2 N_3 S
Avarohana: S N_3 D_2 P M_2 G_3 R_2 S

If mere permutations and combinations of Swaras are used to generate compositions in these Ragas, one would not be able to easily identify which Raga a particular composition follows, if the compositions do not use the Swara 'M'. The characteristic phrases must be accentuated well enough in the composition to enable correct identification of the Raga.

In this paper, we describe an algorithm that learns the characteristics of Ragas and synthesizes sequences conforming to the properties of that particular Raga. The primary motive of this paper is to demonstrate the ability of a probabilistic algorithm to understand the characteristics of a Raga and reproduce sequences of the same Raga that do not delve from the rules of the said Raga.

This paper is organized as follows: Section 2 talks about the related work in this field, Section 3 explains the proposed method, Section 4 discusses the assumptions and constraints that exist in the method proposed, Section 5 talks about the results and Section 6 concludes the paper with possible future extensions.

2 Related Work

Hari V. Saharabuddhe [2] talks about generating automatic and computer assisted notes that correspond to a particular Raga by constructing a Finite State automata for the Raga, with each states in the Finite State Automata being the alaps (or) phrases of the Raga. The paper claims that remembering the last two notes can produce reasonable performance in a majority of ragas. The authors have further stated that preserving information about elongation, grace notes, and/or successor selection frequencies often enhances the quality of performance.

Dipanjan Das *et al.* [3] talks about generating the *'Aarohanas'* and *'Avarohanas'* present in each Raga that conforms to the Hindustani Classical Music for the purpose of discovering and identifying new sequences. This idea implements a Finite State Machine (FSM) which was proposed by H.V. Sahasrabuddhe [2] that generates a sequence of *Swaras* that conforms to the rules of a particular Raga. Dipanjan Das *et al.*, designed a probabilistic finite state machine that generates only the Avarohana and Aarohana of the said Raga. The FSM used in the paper can be looked upon as a bigram model of three most frequent notes along with their probabilities that they follow while moving in either direction so as to generate the ascending and descending sequences of the particular Raga. The FSM for each Raga was constructed manually as the idea deals with only three most frequently occurring notes. Once the FSM created the ascending and descending notes, an algorithm is run on it to generate one instance or sample of the said Raga. This algorithm takes the number of notes and the start note of the Raga as input. Output of the entire process would be a sequence conforming to a particular Raga which was generated upon constructing a sample of the Aarohana and Avarohana of the same. The main drawback that we have identified from this paper is that, a sequence generated using a randomized approach without any defined rules to monitor the occurrence and position of the notes will result in a sequence that does not conform to the characteristics of the Raga.

Subramanian in his paper [4], talks about generating computer music from skeletal notations, along with the gamakkas that produce a smooth transition from one Swara to another. The proposed system uses user defined gamakkam definition files for each Raga. Addition of gamakkas depend on a) The Raga b) The context in which the note occurs and c) The duration of the note. Gamakkams can be modeled in different ways. One method would be to analyze a large number of recordings sung by music experts and extract common features for each note of the Raga. This requires a system that can identify note boundaries and transcribe live music into notation format.

3 The Proposed Method

The method to generate notes for a Raga is given in Fig. 1. The following subsections discuss the proposed method to Raga note generation.

3.1 First Order Markov Model

3.1.1 Learning Phase

The idea of using Markov model was inspired by the work of Simone Hill [5], [6]. In the learning phase, note sequences of existing compositions of the chosen Raga are input to the system in the form of text, and the system constructs a 16*16 (seven basic swaras along with their subtypes) transition matrix (Markov Chain) for the Raga based on this sequence. This table calculates and stores, the probability with which one Swara follows another. By this way, characteristic phrases of the chosen Raga (phrases that occur most frequently in the compositions belonging to that Raga) are assigned maximum probability. The learning part involves the computation of the probability with which the Swaras occur together and the frequency of the individual Swaras.

3.1.2. Synthesis Phase

In the synthesis phase, a sequence of Swaras pertaining to the desired Raga is generated based on the probabilities in the transition matrix of the Markov Chain. Samples are drawn using a Pseudo Random Number generator based on the cumulative probability for transitions from each input Swara. The initial Swara is chosen based on a starting probability distribution estimated using multiple compositions of the same Raga. As a result of sampling, we get a sequence of Swaras of desired length.

3.1.3 Text to Music Phase

In this phase, each of the sixteen Swaras is assigned a frequency relative to a chosen fundamental frequency. The composition generated in the previous phase is fed into this phase which then maps the Swaras in the composition to its corresponding frequency and subsequent generation of music is done using Matlab.

3.2 Hidden Markov Model

Hidden Markov Models (HMMs) are used to model finite discrete or continuous observation sequences when the state transitions cannot be observed directly [7]. HMMs have been employed successfully for speech recognition, speech synthesis, machine translation, parts-of-speech tagging, cryptanalysis, gene prediction, etc. HMM is an extension to the finite state Markov model which makes it more applicable to real-world processes, where the sequence of states cannot be observed.

HMMs have been used for generation of chord harmonizations in Western Music [6]. Its ability to capture harmonies from the training sequence makes it attractive for generation of complex melodies. We propose to use HMM for generation of Carnatic music and provide a comparison of its performance against the method previously discussed, in the Results and Analysis section.

The algorithm takes as input a Raga chosen by the user and a training sequence. The training sequence consists of Swaras which constitute a musical piece conforming to the chosen Raga. The algorithm learns from the training sequence and outputs a sequence of generated Swaras conforming to the chosen Raga.

Steps:

1. Input the training sequence and Raga
2. Learn the HMM model from the input training sequence and parameters
3. Sample from the trained HMM model
4. Generate output sequence of desired length

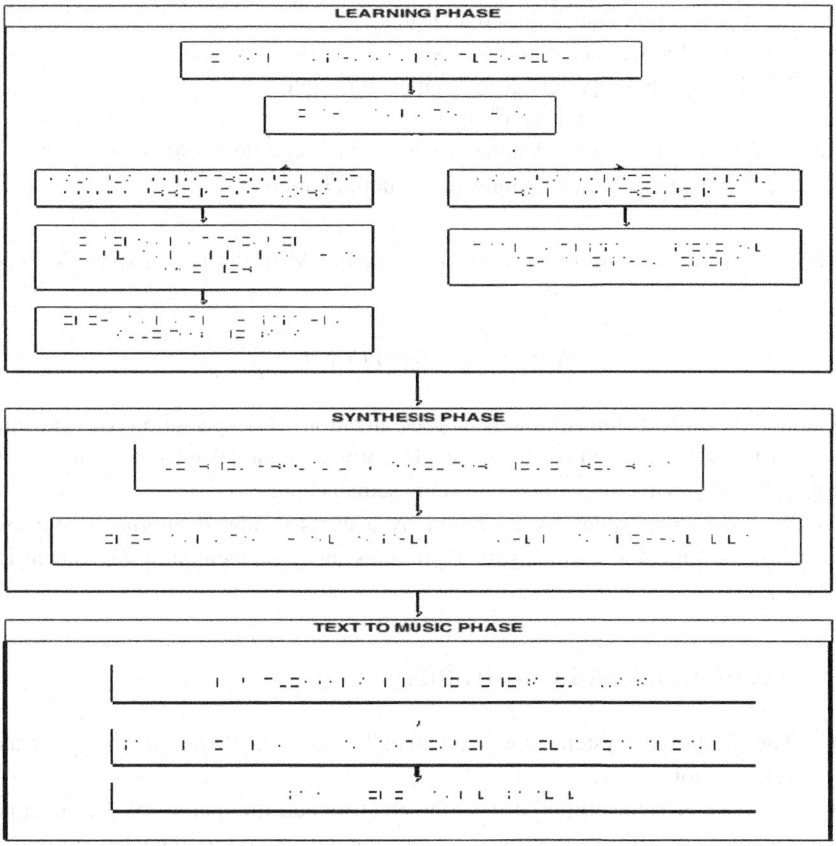

Fig. 1. Block Diagram of the notes synthesis system

The Hidden Markov Model is composed of three probability distributions namely Transition, Emission and Initial probability distributions. The parameters include number of hidden states and visible states, and the vocabulary used(set of symbols emitted).

We use a finite discrete HMM model for generating the sequence of Swaras. The hidden states correspond to the phrases(a group of Swaras). The emitted symbols correspond to the Swaras.

The intuition behind using HMMs is that they are better at capturing this phrasal structure of Swaras from the training examples which is necessary for generating complex harmonies.

HMMs are typically described using a three-tuple representation M = (A, B, Π) [5], where

A - NxN matrix which describes transition probability between states.

B - Emission probability distribution ($B = \{b_1, b_2, \ldots, b_N\}$).

Π - Initial probability distribution over states.
N - Number of hidden and visible states.
a_{ij} - Transition probability between i^{th} state and j^{th} state.
b_i - Emission probability for the ith state.
X_t - Random variable which denotes the state of the system at time t.
Y_t - Random variable which denotes the symbol emitted at time t.

The aim of the learning phase is to find a model M in the model space M* such that

$$M = \text{argmax}_{M \, \varepsilon \, M^*} P(Y|M) \quad (1)$$

Baum-Welch algorithm which is a specialization of Expectation-Maximization algorithm is used for learning the model. It learns that model which maximizes the likelihood of observing the sequence training sequence Y.

Number of hidden states were chosen by a cross-validated approach. We used scikit-learn package for Python which provides an implementation for categorical HMM [8], [9].

4 Assumptions and Constraints

- The proposed system does not handle the gamakkas of the generated composition.
- Since the system proposed is a statistical model, the quality of the generated output is directly proportional to the amount of corpus data fed during the learning phase of the system to fit the model.

5 Results and Analysis

The aim of the experiment was to measure the goodness of the notes generated by the generative model and to validate whether the notes correspond to the Raga given as input.

Six people trained in Carnatic music took part in the experiment. A dataset consisting of songs in .wav format was prepared. The participants were asked to rate each musical composition in the dataset on a scale of 1 to 5, based on their perceived similarity to the corresponding Raga to which the composition originally belonged. Table 1 shows the mean and standard deviation (in brackets) of scores on each Raga for both the methods.

Table 1. Results of the survey that was conducted

Hidden Markov Model	First Order Markov Model	Raga Name
1.75 (0.41)	1.5 (0.45)	Thodi
2.5 (0.63)	2.5 (0.63)	Shankarabharanam
3 (0.45)	2.75 (0.69)	Mohanam
2.25 (0.52)	1.75 (0.69)	Malahari
2.5 (0.71)	2.5 (0.55)	Kalyani

As seen from the above table, HMM performs better than the Bigram model. HMM is able to learn complex and long-range associations in the composition. HMM is also able to capture hierarchy in the compositions for reproducing characteristic phrases in the Raga. The Bigram model is too simple and doesn't allow for the generation of complex melodies.

6 Conclusion and Future work

In this paper, we have compared the performance of First Order Markov Model with that of HMM for synthesis of Carnatic Music. The result indicates that even though HMM performs better than Bigram model, the performance is poor on an absolute scale. Currently, our HMM is trained only on a single observation. It can be extended to use multiple observations. The simple Bigram model could also be extended to a N-gram model. Also, we have only used generative models for synthesis. In the future, we would like to work with discriminative models to test their feasibility for this problem. Our future work would focus on all these improvements.

Acknowledgement. We thank the Center for Technology Development and Transfer, College of Engineering, Guindy, Anna University, for funding this project.

References

1. Sambamoorthy, P.: South Indian Music, vol. 4. Indian Music Publishing (1969)
2. Sahasrabuddhe, H.V.: Analysis and Synthesis of Hindustani Classical Music, University of Poona (1992), http://www.it.iitb.ac.in/~hvs/paper_1992.html
3. Das, D., Choudhury, M.: Finite state models for generation of Hindustani classical music. In: Proceedings of International Symposium on Frontiers of Research in Speech and Music (2005)
4. Subramanian, M.: Generating Computer Music from Skeletal Notations for Carnatic Music Compositions. In: Proceedings of the 2nd CompMusic Workshop (2012)
5. Hill, S.: Markov Melody Generator

6. Steinsaltz, D., Wessel, D.: In Progress,The Markov Melody Engine: Generating Random Melodies With Two-Step Markov Chains. Technical Report, Department of Statistics, University of California at Berkeley
7. Kohlschein, C.: An introduction to hidden Markov models: Probability and Randomization in Computer Science. Aachen University (2006-2007)
8. Pedregosa, F., Varoquaux, G., Gramfort, A., Michel, V., Thirion, B., Grisel, O., Blondel, M., Prettenhofer, P., Weiss, R., Dubourg, V., Vanderplas, J., Passos, A., Cournapeau, D., Brucher, M., Perrot, M., Duchesnay, E.: Scikit-learn: Machine learning in Python. The Journal of Machine Learning Research 12, 2825–2830 (2011)
9. Oliphant, T.E.: Python for scientific computing. Computing in Science & Engineering 9(3), 10–20 (2007)

Smart Card Application for Attendance Management System

Shalini Jain and Anupam Shukla

ABV-Indian Institute of Information Technology,
Department of Computer Science and Engineering,
Gwalior, India
shalinijain58@gmail.com, anupamshukla@iiitm.ac.in

Abstract. This paper presents attendance management system based on the smart cards. This introduces the entities of the system and describes the set up of attendance management system and the role of these entities in the management system for managing attendance. This also represents the graphical representation of the working of entities of the system. This includes how card makes connection with the attendance reader and how verification authority verifies the smart card. And GUI interface for communicating with the card through this application is also shown with some description.

Keywords: smart card, management system, verification, validation.

1 Introduction

Smart cards are electronic cards which consists processor and memory. Hence can store some data on memory and can also perform some computations on that data through processor. Smart card can be either contactless or contact card depending upon how they communicate with the outside world. Contact smart cards communicate with the reader through a physical medium while contactless smart card does not have any physical medium, it transfers data over radio waves through a RF interface.

In this paper, an application for attendance management is developed for managing the records of attendance of group of people belonging to an organization. Group of people can either be the students of a college or employees of an organization. For attending attendance, user makes contact with the attendance reader and after finding a valid card, user insert finger for biometric authentication. Apart from being used for authentication and attendance these cards can also be used for multiple applications like library management (issue and return of books) and canteen access (E-cash application).

In this paper, we will discuss all the entities, which have been taken for systematic working of this application, in section 2. In section 3, set up of the system will be described with smart card reader requirements and details, access points and other setup details. Section 4 will give details of implementation of the system. And at the last section 5 will conclude our work.

M.K. Kundu et al. (eds.), *Advanced Computing, Networking and Informatics - Volume 1*,
Smart Innovation, Systems and Technologies 27,
DOI: 10.1007/978-3-319-07353-8_18, © Springer International Publishing Switzerland 2014

2 Related Work

According to [4], Smart card deployment is increasing pressure for vendor regarding-security features and improvements in computing power to support encryptdecryptal-gorithms with bigger footprints in the smart card chips in the lastdecade. Typical applications described in [4] include subscriber identificationmodule cards (SIM cards), micropayments, commuter cards, and identificationcards.

In [6], gives the new software design that provides the opportunity for developing new multi-application smart cards. It also considers other techniques and software design apart from the conventional cryptographic algorithms. A smart social contact management system is described in [7], provides a better way of managing large amount of social contacts through smart card.

The state of the art of unified read-write operations of the smart card with different data formatsis given in [1]. Based on framework designing and system organizational structures of "card type determining-> function calling ->Data Conversion- >unified read-write", a data reading, writing and reception data management of smart card of different data formats through PC is developed in [1].System like Smart card is a portable media which store sensible data. The information protection is possible with personal identification number (PIN), orthrough finger-print or retina based biometrics.Algorithms and data structures are developed in [2] to solve security problem.

File system of Smart Card in [3] consists three types of files i.e. master file,dedicated file, and elementary File. Master file and dedicated files are knownas catalog files, while elementary files are known as data file. Master file is theroot file of the cards. Each card consists only one MF, all the other sub-files of master file are grandson of master file. All files (including master file) that consistssub-files are considered as dedicated files. If file node is a leaf node, and has nochild nodes, then file is an elementary file. Elementary files are useful for thoseapplications that need to keep data in the elementaryfile. Reference [3] classifiedaccording to data structure, elementaryfile also includes a transparent binary file,the linear Fixed-length record file, circular recording document and linear variablelength record files.

Leong PengChor and Tan Eng Chong describes in [8], a design of a controlaccess method that allows access to an empty laboratory only to groups of authorizedstudents of a size satisfying a prescribed threshold. The scheme employsa shared-secret scheme, symmetric cryptography together with smart-card technology. The access right of a student is carefully encoded and stored on a smartcard. Access records are kept on these smart cards in a distributive fashion andwith duplication.

HoonKo and Ronnie D. Caytiles discuss in [9], Smartcard evolved from verysimple phone cards to business cards made with inferior equipment into complexhigh technology security solutions that can now support a large number of applications. Smartcard usage is rigorously growing over the last decade as theyare being used in telecommunications (GSM), banking services and various otherareas.

3 Entities in The Smart Card System

- Central Authority: This authority will generate the keys in the smart cards, so that card issuer authority can use the cards.
- Card Issuer Authority: Card issuer authority will insert all the required related information in the card and issue that to the respective user.
- Application Creator Authority: This authority has privileged to create different applications within a card. In this work, application creator authority will create an application for attendance management purpose.
- Verification Authority: Verification authority can read only certain data after getting permission. By read information this will verify it, can't make changes in it.
- Card User: who has received a card from card issuer authority, is only permitted to use that card.

These all authorities have different types of access tothe user card based on the cryptographic keys. Theyneed to insert specific key and after successfulauthentication and authorization they got privileges for accessing user card.

4 Set up

We have taken smart card from Infineon SLE 77 series, which consists 136 Kbytes solid flash, 6 Kbytes RAM and 16 bit security controller. These cards support symmetric as well as asymmetric cryptography.

Fig. 1. Infineon Smart Card

In our implementation, we have taken Congent Mini-Gate smart card reader shown below in Fig. 1, which supports fingerprint authentication, can sense contactless smart cards, provides pin combination as a access way and also have data network interface, a keypad and multi color LCD Fig.2.

Fig. 2. Cogent Mini Gate Smart Card Reader

Smart card reader should be smart enough to mark the attendance of the student within a second when he/she makes contact to the Reader with the smart card. When the attendance is marked it should display the name of the student, time of the contact and produce a buzzer sound. The reader should have the fingerprint sensor to take the fingerprint of student and authenticate the student and provide the access. It should be connected with the LAN (or wireless), with the power and should be battery operated in case of power failure. The Reader should be able to update the data with defined frequency to the server. The reader should have sufficient memory to store the attendance of the students when the network is down. Also it should be able to work in offline mode also, i.e., when there is no connection with LAN, it should be able to mark and store the attendance.

Access points where Reader needs to be installed are the classrooms, laboratories, hostel and canteen. The Reader needs to be mounted at a suitable place at the access points. One Reader needs to be installed per access point except those points where students count is more.

Attendance Management system should have the capability of writing/Reading/printing of individual student data into the Card so that it can be issued to the student by the approving authority.

5 Implementation

Working of entities of system is graphically represented in Fig.3. Central authority has privileged to generate the keys in all the blank smart card. After generating keys it will forward cards to card issuer authority. Card issuer authority will create files and insert user specific information in that card. Then check that card is usable or not. If not, then send it to application creator authority that will create necessary application in card and return back to card issuer authority. If card issuer authority finds it usable with particular user, thenissue it to that particular user. Now user can use this card with some accessible attendance reader.

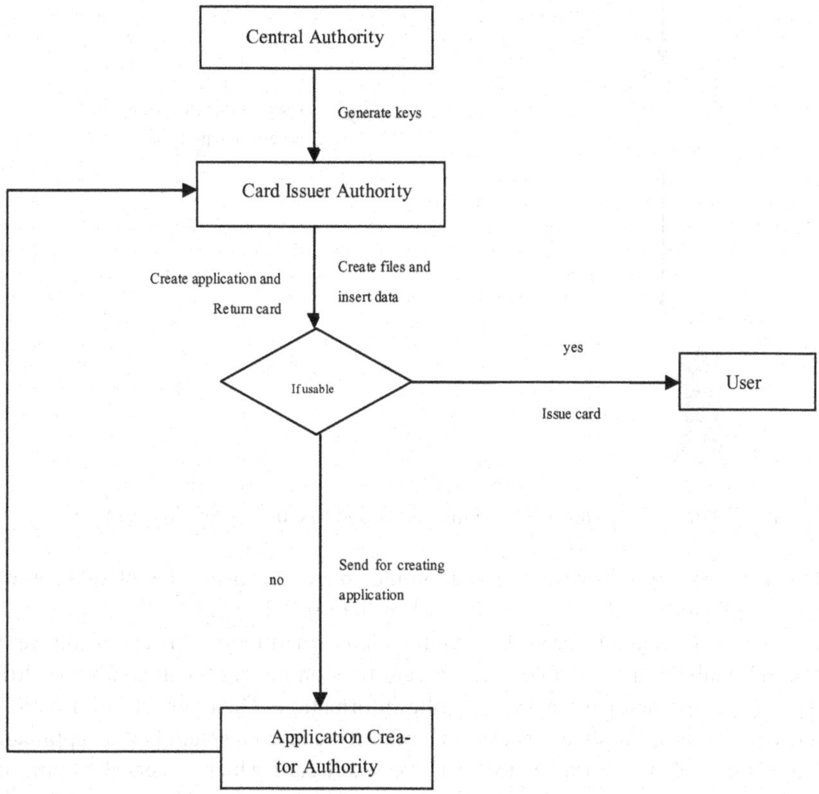

Fig. 3. Block Diagram of working entities of Smart Card

Fig.4 represents how connection takes place between smart card and attendance reader. User makes contact of his card with the specified reader. After successful mutual verification reader make connection with the card. Then it ask user for some security checks like pin code or fingerprint verification. After successful security checks user can perform any accessible operation on the card. If user sends some verification request to the verification authority then verification authority will verify user card. Verification authority can only verify if it submit its key successfully.

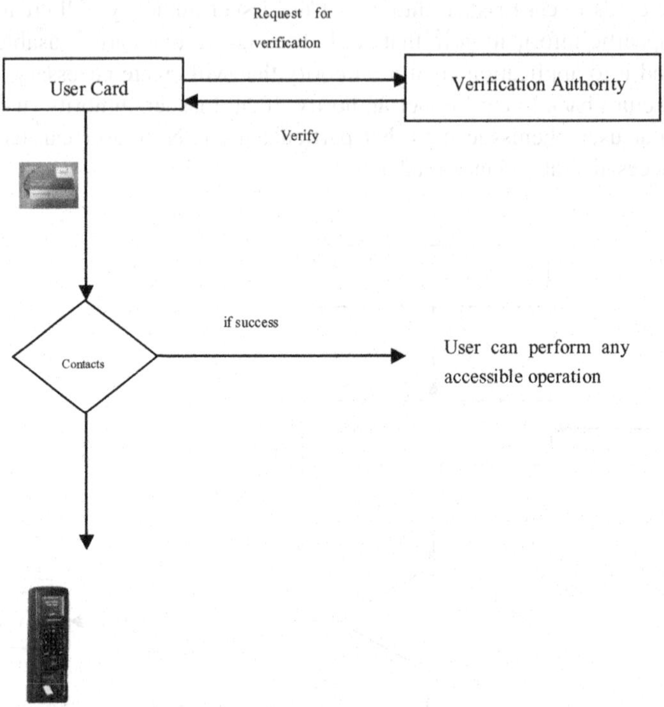

Fig. 4. Smart Card and Reader connectivity and role of Verification Authority

Card issuer authority will write the data on the smart card using the attendance management application and will issue thecard to the user. This application is providing an easiest way to create different files on the blank smart card. This can also reset the card, which makes card reusable. After create files on the card, can also write different types of mentioned personal and public information. Then needed information is printed on the card. Application creator authority can create number of application directory files with in a card. A user can use smart card which is issued to him only because of some security checks either in terms of pin or key or biometric authentication. Only card issuer authority can issue a duplicate card if card has been lost. If anyone want to update of terminate his card than only card issuer authority has permission to do this.

When user wants to make attendance, need to make contact with the attendance reader. After successful mutual authentication reader ask for fingerprint minutia. Then user needs to insert finger for verification purpose. If a wrong or unauthorized user is using it then reader will not take attendance. It will make attendance only if it recognizes a valid fingerprint minutia. Reader will show the date and time of attendance after making it. When attendance of any user has got mark successfully, database on the reader will updated for that user. After a specific time interval reader will send the updated data through the Ethernet to the server (refer Fig.50).

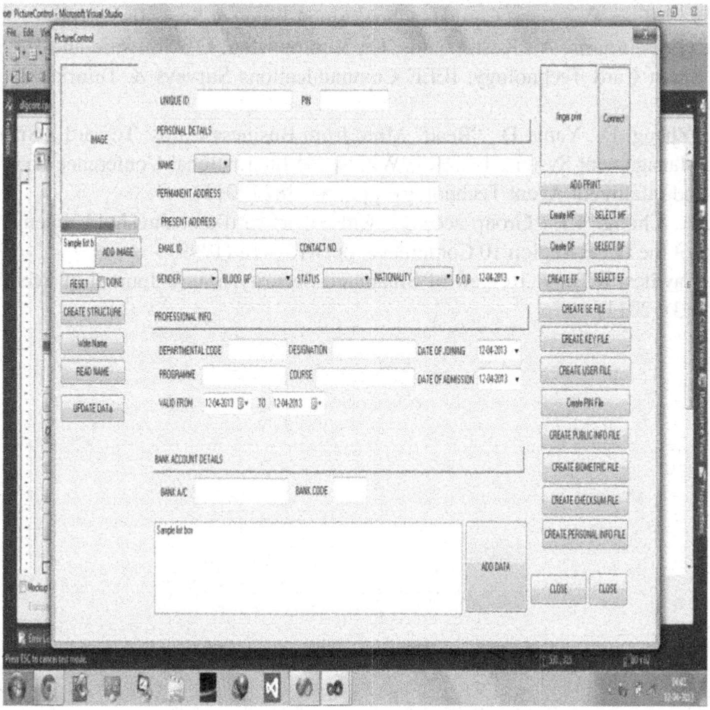

Fig. 5. GUI Interface for dealing with Smart Card

6 Conclusion

In this paper, an application of smart card for attendance management is described that could be used to manage attendance of people, electronically through smart card and smart card reader for reading attendance. There are number of more ways to manage attendance but through smart card, it can be use for many other application like e-cash application and library application. So that user does not need to carry number of cards.

References

1. Wang, R.-D., Wang, W.: Design and implementation of the general read-write system of the smart card. In: 3rd IEEE International Conference on Ubi-Media Computing (2010)
2. Reillo, R.S.: Securing information and operations in a smart card through biometrics. In: Proceedings of IEEE 34th Annual International Carnahan Conference on Security Technology (2000)
3. Yuqiang, C., Xuanzi, H., Jianlan, G., Liang, L.: Design and implementation of Smart Card COS. In: International Conference on Computer Application and System Modeling (2010)
4. Chandramouli, R., Lee, P.: Infrastructure Standards for Smart ID Card Deployment. IEEE Security & Privacy 5(2), 92–96 (2007)

5. Rankl, W., Effing, W.: Smart Card Handbook. John Wiley & Sons, Inc. (2002)
6. Selimis, G., Fournaris, A., Kostopoulos, G., Koufopavlou, O.: Software and Hardware Issues in Smart Card Technology. IEEE Communications Surveys & Tutorials 11(3), 143–152 (2009)
7. Guo, B., Zhang, D., Yang, D.: "Read" More from Business Cards: Toward a Smart Social Contact Management System. In: IEEE/WIC/ACM International Conference on Web Intelligence and Intelligent Agent Technology, pp. 384–387 (2011)
8. Chor, L.P., Chong, T.E.: Group accesses with smart card and threshold scheme. In: Proceedings of the IEEE Region 10 Conference, pp. 415–418 (1999)
9. Ko, H., Caytiles, R.D.: A Review of Smartcard Security Issues. Journal of Security Engineering 8(3) (2011)

Performance Evaluation of GMM and SVM for Recognition of Hierarchical Clustering Character

V.C. Bharathi and M. Kalaiselvi Geetha

Department of Computer Science and Engineering, Annamalai University, India
bharathivc@gmail.com, geesiv@gmail.com

Abstract. This paper presents an approach for performance evaluation of hierarchical clustering character and recognition of handwritten characters. The approach uses as an efficient feature called Character Intensity Vector. A hierarchical recognition methodology based on the structural details of the character is adopted. At the first level similar structured characters are grouped together and the second level is used for individual character recognition. Gaussian Mixture Model and Support Vector Machine are used in first level and second level classifiers and evaluate the accuracy performance of the handwritten characters. Gaussian Mixture Model is used for classification which achieves an overall accuracy of character level 94.39% and Support Vector Machine which achieves an overall accuracy of character level 93.61% is achieved.

Keywords: Handwritten Character Recognition, Character Intensity Vector(CIV), Hierarchical Character Clustering, Support Vector Machine(SVM), Gaussian Mixture Model(GMM).

1 Introduction

Modeling handwriting and human like behaviour is becoming increasingly important in various fields. Handwritten character recognition has been one of the most interesting and challenging research areas in field of image processing and pattern recognition. It contributes enormously performance of an automation process and can improve the interaction between human and computers. The scope of handwritten recognition can be extended to reading letters sent to companies or public offices since there is a demand to sort, search, and automatically answer mails based on document content. Handwritten character recognition is comparatively difficult, as different people have different writing styles. So, handwritten character is still a subject of active research and has wide range of application. Automatic recognition of handwritten information present on documents like cheques, envelopes, forms and other manuscripts.

1.1 Related Work

Choudhary *et al.* [1] proposed binarization technique to extract features and recognize the character using multi-layered feed forward neural network obtained

M.K. Kundu et al. (eds.), *Advanced Computing, Networking and Informatics - Volume 1,* Smart Innovation, Systems and Technologies 27,
DOI: 10.1007/978-3-319-07353-8_19,

recognition accuracy. Vamvakas *et al.* [2] proposed new feature extraction based on recursive subdivisions of the foreground pixels. SVM is used to classify the handwritten character. Lei *et al.* [3] proposed directional string of statistical and structure features implemented and classified by using a nearest neighbor matching algorithm. Das *et al.* [4] and Pirlo and Impedovo [5] handwritten character recognition based on static and dynamic zoning topologies designed the regular grids that superimpose on the pattern images into regions of equal shape using Voronoi based image zoning. [6] proposed quad tree longest run features taken in vertical, horizontal and diagonal directions. SVM is used to classify and recognize the handwritten numerals. Pauplin and Jiang [7] proposed two approaches to improve handwritten character recognition. The first approach dynamic Bayesian network models, observing raw pixel values optimizing the selection and layout using evolutionary algorithms and second approach learning the structure of the models. Reza and Khan *et al.* [8] developed a reliable grouping schema for the similar looking characters using density of nodes is reduced by half from one level to the next and classified the groups in SVM. Bhowmik *et al.* [9] proposed daubechies wavelet transform with four coefficients for feature extraction, then the features are grouped based on similar shapes. SVM classifier is used for classify the groups.

1.2 Outline of the Work

The proposed work initially considers the problem of grouping of alphabets based on their structural details. At the next step, the individual characters are identified at another level of granularity where the misclassifications between them are least possible.

The rest of the paper is organized as follows, Section 2 describes hierarchical character recognition. Feature extraction is described in Section 3. Hierarchical character clustering is presented in Section 4. Gaussian Mixture Model is presented in Section 5. Section 6 describes Support Vector Machine. Section 7 shows the experimental results of our approach and Section 8 concludes this work.

2 Proposed Hierarchical Character Recognition

This section describes all steps to carry out hierarchical character recognition as shown in Fig. 1.

2.1 Dataset

Handwritten dataset is created for uppercase characters. The input images were saved in JPEG/PNG format for further preprocessing as shown in Fig. 2(a).

2.2 Pre-processing

The pre-processing is a set of operation performed on the input document [10]. The various task involved in the document images are

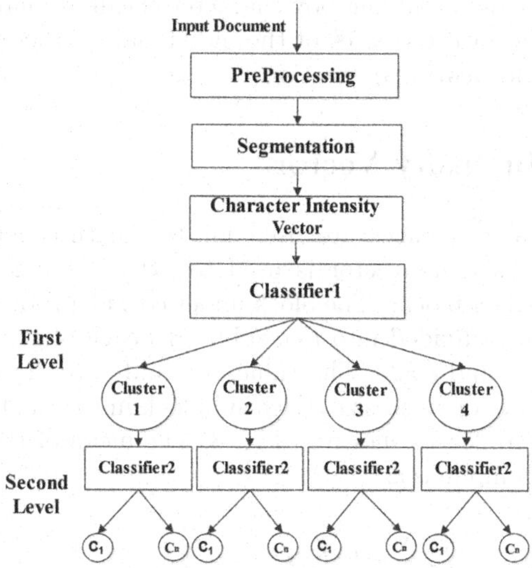

Fig. 1. Hierarchical character recognition

Binarization. Original RGB image is converted to grayscale and then the image is complimented to obtain an image as shown in Fig. 2(b) for further processing.

Morphological Filtering. An image may contain numerous imperfections. Morphological processing removes these imperfections using area open method. It removes small objects fewer than 30 pixels as shown in Fig. 2(c).

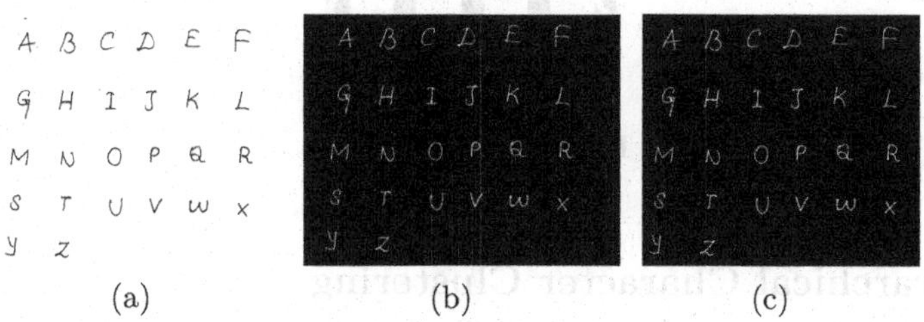

Fig. 2. (a) Input Document (b) Binarization (c) Morphological Filtering

Segmentation. Segmentation is the process of dividing groups of character into multiple segments. Segmentation is done by finding the connected components and measure the properties of labeled region by placing bounding box over the character. To extract the bounding box character is called contours. Extracting the contours which are various sizes of the individual characters are uniformly resized to 100 ×100 for feature extraction.

3 Character Intensity Vector

All segmented character images are used for the feature extraction. In our approach character intensity vector is used to extract the features from the grayscale image based on blocks. The block under consideration of size 100×100. This 100×100 region is divided into r×c subblocks each of size 5×5 number of subblocks(n), where r and c=20. The value of r and c are fixed in such a way that each subblocks are of equal size. Each subblock image of 20×20 pixels with intensity range of 0 to 255 as shown in Fig. 3. The mean of the pixel intensity of each subblock is computed as

$$b_n = \frac{\sum_{x=1}^{r}\sum_{y=1}^{c} count(p_{x,y})}{r \times c} \quad if(p_{x,y} > 127) \tag{1}$$

where b_n is number of subblocks n=1,..,25, $p_{x,y}$ intensity of pixels, r number of rows and c number of columns. So totally 25 subblocks and for each subblock computes the mean of pixel intensity. There by 25 features extracted from each block.

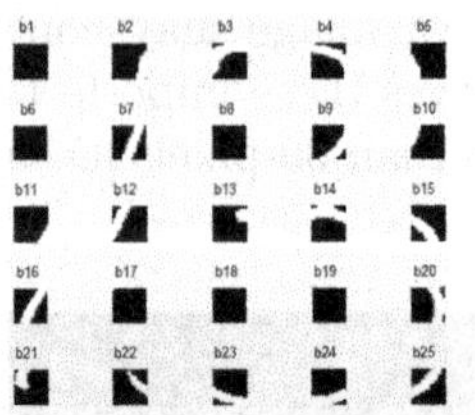

Fig. 3. 5×5 Subblocks

4 Hierarchical Character Clustering

Character clustering is the task of grouping a set of character in such a way that characters in the same group are called clustering. It can be defined as grouping organisms according to their structural similarities. This means that organisms that share similar features are placed in one group. The goal of hierarchical clustering analysis is to find of ultrametric proximities that are reasonably close to the observations. Based on the structural similarities and the centroid of character are grouped are shown in the Table 1.

Table 1. Hierarchical character grouping

Label	Cluster Name	Group Name	Characters
1	Cluster1	Horizontal	A,B,E,F,G,H,R
2	Cluster2	Vertical	I,J,K,L,M,N,P,T
3	Cluster3	No Center	C,D,O,Q,U,V,W
4	Cluster4	Center	S,X,Y,Z

5 Gaussian Mixture Model

A GMM is a probabilistic model for density estimation using a mixture distribution and is defined as a weighted sum of multi variant Gaussian densities[11]. A GMM is a weighte sum of M components densities is given by the form

$$p\left(\frac{x}{\lambda}\right) = \sum_{i=1}^{M} w_i b_i \tag{2}$$

Where x is a dimensional random vector $b_i\,(x)$, i= 1,2..,M is the component densities and w_i i=1,2..,M is the mixture weights. The Gaussian function can be defined of the form

$$b_i\,(x) = \frac{1}{(2\prod)^{\frac{D}{2}} |\sum_i|^{\frac{1}{2}}} \exp\left\{-\frac{1}{2}(x-\mu_i)^{'} \sum\nolimits_i^{-1} (x-\mu_i)\right\} \tag{3}$$

With mean vector μ_i and covariance matrix $\sum_i$. The mixture weight satisfy the constraint that $\sum_{i=1}^{M} w_i = 1$. The complete Gaussian mixture model is parameterized by the mean vectors, covariance matrices and mixture weight from all component densities[12]. These parameters can collectively represented by the notation: $\lambda = \{w_i, \mu_i, \sum_i\}$ for 1=1,2,..,M. The GMM parameters are estimated by the Expecation-Maximization(EM) algorithm using training data by the handwritten dataset. The basic idea of the EM algorithm is, beginning with an initial language model λ, to estimate a new model $\lambda^{'}$ such that $p\left(X/\lambda^{'}\right) \geq \mathrm{p}\,(X/\lambda\,)$.

6 Support Vector Machine

Support Vector Machine (SVM) is a group of supervised learning methods that can be applied to classification and regression. It is a popular technique for classification in visual pattern recognition[13,14]. The SVM is mostly used in kernel learning algorithm. It achieves reasonably vital pattern recognition performance in optimization theory[15]. A classification task are typically involved with training and testing data. The training data are separated by $(x_1, y_1), (x_2, y_2)......(x_m, y_m)$ into two classes, where $x_i \in R_N$ contains n-dimensional feature vector and

$y_i \in \{+1, -1\}$ are the class labels. The aim of SVM is to generate a model which predicts the target value from testing set. In binary classification the hyper plane $w.x + b = 0$ where $w \in R^n, b \in R$ is used to separate the two classes in some space Z. The maximum margin is given by $M = \frac{2}{\|w\|}$. The minimization problem is solved by using Lagrange multipliers $\alpha_i (i = 1, ...m)$ where w and b are optimal values obtained from Eq. 4.

$$f(x) = sgn\left(\sum_{i=1}^{m} \alpha_i y_i \; K(x_i, x) + b\right) \tag{4}$$

The non-negative slack variables ξ_i are used to maximize margin and minimize the training error. The soft margin classifier obtained by optimizing the Eq. 5 and Eq. 6.

$$\min_{w,b,\xi} \; \frac{1}{2} w^T w + C \sum_{i=1}^{l} \xi_i \tag{5}$$

$$y_i (w^T \phi(x_i) + b \geq 1 - \xi_i, \xi_i \geq 0 \tag{6}$$

7 Experimental Results

The experimental are carried out in windows 8 operating system with Intel Xeon X3430 Processor 2.40 GHz with 4 GB RAM. Proposed method is evaluated using the handwritten dataset. Extracted CIV features are fed to GMM and LIBSVM [16] tool to develop the model for each groups and these models are used to recognize the characters.

7.1 Handwritten Dataset

In our experiment 6500 samples of handwritten characters are collected from 50 different writers. Each writer contributed 5 times of each upper case alphabet out of 26. Upper case alphabets collected different age of people written various styles and size. The handwritten documents were scanned at dimensions 1700×2338 pixels, 200 dpi with 24 bit image. Out of 50 different writer, 34 writer samples taken for training and 16 writer samples taken for testing.

7.2 Evaluation Metrics

Precision (P) and Recall (R) are the commonly used evaluation metrics and these measures are used to evaluate the performance of the proposed system. The measures are defined as follows:

$$P = \frac{\text{No. of TP}}{\text{No. of TP + FP}}$$

$$R = \frac{\text{No. of TP}}{\text{No. of TP + FN}}$$

The work used F-score as the combined measure of Precision (P) and recall (R) for calculating accuracy which is defined as follows: $F_{\propto} = 2.\frac{PR}{P+R}$ where $\propto$ is weighting factor and $\propto = 1$ is used.

7.3 Classifier

First Level Classifier. Initially 25 dimensional features are extracted using character intensity vector. All the features are hierarchical grouped as shown in Table. 1 and the clusters are trained and tested in GMM and SVM with RBF kernel. GMM is to identified individual groups recognition accuracy and SVM with RBF kernel with C=500, γ=0.1 to identified individual groups (horizontal, vertical, no center and center) recognition accuracy are shown in Fig. 4.

Second Level Classifier. In the first level classifier construct four models M1(horizontal), M2(vertical), M3(no center) and M4(center) using GMM and

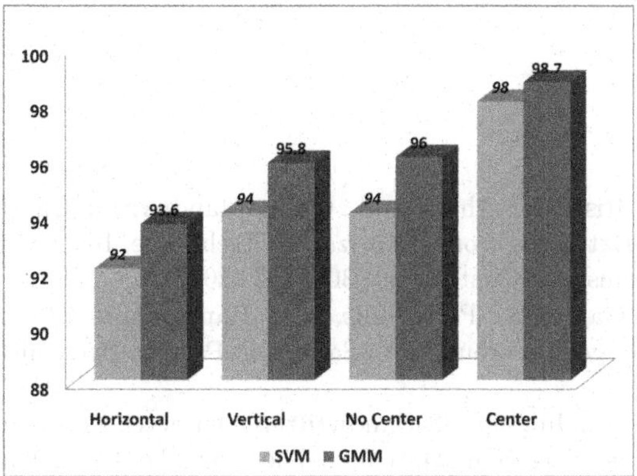

Fig. 4. Recognition accuracy of groups at the first level

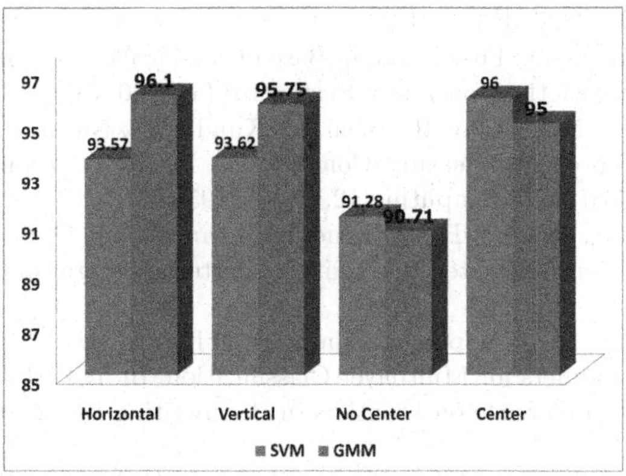

Fig. 5. Recognition accuracy of characters at the second level

SVM. In second level classifier, the testing samples tested in corresponding models to identify the individual character from the cluster. The recognition accuracy of GMM and SVM with RBF kernel of individual characters as shown in Fig. 5.

8 Conclusion

This paper presented an approach the performance evaluation of hierarchical clustering character and recognizing individual character using character intensity vector. Experiments are conducted on handwritten dataset considering upper case character. Based on the structural similarities and centroid of the region character are grouped and recognize the individual character using GMM and SVM with RBF kernel. For group level and character level classification on GMM have obtained overall recognition accuracy 96.02% and 94.39%. For group level and character level classification on SVM have obtained overall recognition accuracy 94.5% and characters 93.61%. From the results GMM outperforms the SVM classifier.

References

1. Choudhary, A., Rishi, R., Ahlawat, S.: Offline Handwritten Character Recognition using Features Extracted from Binarization Technique. In: AASRI Conference on Intelligent Systems and Control, pp. 306–312 (2013)
2. Vamvakas, G., Gatos, B., Perantonis, S.J.: Handwritten Character Recognition Through Two-Stage Foreground Sub-Sampling. Pattern Recognition 43, 2807–2816 (2010)
3. Lei, L., Li-liang, Z., Jing-fei, S.: Handwritten Character Recognition via Direction String and Nearest Neighbour Matching. The Journal of China Universities of Posts and Telecommunications 19(2), 160–165 (2012)
4. Pirlo, G., Impedovo, D.: Adaptive Membership Functions for Handwritten Character Recognition by Voronoi Based Image Zoning. IEEE Transactions on Image Processing 21(9), 3827–3837 (2012)
5. Pirlo, G., Impedovo, D.: Fuzzy Zoning Based Classification for Handwritten Characters. IEEE Transactions on Fuzzy Systems 19(4), 780–785 (2011)
6. Das, N., Reddy, J.M., Sarkar, R., Basu, S., Kundu, M., Nasipuri, M., Basu, D.K.: A Statistical Topological Feature Combination for Recognition of Handwritten Numerals. Applied Soft Computing 12, 2486–2495 (2012)
7. Pauplin, O., Jiang, J.: DBN-Based Structural Learning and Optimisation for Automated Handwritten Character Recognition. Pattern Recognition Letters 33, 685–692 (2012)
8. Reza, K.N., Khan, M.: Grouping of Handwritten Bangla Basic Characters, Numerals and Vowel Modifiers for Multilayer Classification. In: ICFHR 2012 Proceedings of International Conference on Frontiers in Handwriting Recognition, pp. 325–330 (2012)
9. Bhowmik, T.K., Ghanty, P., Roy, A., Parui, S.K.: SVM based Hierarchical Architectures for Handwritten Bangala Character Recognition. International Journal on Document Analysis and Recognition (IJDAR) 12, 97–108 (2009)

10. Bharathi, V.C., Geetha, M.K.: Segregated Handwritten Character Recognition using GLCM Features. International Journal of Computer Applications 84(2), 1–7 (2013)
11. Reynolds, D.A.: Gaussian Mixture Models. MIT Lincoln Laboratory, USA
12. Sarmah, K., Bhattacharjee, U.: GMM based Language Identification using MFCC and SDC Features. International Journal of Computer Applications 85(5), 36–42 (2014)
13. Cristianini, N., Shawe-Taylor, J.: An Introduction to Support Vector Machines and Other Kernel-based Learning Methods. Cambridge University Press (2000)
14. Mitchell, T.: Machine Learning. Hill Computer Science Series (1997)
15. Vapnik, V.: Statistical Learning Theory. Wiley, NY (1998)
16. Chang, C.-C., Lin, C.-J.: LIBSVM: A library for support vector machines. ACM Transactions on Intelligent Systems and Technology 2, 1–27 (2011)

Data Clustering and Zonationof Earthquake Building Damage Hazard Area Using FKCN and Kriging Algorithm

Edy Irwansyah[1] and Sri Hartati[2]

[1] Department of Computer Science, Bina Nusantara University, Jakarta, Indonesia
[2] Deptartment of Computer Science and Electronic, Universitas Gadjah Mada, Yogyakarta, Indonesia
edirwan@binus.ac.id, shartati@ugm.ac.id

Abstract. The objective of this research is to construct the zonation of earthquake building damage hazard area using fuzzy kohonen clustering network (FKCN) algorithm for data clustering and kriging algorithm for data interpolation. Data used consists of the earth data in the form of peak ground acceleration (PGA), lithology and topographic zones and Iris plant database for algorithm validation. This research is comprised into three steps which are data normalization, data clustering and data interpolation using FKCN and kriging algorithm and the construction of zonation. Clusterization produces three classes of building damage hazard data. The first class is consisting of medium PGA,dominantby high compaction lithology in the topography of inland area. The second class with low PGA, dominant low compaction lithology in the lowland topographic zone and the third class with high PGA, dominant by unvery low compactionlithology in swamp topographic zone. Banda Aceh cityas location sample is divided into three building damage hazard zone which is high hazard zone, medium hazard zone and low hazard zone for building damage which is located towards inland area.

Keywords: Clusterization, damage hazard, earthquake, FKCN algorithm.

1 Introduction

USGS recorded that there are four large earthquake events in Indonesia which is Banda earthquake (8.5 Mw) in 1983, Sumatra earthquake (9.1) in 2004,Nias Island earthquake (8.6) in 2005 and Sumatra West Coast earthquake (8.6) in 2012 [21]. The high earthquake intensity has become the main tectonic characteristic of the archipelago of Indonesia that located between three main plates, which is Eurasia at North, Indo-Australia at South and the Pacific plate at Northeast. An earthquake with a certain intensity and magnitude as a response to the movements of plates can result in physical infrastructure damage and people casualties. The physical infrastructure damage characterized as massive is caused by earthquake which is the damage on buildings, caused by poor construction qualities (internal) and environment (external) where the building is built. The recorded figure of building damages due to

M.K. Kundu et al. (eds.), *Advanced Computing, Networking and Informatics - Volume 1*,
Smart Innovation, Systems and Technologies 27,
DOI: 10.1007/978-3-319-07353-8_20, © Springer International Publishing Switzerland 2014

earthquake happened at Banda Aceh city Indonesia in 2004 with a total of damaged buildings of 35 percent compared to total existing building [9] and over 140,000 building units is damaged due to Yogyakarta Indonesia earthquake in 2006 [17].

Currently, research related to the building damage effectdue to earthquake has been implemented using geographical information system (GIS) for hazard zoning of earthquake prone areas [18], [20] and the evaluation of building damage due to earthquake using neuro-fuzzy method[3], [4],[5], [13], [16] and [19]. Neuro-fuzzy algorithm that has been implemented in the researches are mostly using adaptive neuro-fuzzy inference system (ANFIS) and it is not using spatial data and not used to construct zones of hazard. An important value in research of zonation is to facilitate the researcher to understand the important phenomenon in evaluating the building damage due to earthquake in a spatial structure. Specifically, the research related to data clusterization using FKCN algorithm has been proposed by [1],[2], [22] and applied on several area by [12] and [15].

Based on the above facts, the research is to be performed with the objective to constructthe zonation of building damage hazard area due to earthquake usingFKCN algorithm, in order to cluster the data and using krigingal gorithm for data interpolation and construct the zonation..

2 FKCN Algorithm: Development and Its Application

FKCN is an algorithm which is proposed by [2] which is a development Kohonen clustering network (KCN) algorithm which was proposed by Kohonenin 1989 [14]. In principle, FKCN algorithm is an integration of fuzzy c-means (FCM) model of the learning rate and updating strategy (optimization procedure) from a traditional KCN. KCN is a clustering algorithm which is unsupervised where the weights of the clusters are implemented iteratively and sequentially.

Yang *et al.* [22] performed an enhancement of the learning rate of FKCN algorithm by associating the learning rate with a specific fuzzy membership with a specific value. Experiments showed that the modification successfully increased the correct rate from 91.3 percent to 92.7 percent with a difference in operation time which is excellent from 11.42 seconds to only 3.63 seconds.The integration of FCM and KCN in FKCN is performed in order to optimize the algorithm to solve a couple of problems that are faced often such as the capability to generate the output value which is continuous in the fuzzy clustering and to improve the calculation efficiency because of the use of weight updates which is linear [15]. FKCN is an algorithm which is self-organizing, because the "size" of the environment is automatically adjusted during the learning process, and usually FKCN will end in a way such that the function and its objectives are minimized.The learning rate modification from the FKCN algorithm using a method to associate the learning rate of fuzzy member on specific value has been researched by [22]. Experiments showed that the modification successfully increased the correct rate from 91.3 percent to 92.7 percent with a difference in operation time which is excellent, from 11.42 seconds to only 3.63 seconds.

The implementation of FKCN algorithm specially for building damage data due to earthquake until now has not been implemented. Jabbar *et al.*[12] used this algorithm for segmentation of the colors within a diagram.

3 Methodology

The data used in this research is composed of two types of data which is earthquake data that contains data from peak ground acceleration (PGA), lithology and topographic zones and data from IRIS Plants database. The conversion process from lithology data and topographic zoning from map to become data with a class value based on each contribution on each level of building damage. This research consists of three steps which is (1) data normalization, (2) data clustering using FKCN algorithm and (3) data interpolation using kriging algorithm and creation of zones (zonation)

Data normalizes using min-max is a method that resulted in a linier transformation data on the original data where min-max normalizationis mapping a value v from A to v' within a new minimal and maximal range[7]. The formula for min-max normalization can be seen in the following equation.

$$v = \frac{v-minA}{maxA-minA} * (new_{maxA} - new_{minA}) + new_{minA} \quad (1)$$

Where minA, maxA, new maxA and new minA is the minimum value, maximum value from attribute A and maximum and minimum value on the new scale of attribute A respectively.

Earthquake hazard data clustering is implemented using FKCN algorithm. The implementation steps are as follows:

Step 1: Suppose a sample space $X=\{x_1,x_2,\ldots,x_N\}$, $x_i \in R^f$

Where N is the number of patterns and f is the pattern vector dimension; the distance is II_A, c as the number of clusters and error threshold > 0 some small positive constant.

Step 2: Initialization by determining the weight of initial vector $V_0 = (v_{10}, v_{20}, \ldots v_{c0})$, the number of initial cluster (m_0) and t_{max} = iteration limit.

Step 3: For $t=1,2,\ldots,t_{max}$

a) Calculate all learning rate

$$m_t = m_0 - t * \Delta m, \ \Delta m = (m_0 - 1)/t_{max} \quad (2)$$

$$u_{ik,t} = \left(\sum\left(\frac{\text{II}X_k - V_{i,t-1}\text{II}}{\text{II}X_k - V_{j,t-1}\text{II}}\right)^{\frac{2}{m-1}}\right)^{-1} \quad (3)$$

$$\alpha i_{k,t} = (u_{ik,t})^{mt} \quad (4)$$

b) Update all vector weight

$$v_{i,t} = v_{i,t-1} + \frac{\sum_{k=1}^{n} \alpha_{ik,t}(x_k - v_{i,t-1})}{\sum_{j=1}^{n} \alpha_{ij,t}} \quad (5)$$

c) Calculate:

$$E_t = \| V_t - V_{t-1} \| \quad (6)$$

If $E_t <$ stop and go to step 3

The next step is data interpolation using kriging algorithm and plotting the average value of the result of clustering on each grid for zonation. Kriging is one method of the prediction and interpolation in geo-statistics. Interpolation analysis is required because it is impossible to take data from all the existing location. The interpolation technique takes data from some location and resulted in a predicted value for the other locations.

A sample data from location 1,2…,n is $V(x_1)$, $V(x_2)$, … , $V(x_n)$, therefore to predict $V(x_0)$ is [6]:

$$\hat{V}(x_0) = \sum_{i=1}^{n} w_i V(x_i) \quad (7)$$

(Where *wi* is predicted using the following matrix :

$$W = C^{-1}D \quad (8)$$

Where Cis the matrix of semi-variances between data points and D is the vector of semi variances between data points and the target

4 Analysis and Result

4.1 Resulting Cluster Data Validation

FKCN algorithm in this research is implemented using MATLAB R2009a software using a personal computer Pentium® 3.2 GHz. IRIS Plant Database which consists of 150 data where each class consist of 50 data, used to validate the result of the clustering. The validation of clustering algorithm is performed using Iris plants database [6]. Iris data has been generally used by researchers to validate and test the clustering algorithms.The validation is performed by comparing the results obtained using research performed by [2], [8]and [22]especially on cluster center resulting from this research have similarities with cluster center from previous researches.

The number of iteration to obtain cluster center in this research in fact is less which is 3 with the value of the correct number and correct rate which is better although the value of square sum error (SSE) is relatively still large compared to the results from previous researches. In detail, the results of this research compared to the results of previous researches can be seen in Table 1.

4.2 Clusterization of Building Damage Hazard Data

Codification and cluster result that has been validated IRIS Plant database, and further implemented for building damage hazard that consists of data obtained from peak ground acceleration (PGA), lithology and topographic zone data.

Clusterization process divides the research data into 3 (three) classes with different characteristics. Using cluster center data that was further obtained the membership of each data can be determined based on the distance of each data and the cluster center for each class. The first class is a class with average PGA value 0.87813 (medium) with lithology domination that has a high compaction and fine grained.

Table 1. Comparison of IRIS data clusterization using FKCN algorithm

	FKCN (Yang dkk, 1992)	Improved FKCN(Yang dkk, 2008)	FKCN(Irwansyah danHartati, 2013)
Parameter	m=3;ε =0.001; tmax=50;c=3	m=3;ε=0.001;tmax=50; c=3;tu=0.7;td=0.3; mu=0.4; md=3	m=3;ε =0.001; tmax=50;c=3
Correct Number	5480	5560	5580
Correct Rate	91.3%	92.7%	93%
Iteration Number	12	4	3
Hathaway'sCluster Center(Actual clusterCenter) – Hathaway andBezdek, 1994)		5.00 5.93 6.58 3.42 2.77 2.97 1.46 4.26 5.55 0.24 1.32 2.02	
Cluster Center	5.01 5.75 6.60 3.14 2.73 3.01 1.48 4.14 5.36 0.25 1.29 1.91	5.00 5.91 6.58 3.41 2.79 3.00 1.48 4.22 5.44 0.25 1.32 1.98	5.00 5.85 6.74 3.42 2.75 3.04 1.46 4.20 5.66 0.24 1.29 2.09
The square sum of the relativecluster center'serror	0.1073	0.0256	0.1169

Table 2. Data Characteristics on each class of clusterization result

Data Characteristic	***Class 1***	***Class 2***	***Class 3***
Average PGA	0,87813	0,87779	0,87935
Range PGA	0,87748 – 0,87974	0,87670 – 0,87852	0,87855 – 0,88064
Dominant Lithology Dominan	Sandy Clay, Clayey Sand and Clay	Swamp and Sand	Swamp
Main Topograohic Zone	Coastal Plain, Inland Deep and Inland Plain	River and Coastal Deep	River and Coastal Deep

The topographic zone on the first class is mainly from the inner earth with an average PGA value 0.87779 (low) which is dominated by lithology that has not been compacted therefore the low compaction, and located in low topographic zone close to the coast and river side. The third class is a class with an average PGA 0.87935 (high) which is dominated by lithology that has not been compacted on topographic zone swamp which is very close to the coast. The data characteristic on each class resulted from the clusterization using FKCN algorithm can be seen in Table 2.

4.3 Zonation of Earthquake Building Damage Hazard Area

Research on the earthquake building damage hazard using clusterization technique and kriging algorithm, has been conducted by [10] and [11].

Irwansyah and Hartati, 2012[10] with three data variable using self-organizing map (SOM) algorithm and kriging algorithm, concluded that the Banda Aceh city can be divided into three building damage hazard classes with zonation damage class which is relative parallel to the coastline. Irwansyah*et al*[11]used the total optimal group and the data grouping result using *k*-Means algorithm and implementing kriging algorithm to perform data interpolation with the same objective to construct hazard zone in Banda Aceh city. The zonation resulting from the research conducted by [11] treating every data and group data in spatial manner and returning it back to the grid center. The research will produce a building damage zone that has a relatively similar pattern to the zone produced in the research conducted by [10] with two classes of hazard area.

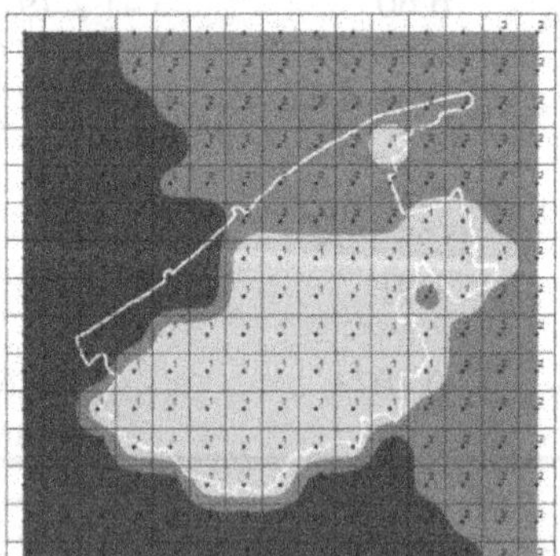

Fig. 1. Buildingdaamage hazard zonation in Banda Aceh (red: high building damage hazard area; orange: mediumbuilding damage hazard area;yellow: low building damage hazard area)

The application of different algorithms in this research produced a different building damage hazard zonation caused by earthquake conducted in previous research. The first zone is the zone of low level of hazard oh building damage with characteristic an average medium PGA, dominant lithology that has been compacted and exists in the topographic zone of the deep inland. The third zone is the opposite, where the PGA value is high with un-compacted lithology and topographically is in swamp area or near the river. The zone with this characteristic is a zone with a high level hazard of building damage. The second zone is the zone with the characteristics in between those two zones. The zonation produced in this research is showing spatial patterns where the coastal area is divided into two zones which are a high building damage hazard zone and a medium building damage caused by earthquake. Low building damage zone is located relatively toward land. (Fig.1).

5 Conclusion

The FKCN algorithm which is implemented in this research using the same parameter successfully increased the correct number and correct rate and at the same time producing cluster center that was produced in previous researches.

Clusterization produces three data classes of building damage hazard which is the first class with an average PGA 0.87813 (medium) with a dominant lithology that has been compacted until high compaction on topographic zones in the Inland Area. The second class with an average PGA 0.87779 (low) with dominant lithology that has been compacted until low compaction and it is in the low topographic zone near the coast and river sided, and the third zone with an average PGA 0.87935 (high) which is dominated by lithology that has not been compacted in swamp topographic zone which is close to coast area.

Banda Aceh City and the surrounding area is divided into three zones (zonation) which is high hazard zone, medium hazard zone caused by earthquake originating from coastal area and low building damage hazard zone which is located relatively further away towards the inland.

References

1. Almeida, C.W.D., Souza, R.M.C.R., Ana Lúcia, B.: IFKCN: Applying Fuzzy Kohonen Clustering Network to Interval Data. In: WCCI 2012 IEEE World Congress on Computational Intelligence, pp. 1–6 (2012)
2. Bezdek, J.C., Tsao, E.C.-K., Pal, N.R.: Fuzzy Kohonen Clustering Networks. Fuzzy Systems 27(5), 757–764 (1992)
3. Carreño, M.L., Cardona, O.D., Barbat, A.H.: Computational Tool for Post-Earthquake Evaluation of Damage in Buildings. J. Earthquake Spectra 26(1), 63–86 (2010)
4. Elenas, A., Vrochidou, E., Alvanitopoulos, P., Ioannis Andreadis, L.: Classification of Seismic Damages in Buildings Using Fuzzy Logic Procedures. In: Papadrakakis, et al. (eds.) Computational Methods in Stochastic Dynamic. Computational Methods in Applied Sciences, vol. 26, pp. 335–344. Springer, Heidelberg (2013)
5. Fallahian, S., Seyedpoor, S.M.: A Two Stage Method for Structural Damage Identification Using An Adaptive Neuro-Fuzzy Inference System and Particle Swarm Optimization. Asian J. of Civil Engineering (Building and Housing) 11(6), 795–808 (2010)
6. Fisher, M.M., Getis, A. (eds.): Handbook of Applied Spatial Analysis –Software Tools, Methods and Applications. Springer, Heidelberg (2010)
7. Han, J., Kamber, M., Pei, J.: Data Mining Concept and Techniques, 3rd edn. Morgan Kaufmann-Elsevier, Amsterdam (2012)

8. Hathaway, R.J., Bezdek, J.C.: Nerf C-Means Non-Euclidean Relation Fuzzy Clustering. Pattern Recognition 27(3), 429–437 (1994)
9. Irwansyah, E.: Building Damage Assessment Using Remote Sensing, Aerial Photograph and GIS Data-Case Study in Banda Aceh after Sumatera Earthquake 2004. In: Proceeding of Seminar on Intelligent Technology and Its Application (SITIA 2010), vol. 11(1), pp. 57–65 (2010)
10. Irwansyah, E., Hartati, S.: Zonasi Daerah BahayaKerusakanBangunanAkibatGempaMenggunakanAlgoritma SOM Dan AlgoritmaKriging. In: Proceeding of Seminar Nasional TeknologiInformasi (SNATI 2012), vol. 9(1), pp. 26–33 (2012) (in Bahasa)
11. Irwansyah, E., Winarko, E., Rasjid, Z.E., Bekti, R.D.: Earthquake Hazard Zonation Using Peak Ground Acceleration (PGA) Approach. Journal of Physics: Conference Series 423(1), 1–9 (2013)
12. Jabbar, N., Ahson, S.I., Mehrotra, M.: Fuzzy Kohonen Clustering Network for Color Image Segmentation. In: 2009 International Conference on Machine Learning and Computing, Australia, vol. 3, pp. 254–257 (2011)
13. Jiang, S.F., Zhang, C.M., Zhang, S.: Two-Stage Structural Damage Detection Using Fuzzy Neural Networks and Data Fusion Techniques. Expert Systems with Application 38(1), 511–519 (2011)
14. Kohonen, T.: New Developments and Applications of Self-Organizing Map. In: Proceeding of the 1996 International Workshop on Neural Networks for Identification, Control, Robotics, and Signal/Image Processing (NICROSP 1996) (1996)
15. Lind, C.T., George Lee, C.S.: Neural Fuzzy System: A Neuro-Fuzzy Synergism to Intelligent System. Prentice-Hall, London (1996)
16. Mittal, A., Sharma, S., Kanungo, D.P.: A Comparison of ANFIS and ANN for the Prediction of Peak Ground Acceleration in Indian Himalayan Region. In: Deep, K., Nagar, A., Pant, M., Bansal, J.C. (eds.) Proceedings of the International Conf. on SocProS 2011. AISC, vol. 131, pp. 485–495. Springer, Heidelberg (2012)
17. Miura, H., Wijeyewickrema, A.C., Inoue, S.: Evaluation of Tsunami Damage in the Eastern Part of Sri Lanka Due To the 2004 Sumatra Earthquake Using High-Resolution Satellite Images. In: Proceedings of 3rd International Workshop on Remote Sensing for Post-Disaster Response, pp. 12–13 (2005)
18. Ponnusamy, J.: GIS based Earthquake Risk-Vulnerability Analysis and Post-quake Relief. In: Proceedings of 13th Annual International Conference and Exhibition on Geospatial Information Technology and Application (MapIndia), Gurgaon, India (2010)
19. Sanchez-Silva, M., Garcia, L.: Earthquake Damage Assessment Based on Fuzzy Logic and Neural Networks. Earthquake Spectra 17(1), 89–112 (2001)
20. Slob, S., Hack, R., Scarpas, T., van Bemmelen, B., Duque, A.: A Methodology for Seismic Microzonation Using GIS And SHAKE—A Case Study From Armenia, Colombia. In: Proceedings of 9th Congress of the International Association for Engineering Geology and the Environment: Engineering Geology for Developing Countries, pp. 2843–2852 (2002)
21. United States Geological Survey-USGS, `http://earthquake.usgs.gov/`
22. Yang, Y., Jia, Z., Chang, C., Qin, X., Li, T., Wang, H., Zhao, J.: An Efficient Fuzzy Kohonen Clustering Network Algorithm. In: Proceedings of 5th International Conference on Fuzzy Systems and Knowledge Discovery, pp. 510–513. IEEE Press (2008)

Parametric Representation of Paragraphs and Their Classification

Dinabandhu Bhandari[1] and Partha Sarathi Ghosh[2]

[1] Department of Information Technology, Heritage Institute of Technology, Kolkata, India
[2] Wipro Technologies, Kolkata, India
{dinabandhu.bhandari,partha.silicon}@gmail.com

Abstract. Automatic paragraph classification is an important task in the field of information retrieval and digital publication. The work presents a novel approach to represent a paragraph of a document using a set of parameters extracted from it and a methodology has been proposed based on multi layer perceptron in designing an automatic paragraph classifier. The proposed framework has been tested on large industrial data and found improved performance compare to conventional rule based approach.

Keywords: Text Classification, Information Retrieval, Machine Intelligence, Multi Layer Perceptron.

1 Introduction

Automatic text classification is a very challenging computational problem to the information retrieval and digital publishing communities, both in academic as well as in industry. Currently, text classification problem has generated lot of interests among researchers. There are several industrial problems that lead to text classification or identification. One such problem that digital publishing companies encounter is paragraph classification. A document have multiple paragraphs of various sections such as title, abstract, introduction, conclusion, acknowledgement, bibliography and many more. When a document is published online or offline, different visualization style is applied on paragraphs based on the section, it belongs to, and its contextual information.

In the era of digitalization, along with the printing format every document is published in the web for easier access to the user over internet. All the paragraphs in documents are not formatted with same style because of better visualization and appearance. Moreover, the style applied on digital documents and on documents meant for printing are different. Sometimes, the style of a paragraph of the same document, published by different publishers, is different. As a business strategy, online publishers display some parts of the documents depending on types of user. Based on the document, online publishers decide which parts of the documents or which paragraphs are to be displayed for a user. The identification of the paragraphs will help to automatically apply the style once the paragraphs are correctly identified. The automatic classification of paragraphs

M.K. Kundu et al. (eds.), *Advanced Computing, Networking and Informatics - Volume 1*, Smart Innovation, Systems and Technologies 27,
DOI: 10.1007/978-3-319-07353-8_21, © Springer International Publishing Switzerland 2014

will reduce the manual interaction and effort. Along with the several business benefits it will also help to build up the repository of references.

At present, the identification of paragraphs is carried out using rule based heuristic algorithms. The heuristics are developed based on individuals' experience and some trivial rules. These systems are document specific and require significant amount of human interactions in their implementation.

There are very few articles, available in literature that focuses on the paragraph classification. In [1], Crossley et. al. have tried to classify the paragraphs taking into consideration the linguistic features such as textual cohesion and lexical sophistication. They correlate these features and proposed a model using statistical techniques. They have also tried to demonstrate the importance of the position of the paragraph in a document while classifying a paragraph. Sporleder & Lapta have used BoosTexter as machine learning tool to identify the paragraphs automatically in different languages and domains based on language modeling features, syntactic features and positional features [2]. They also investigated the relation between paragraph boundaries and discourse cues, pronominalization and information structure on German data for paragraph segmentation. Filippova et. al. proposed a machine-learning approach to identify paragraph boundary utilizing linguistically motivated features [3]. Taboada et. al. distinguished different type of paragraphs based on lexicon and semantic analysis within the paragraph and used naive bayes classifier, SVM and linear regression as machine learning tool [4].

The primary challenge in designing an automatic paragraph classifier is to represent the paragraphs in such a way so that computer can easily process them. At the same time, the representation should maintain the identity of the paragraphs. These motivate us to propose a new framework that represents a paragraph using its characteristics keeping in mind the objective of classification. In this article, an attempt has been made to represent a paragraph using its properties or features. Once the paragraphs are represented or characterized by some parameters, one can always use a tool to design a supervised classifier. One such tool being used here is multi layer perceptron (MLP) as it is a sophisticated modeling technique used in different application areas viz., aerospace, medical science, image processing, natural language processing and many more [5,6,7,8].

A brief discussion on paragraph classification is presented in Section 2. Section 3 discusses the proposed representation technique of a paragraph using its properties and MLP based classifier. Experiments and computer simulation results are presented in Section 4. The final section contains concluding remarks and a short discussion on future scope of work.

2 Classification of Paragraph

A paragraph of a document is a collection of sentences. In certain cases, a word or phrase can also be a paragraph, such as, introduction, abstract, conclusion, acknowledgement. We can define an article or document as a set of paragraphs and there by each paragraph can be considered as an element of the article or document. In this article, we refer a paragraph as an element of a document. Let us consider a document that consists of several elements (paragraphs). As mentioned above, a document is a collection of several paragraphs that are formatted in different styles. To determine the style of a paragraph or to decide the paragraphs that are to be displayed to a user, one has to identify the class or category of the paragraph. It is clear that the problem of identification of the paragraphs in several classes is nothing but to classify the paragraphs in different classes such as 'Title', 'Author', 'Keywords', 'Introduction', 'Sections', 'Reference items (bibitem)'. Mathematically, let D be the document containing $p_1, p_2, ..., p_N$ paragraphs and $C = c_1, c_2, ..., c_M$ $(M \leq N)$ be the member of classes of the paragraphs. In other words, for all $p_i, i = 1, 2, ...N$ can be classified as one of the $C_j, j = 1, 2, ..., M$.

In the next section, we will describe how a paragraph can be characterized by some properties that would be used in classifying a paragraph in different categories. It is to be mentioned here that the present work focusses on the articles of journals and conferences only.

3 Parameterizations of a Paragraph

In the rule based classification technique, various rules are generated based on human experience and the paragraphs are categorized in several classes. Generally, the rules are perceived by human beings depending on several characteristics of a paragraph. There are some obvious protocol that is maintained in arranging the paragraphs in a document. The authors take into consideration the readers' mind to arrange the paragraphs in preparing a document. People gain the experience of presenting a document by reading numerous documents and analyzing the style of presentation. In general, the paragraphs are classified depending on its content. However, the position of the paragraph also plays an important role in deciding its class. Moreover, different publishing houses follow different processes in arranging the paragraphs.

The objective of this work is to develop a methodology that would automatically classify paragraphs emulating the process that human beings adapt by reading numerous documents and analyzing readers' perception. In this context, the primary challenge is to present a paragraph to a machine. More specifically, paragraphs have to be fed into the machine so that machine can learn the patterns of the paragraphs in order to identify an unknown paragraph into a specific class. In this work, the paragraphs are represented using number of properties or features that would be utilized to categorize the paragraphs in different classes correctly. One can extract three different types of features of a paragraph,viz., *Positional*, *Visual* and *Syntactic properties*

Positional properties: The position of a paragraph provides significant information that would be helpful in identifying its class in the document. One can define number of positional properties of a paragraph depending on its position in the document. Some of the properties, used in this work are paraCount(total paragraph present in a document) and paraPos(position of the current paragraph).

Visual properties: Visual properties signifies the look and feel of the paragraph of a document. They are the textual information of a para and are defined based on font, color, alignment, indentation, spacing and many more. This properties are very effective in discriminating a paragraph. Some of the features, in this regard, are: font element such as font-size, bold, italic, underline; formatting element like indentation, spaces.

Syntactic properties: Syntactic properties are the contextual information of a paragraph. A simple set of properties are extracted from the string content, using simple text analysis techniques. Some of the syntactic properties are phraseCount (number of total phrases present into the paragraph), sentenceCount (number of sentences present into the paragraph), articleCount (number of articles present into the paragraph).

Once the parameters are defined to characterize a paragraph, designing the MLP based classifier is straight forward. The basic steps of the MLP-based methodology are:

1. Identification of features or parameters to be used to characterize the paragraphs
2. Extraction of parameter values
3. Preprocessing (Quantification, Normalization) of parameter values and removal of outliers
4. Design of the MLP based classifier and Identify training data set
5. Training the MLP classifier using back propagation technique
6. Testing the model using test data set.

3.1 Identification of Parameters

In this process, the properties or features are identified that characterize a paragraph in order to build the required classifier. It requires sufficient knowledge about the current process of paragraph identification. There are numerous properties or features can be found of a paragraph as discussed earlier. The primary task here is to identify the properties that would explicitly characterize a paragraph. Initially, 50 properties were identified. In our experiment, only 27 properties are used to design the classifier. Few such parameters, with detail description, are provided in Table 1.

3.2 Extraction of Parameter Values

The properties of a paragraph of a document are obtained or extracted using XSLT program. An XSLT style sheet is an XML document and XSLT instructions, which are expressed as XML elements. Element type of an XML tag denotes the feature of the paragraph and the respective value of the XML element indicates the property value. The designed XSLT program extracts all character and paragraph level properties. A few styles, which are used to extract the value from documents are provided in Table 2.

Table 1. Some of the parameters extracted from paragraphs with description

Sl. No	Parameter Name	Description	Type of properties
1	paraPosition	Position of the para into the document.	Positional
2	maxParaPosition	Total number of para into the document.	Positional
3	justifiedVal	Paragraph is left, right or central justified.	Positional
4	leftIndent		Positional
5	fontSize	Overall font size of the paragraph into the document.	Visual
6	isBold	Font faces of the entire paragraph is bold or not.	Visual
7	listType	Para is indicates any list value or not?	Visual
8	stringLength	Total number of word present into paragraph.	Syntatic

Table 2. A selected list of XSLT styles

Sl. No	Style Name	Description (possible values)
1	*fo:font-size*	positive integer values indicating the text size of the string
2	*fo:font-variant*	Indicates whether the string contains small caps or not
3	*fo:font-weight*	bold or normal
4	*fo:font-style*	font style could be normal or italic
5	*style:text-underline-type*	none, single, and double, default is single
6	*style:text-underline-width*	auto, normal, bold, thin, dash, medium, and thick.
7	*style:text-underline-mode*	set to skip-white-space so that whitespace is not underlined.

The feature values obtained using XSLT are of numeric, boolean or text types. For easier processing and to provide numeric inputs to the artificial neural network (ANN), the values are quantified. The boolean values are mapped to 0 and 1. To quantify the text properties, we used some of the predefined look up tables. As an example, an XSLT element **style:text-underline-width**, the possible outcomes are auto, normal, bold, thin, dash, medium, or thick. In such case, one can assign 1 for 'auto', 2 for 'normal', 3 for 'bold' and so on.

3.3 Data Preprocessing

The most important step in a pattern classification technique is preprocessing of the raw data. This task often receives little attention in the research literature, mostly because it is considered too application-specific. However, more effort is expended in preparing data than applying pattern recognition methods. There are two major tasks for the preparation of data. The first task is to quantify the data captured or extracted and the second is to normalize the data. It is expected that each feature, considered in designing the classifier, contributes consistently and equally. To mandate the equal contribution of all the features, it is wise to normalize the quantified data in a specific range, e.g., $[-n, n]$, where, n is a small quantity.

In this experiment, the z-score is used to standardize the data. The z-score is computed following equation (1), using standard deviation normalization to obtain a statistically stable data. In this case, d'_j satisfies normal distribution (identically and independently distributed) with zero mean and unity standard deviation.

$$\left(d_j'\right) = \frac{d_j - \mu\left(d_j\right)}{\sigma\left(d_j\right)}, \forall j \tag{1}$$

3.4 Identification of Outliers and Removal

Often in large data sets, there exist samples (called outliers) that do not comply with the general behavior of the data. In our paragraph identification problem, outliers can be caused by multiple heterogeneous sources of articles received from numerous authors and due to the error introduced during quantification. Outliers affect the learning process of the MLP and thereby reduces the classification performance in terms of accuracy and convergence. It is customary, in any classification algorithm, to identify the outliers and eliminate them from the data set. However, it is also to be noted that the number of eliminated data points should be as minimum as possible. In this work, the z-score values are used to identify the outliers. The standardized samples (d_t') into a compact range, normally between [α, -α] (where α is a small quantity) could minimize the effect from random noise. In our experiment, it has been observed that considering α value as 3, less than 3% of total data points are identified as outliers.

3.5 Training and Testing

Back propagation technique is used here to train the MLP. Training is continued until the Mean Square Error (MSE) is less than some predefined small quantity (ϵ) (in our experiment, ϵ is assumed as 0.001). The initial weights are selected randomly. The learning rate (η) is considered to be varying from 0.1 to 0.001 (decreasing with iteration number) and the moment is set to 0.8.

4 Experimental Result

The experiment is carried out using Java on Microsoft platform (Windows 98) and 2GB CPU RAM. In our experiment, 27 input features are considered for the paragraphs belonging to 9 classes which are widely used and very difficult to classify. The features were extracted from about 20,000 paragraphs of 1000 documents taken from various domains. The paragraphs belong to 30 different classes including the trivial classes. 20% of randomly selected data points were used for training the network and rest of the data points were considered for testing. To validate the effectiveness of the methodology, the experiment was performed 5 times by randomly selecting different training data sets. It has been observed that each run produces consistently similar results. Table 3 shows the performance of the rule based approach and the proposed framework using MLP.

The results clearly shows that the proposed parameter based approach is superior than the conventional rule based approach. The proposed approach performs consistently well for all types of paragraphs. The rule based approach

Table 3. Rule based system Vs. MLP based classifier system

Classes	Rule based system	MLP Classifier based system	
		Before discarding outliers	After discarding outliers
articleTitle	98.551 %	95.775 %	95.775 %
author	98.361 %	96.774 %	96.774 %
affiliation	97.62 %	94.620 %	94.62 %
bibitem	95.835 %	97.942 %	99.629 %
correspondingAuthor	100 %	73.811 %	92.857 %
paragraph	62.26 %	99.071 %	99.07 %
sectiona	79.25 %	90.119 %	93.281 %
sectionb	74.146 %	96.845 %	96.845 %
sectionc	64.89 %	74.600 %	82.54 %
Overall Accuracy	86.101%	97.279%	98.451 %

provides better classification for classes like articleTitle, author, affiliation and correspondingAuthor, due to their distinct visual and syntactic features. The performance of the proposed approach may be improved by defining some appropriate properties or features that would enhance the discrimination ability of the MLP. The performance of the parameter based approach with and without outliers is also been provided in Table 3. As expected, that removal of outliers results in improved performance.

5 Conclusion and Future Scope

An attempt has been made to parameterize the paragraphs of documents using various visual and syntactic features. The parameters are then used to design an MLP based classifier. The performance of the proposed framework is found to produce better result (98.451%) compare to that (86.101%) of the conventional rule based technique. The experiment is carried out to classify paragraphs into 9 important classes though there are more than 30 different classes present in a simple document. The proposed framework has been tested using a set of 27 parameters, one can define and use different parameters to obtain better performance, especially, for the paragraph classes articleTitle, author, affiliation and correspondingAuthor. Though the MLP has been used here in designing the classifier, other classification tool such as support vector machine (SVM) can also be used. The investigation may further be extended by designing a hybrid approach combining the conventional rule based and the proposed parameter based approaches.

Acknowledgement. The authors sincerely acknowledge Dr. S. K Venkatesan of TNQ Books and Journals for his valuable suggestions and for providing relevant data to accomplish the work.

References

1. Crossley, S.A., Dempsey, K., McNamara, D.S.: Classifying paragraph types using linguistic features: Is paragraph positioning important? Journal of Writing Research 3(2), 119–143 (2011)
2. Sporleder, C.: Automatic paragraph identification: A study across languages and domains. In: Proceedings of the Conference on Empirical Methods in Natural Language Processing, pp. 72–79 (2004)
3. Filippova, K., Strube, M.: Using linguistically motivated features for paragraph boundary identification. In: EMNLP, pp. 267–274 (2006)
4. Taboada, M., Brooke, J., Stede, M.: Genre-based paragraph classification for sentiment analysis. In: Proceedings of SIGDIAL 2009, pp. 62–70 (2009)
5. Lieske, S.P., Thoby-Brisson, M., Telgkamp, P., Ramirez, J.M.: Reconfiguration of the neural network controlling multiple breathing patterns: eupnea, sighs and gasps. Nature Neuroscience 3, 600–607 (2000)
6. Collobert, R., Wetson, J.: Fast semantic extraction using a novel neural network architecture. In: Proceedings of the 45th Annual Meeting of the Association of Computational Linguistics, pp. 560–567 (2007)
7. Sebastiani, F.: Machine learning in automated text categorization. ACM Computing Surveys 34, 1–47 (2002)
8. Pomerleau, D.A.: Neural network simulation at warp speed: how we got 17 million connections per second. IEEE 2, 143–150 (1988)

LDA Based Emotion Recognition from Lyrics

K. Dakshina and Rajeswari Sridhar

Department of Computer Science and Engineering,
College of Engineering Guindy, Anna University,
Chennai, India
{dakshinaceg,rajisridhar}@gmail.com

Abstract. Music is one way to express emotion. Music can be felt/heard either using an instrument or as a song which is a combination of instrument and lyrics. Emotion Recognition in a song can be done either using musical features or lyrical features. But at times musical features may be misinterpreting, when the music dominates the lyrics. So a system is proposed to recognize emotion of the song using Latent Dirichlet Allocation (LDA) modelling technique. LDA is a probabilistic, statistical approach to document modelling that discovers latent semantic topics in large corpus. Since there is a chance of more than one emotion occurring in a song, LDA is used to determine the probability of each emotion in a given song. The sequences of N-gram words along with their probabilities are used as features to construct the LDA. The system is evaluated by conducting a manual survey and found to be 72% accurate.

1 Introduction

The recognition of emotion has become in a multi-disciplinary research area that has received great interest. It plays an important role in the improvement of Music information retrieval (MIR), content based searching, human machine interaction, emotion detection and other music related application. The word emotion covers a wide range of behaviours, feelings, and changes in the mental state [1]. Emotion/mood is a feeling that is private and subjective. There are eight basic emotions such as happy, sadness, acceptance, disgust, anger, fear, surprise and anticipation [2]. The recognition of emotion/mood in a song can be done either using the lyrical features or musical features. But, sometimes the musical features like tone may mislead the emotion of the song. Hence, in this work, the emotion of the song is identified by considering only the lyrics. In this paper, Section 2 focuses on literature survey that discusses on the various proposed methods to classify emotion. Section 3 on system design, Section 4 focuses on the implementation details along with the parameters set and used with the justification details and finally Section 5 concludes the paper and future extension for this current work.

2 Literature Survey

Hu, Xiao, Mert Bay, and J. Stephen Downie [3] used social tags on last.fm to derive a set of three primitive mood categories. Social tags of single adjective words on a

M.K. Kundu et al. (eds.), *Advanced Computing, Networking and Informatics - Volume 1*,
Smart Innovation, Systems and Technologies 27,
DOI: 10.1007/978-3-319-07353-8_22, © Springer International Publishing Switzerland 2014

publicly available audio dataset, USPOP, were collected and 19 mood related terms of the highest popularity were selected manually which was later reduced to three latent mood categories using multi-dimensional scaling. This finally scaled mood set was not used by others because three categories were seen as a domain oversimplification.

Yang and Lee [4] performed early work on supplementing audio mood classification with lyric text analysis. A lyric bag-of-words (BOW) approach was combined with 182 psychological features proposed in the General Inquirer to disambiguate categories which was found confusing by the audio-based classifiers and the overall classification accuracy was improved by 2.1%. However, no reliable conclusions were drawn as the dataset was too small.

Dang, Trung-Thanh, and Kiyoaki Shirai [5] constructed a system that automatically classifies the mood of the songs based on lyrics and metadata, and many methods were employed to train the classifier by the means of supervised learning. The classifier was trained using the data gathered from a Live Journal blog where song and mood were tagged appropriately for each entry of the blog. The classifier was trained with the help of various machine learning algorithms like Support Vector Machines, Naive Bayes and Graph-based methods. The accuracy of mood classification methods was not good enough to apply for a real music search engine system. There are two main reasons: mood is a subjective metadata; lyric is short and contains many metaphors which only human can understand. Hence, the authors plan in future to integrate audio information and musical features with lyrical features for improving the accuracy and efficiency of the mood identification system.

Chen, Xu, Zhang and Luo [6] applied the machine learning approach to classify song sentiment using lyric text, in which the VSM is adopted as text representation method. The authors performed error analysis and have identified that the term based VSM is ineffective in representing song lyric. Hence the authors concluded that a new lyric representation model is crucial for song sentiment analysis.

Han, Byeong-jun, Seungmin Ho, Roger B. Dannenberg, and Eenjun Hwang [7] designed music emotion recognition using Support Vector Regression. Several machine learning classification algorithms such as SVM, SVR and GMM are considered for evaluating the system that recognizes the music automatically. There was a tremendous increase in accuracy using SVR when compared to GMM. Many other features can be considered with other classification algorithms such as fuzzy and kNN (k-Nearest Neighbor).

Yang, Dan, and Won-Sook Lee [8] proposed the novel approach of identifying discrete music emotions from lyrics. Statistical text processing tools were applied to the novel application area of identifying emotion in music using lyrics. A psychological model of emotion was developed that covers 23 specific emotions. A feature vector of 182 psychological features was constructed from the song lyrics using a content analysis package. Classification decision trees were used to understand the classification models. Results for each specific emotion were promising.

Latent Dirichlet allocation (LDA) model has been proposed by David Blei [9]. This is an unsupervised model for topic modelling. LDA is a probabilistic, statistical approach to document modeling that discovers latent semantic topics in large collections of text documents [9]. Later LDA has been applied for classification of text, images and music.

Wu, Qingqiang, Caidong Zhang, Xiang Deng, and Changlong Jiang [10] performed topical evolutionary analysis and proposed a text mining model. The authors used the LDA model for the corpus and text to get the topics, and then Clarity algorithm was used to measure more-similar topics in order to identify topic mutation and discovered the topic hidden in the text.

Maowen, Wu, Zhang Cai Dong, Lan Weiyao, and Wu Qing Qiang [11] performed topic analysis using co-occurrence theory and Latent Dirichlet Allocation (LDA) .The text topics of stem cell research literatures from 2006-2011 in PubMed were identified using LDA and Co-occurrence theory. As a result, several stem cell research topics were obtained. The result was analyzed in regards with the topic label, and the relationship between the topics. However the LDA employed suffered with some deficiencies, which were stated for future improvement. LDA has also been used in the Indian context for identification of Raga [12] and emotion from music [13] by observing the Raga lakshana and musical features respectively.

Based on the literature survey for emotion recognition, in this paper an algorithm has been proposed for Tamil lyric emotion recognition based on constructing a LDA model. The use of LDA is justified as it is a probabilistic model. The emotion of a particular lyric is subjective and hence one lyric can be thought of as having multiple emotions with varied probability.

3 System Design

In this section, the system architecture and the steps involved in the design process are considered.

3.1 System Architecture

The block diagram of the proposed work for recognizing the emotion is given in Fig. 1. The emotion recognition process is carried out by initially collecting the Tamil lyric texts and pre-processing it. The frequency and probability for the pre-processed Tamil lyric document is calculated and *N*-gram model is constructed. The constructed *N*-gram model is used as features in LDA. The designed LDA model is tested with a new Tamil lyric document.

3.2 Steps in Emotion Recognition

The Steps involved in recognizing emotions is discussed in the following subsections.

Data Collection and Preprocessing. Collecting data is the first part of the training phase of emotion recognition process. Sufficient data must be collected in both the

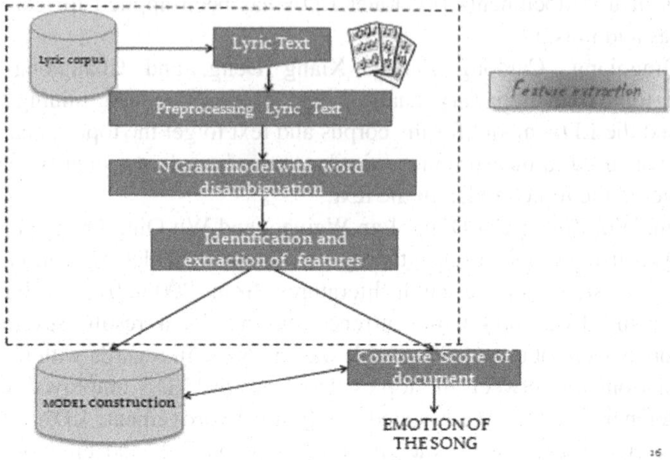

Fig. 1. System Architecture

training phase and testing phase for effective measurements. The accuracy of the recognition process largely depends on the training samples. The system must be trained efficiently with large number of training samples. A large number of lyric samples are collected for different emotions of the songs. A subset of the collected samples is used for training. For testing purpose, any sample can be used. The pre-processing and handling of the Tamil lyrics is initially performed to remove stop words for recognising emotion. Emotion of the song is very sensitive, so unnecessary words which do not contribute to the emotion recognition should be removed during the pre-processing of the lyrics and the POS tagger is used for collecting relevant parts of speech and avoids word disambiguation problems arising due to polysemy.

Feature Extraction. Use of features is the basis of emotion recognition process. Different features can be used for emotion recognition process. So the selection of features plays an important role in the recognition process. In the proposed system, the sequence of words is used as features. The probability of a word following the other is modelled as unigram, bigram and the Trigram. The probability of occurrence of these word sequences are used as features for recognizing emotion.

N-gram Model Construction. The frequency and probability of each word in a document after pre-processing is computed and Uni-gram is constructed. Similarly, from the Uni-gram model, the Bi-gram and Tri-gram model are constructed, for Bi-gram and Tri-gram model the frequency and probability of two co-occurring and three co-occurring words are taken respectively.

LDA Construction and Emotion Recognition. LDA strongly asserts that each word has some semantic information, and documents having similar topics will use same set of words. Latent topics are thus identified by finding groups of words in the corpus that frequently occur together within documents. In this way, LDA depicts each document as a random mixture containing several latent topics, where each and every topic has its own specific distribution of words [14]. In this paper, techniques and results are shown and it is suggested that LDA aids not only in the text domain, but also in lyric domain.

Initially a training set is considered and supervised learning is given for the training set. The LDA dirichlet parameters are derived from the training set and the same is used for the test set based on which the mood/emotion has to be determined.

Distribution Over Emotion and Features. Fig. 2 shows the influence of the dirichlet parameters of LDA over emotion recognition. If a feature appears in a emotion, it is indicated that the lyric document in which the feature is present, contains the emotion.

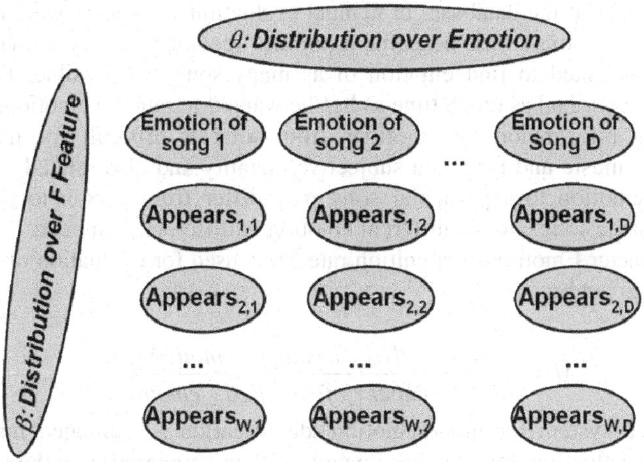

Fig. 2. Distribution of emotion over features

4 Implementation and Evaluation

4.1 LDA Parameters

The LDA model is constructed using the alpha (α) and beta (β) parameters derived from the training set. The calculation of alpha and beta are given below. Alpha (α) is the dirichlet prior on the per-document topic distributions. It is given by

$$\alpha = \frac{document\_count \times topic\_count}{count} \tag{1}$$

Beta (β) is the dirichlet prior on the per-topic word distributions.

$$\beta = \frac{probability\_of\_feature\_occuring\_in\_a\_topic}{topic\_count} \quad . \tag{2}$$

Here count is the number of emotions per document and topic count = 6.

4.2 Emotion Recognition

The N gram features are used in identifying the emotion of the Tamil song lyrics. The probability of these features are calculated and appended in a document. The probability found is used in the calculation of the LDA parameters stated in section 4.1.

4.3 Evaluation

Evaluation of the system carried out with the help of 5 randomly selected volunteer to validate the data in the database. In manual evaluation volunteers were asked to read the lyrics of the songs in a peaceful room and recognize emotion manually. Each volunteer was asked to find emotion of as many songs as possible. Between each songs, the listener had as much time as he/she wanted to rate the emotion.

Extracting information on emotions from song is difficult for many reasons, because both music and lyric is a subjective quantity and also related to culture. So, the feel of emotion to a particular song may differ from person to person. While evaluation, same song evoked different emotion to different volunteer.

The parameter Emotion Recognition rate M, is used for evaluation of the proposed system. It is given by

$$M = \frac{correctly\_classified\_emotion}{total\_number\_ofclassified\_emotions} \tag{3}$$

The proposed system for mood/emotion identification is evaluated manually by 5 people. The evaluation done by these people is then compared with the system which

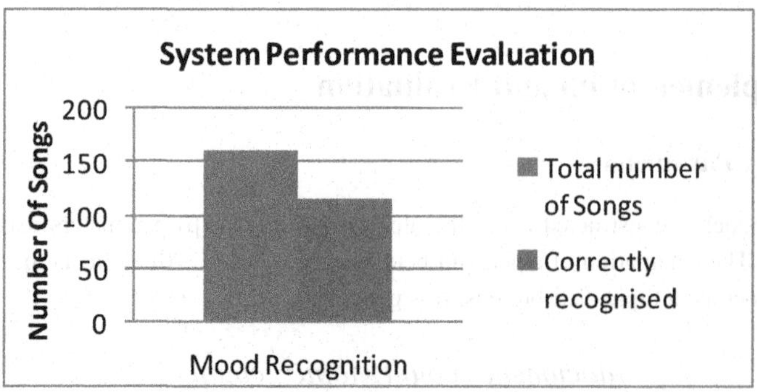

Fig. 3. Overall System Performance

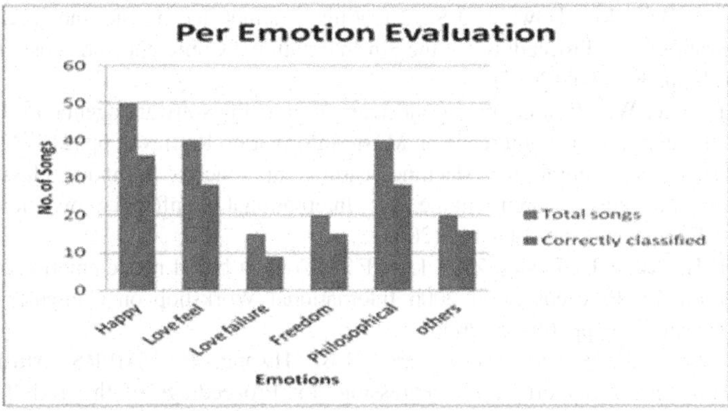

Fig. 4. Evaluation of individual emotions

identifies the mood/emotion for a given song. The different possible emotions for a particular song with probability are given. Since a probability model is used, not always the emotion with correct probability is recognized by the system. Hence, our system identified the emotions correctly for more than 100 songs, upon a total of 160 songs and is given in Fig. 3 and Fig. 4.

5 Conclusion and Future Work

A system where the emotion of the song is identified based on the lyrics of the Tamil song is being proposed and implemented. The lyrical features considered in this work is the probability of one word occurring after another. The emotion is identified by constructing an LDA model using the probability of word sequence as features. The use of LDA technique improves the accuracy of recognition process greatly and identifies the various emotions that are present in a song with their respective probability. However, although the proposed system recognizes emotion of a song with good level of accuracy, consideration of more features will improve the accuracy of the emotion recognition system.

In future, emotion can be used as features for designing a song Search system. This search system could be designed to retrieve similar songs that match the input lyric's emotion.

References

1. http://en.wikiversity.org/wiki/Motivation_and_emotion/Book/2013/Pet_ownership_and_emotion
2. http://www.translate.com/english/lyrics-in-singular-form-lyric-are-a-set-of-words-that-make-up-a-song-usually-consisting-of-verses/

3. Xiao, H., Bay, M., Downie, J.S.: Creating a simplified music mood classification groundtruth set. In: Proceedings of the 8th International Conference on Music Information Retrieval, pp. 309–310 (2007)
4. Yang, D., Lee, W.: Disambiguating music emotion using software agents. In: Proceedings of the 5th International Conference on Music Information Retrieval, pp. 218–223 (2004)
5. Trung-Thanh, D., Shirai, K.: Machine learning approaches for mood classification of songs toward music search engine. In: International Conference on Knowledge and Systems Engineering, pp. 144–149 (2009)
6. Chen, R.H., Xu, Z.L., Zhang, Z.X., Luo, F.Z.: Content based music emotion analysis and recognition. In: Proceedings of 2006 International Workshop on Computer Music and Audio Technology, pp. 68–75 (2006)
7. Byeong-Jun, H., Ho, S., Dannenberg, R.B., Hwang, E.: SMERS: Music emotion recognition using support vector regression. In: Proceedings of the 10th International Society for Music Information Conference, pp. 651–656 (2009)
8. Yang, D., Lee, W.-S.: Music emotion identification from lyrics. In: 11th IEEE International Symposium In Multimedia, pp. 624–629 (2009)
9. Blei, D.M., Ng, A.Y., Jordon, M.I.: Latent Dirichlet Allocation. Journal of Machine Learning Research 3, 993–1022 (2003)
10. Qingqiang, W., Zhang, C., Deng, X., Jiang, C.: LDA-based model for topic evolution mining on text. In: 6th International Conference in Computer Science & Education, pp. 946–949 (2011)
11. Wu, M., Dong, Z.C., Weiyao, L., Qiang, W.Q.: Text topic mining based on LDA and co-occurrence theory. In: 7th International Conference on Computer Science & Education, pp. 525–528 (2012)
12. Sridhar, R., Geetha, T.V., Subramanian, M., Lavanya, B.M., Malinidevi, B.: Latent Dirichlet Allocation Model for raga identification of carnatic music. Journal of Computer Science, 1711–1716 (2011)
13. Arulheethayadharthani, S., Sridhar, R.: Latent Dirichlet Allocation Model for Recognizing Emotion from Music. In: Kumar M., A., R., S., Kumar, T.V.S. (eds.) Proceedings of ICAdC. AISC, vol. 174, pp. 475–481. Springer, Heidelberg (2013)
14. Hu, D., Saul, L.: Latent Dirichlet Allocation for text, images, and music. University of California (2009)

Efficient Approach for Near Duplicate Document Detection Using Textual and Conceptual Based Techniques

Rajendra Kumar Roul, Sahil Mittal, and Pravin Joshi

BITS Pilani K. K. Birla Goa Campus, Zuarinagar, Goa-403726, India
rkroul@goa.bits-pilani.ac.in,
{sahilindia33,pravinjoshi95}@gmail.com

Abstract. With the rapid development and usage of World Wide Web, there are a huge number of duplicate web pages. To help the search engine for providing results free from duplicates, detection and elimination of duplicates is required. The proposed approach combines the strength of some "state of the art" duplicate detection algorithms like Shingling and Simhash to efficiently detect and eliminate near duplicate web pages while considering some important factors like word order. In addition, it employs Latent Semantic Indexing (LSI) to detect conceptually similar documents which are often not detected by textual based duplicate detection techniques like Shingling and Simhash. The approach utilizes hamming distance and cosine similarity (for textual and conceptual duplicate detection respectively) between two documents as their similarity measure. For performance measurement, the F-measure of the proposed approach is compared with the traditional Simhash technique. Experimental results show that our approach can outperform the traditional Simhash.

Keywords: F-measure, LSI, Shingling, Simhash, TF-IDF.

1 Introduction

The near duplicates not only appear in web search but also in other contexts, such as news articles. The presence of near duplicates has a negative impact on both efficiency and effectiveness of search engines. Efficiency is adversely affected because they increase the space needed to store indexes, ultimately slowing down the access time. Effectiveness is hindered due to the retrieval of redundant documents. For designing robust and efficient information retrieval system, it is necessary to identify and eliminate duplicates.

Two documents are said to be near duplicates if they are highly similar to each other [3]. Here the notion of syntactic similarity and semantic similarity between two documents has to be carefully considered. A set of syntactically similar documents may not necessarily give positive result when tested for semantic similarity and vice versa. So two independent strategies have to be employed to detect and eliminate syntactically and semantically similar documents. Most of the traditional duplicate

M.K. Kundu et al. (eds.), *Advanced Computing, Networking and Informatics - Volume 1*,
Smart Innovation, Systems and Technologies 27,
DOI: 10.1007/978-3-319-07353-8_23,

detection algorithms [1],[2] have considered only the aspect of syntactic similarity. Here, apart from performing textual near duplicate detection, an approach to detect and eliminate semantically (conceptually) similar documents has been proposed. Another aspect of duplicate detection algorithms is the precision-recall trade off. Often an algorithm based on mere presence or absence of tokens in documents to be compared performs low on precision but yields a high recall considering only textually similar documents. For example if shingling [1] is implemented with *one* as the shingle size, it yields high recall but low precision. On the other hand, when techniques based on co-occurrence of words in document are used, taking into account even the order of words, its precision increases by a reasonable amount but its recall decreases. Thus, it is essential to give a right amount of importance to co-occurrence of words in documents but at the same time taking care of the recall. In the proposed approach, F-measure [9] has been used as a performance measure to give optimum precision-recall combination.

The outline of the paper is as follows: Section 2reviews previous research work on detection of duplicate documents.Section 3 proposesthe approach to effectively detect near duplicate documents. Section 4demonstrates the experimental results which are followed by conclusion and future work covered in Section 5.

2 Related Work

One of the major difficulties for developing approach to detect near duplicate documents was the representation of documents. Broder [1] proposed an elegant solution to this by representing a document as a set of Shingles. The notion of similarity between two documents as the ratio of number of unique Shingles common to both the documents to the total number of unique Shingles in both the documents was defined as resemblance. The approach was brute force in nature and thus not practical. Charikar [2] proposed another approach which required very low storage as compared to the Shingling. The documentswere represented as a short string of binaries (usually 64 bits) which is called fingerprint.Documents are then compared using this fingerprint which is calculated using hash values. Both of the above approaches became the "state of the art" approaches for detection of near duplicate documents. Henzinger [3] found that none of these approaches worked well for detecting near duplicates on the same site. Thus, they proposed a combined approach which had a high precision and recall and also overcame the shortcomings of Shingling and Simhash. Bingfeng Pi *et al.* [6] worked on impact of short size on duplicate detection and devised an approach using Simhash to detect near duplicates in a corpus of over 500 thousand short messages. Sun, Qin *et al.* [5] proposed a model for near duplicate detection which took query time into consideration. They proposed a document signature selection algorithm and claimed that it outperformed Winnowing – which is one of the widely used signature selection algorithm.Zhang *et al.* [7] proposed a novel approach to detect near duplicate web pages which was to work with a web crawler. It used semi-structured contents of the web pages to crawl and detect near duplicates. Figuerola *et al.* [8] suggested the use of fuzzy hashing to generate the fingerprints of the web documents. These fingerprints were then used to estimate the similarity between two documents. Manku *et al.* [4] demonstrated practicality of using Simhash to identify near duplicates in a corpus containing multi-billion documents. They also showed that fingerprints with length 64 bits are appropriate to detect near duplicates. Martin Theobald *et al.* [10] took the idea of Shingling forward to include the stopwords to form short chains of adjacent content terms to create robust document signatures of the web pages.

In this paper, Shingling along with Simhash is used to detect textually similar documents. LSI [13] has also been used for detecting conceptually similar documents. Gensim [11], a Python toolkit has been used for experimental purpose and, it has been found out that the F-measure of the proposed approach is better than the traditional Simhash technique.

3 Proposed Approach

Input: A preprocessed set of documents <D_1, D_2, D_3,…,D_n>, where n is the total number of documents in the corpus.
Output: A set of documents <D_1, D_2, D_3,…, D_m> where, m<=n, which are free of near duplicates.
The proposed approach is described in Fig. 1.

1. Stop words are removed from the documents by using preProcess()method.
2. TF-IDF [12] is calculated for each token of each document using getTFIDF() method and stored in Tfidf array. TF-IDF weights are normalized by the length of the document so that length of the document does not have any impact on the process.
3. Each document is then represented by a set of shingles of size *three*(experimentally determined). These shingles are stored in Shingles array.
4. Each shingle hashed value is stored in HashValues array. Hashing the shingle itself (and not the token) takes into account effect of *word order* and not mere presence or absence of words as in case of traditional Simhash. Thus two documents, with same set of words used in different contexts, will not be considered near duplicates unless their word order matches to a large extent. If only single words were selected as features, two documents with largely same words, in different contexts and with different meanings are considered duplicates.
5. A 'textual fingerprint' (64 bit) textFing is generated for every document.
6. 'Conceptual fingerprints' lsiFing are computed for each document in the corpus. This is done using SVD (Singular Value Decomposition) matrices.
7. The list of textual fingerprints is sorted so that similar fingerprints (fingerprints which differ from each other in very less number of bits) come closer to each other.
8. The thresholds used for filtering the near duplicates are as follows:

   i. th1–This is a threshold that must be exceeded by textual similarity of a pair of documents to be considered for further detection. Here the idea that a pair of conceptually similar documents will also invariably exhibit some amount of textual similarity as well is used by keeping the threshold reasonably low.
   th2-This is the textual similarity threshold. If the textual similarity of the pair of documents exceeds th2, they are considered as textual near duplicates. This is set to a quite higher value as compared to th1 based on experimental results.

```
1. for each di in <D1, D2,...,Dn>
       preProcess(di)
       Tfidf[i] = getTFIDF(di)/len(di)
       Shingles[i] = getShingles(di, 3)//3 is shingle size
2. for each si in Shingles
       //HashValues[i] is a list of all hashes of i^th doc
       HashValues[i] = SHA1(si)//Hashing the shingles
3. textFing = Initialize2DArray(n, 64)//64 bit arrays
for i in range(0, n)
       for each value in HashValues[i]
              for j in range(0, 64)
                     ifHashValues[i][j]==1
                            textFing[i][j]+=Tfidf[i][j]

                     else
                            textFing[i][j]-=Tfidf[i][j]
for i in range(0, n)//converting to binary
       for j in range(0, 64)
              iftextFing[i][j] > 0
                     textFing[i][j] = 1
              else
                     textFing[i][j] = 0
4. for each di in <D1, D2,...,Dn>
       termDocMat = getTermDocMat(di)
       U, s, V = getSVD(termDocMat)
        lsiFing = s*(transpose(V))//lsi fingerprint
5. for i in range(0, 64)
       //rotate all textual fingerprints by 1 bit
       rotate(textFing)
       sort(textFing)
       for j in range(0, n-1)
       textSim = 64 - HamDist(textFing[j],textFing[j+1])
       concSim = cosine(j^th doc, j+1^th doc)
       if(textSim>th1)
              if(textSim<th2)
              //cosine similarity calculated using lsiFing
                     if(concSim>th3)
                            //add edge to Conceptual Graph
                            grConc.add(jth doc, j+1th doc)
              elif(textSim>th2 &&textSim<th3)
              grText.add(j^th doc, j+1^th doc)
              else
              grExact.add(j^th doc, j+1^th doc)
6. removeDups(grText, grConc, grExact)
```

Fig. 1. The proposed approach

ii. th3 - This threshold is used to detect exact duplicates. Thus it is set to a very high value.
iii. th4 – This is the conceptual similarity threshold. If the cosine similarity [9] of two documents being checked for conceptual similarity exceeds th4, they are considered conceptual near duplicates.

9. Each pair of adjacent documents in the sorted list is checked for near duplicates as follows:

i. If the documents under consideration are textual near duplicates ((64 - hamming distance) exceeds th2), or exact duplicates ((64 - hamming distance) exceeds th3), they are added to graph of Textual and Exact duplicates respectively.
ii. If none of the conditions mentioned above are satisfied and if the (64 - hamming distance) of documents under consideration exceed th1, the documents are checked for conceptual similarity. If their cosine similarity exceeds a threshold, th4, they are added to the graph of conceptual duplicates.

10. Step (7) and (8) are repeated 64 times, each time rotating each fingerprint by one bit so as to cover all the possible near duplicate pairs. Here, the approach takes advantage of the property of the fingerprints that similar documents differ by very low number of bits in their respective fingerprints. Thus, every document need not be checked with every other document in the corpus for textual similarity.

11. Select a document from each set of duplicate documents (unique document, kept in the corpus) and remove others from input list. Remove Dups method returns the set of documents which are free of near duplicates.From each set of near duplicate documents, it keeps the longest document with a view to lose minimum information. Here, other techniques can be applied like computing each document's relevance to the query and outputting the one with the highest value.

4 Experimental Results

For experimental purpose, *20 Newsgroup Dataset* [14] is used. The *20 Newsgroup Dataset* is a collection of approximately 20,000 newsgroup documents, partitioned (nearly evenly) across 20 different newsgroups. The *20 Newsgroup* collections are widely used for experiments in areas of text processing and clustering. It offers a group of documents which already contains some duplicates, making it suitable for testing near duplicate detection algorithms. The proposed duplicate detection approach was implemented using Python and two kinds of experiments were majorly performed on dataset as described below.

4.1 Determination of Optimum Values for Parameters Involved in Algorithm

For figuring out the optimal combination of parameters, algorithm was run for a range of different thresholds and Shingle sizes. The thresholds th1, th3 and th4 are fixed to a reasonable value. As th1 is the minimum similarity threshold which two documents must pass to be considered for duplicate detection, it can be set to an intuitively low value. Similar argument holds true forth3 and th4. So, only the Shingle size and th2 need to be determined experimentally. Fig. 2 described the precision, recall and F-measure with respect to Shingle size (where threshold is set to optimum). As the graph clearly shows, the Shingle size = 3 gives the best results with highest F-measure of 0.88. When Shingle size is very low, the algorithm tends to neglect the effect of word order while testing two documents for duplicates. On the other hand, when Shingle size is very high, algorithm can only find duplicates which are exact copies of each other and lacks ability to find *near duplicates*. A balance between these two extremes is obtained using a sequence of tokens of length 3 each as suggested by the results. Fig. 3 shows the graphs of precision, recall and F-measure with respect the value of th2. Here, it can be seen that as the value of th2 increases, precision increases and recall decreases. When th2 value is increased, two documents are considered to be near duplicates only when they have a large chunk of text in common and thus this does not cover all near duplicate documents, leading to low recall. Whereas, when th2 value is very low, almost every document is considered as a near duplicate covering all the actual near duplicate documents but performing poorly on precision. F-measure is used to determine the optimum th2 value which turns out to be 46.

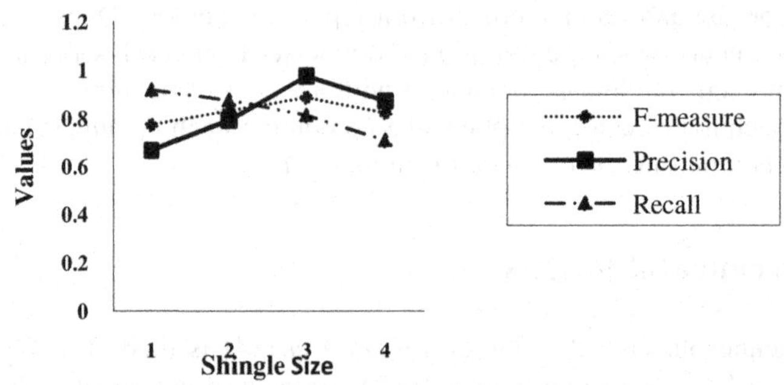

Fig. 2. F-measure, Precision and Recall v/s Shingle size

4.2 Comparison of Proposed Approach with Traditional Simhash Implementation

Simhash-0.1.0, a Python library is used for comparing the proposed approach with Simhash. To demonstrate the work and for comparison purposes, a corpus of 128 documents from *20 Newsgroup*dataset out of which 48 documents were near duplicates (as mentioned in [14]) has been considered. Both the algorithms were run for same set of conditions (the corpus and the thresholds).

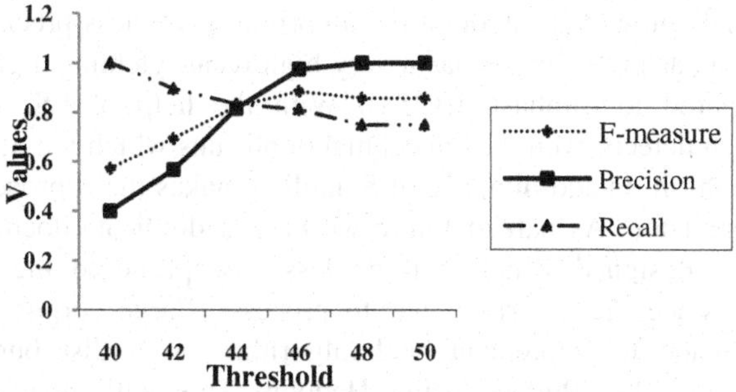

Fig. 3. Threshold (th2) v/s F-measure, Precision and Recall

The results are shown in Fig. 4. Simhash could only identify 7 documents as near duplicates whereas the proposed approach could identify 40 near duplicate documents. Simhash performs excellently on precision but poorly for recall. This is because, Simhash, when comparing two documents, totally relies on syntactic occurrences of words in the documents and does not take into account the context of occurrence of words. The use of Shingling along with Simhash and LSI, in the proposed approach, takes into account not only the syntactic occurrence of words in the documents, but also the context or the co-occurrence of terms. This leads to not only high precision but high recall as well, because both – syntactic and semantic near duplicates are identified.

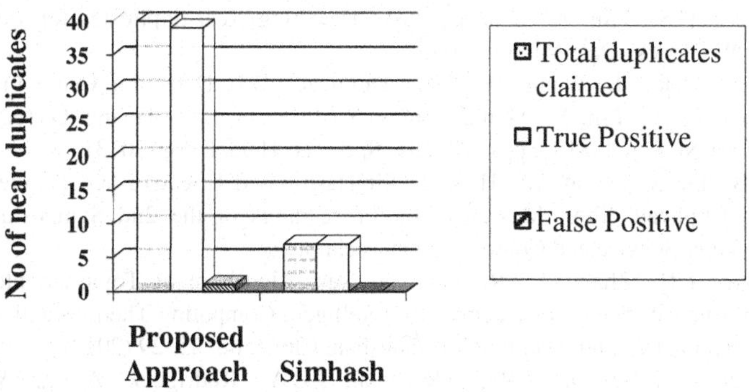

Fig. 4. Comparison of Proposed Approach and Simhash based on the No. of duplicates detected

5 Conclusion and Future Work

In this paper, the proposed approach uses a combination of Shingling and Simhash techniques to detect textual near duplicates. For detecting conceptual near duplicates, LSI is used to generate fingerprints of the documents. Optimal Shingle size and threshold is determined by experiment using F-measure as the performance measure. Experimental result show that although the algorithm gives less precision (0.975) in comparison to Simhash (1.0) but it has a very high recall yielding a good F-measure (0.886) as compared to Simhash (0.254). With the help of LSI technique, the proposed approach detects even the conceptual duplicates which are often missed out by traditional Simhash. In addition, use of Shingling makes the approach more robust in terms of its precision. As part of future work, near duplicate document detection algorithm can be designed which will be less susceptible to the length of the document. Also, since the algorithm has to process a large corpus of documents, efforts can be made to implement such algorithm on a distributed computing framework such as Map Reduce using Hadoop. This will greatly increase the efficiency of the algorithm.

References

1. Broder, A.Z.: Identifying and filtering near-duplicate documents. In: Giancarlo, R., Sankoff, D. (eds.) CPM 2000. LNCS, vol. 1848, pp. 1–10. Springer, Heidelberg (2000)
2. Charikar, M.S.: Similarity estimation techniques from rounding algorithms. In: STOC 2002: Proceedings of the 34th Annual ACM Symposium on Theory of Computing, pp. 380–388. ACM, New York (2002)
3. Henzinger, M.: Finding near-duplicate web pages: a large-scale evaluation of algorithms. In: SIGIR 2006: Proceedings of the 29th Annual International ACM SIGIR Conference on Research and Development in Information Retrieval, pp. 284–291. ACM, New York (2006)
4. Manku, G.S., Jain, A., Sharma, A.D.: Detecting Near-duplicates for web crawling. In: WWW / Track: Data Mining (2007)
5. Sun, Y., Qin, J., Wang, W.: Near Duplicate Text Detection Using Frequency-Biased Signatures. In: Lin, X., Manolopoulos, Y., Srivastava, D., Huang, G. (eds.) WISE 2013, Part I. LNCS, vol. 8180, pp. 277–291. Springer, Heidelberg (2013)
6. Pi, B., Fu, S., Wang, W., Han, S.: SimHash-based Effective and Efficient Detecting of Near-Duplicate Short Messages. In: Proceedings of the 2nd Symposium International Computer Science and Computational Technology
7. Zhang, Y.H., Zhang, F.: Research on New Algorithm of Topic-Oriented Crawler and Duplicated Web Pages Detection. In: Intelligent Computing Theories and Applications 8th International Conference, ICIC, Huangshan, China, pp. 25–29 (2012)
8. Figuerola, C.G., Díaz, R.G., Berrocal, J.L.A., Rodríguez, A.F.Z.: Web Document Duplicate Detection using Fuzzy Hashing. In: Trends in Practical Applications of Agents and Multiagent Systems, 9th International Conference on Practical Applications of Agents and Multiagent Systems, vol. 90, pp. 117–125 (2011)
9. Tan, P.N., Kumar, V., Steinbach, M.: Introduction to Data Mining. Pearson
10. Theobald, M., Siddharth, J., Paepcke, A.: SpotSigs: Robust and Efficient Near Duplicate Detection. In: Large Web Collections in (SIGIR 2008), pp. 20–24 (2008)
11. Rehurek, R., Sojka, P.: Software Framework for Topic Modeling with Large Corpora. In: Proceedings of LREC workshop New Challenges for NLP Frameworks, pp. 46–50. University of Malta, Valleta (2010)

12. Robertson, S.: Understanding Inverse Document Frequency: On theoretical arguments for IDF. Journal of Documentation 60(5), 503–520
13. Golub, G.H., Reinsch, C.: Singular value decomposition and least square solutions. Numerische Mathematik 10. IV 5(14), 403–420 (1970)
14. Celikik, M., Bast, H.: Fast error-tolerant search on very large texts. In: SAC 2009 Proceedings of the ACM Symposium on Applied Computing, pp. 1724–1731 (2009)

Decision Tree Techniques Applied on NSL-KDD Data and Its Comparison with Various Feature Selection Techniques

H.S. Hota[1] and Akhilesh Kumar Shrivas[2]

[1] Guru Ghasidas Central University, Chhattisgarh, India
[2] Dr. C.V. Raman University, Chhattisgarh, India
{profhota,akhilesh.mca29}@gmail.com

Abstract. Intrusion detection system (IDS) is one of the important research area in field of information and network security to protect information or data from unauthorized access. IDS is a classifier that can classify the data as normal or attack. In this paper, we have focused on many existing feature selection techniques to remove irrelevant features from NSL-KDD data set to develop a robust classifier that will be computationally efficient and effective. Four different feature selection techniques :Info Gain, Correlation, Relief and Symmetrical Uncertainty are combined with C4.5 decision tree technique to develop IDS . Experimental works are carried out using WEKA open source data mining tool-and obtained results show that C4.5 with Info Gain feature selection technique has produced highest accuracy of 99.68% with 17 features, however result obtain in case of Symmetrical Uncertainty with C4.5 is also promising with 99.64% accuracy in case of only 11 features . Results are better as compare to the work already done in this area.

Keywords: Decision Tree (DT), Feature Selection(FS), Intrusion Detection System(IDS).

1 Introduction

The ever increasing size of data in computer has made information security more important. Information security means protecting information and information systems from unauthorized access. Information security becomes more important as data are being accessed in a network environment and transferred over an insecure medium. Many authors have worked on this issue and applied feature selection technique on NSL-KDD data set for multiclass problem. Mukharjee, S. *et al.*[9] have proposed new feature reduction method: Feature Validity Based Reduction Method (FVBRM) applied on one of the efficient classifier Naive Bayes and achieved 97.78%accuracy on reduced NSL-KDD data set with 24 features. This technique gives better performance as compare to Case Based Feature Selection (CFS), Gain Ratio (GR) and Info Gain Ratio (IGR) to design IDS. Panda, M.*et al.*[10] have suggested hybrid technique with combination of random forest, dichotomies and ensemble of balanced nested dichotomies (END) model and achieved detection rate 99.50% and low false alarm

M.K. Kundu et al. (eds.), *Advanced Computing, Networking and Informatics - Volume 1*, Smart Innovation, Systems and Technologies 27,
DOI: 10.1007/978-3-319-07353-8_24,

rate 0.1% which are quite encouraging in comparison to all other models. Imran, H. M.*et al.*[11] have proposed hybrid technique of Linear Discriminant Analysis (LDA) algorithm and Genetic Algorithm (GA) for feature selection . They proposed features selection technique applied on radial basis function with NSL-KDD data set to develop a robust IDS. The different feature subsets are applied on RBF model which produces highest accuracy of 99.3% in case of 11 features. Bhavsar, Y. B.*et al.* [12] have discussed different support vector machine (SVM) kernel function as Gaussian radial basis function (RBF) kernel, polynomial kernel, sigmoid kernel to develop IDS. They have compared accuracy and computation time of different kernel function as classifier, authors suggested Gaussian RBF kernel function as the best kernel function ,which achieved highest accuracy of 98.57% with 10- fold cross validation. Amira, S.A.S.*et al.* [8] have proposed Minkowski distance technique based on genetic algorithm to develop IDS to detect anomalies. The proposed Minkowski distance techniques applied on NSL-KDD data produces higher detection rate of 82.13% in case of higher threshold value and smaller population. They have also compared their results with Euclidean distance. In a recent work by Hota, H. S. *et al.*[15] , a binary class based IDS using random forest technique combined with rank based feature selection was investigated. Accuracy achieved in case of above technique was 99.76% with 15 features.

It is observed from the literature that feature selection is an important issue due to high dimensionality feature space of IDS data. This research work explores many existing feature selection techniques to be applied on NSL-KDD data set to reduce irrelevant features from data set, so that IDS will become computationally fast and efficient. Experimental work is carried out using WEKA (Waikato Environment for Knowledge Analysis)[13] and Tanagra [14] open source data mining tool and obtained results revealed that C4.5 with Info Gain feature selection technique and C4.5 with Symmetrical Uncertainty feature selection technique produces highest accuracy respectively with 17 and 11 features and is highest among the entire research outcome reviewed so far.

2 Methods and Materials

Various methods (Techniques) based on data mining techniques and benchmark data (material) used in this research work are explained in detail as below:

2.1 Decision Tree

Decision tree [2] is probably the most popular data mining technique commonly used for classification. The principle idea of a decision tree is to split our data recursively into subsets so that each subset contains more or less homogeneous states of our target variable. At each split in the tree, all input attributes are evaluated for their impact on the predictable attribute. When this recursive process is completed, a decision tree is formed which can be converted in simple If –Then rules. We have used various decision tree techniques like C4.5, ID3 (Iterative Dichotomizer 3), CART (Classification and Regression Tree), REP Tree and decision list.

Among the above decision tree techniques, C4.5 is more powerful and produces better results for many classification problems. C4.5 [1] is an extension of ID3 that accounts for unavailable values, continuous attribute value ranges, pruning of decision trees and rule derivation. In building a decision tree, we can deal with training set that have records with unknown attributes values by evaluating the gain, or the gain ratio, for an attribute values are available. We can classify the records that have unknown attribute value by estimating the probability of the various possible results. Unlike CART, which generates a binary decision tree, C4.5 produces tree with variable branches per node. When a discrete variable is chosen as the splitting attribute in C4.5, there will be one branch for each value of the attribute.

2.2 Feature Selection

Feature selection [5]is an optimization process in which one tries to find the best feature subset from the fixed set of the original features, according to a given processing goal and feature selection criteria. A solution of an optimal feature selection does not need to be unique. Different subset of original features may guarantee accomplishing the same goal with the same performance measure. An optimal feature set will depend on data, processing goal, and the selection criteria being used. In this experiment, we have used following ranking based feature selection techniques:

Information gain [4] is a feature selection technique which measure features of data set based on its ranking. This measure is based on pioneering work by Claude Shannon on information theory, which studied the value or information content of messages. Let node *N* represent or hold the tuples of partition D. The attribute with the highest information gain is chosen as the splitting attribute for node *N*. This attribute minimizes the information needed to classify the tuples in the resulting partitions and reflects the least randomness or impurity in these partitions.

Correlation based feature selection (CFS) [7] is used to determine the best feature subset and is usually combined with search strategies such as forward selection, backward elimination, bi-directional search, best-first search and genetic search. Among given features, it finds out an optimal subset which is best relevant to a class having no redundant feature. It evaluates merit of the feature subset on the basis of hypothesis: Good feature subsets contain features highly correlated with the class, yet uncorrelated to each other. This hypothesis gives rise to two definitions. One is feature class correlation and another is feature -feature correlation. Feature-class correlation indicates how much a feature is correlated to a specific class while feature-feature correlation is the correlation between two features.

ReliefF [6] is an extension of the Relief algorithm that can handle noise and multiclass data sets. ReliefF uses a nearest neighbor implementation to maintain relevancy scores for each attribute. It defines a good discriminating attribute as the attribute that has the same value for other attributes in the same class and different from attribute values in different classes. ReliefF algorithm is stronger and can handle incomplete and noisy data as well as multiclass dataset.

Symmetrical Uncertainty [7] is another feature selection method that was devised to compensate for information gain's bias towards features with more values. It capitalizes on the symmetrical property of information gain. The symmetrical uncertainty between features and the target concept can be used to evaluate the goodness of features for classification.

2.3 NSL-KDD Data

Many benchmark data sets related to IDS are available in repository sites .One of the data set publicly available for the development and evaluation of IDS is NSL-KDD data set [3]. This data set is inherent data set of KDD99. One of the problem with KDD99 data set is the huge number of redundant records, which causes the learning algorithms to be biased towards the frequent records, and thus prevent them from learning infrequent records which are usually more harmful to networks such as U2R and R2L attacks. In addition, the existence of these repeated records in the test set will cause the evaluation results to be biased by the methods which have better detection rates on the frequent records. Data set has total 25192 samples consisting 41 features with four different types of attack and normal data. Detail of samples according to attack class is shown in Table 1.

Table 1. Different attacks and Normal data along with sample size

Class type	Number of instance
Normal	13449
DoS	9234
R2L	209
U2R	11
Probe	2289
Total	25192

From the above table it is clear that data set is highly unbalanced or there is no uniform distribution of samples. Number of samples of DoS type attack is high (9234 samples) while on the other hand U2R type of attack has only 11 samples. This unbalanced distribution of samples may create problem during training of any data mining based classification model.

In order to verify efficiency and accuracy of IDS model, data set is divided into various partitions based on training and testing samples.

3 Experimental Work

The experimental work is carried out using WEKA [13] and TANAGRA [14] open source data mining software. This entire experimental work is divided into two parts: First building multiclass classifier and second applying feature selection technique. Multiclass classifier for IDS is developed based on various decision tree techniques and finally a best model (C4.5) with reduced feature subset is selected.

As explained NSL-KDD data set is divided into three different partitions as 60-40%, 80-20%, and 90-10% as training–testing samples. Decision tree based models are first trained and then tested on various partitions of data set. Accuracy of different models with different partitions is shown in Table 2.Accuracy of model is varying from one partition to other partition, highest accuracy of 99.56% was achieved in case of C4.5 with 90-10% partition with all available features.

Table 2. Accuracy of different partitions of data set

Model	60-40% partition	80-20% partition	90-10% partition
C4.5	99.39	99.52	**99.56**
ID3	97.75	97.92	97.86
CART	99.29	99.33	99.05
REP Tree	98.55	98.86	99.12
Decision Table	98.44	98.70	98.69

Table 3. Obtained rank in case of various ranking based feature selection techniques

Feature selection Technique	Features with rank (In descending order)
Info Gain	5,3,6,4,30,29,33,34,35,38,12,39,25,23,26,37,32,36,31,24, 41,2,27,40,28,1,10,8,13,16,19,22,17,15,14,18,11,7,9,20,21
Correlation	29,39,38,25,26,33,34,4,12,23,32,3,35,40,27,41,28,36,31,3 7,2,30,8,1,22,19,10,24,14,15,6,17,16,13,18,11,5,9,7,20,21
ReliefF	3,29,4,36,32,38,12,33,2,34,39,23,26,35,40,30,31,8,24,25, 37,27,41,28,10,22,1,6,14,11,13,15,19,18,16,5,17,9,7,20,21
Symmetrical Uncertainty	4,30,5,25,26,39,38,6,29,12,3,35,34,33,23,37,36,32,31,40, 2,41,27,24,1,28,10,22,8,13,16,14,17,19,11,15,9,18,7,20,21

Feature selection is an optimization process in which we can find the best feature subset from original data set. Best model obtained in case of 90-10% partition using C4.5 technique is selected for applying feature selection. Various ranking based feature selection techniques like Info Gain , Correlation, ReliefF and Symmetrical Uncertainty are applied on C4.5 model. Table 3 shows rank obtained after applying various feature selection techniques in descending order (Left to right) ,these features are then removed one by one and applied on C4.5 model. Rank obtained in case of different feature selection techniques are different, hence results obtained in case of all these techniques will be different in terms of accuracy. Table 4 shows accuracy of model with different feature selection techniques with reduced subset of features. C4.5 with Info Gain and C4.5 with Symmetrical Uncertainty are producing best accuracy as 99.68% (With 17 features) and 99.64% (With 11 feature) respectively.

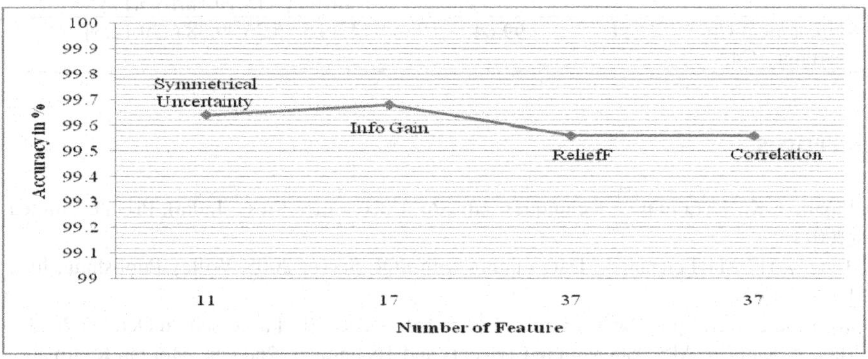

Fig. 1. Accuracy graph in case of different feature selection techniques with C4.5

Both the models are competitive and can be accepted for the development of IDS. Fig. 1 shows pictorial view of accuracy obtained in case of different feature selection techniques combined with C4.5 model.

4 Conclusion

Information security is a crucial issue due to exchanging of huge amount of data and information in day to day life .Intrusion detection system (IDS) is a way to secure our host computer as well as network computer from intruder. Efficient and computationally fast IDS can be developed using decision tree based techniques. On the other hand irrelevant features available in the data set must be reduced. In this research work ,an attempt has been made to explore various decision tree techniques with many existing feature selection techniques. Four different feature selection techniques are tested on CART, ID3, REP, Decision table and C4.5 .Two rank based feature selection techniques: Info Gain and Symmetrical Uncertainty along with C4.5 produces 99.68% and 99.64% accuracy with 17 and 11 features respectively. Results obtained in these two cases are quite satisfactory as compare to other research work already done in this field.

In future a feature selection technique based on statistical measures will be developed and will be tested on more than one benchmark data related to intrusion; also proposed feature selection technique will be compared with all other existing feature selection techniques.

Table 4. Selected features in case of C4.5 model using various feature selection techniques

Feature Selection Technique	No. of features	Accuracy	Selected Features
Info Gain-C4.5	**17**	**99.68**	{5,3,6,4,30,29,33,34,35,38,12,39,25,23,26,37,32}
Correlation-C4.5	37	99.56	{29,39,38,25,26,33,34,4,12,23,32,3,35,40,27,41,28,36,31,37,2,30,8,1,22,19,10,24,14,15,6,17,16,13,18,11,5}
ReliefF-C4.5	37	99.56	{3,29,4,36,32,38,12,33,2,34,39,23,26,35,40,30,31,8,24,25,37,27,41,28,10,22,1,6,14,11,13,15,19,18,16,5,17}
Symmetrical Uncertainty-C4.5	**11**	**99.64**	{4,30,5,25,26,39,38,6,29,12,3}

References

1. Pujari, A.K.: Data mining techniques, 4th edn. Universities Press (India), Private Limited (2001)
2. Tang, Z.H., MacLennan, J.: Data mining with SQL Server 2005. Willey Publishing, Inc., USA (2005)
3. Web sources, `http://www.iscx.info/NSL-KDD/` (last accessed on October 2013)
4. Han, J., Kamber, M.: Data Mining Concepts and Techniques, 2nd edn. Morgan Kaufmann, San Francisco (2006)

5. Krzysztopf, J.C., Pedrycz, W., Roman, W.S.: Data mining methods for knowledge discovery, 3rd edn. Kluwer Academic Publishers (2000)
6. Zewdie, M.: Optimal feature selection for Network Intrusion Detection System: A data mining approach. Thesis, Master of Science in Information Science, Addis Ababa University, Ethiopia (2011)
7. Parimala, R., Nallaswamy, R.: A study of a spam E-mail classification using feature selection package. Global Journals Inc. (USA) 11, 44–54 (2011)
8. Aziz, A.S.A., Salama, M.A., Hassanien, A., Hanafi, S.E.-O.: Artificial Immune System Inspired Intrusion Detection System Using Genetic Algorithm. Informatica 36, 347–357 (2012)
9. Mukherjee, S., Sharma, N.: Intrusion detection using Bayes classifier with feature reduction. Procedia Technology 4, 119–128 (2012)
10. Panda, M., Abrahamet, A., Patra, M.R.: A hybrid intelligent approach for network intrusion detection. Proceedia Engineering 30, 1–9 (2012)
11. Imran, H.M., Abdullah, A.B., Hussain, M., Palaniappan, S., Ahmad, I.: Intrusion Detection based on Optimum Features Subset and Efficient Dataset Selection. International Journal of Engineering and Innovative Technology (IJEIT) 2, 265–270 (2012)
12. Bhavsar, Y.B., Waghmare, K.C.: Intrusion Detection System Using Data Mining Technique: Support Vector Machine. International Journal of Emerging Technology and Advanced Engineering 3, 581–586 (2013)
13. Web sources, `http://www.cs.waikato.ac.nz/~ml/weka/` (last accessed on October 2013)
14. Web sources, `http://eric.univ-lyon2.fr/~ricco/tanagra/en/tanagra` (last accessed on October 2013)
15. Hota, H.S., Shrivas, A.K.: Data Mining Approach for Developing Various Models Based on Types of Attack and Feature Selection as Intrusion Detection Systems (IDS). In: Intelligent Computing, Networking, and Informatics. AISC, vol. 243, pp. 845–851. Springer, Heidelberg (2014)

Matra and *Tempo* Detection for INDIC *Tala*-s

Susmita Bhaduri, Sanjoy Kumar Saha, and Chandan Mazumdar

Centre for Distributed Computing
Department of Computer Science and Engineering
Jadavpur University, Kolkata, India
{susmita.sbhaduri,chandan.mazumdar}@gmail.com,
sks_ju@yahoo.co.in

Abstract. In the context of Indian classical music, *matra* and Tempo are two important parameters to represent the rhythmic pattern of a composition. Detection of such parameters enables organized archiving and retrieval of music data. In this work, a simple methodology for detecting *matra* and Tempo is presented. It is based on the extraction of basic beat pattern that gets repeated in the signal. Such pattern is identified by processing the amplitude envelope of a music signal. *matra* and Tempo are detected from the extracted beat pattern. Experiment with number of audio clip of *tablaa* signal for various *tala* and Tempo indicates the effectiveness of the proposed methodology.

Keywords: *matra* Detection, Tempo Detection, Indian Classical Music.

1 Introduction

tala is the term used in Indian classical music for the rhythmic pattern of any composition and for the entire subject of rhythm, roughly corresponds to metre in Western music. They are rhythmic cycles that group long measures (*anga*-s). In theory, there are 360 *tala*-s which range from 3 to 108 *matra*-s, although only 30 to 40 are in use today [1].

Musical metadata, or information about the music, is of growing importance in a rapidly expanding world of digital music analysis and consumption. To make the desired music easily accessible to the consumer, it is important to have meaningful and robust descriptions of music that are amenable to search.

An efficient music classification system can serve as the foundation for various applications like music indexing, content based music information retrieval, music content description, and music genre classification. With Electronic Music Distribution, as the music catalogues are becoming huge, it is becoming almost impossible to manually check each of them for choosing the desired genre song. In order to label the data set, one crude option is labelling them manually which is of huge cost. Thus, automatic genre discrimination is becoming more and more popular area of research. In this context, Music Information Retrieval for INDIC music based on *matra,tala* and Tempo, can act as an elementary step for content based music retrieval system and organizing the digital library for INDIC music.

M.K. Kundu et al. (eds.), *Advanced Computing, Networking and Informatics - Volume 1*, Smart Innovation, Systems and Technologies 27,
DOI: 10.1007/978-3-319-07353-8_25,

In this work, we have dealt with the signals of *tablaa*, an Indian drum instrument. *matra* and Tempo are automatically detected from the signal to enable organized archiving of such audio data and fast retrieval based on those fundamental rhythmic parameters. The rest of the paper is organized as follows. Definition and details of *matra* and *tala* are presented in Section 2. A survey of past work is placed in Section 3. Section 4 elaborates the proposed methodology. Experimental results are presented in Section 5. The paper is concluded in Section 6.

2 *Matra* and *Tala* in INDIC Music

Each *tala* consists of a fixed number of beats. The beats have different degrees of emphasis within a *tala* and those are represented with hand claps, hand waves and movements of the fingers. A *bol* is an onomatopoeic syllable representing the drum sounds. Each *tala* has a *theka*, a set of *bol*-s that delineate the standard groove representing the *tala*. Each repeated cycle of a *taala* is called an *avartan* [1]. Corresponding *bols* in each cycle of a *taala* have very similar amplitude.

The details of the *tala*-s include number of *matra* or beat pattern, *bol* and *anga* (or long measures and *tali*-s) for a single cycle of each *tala* [1]. *matra*/beat pattern for few common *tala* of Indian classical music are described in the Tables 1 - 4. Different rows of the tables depict the following.

- *tali*-s or important beats at the start of each *anga* is depicted in the first row.
- The numbers in the second row are *matra*.
- In the third row *bol*-s like *dhi, dha, dhin* are shown.
- In the last row *anga* or long measures (set of *matra*-s) are presented.

Table 1. Description of *tintal*

tali	+	-	-	-	2	-	-	-	0	-	-	-	3	-	-	-
matra	1	2	3	4	5	6	7	8	9	10	11	12	13	14	15	16
bol	dha	dhin	dhin	dha	dha	dhin	dhin	dha	na	tin	tin	na	tete	dhin	dhin	dha
anga	1	-	-	-	2	-	-	-	3	-	-	-	4	-	-	-

tintal is described in Table 1. It has 16 *matra*-s with 4-4-4-4 beat pattern divided in 4 *anga*-s. *kaharba*, as dpicted in Table 2 has 8 *matra*-s with 4-4 beat pattern divided in 2 *anga*-s. *dadra* (see Table 3) consists of 6 *matra*-s with 3-3 beat pattern divided in 2 *anga*-s. For *jhaptal* as shown in Table 4, 10 *matra*-s are there with 2-3-2-3 beat pattern divided in 4 *anga*-s.

Table 2. Description of *kaharba*

tali	+	-	-	-	0	-	-	-
matra	1	2	3	4	5	6	7	8
bol	dha	ge	na	ti	na	ka	dhi	na
anga	1	-	-	-	2	-	-	-

Table 3. Description of *dadra*

tali	+	-	-	0	-	-
matra	1	2	3	4	5	6
bol	dha	dhi	na	na	ti	na
anga	1	-	-	2	-	-

3 Past Work

Previous work in this context includes the areas like Recognition of *tablaa bol*-s and *tablaa* strokes as well as *tablaa* acoustics.

Tonal Quality of *tablaa* is first discussed by C. V. Raman [2]. The importance of the first three to five harmonics which are derived from the drumhead's vibration mode was highlighted. Bhat [3] developed a mathematical model of the membrane's vibration modes that could well be applied to the *tablaa*.

Malu and Siddharthan [4] confirmed C.V. Raman's observations on the harmonic properties of Indian drums, and the *tablaa* in particular. They attributed the presence of harmonic overtones to the "central loading"(black patch) in the center of the *dayan* (the *gaab*). Chatwani [5] developed a computer program based on linear predictive coding (LPC) analysis to recognize spoken *bol*-s. An acoustic and perceptual comparison of *tablaa bol*-s (both, spoken and played) was done by Patel *et al.* [6]. It was found that spoken *bol*-s do indeed have significant correlations in terms of acoustical features (e.g. spectral centroid, flux) with played *bol*-s. It enables even an untrained listeners to match syllables to the corresponding drum sound. This provides strong support for the symbolic value of *tablaa bol*-s in the North Indian drumming tradition.

Gillet *et al.* [7] worked on *tablaa* stroke recognition. Their approach follows three steps: stroke segmentation, computation of relative durations (using beat detection techniques), and stroke recognition. Transcription is performed with a

Table 4. Description of *jhaptal*

tali	+	-	2	-	-	0	-	3	-	-
matra	1	2	3	4	5	6	7	8	9	10
bol	dhi	na	dhi	dhi	na	ti	na	dhi	dhi	na
anga	1	-	2	-	-	3	-	4	-	-

Hidden Markov Model (HMM). The advantage of their model is its ability to take into account the context during the transcription phase. A *bol* recognizer with cepstrum based features and and HMM has been presented in [8]. Essl *et al.* [9] applied the theory of banded wave-guides to highly inharmonic vibrating structures. Software implementations of simple percussion instruments like the musical saw, glasses and bowls along-with the model for *tablaa* were presented. The work of Gillet *et al.* [7] was extended in the work of Chordia [10] by incorporating different classifiers like neural network, decision trees, multivariate Gaussian model. A system that segments and recognizes *tablaa* strokes has been implemented.

Conventional works in rhythm detection relies on frame (audio segment of small duration) based autocorrelation techniques which is relatively costly. In this work, we present a simple scheme that relies on perceptual features like the amplitude envelope of recorded clips of electronic *tablaa.* This proposed system detects the number of *matra*-s and Tempo (in Beats per Minute *i.e.* BPM) of a *tala.* Detection is made based on the distribution of amplitude peaks and matching of repetitive pattern present in the clip with the standard patterns for *tala*-s of INDIC music.

4 Proposed Methodology

tabla is a two-part instrument. The left part is called *bayan* and the right part *dayan.* In *tablaa* there is a central loading (black patch) at the center of the *dayan.* When played it gives rise to various harmonic overtones and that forms the characteristics of a *tablaa* signal. Depending on the *bol*-s used in the *theka* of a *tala,* number of harmonics and their amplitude envelope varies. This results into different repetitive pattern for different *tala*-s. Based on this understanding, the proposed methodology revolves around the detection of onset pattern in the signal. Thus, onset detection is the fundamental step. As suggested in [11], [12], it follows the principle that an onset detection algorithm should follow the human auditory system by treating frequency bands separately and finally combining the results. The major steps to identify the *matra* and Tempo as follows.

- Extraction of peaks from amplitude envelope
  - Decompose the signal into different frequency bands.
  - Create the envelope for each band and sum them up.
  - Extract the amplitude peaks
- Detect *matra* and Tempo.

MIRtoolbox [13] is used to create the amplitude envelope from the audio clip and subsequent extraction of peaks. As the peaks correspond to the beats, by analysing the inherent patterns in the peaks, *matra*-s and Tempos are detected.

4.1 Extraction of Peaks from Amplitude Envelope

One common way of estimating the rhythmic pulsation or *matra,* is based on auditory modelling. The audio signal is first decomposed into auditory channels using Equivalent Rectangular Bandwidth (ERB) filterbank, meaning that the width of each band is determined by a particular psycho-acoustical law. A bank of 20 filters is used that matches the actual process of human perception, corresponding to the distribution of frequencies into critical bands in the cochlea [11], [12]. Fig. 1(b) shows the decomposed signal corresponding to the *tablaa* signal in Fig. 1(a).

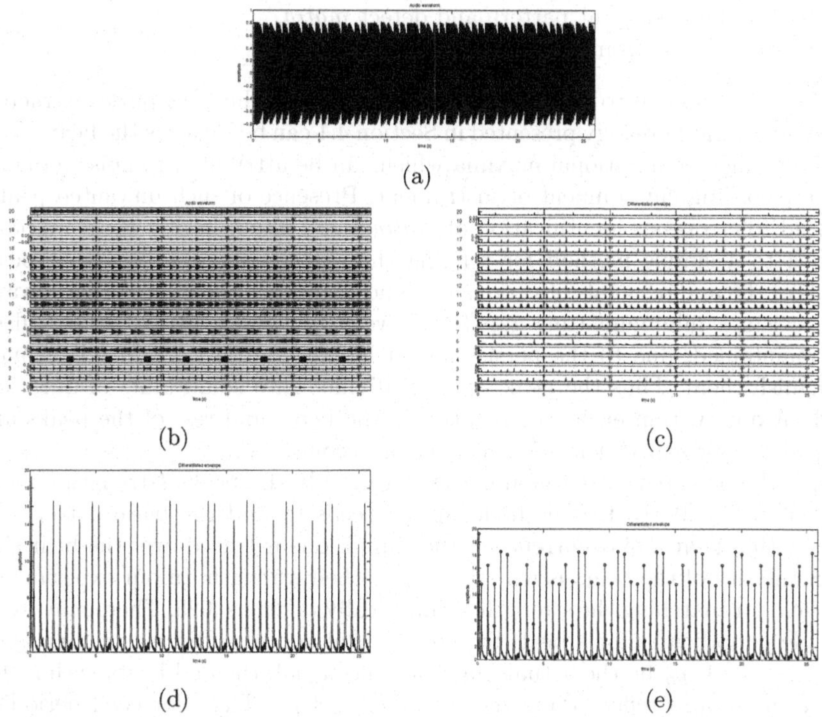

Fig. 1. Peak extraction from amplitude envelope: (a) Time varying signal of *tablaa*, (b) Decomposed signal (c) Differentiated envelope for the bands (d) Overall differentiated signal, (e) Detected peaks

For each band, the differential envelope is then obtained by discarding the negative part through half wave rectification. Such differential envelope is shown in Fig. 1(c). All such 20 envelopes are summed up to reconstruct the overall differential signal which reflects the global outer shape of original signal (see Fig. 1(d)). Peaks are detected from this summed up envelope by considering only the local maxima. Detected peaks are circled in Fig. 1(e).

4.2 Detection of *Matra* and Tempo

A *tala* has a specific no of *matra*-s which represents a basic beat pattern. In an audio clip of specific *tala*, corresponding beat pattern is repeated. The number of peaks (may be of different magnitude) in the amplitude envelope of a *tala* represents the no of *matra*-s of the corresponding basic beat pattern. The task of detecting the *matra*-s of an audio clip maps on to finding the basic beat pattern and thereafter finding out the number of beats within it. Once the beats are identified, Tempo can be detected very easily as it stands for number of beats in a minute. The broad steps are as follows.

- Detect beats from extracted amplitude peaks
- Identify the basic beat pattern and detect *matra*
- Measure Tempo from detected beats

A beat corresponds to high amplitude in the signal. Thus, the peaks extracted following the methodology presented in Section 4.1 can be taken as the beats. But those may include additional maxima which can be attributed to noise induced by the recording environment or instrument. Presence of such unwanted peaks will lead to erroneous identification of *matra* and Tempo. In our work, we have adhered to a simple method for refining the extracted peaks. Studying peak signals of the audio signals of various *tala* and Tempo, it is observed that always there exists a minimum time gap (t) between two consecutive beats. Unless it is maintained, due to the persistence of human auditory system, it cannot distinguish them. Thus the peak signal is divided into small units of duration t and maxima within each unit is taken as the beats and rest of the peaks are dropped. In our experiment t is taken as 0.1 second.

With the detected beat signal consisting of only the peaks (strength is normalized within $[0,1]$), the methodology proceeds to find the basic pattern. It starts with a beat and occurrence of the similar one is traced over the beat signal. Similarity of the beats is with respect to their strength. In our experiment, threshold for matching the strength is empirically taken as .025. There may exist multiple periodicities for a beat. Suppose, a beat repeats after every interval of $p_1, p_2, \cdots, p_n$. If p_a be the actual *matra* for the signal, then all beats within the interval must also reflect the same periodicity of p_a. Thus, for each periodicity $p_i \in \{p_1, p_2, \cdots, p_n\}$, repetitiveness of the intermediate beats are verified to determine p_a. Once actual periodicity is obtained, then number of beats in the that interval denotes *matra*.

From the beat signal Tempo is extracted simply as $\frac{N}{T}$ where, N stands for number of beats in the signal and T is the signal duration. Normally it is expressed in beats per minute (BPM).

5 Experimental Results

In our experiment we have worked with four different *tala*-s namely, *dadra*, *kaharba*, *tintal* and *jhaptal*. They are of different *matra*. Audio clips of different *tala* are obtained by recording the signals of electronic *tablaa* instrument. Audio clips are of duration 10 to 20 seconds. Different signals for each *tala* are generated for various Tempo. The detailed description of the data used is shown in Table 5.

Table 5. Description of data

tala	*matra*	Tempo (in BPM)	No. of Clips
dadra	6	120	30
		150	30
		200	30
kaharba	8	120	13
		150	30
		200	30
tintal	16	120	11
		130	11
		150	11
jhaptal	10	120	27

Table 6. Performance of *matra* detection (in %)

tala-s	Accuracy
dadra	100
kaharba	100
tintal	100
jhaptal	100

Table 7. Performance of Tempo detection (in %)

tala-s	Accuracy
dadra	87.77
kaharba	100.00
tintal	81.80
jhaptal	100.00

Thus, the data reflects variation in terms of duration, Tempo and *tala* to establish the applicability of proposed methodology in identifying *matra* and Tempo.

Table 6 and 7 show the accuracy in detecting *matra* and Tempo respectively. In Tempo detection, a tolerance of $\pm 2\%$ has been considered for matching as it is more affected by the presence of noise. In general, the experimental result indicates the effectiveness of proposed methodology.

6 Conclusion

In this work, a simple methodology to detect two important rhythmic parameters like *matra* and Tempo of music signal is proposed. The methodology works with the perceptual features based on amplitude envelope of recorded audio clips of electronic *tablaa* of different *tala* and Tempo. Then from the envelope the peaks are extracted and refined subsequently to get the beat signal. Tempo can be readily obtained from it. From the beat signal, the basic repetitive beat pattern is identified which enables the detection of *matra*. Experiment with number of *tala* of various Tempo indicates that the performance of proposed methodology is satisfactory. The dataset may be enhanced to include other *tala*-s and the methodology can be extended to develop a framework for *bol* driven identification of *theka* for Indian Classical Music system.

References

1. Sarkar, M.: A real-time online musical collaboration system for indian percussion, 33–36 (2007)
2. Raman, C.V., Kumar, S.: Musical drums with harmonic overtones. Nature, 453–454 (1920)
3. Bhat, R.: Acoustics of a cavity-backed membrane: The indian musical drum. Journal of the Acoustical Society of America 90, 1469–1474 (1991)
4. Malu, S.S., Siddharthan, A.: Acoustics of the indian drum. Technical report, Cornell University (2000)
5. Chatwani, A.: Real-time recognition of tabla bols. Technical report, Senior Thesis, Princeton University (2003)
6. Patel, A., Iversen, J.: Acoustic and perceptual comparison of speech and drum sounds in the north indian tabla tradition: An empirical study of sound symbolism. In: Proc. 15th International Congress of Phonetic Sciences, ICPhS (2003)
7. Gillet, O., Richard, G.: Automatic labelling of tabla signals. In: Proceedings of the 4th International Conference on Music Information Retrieval, ISMIR 2003 (2003)
8. Samudravijaya, K., Shah, S., Pandya, P.: Computer recognition of tabla bols. Technical report, Tata Institute of Fundamental Research (2004)
9. Essl, G., Serafin, S., Cook, P., Smith, J.: Musical applications of banded waveguides. Computer Music Journal 28, 51–63 (2004)
10. Chorida, P.: Segmentation and recognition of tabla strokes. In: Proceedings of 6th International Conference on Music Information Retrieval, ISMIR 2005 (2005)
11. Scheirer: Tempo and beat analysis of acoustic musical signals. Journal of the Acoustical Society of America 103, 588–601 (1996)
12. Klapuri, A.: Sound onset detection by applying psychoacoustic knowledge. In: Proc. Int. Conf. on Acoustics, Speech, and Signal Processing, pp. 3089–3092 (1999)
13. Lartillot, O., Toiviaine, P.: A matlab toolbox for musical feature extraction from audio. In: Proc. Int. Conference on Digital Audio Effects (2007)

Modified Majority Voting Algorithm towards Creating Reference Image for Binarization

Ayan Dey[1], Soharab Hossain Shaikh[1], Khalid Saeed[2], and Nabendu Chaki[3]

[1] A. K. Choudhury School of Information Technology, University of Calcutta, India
[2] Faculty of Physics and Applied Computer Science,
AGH University of Science and Technology,
Poland
[3] Department of Computer Science & Engineering, University of Calcutta, India
deyayan9@gmail.com, {soharab.h.shaikh,nabendu}@ieee.org,
saeed@agh.edu.pl

Abstract. The quantitative evaluation of different binarization techniques to measure their comparative performance is indeed an important aspect towards avoiding subjective evaluation. However, in majority of the papers found in the literature, creating a reference image is based on manual processing. These are often highly subjective and prone to human error. No single binarization technique so far has been found to produce consistently good results for all types of textual and graphic images. Thus creating a reference image indeed remains an unsolved problem. As found in the majority voting approach, a strong bias, due to poor computation of threshold by one or two methods for a particular image, has often had an adverse effect in computing the threshold for the reference image. The improvement proposed in this paper helps eliminate this bias to a great extent. Experimental verification using images from a standard database illustrates the effectiveness of the proposed method.

Keywords: Image binarization, global thresholding, reference image, quantitative evaluation.

1 Introduction

Image binarization is a fundamental step in image processing. It is required in most of the applications related to computer vision and pattern recognition. Use of binary image reduces computational load for all applications. Binarization takes a major role to identify region of interest in medical images. Applications of binarization include optical character recognition, document analysis, material inspection etc. One recent application of binarization for finger vein images has been found in [3]. An adaptive binarization algorithm using ternary entropy-based approach has been proposed in [4]. The method classifies noise into two categories which are processed by binary morphological operators, shrink and swell filters, and graph searching strategy. In [5] a method for binarization based on text blòck extraction has been presented which uses a combination between a preprocessing step and a localization step. In [6] a local

M.K. Kundu et al. (eds.), *Advanced Computing, Networking and Informatics - Volume 1*,
Smart Innovation, Systems and Technologies 27,
DOI: 10.1007/978-3-319-07353-8_26, © Springer International Publishing Switzerland 2014

thresholding algorithm for handwritten historical document images has been presented. A local thresholding approach is described in [7] using edge information. The document image binarization technique presented in [8] separates text from background in badly illuminated document images based on background estimation by using morphological closing operation. A comprehensive survey of image thresholding techniques for binarization can be found in [11]. Effect of different binarization techniques on different category of images from standard database is analyzed in [19].

In the process of binarization of a gray-scale image, it is required to select a particular gray-level known as the threshold. Based on this threshold value, the gray-scale image is converted to a bi-level image where the object is represented with gray value 0 (black) and the background with 1 (white) or vice versa.

Let us consider a digital image of size *NxN*. Let (x, y) be the spatial coordinates of a pixel in the image and $G = \{0, 1,\ldots, L-1\}$ be a set of positive integers representing the gray-values, where L is 256, the total number of possible gray-levels. $0 \leq G \leq 255$ supposing 8-bit gray-scale image. An image mapping function f is defined by f: $NxN \rightarrow G$. The intensity of a pixel having the spatial coordinates (x, y) is denoted as $f(x, y)$. Let $t \in G$ be a threshold and $B = \{0, 1\}$ be a pair of binary gray levels. The thresholding function f at gray level t is a binary image function f_t: $NxN \rightarrow B$ such that,

$$f_t(x,y) \quad = 0 \text{ if } f(x,y) < t$$
$$\quad = 1 \text{ if } f(x,y) \geq t$$

A global thresholding method computes a single threshold for the entire image whereas in local thresholding technique, an image is partitioned into several segments and separate threshold values are determined for each sub-image in a localized spatial region. Performance evaluation is important whenever a new binarization method is proposed. This calls for the requirement of preparing a reference image with which different thresholding techniques should be compared to produce a quantitative result. However, as pointed out in [2] that there is a gap in the literature for preparing a reference image for comparing graphic images. In most of the cases for quantitative comparison, a reference image has been prepared manually which involves human interference. This approach is highly subjective and error-prone, so the quality of the reference image varies with a large extent from person to person. Therefore, an alternative technique is required for an objective measure of the reference image.

In [1-2] a method using majority voting scheme has been proposed. In the context of the present study the authors present a variation of the majority voting scheme considering a deviation parameter. The results are comparable to what is obtained from majority voting [1-2] and in some cases the proposed modified version outperforms the majority voting approach. The following binarization methods have been considered in the majority voting scheme: Otsu [17], Iterative-thresholding [12], Balanced histogram [10], Kapur [14], Johannsen [15], Ridler [16], and Kittler [13].

2 The Proposed Methodology

The proposed method is a modified version of the majority voting approach found in [1- 2] for creating a reference image for quantitative evaluation of different binarization methods. In the present study graphic images have been considered for performance evaluation from USC-SIPI database [18].

The basic concept behind this approach is to find an odd number of binarization methods among all global methods for binarization. Majority voting [2] is used subsequently for the selected subset of methods. The selection of global methods is done by calculating the percentage deviation from the average global thresholds. The percentage deviation for each method from an initial average of all the methods is computed as described in the proposed algorithm. Method having maximum deviation is discarded to eliminate any undesired bias. In fact, more than one method may be discarded at this stage if each of such methods deviates significantly from the initial average. A deviation parameter (DP) has been computed to implement this. If the remaining methods for a particular image are found to be in an odd number, then these methods would be used for creating a reference image using majority voting. Otherwise, *DP* will be decremented in step 1 until an odd number of global methods are selected or the value of *DP* becomes 0. With an odd number of methods left, the majority voting is applied as explained above. If *DP* becomes zero then the results of the binarization method which is closest to the average threshold would be considered as the reference image. In this paper, a total seven binarization methods have been taken into consideration for experimental evaluation.

2.1 Proposed Method

Let there be n global binarization methods $\{M_1, M_2, M_3, ..,M_n\}$ and t_i denotes threshold for method M_i.

Step 1: The average threshold is calculated as

$$t_{avg}= [\sum^{n}_{i=1} t_i]/n.$$

Step 2: The percentage deviation has been calculated as follows:

$$dev_i = [|t_i- t_{avg}|/t_{avg}]*100\%.$$

Step 3: A deviation parameter (DP) is computed as:

$$DP = max\{dev_1, dev_2,..., dev_{n-1}, dev_n\} - 1.$$

Step 4: Select m methods from the initial n methods such that

$$m \leq n\ (dev_i \leq DP\ for\ 1 \leq i \leq m\ )$$

Step 5: If mod (m, 2) *equals to* 0, then

If DP > 1 then

DP = DP - 1;

Go to Step 4;

Else

Select M_p as the reference image where
$dev_p = min\ \{dev_1, dev_2, \ldots, dev_{n-1}, dev_n\}$;

Endif

Else

Compute reference image using majority voting by *m* methods selected in Step 4;

Endif

3 Experimental Verification

Fig. 1 shows a set of test images and the results of the proposed method. The first column represents the original gray-scale images.

Table 1. Performance Evaluation

Images	Proposed			Majority Voting		
	ME	RAE	PSNR	ME	RAE	PSNR
Image-1	62.05	11.09	50.24	62.15	10.7	50.23
Image-2	21.45	41.24	54.85	36.64	66.36	52.53
Image-3	88.28	4.24	48.71	88.28	4.24	48.71
Image-4	89.19	41.99	48.66	89.19	23.57	48.66
Image-5	98.27	12.93	48.24	98.27	12.93	48.24
Average	71.848	22.3	50.14	74.91	23.56	49.67

The manually generated reference images have been presented as the second column. The third column shows the results of majority voting and the results of the proposed method have been presented in the fourth column.A number of performance evaluation techniques have been presented in [9], [11]. In the context of the present paper the authors have taken Misclassification error (ME), Relative Foreground Area error (RAE) and peak signal to noise ratio (PSNR) as in [1], [11] as the performance evaluation metrics.

A reference image has been created manually for the sake of performance evaluation. Performance has been evaluated using images from the USC-SIPI [18] image database. For each image, two reference images have been created; one using the majority voting scheme and another using the proposed method. These two reference images have been compared with the manually generated reference image to calculate the values of different metrics. The results of the experiments have been presented in Table 1.

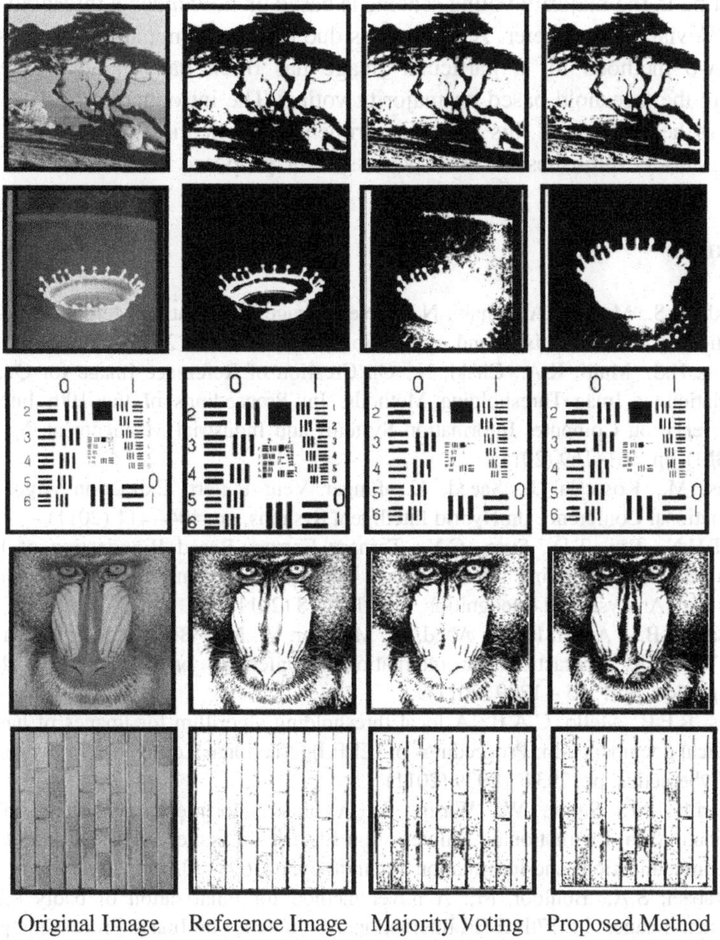

Fig. 1. Test images (shown in each row)

The results presented in Table 1 shows that for some of the images the results of majority voting as well as that of the proposed method are comparable and for some cases the proposed method generates a reference image that is more close to the manually generated reference image.

4 Conclusions

No single binarization technique so far has been found to produce consistent results for all types of textual and graphic images. A method for creating a reference image for quantitative evaluation of binarization method has been proposed in this paper. Besides considering seven distinct binarization techniques for majority voting, a deviation parameter has been calculated. Selection of the binarization methods participating in the majority voting scheme is based on how much a threshold deviates from the deviation parameter. A strong bias due to poor computation of threshold by one or two methods for a particular image has often had an adverse effect in computing the threshold based on majority voting. The introduction of the deviation parameters helps remove this bias. Experimental verification over images form a standard dataset establishes the effectiveness of the proposed method.

References

1. Shaikh, H.S., Maiti, K.A., Chaki, N.: A New Image Binarization Method using Iterative Partitioning. Machine Vision and Application 24(2), 337–350 (2013)
2. Shaikh, H.S., Maiti, K.A., Chaki, N.: On Creation of Reference Image for Quantitative Evaluation of ImageThresholding Methods. In: Proceedings of the 10th International Conference on Computer Information Systems and Industrial Management Applications (CISIM), pp. 161–169 (2011)
3. Waluś, M., Kosmala, J., Saeed, K.: Finger Vein Pattern Extraction Algorithm. In: International Conference on Hybrid Intelligent Systems, pp. 404– 411 (2011)
4. Le, T.H.N., Bui, T.D., Suen, C.Y.: Ternary Entropy-Based Binarization of Degraded Document Images Using Morphological Operators. In: International Conference on Document Analysis and Recognition, pp. 114–118 (2011)
5. Messaoud, B.I., Amiri, H., El. Abed, H., Margner, V.: New Binarization Approach Based on Text Block Extraction. In: International Conference on Document Analysis and Recognition, pp.1205 – 1209 (2011)
6. Neves, R.F.P., Mello, C.A.B.: A local thresholding algorithm for images of handwritten historical documents. In: Proceedings of IEEE International Conference on Systems, Man, and Cybernetics, pp. 2934 – 2939 (2011)
7. Sanparith., M., Sarin., W., Wasin, S.: A binarization technique using local edge information. In: International Conference on Electrical Engineering/Electronics Computer Telecommunications and Information Technology, pp. 698–702 (2010)
8. Tabatabaei, S.A., Bohlool, M.: A novel method for binarization of badly illuminated document images. In: 17th IEEE International Conference on Image Processing, pp. 3573–3576 (2010)
9. Stathis, P., Kavallieratou, E., Papamarkos, N.: An evaluation technique for binarization algorithms. Journal of Universal Computer Scienc 14(18), 3011–3030 (2008)
10. Anjos, A., Shahbazkia, H.: Bi-Level Image Thresholding - A Fast Method. Biosignals 2, 70–76 (2008)
11. Sezgin, M., Sankur, B.: Survey over image thresholding techniques and quantitative performance evaluation. Journal of Electronic Imaging 13(1), 146–165 (2004)
12. Gonzalez, R., Woods, R.: Digital Image Processing. Addison-Wesley (1992)
13. Kittler, J., Illingworth, J.: Minimum error thresholding. Pattern Recognition 19, 41–47 (1986)
14. Kapur, N.J., Sahoo, K.P., Wong, C.K.A.: A new method for gray-level picture thresholding using the entropy of the histogram. Computer Vision, Graphics, and Image Processing 29(3), 273–285 (1985)
15. Johannsen, G., Bille, J.: A threshold selection method using information measures. In: 6th International Conference on Pattern Recognition, pp. 140–143 (1982)

16. Ridler, T., Calvard, S.: Picture thresholding using an iterative selection method. IEEE Transaction on Systems Man Cybernetics 8, 629–632 (1978)
17. Otsu, N.: A threshold selection method from gray-level histogram. IEEE Transactions on Systems, Man and Cybernetics 9, 62–66 (1979)
18. USC-SIPI Image Database, University of Southern California, Signal and Image Processing Institute, http://sipi.usc.edu/database/
19. Roy, S., Saha, S., Dey, A., Shaikh, H.S., Chaki, N.: Performance Evaluation of Multiple Image Binarization Algorithms Using Multiple Metrics on Standard Image Databases, ICT and Critical Infrastructure. In: 48th Annual Convention of Computer Society of India, pp. 349–360 (2013)

Multiple People Tracking Using Moment Based Approach

Sachin Kansal

Indian Institute of Technology Delhi, Delhi, India
sachinkansal87@gmail.com

Abstract. This paper has the capability to detect multiple people in indoor and outdoor environment. In this paper we have used single camera. In this paper we proposed a technique in which it performs multiple face detection, from this it extracts the people's torso regions and stores the HSV range of each person. After this when person's face is not in front of the camera it will track all those people's using moment based approach i.e. it will compute the area of exposed torso region and centre of gravity of the segmented torso region of each person. In this paper we consider torso region HSV range as the key feature. From this we calculate the tracking parameters for each person. In this paper speech thread module is implemented to have interaction with the system. Experiment results validate the robust performance of the proposed approach.

Keywords: Face Detection, HSV Range, Tracking and following, Color features, Moment Calculation, Speech Generation.

1 Introduction

Multiple people tracking calculation in the dynamic indoor and outdoor environment, i.e., having uncontrollable lighting conditions is an important issue to be addressed now-a-days. There are many methods which are existing to track people by, and some are rely on proximity sensors such as visual methods [1-3], in laser range finder [4][5], we also get information from sonar sensors [6] . Some use concept of localization by microphones from any sound source [7]. Single person tracking has been implemented using face detection technique [8]. As none of the above techniques are capable of multiple people's tracking in the real time environment in an efficient fashion. In these techniques we use single camera (Logitech C510). In this paper multiple people are tracked by extracting the motion parameters. Here we use (Right, Left) by the centre of gravity and (Away, toward) using area of segmented of the torso region. For every sample frame we compute all the parameters.

This paper is organized as follows. Section1 briefly introduces the work which had been implemented. Section 2 depicts method of framework. Section 3 introduces detailed approach of multiple people's tracking. Section 4 introduces performance of multiple people's tracking with different distances from the camera, having different ambience of lightning condition and time. Finally, section 5 depicts the conclusion and future work.

M.K. Kundu et al. (eds.), *Advanced Computing, Networking and Informatics - Volume 1*,
Smart Innovation, Systems and Technologies 27,
DOI: 10.1007/978-3-319-07353-8_27, © Springer International Publishing Switzerland 2014

2 Methodology

For multiple people tracking have to identify the person's face, i.e. target person(s) for tracking. Then target person by detecting the torso region and extracting the HSV range of each person separately. After this we calculate moments (zero and first order for area and centre of gravity respectively). At last we calculate all tracking parameters. We had also done the calibration of the camera in order to have better accuracy while handling the threshold values. The framework of the multiple people(s) tracking is shown in Fig. 1.

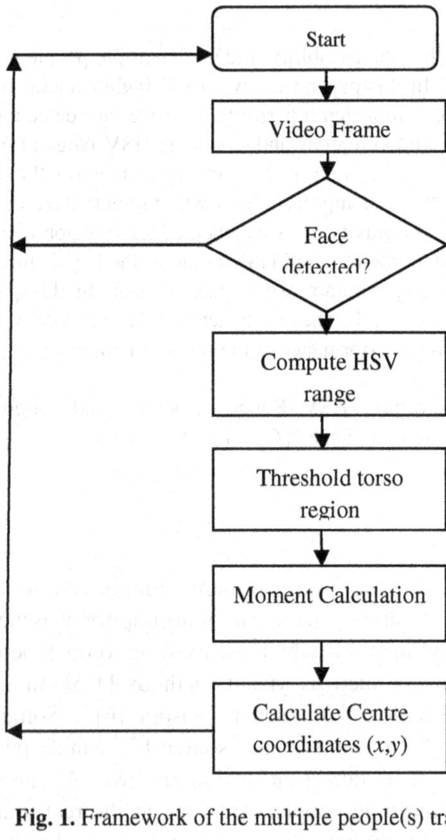

Fig. 1. Framework of the multiple people(s) tracking

2.1 Robust Multiple People Tracking

As the for multiple people tracking it needs a very high real time processing speed and it due to this it has dynamic changing of the background throughout. In order to extract torso region of the person(s), we implemented HSV range as feature in the dynamic environment.

2.1.1 Face Detection

This module introduces the Viola Jones face detection technique and it comprises of Ada boost for learning and Haar cascade as classifier for face detection.

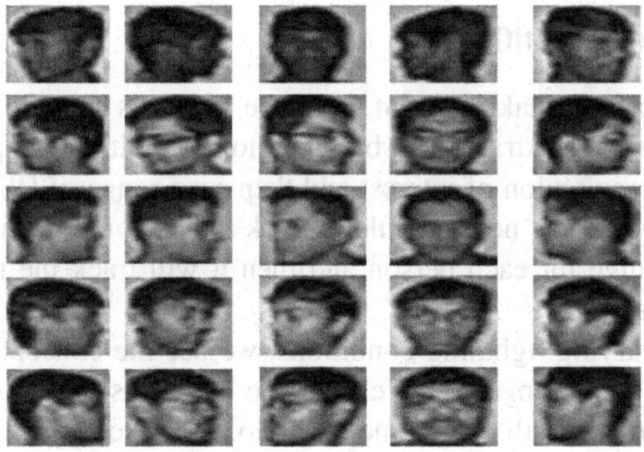

Fig. 2. Frontal faces database

2.1.2 AdaBoost Learning Algorithms

In this learning algorithm we have a selected small number of features from a larger set and thus we get extremely efficient classifiers. In this algorithm within a sub image of the image we have the features vectors greater than total sum number of pixels in order to have fast classification; the method used in learning algorithm should exclude the majority critical features and concentrate on small set of features.

Fig. 3. Non – Face Images Database

In each frame we detect the face of multiple people(s) to have torso based information as features for further processing. This provides information exchange to next subsequent module. As stated above, we use Haar-like features. In this training has been done for hundreds of samples, sample space can be a train, toy or a face i.e. termed as positive samples. Simultaneously we also take negative samples which are of same size say 50×50. As we train the classifier and we implement to the ROI (region of interest). In this it results "1" if it is likely to detect the object else it will show "0".

2.2 HSV Range Identification

This Module works by taking input from the previous module (ROI from the face detection module). We extract torso based region of multiple people(s). Calculation of HSV range on some region of interest and then set a range of HSV from it in order to provide the feature to the next module to track the person. For multiple People(s), we compute HSV range for each person and then it will track the multiple people(s) in real time environment.

As we have varying lightning conditions, we use the concept HSI as color space. After extracting HSV range we threshold the torso based region from the captured image by the camera in the real time scenario. We also implemented operation like morphological closing operator which removes noisy. Segmentation of torso based region of a single person is shown in Fig. 4.

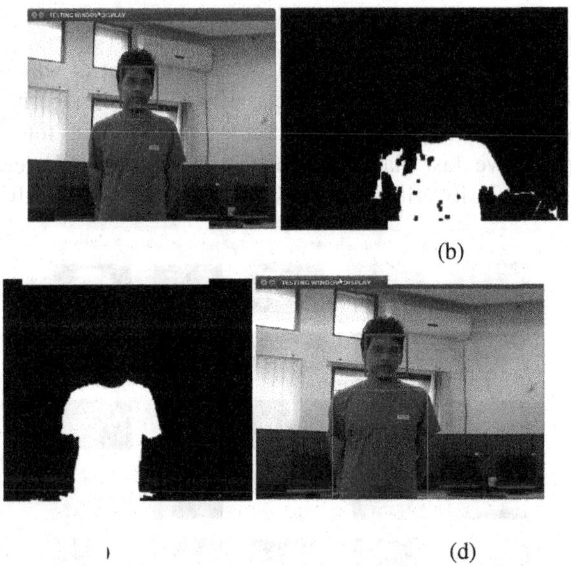

Fig. 4. Image Segmentation (a) Original image, (b) Binary image of torso color, (c) After closing operation, (d) Tracking module detection

In our approach a facial image is segmented (after region of interest is located) into small grid. If the input image is of size M x N then single module is about of size M/2×N/2. The main concept behind the segmentation is to extract main contributed facial features and to reduce the search space. For example, if one expresses smile by stretching both lips corners while the other expresses only stretching one lip corner.

At the time of classification all face models will be compared and maximum similarity will be detected. If the face image is divided into very small regions, then the global information of the face may be lost and the accuracy will be deteriorated.

2.3 Moment Calculation

Initially we extract HSV range and then even face is not detected it will track multiple people(s). We compute moment for each frame. Steps for calculating moment are given below:
Step 1: Camera grabbed the frame in a real time environment.
Step 2: In each frame we extract HSV range.
Step 3: After this step we extract torso region.
Step 4: After this step we threshold the torso region.
Step 5: Calculation of moments from calculated threshold image.
Step 6: From the moments we compute centre coordinates (Xc, Yc) of multiple people for each captured frame.

Definition of moment is given below:

2.3.1 Zero Moment Definition

$$M00 = \sum x \sum y\ F(x, y)$$

2.3.2 First Order Moment Definition

$M10 = \sum x \sum y\ x\ F(x, y)$;
$M01 = \sum x \sum y\ y\ F(x, y)$
Where; F(x, y) is intensity of the pixel (x, y).

2.3.3 Centre of Gravity

$Xc = M10 / M00$
$Yc = M01 / M00$

2.4 Calculation of Away and Toward Parameter

Steps are as follows:
Step 1: We calculate the area from the moments.
Step 2: Erosion and Dilation also implemented to remove the noise.

2.4.1 Towards Calculation

If we move towards the camera, area increased by some factor (say x).
From this we can easily compute this motion parameter.

2.4.2 Away Calculation

If we move towards the camera, area increased by some factor (say x).
From this we can easily compute this motion parameter.

2.4.3 Left Calculation

Step 1: We calculate the area from the moments.
Step 2: Calculate centre of gravity (Xc, Yc).
Xc = M10 / M00
Yc = M01 / M00
Step3: Compare the centre of gravity $(Xc, Yc)^1$ with next fame centre of gravity $(Xc, Yc)^2$.
Step4: By taking difference between two centre of gravity, $(Xc, Yc)^2$ and $(Xc, Yc)^1$.
Step5: We apply threshold and compare which parameter is changing.

2.4.4 Right Calculation

Step 1: We calculate the area from the moments.
Step 2: Calculate centre of gravity (Xc, Yc).
Xc = M10 / M00
Yc = M01 / M00
Step3: Compare the centre of gravity $(Xc, Yc)^1$ with next fame centre of gravity $(Xc, Yc)^2$.
Step4: By taking difference between two centre of gravity, $(Xc, Yc)^2$ and $(Xc, Yc)^1$.
Step5: We apply threshold and compare which parameter is changing.

2.5 Calibration Techniques

Let object is d (meters) from the camera. From this we calibrate in order to map between distance and pixel. In this technique we also provide threshold condition to the tracking system i.e. for a slight change of movement by the people will not affect the tracking system.

3 Experiment Results

3.1 Single Person Tracking

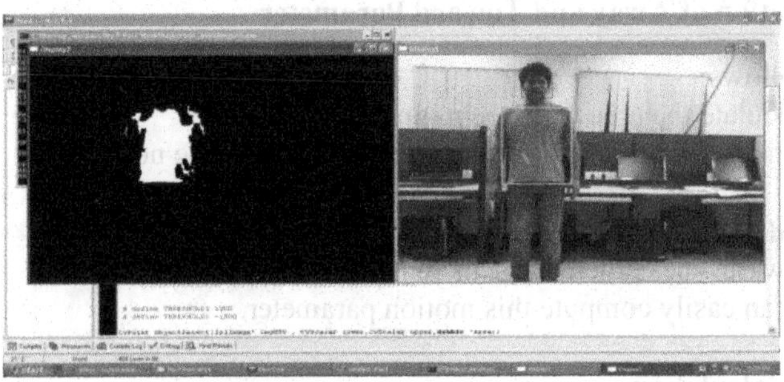

Fig. 5. Snapshot of single person tracking result

3.2 Multiple Person(s) Tracking

3.2.1 Initial Capturing Image from Camera

Fig. 6. Snapshot of capturing image

3.2.2 Multiple People(s) Tracking (Frontal Side)

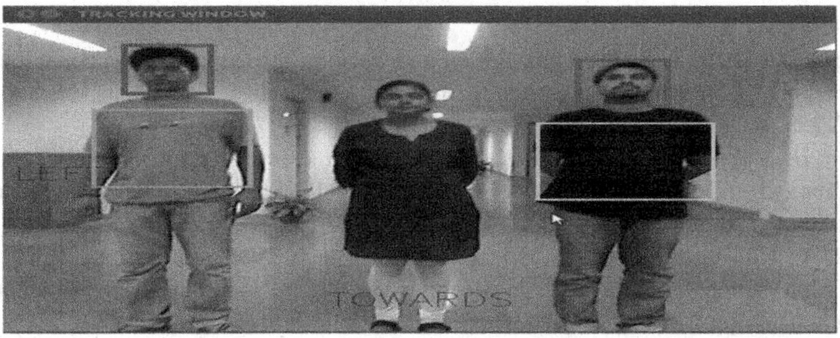

Fig. 7. Snapshot of frontal side multiple people tracking

3.2.3 Multiple People(s) Tracking (Back Side)

Fig. 8. Snapshot of back side multiple people tracking

4 Discussion

Image resolution is 640×480.We have done variation in the distance between people and the camera from 1.3 m to 3.4 m. We are streaming the frames at a speed of 5 fps (frames per second). We are having about 98% average corrected rate and having 2% average false positive rate. It will also track the multiple people even their faces are not detected all the time

5 Conclusion

Calculating zero and first order moments results in all motion parameters. So that multiple people(s) can be tracked in a real time scenario in an effective fashion.

In the future work we will implement multiple people tracking by using Open MPI which provides parallel computing when number of person increases for tracking.

References

1. Song, K.T., Chen, W.J.: Face recognition and tracking for human-robot interaction. IEEE International Conferences on Systems, Man and Cybernetics, The Hague, Netherlands 3, 2877–2882 (2004)
2. Kwon, H., Yoon, Y., Park, J.B., Kak, A.C.: Person tracking with a mobile robot using two uncalibrated independently moving cameras. In: IEEE International Conferences on Robotics and Automation, Barcelona, Spain, pp. 2877–2883 (2005)
3. Food, A., Howard, A., Mataric, M.J.: Laser based people tracking. In: Proceedings of the IEEE International Conferences on Robotics & Automation (ICRA), Washington, DC, United States, pp. 3024–3029 (2002)
4. Montemerlo, M., Thun, S., Whittaker, W.: Conditional particle filters for simultaneous mobile robot localization and people tracking. In: Proceedings of the IEEE International Conference on Robotics & Automation (ICRA), Washington, DC, USA, pp. 695–701 (2002)
5. Shin, J.-H., Kim, W., Lee, J.-J.: Real-time object tracking and segmentation using adaptive color snake model. International Journal of Control, Automation, and Systems 4(2), 236–246 (2006)
6. Scheutz, M., McRaven, J., Cserey, G.: Fast, reliable, adaptive, bimodal people tracking for indoor environments. In: Proc. of the 2004 IEEE/RSJ Int. Conf. on Intelligent Robots and Systems (IROS 2004), Sendai, Japan, vol. 2, pp. 1347–1352 (2004)
7. Fritsch, J., Kleinehagenbrock, M., Lang, S., Fink, G.A., Sagerer, G.: Audiovisual person tracking with a mobile robot. In: Proceedings of International Conference on Intelligent Autonomous Systems, pp. 898–906 (2004)
8. Kansal, S., Chakraborty, P.: Tracking of Person Using monocular Vision by Autonomous Navigation Test bed (ANT). International Journal of Applied Information Systems (IJAIS) 3(9) (2012)

Wavelets-Based Clustering Techniques for Efficient Color Image Segmentation

Paritosh Bhattacharya[1], Ankur Biswas[1], and Santi Prasad Maity[3]

[1] Dept of CSE, National Institute of Technology, Agartala, India
[2] Dept of CSE, Tripura Institute of Technology, Agartala, Tripura, India
[3] Dept of IT, Bengal Engineering and Science University, Shibpur, India
abiswas.agt@gmail.com

Abstract. This paper introduces efficient and fast algorithms for unsupervised image segmentation, using low-level features such as color and texture. The proposed approach is based on the clustering technique, using 1. Lab color space, and 2. the wavelet transformation technique. The input image is decomposed into two-dimensional Haar wavelets. The features vector, containing the information about the color and texture content for each pixel is extracted. These vectors are used as inputs for the k-means or fuzzy c-means clustering methods, for a segmented image whose regions are distinct from each other according to color and texture characteristics. Experimental result shows that the proposed method is more efficient and achieves high computational speed.

1 Introduction

Image segmentation has been a focused research area in the image processing, for the last few decades. Many papers has been published, mainly focused on gray scale images and less attention on color image segmentation, which convey much more information about the object or images. Image segmentation is typically used to locate objects and boundaries in images. More precisely, image segmentation is the process of assigning a label to every pixel in an image such that the pixels with the same label share certain visual characteristics. The color image segmentation plays crucial role in image analysis, computer vision, image interpretation, and pattern recognition system.

The CIELAB color space is an international standard where the Euclidean distance between two color points in the CIELAB color space corresponds to the difference between two colors by the human vision system [1]. This property CIELAB color space is attractive and useful for color analysis. The CIELAB color space has shown better performance in many color image applications [2,4]. Because of these the CIELAB color space has been chosen for color clustering.

Clustering is an unsupervised technique that has been successfully applied to feature analysis, target recognition, geology, medical imaging and image segmentation [5,7]. This paper considers the segmentation of image regions based on two clustering methods: 1. color feature using fuzzy c-means or k-means, 2. color and texture feature using fuzzy c-means or k-means through Haar wavelet transformation.

M.K. Kundu et al. (eds.), *Advanced Computing, Networking and Informatics - Volume 1*,
Smart Innovation, Systems and Technologies 27,
DOI: 10.1007/978-3-319-07353-8_28, © Springer International Publishing Switzerland 2014

The rest of the paper is organized as follows: In section 2 and 3, a brief description of texture characterization using Haar wavelet transformation is presented. The section 3 describes the. The section 4, presents the proposed image segmentation technique using color and the section 5, presents the segmentation technique using color and texture. The experimental results are presented in section 6. The future scope and conclusions are presented in section 7.

2 Haar Wavelet Transformation

The Wavelets are useful for hierarchically decomposing functions in ways that are both efficient and theoretically sound [8]. Through Wavelet transformation on the color image, four sub images like low resolution copy of original image and three-band pass filtered images in specific directions: horizontal, vertical and diagonal, will be produced. In this paper, we use Haar wavelets to compute feature signatures because they are the fastest and simplest to compute and have been found to perform well in practice [9]. The Haar wavelet is a certain sequence of rescaled 'square-shaped' functions which together form a wavelet family or basis. The Haar wavelet transformation technique is explained in the paper [13].

3 Clustering Technique on Images

Clustering is the most important unsupervised learning problem which is the process of organizing objects into groups whose members are similar in some way. A cluster is therefore a collection of objects which are 'similar' between them and are 'dissimilar' to the objects belonging to other clusters. The two basic clustering techniques are K-means (an exclusive clustering algorithm) and Fuzzy C-means (an overlapping clustering algorithm). [5,6,3,10].

K-Means Algorithm

K-means algorithm was originally introduced by McQueen in 1967 [11]. The K-means algorithm is an iterative technique that is used to partition an image into K clusters.

Let $X = \{x_1, x_2, ..., x_n\}$ represent a set of pixels of the given image, where n is the number of pixels. $V = \{v_1, v_2, ..., v_k\}$ is the corresponding set of cluster centres, where k is the number of clusters. The aim of K-means algorithm is to minimize the objective function J(V), in this case a squared error function:

$$J(V) = \sum_{i=1}^{k} \sum_{j=1}^{k_t} \left\| x_{ij} - v_j \right\|^2 \qquad (1)$$

Where, $\left\| x_{ij} - v_j \right\|$ is the Euclidean distance between x_{ij} and v_j.k_i is the number of pixels in the cluster i.

The difference is typically based on pixel colour, intensity, texture, and location, or a weighted combination of these factors. In our study, we have considered pixel intensity. The ith cluster centre v_i can be calculated as:

$$v_i = \frac{1}{k_i}\sum_{j=1}^{k_t} x_{ij} \qquad (2) \qquad \text{for } i = 1, \dots, k.$$

The basic algorithm is:

i) Randomly select *k* cluster centres.
ii) Calculate the distance between all of the pixels in the image and each cluster centre.
iii) A pixel is assigned to a cluster based on the minimum distance.
iv) Recalculate the centre positions using equation (2).
v) Recalculate the distance between each pixel and each centre.
vi) If no pixel was reassigned, then stop, otherwise repeat step (iii).

This algorithm is guaranteed to converge, but it may not return the optimal solution. The quality of the solution depends on the initial set of clusters and the value of k.

Fuzzy C-means (FCM) Algorithm

The FCM algorithm was originally introduced by Bezdek in 1981 [12,13]. It is an iterative algorithm. FCM can be used to build clusters (segments) where the class membership of pixels can be interpreted as the degree of belongingness of the pixel to the clusters.

Let X = $\{x_1, x_2 \dots x_n\}$ represent a set of pixels of the given image, where n is the number of pixels and V = $\{v_1, v_2 \dots v_c\}$ is the corresponding set of fuzzy cluster centers, where c is the number of clusters. The main aim is to minimize the objective function J(U,V), which is a squared error clustering criterion defined as :

$$J(U,V) = \sum_{i=1}^{n}\sum_{j=1}^{c} \mu_{ij}{}^{m} \left\|x_i - v_j\right\|^2 \quad (3)$$

Where, $\left\|x_{ij} - v_j\right\|$ is the Euclidean distance between x_{ij} and v_j. μ_{ij} is the membership degree of pixel x_i to the cluster centre v_j and μ_{ij} has to satisfy the following conditions:

$$\mu_{ij} \in [0,1],\, , \forall i = 1,\dots n, \forall j = 1,\dots c \quad (4)$$

$$\sum_{j=1}^{c} \mu_{ij} = 1\,, \quad \forall i = 1,\dots n \qquad (5)$$

$U = (\mu)_{ij_{n*c}}$ is a fuzzy partition matrix. Parameter m is called the “fuzziness index”; it is used to control the fuzziness of membership of each pixel. The value of m should be within the range $m \in [1, \infty]$. m is a weighting exponent that satisfies $m > 1$ and controls the degree of “fuzziness” in the resulting membership functions: As m approaches unity, the membership functions become more crisp, and approach binary functions. As m increases, the membership functions become increasingly fuzzy.

The FCM algorithm is as stated as follows:

i) Initialize the cluster centres V = {v_1, $v_{2....,}$ v_c}, or initialize the membership matrix μ_{ij} with random value such that it satisfies conditions (4) and (5). Then calculate the cluster centres.

ii) Calculate the fuzzy membership μ_{ij} using:

$$\mu_{ij} = \frac{1}{\sum_{k=1}^{c}(\frac{d_{ij}}{d_{ik}})^{\frac{2}{m-1}}} \tag{6}$$

Where, $d_{ij} = ||x_i - v_j||$, $\forall i = 1,...,n$, $\forall j = 1,...,c$.

iii) Compute the fuzzy centres v_j using:

$$v_j = \frac{\sum_{i=1}^{n}(\mu_{ij})^m x_i}{\sum_{i=1}^{n}(\mu_{ij})^m} \tag{7}$$

iv) Repeat step (ii) to (iii) until the minimum J value is achieved.

4 Proposed Image Segmentation Technique Using Color Feature

In the present study an improved image segmentation is achieved by smoothing the image and converting into L*a*b* color space before applying the KM and FCM algorithms, as explain in subsequent sequence

A. K-Means Clustering (KM)

- Read color image.
- Smoothing the image using average filter.
- Convert the image into L*a*b* color space.
- Segment the feature vector using K-Means Clustering.
- Label every pixel in the image using the Results from K-Means Clustering.
- Create segment image.

B. Fuzzy–C-Means Clustering (FCM)

- Read color image.
- Smoothing the image using average filter.
- Convert image into L*a*b* color space.
- Segment the feature vector using FCM.
- Assign the pixels to the clusters as follows:
  A pixel x is assigned to cluster i such that the value of the membership function of x for i is maximum.
- Label every pixel in the image using the results from FCM.
- Create segment image.

5 Proposed Segmentation Technique Using Color and Texture

The Image Segmentation, using color and texture features, is achieved by using wavelet transformation, as explain in subsequent sequence

Wavelet-Based K-Means Clustering (WKM)

- Read color image.
- Smoothing the image using average filter.
- Split the image into Red, Green, and Blue Components.
- Decompose each Red, Green, Blue Component Using Haar Wavelet transformation at 1st level to get approximate coefficient and vertical, horizontal and diagonal detail coefficients.
- Combine approximate coefficient of Red, Green, and Blue Component.
- Similarly combine the horizontal and vertical coefficients of Red, Green, and Blue Component.
- Assign suitable weights to approximate, horizontal and vertical coefficients.
- Segment the approximate, horizontal and vertical coefficients using K-Means Clustering.
- Label every pixel in the image using the results from K-Means Clustering.
- Create segment image.

Wavelet-Based Fuzzy-C–Means Clustering (WFCM)

- Read color image.
- Smoothing the image using average filter.
- Split the image into Red, Green, and Blue Components.
- Decompose each Red, Green, Blue Component using Haar Wavelet transformation at 1st level to get approximate coefficient and vertical, horizontal and diagonal detail coefficients.
- Combine approximate coefficient of Red, Green, and Blue Component.
- Similarly combine the horizontal and vertical coefficients of Red, Green, and Blue Component.
- Assign suitable weights to approximate, horizontal and vertical coefficients.
- Segment the approximate, horizontal and vertical coefficients using FCM Clustering.
- Assign the pixels to the clusters as follows:

A pixel x is assigned to cluster i such that the value of the membership function of x for i is maximum.

- Label every pixel in the image using the results from FCM.
- Create segment image.

6 Experimental Results

The algorithms are implemented on Matlab. The four algorithms, viz., K-Means, Wavelet-based Fuzzy C-Means, Fuzzy C-Means and, Wavelet-based K-Means are implemented on image databases. The K-means and FCM algorithm based on color segmentation are described in section IV of A and B, has been tested on L*a*b* color

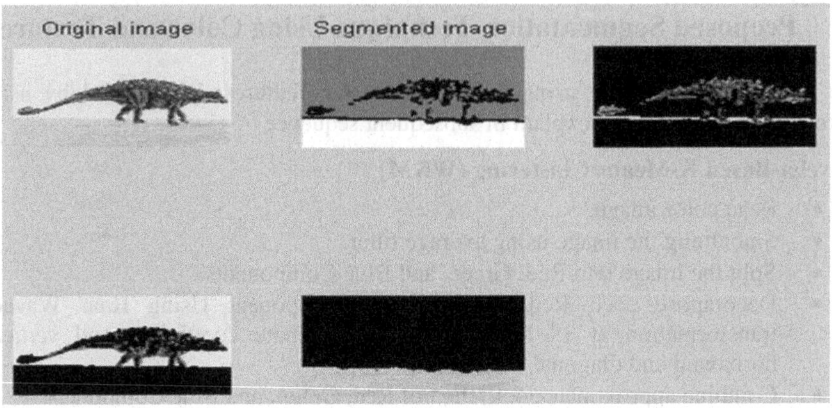

Fig. 1. Segmentation using K-means

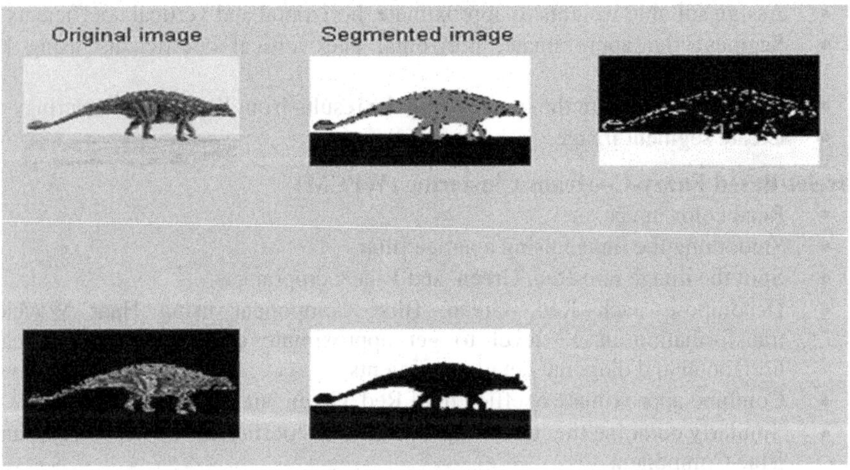

Fig. 2. Segmentation using Wavelet based Fuzzy C-Means

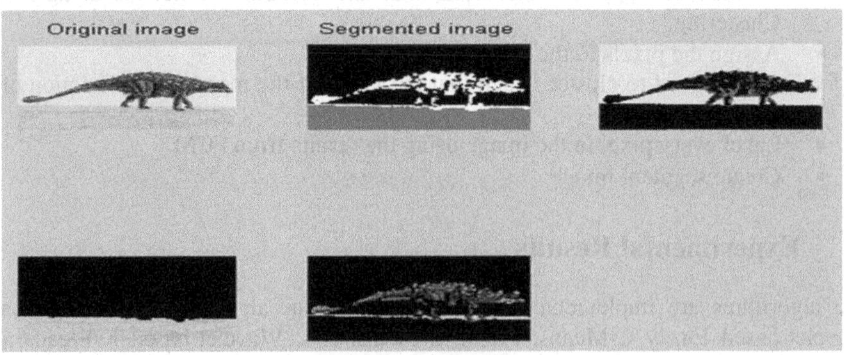

Fig. 3. Segmentation using Fuzzy C-Means

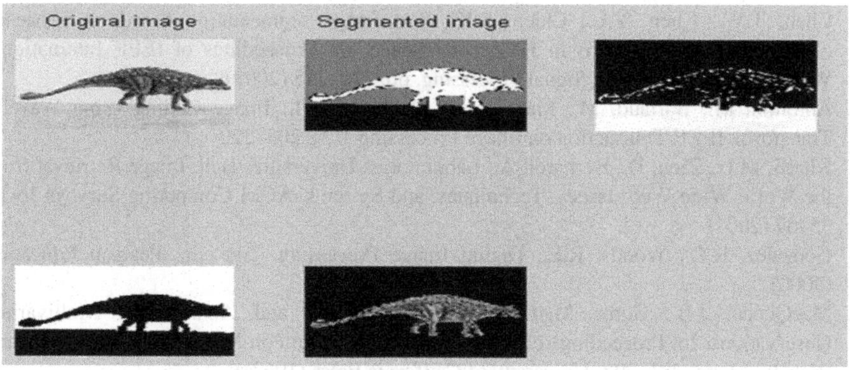

Fig. 4. Segmentation using Wavelet based K-Means

space and their results are shown in Fig. 1 and Fig. 3. Color and texture based image segmentation are described using Wavelet-based K-Means (WKM) and Wavelet-based Fuzzy-C–Means Clustering (WFCM) in section V of C and D, and their tested results are shown in Fig. 4 and Fig. 2.

7 Future Scope and Conclusion

In this paper we have presented algorithms for segmentation of color images using Fuzzy C-Means (FCM), K-Means, Wavelet based K-Means (WKM) Wavelet Based Fuzzy C-Means (WFCM), described in Section 4 and Section 5, have been developed. The experimental results in Fig. 1 to 4, By observing the results it can be said that proposed method can be successfully applied on different types of images. Using wavelet and clustering technique for color image segmentation, it not only enhances but also execution time of each image takes less time than single clustering technique.

References

1. Wyszecki, G., Stiles, W.S.: Color science: concepts and methods, quantitative data and formulae, 2nd edn. John Wiley and Sons (2000)
2. Jin, L., Li, D.: A Switching vector median based on the CIELAB color space for color image restoration. Signal Processing 87, 1345–1354 (2007)
3. Hartigan, J.A., Wong, M.A.: A K-means clustering algorithm. Appl. Stat. 28(1), 100–108 (1979)
4. Kwak, N.J., Kwon, D.J., Kim, Y.G., Ahn, J.H.: Color image segmentation using edge and adaptive threshold value based on the image characteristics. In: Proceedings of International Symposium on Intelligent Signal Processing and Communication Systems, pp. 255–258 (2004)
5. Klir, G.J., Yuan, B.: Fuzzy Sets and Fuzzy Logic-Theory and Applications. PHI (2000)
6. Bezdek, J.C., Ehrlich, R., Full, W.: FCM: The Fuzzy c-Means clustering algorithm. Computers and Geosciences 10, 191–203 (1984)

7. Chen, T.W., Chen, Y.L., Chien, S.Y.: Fast image segmentation based on K-Means clustering with histograms in HSV color space. In: Proceedings of IEEE International Workshop on Multimedia Signal Processing, pp. 322–325 (2008)
8. Antonini, M., Barlaud, M., Mathieu, P., Daubechies, I.: Image Coding using Wavelet Transform. IEEE Transaction on Image Processing 1(2), 205–220
9. Kherfi, M.L., Ziou, D., Bernardi, A.: Laboratories Universities Bell. Image Retrieval from the World Wide Web: Issues, Techniques, and Systems. ACM Computing Surveys 36(1), 35–67 (2004)
10. Gonzalez, R.C., Woods, R.E.: Digital Image Processing, 2nd edn. Pearson Education (2000)
11. MacQueen, J.B.: Some Methods for classification and Analysis of Multivariate Observations. In: Proceedings of 5th Berkeley Symposium on Mathematical Statistics and Probability, pp. 281–297. University of California Press (1967)
12. Bezdek, J.C.: Pattern Recognition and Fuzzy Objective Function Algorithms. Plenum Press (1981)
13. Castillejos, H., Ponomaryov, V.: Fuzzy Image Segmentation Algorithms in Wavelet Domain. In: 2011 8th International Conference on Electrical Engineering Computing Science and Automatic Control (2011)

An Approach of Optimizing Singular Value of YCbCr Color Space with q-Gaussian Function in Image Processing

Abhisek Paul[1,*], Paritosh Bhattacharya[1], and Santi Prasad Maity[2]

[1] Department of Computer Science and Engineering,
National Institute of Technology, Agartala, India
[2] Department of Information Technology,
Bengal Engineering and Science University, Shibpur, India
abhisekpaul13@gmail.com

Abstract. To increase performance of Radial Basis Functions (RBFs) in Artificial Neural Network q-Gaussian Radial Basis Function (q-GRBF) is introduced to optimize singular value of Y, Cb and Cr color image components. Various radial basis functions such as Gaussian Radial Basis Function (GRBF), Multi Quadratic Radial Basis Function (MCRBF), Inverse Multi Quadratic Radial Basis Function (IMCRBF) and Cosine Radial Basis Function (CRBF) are also introduced and compared with singular values of Y, Cb and Cr component of color images. Simulation and analysis shows that q-Gaussian Radial Basis Function gives lesser error and better result compared to the other radial basis functions in artificial neural network.

Keywords: Gaussian RBF, Multi Quadratic RBF, Inverse Multi Quadratic RBF, Cosine RBF, q-Gaussian RBF, Radial Basis Function, Artificial Neural Network

1 Introduction

Radial Basis Functions (RBFs) of Artificial Neural Network (ANN) are utilized in various areas such as pattern recognition, optimization etc. In this paper we have chosen Radial Basis Function such as Gaussian, Multi Quadratic Radial, Inverse Multi Quadratic, Cosine and q-Gaussian. Singular values of colour images are calculated and compared with such RBFs in this paper. Colour image is taken as example and their YCbCr colour space components are being introduced for the analysis. Gaussian, Multi Quadratic Radial, Inverse Multi Quadratic, Cosine and q-Gaussian are computed and compared with normal methods. In section 2 architecture of RBF neural network is described. Section 3 shows the analysis. In section 4 simulations is given and finally in Section 5 we have given the conclusion [1-4], [7].

* Corresponding author.

M.K. Kundu et al. (eds.), *Advanced Computing, Networking and Informatics - Volume 1*, Smart Innovation, Systems and Technologies 27,
DOI: 10.1007/978-3-319-07353-8_29, © Springer International Publishing Switzerland 2014

2 Architecture of Radial Basis Function Neural Network

In Fig. 1 RBF neural network architecture consist of three layers: the input layer, hidden layer and output layer. In Fig. 1 inputs are X = $\{x_1,... x_d\}$ which enter into input layer. Radial centres and width are C = $\{c_1........., c_n\}^T$ and σ_i respectively. In hidden layer $\Phi = \{ \Phi_1,......, \Phi n \}$are the radial basis functions. Centres are of $n \times 1$ dimension when the number of input is n. The desired output is given by y which is calculated by proper selection of weights. $\Omega = \{w_{11},..., w_{1n},...., w_{m1},.... w_{mn}\}$ is the weight. Here, w_j is the weight of ith centre [2-3].

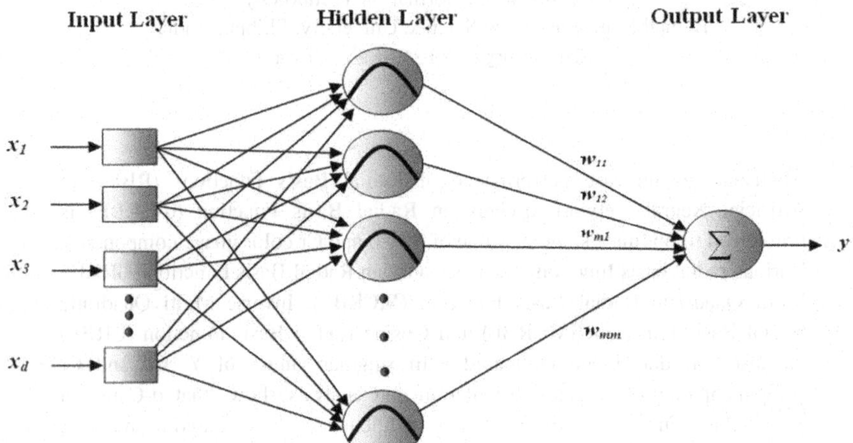

Fig. 1. Radial basis function neural network architecture

$$y = \sum_{i=1}^{m} w_i \phi_i \tag{1}$$

$$\phi(x) = \| x - c \| = r \tag{2}$$

Radial basis functions like Linear, Cubic, Thin plane spline and Gaussian are given in the in the Eqn.2, Eqn.3, Eqn4, and Eqn.5 respectively.

Gaussian: $$\phi(x) = \exp(-\frac{(r)^2}{2\sigma^2}) \tag{3}$$

Multi Quadratic: $$\phi(x) = (r^2 + \sigma^2)^{1/2} \tag{4}$$

Inverse Multi Quadratic: $$\phi(x) = (r^2 + \sigma^2)^{-1/2} \tag{5}$$

Cosine: $$\phi(x) = \frac{\sigma}{\sqrt{r^2 + \sigma^2}} \tag{6}$$

q-Gaussian: $$\phi(x) = \exp(-\frac{r^2}{(3-q)\sigma^2}) \tag{7}$$

3 Analysis

We have taken one colour image. After that we extracted the red, green and blue colour components from the colour images Fig. 2. After that we calculated Y, Cb and Cr colour component [1] from the Eq.8, Eq.9 and Eq.10. Relation between Y, Cb and Cr colour components and R, G and B colour component is given in Eq.6. As we have applied. Gaussian, Multi Quadratic Radial, Inverse Multi Quadratic, Cosine and q-Gaussian RBF in artificial neural network, RBF needs optimal selections weight and centres. We have calculated all the method with pseudo inverse technique [5]. As in put we have chosen House image. Matrix sizes of all the input images are of 128×128 pixels for every analysis and experiment.

$$Y = 0.299R + 0.587G + 0.114B \tag{8}$$

$$Cb = 0.172R - 0.339G + 0.511B + 128 \tag{9}$$

$$Cr = 0.511R - 0.428G - 0.0838B + 128 \tag{10}$$

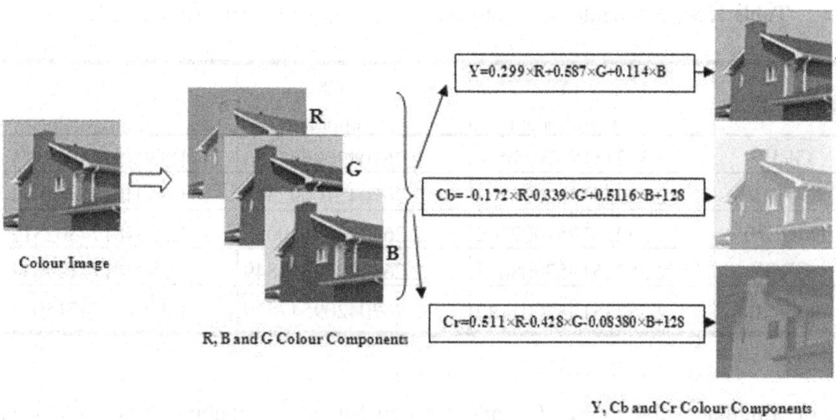

Fig. 2. Colour image with RGB and YCbCr Components

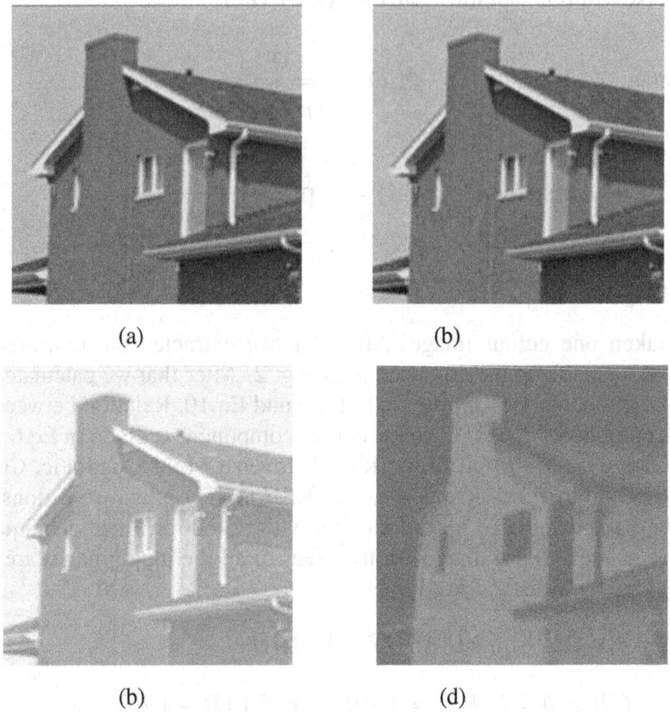

Fig. 3. House image (a) Original Color Image, (b) Y Component, (c) Cb Component, (d) Cr Component

Table 1. Mean Singular value of House image with different RBF methods

RBF methods	Y Component	Cb Component	Cr Component
GRBF	2.352351957037448	2.26704209518650	1.36048170627175
MQRBF	2.352351960698278	2.26704210092544	1.36048143482934
IMQRBF	2.352351957007302	2.26704209550038	1.36048174302319
CRBF	2.352351957008633	2.26704209548849	1.36048174199634
q-GRBF	2.352351957007182	2.26704209547479	1.36048174215049

Mean Singular value of Y, Cb and Cr component of image House in normal method are 2.352351957007144, 2.267042095475595 and 1.360481741975729 respectively.

Table 2. Mean Square Error (MSE) of Singular value of House image with different RBF methods

RBF methods	Y Component	Cb Component	Cr Component
GRBF	$5.587050298754\times 10^{-17}$	$2.390966601750\times10^{-16}$	$3.759190991746\times10^{-15}$
MQRBF	$7.903088842942\times 10^{-13}$	$1.199463110908\times10^{-11}$	$3.053300779145\times10^{-9}$
IMQRBF	$8.704202668210\times 10^{-20}$	$1.583915586350\times10^{-18}$	$2.695149918667\times10^{-15}$
CRBF	$8.234144691833\times 10^{-20}$	$1.807522296961\times10^{-18}$	$2.064707816520\times10^{-15}$
q-GRBF	$1.010905951693\times 10^{-21}$	$1.264231758406\times 10^{-20}$	$6.797597647388\times10^{-17}$

4 Simulation

Colour images with YCbCr colour components are taken for the simulation. In this paper we have experimented through House colour image which the size of 128×128 pixels. We have simulated singular value of matrixes of these images with normal method, Gaussian, Multi Quadratic Radial, Inverse Multi Quadratic, Cosine and q-Gaussian methods. We have used MATLAB 7.6.0 software [6] for the analysis and simulation process. In Fig. 3, House image and its corresponding YCbCr colour component images are shown. Mean singular values of matrixes of all these images with Gaussian RBF, Multi Quadratic RBF and Inverse Multi Quadratic RBF, Cosine RBF and q-Gaussian RBF are calculated, simulated. Comparative singular values are shown in Table.1. Corresponding Mean Square Error of singular values of matrixes are shown in Table 2. In Fig. 4 graphically represents the Mean sequence error of GRBF, MCRCF, IMCRBF, CRBF, q-GRBF.

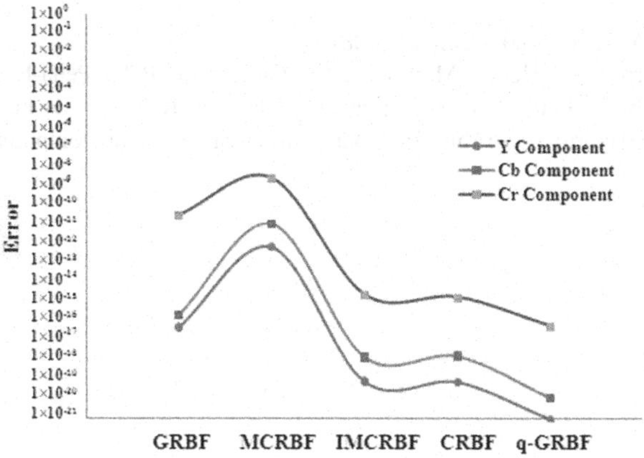

Fig. 4. Comparison of Error of Eigen value with YCbCr color components in House image

5 Conclusion

In this paper Gaussian, Multi Quadratic Radial, Inverse Multi Quadratic, Cosine and q-Gaussian Radial Basis Functions are utilized for the approximation of singular values of image and their corresponding YCbCr colour component matrixes. Simulation results give better result and lesser Mean Square Error for q-Gaussian RBF method. So, it can be conclude that q-Gaussian RBF method could be used for the approximation compared to the other relative methods in Artificial Neural Network.

Acknowledgements. The authors are so grateful to the anonymous referee for a careful checking of the details and for helpful comments and suggestions that improve this paper.

References

1. Noda, H., Niimi, M.: Colorization in YCbCr color space and its application to JPEG images. Pattern Recognition 40(12), 3714–3720 (2007)
2. Scholkopf, B., Sung, K.-K., Burges, C.J.C., Girosi, F., Niyogi, P., Poggio, T., Vapnik, V.: Comparing support vector machines with Gaussian kernels to radial basis function classifiers. IEEE Transactions on Signal Processing 45, 2758–2765 (1997)
3. Mao, K.Z., Huang, G.-B.: Neuron selection for RBF neural network classifier based on data structure preserving criterion. IEEE Transactions on Neural Networks 16(6), 1531–1540 (2005)
4. Luo, F.L., Li, Y.D.: Real-time computation of the eigenvector corresponding to the smallest Eigen value of a positive definite matrix. IEEE Transactions on Circuits and Systems 141, 550–553 (1994)
5. Klein, C.A., Huang, C.H.: Review of pseudo-inverse control for use with kinematically redundant manipulators. IEEE Transactions on System, Man, Cybernetics 13(3), 245–250 (1983)
6. Math Works. MATLAB 7.6.0, R2008a (2008)
7. Fernandez-Navarroa, F., Hervas-Martineza, C., Gutierreza, P.A., Pena-Barraganb, J.M., Lopez-Granadosb, F.: Parameter estimation of q-Gaussian Radial Basis Functions Neural Networks with a Hybrid Algorithm for binary classification. Neurocomputing 75(1), 123–134 (2012)

Motion Tracking of Humans under Occlusion Using Blobs

M. Sivarathinabala and S. Abirami

Department of Information Science and Technology,
College of Engineering,
Anna University, Chennai
sivarathinabala@gmail.com, abirami_mr@yahoo.com

Abstract. In today's scenario, Video Surveillance plays a major role in building intelligent systems. This involves the phases such as Motion detection, Object classification and Object tracking. Among these, Object tracking is an important task to identify/detect the objects and track its motion correspondingly. After object identification, the location of the objects is crucial to understand the nature of the moving objects. There arises a need for tracking the occluded objects also when multiple objects are under surveillance. In this paper, a new tracking mechanism has been proposed to track the objects under surveillance though they occlude. Initially, background has been modelled with the Adaptive background modelling using GMM (Gaussian Mixture Model) to obtain the foreground as blobs. Later, Objects represented using Contours are integrated with simple particle filters to obtain a new state which would track the object/person effectively. Using the path estimated by particle filters, Occlusion of blobs gets determined based on the interference of their radii lying in that path. Performance of this system has been tested over CAVIAR and User generated data sets and results seem to be promising.

Keywords: Video Surveillance, Object Tracking, Background Modelling, Blob Tracking, Particle Filters.

1 Introduction

Video Surveillance is one of the major research field in video analytics. The aim of video surveillance is to collect information from videos by tracking the people involved in them. For Security purpose, Video Surveillance has been employed in all public places. Over a long period of time, a human guard would be unable to watch the behavior of the persons from the videos. To avoid the manual intervention, automatic video surveillance has been emerged where people/objects could be identified and tracked effectively.

Numerous tracking approaches have been proposed in the literature to track objects [1,4,6,12].Tracking an object/human can be complex due to the existence of noise over the images, partial and total occlusion, distorted shapes and complex object motion. An Object Tracking mechanism starts with the Object detection stage which

M.K. Kundu et al. (eds.), *Advanced Computing, Networking and Informatics - Volume 1*,
Smart Innovation, Systems and Technologies 27,
DOI: 10.1007/978-3-319-07353-8_30, © Springer International Publishing Switzerland 2014

consists of Background modeling, Segmentation, Object representation and Feature extraction. Background is the reference frame that is used to isolate moving object of interest. Background modeling is required to differentiate/segment the foreground pixels from the background. Median based models, Gaussian based and Kernel density based models have been used to model the background [6,7,5]. Adaptive background modeling ie., Gaussian based modeling[12] has been effectively used to model the background in literatures. Adaptive Background modeling updates the background frame frequently in order to adapt to the changes in the environment. Next to this, segmentation has been done to extract the moving regions separately. After segmentation, segmented blob could be considered as an object. In general, Object can be represented by its points, ellipse or bounding box, contour or silhouette [1] and color or textural features of the object could be extracted.

Next to Object detection, represented objects could be tracked (Object Tracking) by estimating the position of the object in every frame over time. Particle filters and Kalman filters [13] are widely used for tracking the movement of the interested objects in previous works. However, the Kalman filter [13] does not follow the Gaussian distribution and results in poor estimation of state variables. While tracking the movement of the objects, Occlusion, a challenging issue may arise. Occlusion means hiding the target object/person either partially or fully. This occurs when two persons cross each other or person crossing an object. A particular person/object may fail to get tracked under occlusion. Even though many works have been reported in literatures, this research made a special attempt to track the blob of few objects even under occlusion. Here, Particle filter has been integrated along with the blob tracking approach to track the particular object/person effectively under occlusion conditions.

2 Related Works

Object Tracking includes the following stages: Background modelling, Segmentation, Object representation, Feature selection and tracking using filters. Major works in the field of video analytics have been devoted in the object tracking phase and they are addressed in this section.Mixture of Gaussians (MOG) background model [8] is considered to be more effective in modelling multi-modal distributed backgrounds. Parisa Darvish *et al.* [11] proposed a region based method for background subtraction. This method has been done based on color histograms, texture information and successive division of candidate rectangular image regions to model the background. This method provides the detail of the contour of the moving shapes. Stauffer *et al.* [7] proposes Gaussian mixture background model based on the mixture modelling of pixel sequences by considering the time continuity of the pixels, but this does not combine the information in the spatial neighbourhood of the pixels. Several techniques for foreground segmentation [1,13,5] are available, among them background subtraction [4] seems to be the simpler method. In background subtraction, current frame has been subtracted from the previous frame. This segmentation approach has been suited for static environments. Hajer Fradi *et al.* [12] proposed a segmentation approach, background subtraction based on incorporating a uniform motion model into GMM. Tracking can be defined as the problem of estimating trajectories of the particular object. Hamidreze Jahandide *et al.* [2] proposed a hybrid model for predicting the object position and color. In this, motion and appearance model has been constructed and Kalman filter has been applied to both the components. This algorithm seems to be suitable for various illumination conditions. Apart from this, tracking under occlusion has also been considered as a major problem. Nam-Gyu-Cha *et al.* [10] proposed a new method to avoid self occlusion that estimates the occlusion states by using Markov Random Field (MRF). Yaru Wang *et al.* [13] designed particle filter based weight tracker to update multiple cues. Here, the author proposed that Particle filter has the ability to deal with non-linear, non-gaussian and multimodal problems. Tracking the particular person/object under illumination conditions, occlusion and dynamic environments is still a challenging research.

3 Object Tracking

In this research, to detect/track the objects, frames are extracted from input videos. Temporal information which reveals the moving region details are obtained from those sequence of frames. A suitable background model has been selected using GMM mixture of Gaussians and the background information gets updated constantly. Later, suitable features are selected to represent every object and by employing blob and particle filter, trajectories are identified for those objects.

3.1 Background Modelling

In Object Detection, Background modelling is the first and foremost step. This can be achieved by understanding and representing the background scene from the videos. To obtain the background model, Gaussian Mixture model (GMM)[12] has been employed here. For outdoor environments, Single Gaussian model is not suitable. Therefore to suit outdoor and indoor environments, mixture of Gaussians has been used here to model the background. In GMM, each pixel in an image has been modelled by mixture of K Gaussians. The probability of the current pixel value has been determined using

$$P(X_t) = \sum_{i=1}^{k} \omega_{i,t}\eta(X_t, \mu_{i,t}, \Sigma_{i,t}) \quad (1)$$

Where k denotes the number of distributions X_t denotes each pixel value of tth image frame. $\omega_{i,t}$ denotes the weight factor for corresponding Gaussian distribution. $\mu_{i,t}$ denotes the mean value of ith Gaussian. $\Sigma_{i,t}$ represents the co-variance matrix of ith Gaussian. η represents the Gaussian probability density function. Threshold is predefined here through which binary motion detection mask has been calculated. For each frame, the difference of pixel value from mean has been calculated to derive pixel color match component. If the match component is zero, then the Gaussian components for each pixel has been updated else if it is one, Updation of weight, mean and standard deviation has been done. This enables the new background to be get detected and updated subsequently.

3.2 Segmentation

Once background has been modelled, it needs to get subtracted from the frames to obtain the foreground image. Background subtraction [3] has been performed by taking the difference between the current image and the reference background image in a pixel by pixel fashion. Here, threshold value has been fixed. When the pixel difference is above threshold value, it is considered as a foreground image. Blobs or pixels segmented as foreground are necessary objects which need to get represented to track its motion. This object could be either a particular thing/vehicle/person etc.

4 Object Representation and Feature Selection

Object representation varies depending upon the environment where we apply and their appearance and shape. Object/Shape representation is required to track the object/person in the non-rigid motion. Here in this paper, the target person/object has been represented in contour. Next to this, feature selection is an important task in object tracking. In general, color, texture, object edges and contour are usually considered as features of objects [1]. Here contour of the object/person, average of individual RGB components (Color histogram information) are selected as features for tracking a particular person. The probability density of appearance termed as color has been computed as a feature from the region of rectangle. Next to the foreground segmentation, foreground image has been binarized and contour points are obtained using Moore's neighbourhood algorithm. Using those points, Contour has been formed [4] using 4 -neighbourhood masking algorithm. From the centre point of the contour, Circle has been drawn and radius has been calculated from the contour. Blob appearing in successive frames have been analysed if there have been any similarity in their shape and color.

5 Proposed Tracking Algorithm

Once the shape of the object has been represented using contour, its motion needs to get tracked. In this paper, our motivation is to track the objects even if they are in occluded state also. In general, Occlusion [9] can be classified into three categories: self occlusion, partial occlusion and total occlusion. In this Paper, to handle self and partial occlusion problems, a combination of blob tracking method and particle filter approach, has been employed by using Contour as shape representation and color as feature for tracking. In addition to this, blob splitting and merging approach has been attempted to identify occlusion.

5.1 Blob Tracking Approach

In blob tracking, initially, a centre point of the contour has been considered as centroid, and assuming a threshold radius from the centroid, circle has been drawn inside the contour. In each and every frame, the circle has been tracked continuously. If the circles get overlapped in consecutive frames, then the blob has been considered as the same or existing object and the person has been moving in contiguous frames. Figure 1(a) shows the movement of the human blob (Person 1) getting tracked continuously with the interference of their circles. Area of the circle has been computed for further processing.

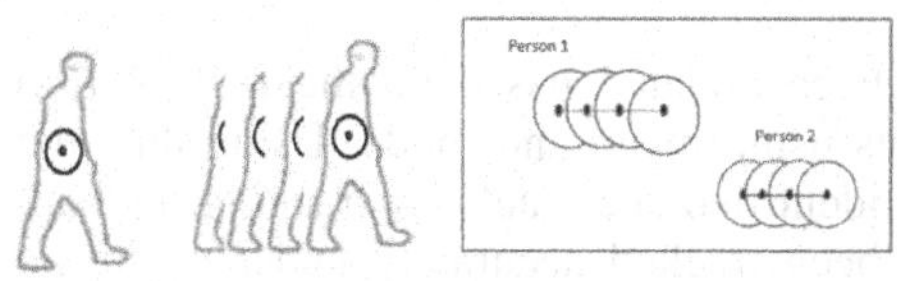

Fig. 1. (a) Bolb creation of two persons moving in the same direction

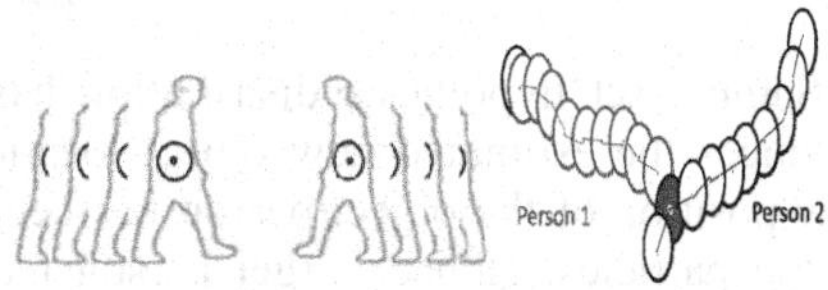

Fig. 1. (b) Crossover of bolbs during motion of two persons in opposite directions

Fig. 1(b) shows the existence of two different objects in the same frame. Both the objects are tracked simultaneously using the circles lying inside their contours. Fig. 1(b) shows the existence of two separate blobs being identified; hence two persons need to get tracked separately. According to this figure, two persons are walking towards each other. Initially, Person 1 and Person 2 are considered as new objects and they are tracked simultaneously. At some point, crossover has been happened between them. When one circle overlaps with two other circles of different directions, blob merging occurs. Blob merging represents the first occurrence of occlusion. Using particle filters, a new tracking hypothesis has been created to track the occluded persons (Person 1 and 2 in this case).

5.2 Particle Filter Tracking Approach

Simplified Particle filter has been used here to create the new tracking hypothesis. Color Distribution has been considered as feature for particle filter. Particle filter tracks the particular object in static as well as dynamic environments. The particles have been generated from the blob and each sample in the particle filter possess color feature. In Particle filter [1], Conditional state density p(Xt|Zt) for the set of samples has been defined by,

$S=\{S^{(n)},\pi(^{n})\}$ Where $S^{(n)}$ represents the sample set and $\pi(^{n})$ represents the sampling probability and it has been represented as,

$$\pi^{(n)} = \frac{1}{\sqrt{2\pi\sigma^2}} e^{\frac{d(n)}{2\sigma^2}} \tag{2}$$

At each sample locations, hypothesis distribution has been computed. The hypothesis has been compared with the target object distribution using Bhattacharya distance and the new weights have been calculated. Here, d(n) is the Bhattacharya distance between target color distribution and sample color distribution .

$$d^{(n)} = 1 - \rho[p^{(n)}, q] \tag{3}$$

$$\rho[p^{(n)}, q] = \sum_{u=1}^{m} \sqrt{p_u{}^{(n)} q_u} \tag{4}$$

Equation (4) gives the distance measurement between two distributions p and q. ρ is the Bhattacharya co-efficient, m is the histogram bin number, q is the target color distribution, p(n) is the sample hypothesis color distribution.

$$E(s) = \sum_{n} \pi^{(n)} s^{(n)} \tag{5}$$

Sample set hypothesis distribution has been updated based on mean E(s). E(s) produces the estimated new object locations using mean and mode. Using this, the new position of the object/person has been identified accurately by tracking the path of the particles. Hence, target person has been tracked accurately under occlusion state also.

6 Results and Discussion

The implementation of this object tracking system has been done using MATLAB (Version2013a). MATLAB is a high performance language for technical computing. The input videos are taken from CAVIAR dataset [12] and user generated datasets. The proposed algorithm have been applied and tested over many different test cases and two of the scenarios have been shown here. CAVIAR dataset video has been considered in test case 1 whereas user generated data set has been considered for test case 2.

Test case 1: CAVIAR dataset shows the videos which consists of a person walking along the corridor as shown in Fig. 2(a) and Fig. 2 (b) shows the input frame and the foreground image segmented from the background. Fig. 3(a) shows the input frame and Fig. 3(b) show the humans represented by circle. Fig. 4(a) and 4(b) and 4(c) depict the movement of humans through particle filter.

Fig. 2. a) An input image and b) its segmented frame

Fig. 3. a) An input frame and b) object representation

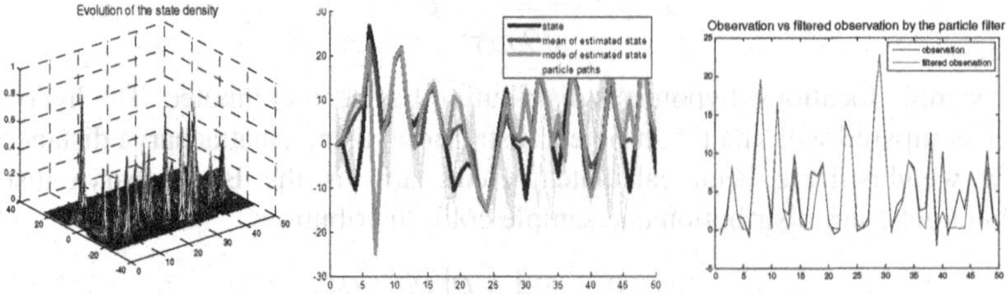

Fig. 4. a). Evolution of State density using Particle Filter, b)Estimated State using Particle Filter and c) Filtered Observation, all in sequence

Test case 2:

User generated videos has been considered in the test case 2. But in this scenario, a person walks in the terrace and returns back to his initial position and at the same time, another person walks on the same path as shown in figures 5(a). Our proposed algorithm tracks the multiple person successfully even under occlusion. Figure 6(a), 6(b) shows the circle representation on human Figure 7(a), 7(b) and 7(c) depict the movement of humans through particle paths of Estimated state and Filtered observation using Particle Filter.

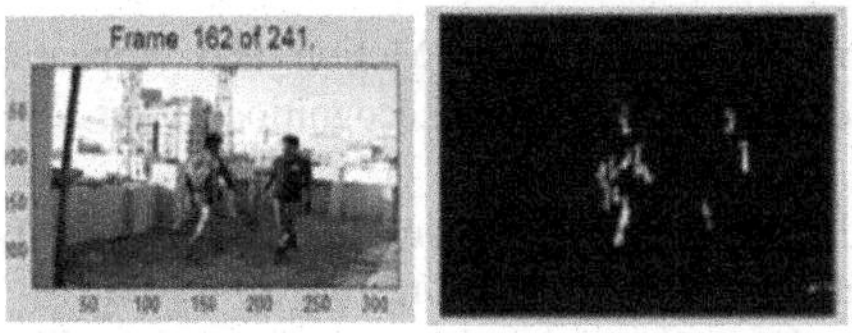

Fig. 5. a) Input image and its Segmented Frame

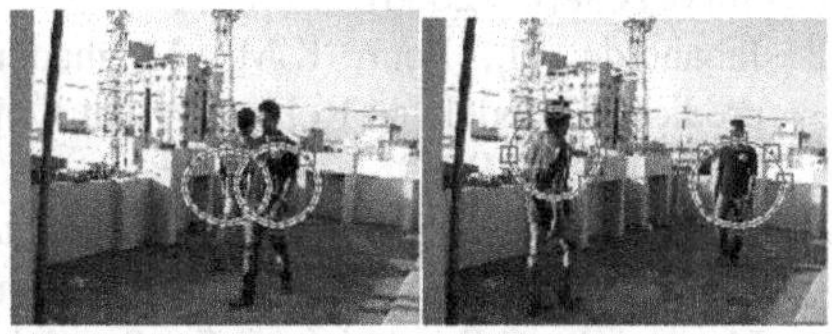

Fig. 5. b) The blob spliting and merging using the circles drawn

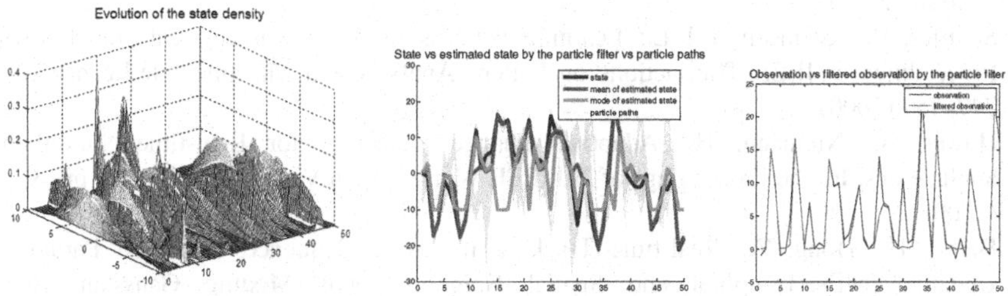

Fig. 6. a)Evolution of State Density using Particle Filter, b). Estimated State using Particle Filter and c)Filtered observation using Particle Filter, all in sequence

Performance of this system has been measured using Occlusion detection rate. This can be defined as the number of occluded objects tracked to the total number of occluded objects. The Proposed algorithm detects the person when he/she is occluded partially and the detection rate has been tabulated for two different datasets as in Table 1.

Table 1. Occlusion Detection Rate

Dataset Used	Two Persons
CAVIAR Dataset	94%
User Generated Dataset	91%

7 Conclusion

In this research, an automated tracking system has been developed using blob and particle filters to detect partial occlusion states. The proposed algorithm has been tested over CAVIAR and user generated datasets and the results are promising. This system has the ability to track the objects even if there is an object crossover also. In future, this system could be extended along with the detection of full occlusion and multiple objects tracking too.

References

1. Yilmaz, A., Javed, O., Shah, M.: Object Tracking: A Survey. ACM Computing Surveys 38(4) (2006)
2. Jahandide, H., Pour, K.M., Moghaddam, H.A.: A Hybrid Motion and Appearance prediction model for Robust Visual Object Tracking. Pattern Recognition Letter 33(16), 2192–2197 (2012)
3. Bhaskar, H., Maskell, L.M.S.: Articulated Human body parts detection based on cluster background subtraction and foreground matching. Neurocomputing 100, 58–73 (2013)
4. Manjunath, G.D., Abirami, S.: Suspicious Human activity detection from Surveillance videos. International Journal on Internet and Distributed Computing Systems 2(2), 141–149 (2012)
5. Gowshikaa, D., Abirami, S., Baskaran, R.: Automated Human Behaviour Analysis from Surveillance videos: a survey. Artificial Intelligence Review (April 2012), doi:10.1007/s 10462-012-9341-3
6. Stauffer, C., Grimson, E.E.L.: Learning patterns of activity using real-time tracking. Proceedings of IEEE Transactions on Pattern Analysis and Machine Intelligence 22(8), 747–757 (2000)
7. Huwer, S., Niemann, H.: Adaptive Change Detection for Real-time Surveillance applications. In: the Proceedings of 3rd IEEE Workshop on Visual Surveillance, pp. 37–45 (2000)
8. Haibo, H., Hong, Z.: Real-time Tracking in Image Sequences based-on Parameters Updating with Temporal and Spatial Neighbourhoods Mixture Gaussian Model. Proceedings of World Academy of Science, Engineering and Technology, 754–759 (2010)
9. Lu, J.-G., Cai, A.-N.: Tracking people through partial occlusions. The Journal of China Universities of Post and Telecommunications 16(2), 117–121 (2009)
10. Cho, N.G., Yuille, A.L., Lee, S.W.: Adaptive Occlusion State estimation for human pose tracking under self-occlusions. Pattern Recognition 46(3) (2013)
11. Varcheie, P.D.Z., Sills-Lavoie, M., Bilodeau, G.-A.: A Multiscale Region-Based Motion Detection and Background Subtraction Algorithm. The Proceedings of Sensor Journal 10(2), 1041–1061 (2010)
12. Fradi, H., Dugelay, J.-L.: Robust Foreground Segmentation using Improved Gaussian Mixture Model and Optical flow. In: The Proceedings of International Conference on Informatics, Electronics and Vision, pp. 248–253 (2012)
13. Wang, Y., Tang, X., Cui, Q.: Dynamic Appearance model for particle filter based visual tracking. Pattern Recognition 45(12), 4510–4523 (2012)

Efficient Lifting Scheme Based Super Resolution Image Reconstruction Using Low Resolution Images

Sanwta Ram Dogiwal[1], Y.S. Shishodia[2], and Abhay Upadhyaya[3]

[1] Department of Computer Science, Skit, Ramnagaria, Jaipur, India
[2] Department of Computer Science, JaganNath University, Jaipur, India
[3] University of Rajasthan, Jaipur, India
dogiwal@gmailcom, pvc@jagannathuniversity.org, abhayu@rediffmail.com

Abstract. Super resolution (SR) images can improve the quality of the multiple lower resolution images. it is constructed using raw images like noisy, blurred and rotated. In this paper, Super Resolution Image Reconstruction (SRIR) method is proposed for improving the resolution of lower resolution (LR) images. Proposed method is based on wavelet lifting scheme with Daubechies4 coefficients. Experimental results prove the effectiveness of the proposed approach. It is observed from the experiments that the resultant reconstructed image has better resolution factor, MSE and PSNR values.

Keywords: Super Resolution, SRIR, LR, Wavelets Lifting Scheme, Daubechies.

1 Introduction

Images [1] are obtained for many areas like a remote sensing, medical images, microscopy, astronomy, weather forecasting. In each case there is an underlying object or science we wish to observe, the original or true image is the ideal representation of the observed scene [1], [2]. There is uncertainty in the measurement occurring as noise, bluer, rotation and other degradation in the recorded images. There for remove these degradation using Super Resolution technology [5]. In super resolution technology in image processing area to get a High Resolution (HR) image. The central aim of Super Resolution (SR)[5] is to enhance the spatial resolution of multiple lower resolution images. HR means pixel density within the image is high and indicates more details about original scene. The super resolution technique is an efficient lossy and low cost technology. In this paper we are using Wavelet Transform (WT) technique to get an HR image from Low Resolution (LR) images by involving image blurring, registration, deblurring, denoising and interpolation.

The mathematical modeling of SRIR is done from several low resolution images. Low resolution means the lesser details of image. The CCD discretizes the images

M.K. Kundu et al. (eds.), *Advanced Computing, Networking and Informatics - Volume 1*, Smart Innovation, Systems and Technologies 27,
DOI: 10.1007/978-3-319-07353-8_31, © Springer International Publishing Switzerland 2014

and produces digitized noisy, rotated and blurred images. The images are not sampled according to the Nyquist criterion by these imaging systems. As a result, image resolution diminishes due to reduction in the high frequency components.

2 Super Resolution Image Reconstruction

We give the model for super resolution reconstruction [6] for set of low resolution images. The problem has been stated by casting a low resolution restoration frame work are P observed images$\{y_m\}_{m=1}^{P}$, each of size M_1xM_2 which are decimated, blurred and noisy versions of a single high resolution image Z of size N_1xN_2 where $N_1=qM_1$ and$N_2=qM_2$ After incorporating the blur matrix, and noise vector, the image formation model is expressed as $Y_m = H_m DZ + \eta_m$ where m =1, …, P

Here D is the decimation matrix of size $M_1M_2xq^2M_1M_2$, H is the PSF of size $M_1M_2xM_1M_2$, η_m is M_1M_2x1 noise vectors and P is the number of low resolution observations. Stacking P vector equations from different low resolution images into a single matrix vector gives

$$\begin{bmatrix} Y_1 \\ \cdot \\ \cdot \\ \cdot \\ Y_2 \end{bmatrix} = \begin{bmatrix} DH_1 \\ \cdot \\ \cdot \\ \cdot \\ DH_P \end{bmatrix} Z + \begin{bmatrix} \eta_1 \\ \cdot \\ \cdot \\ \cdot \\ \eta_P \end{bmatrix}$$

3 Lifting Schemes

The wave lifting scheme [10] is a method for decomposing wavelet transform into a set of stages. The forward lifting wavelet transform divides the data being processed into an even half and odd half. The forward wavelet transform expressed in the lifting scheme is shown in Fig. 1.

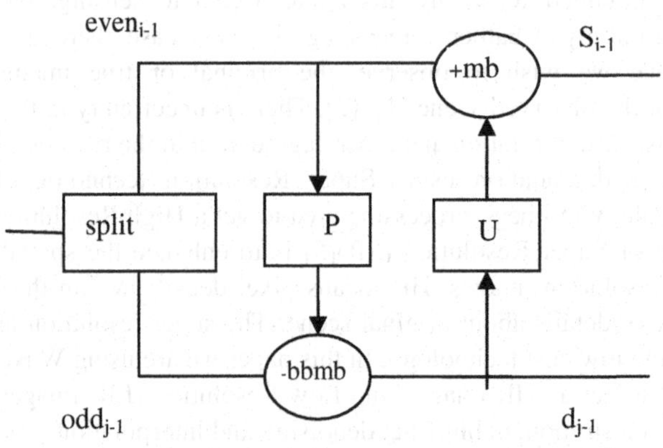

Fig. 1. The three step lifting scheme

The predict step calculates the wavelet function in the wavelet transform. In split step sort the entries into the even and odd entries. Generally prediction procedure p and then compute d_{j-1} = odd_{j-1} − p($even_{j-1}$). Update U for given entry, the prediction is made for the next entry has the small values and difference is stored. Updates are $S_{j-1}[n] = S_j[2n] + d_{j-1}[n]/2$ and $S_{j-1} = even_{j-1} + U(d_{j-1})$.This is high pass filter. The update step calculates the scaling function, which results in a smoother version of the data. This is low pass filter.

3.1 Lifting Scheme of the Daubechies4 Wavelet Transforms

The two steps of Lifting Scheme wavelet transform are update and predict. Here a new step is included named normalization; see Fig. 2.

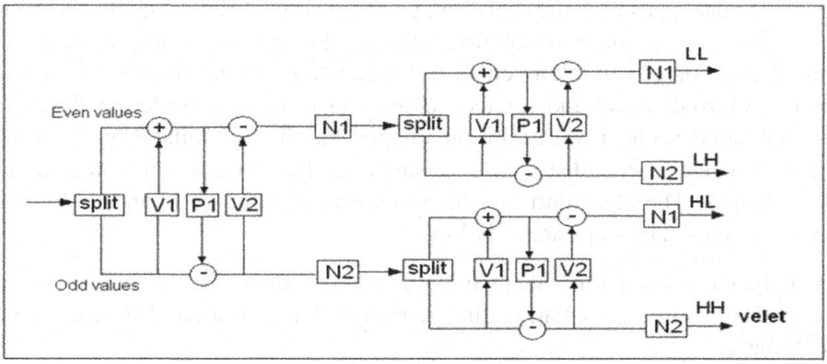

Fig. 2. Two stages Daubechies4 forward lifting wavelet transform

The input data in the split step is separated into even and odd elements. The even elements are kept in S_0toS_{half-1}, the first half of an N element array section. The odd elements in S_{half}, toS_{n-1},, the second half of an N element array section. The term LL refers to low frequency components and LH, HL, HH represents the high frequency components in the horizontal, vertical and diagonal directions respectively.

The forward step are Update1 (UI): for n=0 to half-1

S[n]=S[n]+√3S[half+n], Predict (PI): S[half]=S[half]-$\frac{\sqrt{3}}{4}$S[0]-$\frac{\sqrt{3-2}}{4}$S[halh-1}

For n=1 to half-1

S[half]=S[half]-$\frac{\sqrt{3}}{4}$S[n]-$\frac{\sqrt{3-2}}{4}$S [n-1], Update2 (U2): For n=0 to half-2

S[n]=S[n]-S[half+n+1], S[half-1]=S[half-1]-S[half]

Normalize (N): for n=0 to half-1

S[n]=$\frac{\sqrt{3-1}}{\sqrt{2}}$S[n] and S[n+half]=$\frac{\sqrt{3+1}}{\sqrt{2}}$S[n+half]

The backward transform, the subtraction and addition operation are interchanged.

3.2 SPHIT Technique

Set partitioning in hierarchical Trees (SPIHT) [3-12] is an image compression technique that exploits the inherent similarities across the sub bands in a wavelet decomposition of an image. In this scheme the code is most important wavelet transform coefficient first, and transmits the bits so that an increasing refined copy of the original image can be obtained progressively.

In the decomposition, SPIHT allows three lists of coordinates of coefficients. In the order, they are the List of Insignificant Pixels (LIP), the List of Significant Pixels (LSP), and the list of Insignificant Sets (LIS). At a certain threshold, a coefficient is considered significant if its magnitude is greater than or equal to threshold. The LIP, LIS, and LSP concept can be defined based on the idea of significance.

4 Proposed Super Resolution Reconstruction

The growing requirement for high resolution images resulted in the need for super resolution image reconstruction. Our goal in creating a super resolution image is to consider different types of same scene image of low resolution as inputs and combine them to generate a high resolution image through a series of steps. The implementation consists of taking either a source image for developing low resolution images like blurred, noisy and rotated versions or directly considering the different versions of same scene low resolution images based on availability. In proposed method we suggest a general structure for super resolution reconstruction using lifting wavelet schemes. The algorithms are based on breaking the problem into three steps to work on images. The steps are as below.

Step 1: Image registration, means mis-registered images (down sampled, wrapped, blurred, noisy, shifted) estimating the geometrical registration difference between different images.

Step 2: Restoration of the registered images and fusion using proposed lifting based de-noising.

Step 3: Interpolation of the resulting image from step 2 to obtain a super resolution image.

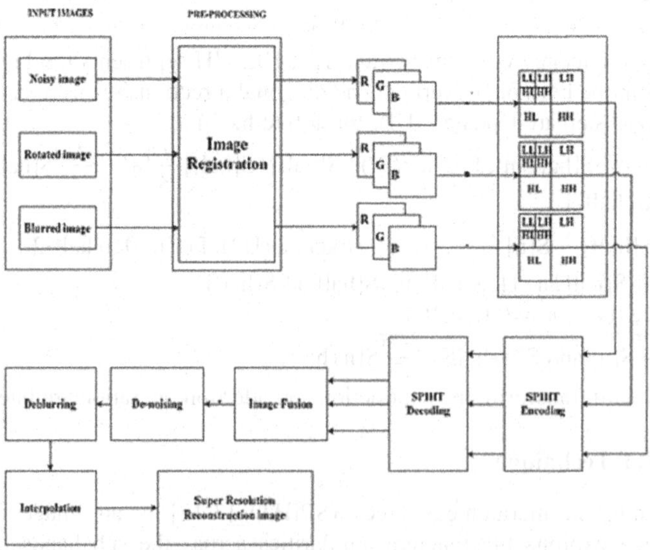

Fig. 3. Proposed diagram for Super Resolution

This working diagram fig. 3 is the way of doing our work. This is the step by step procedure that how we implement the images and how we evaluate the working parameters. In this approach three input low resolution blurred, noise, rotated images are considered. These images are registered using FFT based algorithms [7]. The registered low resolution images are decomposed using lifting schemes to specified levels. At each level we will have one approximation i.e. LL sub band and three detail sub band, i.e. LH, HL, HH coefficients. Each low resolution image is encoded using SPHIT scheme. The decomposed images are fused are using the fusion rule and inverse lifting scheme is applied to obtained fused image. The fused image is decoded using DSPHIT [8]. Restoration is performing in order to remove the bluer and noise from the image. We obtained the super resolution image using wavelet based interpolation.

5 Measurement of Performance

Mean Square Error (MSE) and Peak Signal to Noise Ratio (PSNR) are the two objective image quality measures used here to measure the performance of super resolution algorithm. *MSE* gives the cumulative squared error between the original image and reconstructed image and *PSNR* gives the peak error.TheMSE andPSNRof the reconstructed image are [11]

$$MSE = \frac{\sum[f(i,j)-F(i,j)]^2}{N^2}$$

Where f (i, j) is the source image and F(I, J) is the reconstructed image, containing N xN pixels, and for 8bits representations.

$$\text{PSNR}=20log_{10}(\frac{2^8}{MSE})$$

6 Results

In simulation, the proposed method was tested on several images. Results on two such images are shown in Fig. 4, Fig. 5 and Table 1. In Fig. 4 and Fig. 5a high resolution source image (512, 512) is a considered, from source image three noisy, blurred and rotated mis-registered LR images (256, 256) are created. We have tested using rotated version of source image by an angle of 15 degrees. It was noted that rotation was checked for different angles and results coalesce, the noisy image derived from source image by adding an additive white Gaussian noise with variance 0.05, the blurred version of source image developed by convolution with an impulse response of [1 2 2 2 2 2 2 2 2 2 1]. The results were verified for many impulse responses. The images were Pre-processed in image registration with arguments such that it applies to only rotational, translational and scaled images. By overlaying with reference images provided, the orientation of rotated image aligns similar to that of reference image. In figure (b) super resolution reconstructed images after reducing blur by blind deconvolution and iterative blind deconvolution respectively and are interpolated to twice the samples with the increase in the image size(512, 512). Experimental results show that proposed method produces better results as compared to simple DWT method. For example, MSE and PSNR values are improved from 36.682 and 32.246 to 36.166 and 32.548 respectively forLand Image. Table 1 shows the comparison between lifting scheme (proposed) based super resolution reconstruction with DWT based super resolution reconstruction Daubechies (D4) wavelets.

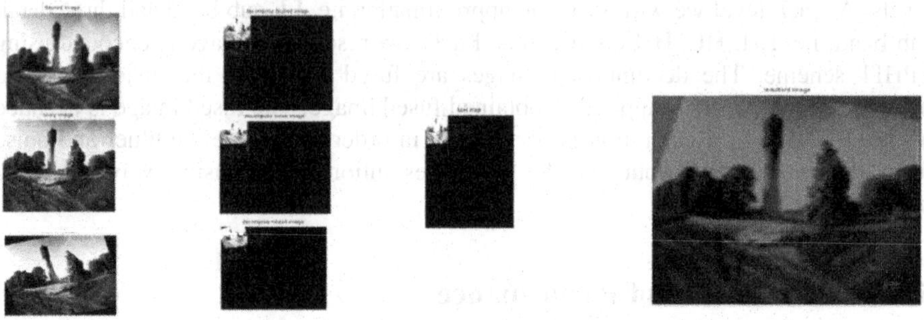

(a) (b)

Fig. 4. (a) Three LR images (noisy, blurred and rotated) and three decomposed images (noisy, blurred and rotated) and are fused using LWT (b) is the super resolution reconstructed image

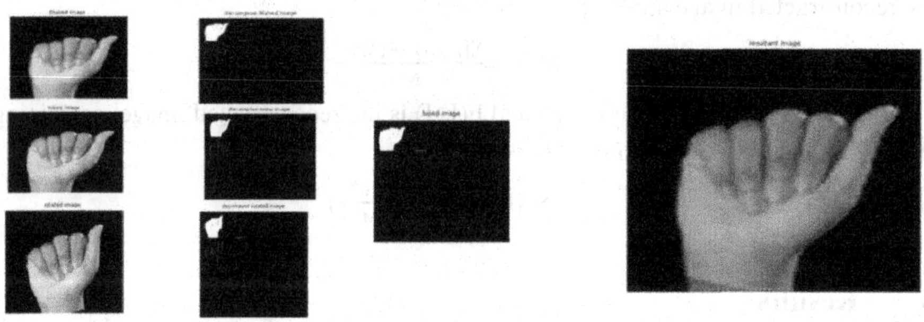

(a) (b)

Fig. 5. (a) Three LR images (noisy, blurred and rotated) and three decomposed images (noisy, blurred and rotated) and are fused using LWT (b) is the super resolution reconstructed image

Table 1. Performance of Super Resolution Reconstruction using WaveletLifting Schemes

Name of Image	Source Image	MSE	PSNR
DWT	Land	36.682	32.246
DWT	Hand	17.544	35.689
PROPOSED	Land	36.166	32.548
PROPOSED	Hand	15.787	36.148

7 Conclusion

Super resolution images enhance the quality of the multiple lower resolution images like noisy, blurred and rotated. Super resolution images increase the recognition rate in various applications like health domain, security applications etc. This paper proposed a technique for improving quality of lower resolution images namely Super Resolution Image Reconstruction (SRIR). Proposed method is based on wavelet lifting scheme. Experiments are performed with various input images like noisy, blurred and rotated images. Experimental results show that proposed method for construction of super resolution image performs well for super resolution image construction. In future, work can be performed feature extraction for LR imagesusing super resolution images with the implementation of Gabor transform.

References

1. Gonzalez, R.C., Woods, R.C.: Digital Image Processing. Prentice Hall (2002)
2. Jayaraman, S., Esakkirajan, S., Veerakumar, T.: Digital Image Processing.Tata McGraw Hill Education Pvt. Ltd. (2009)
3. Morales, A., Agili, S.: Implementing the SPIHT Algorithm in MATLAB. In: Proceedings of ASEE/WFEO International Colloquium (2003)
4. Hu, Y., et al.: Low quality fingerprint image enhancement based on Gabor filter. In: 2nd International Conference on Advanced Computer Control (2010)
5. Chaudhuri, S.: Super-resolution imaging. The Springer International Series in Engineering and Computer Science, vol. 632 (2001)
6. Kumar, C.N.R., Ananthashayana, V.K.: Super resolution reconstruction of compressed low resolution images using wavelet lifting schemes. In: Second International Conference on Computer and Electrical Engineering, pp. 629–633 (2009)
7. Castro, E.D., Morandi, C.: Registration of translated and rotated images using finite Fourier transform. IEEE Transactions on Pattern Analysis and Machine Intelligenc 9(5), 700–703 (1987)
8. Malý, J., Rajmic, P.: DWT-SPIHT Image Codec Implementation.Department of telecommunications, Brno University of Technology, Brno, CzechRepublic
9. Solomon, C., Breckon, T.: Fundamentals of Digital Image Processing: A practical approach with examples in Matlab. Wiley (2011)
10. Jiji, C.V., Joshi, M.V., Chaudhuri, S.: Single-frame image super-resolutionusing learned wavelet coefficients. International Journal of Imaging Systems andTechnology 14(3), 105–112 (2004)
11. Jensen, A., la Cour-Harbo, A.: Ripples in mathematics: The discrete wavelets transform. Springer (2001)
12. Ananth, A.G.: Comparison of SPIHT and Lifting Scheme Image CompressionTechniques for Satellite Imageries. International Journal of Computer Applications 25(3), 7–12 (2011)
13. Wang, K., Chen, B., Wu, G.: Edge detection from high-resolution remotely sensed imagery based on Gabor filter in frequency domain. In: 18th International Conference on Geoinformatics (2010)

14. Reddy, B.S., Chatterji, B.N.: An FFT-based technique for translation,rotation, and scaleinvariant image registration. IEEE Transactionson Image Processing 5(8), 1266–1271 (1996)
15. Zhu, Z., Lu, H., Zhao, Y.: Multi-scale analysis of odd Gabor transform for edge detection. In: First International Conference on Innovative Computing, Information and Control (2006)
16. Zhang, D., Jiazhonghe: Face super-resolution reconstruction and recognition from low – resolution image sequences. In: 2nd International Conference on Computer Engineering and Technology, pp. 620–624 (2010)

Improved Chan-Vese Image Segmentation Model Using Delta-Bar-Delta Algorithm

Devraj Mandal, Amitava Chatterjee, and Madhubanti Maitra

Department of Electrical Engineering
Jadavpur University, Kolkata - 700032, India
devraj89@gmail.com, cha_ami@yahoo.co.in,
Madhubanti.Maitra66@hotmail.com

Abstract. The level set based Chan-Vese algorithm primarily uses region information for successive evolutions of active contours of concern towards the object of interest and, in the process, aims to minimize the fitness energy functional associated with. Orthodox gradient descent methods have been popular in solving such optimization problems but they suffer from the lacuna of getting stuck in local minima and often demand a prohibited time to converge. This work presents a Chan-Vese model with a modified gradient descent search procedure, called the Delta-Bar-Delta learning algorithm, which helps to achieve reduced sensitivity for local minima and can achieve increased convergence rate. Simulation results show that the proposed search algorithm in conjunction with the Chan-Vese model outperforms traditional gradient descent and recently proposed other adaptation algorithms in this context.

Keywords: Image segmentation, Chan-Vese segmentation model, gradient descent search, level set method, Delta-Bar-Delta algorithm.

1 Introduction

Active region based image segmentation is generally carried out by Chan-Vese (C-V) model [1] by wrapping around [1], [7] an initial active contour [8] along the steepest descent direction of energy employing gradient descent search (GDS) method [3]. The Chan-Vese model for image segmentation is well developed and well-cited enough. The formulation of Chan-Vese model usually comes up with the requirement of solving the partial differential equation (PDE) for obtaining the route of contours evolved during the process of computation employing level set formulation [1], [7]. Consequently, the associated energy gradients are derived by using the Euler-Lagrange equations. In this respect, the GDS method is a convenient tool as they can be utilized for minimization of non-convex functionals and easy to implement as they involve calculation of first order derivatives. But it generally converges to the first local minimum it encounters and its rate of convergence is often very slow. For the sake of brevity, in this work we presume the findings of the related previous works to be truthful i.e. the C-V energy fitting functional [1] is non-convex and non-unique in nature and may also have many

M.K. Kundu et al. (eds.), *Advanced Computing, Networking and Informatics - Volume 1*, Smart Innovation, Systems and Technologies 27,
DOI: 10.1007/978-3-319-07353-8_32,

local minima. Thus to alleviate the problem of getting stuck to local minima, this work proposes a modified gradient search technique that guarantees better and faster convergence of the C-V algorithm towards its global minimum to get accurate segmentation results compared to the well established heuristic searches. The present work mainly rests on the efficacy of a variant of GDS, namely, the Delta-Bar-Delta rule (DBR), proposed by Jacobs [5], which utilizes a learning parameter update rule, in each iteration, in addition to the weight update rule. This method bears similarity with the RPROP [3] method, although the update rules are quite distinct in nature. In this work, we propose a modified version of DBR algorithm, namely MDBR algorithm, which utilizes a modified version of DBR algorithm to update learning rate parameters and momentum method to update weights, to achieve even faster convergence. The proposed Chan-Vese-MDBR algorithm has been utilized to segment both scalar and vector valued images and its superiority has been firmly established in comparison with other popular search methods used for level set based image segmentation algorithms e.g. the well established basic GDS model and recently proposed momentum (MOMENTUM), resilient backpropagation (RPROP) and conjugate gradient (CONJUGATE) based learning methods.

The outline of this paper is as follows: In Section 2 we describe the basic C-V model. In Section 3 we describe the fundamental search procedure of the GDS method. In Section 4 our proposed modified GDS algorithm has been presented. In Section 5, segmentation results of various gray-scale and colour images employing MDBR have been illustrated. Performance comparisons of MDBR with some other tools have also been highlighted in this section. Section 6 concludes the present work.

2 The C-V Algorithm

The C-V model [1] assumes that the image is formed by two regions of approximately homogenous intensities. The evolving contour C, tries to segment the image domain Ω into two regions i.e., Ω_1 and Ω_2 (inside and outside C). The fitting energy associated with the C-V model in the level-set representation [1], [7] is given by

$$F_{CV}(c_1, c_2, \phi) = \lambda_1 \int_{\Omega} |u_0 - c_1|^2 H(\phi) dxdy + \lambda_2 \int_{\Omega} |u_0 - c_2|^2 (1 - H(\phi))\, dxdy + \mu \int_{\Omega} \delta_0(\phi) |\nabla \phi| dxdy + \nu \int_{\Omega} H(\phi) dxdy \quad (1)$$

where, ϕ is the level set function, H and δ_0 [1] are the Heaviside and one-dimensional Dirac functions and the constants c_1 and c_2 [1] are, respectively, the mean intensity values within the regions Ω_1 and Ω_2. H_ε and δ_ε are the regularized versions of H and δ_0 respectively. The other terms in Eq. (1) denote the length and area of the curve C [1]. The $F_{CV}(c_1, c_2, \phi)$ can be minimized with respect to ϕ by solving the gradient flow equation [9].

$$\frac{\partial \phi}{\partial t} = -\frac{\partial F_{CV}(c_1, c_2, \phi)}{\partial \phi} \tag{2}$$

The PDE that needs to be solved to evolve the level set function [1] is given by:

$$\frac{\partial \phi}{\partial t} = \delta_\varepsilon(\phi)\left[-\lambda_1 (u_0 - c_1)^2 + \lambda_2 (u_0 - c_2)^2 + \mu.div\left(\frac{\nabla\phi}{|\nabla\phi|}\right) - \nu\right] \tag{3}$$

Then the new level set can be obtained by Eq. (4), where step is the length of the step in the GDS method.

$$\begin{aligned}\phi^{new} &= \phi^{old} - step * \frac{\partial F_{CV}(c_1 c_2, \phi)}{\partial \phi}\\ &= \phi^{old} + step * \frac{\partial \phi^{old}}{\partial t}\end{aligned} \tag{4}$$

The C-V model can also be extended to segment vector-valued images [2] and here also, the level set is evolved by using the basic GDS method.

3 The GDS Method and the Delta-Bar-Delta Algorithm

In the steepest descent implementation of the traditional adaptation, the GDS moves in the negative direction of the gradient, locally minimizing the cost function [3]. The basic GDS algorithm in the form of a standard line search optimization method is given by Eq. (5,6) [3], where x_k is the current solution, s_k is the next step consisting of length α_k and direction $\hat{p}_k$. For GDS, $\hat{p}_k$ is a descent direction, given by, $\hat{p}_k = -\nabla f_k$, where f is the cost function. The Eq. (5,6) are similar to the Eq. (3,4) for the C-V model:

$$x_{k+1} = x_k + s_k \tag{5}$$

$$s_k = \alpha_k \hat{p}_k \tag{6}$$

The modified GDS method highlighted in our work shows another way in which the descent direction and length can be optimally calculated for faster and better convergence than the basic GDS model. Some new methods such as momentum (MOMENTUM), Resilient Backpropagation (RPROP) [3] and conjugate gradient (CONJUGATE) [4] have already been proposed which show significant improvement over the basic GDS algorithm.

Jacobs developed an improved learning algorithm, called the Delta-Bar-Delta (DBR) algorithm, a modified version of the Delta-Delta algorithm, where the learning rule comprises both a weight update rule and a learning rate update rule [5]. The learning rate for a weight is increased by a fixed amount u, if the current derivative of the weight, $\delta(n)$, and the exponential average of the weights previous derivatives, $\bar{\delta}(n)$, are of the same sign, or, reduced in linear proportion, d, to its present value if, they are of opposite signs. The exponentially averaged derivative $\bar{\delta}(n)$ is calculated as [5]

$$\bar{\delta}(n) = (1 - \theta) * \frac{\partial \phi}{\partial t} + \theta * \bar{\delta}(n-1) \tag{7}$$

where, θ represents the base. The weight update rule employs the conventional steepest-descent algorithm where each weight is associated with its own learning rate parameter.

4 The MDBR Algorithm for Level Set Based Image Segmentation

In our algorithm, we have utilized a modified version of the DBR algorithm, named as MDBR algorithm. MDBR algorithm utilizes a modified version of DBR algorithm, where the learning rate is increased or decreased from its previous value by using a multiplying factor $u(> 1)$ and $d(< 1)$. The weight update rule is implemented using the momentum ω, in a bid to achieve faster convergence. This entire algorithm used for level set based image segmentation is given next.

The MDBR algorithm:

1. Construct the initial level set function $\phi(0)$ for iteration $n = 0$.
2. Initialize each learning rate $\eta_{ij}(0)$ and the constants ω, u, d and θ.
3. Calculate the values of the constants $c_1(\phi(n))$ and $c_2(\phi(n))$ respectively.
4. Solve the PDE in ϕ to calculate the partial derivative $\frac{\partial \phi(n)}{\partial t}$ i.e. $\delta(n)$ by Eq. (3).
5. Compute the exponential averaged gradient $\bar{\delta}(n)$ by Eq. (7).
6. Compute the new learning rate $\eta_{ij}(n)$ by Eq. (9).

$$\begin{aligned}
\eta_{ij}(n) &= \eta_{ij}(n-1) * u, \text{ if } \tfrac{\partial \phi_{ij}(n)}{\partial t} * \bar{\delta_{ij}}(n-1) > 0 \\
\eta_{ij}(n) &= \eta_{ij}(n-1) * d, \text{ if } \tfrac{\partial \phi_{ij}(n)}{\partial t} * \bar{\delta_{ij}}(n-1) < 0 \\
\eta_{ij}(n) &= \eta_{ij}(n-1), \text{ otherwise}
\end{aligned} \tag{8}$$

7. Update the level set ϕ by Eq. (9,10) using the new obtained learning rate η_{ij} and the momentum ω of the current solution:

$$s_n = (1 - \omega).\eta(n) * \frac{\partial \phi(n)}{\partial t} + \omega.s_{n-1} \tag{9}$$

$$\phi(n+1) = \phi(n) + s_n \tag{10}$$

8. The level set may have to be reinitialized locally. This step is optional and, if employed, is generally repeated after few iterations of the curve evolution.
9. Compare the level set functions $(\phi(n), \phi(n+1))$. If the solution is not stationary, then repeat from Step 3, otherwise stop contour evolution and report the segmentation result.

5 Implementation and Results

Our proposed algorithm has been used to segment different sample images both for scalar and vector-valued cases. The performances of our proposed method are compared with the traditional GDS algorithm, and the recently proposed MOMENTUM, RPROP [3] and CONJUGATE methods [4] based adaptation of level sets for image segmentation. The lower the fitness function reached indicates the superiority of a learning algorithm in approaching the optimum and the lower the computation time taken indicates better convergence rate achieved. In addition, the segmentation results are also compared with the ground truth to calculate the segmentation performance by using the Dice Coefficient (DC) [6] given as:

$$\mathrm{DC} = [2 * (\mathrm{ASI} \cap \mathrm{MSI})] \,/\, [|\mathrm{ASI}| + |\mathrm{MSI}|] \tag{11}$$

where, $|\mathrm{ASI}|=$ cardinality of segmented image (segmentation output from the C-V model), $|\mathrm{MSI}|=$ cardinality of manually segmented image (ground truth) and $(\mathrm{ASI} \cap \mathrm{MSI})=$ degree of correlation between the two image matrices. The closer the value of DC is to 1 [6], the better is the segmentation performance. For all our implementations, the values of the constants chosen are, $\omega = 0.1$, u=1.1, d=0.9 and θ=0.7.

Fig. 1 and Fig. 2 show the segmentation performance of a sample gray image by the above mentioned competing algorithms. The implementation is done such that, if the fitness function remains unchanged for 50 iterations, the algorithm stops. From Fig. 2, the MDBR based method is shown to be the fastest while it also achieves accurate segmentation results. Table 1 presents the corresponding comparative performances, in quantitative form. From Table 1 it can be easily observed that MDBR algorithm achieves the lowest fitness function value, consumes least computation time and reaches the minimum fitness function value within the fewest number of iterations. The DC value achieved is also quite competitive, greater than 0.99, compared to the other adaptation algorithms. Next, Fig. 3 and Fig. 4 show corresponding segmentation performances for a colour image. Here also, MDBR could successfully segment the given image within the quickest time, using fewest numbers of iterations. The efficiency of our proposed MDBR algorithm can be further illustrated by several examples of gray-scale and colour image segmentations shown in Fig. 5 and Fig. 6.

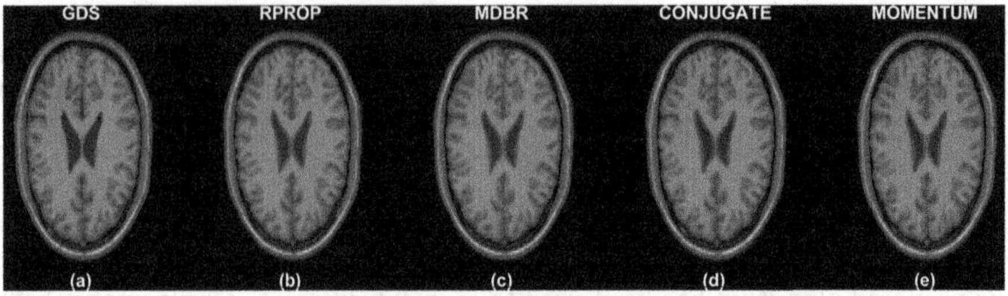

Fig. 1. Sample gray image segmentation by GDS and its variants

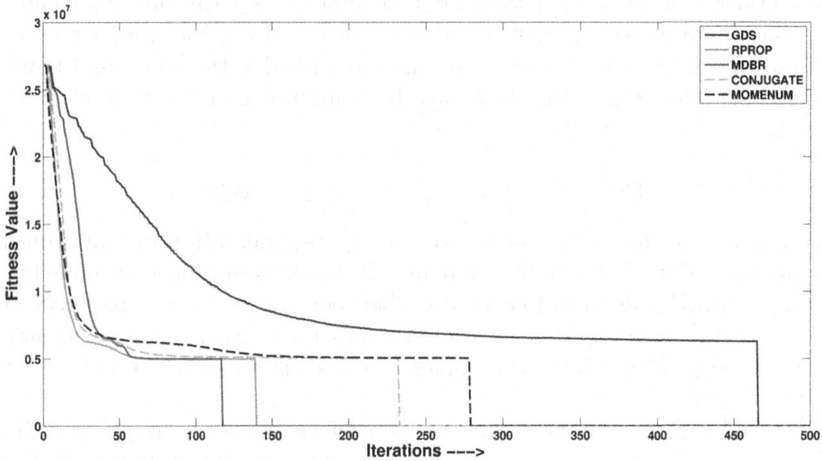

Fig. 2. The plot of the fitness function with number of iterations n for the image segmented in Fig. 1

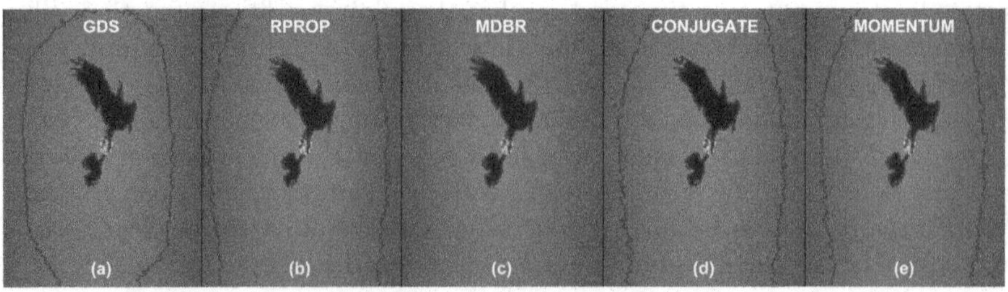

Fig. 3. Sample color image segmentation by GDS and its variants

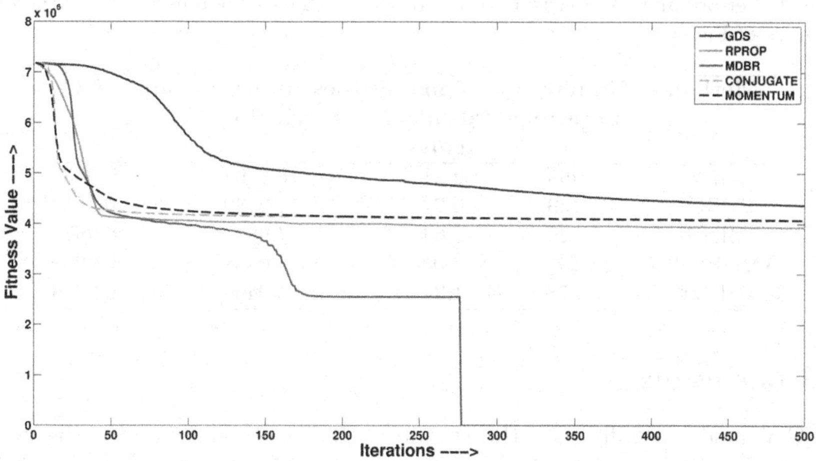

Fig. 4. The plot of the fitness function with number of iterations n for the image segmented in Fig. 3

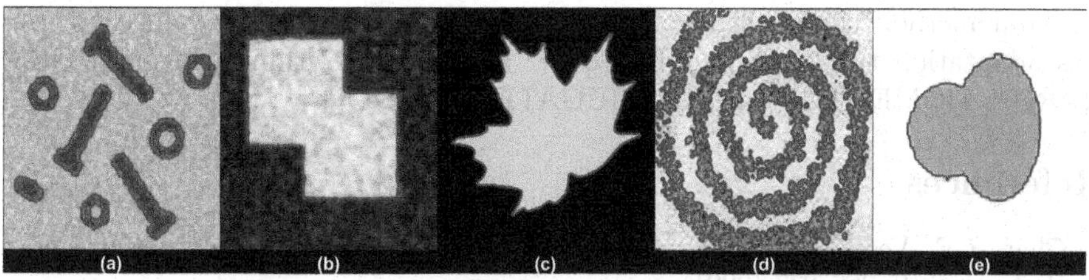

Fig. 5. (a)-(e) Grayscale image segmentation and their segmentation contours marked in red. The DC values, number of iterations and computation time (in seconds) for each image are: (a) 0.9610, 148, 0.94, (b) 0.9979, 117, 0.61, (c) 0.9995, 144, 0.74, (d) 0.9607, 154, 0.95, and (e) 0.9996, 113, 0.58

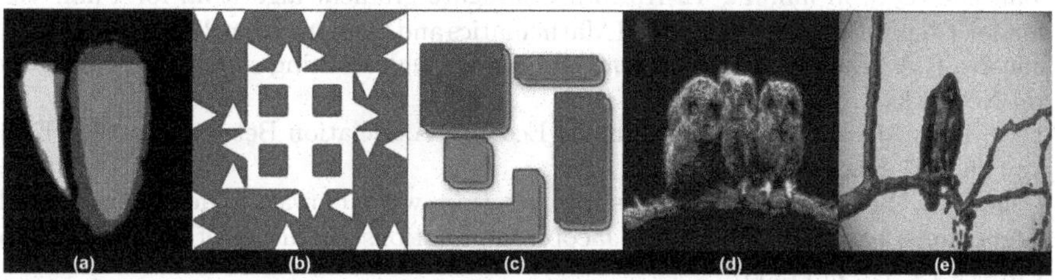

Fig. 6. (a)-(e) Colour image segmentation and their segmentation contours marked in red. The DC values, number of iterations and computation time (in seconds) for each image are: (a) 0.9995, 168, 1.68, (b) 0.9970, 113, 1.16, (c) 0.9004, 126, 1.19, (d) 0.9652, 225, 2.52, and (e) 0.9644, 239, 2.51

Table 1. Performance comparison for the different segmentation methods for the sample image of Fig. 1

Method	Number of iterations	Time taken in seconds	Fitness function value reached (x10^6)	DC
GDS	465	2.21	6.2863	0.9531
RPROP	139	0.74	4.9889	0.9958
MDBR	125	0.64	4.9822	0.9918
CONJUGATE	232	2.08	5.0438	0.9956
MOMENTUM	278	1.34	5.0468	0.9959

6 Conclusions

The C-V model usually uses the standard GDS method to evolve the active contour to achieve proper segmentation result. In this work, we have proposed an advanced adaptation algorithm associated with level sets, named as MDBR algorithm, a modified version of the Delta-Bar-Delta algorithm, so as to achieve reduced sensitivity for local optima and higher convergence speed in segmenting an image. Extensive segmentations of both scalar and vector-valued images show that our method can consistently achieve lower fitness function value and uses less adaptation time, compared to other learning algorithms like basic GDS, RPROP, MOMENTUM and CONJUGATE methods.

References

1. Chan, T.F., Vese, L.A.: Active contours without edges. IEEE Transactions on Image Processing 10, 266–277 (2001)
2. Chan, T.F., Sandberg, B.Y., Vese, L.A.: Active contours without edges for Vector-Valued Images. Journal of Visual Communication and Image Representation 11, 130–141 (2000)
3. Andersson, T., Läthén, G., Lenz, R., Borga, M.: Modified Gradient Search for Level Set Based Image Segmentation. IEEE Transactions on Image Processing 22, 621–630 (2013)
4. Jian-jian, Q., Shi-hui, Y., Ya-Xin, P.: Conjugate gradient algorithm for Chan-Vese model. Communication on Applied Mathematics and Computation 27, 469–477 (2013)
5. Jacobs, R.A.: Increased rates of convergence through learning rate adaptation. Neural Networks 1, 295–307 (1988)
6. Dice, L.R.: Measures of the Amount of Ecologic Association Between Species. Ecology 26, 297–302 (1945)
7. Osher, S., Sethian, J.A.: Fronts propagating with curvature-dependent speed: Algorithms based on Hamilton-Jacobi formulations. Journal of Computational Physics 79, 12–49 (1988)
8. Kass, M., Witkin, A., Terzopoulos, D.: Snakes: Active contour models. International Journal of Computer Vision 1, 321–331 (1988)
9. Li, C., Huang, R., Ding, Z., Gatenby, J.C., Metaxas, D.N.: A Level Set Method for Image Segmentation in the Presence of Intensity Inhomogeneities With Application to MRI. IEEE Transactions on Image Processing 20, 2007–2016 (2011)

Online Template Matching Using Fuzzy Moment Descriptor

Arup Kumar Sadhu[1], Pratyusha Das[1], Amit Konar[1], and Ramadoss Janarthanan[2]

[1] Electronics &Telecommunication Engineering Deptment,
Jadavpur University, Kolkata, India
[2] Computer Science & Engineering Deptment,
TJS Engineering College,
Chennai, India
{arup.kaajal,pratyushargj}@gmail.com,
konaramit@yahoo.co.in, srmjana_73@yahoo.com

Abstract. In this paper a real-time template matching algorithm has been developed using Fuzzy (Type-1 Fuzzy Logic) approach. The Fuzzy membership-distance products, called Fuzzy moment descriptors are estimated using three common image features, namely edge, shade and mixed range. Fuzzy moment description matching is used instead of existing matching algorithms to reduce real-time template matching time. In the proposed matching technique template matching is done invariant to size, rotation and color of the image. For real time application the same algorithm is applied on an Arduino based mobile robot having wireless camera. Camera fetches frames online and sends them to a remote computer for template matching with already stored template in the database using MATLAB. The remote computer sends computed steering and motor signals to the mobile robot wirelessly, to maintain mobility of the robot. As a result, the mobile robot follows a particular object using proposed template matching algorithm in real time.

Keywords: Arduino, Real-Time Template Matching, Fuzzy Moment Descriptor.

1 Introduction

Image matching belongs to computer vision and it is the fundamental requirement for all intelligent vision process [1]. Examples of image matching systems are vision-based autonomous navigation [2], automated image registration [3], object recognition [4], image database retrieval [2], 3D scene reconstruction [2]. Technically, the image matching process generally consists of three components Feature detection, Feature description, feature selection and optimal matching.

Feature detection [6] helps to detect an object by making local decisions at every image point. The local decision is about whether there exists a given image feature of a given type at that point or not. Lately, local features (edges, shades and mixed ranges) have been extensively employed due to their uniqueness and capability to handle noisy

M.K. Kundu et al. (eds.), *Advanced Computing, Networking and Informatics - Volume 1*, Smart Innovation, Systems and Technologies 27,
DOI: 10.1007/978-3-319-07353-8_33, © Springer International Publishing Switzerland 2014

images [6]. *Feature description* [6] represents the detected features into a compact, robust and stable structure for image matching. *Feature selection* [6] selects the best set of features for the particular problem from the pool of features [6].In *Optimal matching* [6] similarity between the features detected in the sensed image and those detected in the reference image is established. Local feature based approaches typically produces a large number of features needed to match [2], [7]. Such methods have very high time complexity and are not suitable for real-time application.

To implement the algorithm on real robots, an Arduino [8] development board is used to develop a mobile robot. Arduino assists to execute steering and motor control signals generated on the remote computer by Matlab, after locating the target template. Control signals are communicated wirelessly (using XBeemodule [8]) from a remote computer to the mobile robot. To locate the target template a dedicated wireless camera is mounted on the roof of the mobile robot. The Camera sends video information to the remote computer using a Bluetooth connection. Now, on the remote computer using Matlab template position is found using template matching algorithm and then steering and motor control signals are sent to the mobile robot using XBee.

In this paper a real-time application of a template matching algorithm has been shown on a real mobile robot is shown. Template matching, steering and motor control signals are generated as explained in the previous paragraph. Initially the grabbed RGB image from wireless camera is converted to gray image in a remote computer. The gray image has been partitioned into several non-overlapped blocks of equal dimensions as per our requirement. Each block containing three regions of possible characteristics, namely, 'edge', 'shade' and 'mixed range' [1] are identified. The sub-classes of edges based on their slopes in a given block are also estimated. The degree of fuzzy membership of a given block to contain the edges of typical sub-classes, shades and mixed range is measured subsequently with the help of a few pre-estimated image parameters like average gradient, variance and the difference of the maximum and the minimum of gradients. Fuzzy moment [1], which is formally known as the membership-distance product of a block b[i,w], with respect to a block b[j,k], is computed for all $1 \leq i, w, j, k \leq n$. A feature called 'sum of fuzzy moments [1]' that keeps track of the image characteristics and their relative distances is used as an image descriptor. The descriptors of an image are compared subsequently with the same ones of the test image. Euclidean distance is used as a measure to determine the distance between the image descriptors of two images. For checking that the test image is matched or not we set a threshold and check the Euclidean distance of the image description of the reference image with all test images is greater or less than the threshold value. If matched, according to the position of the test image the algorithm sends the steering signal to the motors.

The rest of the paper is separated into the following sections. Section 2 describes the required tools. Section 3 gives a brief explanation of the fuzzy template matching algorithm. Section 4 discusses about the experimental results and Conclusion and future work are proposed in Section 5.

2 Required Tools

The proposed methodology needs two basic tools. One is the mobile robot equipped with a wireless camera. Another one is the Fuzzy Moment Descriptor (FMD)[2]. Both of them are discussed in details below.

2.1 Arduino Based Mobile Robot with Camera

In this section a brief introduction about the mobile robot developed on the Arduino platform [8] is given below. This mobile robot is developed at Artificial Intelligence Laboratory, Jadavpur University by our research group. The mobility of the robot is maintained by a DC motor. Steering system (Ackerman steering [5]) is controlled by a DC servomotor. Also, it comprises of two ultrasonic sensors [8] to detect obstacle position, an encoder [8] to measure speed and distance covered by the robot and magnetic compass [8] to measure the magnetic field intensity at a point. It can detect surface texture and material using an infrared sensor. A wireless camera is mounted on the roof of the mobile robot. The camera records video and sends the video to the remote computer to process using MATLAB for template matching.

2.2 Template Matching Using Fuzzy Moment Descriptor

In this section template matching using fuzzy moment descriptor [2] is shown. The definition of 'edge', 'shade', 'mixed range', 'edge angle', 'fuzzy membership distribution', 'gradient', 'gradient difference', 'gradient average', 'variance' are taken from [6].The definition of Image features such as edge [6], shade [6] and mixed-range [6] and their membership distribution are used to calculate fuzzy moment distributions. The *variance* [6] (σ^2) of gradient [6] is defined as the arithmetic mean of square of deviation from mean. It is expressed formally as

$$\sigma^2=\sum\left(G - G_{avg}\right)^2 P(G) \tag{1}$$

Where, G denotes the gradient values of the pixels, and P(G)[1] represents the probability of the particular gradient G in that block.

2.3 Fuzzy Membership Distributions

Table 1summarizes the list of membership functions used in this paper. The membership values of a block *b*[*j,k*] containing edge, shade and mixed-range can be easily estimated if the parameters and the membership curves are known. The constant η, ρ, θ, ϕ, $\alpha, \beta, \lambda, \delta$, c, d, e, f, *a* and *b* are estimated using Artificial Bee Colony Optimization Algorithm [9].

Table 1. Membership functions for features

Parameter x=	Mixed-range membership	Edge membership	Shade membership
G_{avg}	$\frac{\eta x^2}{(\rho+\theta x^2+\phi x^3)}$ $\eta,\rho,\theta,\phi>0$	$1-e^{-bx}$ $b>0$	e^{-ax^2} $a>0$
G_{diff}	$\frac{\alpha x^2}{(\beta+\lambda x^2+\delta x^3)}$ $\alpha,\beta,\lambda,\delta>0$	$1-e^{-bx^2}$ $b>0$	e^{-ax^4} $a>0$
σ^2	$\frac{c x^2}{(d+e x^2+f x^3)}$ $c,d,e,f>0$	$1-e^{-bx^2}$ $b>0$	e^{-ax} $a>>0$

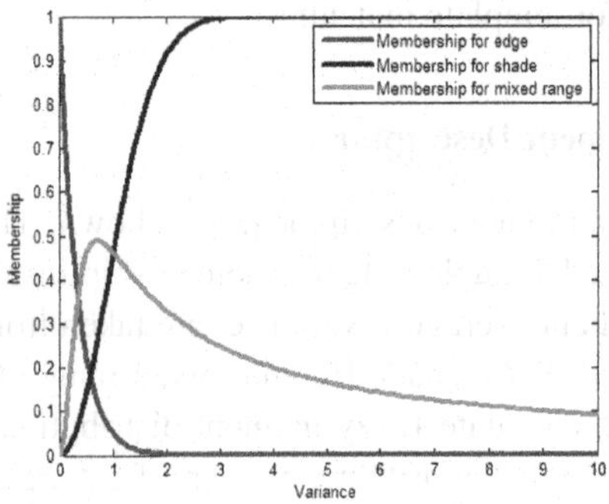

Fig. 1. Membership distribution of edge, shades or mixed range against variance

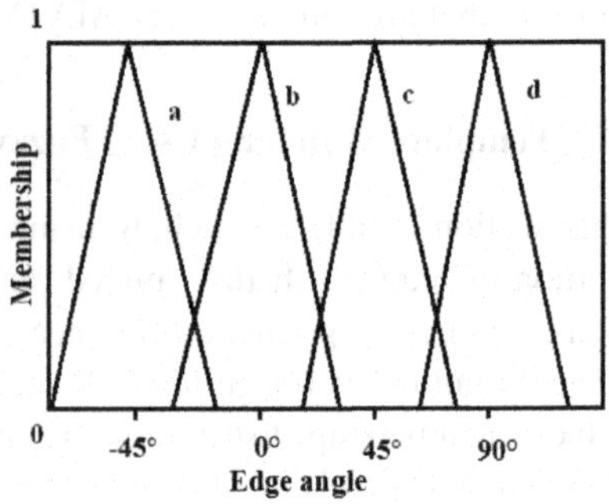

Fig. 2. The membership function of a block $b[j,k]$ containing edge with angle (a) $\alpha=-45^\circ$ (b) $\alpha=0^\circ$ (c) $\alpha=45^\circ$ and (d) $\alpha=90^\circ$

The evaluated values of the constants to calculate membership of edge, shade and mixed-range with respect to Variance (σ^2) are a=3, b=0.618, c=0.9649, d=0.1576, e=0.9706, f=0.9572.

The fuzzy production rules, described below, are subsequently used to estimate the degree of membership of a block $b[j,k]$ to contain edge (shade or mixed range) by taking into account the effect of all the three parameters together.

2.4 Fuzzy Production Rules

A fuzzy production rule is an If-Then relationship representing a piece of knowledge in a given problem domain. The If-Then rules represent logical mapping functions from the individual parametric memberships to the composite membership of a block containing shade. The $\mu_{shade}(b_{jk})$ now can be estimated by obtaining the membership values $\mu_{shade}(b_{jk})$w.r.t. G_{avg}, G_{diff} and σ^2, respectively, by consulting with the membership curves in Fig. 1.and Fig. 2.and applying the fuzzy AND (minimum)operator over these membership values. The single valued membership describes the degree of membership of the block *b*[*j*,*k*] to contain shade. The membership of the other features can be calculated same as shade just by replacing '*shade*' by appropriate features.

2.5 Fuzzy Moment Descriptors

In this section, we will describe fuzzy moments for different features and evaluate image descriptors based on those moments. *Fuzzy shade moment*$[M_{iw}^{jk}]_{shade}$ is estimated by taking the product of the membership value $\mu(b_{jk})_{shade}$ (of containing shade in the block *b*[*j*,*k*]) and normalized Euclidean distance $d_{iw,jk}$ of the block *b*[*j*,*k*] with respect to *b*[*i*,*w*] is given below,

$$[M_{iw}^{jk}]_{shade} = b_{iw,jk}\, X\, \mu_{shade}(b_{jk}) \tag{6}$$

The fuzzy mixed range and edge moments with edge angle α are also estimated using equation (6) with only replacement of the term 'shade' by required features. The *fuzzy sum of moments* (FSM), for shade *S*[*i*,*w*], with respect to block *b*[*i*,*w*] is defined as the sum of shade moments of the blocks where shade membership is the highest among all other membership values and is given below.

$$S_{iw} = \sum_{\exists jk} b_{iw,jk}\, X\, \mu_{shade}(b_{jk}) \tag{7}$$

Where,$\mu_{shade}(b_{jk}) \geq Max[\mu_x(b_{jk})]$,$x \in$ set of features.The FSM of the other features can be defined analogously following(7). After estimating fuzzy membership values for edges with edge angleα, shades and mixed range, the predominant membership value for each block and the predominant feature are saved. The FSMs with respect to the predominant features are evaluated for each blocking the image. For each of six predominant features, shade, mixed range and edges with edge angles, there are six sets of FSMs. Each set of FSM (for example the FSM for shade) is stored in a one-dimensional array and is sorted in a descending/ascending order. These sorted vectors are used as descriptors for the image. The Euclidean distance or image distance between reference image, r and an unknown test image, *y* is given by D_{ry}.

$$D_{ry} = \sum_k \beta_k * [E_{i,j}]_k \tag{8}$$

Where, the suffix i and j in $[E_{i,j}]_k$ corresponds to the set of vectors V_i for image*r* and V_j for image y, for $1 \leq i, j \leq 6$.*k* indicates the kth sorted FSM descriptor. In this paper the image descriptors for a known reference image are evaluated and saved prior to the matching process. The descriptors for the test image, however, are evaluated in real time when the matching process is invoked. A threshold (ε) is introduced in this situation to identify whether the test image is reference image or not. If $D_{ry} \geq \varepsilon$, then the test image is not reference image. Otherwise the test image is reference image.

3 Tracking Algorithm

In this section the real-time tracking algorithm is shown. The average time to track the test object is 5.36secin Intel(R) Core(TM) i7-3770 CPU with clock speed of 3.40GHz using Matlab R2012b. The time complexity of the image matching algorithm is calculated as per [1]. However, for online object tracking purpose, propagation delays are added to the time complexity of fuzzy image matching algorithm. Following this algorithm mobile robot tracks the target image. The real-time template matching algorithm is shown below.

Algorithm

Input: Image of the object to be tracked, select a threshold $\varepsilon \to 0$.

Output: Steering and motor signals.

Initialization:

1.Convert the reference image from rgb to gray
2.Calculate the size (MXN) of the reference image.
3.Calculate the gradient magnitude and gradient Direction of the reference image.
4. Calculate three parameters 'G_{avg}','G_{diff}','variance' for the blocks using equations given in Table 1.
5. Use fuzzy production rule for a block containing edge.
6. Calculate the membership of 'edge', 'shade' with respect to G_{avg}', 'G_{diff}', 'variance'
7. Calculate the membership for edge angles for all blocks.
8. Calculate the fuzzy sum moment for G_{avg}', 'G_{diff}', 'variance'using(7) for all blocks.

Begin:

1. Take the test frame from the video signal.
2.Calculate the fuzzy moment descriptor for the test image usingstep1tostep8 in the initialization.
3. Calculate the Euclidian distance, D_{ry} between Reference image,r and test image, y using (8).
4.**If** the $D_{ry} < \varepsilon$

Then reference image is present in the test image.

Do

Begin

a. Calculate the x-coordinate reference image on the current frame.

```
        If the previous coordinate < present coordi-
        nate
        Move right with a constant velocity.
  Else if previous coordinate > present coordinate
  Move left with a constant velocity.
     Else
  go straight with a constant velocity.
  End
  Else reference image≠test image.
5: Repeat step1 to step4 up to the time we want to
trackthe object.
End
```

4 Experiment on Mobile Robot

In this section template matching algorithm is implemented on a mobile robot. An uneven shaped object is chosen for the experiment. Fig. 3 shows the experimental setup and with tracking object. In Fig. 3.(a) mobile robot is identifying the target in front of it and following it. If the target is on the left or right of the robot and mobile robot adjust its steering and follow the target template. When the object is moving towards the left, the robot turns left turn as shown in Fig. 3(b) to follow the target. Fig. 3(c) shows the right turn taken by the robot to follow the target.

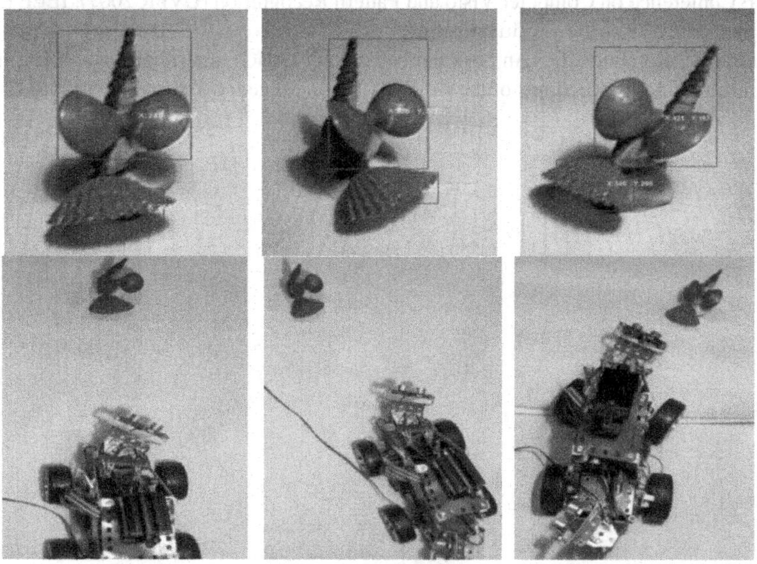

(a) Front tracking (b) Left tracking (c) Right tracking

Fig. 3. Screen shot of tracking targets and tracking by car

5 Conclusion and Future Work

In this paper, we have proposed a real time template matching algorithm for a mobile robot. The image matching algorithm involves the fuzzy moment descriptors which makes the algorithm much faster and accurate as compared to the existing algorithms. The algorithm was also implemented in hardware system (here in Arduino base mobile robot).Though the real-time response is not satisfactory but it is efficient in terms of accuracy.To improve response time the fuzzy moment descriptors along with kernel projection and it can be implemented.

References

1. Biswas, B., Konar, A., Mukherjee, A.K.: Image matching with fuzzy moment descriptors. Engineering Applications of Artificial Intelligenc 14(1), 43–49 (2001)
2. Konar, A.: Computational Intellingence: Principles, Techniques, and Applications. Springer (2005)
3. Hel-Or, Y., Hel-Or, H.: Real-time pattern matching using projection kernels. IEEE Transactions onPattern Analysis and Machine Intelligenc 27(9), 1430–1445 (2005)
4. Baudat, G., Anouar, F.: Feature vector selection and projection using kernels. Neurocomputing 55(1), 21–38 (2003)
5. Everett, H.R.: Sensors for mobile robots: theory and application. AK Peters, Ltd. (1995)
6. Gonzalez, R.C., Woods, R.E.: Digital Image Processing. Addison-Wesley (1992)
7. Wang, Q., You, S.: Real-time image matching based on multiple view kernel projection. In: IEEE Conference on Computer Visio and Pattern Recognition (CVPR 2007). IEEE (2007)
8. [Online] http://www.arduino.cc
9. Basturk. B., Karaboga, D.: An artificial bee colony (ABC) algorithm for numeric function optimization. In: Proceedings of the IEEE Swarm Intelligence Symposium (2006)

Classification of High Resolution Satellite Images Using Equivariant Robust Independent Component Analysis

Pankaj Pratap Singh[1] and R.D. Garg[2]

[1] Department of Geomatics Engineering,
[2] Department of Civil Engineering
Indian Institute of TechnologyRoorkee
Roorkee, Uttarakhand 247667, India
pankajps.iitr@gmail.com, garg_fce@iitr.ernet.in

Abstract. Classification approach helps to extract the important information from satellite images, but it is quite effective on extracting the information from mixed classes. The problem of classes is well known in the area of satellite image processing due to similar spectral resolution among the few classes (objects). It is quite obvious in multispectral data which is having little variation in spectral resolution with heterogeneous classes. In earlier times, Neural network based classification has been used widely, but at a cost of high time and computation complexity. To resolve the problem of the mixed classes due to the spectral behavior in sufficient time, a novel method Equivariant Robust Independent Component Analysis (ERICA) is proposed. This algorithm separates the objects from mixed classes, which shows similar spectral behavior. It can easily predict the objects without using of pre-whitening technique. Therefore, pre-whitening is not playing an important role in convergence of the algorithm. Due to Quasi-Newton based iteration in this algorithm helps to converge to a saddle point with locally isotropic convergence, regardless of the spatial and spectral distributions of satellite images. Hence, this proposed ERICA gives major contribution for classification of satellite images in healthy trees, buildings and road areas. Another important one in the image is shadow information, which helps to show elevated factor due to high rising buildings and flyovers in emerging suburban areas. The experimental results of the remote sensing data clearly indicates that the proposed ERICA has better classification accuracy and convergence speed, and is also appropriate to solve the image classification problems.

Keywords: Equivariant Robust Independent Component Analysis, Information extraction, Image classification, Mixed class, Performance index.

1 Introduction

Satellite Images play a vital role in daily life applications such as classified objects, road network generation and in area of research such as information extraction. Images are often contaminated by factors such as mixed classes problem from multispectral sensors i.e. those having similar spectral characteristics. In absence of

M.K. Kundu et al. (eds.), *Advanced Computing, Networking and Informatics - Volume 1*,
Smart Innovation, Systems and Technologies 27,
DOI: 10.1007/978-3-319-07353-8_34, © Springer International Publishing Switzerland 2014

edge information is become a quite difficult to segregate the objects in images and the noise occurrence in form of mixed classes also mislead for correct classified region. Since, the attenuation of noise and preservation of information are usually two contradictory aspects of image processing. Thus, the noise reduction for image enhancement is a necessary step in image processing. Depending upon the spectral resolution of satellite imagery and the areas of restoration, various noise reduction methods make several assumptions. Therefore, a single method is not useful for all applications. Zhang, Cichocki and Amari has been developed the state-space approach towards the problems related to blind source separation and also the recovery of the original classes by incorporating deconvolution algorithm [2], [6], [14-15]. Nonnegative constraints were imposed to solve the problem of mixed signal. These constraints have utilized nonnegative factors in sparse matrix form to separate correlated signals from mixed signals [7]. Least mean square (LMS) type traditional adaptive algorithm utilized activation function during the learning process. Blind least mean square (BLMS) is the other version of the LMS algorithm with the calculated blind error, which has to be tolerating on improving the performance of BLMS [5].

To extract the roof of buildings automatic or semi-automatic, an image processing algorithm with artificial intelligence methods were proposed [3], [11]. Li *et al.* (2010) proposed an approach for automatic building extraction on considering a snake model with the improved region growing and mutual information [10]. All smaller details (e.g. chimneys) which hinder proper segmentation were removed from each separate multispectral band by applying a morphological opening with reconstruction, followed by a morphological closing with reconstruction [13].In 2013, a hybrid approach is proposed for classification of high resolution satellite imagery. It is comprised of improved marker-controlled watershed transforms and a nonlinear derivative method to derive an impervious surface from emerging urban areas [12].

2 A Proposed Framework for Satellite Image Classification

In this section, an ERICA technique for classification of high-resolution satellite images (HRSI) is proposed. Initially, a HRSI generates components by using ERICA to discriminate mixed classes i.e. having similar spectral characteristics. It is asymptotically equivariant in the presence of Gaussian noise. This algorithm separates the objects from m mixture of n classes. This algorithm is a quasi-Newton iteration that will converge to a saddle point with locally isotropic convergence, regardless of the distributions of sources. This algorithm is capable to converge without preprocessing steps such as low and high-pass filtering on input satellite images, which is defined in the form of higher order cumulants. It is quite effective in the presence of a large number of samples [8]. Hence, the proposed ERICA approach is quite adaptable and extendable algorithm for separating the different classes in an image. At the moment of source separation, the gradient of $\psi(G)$ disappears suddenly (at the expression of G = I or any permutation matrix). The gradient of $\psi(G)$with respect to G is obtained by using the cumulants properties as follows,

$$\frac{\partial \psi(G_r)}{\partial G_r} = S_x{}^{\beta} C_{x,s}^{\beta,1} - G^{-T} \tag{1}$$

While, the above equation depends only on the outputs and its solution will go ahead to an equivalent approximate of matrix G [1].

2.1 Equivariant Robust Independent Component Analysis (ERICA)

In this section, an algorithm is explained that help to converge towards the disjoint solution. Since source, partition is not at least or utmost of ψ(G) but a saddle point, gradient-based approaches cannot be utilized as the predictable gradient or the innate gradient to adapt the segregation system [2, 4, 6]. Instead, the BSS solution is suggested by using a preconditioned process, which uses the second-order information existing at partition. In order to get the zeros of the gradient, the exploitation of a preconditioned iteration form is recommended by Kelley (1995) [9], such as

$$vecG_r^{(m+1)} = vecG_r^{(m)} - \mu^{(m)}(\hat{h\psi})^{-1} vec(\frac{\partial \psi}{\partial G_r}) \tag{2}$$

Where $\hat{h}\psi$is defined as an estimation of the true Hessian matrix in the region of the partition. This Hessian approximation is proposed as follows,

$$\hat{h\psi}(G_r) = k_M((G_r^{-1})^T \otimes G_r^{-1}), \tag{3}$$

This shows the difference only in the diagonal terms of the true Hessian matrix. Furthermore, on the separation of the classes, the aforesaid difference will become negligible, due to the Hessian approximation. It keeps true hessian for the Eigen values with a single module. The true hessian plays an important role to avoid the possibility of converging towards non-separating solutions. On Substituting (4) into the iteration (3) and then the following expression in the following algorithm:

$$G_r^{(m+1)} = G_r^{(m)} - \mu^{(m)}(C_{x,x}^{1,\beta} S_x^{\beta} - I)G_r^{(m)}, \tag{4}$$

the separating system is obtained as in the following Eq. (6),

$$B_s^{(m+1)} = B_s^{(m)} - \mu^{(m)}(C_{x,x}^{1,\beta} S_x^{\beta} - I)B_s^{(m)}, \tag{5}$$

This expression denoted as the CII (Cumulant-based iterative inversion) term, which is significantly also known as Quasi-Newton method, due to the recursion property. The saddle point can be found by the results of the comprehensive and cumulant-based iterative inversion (GCII) algorithm,

$$B_s^{(m+1)} = B_s^{(m)} - \mu^{(m)}\left(\sum_{\beta\in\Omega} w_\beta C_{x,x}^{1+\beta} S_x^{\beta} - I\right)B_s^{(m)}, \tag{6}$$

Where, $\mu^{(m)}$ signifies the optimal step-size value to achieve the separation of classes in the manner of local convergence and $\sum_{\beta\in\Omega} w_\beta = 1$

Therefore, the advanced version of the algorithm utilizes many cumulants matrices to make it extra robust in the sense of deducting the probability, which emphasizes mainly on a specific cumulant. These deductions come due to the occurrence of a few bad choices of cumulant sequence whose outcomes in near zero values of the weighted sum of cumulants for few sources. Additionally, the statistical information utilizes in best way by using many cumulants matrices in the proposed algorithm. However, the variance of the cumulant estimation is inversely proportional to the selected weighting factor.

The proposed ERICA approach is used to classify the existing objects from the HRSI on resolving the mixed class problem with preserving mutual exclusion among classes. The mandatory input argument is a HRSI image as converted in Matrix form 'x', which is observed as in the dimension (sources x, samples s). Each column and row of this matrix corresponds to a different spectral resolution and pixel respectively. The other optional input arguments are the number of independent components (e), which has to extract (default e=1) simultaneously vector with the weighting coefficients (w). This iterative process involvesa sensitivity parameter to stop the convergence.

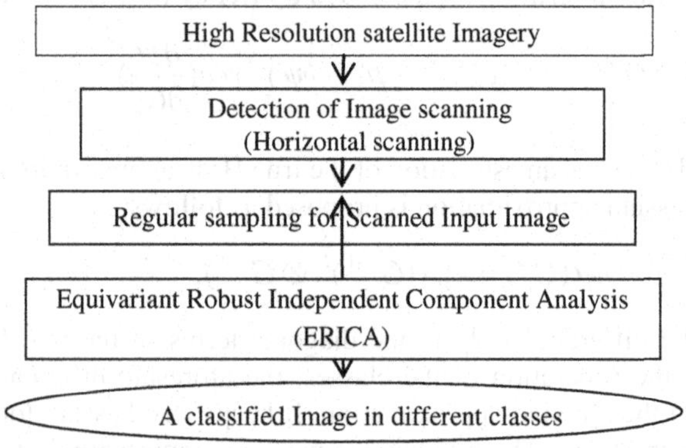

Fig. 1. A Proposed Framework for satellite image classification using ERICA method

Fig.1 shows a proposed Framework for satellite image classification using ERICA method. Initially a HRSI is taken as input to complete the scanning process in horizontal and vertical direction with regular sampling to scan one by one pixel. Now, the proposed ERICA approach is asymptotically in nature to converge the desired saddle points. These saddle points help to approach for separation of classes with the existing cumulants based method in the proposed approach.Finally,ERICA approach resolves the mixed class problem onsegregatingthe different objects in satellite image.

3 Results and Discussion

The classification results for emerging suburban area are shown in Fig.2. Fig.2(a) and Fig. 2(b) denoted as ES1 & ES3 classified satellite images respectively having good PI values, which show that the separation of classes are efficiently achieved. Fig. 2(a) and Fig. 2(b) shows the roads and vegetation area with light grey and light bluish respectively. In Fig.2(b) Non-metallic building 'roof can be easily seen as dark sky blue color with respect to Metallic structure of building 'roof is in dark grey color. Roads are in grey color due to excessive quantities of asphalt andthe snow-white color reflects the shadow effect as shown in Fig. 2(b). The proposed approach shows the importance in classification of satellite images in compare to Maximum likelihood

Classifier (MLC), Biogeography Based Optimization (BBO) approaches, due to better separation among classified classes.

3.1 Performance Index and Accuracy Assessment

The presented qualitatively performance analysis of the proposed ERICA algorithm is evaluated with the help of the performance index (PI). This PI is defined for the processing unit at the kth extraction of the (k, l) element in the matrix of the global system which comprise a separating and mixing matrix. If the occurrence of performance index value is near about to zero or smaller, it means the perfect separation is achieved at the kth extraction-processing unit.

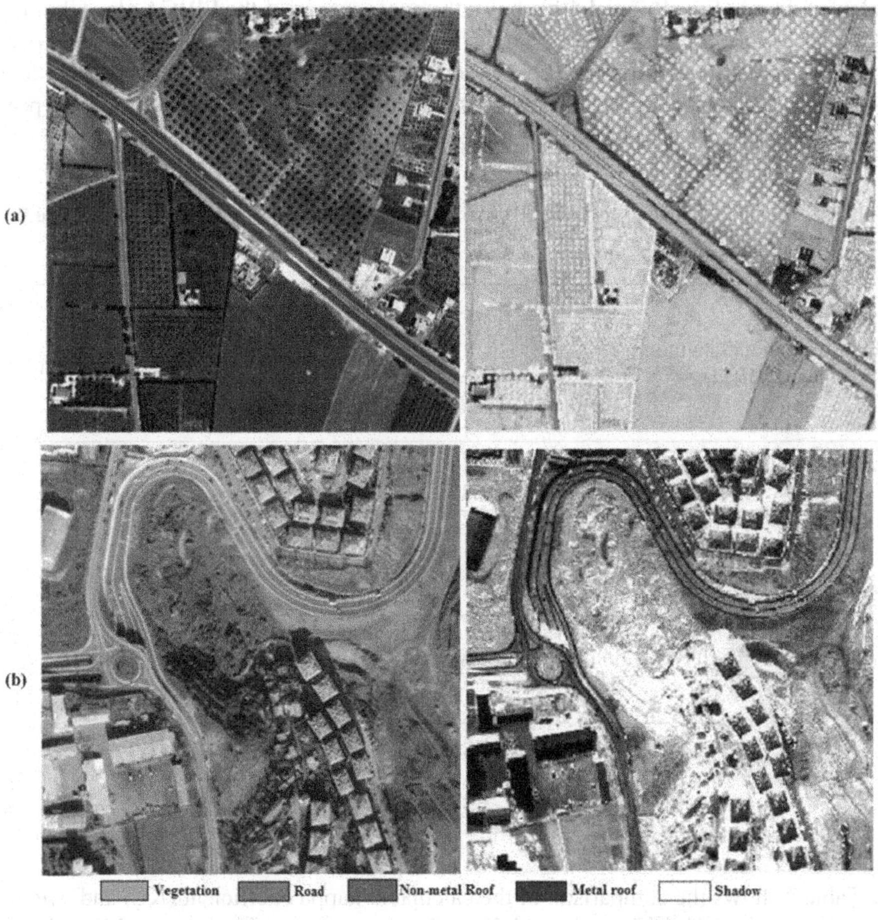

Fig. 2. ERICA based classified results of the emerging suburban area (a-b): (I) Input image; (II) Result of our proposed method

Table 1. Performance measurement of proposed ERICA algorithm for classification of high resolution satellite images

High Resolution Satellite Images (HRSI)		Iteration	Convergence	Time elapsed (Sec)	Performance Index(PI)
Emerging suburban areas	ES1	1 - 25	0.57563 - 0.00008	15.14	0.0325
	ES3	1 - 41	0.20248 - 0.00011	23.10	0.0513

Table 1 shows the performance measurement of proposed ERICA algorithm for classification of high resolution satellite images. It shows the no. of iteration to converge the algorithm in the calculated time and finally shows the performance index in calculated time. Fig. 4(a) and 4(b) shows the convergence of the ERICA algorithm for ES1 and ES3 HRSI images, which took 15.14 and 23.20sec time as the resulted output of classified image respectively. In addition, the calculated P.I. value 0.0325 and 0.058, for ES1 and ES3 HRSI images respectively which significance shows the perfect separation in objects of satellite image of emerging suburban area.

Table 2. Kappa coefficient and overall accuracy of classification of high resolution satellite images

High Resolution Satellite Images (HRSI)		Overall Accuracy (OA)	Kappa Coefficient (K)		
		Proposed ERICA	Proposed ERICA	Maximum Likelihood Classifier (MLC)	Biogeography Based Optimization (BBO)
Emerging suburban areas	ES1	95.25%	0.8756	0.7746	0.6726
	ES3	96.27%	0.8856	0.7836	0.6816

Table 3. Producer's accuracy of classification of high resolution satellite images

High Resolution Satellite Images (HRSI)		Producer's Accuracy (PA) (%)			
		Trees	Buildings	Roads	Shadow
Emerging suburban areas	ES1	98.71	89.91	98.81	-
	ES3	-	93.43	90.79	97.68

Table 2 shows the comparison of the calculated kappa coefficient (KC) and overall accuracy (OA) of classification of high resolution satellite images among MLC, BBO and the Proposed ERICA approaches.

Table 4. User's accuracy of classification of high resolution satellite images

High Resolution Satellite Images (HRSI)		User's Accuracy (UA) (%)			
		Trees	Buildings	Roads	Shadow
Emerging suburban areas	ES1	98.41	89.09	97.81	-
	ES3	-	93.43	90.79	97.68

The proposed ERICA method shows the better KC value in comparison to MLC and BBO methods. It shows the number of iteration to converge the algorithm in the calculated time and finally shows the performance index in calculating time. Table 3 and Table 4 show the producer's accuracy (PA) and user's accuracy (UA) of classification of high resolution satellite images respectively.

4 Conclusion and Future Scope

In this paper, an automated satellite image classification achieved a good level of accuracy with a proposed ERICA approach. Therefore, the classified image helps to identify most relevant information. This information shows the existing objects such as buildings, road, shadow area and vegetation, in emerging urban areas. The Shadow area reflects the elevation factors, which signified as a similar kind of pattern in shadow due to the availability of more buildings. The uncorrelated sources can be removed to extract only relevant regions with the help of deflation procedure in ERICA approach. This proposed approach is quite susceptible towards mixed classes due to spectral similarities among classes. Performance evaluation also promises for a better satellite image classification. Hence, this proposed ERICA gives major contribution for classification of satellite images of healthy trees, buildings and road area. The experimental results of the remote sensing data significantly specified that the proposed ERICA provides better classification accuracy and convergence speed.

References

1. Almeida, L.B., Silva, F.M.: Adaptive Decorrelation. In: Aleksander, I., Taylor, J. (eds.) Artificial Neural Networks, vol. 2, pp. 149–156. Elsevier Science Publishers (1992)
2. Amari, S.: Natural gradient work efficiently in learning. Neural Comput. 10(2), 251–276 (1998)
3. Benediktsson, J.A., Pesaresi, M., Arnason, K.: Classification and feature extraction for remote sensing images from urban areas based on morphological transformations. IEEE Transactions on Geoscience and Remote Sensing 41(9), 1940–1949 (2003)
4. Cardoso, J.F., Laheld, B.: Equivariant adaptive source separation. IEEE Trans. Signal Process. 44(12), 3017–3030 (1996)
5. Choi, S., Cichocki, A., Amari, S.: Flexible independent component analysis. Journal of VLSI Signal Processing 26(1/2), 25–38 (2000)

6. Cichocki, A., Unbehauen, R., Rummert, E.: Robust learning algorithm for blind separation of signals. Electronics Letters 30(17), 1386–1387 (1994)
7. Cichocki, A., Georgiev, P.: Blind Source Separation Algorithms with Matrix Constraints. IEICE Trans. on information and Systems, Special Section on Special issue on Blind Signal Processing E86-A(1), 522–531 (2003)
8. Cruces, S., Castedo, L., Cichocki, A.: Novel Blind Source Separation Algorithms Using Cumulants. In: IEEE International Conference on Acoustics, Speech, and Signal Processing, V, Istanbul, Turkey, pp. 3152–3155 (2000)
9. Kelley, C.T.: Iterative methods for linear and nonlinear equations. In: Frontiers in Applied Mathematics, vol. 16, pp. 71–78. SIAM, Philadelphia (1995)
10. Li, G., Wan, Y., Chen, C.: Automatic building extraction based on region growing, mutual information match and snake model. In: Zhu, R., Zhang, Y., Liu, B., Liu, C. (eds.) ICICA 2010. CCIS, vol. 106, pp. 476–483. Springer, Heidelberg (2010)
11. Segl, K., Kaufmann, H.: Detection of small objects from high-resolution panchromatic satellite imagery based on supervised image segmentation. IEEE Transactions on Geoscience and Remote Sensing 39(9), 2080–2083 (2001)
12. Singh, P.P., Garg, R.D.: A Hybrid approach for Information Extraction from High Resolution Satellite Imagery. International Journal of Image and Graphics 13(2), 340007(1-16) (2013)
13. Yang, J.H., Liu, J., Zhong, J.C.: Anisotropic diffusion with morphological reconstruction and automatic seeded region growing for colour image segmentation. In: Yu, F. (ed.) Proceedings of the International Symposium on Information Science and Engineering, Shangai, China, pp. 591–595 (2008)
14. Zhang, L., Amari, S., Cichocki, A.: Natural Gradient Approach to Blind Separation of Over- and Under-complete Mixtures. In: Proc. of Independent Component Analysis and Signal Separation, Aussois, France, pp. 455–460 (1999)
15. Zhang, L., Amari, S., Cichocki, A.: Equi-convergence Algorithm for blind separation of sources with arbitrary distributions. In: Mira, J., Prieto, A.G. (eds.) IWANN 2001. LNCS, vol. 2085, pp. 826–833. Springer, Heidelberg (2001)

3D Face Recognitionacross Pose Extremities

Parama Bagchi[1], Debotosh Bhattacharjee[2], and Mita Nasipuri[2]

[1] Dept. Of Computer Science & Engineering, MCKV Institute of Engineering Howrah, West Bengal, India
[2] Dept. Of Computer Science & Engineering, Jadavpur University, Kolkata
{paramabagchi,mitanasipuri}@gmail.com, debotosh@ieee.org

Abstract. In this paper, a mathematical model for 3D face image registration has been proposed with poses varying from 0 to ± 90° across yaw,pitch and roll. The method which has been proposed in this paper consists of two major steps:- 3D face registration and comparison of the registered image with any neutral frontal posed model in order to measure the accuracy of registration followed by recognition. In this 3D registration and recognition model, a 3D image is transformed from any different pose to frontal pose. After applying the algorithm on the Bosphorus databases, our proposed method registers the images with poses ranging from 0 to 20° with an average rotational error ranging between 0.003 to 0.009. For poses with an orientation of 40°to 45°, the average rotational error was 0.003 to 0.009 and for poses with90° the average rotational error was 0.004. Features are extracted from the registered images in the form of face normals. The experimental results which were obtained, on the registered facial images, from the 3D Bosphorus face database, illustrate that our registration scheme has attained a recognition accuracy of 93.33%.

Keywords: Hausdorff'sdistance, registration, recognition, translation, scaling.

1 Introduction

All registration algorithms attempt to solve the problem of finding a set of transformation matrices that will align all the data sets into a single coordinate system. In the problem of 3D face recognition, three dimensional registrations is an important issue. For 3D face which is oriented across any angle, for it to be correctly recognized, it has to be correctly registered. The method which has been used in this paper is that, the algorithm takes as input a 3D face and returns the registered images for poses from 0 to ± 90°. We need to specify here that, the proposed method discussed is a rough registration methodology. We have used Hausdorff distance calculation as a metric, to test the performance of our registration method which has been discussed in Section 2. Here, we are going to discuss about some of the related works based on 3D face registration across pose, and discuss the advantages of our method over the existing methods on 3D face registration.In [1], the authors used TPS plane wrapping and Procrustes analysis for rough alignment with ICP algorithm for registration, but in reality ICP is an approach which runs correctly only up to angles of 20 to 30 degrees on different 3D face databases. In [2], the authors have found out the angles which

M.K. Kundu et al. (eds.), *Advanced Computing, Networking and Informatics - Volume 1*, Smart Innovation, Systems and Technologies 27,
DOI: 10.1007/978-3-319-07353-8_35, © Springer International Publishing Switzerland 2014

been used for registration of 3D face images, but the process can only detect pose angles up to 40°. In [3], the authors used Hausdorff's distance as a measure for registration but no angles of pose corrections were mentioned. In [4], only a part of the registration work was done i.e. identifying only the nose tip region across different poses like yaw, roll etc. In [5], facial landmark localization was done using dense SURF features and also angles ranging from -90° to +90° were considered. But, the work composed only of 3 databases namely XM2VTS, the BioID and the Artificial driver datasets out of which both the XM2VTS and the BioID face database comprises of very near frontal faces. In [6], the authors presented a 3D facial recognition algorithm based on the Hausdorff distance metric. But, in addition the authors assumed that there should be a good initial alignment between probe and template datasets. The novelty of the present approach, stated in this paper is that, an attempt has been made to register 3D face scans taken from the Bosphorus databases using Hausdorff's distance, for poses ranging from 0 to 90 degrees. The distinctiveness of the present approach is that, a mathematical model to register a 3D face scan has been set up, which uses Hausdorff's distance metric and considers all pose extremities. The paper has been organized as follows: In Section 2, a description of the proposed method has been described. Our significant contribution and a comparative analysis been discussed in Section 3. Experimental results are discussed in Section 4, and finally in the Section 5, conclusion and future scope have been enlisted.

2 An Overview of the Proposed System

The different steps of the present proposed algorithm are shown in Fig. 1.

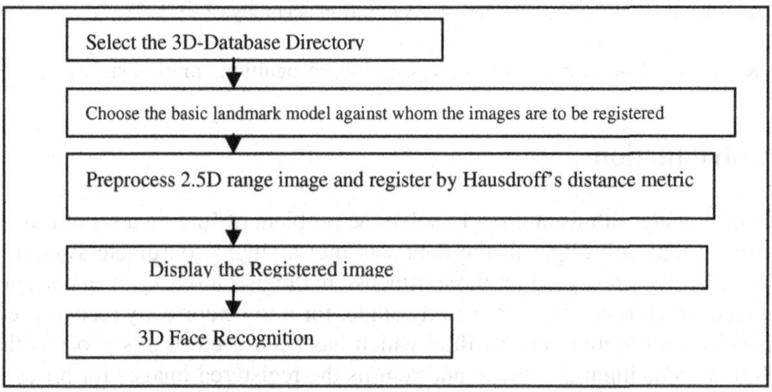

Fig. 1. Snapshot of menu screen for the present proposed method

The steps used in the registration technique are described below:

2.1 Select the 3D-Database Directory

The 3D database directory namely the Bosphorus database used in the registration process consists of .bnt file formats. If a user, clicks on the link in Fig. 1 termed as

"Select the 3D-Database Directory", the user is prompted to select the directory where the *.bnt files reside.

Choose the basic landmark model against whom the images are to be registered:- In this example, when the user clicks on the second button, then he is prompted to choose the basic landmark model as shown in Fig.2(b), against whom the 3D face model as selected in Step 2.1 and shown in Fig.2(a), isto be registered. The figure shown in Fig. 2(a) is called a 2.5D range because, it contains depth information at each of it's pixel position. The preprocessing stages which have been used to reduce noise in the 3D face images are subdivided into two parts as follows:-

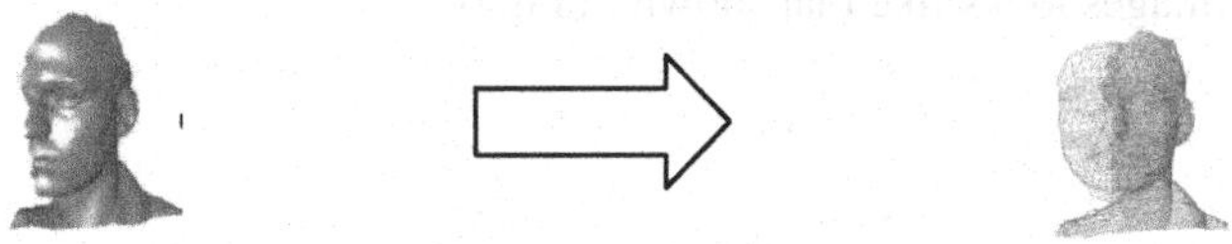

Fig. 2a. A 2.5D range image rotated about 45° about y-axis

Fig. 2b. Registered image corresponding to Fig. 2a)

2.1.1 Cropping

Fig.3 shows a 2.5D range image of a frontal pose taken from the Bosphorus database after being cropped by fitting an ellipse to it.

Fig. 3. Diagram of a cropped 3D range image from Bosphorus database

2.1.2 Image Smoothing

Every range image is subjected to noise caused by outliers. So, the 3D face images have been smoothed by weighted median filters, because weighted median filters have a very good smoothing effect while restoring the entire information. When weighted median filter was applied to the frontal image as shown in Fig.3, the smoothed meshes were obtained as in Fig.4.

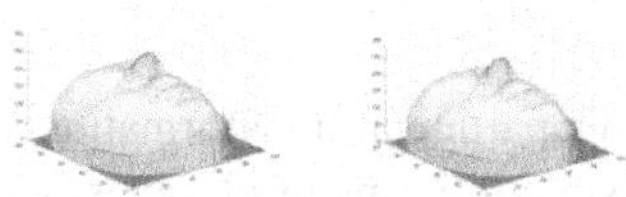

(a) (b)

Fig. 4. Frontal range images from Bosphorus (a, b) after smoothing by weighted median filter

2.2 Preprocess 2.5D Range Image and Register by Hausdorff's Distance Metric

In this step, the user selects the 2.5D range image in rotated pose. Once the user selects the rotated 2.5D range image across X, Y and Z axes, the corresponding mesh images would look like as is shown in Fig.5.

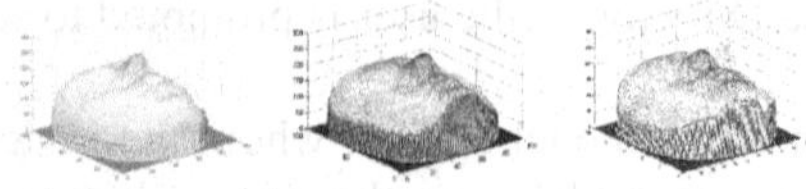

Fig. 5. (a) Frontal pose (b) Rotated about y-axis (c) Rotated about x-axisfrom the Bosphorus database

The images that the user chooses are cropped and smoothed by weighted median filter, the same approach used for the frontal images. After the images are smoothed the final mesh images looks like that shown in Fig.6.

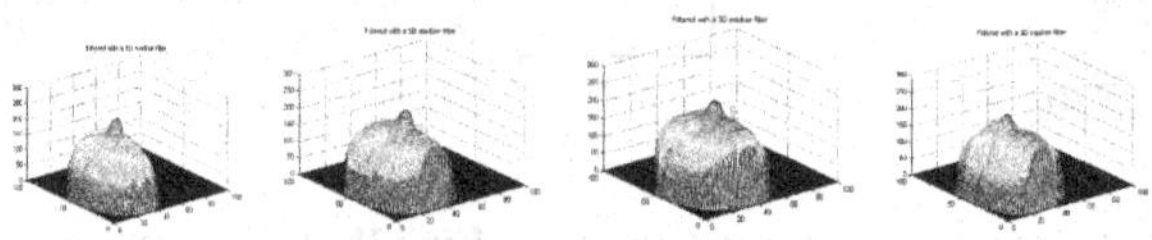

(a) (b) (c) (d)

Fig.6.(a) Frontal pose (b) Rotated about y-axis (c) Rotated about x-axis (d) Rotated about z axis

2.3 Registrationby Hausdorff's Distance Metric

A face recognition system is generally made up of two key parts: - registration and comparison. The accuracy of registration will greatly impact the result of the comparison. The Hausdorff distance was originally designed to match binary edge images [8][10]. Modifications to the Hausdorff distance permit it to handle noisy edge positions, missing edges from registration, and spurious edges from clutter noise [8][9]. Given two sets of facial range data $S = \{s_1, s_2,.......s_k\}$and $M = \{m_1, m_2,......., m_n\}$, the task of 3D image registration is to find the transformation i.e. translation, rotation and scaling which will optimally align the regions of S with those of M. For a transformation group G, it can be formalized as an optimization problem:

$$\min_{g \in G} \partial (M, g(S)) \quad (1)$$

In other words, we should find a particular transformation $g \in G$, which matches M with g(S) as "closely" as possible in terms of closeness evaluation function $\partial(.)$. The method of registration technique used in the present approach consists of the following two steps:

2.3.1 Coarse Registration

Algorithm 1 lists the method used for the registration purpose of the 3D unregistered point-set S to the 3D frontal point-set R.

Algorithm1. Proposed Method

Input:

```
Rot← Rotation matrix i.e. across yaw, pitch or roll
S←3D mesh of unregistered image
R← 3D mesh of frontal image
counter← 1
Output:-S←3D mesh of registered image
```

```
Do
   1.if counter equals 1 then
   2.dist= Call_hausdorff(S, R);
   3.else
   4. S ← Rotate and translate S by the matrix Rot
   5. Update_dist = Call_hausdorff(S, R);
   6. if((Update_dist-dist)/dist>=0.2 || (Update_dist-dist) /dist>=0.1)
   7. break;
    8.end if
    9. counter=counter+1;
   while(1)
   10.Display the final registered 3D image S
   11. End of Algorithm
```

```
Sub Function [dist] = Call_hausdorff (A, B)
dist = max (calculate (A, B),calculate (B A))
End Function
```

```
Sub Function [dist] = calculate (A, B)
1. for k=1: size(A)
2.        D = (A-B).* (A-B);
3.     dist = min (dist);
4.   end   for
5.  dist =max (dist);
End Function
```

As is quite evident from the Algorithm 1, the present proposed method takes as input a 3D range image oriented across any pose from 0° to 90°, rotates and translates it and the process is continued till the difference between the original 3D mesh image and the registered mesh image crosses a maximum threshold of 0.1. We have fixed the threshold value 0.1 or 0.2, by making a rough estimate on the basis of the database used in the method. The function Call_Hausdorff()[11] calculates the maximum distance between the two point clouds. To calculate the percentage rotational error between the neutral image in frontal pose and the registered image, we calculate the RMS(Root Mean Square) error between the frontal and the registered images. Variable counter is used to check the decision control for initial and final conditional statements.

2.3.2 Accuracy of Proposed Registration Technique

In order to evaluate how correctly the 3D faces have been registered to the frontal image, we employ Algorithm 2.

Algorithm 2. Evaluation of face registration

Input:

S←3D mesh of registered image
R← 3D mesh of frontal image

1.Find the coordinates of the nose-tip by depth map analysis[2]of R and store in xR and yR.
2.Find the coordinates of the nose-tip by depth map analysis[2]of S and store in xS and yS.
3. percen_errorx= abs(xS/xR)
4. percen_errory= abs(yS/yR)
5.Check if (percen_errorx<0.01 &&percen_errory<0.01)
6. The 3D image is perfectly registered
7. Else
8. The 3D image is not perfectly registered
9. End if

The accuracy of registration as illustrated in Algorithm1 is measured by calculating the rotational error, in between the nose-tips of the 3D image in frontal pose and the unregistered image.

2.4 Feature Extraction and Classification by Normal Points

The final stage, in the face recognition process is, feature extraction and classification. PCA vector extraction has been the most common way of feature extraction, but here we have extracted the corresponding normal values, from the registered face images.

3 Our Contribution and a Comparative Analysis with other 3D Registration Systems

1. An original 3D face registration system that combines all pose orientations i.e. across yaw, pitch and roll have been considered.
2. A thorough performance evaluation of the registration system has been done the widely available dataset namely the Bosphorus Dataset.
3. All possible poses including large poses, ranging from 0° to ± 90° degrees have been considered, which is a very novel contribution by itself.

4. Finally, the recognition of the registered images by normal points has also proved, to be very much effective.

Table 1 shows the Comparison of proposed approach with existing ones.

Table 1. Comparison of proposed approach with existing ones

Sl. No	Name of the Registration Method	Databases used for Analysis	Angles of Pose Orientations	Registration Accuracy
1.	Automatic 3D verification from range data by modified Hausdorff's distance[2]	1.3D RMA	1.straight forward 2. left or right 3. upward or downward, no pose angle mentioned	Best EER rate was 3.24%
2.	Pose and Expression Independent Facial Landmark Localization Using Dense-SURF and the Hausdorff Distance[5]	1.XM2VTS 2.BioID 3.Artificial Driver	1. Frontal poses 2. Expression poses 3.Head poses range from ±90° but in 10° intervals	No registration technique was mentioned only landmark localization was mentioned as 99.96%
3.	Adaptive Registration for Registration Based 3D Registration[12]	1.FRGCV 2 2.UMB-DB	1.Neutral Faces 2.Neutral faces, Registrations	Less limited registration approach, recognition rate was 65.80%
4.	Spherical Harmonic Features for 3-D Face Recognition[10]	1.SHREC2007 2.Bosphorus 3.FRGC v2.0	1. 35 expression faces 2. 13 yaw, pitch, cross rotations 3. 4 occlusions Pose angles were up to maximum of 45°	No registration accuracy was mentioned, but recognition rate was 90%
5.	Our Proposed Approach	1.Bosphorus	All possible poses across X,Y and Z axes from 0° to ± 90°	Minimum Registration error was 0.001

4 Experimental Results

The experimental results of the registration performed by the proposed algorithm have been tested on the Bosphorus database (Refer Table 2).The experimental setup consistedof apersonal machine with a system configuration of Intel Core2Duo Processor. The speed of the processor was 2.20GHz with 1GB RAM and 250GB hard disk.

All experimentation was done in Matlab R2011b version.

Table 2. Performance of Registration on the Bosphorus Database

Pose-Orientations	**Registration on the Bosphorus Using Hausdorff's Distance**		
	Angles of pose orientation	***Total no of registered Range images***	***Rotational error***
Pose orientation across X & Y axes	10°	11	0.009
	20°	11	0.003
	30°	11	0.009
	45°(L & R)	22	0.003
	90°(L &R)	22	0.004

The 3D registered images are now fed to the recognition system. Classification was performed on the extracted normal features, taken from the reconstructed face images using MultiLayer Perceptron.

4.1 Recognition of Registered Images

Initially, we extracted features using PCA which was the traditional method. The rank-1 recognition rate on the registered images was 66%, when we used approach feature extraction by PCA.

Table 3. Rank-1 Recognition Rate

Our proposed method	
Feature extraction	PCA approach
Recognition rate	66.66%

As is inevitable from the above Table 4, the PCA based approach works well. The rank-2 recognition rate, for the current proposed method was 93.3%, when we used approach 2, as shown inTable 4. The rank-2 recognition rate is better than rank-1 (as shown inTable 3) because face normals are less prone to pose variations.

Table 4. Rank-2 Recognition Rate

Our proposed method	
Feature extraction	Feature Extraction By Face Normals
Recognition rate	93.33%

5 Conclusion and Future Scope

This paper is a robust approach, to register 3D faces across large pose extremities using Hausdorff's distance metric[7]. The technique used in the paper is very different from existing works in literature because, it registers poses across varying angles of pose i.e. from 0° to ± 90°.The paper tries to address, the problem of pose with respect to 3D face images and tries to find out a solution for it. Much of the work which has been done based on Hausdorff distance has been only based on feature detection as well as correction. Maximum of the state of the art work have first registered the 3D image models and then corrected the registration using Hausdorff distance. Our approach is a very different because we have based our registration technique on Hausdorff's distance metric. As part of the future work, a more robust model which would increase the accuracy of the present registration and subsequently recognition technique are planned to be implemented across many more 3D face databases.In future, we attempt to draw up a more comprehensive analysis of the results of registration as well as recognition on other databases also.

Acknowledgement. The work has been supported by the grant from Department of Electronics and Information Technology (DEITY), Ministry of Communication & Information Technology (MCIT),Government of India.

References

1. Mian, A., Bennamoun, M., Owens, R.: Automatic 3D Detection, Normalization and Recognition. Clarendon, Oxford (1892)
2. Pan, G., Wu, Z., Pan, Y.: Automatic 3D Face Verification From Range Data. In: Proceedings of IEEE Conference on Acoustics, Speech, and Signal Processing, pp. 193–196 (2003)
3. Mian, A.S., Bennamoun, M., Owens, R.: An Efficient Multimodal 2D-3D Hybrid Approach to Automatic Face Recognition. IEEE Transactions on Pattern Analysis and Machine Intelligence 29(11), 1927–1943 (2007)
4. Anuar, L.H., Mashohor, S., Mokhtar, M., Adnan, W.A.W.: Nose Tip Region Detection in 3D Facial Model across Large Pose Variation and Facial Expression. International Journal of Computer Science Issues 7(4), 4 (2010)
5. Sangineto, E.: Pose and Expression Independent Facial Landmark Localization Using Dense-SURF and the Hausdorff Distance. IEEE Transactions on Pattern and Machine Intelligence 35(3), 624–638 (2013)
6. Russ, T.D., Koch, M.W., Little, C.Q.: A 2D Range Hausdorff Approach for 3D Face Recognition. In: IEEE Computer Society Conference on Computer Vision and Pattern Recognition Workshops (2005)
7. Aspert, N., Cruz, D., Ebrahimi, T.: Mesh:measuring errors between surfaces using the hausdorff distance. In: IEEE Conference in Multimedia and Expo, pp. 705–708 (2002)
8. Radvar-Esfahlan, H., Tahan, S.: Nonrigid geometric metrology using generalized numerical inspection fixtures. Precision Engineering 36(1), 1–9 (2012)
9. Alyuz, N., Gokberk, B., Akarun, L.: Adaptive Registration for Registration Based 3D Registration. In: BeFIT 2012 Worksop (2012)
10. Liu, P., Wang, Y., Huang, D., Zhang, Z., Chen, L.: Learning the Spherical Harmonic Features for 3D Face Recognition. IEEE Transactions on Image Processing 22(3), 914–925 (2013)
11. Dubuisson, M.P., Jain, A.K.: A Modified Hausdorff Distance for Object Matching. In: International Conference on Pattern Recognition, pp. 566–568 (1994)
12. Alyuz, N., Gokberk, B., Akarun, L.: Adaptive Registration for Registration Based 3D Registration. In: BeFIT 2012 Worksop (2012)

Indexing Video Database for a CBVCD System

Debabrata Dutta[1], Sanjoy Kumar Saha[2], and Bhabatosh Chanda[3]

[1] Tirthapati Institution, Kolkata, West Bengal, India,
[2] Computer Science and Engineering Department, Jadavpur University, Kolkata, India
[3] ECS Unit, Indian Statistical Institute, Kolkata, India
debabratadutta2u@gmail.com, sks_ju@yahoo.co.in,
chanda@isical.ac.in

Abstract. In this work, we have presented a video database indexing methodology that works well for a content based video copy detection (CBVCD) system. Video data is first segmented into cohesive units called shots. A clustering based method is proposed to extract one or more Representative frames from the shots. On such collection of representatives extracted from all the shots in the video database, triangle inequality based image database indexing scheme is applied. Thus, video indexing is mapped to the task of image indexing. For a shot, following the proposed methodology primarily candidate shots corresponding to the matched representative frames are retrieved. Only on such small number of candidates the rigorous video sequence matching technique can be applied to make final decision by the CBVCD system or video retrieval system. Experimental result with a CBVCD system indicates significant gain in terms of speed, reduces false alarm rate without much compromise in terms of correct recognition rate in comparison to exhaustive search.

Keywords: Video Database, Indexing, CBVCD.

1 Introduction

With the development of different multimedia tools and devices there has been enormous growth in the volume of digital video data. Search and retrieval of desired piece of video data from a large database has become an important area of research. In the applications like content based video retrieval (CBVR) or content based video copy detection (CBVCD) system a query video data is given by the user. It is to be matched with the video data stored in the database. In case of a CBVR system, we are mostly interested in finding few top order matches. But in case of CBVCD detection, the task is to decide whether or not the query is a copied version of any reference video data stored in the database. Further challenge for a CBVCD system is that a copied video data may undergo different photometric and post-production transformations making it different from the corresponding reference video. In both the applications, exhaustive video sequence matching is prohibitive. As a result indexing scheme becomes essential. In this work, we propose an indexing scheme and also present its application for a CBVCD system.

M.K. Kundu et al. (eds.), *Advanced Computing, Networking and Informatics - Volume 1*, Smart Innovation, Systems and Technologies 27,
DOI: 10.1007/978-3-319-07353-8_36, © Springer International Publishing Switzerland 2014

The paper is organized as follows. The brief introduction is followed by the review of past work on indexing in Section 2. Details of the proposed methodology are presented in Section 3. Experimental results and concluding remarks are put in Section 4 and Section 5 respectively.

2 Past Work

A video database can be indexed following different approaches outlined in the work of Brunelli *et al.* [1]. Video data may be annotated manually at various level of abstraction. Abstraction may range from the video title to content details. The database may be organized according to the annotations to support indexing. Such manual annotation is labour intensive and subjective. To overcome the difficulties there was the development of alternate approach in the form of content-based automated indexing. A framework for automated video indexing as discussed in [1] is to segment the video data into shots and to identify the representative frames from the shots. Subsequently, the collection of those representatives is considered as the image database and experience of content based indexing of image database can be applied. Similar trend in video browsing and retrieval still persists as indicated in [2].

Zhang *et al.* [3], in their approach have classified the video shots into groups following a hierarchical partition clustering. Bertini *et al.* [4] have presented a browsing system for news video database. A temporal video management system has been presented in [5] that relies on tree based indexing. Spatio-temporal information data has been widely used for video retrieval [6,7]. Ren and Singh [8] proposed R-string representation to formulate spatio-temporal data into binary string. Non-parametric motion analysis has also been used for video indexing [9]. But, presence of noise and occlusion may lead to failure [10].

As discussed in [1], [4], the major trend for video indexing is to break it into units and to map the video database into image database by taking the representative frames of the segmented units. So it is worth to review the techniques for indexing the image database. A comprehensive study on high dimensional indexing for image retrieval has been presented in [11]. Tree based schemes are widely used. Zhou *et al.* [12] have classified those according to the indexing structure. K-D tree [13], M tree and its variants [14], TV tree [15], are few examples of such indexing structures. Hashing based techniques [16], [17] are also quite common for indexing an image database. Concept of bag of words has been deployed in number of works [18], [19].

Most of the video indexing system has focused on specific domain and thereby concentrated on designing the descriptors accordingly. Details of the underlying indexing structure have not been elaborated. It appears that the technique used for image database is adopted to cater the core need of indexing the video database.

3 Proposed Methodology

In our early work [20], a content-based video copy detection (CBVCD) system has been proposed. The system is robust enough against various photometric and post-production attacks. Each frame in the video data goes through preprocessing to reduce the effect of attacks and then features are extracted. In order to decide whether a shot in the query video is a copied version of any of the reference video shots or not, an exhaustive video sequence matching is carried out following multivariate Wald-Wolfowitz hypothesis test. Such linear search leads to large number of test which is prohibitive. In this effort, our target is to propose a methodology to reduce the number of video sequence matching.

As presented in Section 2, common approach for video indexing is to map the problem to image database indexing. The major steps are like breaking the video data into structural units called shots, extraction of representative frames of the shots to form an image database and finally well established image database indexing technique is applied on the image database formed. Proposed methodology also follows the same approach.

In this work, it is assumed that the video data is already segmented into shots following the technique presented in [21]. A clustering based shot level representative frame detection scheme is devised and subsequently indexing is done based on *triangle inequality* property [22].

3.1 Selection of Shot Representative

A shot is the collection of consecutive frames captured in a single camera session. Mostly, the frames in a shot are visually very similar. Thus, it is good enough to represent the shot by a single frame and redundancy is also removed. But, due to the motion of camera and/or object, presence of lots of activity in a shot it may not be judicious to represent the content by a single representative. Thus, multiple frames may have to be chosen for proper reflection of the content. In this context lot of works have been done [23], [24]. Here we present a simple scheme based on the following steps.

- Verify the uniformity of the shot
- For a uniform shot, select one representative frame
- For non-uniform shot, select multiple representatives based on clustering

Each frame in the shot is first represented by an edge based visual descriptor. A frame is divided into a fixed number of blocks. Normalized count of edge pixels in the blocks arranged in raster scan order forms the multi-dimensional feature vector. We have partitioned the frame into 16 blocks and same is the dimension of the vector. The features thus computed are neither too local nor global.

In order to verify the uniformity of a shot content, similarity of each frame with respect to the first one in the shot is computed. Let, s_i be the similarity value for the i-th frame in the shot. If $min\{s_i\}/max\{s_i\}$ is smaller than a threshold then the shot is taken as non-uniform one and it qualifies for multiple representatives. Similarity between two frames is measured by the Bhattacharya distance between the corresponding feature vectors and threshold is empirically determined as 0.9. For a uniform shot, frame for which the similarity with the first one is closest to $(min\{s_i\}+max\{s_i\})/2$ is taken as the representative.

For a non-uniform shot, K-means clustering is applied to put the frames into different clusters. Number of clusters, n_c is varied starting from 2 and gradually increased till optimal value for n_c is determined. Goodness of the clusters is measured based on Dunn index [25]. Once optimal numbers of clusters are formed, from each cluster the frame nearest to the centre is chosen as the representative. Thus, a shot will have number of representatives same as the number of optimal clusters.

3.2 Indexing the Database of Representative Frames

Indexing scheme is applied on the image database obtained after collecting the representative frames of all the shots in the video sequences. Based on the *triangle inequality* approach, a value for each database image is assigned which corresponds to the lower bound on the distance between database image and a query image. On that value, a threshold is applied to discard the database images which are away from the query image.

Let $I = \{i_1, i_2, \ldots, i_n\}$ and $K = \{k_1, k_2, \ldots, k_m\}$ denote the database of images and collection of key images chosen from I respectively. As per *triangle inequality*, $dist(i_p,Q) + dist(Q, k_j) \geq dist(i_p, k_j)$ is true where $i_p \in I$, $k_j \in K$, Q is a query image and $dist(..)$ stands for a distance measure. The equation can be rewritten as $dist(i_p,Q) \geq | dist(i_p, k_j) \quad dist(Q, k_j) |$. Considering all $k_j \in K$, the lower bound on dist(i_p,Q) can be obtained from $dist(i_p,Q) \geq max_j(| dist(i_p, k_j) \quad dist(Q, k_j) |)$ where $j \in \{1, 2, \ldots ,m\}$. Thus, the major steps are as follows.

- Select the key images K from the image database I
- pre-compute the distance matrix to store $dist(i_p, k_j)$ for all $i_p \in I$ w.r.t all $k_j \in K$

All these steps are offline. In order to select the key images, images in the database are partitioned into clusters following k-means clustering algorithm and number of optimal clusters is decided based on Dunn-index. For each cluster, image nearest to the cluster centre is taken as key image. In order to prepare the distance matrix, n ×m of computation is required. In our experiment, $dist(i_p, k_j) = 1 \quad bhatt\_dist(i_p, k_j)$ where $bhatt\_dist(i_p, k_j)$ provides the similarity between i_p and k_j.

For a CBVCD system, our point of interest is to find out the images from the database for which the lower bound does not exceed a threshold t. For searching against a query image(Q) the steps to be carried out in online mode are as follows.

- Compute $dist(Q, k_j)$ for all $k_j \in K$
- Retrieve all i_p such that $| dist(i_p, k_j) \quad dist(Q, k_j) | \leq t$ *for all* k_j *in* K

The first step requires m number of computation whereas the second step involves n × m simple operations if searched linearly. The comparison can be reduced further based on the actual implementation. In our work, with respect to each k_j a separate structure stores $dist(i_p, k_j)$ in an order. It enables binary search to retrieve the i_ps satisfying the condition $dist(Q, k_j) - t \leq dist(i_p, k_j) \leq dist(Q, k_j) + t$. Thus, the overhead of accessing the index is also reduced significantly. In our experiment value of t is empirically determined and taken as 0.01.

Now, for video copy detection, representative frames are extracted from the shot to be tested. Each such representative frame is used as the query image (Q) to retrieve the desired images from the database. With the shots corresponding to all the retrieved images, the final hypothesis test is carried out to decide the outcome. Thus the final test is carried out with only a small subset of the shots in the reference database making the CBVCD system faster.

4 Experimental Result

In order to carry out the experiment we have worked with a collection of video data taken mostly from TRECVID 2001 dataset and a few other recordings of news and sports program. The sequences are segmented into shots following the methodology presented in [21]. The dataset contains 560 shots. Performances of the proposed methodology to extract the representative frames of the shot and effectiveness of indexing are studied through experiments.

All the shots are manually ground-truthed and marked as single or multiple depending on whether a shot requires a single frame or more than one frame as its representative. For shots of *multiple* types, different homogeneous sub-shots are also noted. Performance of the proposed scheme for extracting the representative frames is shown in Table 1. It shows that the homogeneous shots are correctly identified as single and it fails only for a few cases of shots of type *multiple*. There may be other type of errors like *over-splitting* and *under-splitting* in case the number of extracted frame is more or less than the expected count respectively. In our experiment, only five shots are under-splitted extracted frames is one less than the desired count and over-splitting occurred for three shots.

To study the performance of the indexing system, we have considered the application of CBVCD. Index based search has been incorporated in the CBVCD system presented in [20]. In linear method, a query video shot is verified with all the shots in the reference video database using hypothesis test based sequence matching technique. But in case of indexed method, following the proposed scheme a subset of shots whose representative frames match with those of query shot are retrieved. Only with the retrieved shots final verification is made. Adoption of indexing scheme is expected to make the process faster but it may affect the detection performance. The performance of a CBVCD can be measured in terms of correct recognition rate (CR) and false alarm rate (FR). These are measured as $CR=(n_c/\ n_a)\ \text{x}\ 100\ \%$ and $FR = (n_f/n_q)\text{x}100\%$where n_c, n_a, n_f , n_q are number of correctly detected copies, number of actual copies, number of false alarm and number of queries respectively.

Table 1. Performance of Extraction of Representative Frame from Shot

Shot type	Number of shots	Detected as	
		Single	Multiple
Single	489	489	0
Multiple	71	4	67

Table 2. Performance of CBVCD system

Attack type	No. of query	Linear method		Indexed method	
		CR(in%)	FR(in%)	CR(in%)	FR(in%)
No attack	560	100.00	18.57	100.00	15.18
Brightness change	300	100.00	23.67	99.67	11.33
Contrast change	300	100.00	22.67	99.67	11.00
Noise Corruption	200	99.00	19.00	98.50	10.50
Flat file	100	99.00	18.00	98.00	13.00
Letter box	100	98.00	16.00	98.00	14.00
Pillar	100	98.00	16.00	97.00	14.00
Logo insertion	100	98.00	18.00	94.00	15.00
Overal	1760	99.49	19.82	98.98	8.47

Table 2 shows the performance of linear and indexed method for the CBVCD system. In case of indexed method the sequence matching is carried out in a reduced search space. As a result, exclusion of similar sequence is possible, particularly for the transformed version. But, for a CBVCD system similarity is not synonymous with copy. It is well reflected in Table 2 that reduction of search space has reduced both CR and FR. But, CR is reduced marginally and FR (it is more sensitive and should have low value) is reduced significantly. Even under different photometric (change in brightness, contrast, corruption by noise) and post-production (insertion of logo, change in display format – flat file, letter box, pillar) attacks, the indexed method reduces FR substantially without much compromising the correct recognition rate.

As the indexing scheme reduces the search space, the copy detection process as a whole becomes faster. Experiment with 1760 query shots has revealed that on an average the system becomes five times faster.

5 Conclusion

In this work a simple but novel video indexing scheme is presented. A clustering based scheme is proposed which dynamically determines the number of representative frames and extracts them. A triangle inequality based indexing scheme is adopted for the image database formed by collecting the representative frames for all the shots. For a shot given as the query, candidate shots are retrieved based on the proposed methodology. On the retrieved candidate shots, video sequence matching technique can be applied to fulfill the requirement of a CBVCD system. Experiment indicates the effectiveness of the proposed methodology in extracting the representative frames. Applicability of the indexing scheme in CBVCD system is also well established as it reduces false alarm rate drastically without making much compromise on correct recognition rate and it speeds up the process significantly.

References

1. Brunelli, R., Mich, O., Moden, C.M.: A survey on the automatic indexing of video data. Journal of Visual Communication and Image Representation 10, 78–112 (1999)
2. Smeaton, A.F.: Techniques used and open challenges to the analysis, indexing and retrieval of digital video. Information Systems 32, 545–559 (2007)
3. Zhang, H.J., Wu, J., Zhong, D., Smoliar, S.W.: An integrated system for content based video retrieval and browsing. Pattern Recognition 30(4), 643–658 (1997)
4. Bertini, M., Bimbo, A.D., Pala, P.: Indexing for reuse of tv news shot. Pattern Recognition 35, 581–591 (2002)
5. Li, J.Z., OZsu, M.T., Szafron, D.: Modeling video temporal relationships in an object database systems. In: Proc. SPIE Multimedia Computing and Networking, pp. 80–91 (1997)
6. Pingali, G., Opalach, A., Jean, Y., Carlbom, I.: Instantly indexed multimedia databases of real world events. IEEE Trans. on Multimedia 4(2), 269–282 (2002)
7. Ren, W., Singh, S., Singh, M., Zhu, Y.S.: State-of-the on spatio-temporal information-based video retrieval. Pattern Recognition 42 (2009)
8. Ren, W., Singh, S.: Video sequence matching with spatio-temporal constraint. In: Intl. Conf. Pattern Recog., pp. 834–837 (2004)
9. Fablet, R., Bouthmey, P.: Motion recognition using spatio-temporal random walks in sequence of 2d motion-related measurements. In: Proc. Intl. Conf. on Image Processing, pp. 652–655 (2001)
10. Fleuret, F., Berclaz, J., Fua, P.: Multicamera people tracking with a probabilistic occupancy map. IEEE Trans. on PAMI 20(2), 267–282 (2008)
11. Fu Ai, L., Qing Yu, J., Feng He, Y., Guan, T.: High-dimensional indexing technologies for large scale content-based image retrieval: A review. Journal of Zhejiang University-SCIENCE C (Computers & Electronics) 14(7), 505–520 (2013)
12. Zhou, L.: Research on local features aggregating and indexing algorithm in large-scale image retrieval. Master Thesis, Huazhong University of Science and Technology, China 10–15 (2011)
13. Robinson, T.J.: The k-d-b tree: A search structure for large multidimensional dynamic indexes. In: Proc. ACM SIGMOD Intl. Conf. on Management of Data, pp. 10–18 (1981)
14. Skopal, T., Lokoc, J.: New dynamic construction techniques for m-tree. Journal of Discrete Algorithm 7(1), 62–77 (2009)
15. Lin, K.I., Jagadish, H.V., Faloutsos, C.: The tv-tree: An index structure for high-dimensional data. VLDB Journal 3(4), 517–542 (1994)

16. Zhuang, Y., Liu, Y., Wu, F., Zhang, Y., Shao, J.: Hypergraph spectral hashing for similarity search of social image. In: Proc. ACM Int. Conf. on Multimedia, pp. 1457–1460 (2011)
17. Heo, J.P., Lee, Y., He, J., Chang, S.F., Yoon, S.E.: Spherical hashing. In: Proceedings of IEEE Conference on Computer Vision and Pattern Recognition, pp. 2957–2964 (2012)
18. Avrithis, Y., Kalantidis, Y.: Approximate gaussian mixtures for large scale vocabularies. In: Proc. European Conf. on Computer Vision, pp. 15–28 (2012)
19. Jegou, H., Douze, M., Schmid, C.: Product quantization for nearest neighbor search. IEEE Trans. PAMI 33(1), 117–128 (2011)
20. Dutta, D., Saha, S.K., Chanda, B.: An attack invariant scheme for content-based video copy detection. Signal Image and Video Processing 7(4), 665–677 (2013)
21. Mohanta, P.P., Saha, S.K., Chanda, B.: A model-based shot boundary detection technique using frame transition parameters. IEEE Trans. on Multimedia 14(1), 223–233 (2012)
22. Berman, A.P., Shapiro, L.G.: A flexible image database system for content-based retrieval. Computer Vision and Image Understanding 75(1/2), 175–195 (1999)
23. Ciocca, G., Schettini, R.: An innovative algorithm for key frame extraction in video summarization. Real Time IP 1, 69–98 (2006)
24. Mohanta, P.P., Saha, S.K., Chanda, B.: A novel technique for size constrained video storyboard generation using statistical run test and spanning tree. Int. J. Image Graphics 13(1) (2013)
25. Dunn, J.C.: Well separated clusters and optimal fuzzy partitions. Journal of Cybernetica 4, 95–104 (1974)

Data Dependencies and Normalization of Intuitionistic Fuzzy Databases

Asma R. Shora and Afshar Alam

Department of Computer Science, Jamia Hamdard University,
Hamdard Nagar, New Delhi, India
{asmarasheid,mailtoafshar}@gmail.com

Abstract. Intuitionistic Fuzzy sets can be considered as a generalization of Fuzzy sets. It is an emerging branch of research on soft computing. Intuitionistic Fuzzy logic adds the indeterminacy factor to the Fuzzy logic techniques and is thus capable to solve multi-state logical problems. It can help machines make complex decisions, involving degrees of uncertainty and imprecision. In order to facilitate efficient retrieval and updating, the data stored in the Intuitionistic Fuzzy databases has to have an efficient information base, which can be ensured by proper organization of data. In this paper, we propose Intuitionistic Fuzzy Normalization. This process decomposes the Intuitionistic fuzzy relation into sub relations, in order to provide an efficient storage mechanism. We define data dependencies and their properties and use the same for Normalizing Intuitionistic fuzzy databases.

Abbreviations: IF (Intuitionistic Fuzzy), IFS (Intuitionistic Fuzzy Set), IFDB (Intuitionistic Fuzzy database), IFFD (Intuitionistic Fuzzy Functional dependency), NF – IF or NF (IF) (Intuitionistic Fuzzy normal form)

Keywords: Intuitionistic Fuzzy normal forms, Intuitionistic Fuzzy key, BCNF (IF), Soft Computing.

1 Introduction

Intuitionistic Fuzzy Sets were introduced by Krassimir Atanasov [4]. Intuitionistic Fuzzy logic helps us deal with imprecise and incomplete information, thus providing us with a better tool to process human logic. In most of the situations, human decisions are way more complicated than simple YES/NO decisions and are based on information that is imprecise and uncertain in nature rather than being crisp. Intuitionistic Fuzzy logic beautifully extrapolates the concepts and operations of Crisp and Fuzzy logic. We can translate and quantify human linguistic phrases and variables and process them using Intuitionistic Fuzzy logic. The variable may define any attribute or quality of a living or a non-living entity. We can even assign the degree or extent of the same. Suppose we wish to represent how severe or 'moderate' or 'low' a patient experiences a particular symptom.

M.K. Kundu et al. (eds.), *Advanced Computing, Networking and Informatics - Volume 1*, Smart Innovation, Systems and Technologies 27,
DOI: 10.1007/978-3-319-07353-8_37, © Springer International Publishing Switzerland 2014

1.1 An Example

Following instance is a part of an Intuitionistic Fuzzy medical diagnosis proposed by us [5]. We use a relation which quantifies the symptoms. Let us consider the following Tolerance relation.

Table 1. Symptoms of a disease quantified using IF database [5] (MV, NMV, HESITATION)

Symptom	Mild	Moderate	Severe	Absent
Mild	(1.0,0.0,0.0)	(0.8,0.2,0.0)	(0.9,0.1,0.0)	(0.2,0.7,0.1)
Moderate	(0.8,0.2,0.0)	(1.0,0.0,0.0)	(0.7,0.2,0.1)	(0.4,0.5,0.1)
Severe	(0.9,0.1,0.0)	(0.7,0.2,0.1)	(1.0,0.0,0.0)	(0.77,0.03,0.2)
Absent	(0.2,0.7,0.1)	(0.4,0.5,0.1)	(0.77,0.03,0.2)	(1.0,0.0,0.0)

Based on these values, a decision can be formed as to which extent has the disease spread and what treatment should be given to the patient. The IF- DSS [5] works on the linguistic variables which are quantified and represented through IF databases.

2 Intuitionistic Fuzzy Sets and Human Logic

Formally, an Intuitionistic Fuzzy set can be defined as follows: Let a set E be fixed. An IFS A in E is an object of the following form: $A = \{(x, \mu_A(x), \nu_A(x))\ x \in E\}$, When $\nu_A(x) = 1 - \mu_A(x)$ for all $x \in E$ is ordinary fuzzy set. In addition, for each IFS A in E, $\pi_A(x) = 1 - \mu_x - \nu_x$, where $\pi_A(x)$ is called the degree of indeterminacy of x to A, or called the degree of hesitancy of x to A. The following figure shows a gradual curve for the membership function of an Intuitionistic Fuzzy set and a discrete yes/no logic.

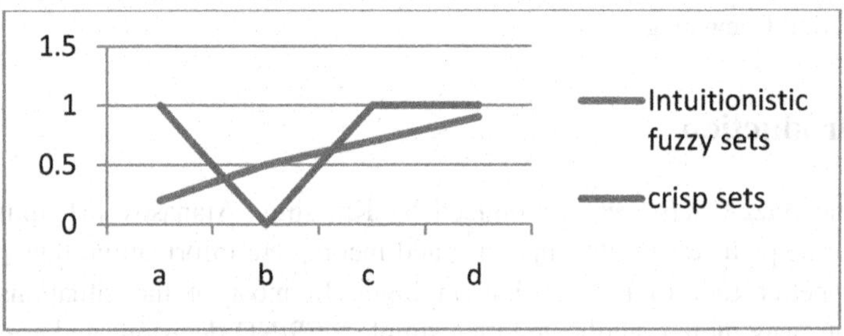

Fig. 1. Crisp sets vs Intuitionistic Fuzzy sets: This figure represents a smooth transition from membership to non-membership degrees

An Intuitionistic fuzzy database [7] is a set of relations where each pair of such relation R is a subset of the cross product: 2D1 × 2D2 × ………2Dm, where 2Di = P (Di) and P (Di) is the power set of Di, here R is called the Intuitionistic Fuzzy database relation.

An Intuitionistic fuzzy set [4], [18] is often considered a generalization of fuzzy set [6].Suppose, we have an Intuitionistic Fuzzy set,S_1 {(4, 0.5, 0.3), (5, 0.8, 0.1), (6, 0.7, 0.2)} and a Fuzzy set, S_2 {(4, 0.5), (5, 0.8), (6, 0.7)}. S_1 and S_2 represent the same information but differently. Set S_1 denotes that element '4' belongs to S_1 to a degree of membership, 0.5 and the degree of non membership, 0.3. In S_2, a part of this information is lost. It only denotes that the degree of membership is 0.5 and the fact that the possibility of not belonging (to S_2 is 0.3) is lost. The degree of non membership i.e. the possibility of non occurrence of an event is equally important as the possibility of occurrence of events in our daily decision making process.

Since, both fuzzy and Intuitionistic Fuzzy logic simulate human logic closely, there are evidences that support the fact that Intuitionistic Fuzzy logic is even more closer to human decision making logic [12], [19].

3 Normalization

It is a reversible process of breaking down a large complex database into simpler tables, so as to achieve the elimination of data redundancy along with data security and consistency. It is an important part of modern Relational database design [8]. IF Database contains both crisp and IF attributes. Further, there are IF dependencies and other IF data constraints have to be taken care of by the IF Normalization procedure.

Normalization requires us to have the knowledge base and the rules. Some constraints are set on the knowledge base and the data to be entered (or already present) must agree to the same. Breaking the database helps the smaller tables represent one unit of information thus organizing and optimizing our data. Normalized data is easier and faster to access. It helps minimize redundancy and non-key dependencies. The Intuitionistic Fuzzy Relational databases are normalized on the lines of grade of membership and non-membership values of data thus making it a challenge to decompose such complex datasets.

4 Related Work

The research [20] on Fuzzy and Intuitionistic Fuzzy sets has gained a lot of importance over the past few years. Fuzzy and Intuitionistic Fuzzy DBMS has also been in focus [7], [16], [21], [22]. Database schema design is the most important thing to be considered when it comes to RDBMS and so is the case with Fuzzy/ IF RDBMS. Work done on Fuzzy dependencies [10], [11] has proved to be helpful for exploring Fuzzy normalization techniques. An alternative approach [17] to define Functional dependencies has also been proposed for handling transitive dependencies. The need for fuzzy dependency preservation led to the need for decomposing Fuzzy databases [14]. Similarly, Intuitionistic Fuzzy relations and dependencies have been studied in detail [13], [15], [16]. As a result, Intuitionistic Fuzzy normal forms were defined [1], [2], [3] on the basis of dependencies among data. These dependencies are also Intuitionistic Fuzzy in nature as they have a grade of membership and non-membership associated with them.

5 Proposed Intuitionistic Fuzzy Normalization

We propose a model for the Intuitionistic Fuzzy Normalization, which takes as input a dataset that may be a combination of both Intuitionistic fuzzy and crisp data. The function of the (abstract) Intuitionistic Fuzzy interface is to convert the raw dataset into an IF database - a collection of Intuitionistic Fuzzy and crisp data.

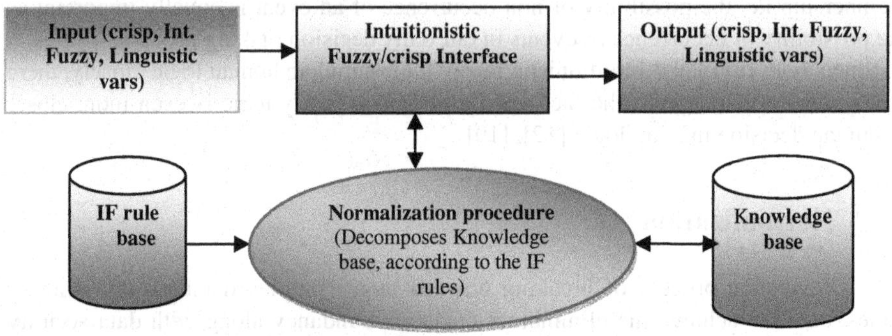

Fig. 2. Proposed Intuitionistic Fuzzy Normalization Model: This figure represents only a part of our data model. There are other procedures apart from the Normalization procedure that work on the knowledge / IF rule base.

The core process is the IF Normalization process, which may have more than one sub – processes. The data stores being used by the normalization procedure are: the Knowledge base and the IF rule base. The knowledge base contains processed data, and is the actual data warehouse for our data model. e. g., the medical database acts as the Knowledge base for storing complete information about the patients that includes- their personal information, as well as the symptoms and other medical history. There are a set of constraints that are applied on the Knowledge base in order to preserve the integrity of our data. The Intuitionistic Fuzzy rule base contains every single rule that is applied by the Normalization procedure on the Knowledge base to achieve the desired results.

We are working on a variant of the above proposed IF normalization model which helps us improve the Knowledge base as well. Discussing the technicalities of the same is beyond the scope of this paper. The idea is that the IF rule base is being updated every time it learns a new pattern or a new query and we end up creating a machine learning model, which normalizes the knowledge base and at the same time learns and upgrades its IF rules using some machine learning tools.

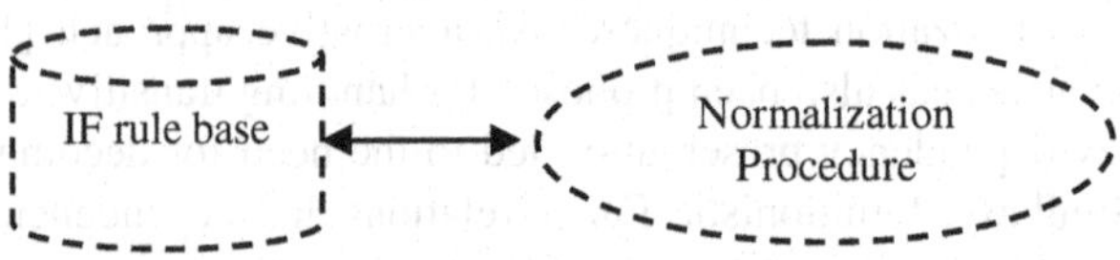

Fig. 3. Modified version of the proposed Intuitionistic Fuzzy Normalization Model

6 Intuitionistic Fuzzy Normal Forms

The data dependencies in a RDBMS can lead to an undesirable situation formally known as anomalies, e.g. appending, updating, deletion and sometimes searching anomalies. The Intuitionistic Fuzzy dependencies can prove harmful to the integrity of our database especially when the vital operations are being performed on the data. The purpose of Normalization is to avoid these anomalies by decomposing tables so that there are only desirable dependencies among data.

6.1 Intuitionistic Fuzzy Equality

One of the most effective techniques for comparing and determining Fuzzy dependencies is the Equality measure [9]. Equality between any two or more sets can always be expressed by comparing the elements of each set with another. The equality measure plays an important role in defining dependencies among the attributes and tuples.

Equality measure can be of types:

a) Syntactic
b) Semantic

The **Syntactic** equality is used for crisp comparisons, where information is complete and the sets to be compared are absolutely comparable (element by element).

The **Semantic** equality measure is relevant to us however. It is used to compare sets on the basis of approximation or similarity.

We define the Intuitionistic fuzzy Equality and use the same to define Intuitionistic Fuzzy dependencies. We denote it as IEQ (t_1, t_2), where $t_1,t_2 \in U$ and A_1 through A_n belong to an Intuitionistic Fuzzy relation R.

$$IEQ(t_1,t_2) = \{ (t_1,t_2), \mu_{IEQ}(t_1,t_2), \nu_{IEQ}(t_1,t_2) \} \tag{1}$$

$$\text{where, } \mu_{IEQ}(t_1,t_2) = \min\{\mu_{IEQ}(t_1[A_1]), t_2[A_1]),\ldots,\mu_{IEQ}(t_1[A_n]), t_2[A_n]) \} \tag{2}$$

$$\nu_{IEQ}(t_1,t_2) = \max\{\nu_{IEQ}(t_1[A_1]),t_2[A_1]),\ldots.,\nu_{IEQ}(t_1[A_n]), t_2[A_n]) \} \tag{3}$$

6.1.1 Properties of Intuitionistic Fuzzy Equality Measure

1) IEQ is **Reflexive**, i.e. IEQ(t, t) = 1 for all $t \in U$
2) IEQ is **Symmetric** i.e. IEQ (t_1, t_2) = IEQ (t_2, t_1) , for all $a, b \in U$

6.2 Intuitionistic Fuzzy Functional Dependency

Normalization process decomposes a relation based on the dependencies. One such dependency is the functional dependency among the attributes of a relation. Since, functional dependency is based upon the concept of *equality* of values, we need to compare data (from both domain and the range of a function) e.g. f(x) = f(y) => g(x) =g(y). Hence, we use the equality measure defined in section 6.1.

Definition: An Intuitionistic Fuzzy dependency (IFFD): $X \rightarrow_{p,q} Y$ exists in an Intuitionistic Fuzzy relation R if for all tuples t_1 and t_2 of R:

$$\mu_{IEQ}(t_1[X], t_2[X]) <= \mu_{IEQ}(t_1[Y], t_2[Y]) \quad (4)$$

$$\nu_{IEQ}(t_1[X], t_2[X]) > \nu_{IEQ}(t_1[Y], t_2[Y]) \quad (5)$$

where, $X, Y \in R$, $0 > \mu_{IEQ}(t) > 1$ and $0 > \nu_{IEQ}(t) > 1$, for all $t \in R$.

6.2.1 Reduction of the Definition to Fuzzy Functional Dependency

The above definition can be reduced to accommodate the definition of a Fuzzy dependency as follows.

Proof: Fuzzy dependency is a special case of IFFD, where q=0 and $X \rightarrow_{p,q} Y$ reduces to $X \rightarrow_p Y$. i.e. Fuzzy functional dependency $X \rightarrow_p Y$ exists in an Intuitionistic Fuzzy relation R if for all tuples t_1 and t_2 of R:

$$\mu_{IEQ}(t_1[X], t_2[X]) <= \mu_{IEQ}(t_1[Y], t_2[Y])$$
$$\nu_{IEQ}(t_1[X], t_2[X]) > \nu_{IEQ}(t_1[Y], t_2[Y])$$

Again, $\nu_{IEQ}(t) = 0$, which leaves us with the inequality:

$\mu_{IEQ}(t_1[X], t_2[X]) <= \mu_{IEQ}(t_1[Y], t_2[Y])$

We have, $X \rightarrow_p Y$ exists in an Intuitionistic Fuzzy relation R if for all tuples t_1 and t_2 of R

$$\mu_{IEQ}(t_1[X], t_2[X]) <= \mu_{IEQ}(t_1[Y], t_2[Y])$$

which is the definition [9] for Fuzzy functional dependency.

6.2.2 Inference Rules for Intuitionistic Fuzzy Functional Dependency

Suppose we have a relation R, and t_1, t_2 R, the following inference rules hold in R:

Rule 1:(Reflexivity)If $Y \subseteq X$, then $X \rightarrow_{p,q} Y$ in R
Proof: Since, $Y \subseteq X$,

$$\mu_{IEQ}(t_1[X], t_2[X]) <= \mu_{IEQ}(t_1[Y], t_2[Y])$$
$$\nu_{IEQ}(t_1[X], t_2[X]) > \nu_{IEQ}(t_1[Y], t_2[Y])$$

hence, $X \rightarrow_{p,q} Y$ is **trivial**
Rule 2: (Augmentation) If $X \rightarrow_{p,q} Y$ holds, then $XZ \rightarrow_{p,q} YZ$ in R
Proof: Since, $X \rightarrow_{p,q} Y$ holds,

$$\mu_{IEQ}(t_1[X], t_2[X]) <= \mu_{IEQ}(t_1[Y], t_2[Y])$$
$$\nu_{IEQ}(t_1[X], t_2[X]) > \nu_{IEQ}(t_1[Y], t_2[Y])$$

Hence,
$\min\{\mu_{IEQ}(t_1[X], t_2[X], \mu_{IEQ}(t_1[Z], t_2[Z])\} <= \min\{\mu_{IEQ}(t_1[Y], t_2[Y], \mu_{IEQ}(t_1[Z], t_2[Z])\}$
and
$\max\{\nu_{IEQ}(t_1[X], t_2[X], \mu_{IEQ}(t_1[Z], t_2[Z])\} > \max\{\nu_{IEQ}(t_1[Y], t_2[Y]), \mu_{IEQ}(t_1[Z], t_2[Z]\}$
that is: $XZ \rightarrow_{p,q} YZ$ holds in R

Rule 3: (Transitivity)If X $->_{p,q}$ Y and Y $->_{p,q}$ Z , X $->_{p,q}$ Z holds in R

Proof: Since, X $->_{p,q}$ Y and Y $->_{p,q}$ Z

$$\mu_{IEQ}(t_1[X], t_2[X]) <= \mu_{IEQ}(t_1[Y], t_2[Y]) \quad (6)$$

$$\nu_{IEQ}(t_1[X], t_2[X]) > \nu_{IEQ}(t_1[Y], t_2[Y]) \quad (7)$$

and

$$\mu_{IEQ}(t_1[Y], t_2[Y]) <= \mu_{IEQ}(t_1[Z], t_2[Z]) \quad (8)$$

$$\nu_{IEQ}(t_1[Y], t_2[Y]) > \nu_{IEQ}(t_1[Z], t_2[Z]) \quad (9)$$

from (6) and (8), we have:

$$\mu_{IEQ}(t_1[X], t_2[X]) <= \mu_{IEQ}(t_1[Z], t_2[Z]) \quad (10)$$

from (7) and (9), we have:

$$\nu_{IEQ}(t_1[X], t_2[X]) > \nu_{IEQ}(t_1[Z], t_2[Z]) \quad (11)$$

(10) and (11) implies: **X $->_{p,q}$ Z holds in R**

The following inference rules follow from the above proved rules:

Rule 4: (Union) if X $->_{p,q}$ Y and X $->_{p,q}$ Z hold , then , X $->_{p,q}$ YZ also holds.
Rule 5: (Decomposition) if X $->_{p,q}$ YZ holds, then X $->_{p,q}$ Y and X $->_{p,q}$ Z hold.
Rule 6: (Pseudo- transitivity) if X $->_{p,q}$ Y and YW $->_{p,q}$ Z hold, then XW $->_{p,q}$ Z also holds.

6.3 Normalization Process

Based on the data dependencies, we introduce the normalization method as follows: Suppose we have an IF relation R (A, B, C, D, E, F) with the following IFFDs:

$$A \rightarrow_{p,q} B \quad (I)$$

$$AC \rightarrow_{r,s} DE \quad (II)$$

$$D \rightarrow_{t,u} C \quad (III)$$

After calculating the IFFD closure for R, we know that the relation R has two overlapping candidate keys:

(A, C)
(A, D)

We choose AC to be the key for solving this very example.

1- NF (IF) check

1 NF (IF) states: All the tuples present in the IF database must have a unique identifier. This condition holds in R, therefore 1NF (IF) holds.

2- NF (IF) check

2- NF (IF) states: The relation must be in 1NF-IF and every non key attribute must be fully dependent on the IF key. Since, there is a partial dependence of a non –key attribute B on the part of the key **A**, R is not in 2- NF (IF) i.e., $A \rightarrow_{p,q} B$. Thus, we need to decompose the relation R into two relations:

R (<u>A</u>, <u>C</u>, D, E, F) and R1 (<u>A</u>, B) using (I)

3- NF (IF) check

3- NF (IF) states: The relation must be in 2-NF (IF) and all the IF- determinants in the relation must be all keys and if not keys, the determined attributes must be a part of key i. e., Non transitive IF dependence.

The determinants in R and R1 are all keys and if not keys, the determined attributes are a part of key. This is the condition that rules out transitivity. Hence, there is no need to carry out the transitivity check

BCNF – IF check

BCNF – IF states: An Intuitionistic Fuzzy relational database is in BCNF-IF iff:

*a) It is in 3NF – **IF***

b) For every determinant K ∈ U, such that
*$K_{ab} \rightarrow X$, in R where X ∈ U, implies K is a **pq** - candidate key for relation R.*

The dependency $A \rightarrow_{p,q} B$ doesn't violate any BCNF (IF) rules as A is a pq –key for the relation **R1**. Similarly, $AC \rightarrow_{r,s} DE$ is acceptable. But the dependency $D \rightarrow_{t,u} C$ creates a violation as D is not a key attribute. So, we need to decompose the relation **R** again, so as to achieve **BCNF (IF).**The final decomposition is as follows:

R (<u>A</u>, <u>D</u>, E, F), R1 (<u>A</u>, B) and **R2 (<u>D</u>, C)** using (III)

As we have seen with the crisp databases, BCNF is always stronger than 3NF. Same is the case with Intuitionistic Fuzzy relational databases. Both 3NF –IF and BCNF- IF behave the same till there is only one candidate key. Once we have more than one overlapping candidate keys, there arise chances of redundancy once again. BCNF (IF) takes care of this situation.

7 Conclusion

Normalization is a complex task in case of Intuitionistic Fuzzy databases. The data is not in a standard or conventional form and handling functional dependencies becomes even more difficult. There are challenges that we overcome in this procedure e.g. isolating the Intuitionistic Fuzzy part from the crisp one, then search for a pq-candidate key, identifying dependencies, transitivity, etc.

Further, the violation of constraints in the IF relations needs to be figured out. This helps us optimize the knowledge base in use. So that the underlying interfaces and procedures find it easy and quick to gather data and apply relevant inference rules on the same. Since the knowledge bases are huge and complex entities, we need pre-defined constraints and normalization techniques to keep our data maintained in order. Knowledge bases have to be normalized so that the Intuitionistic Fuzzy retrieval model we are using can be effectively implemented.

References

1. Alam, A., Ahmad, S., Biswas, R.: Normalization of Intuitionistic Fuzzy Relational Database. NIFS 10(1), 1–6 (2004)
2. Hussain, S., Alam, A., Biswas, R.: Normalization of Intuitionistic Fuzzy Relational Database into second normal form- 2NF (IF). International J. of Math. Sci. & Engg. Appls. 3(III), 87–96 (2009)
3. Hussain, S., Alam, A.: Normalization of Intuitionistic Fuzzy Relational Database into third normal form- 3NF (IF). International J. of Math. Sci. & Engg. Appls. 4(I), 151–157 (2010)
4. Attanasov, K.T.: Intuitionistic Fuzzy sets. Physica Verlag (1999)
5. Shora, A.R., Alam, M.A., Siddiqui, T.: Knowledge-driven Intuitionistic Fuzzy Decision Support for finding out the causes of Obesity. International Journal on Computer Science and Engineering 4(3) (2012)
6. Buckles, B.P., Petry, F.E.: A fuzzy representation of data for relational databases. Fuzzy sets and Systems 7(3), H213–H226 (1982)
7. De, S.K., Biswas, R., Roy, A.R.: Intuitionistic Fuzzy Database. In: Second International Conference on IFS, pp. 34–41 (1998)
8. Codd, E.F.: Recent Investigations into Relational Data Base Systems. IBM Research ReportRJ1385 (1974)
9. Raju, K.V.S.V.N., Majumdar, A.K.: Fuzzy Functional dependencies and lossless join decomposition of Fuzzy Relational database systems. ACM Transactions on Database Systems 13(2), 129–166 (1988)
10. Jyothi, S., Babu, M.S.: Multivalued dependencies in Fuzzy relational databases and loss less join decomposition. Fuzzy Sets and Systems 88, 315–332 (1997)
11. Vucetic, M., Hudecb, M., Vujosevica, M.: A new method for computing fuzzy functional dependencies in relational database systems. Expert Systems with Application 40(7), 2738–2745 (2013)
12. Shora, A.R., Alam, M.A., Biswas, R.: A comparative Study of Fuzzy and Intuitionistic Fuzzy Techniques in a Knowledge based Decision Support. International Journal of Computer Applications 53(7) (2012)
13. Kumar, D.A., Al-adhaileh, M.H., Biswas, R., Al-adhaileh, M.H., Biswas, R.: A Method of Intuitionistic Fuzzy Functional Dependencies in Relational Databases. European Journal of Scientific Research 29(3), 415–425 (2009)
14. Shenoi, S., Melton, A., Fan, L.T.: Functional dependencies and normal forms in the fuzzy relational database model. Information Sciences 60(1-2), 1–28 (1992)
15. Deschrijver, G., Kerre, E.E.: On the composition of Intuitionistic fuzzy relations. Fuzzy Sets and Systems 136, 333–361 (2003)
16. Umano, M.: FREEDOM-O. In: Gupta, M.M., Sanchez, E. (eds.) A Fuzzy Database System. Fuzzy Information and Decision Processes, pp. 339–349. North Holland, Amsterdam (1982)
17. Hamouz, S.A., Biswas, R.: Fuzzy Functional Dependencies in Relational Databases. International Journal of Computational Cognition 4(1) (2006)
18. Atanassov, K.T.: Intuitionistic Fuzzy Sets Past, Present and Future. In: 3rd Conference of the European Society for Fuzzy Logic and Technology (2003)

19. Boran, F.E.: An integrated intuitionistic fuzzy multicriteria decision making method for facility location selection. Mathematical and Computational Applications 16(2), 487–496 (2011)
20. Beaubouefa, T., Petry, E.F.: Uncertainty modeling for database design using Intuitionistic and rough set theory. Journal of Intelligent & Fuzzy Systems 20 (2009)
21. Yana, L., Ma, Z.M.: Comparison of entity with fuzzy data types in fuzzy object-oriented databases. Integrated Computer-Aided Engineering, 199–212 (2012)
22. Bosc, P., Pivert, O.: Fuzzy Queries Against Regular and Fuzzy Databases. Flexible Query Answering Systems, 187–208 (1997)

Fuzzy Logic Based Implementation for Forest Fire Detection Using Wireless Sensor Network

Mamata Dutta[1], Suman Bhowmik[2], and Chandan Giri[1]

[1] Department of Information Technology
Indian Institute of Engineering Science and Technology
Shibpur, West Bengal, India
[2] Department of Computer Science
College of Engineering and Management
Kolaghat, West Bengal, India
{mamata.mithi,suman.bhowmik,chandangiri}@gmail.com

Abstract. The detection and prevention of forest fire is a major problem now a days. Timely detection allows the prevention units to reach the fire in its initial stage and thus reduce the risk of spreading and the harmful impact on human and animal life. Because of the inadequacy of conventional forest fire detection on real time and monitoring accuracy the Wireless Sensor Network (WSN) is introduced. This paper proposes a fuzzy logic based implementation to manage the uncertainty in forest fire detection problem. Sensor nodes are used for detecting probability of fire with variations during different time in a day. The Sensor nodes sense *temperature, humidity, light intensity, CO_2 density and time* and send the information to the base station. This proposed system improves the accuracy of the forest fire detection and also provides a real time based detection system as all the input variables are collected in real time basis.

Keywords: Wireless Sensor Network (WSN), Fuzzy Logic, Forest Fire Detection.

1 Introduction

Forest fire is a major problem due to the destruction of forests and generally wooden nature reserves [2]. A WSN is a group of specialized transducers and associated controller units which make up a system of wireless sensor nodes deployed in some geographical area, that autonomously monitor physical or environmental conditions and send the collected information to a main controller for certain action to be taken [1]. One major constraint of sensor nodes is limited battery power. This proposed system is self sufficient in maintaining a regular power supply that is provided by the solar systems.

This work proposes a real time forest fire detection method using fuzzy logic based implementation for Wireless Sensor Network (WSN). In this proposed

DOI: 10.1007/978-3-319-07353-8_38, © Springer International Publishing Switzerland 2014

model a number of sensor nodes are densely deployed in a forest. These sensor nodes collect the variations of temperature, humidity, light intensity, CO_2 density, in its vicinity throughout its lifetime and send to the nearby cluster head to forward aggregated data to the central sink node. Use of fuzzy logic can make real time decisions without having specific information about the event [10]. Since this technique deals with linguistic values of the controlling variables in a natural way instead of logic variables, it is highly suitable for applications with uncertainties. The sensed data is fed to the fuzzy inference engine to infer the possibility of forest fire.

The rest of the paper is organised as follows. Section 2 discusses the related works done so far in this area. A short discussion on the problem statement is given in Section 3. In Section 4 our proposed solution for forest fire detection is explained in detail. Section 5 gives a brief overview of network topology for energy efficient WSN. Section 6 evaluates the performance of the proposed model followed by a conclusion in Section 7.

2 Related Works

In [1], the authors discussed the general causes of increase in frequency of forest fires and describes the architecture of wireless sensor network and a scheme for data collection in real-time forest fire detection. The authors proposed two algorithms in [2] for forest fire detection. The proposed algorithms are based on information fusion techniques. The first algorithm uses a threshold method and the other uses Dempster- Shafer theory. Both algorithms reported false positives when the motes were exposed to direct sunlight. However, if the motes are covered to avoid direct sunlight exposure, the number of false positives may be reduced.

In [3], the authors have presented how to prevent the forest fires using wireless sensor network by the design of a system for monitoring temperature and humidity in the environment.The respective values of temperature and humidity in the presence of fire is required. Authors in [4] estimated total carbon release and carbon dioxide, carbon monoxide, and methane emissions through the analysis of fire statistics from North America and satellite data from Russia.

The authors in [5] proposed an improved approach to track forest fires and to predict the spread direction with WSNs using mobile agents.

In [6], the authors have discussed the causes of environmental degradation in the presence of forest and rural fires. The authors have developed a multi sensor scheme which detects a fire, and sends a sensor alarm through the wireless network to a central server. In [7], Zhang et al. discussed a wireless sensor network paradigm based on a ZigBee technique. Environmental parameters such as temperature, humidity, light intensity in the forest region monitored in real time. The authors in [9] have introduced a wireless sensor network paradigm for real-time forest fire detection where neural network is applied for data processing. Compared with the the traditional satellite-based detection approach the wireless sensor network can detect and forecast forest fire more promptly. In [11], the necessity of intelligent decision making (IDM) is discussed.The authors have focussed on predefined sensitivity levels that are needed to activate the necessary actions. They proposed fire detection to illustrate the IDM capability of the system. A fuzzy logic based algorithm is developed and simulated to obtain the results.

Effects of weather, terrain, fuels on fire severity and compared using remote sensing of the severity of two large fires in south-eastern Australian forests are discussed in [12]. The probability of contrasting levels of fire severity (fire confined to the understoryed vs. tree canopies consumed) was analysed using logistic regression.

The following section describes the problem before going to the proposed solution strategy.

3 Problem Statement

This paper aims to design a system for early detection of forest fires. It is very important to predict the direction in which the forest fires are going to spread as they can spread quickly[4]. Availability of air, heat and fuel are the main parameters that initiate a fire in a forest. The moisture content of the combustible material plays an important role in assessment and prediction of forest fire. The moisture content is related with relative humidity in the atmosphere, wind, temperature of the air and similar factors while relative humidity affects water evaporation. The physical properties of the combustible materials vary indirectly by air temperature. Temperature, humidity, light intensity also vary with time and weather. Hence, in this work we have assumed the parameters like temperature, humidity, light intensity, CO_2 density, time for forest fire detection using fuzzy logic based implementation.

4 Proposed Solution for Forest Fire Detection

The proposed solution for forest fire detection uses the concept of Fuzzy Logic System (FLS). The necessary steps used for detecting the probability of fire using FLS are fuzzification, fuzzy rules, fuzzy inference system and defuzzification process [8]. The steps are as follows:

Step 1: In the first step each crisp input is transformed into fuzzy input which is called **fuzzification**. The inputs measured by the sensor nodes in our proposed solution are the crisp inputs such as *temperature, humidity, light intensity, CO_2 density and time.* For each of the crisp inputs a specific range is defined. The main reason to choose these input variables are that the actual heat, moisture i.e humidity, light intensity and smoke i.e CO_2 density are the main parameters in a forecast of forest fire. Crisp inputs that are passed into the control system for processing have its own group of membership functions. The group of the membership functions for each FIS (Fuzzy Inference System) variable i.e the input fuzzy variables and the output fuzzy variables is defined

Table 1. Fuzzy variables

Variable	Range	Linguistic Values
Temperature	0 °C to 590 °C [12]	VeryHigh(VH),High(H),Moderate(M),Low(L)
Humidity	0 to 100 ppm [3]	High(H),Optimum(O),Low(L),Very Low(VL)
Light Intensity	0 to 10000 lux [12]	Very High(VH),High(H),Moderate(M),Low(L)
CO_2 Density	500 to 5000 ppm	High(H),Normal(N),Low(L)
Time	0 to 24 hours	Afternoon,Noon,Beforenoon
Fire Probability	0 to 100	Very High(VH),High(H),Medium(M),Low(L),Very Low(VL)

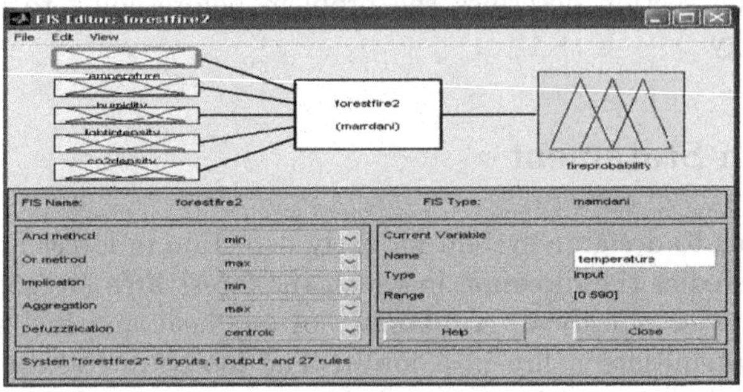

Fig. 1. FIS editor of forest fire detection

in Table 1. Each of these linguistic values represent one fuzzy set and can be defined by a membership function (MF).

Temperature, relative humidity and CO_2 density are the most important weather factors that can be used to predict and monitor fire behaviour. The danger from firebrands was lower if the ambient air temperature was below 15 °C. The corresponding temperature values in forest in the presence of fire are 100 °C - wood is dried, 230 °C - releases flammable gases, 380 °C - smoulder, 590 °C - ignite [12]. So we take the upper range of temperature value as 590 °C. There is a medium probability for a spotfire occurring when the relative humidity is below 40 percent. If the amount of CO_2 in the air of a forest is above 500ppm then only we can predict that fire has ignited [4]. The minimum and extreme light intensity in the forest when fire occurs are 500lux and 10000lux [12]. The variations in the values of these input variables in different time in day as well as seasonal variations can affect the detection process. For example the normal temperature value in the noon is much higher than the value in night. So if the sensors give same temperature value for both in day and night then it provides more information of occurrence of fire in the night. Similar situations arise for light intensity, CO_2 density. Likewise the different values of temperature in different seasons like summer, winter, spring can affect the inference system. But in this work seasons are not included.

The proposed Mamdani fuzzy inference system of forest fire detection including all input and output variables is shown in Fig. 1. The proposed membership

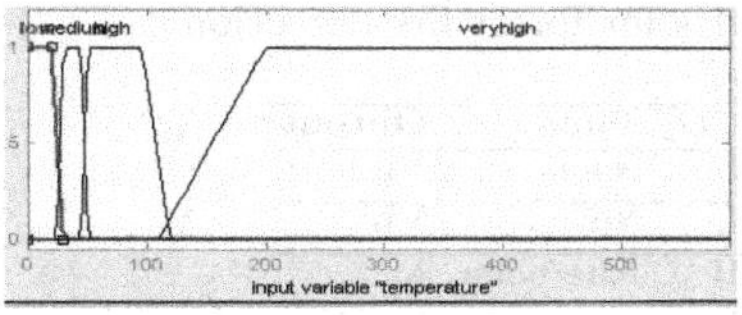

Fig. 2. Membership function for temperature

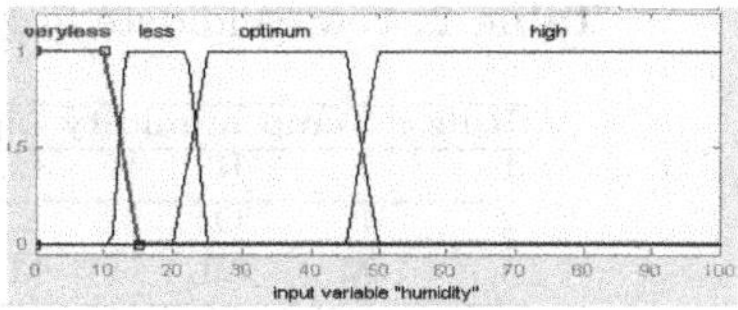

Fig. 3. Membership function for humidity

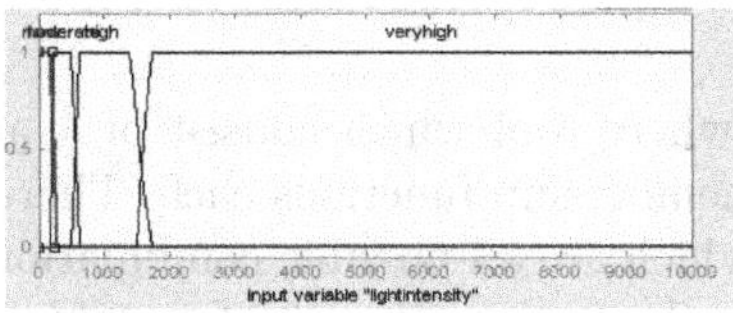

Fig. 4. Membership function for light intensity

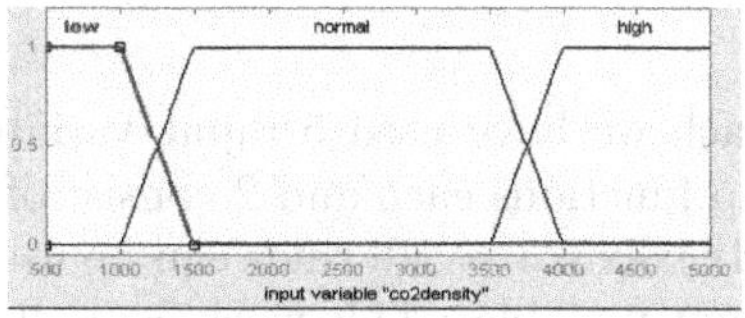

Fig. 5. Membership function for CO_2 density

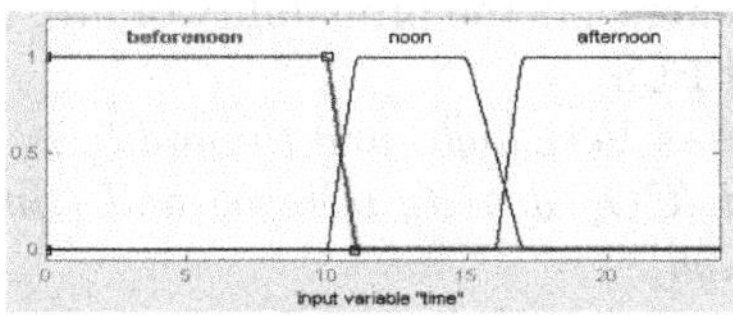

Fig. 6. Membership function for time

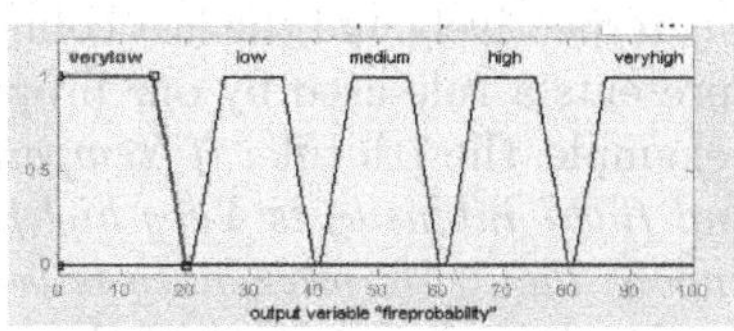

Fig. 7. Membership function for fire probability

functions for all the input and output linguistic variables are shown in figures from Fig. 2 to Fig. 7.

From Fig. 2, it is noted that if the temperature is 110 °C then the membership grade is calculated from the high and very high membership function and the rules associated with these two membership functions determines the output.

Step 2: In this step fuzzy reasoning is used to map input space to the output space. This process is called **fuzzy inference process**. The fuzzy inference process involves membership functions, logical operations and If-Then rules. Here we use Mamdani fuzzy inference method [8] for decision making. A set of fuzzy rules are defined which are a collection of linguistic statements that describe how the fuzzy inference system should make a decision regarding classifying an input or controlling an output. A series of IF-Operator-THEN rules evaluate the rule. We may assume that temperature and light intensity are very high, humidity is very low, CO_2 density is high and time is before noon. The output achieved by these assumptions is addressed to very high, i.e, for these inputs probability of fire is very high. Consequently new rules can be produced. In our proposed

Table 2. Few candidate fuzzy rules for forest fire detection

Rule#	Temp	humidity	Light	CO_2	Time	Output
1	L	H	L	L	Noon	VL
2	L	O	L	L	Noon	VL
3	M	VL	M	H	Afternoon	M
4	M	H	L	L	Noon	VL
5	H	O	H	N	Noon	M
6	H	L	H	N	Noon	H
7	VH	VL	VH	H	Beforenoon	VH
8	VH	O	VH	H	Afternoon	VH

approach we have used 5 input variables where 3 of which consist of 4 membership functions each and 2 consist of 3 membership functions each. Therefore total 4x4x4x3x3 = 576 rules can be used, which are all possible combinations of the input variables. This work uses all the rules in detecting the forest fires.

Table 2 shows some candidate fuzzy rules and represents the combinations of parameters that are used for different linguistic states to form fuzzy If-Then rules. Table 2 consists of six columns. The first five columns indicate the different conditions of inputs i.e the *temperature, humidity, light intensity, CO_2 density and time* respectively and the last column represents the probability of fire. Each row represents a rule used by our proposed FLS.

For example, the 7th rule: *If (temperature is Very high) and (humidity is Very low) and (light intensity is Very high) and (CO_2 density is high) and (time is Beforenoon) then (fire probability is very high).*

Thus the changes in values of the parameters in different time in a day can change the probability of fire accordingly. After defining the rules all the fuzzified inputs are combined according to the fuzzy rules to establish a rule strength. Here we use the "AND" operator is used for combining the fuzzified inputs. Then we find the consequence of the rule by combining the rule strength and the output membership function. Then the outputs of all the fuzzy rules are combined to obtain one fuzzy output distribution.

Step 3: In the final step a crisp value i.e a numeric digit is determined from output fuzzy set as a representative value. This process is called **defuzzification** [8]. In our work we have used "centroid of area" method.

5 Network Topology for Energy Efficient WSN

In this work the proposed WSN is based on the work as presented in [13]. Authors in [13] presented a clustering mechanism based on fuzzy communication model. The nodes in the network are randomly deployed and several clusters are determined. Each cluster contains a cluster head that communicates with the neighbouring cluster or base station. The cluster head collects data from the other member nodes in the cluster and sends directly (when it is close to base station) or indirectly (via connector nodes) to the base station. Based on this clustering method same authors also reported one k-fault tolerant topology

Table 3. conditions of fire probability

Sl No.	Output	Fire probability
1	0-20	Very Low
2	21-40	low
3	41-60	medium
4	61-80	high
5	81-100	very high

control algorithm [14] which can be used in an efficient way to send the different parameters sensed by the sensors to the base station through cluster heads and connector nodes. This communication model extends the life of the network by saving the energy of the sensors at a particular instant by using only few sensors that are active for communication.

6 Simulation and Results

In order to verify the proper functioning and the effectiveness of the proposed system with respect to the environmental conditions, MATLAB simulation is carried out. The scale of fire probability gives the results which are simply crisp numbers from 0 to 100. As shown in Fig. 7 the FIS variable fire probability has five membership functions which are distributed in the range [0, 100]. The decision of whether fire has occurred and the intensity of the fire can be made by the crisp value obtained from the defuzzification process as shown in Table 3.

Simulation results are presented in Table 4. For example an input set is taken as [80 12 1500 3000 20], where the first, second, third, fourth and fifth elements of input matrix represents *Temperature, Humidity, Light Intensity, Carbon Dioxide Density and Time* respectively. The output obtained is 91. Hence the probability of fire is *Very High.* Here temperature is 80 °C, humidity is 12ppm, light intensity is 1500lux, CO_2 density is 3000ppm and time is 6 pm and trapezoidal MFs are used for each FIS variable. So when temperature is 80 °C, it lies between *high* and *very high.* Membership grades are 1 for *high* and 0.05 for *very high.* Similarly, 12ppm humidity lies between *very low* and *low* and for *very low* sensitivity of humidity has a greater weight than *low.* In case of light intensity for 1500 ppm the membership grade of *high* is greater than *very high.* CO_2 density has higher weight in *high* compared to the *normal* level sensitivity. The time can be easily evaluated as 6pm that is 20 hours lies only in the *afternoon* region.The probability of fire is 91.3846% calculated from fuzzy logic toolbox in Matlab. For different combination of inputs, probability of fire can be calculated easily. Table 4 shows the simulation results evaluated from a variety of values of all the input variables. As an example in 1st row of Table IV temperature is 20 °C, humidity is 90ppm, light intensity is 150lux, CO_2 density is 800ppm and time is 6 pm; output is 8.55% i.e the probability of fire is *very low.* On the other hand in 6th row temperature is 85 °C, humidity is 60ppm, light intensity is 2000lux, CO_2 density is 2000ppm and time is 6 am; output is 70.50% i.e the probability of fire is *high.* Hence it is observed that the proposed approach can effectively detects the forest fire for different environmental conditions.

Table 4. Experimental output probability values for different situations

Rule#	Temp	Humidity	Light	CO_2	Time	Output(%)
1	20	90	150	800	18	8.55
2	20	40	170	900	12	8.55
3	40	60	150	900	14	8.55
4	75	10	1000	4000	19	90.00
5	80	12	1500	3000	20	91.38
6	85	60	2000	2000	6	70.50
7	300	6	5000	4500	8	91.38
8	500	40	5000	4500	20	91.94

7 Conclusion

This paper, investigates the use of fuzzy logic in determining the probability of forest fire using multiple sensors. Some vagueness related to the different environmental conditions can easily be handled by this proposed method. It gives accurate and robust result with variation of temperature, humidity, etc as all the input variables are defined by real time data.

References

1. Abraham, A., Rushil, K.K., Ruchit, M.S., Ashwini, G., Naik, V.U.: G, N.K.: Energy Efficient Detection of Forest Fires Using Wireless Sensor Networks. In: Proceedings of International Conference on Wireless Networks (ICWN 2012), vol. 49 (2012)
2. Diaz-Ramirez, A., Tafoya, L.A., Atempa, J.A., Mejia-Alvarez, P.: Wireless Sensor Networks and Fusion Information Methods for Forest Fire Detection. In: Proceedings of 2012 Iberoamerican Conference on Electronics Engineering and Computer Science, pp. 69–79 (2012)
3. Lozano, C., Rodriguez, O.: Design of Forest Fire Early Detection System Using Wireless Sensor Networks. The Online Journal on Electronics and Electrical Engineering 3(2), 402–405
4. Kasischke, E.S., Bruhwiler, L.P.: Emissions of carbon dioxide, carbon monoxide, and methane from boreal forest fires in 1998. Journal of Geophysical Research 108(D1) (2003)
5. Vukasinovic, I., Rakocevic, G.: An improved approach to track forest fires and to predict the spread direction with WSNs using mobile agents. In: International Convention MIPRO 2012, pp. 262–264 (2012)
6. Lloret, J., Garcia, M., Bri, D., Sendra, S.: A Wireless Sensor Network Deployment for Rural and Forest Fire Detection and Verification. In: Proceedings of International Conference on IEEE Sensors, pp. 8722–8747 (2009)
7. Zhang, J., Li, W., Han, N., Kan, J.: Forest fire detection system based on a ZigBee wireless sensor network. Journal of Frontiers in China 3(3), 359–374 (2008)
8. Jang, J.-S.R., Sun, C.-T., Mizutani, E.: Neuro-fuzzy and soft computing. PHI Learning
9. Yu, L., Wang, N., Meng, X.: Real-time Forest Fire Detection with Wireless Sensor Networks. In: Proceedings of International Conference on Wireless Communications, Networking and Mobile Computing, vol. 2, pp. 1214–1217 (2005)

10. Sridhar, P., Madni, A.M., Jamshidi, M.: Hierarchical Aggregation and Intelligent Monitoring and Control in Fault-Tolerant Wireless Sensor Networks. International Journal of IEEE Systems 1(1), 38–54 (2007)
11. Bolourchi, P., Uysal, S.: Forest Fire Detection in Wireless Sensor Network Using Fuzzy Logic. In: Fifth International Conference on Computational Intelligence, pp. 83–87 (2013)
12. Bradstock, R.A., Hammill, K.A., Collins, L., Price, O.: Effects of weather, fuel and terrain on fire severity in topographically diverse landscapes of south-eastern Australia. Landscape Ecology 25, 607–619 (2010)
13. Bhowmik, S., Giri, C.: Energy Efficient Fuzzy Clustering in Wireless Sensor Network. In: Proceedings of Ninth International Conference on Wireless Communication & Sensor Networks (2013)
14. Bhowmik, S., Mitra, D., Giri, C.: K-Fault Tolerant Topology Control in Wireless Sensor Network. In: Proceedings of International Symposium on Intelligent Informatics (2013)

Fuzzy Connectedness Based Segmentation of Fetal Heart from Clinical Ultrasound Images

Sridevi Sampath and Nirmala Sivaraj

[1] Dhirajlal Gandhi College of Technology, Salem, Tamilnadu, India
[2] Muthayammal Engineering College, Rasipuram, Tamilnadu, India
{balasriya,nirmala.ramkamal}@gmail.com

Abstract. *Congenital Heart Disease(CHD)* is one among the most imperative causes of neonatal morbidity and mortality. Nearly 10 percentile of contemporary infant mortality in India is accounted for CHD. It is optimal to use ultrasound imaging modality for scanning the well-being of growing fetus owing to its non-invasive nature. But then several issues such as the manifestation of speckle noise, poor quality of ultrasound images with low signal to noise ratio and rapid movements of anatomically small fetal heart makes ultrasound prenatal diagnosis of cardiac defects as a most challenging task, which can only be done flawlessly by experienced radiologists. This paper demonstrates testing the method of *fuzzy connectedness* based image segmentation to detect the fetal heart structures from ultrasound image sequences. The proposed work involves *Probabilistic Patch Based Maximum Likelihood Estimation(PPBMLE)* based image denoising technique as a pre-processing step to remove the inherent speckle noise present in ultrasound images. The second step is use of Fuzzy connectedness based image segmentation algorithm with predefined seed points selected manually inside the fetal heart structure. The results of Matlab based simulation on fetal heart ultrasound dataset proves that the combination of above mentioned image processing techniques was predominantly successful in delineating the fetal heart structures. Quantitative results of the proposed work is apparently shown to illustrate the efficacy of the PPBMLE preprocessing technique.

Keywords: Speckle suppression, Fuzzy connectedness, Image segmentation, Probabilistic patch based weights, Maximum Likelihood Estimation.

1 Introduction

CHD is a birth defect, it can be categorized as structural and functional heart defects which occurs during conception of fetus. Approximately 33 percentile to 50 percentile of these defects are critical, requiring early intervention in the first year of neonatal life [1]. Diagnosing CHD from fetal ultrasound image requires unique set of progressive anatomy identification skills and exposure [2]. Earlier diagnosis of CHD aids in maximizing the chance of neonatal mortality. Thus

M.K. Kundu et al. (eds.), *Advanced Computing, Networking and Informatics - Volume 1,* Smart Innovation, Systems and Technologies 27,
DOI: 10.1007/978-3-319-07353-8_39,

there exists a significant need for computerized algorithm to detect the fetal heart structures from the ultrasound image sequences.

Owing to the favorable merit of the non-invasive nature of ultrasound modality, it is most commonly used in inferring health status of growing fetus in the mothers' womb. The inherent demerits of fetal heart ultrasound images which complicates the diagnostic interpretation are low signal to noise ratio, anatomic complexity and dynamic nature of fetal heart. Speckle noise is inherent in clinical ultrasound images which makes it very difficult to interpret fine diagnostic facets and limits the detectability of low contrast lesions approximately by a factor of eight [10]. The biological structures in ultrasound images seems to appear with missing boundaries for the reason of poor contrast. Only because of these demerits, application of computerized algorithms are difficult for ultrasound image analysis. As the object boundaries are most important aspects for interpretation of images, delineation of the specific objects is utterly useful in realizing computerized tool to automatically recognize specific ultrasound fetal heart structures.

The prominent objective of Ultrasound speckle suppression technique is to better preserve the image edge structures while suppressing the undesired visual effects of speckle pattern. Making suitable assumption about the statistical behavior of speckle noise in ultrasound images helps in contriving effective speckle suppression technique. Numerous statistical models have been reported in the literature to model the speckle pattern. One such statistical noise model is Nakagami-Rayleigh joint probability density function. At this juncture, it is optimal to use a despeckling algorithm which follows the assumption of modelling the speckle pattern in terms of Nakagami-Rayleigh joint probability density function namely Probabilistic patch based Weighted Maximum Likelihood Estimation (PPBMLE) proposed by Charles et al [4]. Maximum Likelihood estimation is a parameter estimation technique holding substantial importance in statistical estimation. This patch based filtering is based on nonlocal means method, as pixel values far apart can be averaged with Patch similarity based weights together to generate best estimate of the noise free pixel.

Though multitude of segmentation techniques were reported in the literature, detecting the cardiac structures from fetal ultrasound image remains still difficult. Lassige et al [8] utilized the level set segmentation technique to delineate the septal defects present in fetal heart chambers. Enhanced version of level set segmentation algorithm utilizing shape prior information is used by Dindoyal et al [9] to segment and extract the fetal heart chambers. The proposed method involves the investigation of Fuzzy connectedness based image segmentation for extracting the fetal heart structures. This method is basically a region growing method of object segmentation [7] by considering the pixel properties of spatial relationship and intensity similarities, which needs users involvement to select seed points. The segmentation results of this method gives high valued fuzzy connectedness map defined by the strength of hanging togetherness of pixels and appropriate object extraction can be performed with reference to the seed points. This paper is organized as follows. Section 2 describes the methodology used in the proposed work with PPBMLE based preprocessing and FC based image segmentation scheme. Section 3 presents the quantitative validation for FC based segmentation scheme with and without PPBMLE preprocessing technique. Section 4 describes the results of the proposed work implemented in both phantom ultrasound image and clinical ultrasound images.

2 Methodology

The proposed work involves the extraction of Region of interest (ROI), use of PPBMLE based despeckling to remove the speckle noise and Fuzzy connectedness based image segmentation to detect the structure of fetal heart from four chamber view of ultrasound plane. The scheme of the proposed work is illustrated in the flowchart shown in Fig.1.

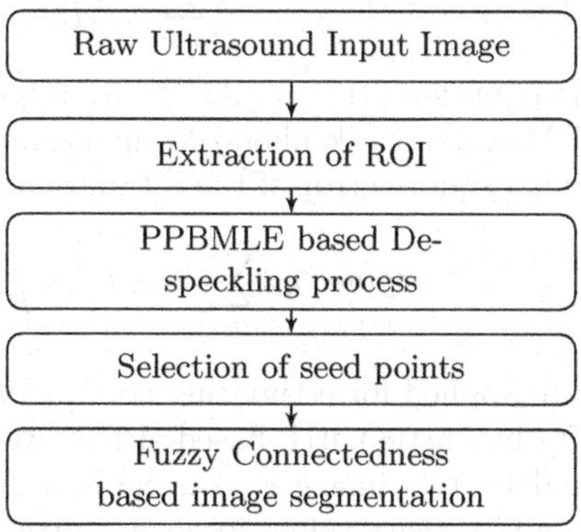

Fig. 1. Flowchart for proposed scheme

2.1 Probabilistic Patch Based Weighted Maximum Likelihood Despeckling

The inherent speckle pattern in Ultrasound image obscures the obstetrical screening of diagnostic information. According to Goodman's statistic details [8], the amplitude of envelop signal emerged from ultrasound transducer appears in the form of backscattered echoes reflected from biological tissues and hence gets multiplicatively corrupted with speckle noise. The amplitude A_s can be well modelled by independent and identically distributed joint probability density function such as Nakagami-Rayleigh distribution [12]. It can be defined as

$$p(A_s|R_s^*) = \frac{2L^L}{\Gamma(L)R_s^{*L}} A_s^{2L-1} exp^{\frac{-LA_s^2}{R_s^*}} \tag{1}$$

where A_s- Pixel amplitude, R_s^*- Reflectivity image and $L-$Shape parameter. With decreased signal to noise ratio of fetal ultrasound image, the noise free signal amplitude can be estimated by extracting priori information of image from statistical parameters of the noise distribution. Weighted Maximum Likelihood estimator (WMLE) accomplishes the good estimate of the true intensity of image signal by utilizing the appropriate weights $w(s,t)$ updated with respect to Probabilistic patch based (PPB) similarity measure.

Assuming image region being defined by regular grid S of 3x3 neighbourhood ? with the cardinality of 9 pixels, let v_s be a pixel at the site s??. The noise free pixel intensities $\widehat{R_s}$ can be estimated from reflectivity image by the relationship defined as

$$\widehat{R_s} = \frac{\sum_t w(s,t)A_t^2}{\sum_t w(s,t)} \tag{2}$$

At every pixel position of neighbourhood, the Maximum Likelihood Estimator (MLE) presumes a robust estimate $\widehat{R_s}$ from the underlying reflectivity image R_s^* given by

$$\widehat{R_s}^{(MLE)} \cong argmax_{Rs} \sum_{S \in \Omega} log \frac{v_t}{R_s} \tag{3}$$

MLE framework given in equation (3) generalized with appropriate PPB weights is defined by Weighted Maximum Likelihood Estimation (WMLE) method. It intends to reduce the mean square error of the estimate and is given by

$$\widehat{R_s}^{(MLE)} \cong argmax_{Rs} \sum_{S \in \Omega} w(s,t) log \frac{v_t}{R_s} \tag{4}$$

The similarity measure applied for estimating true intensity from noisy pixel intensities is called as Probabilistic patch based (PPB) weights. The weights of the estimate is expressed by two image region patches Δ_s and Δ_t with close probability assuming equal intensity values for center pixel. The weights defined by patch based similarity is denoted by

$$w(s,t)^{(PPB)} \cong p(R_{\Delta_s}^* = R_{\Delta_t}^* v)^{(1/h)} \tag{5}$$

where $R_{\Delta_s}^*$ and $R_{\Delta_t}^*$ denotes sub-image extracted from the parameter image R^* in the corresponding windows of Δ_s and Δ_t and h is a scalar parameter. The scalar parameter depends on the size of weight.

2.2 Fuzzy Connectedness Based Image Segmentation

To solve the problem of image analysis, the notion of pixel connectivity gives significant decisive statistics to infer about the nature of image. The framework of Fuzzy Connectedness (FC) based segmentation has been proven as quite successful for image analysis [7]. FC declares global fuzzy relation over the entire fuzzy spels by assigning a strength of connectedness to every pair of spels in an image to identify objects using dynamic programming. The prominent goal of this scheme is to grow the region of an object of interest by capturing the specific intensity patterns starting from the set of seed point spels, which can been selected manually. A seed spel is selected in such a way, as it belongs to

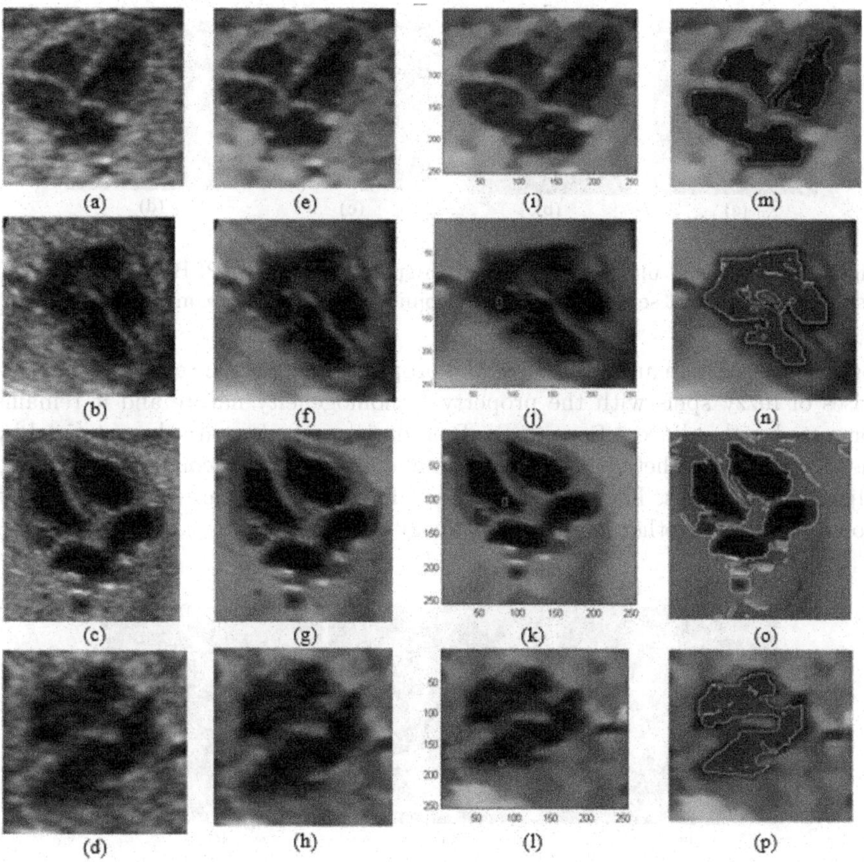

Fig. 2. (a,b,c,d) Fetal heart ROI extracted from clinical raw ultrasound images; (e,f,g,h) Despeckled images using PPBMLE; (i,j,k,l) Selection of seed points to aid segmentation; (m,n,o,p) Fuzzy connectedness based segmentation resulting delineation of fetal heart structures

an object of interest which needs to be delineated. Here, the user needs to select the seed spel from the fetal heart structure.

Considering a sample image consisting of numerous fuzzy subsets, they are characterized by the strength of pixel elements assigned in the range of 0 to 1 in terms of fuzzy membership values or Degree of membership $\mu_F(s,t)$ with as maximum gray scale intensity value of k = 255. It can be denoted by

$$\mu_F(s,t) = C\left\langle F(s,t)\right\rangle/k \tag{6}$$

The concept of Fuzzy connectedness effectively captures fuzzy hanging togetherness of objects in the image [7]. It is realized by the property of hanging togetherness among spels in a scene C and is called as Local hanging togetherness and it can also be extended over entire image to define Global hanging

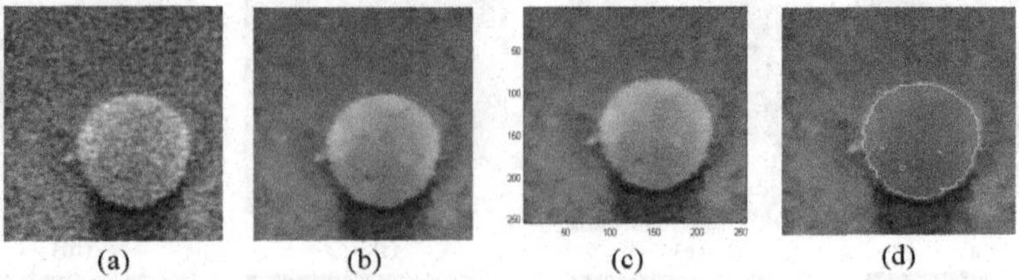

Fig. 3. (a) Phantom ultrasound image with speckle noise; (b) PPBMLE despeckling result; (c) Seed point selection; (d) Fuzzy connectedness based segmented image

togetherness. These are defined by spel properties. For instance, an object comprises of fuzzy spels with the property of homogeneity nature and it remains connected with adjacent fuzzy spels. Boundaries and edges are characterized by the fuzzy spels of heterogeneity nature and it remains unconnected with the adjacent fuzzy spels. The concept of fuzzy affinity relations are utilized to define global hanging togetherness called as Fuzzy connectedness.

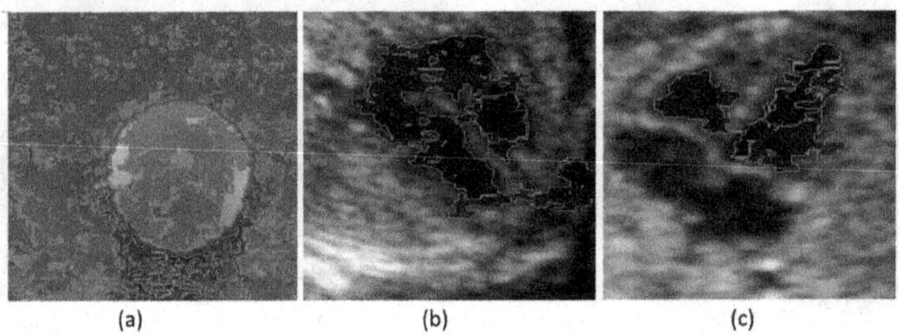

Fig. 4. (a) Segmented Phantom ultrasound image without the use of PPBMLE based despeckling; (b,c) Segmented clinical ultrasound fetal heart images without the use of PPBMLE based despeckling

The strength of fuzzy connectedness in contiguous region of a scene between any two spels say s and t is defined as fuzzy spel affinity $\mu_k(s,t)$ relation given in equation(7) and it can be acheived by a combination of three prominent parameters. The first parameter is the unit of co-ordinate space adjacency $\mu_\alpha(s,t)$, which is a measure of topological spatial location and it is denoted by equation(8). The second parameter is the unit of intensity space adjacency $\mu_\phi(s,t)$, which is a measure of homogenous nature of pixel intensity and it is denoted by equation(9). The third parameter is the unit of gradient intensity space adjacency $\mu_\psi(s,t)$, which is a measure to identify object boundaries and is defined in equation(10).

$$\mu_k(s,t) = \mu_\alpha(s,t)[\omega_1\mu_\phi(f(s),f(t))] + [\omega_2\mu_\psi(f(s),f(t))] \tag{7}$$

$$\mu_\alpha(s,t) = \begin{cases} \frac{1}{1+k\sqrt{\sum_{i=1}^{2}(s_i-t_i)^2 \le 2}}, & \text{if} \sum_{i=1}^{2}(s_i - t_i)^2 \le 2 \\ 0, & \text{otherwise} \end{cases} \tag{8}$$

$$\mu_\phi(s,t) = e^{-1/2[(\frac{1}{2}(f(s)+f(t))-m_i)/v_i]} \tag{9}$$

$$\mu_\psi(s,t) = e^{-1/2[(\frac{1}{2}(f(s)+f(t))-m_g)/v_g]} \tag{10}$$

The value of affinity nears 0 for non-adjacent spels and ranges from 0 to 1 for adjacent spels. Fuzzy connectivity details among spels are represented as connectivity map where the object of interest is extracted by thresholding the image at specified threshold. The computation of fuzzy connectivity image is carried out by initiating from the selected one or more seed points within the object of interest. The statistical measures like mean and standard deviation of the selected seed points are computed separately for extracting the features of the specified object of interest. The fuzzy k-component based object extraction with specified threshold value is defined as

$$\mu_k(s,t) = \begin{cases} 1, & \text{if} \mu_k(s,t) \ge \theta \in [0,1] \\ 0, & \text{otherwise} \end{cases} \tag{11}$$

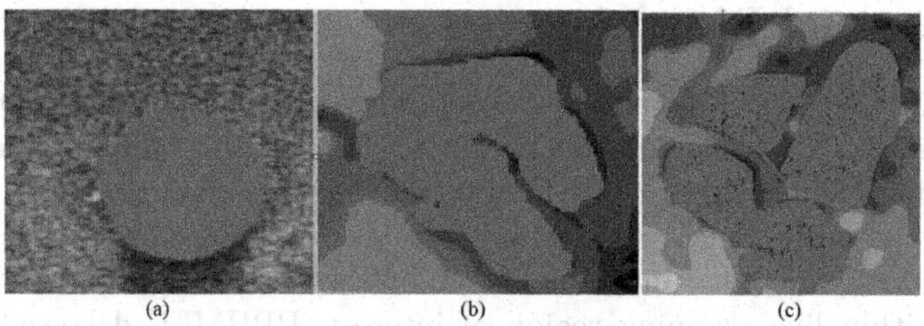

Fig. 5. (a) Manually segmented ground truth of phantom images; (b and c) Manually segmented ground truth of clinical ultrasound images

3 Quantitative Validation of Segmentation

The performance of Fuzzy connectedness based image segmentation scheme is validated using Dice similarity coefficient (DSC) and Tanimoto Coefficient (TC). The DSC index is used to measure the spatial overlap among two segmentation regions, namely manual segmentation region D_1 and algorithmic segmentation region D_2.

$$DSC(D_1, D_2) = \frac{|D_1 \cap D_2|}{|D_1 \cup D_2|} \tag{12}$$

Table 1. Quantitative results for validating segmentation

Image segmented after PPBMLE technique			
S.No	Name of Image	DSC	TC
1	Phantom Image	0.998	0.993
2	First Ultrasound image	0.985	0.974
3	Second Ultrasound image	0.974	0.961

Image segmented Without PPBMLE technique			
S.No	Name of Image	DSC	TC
1	Phantom Image	0.325	0.283
2	First Ultrasound image	0.246	0.147
3	Second Ultrasound image	0.126	0.098

The TC coefficient is also used to measure the overlap of two regions namely D_1 and D_2. TC is defined as the ratio of pixel quantity in both manually segmented image and algorithmically segmented image to the quantity of pixels in union of both.

$$TC(D_1, D_2) = \frac{D_1 \cdot D_2}{\|D_1\|^2 + \|D_2\|^2 - D_1 \cdot D_2} \tag{13}$$

4 Results and Discussion

This section describes the results of proposed work. The profound intention of the proposed work is to clearly delineate the fetal heart chamber. Various image processing techniques discussed above are experimented in both ultrasound phantom image and clinical ultrasound four chamber view of the heart image. Fig.2 shows four different clinical ultrasound fetal heart images at various stages of the algorithm like cropping region of interest, PPBMLE despeckling, Seed points selection and Fuzzy connectedness based fetal heart delineation. Fig.(3) shows the application of the PPBMLE and FC segmentation adopted for processing the phantom ultrasound image. To illustrate the efficacy and to highlight the role of PPBMLE preprocessing technique in image segmentation, Fig.(4) shows the ultrasound images improperly delineated structures without the use of PPBMLE technique. Fig.(5) shows the ground truth of manually segmented images which is used for quantitative evaluation of segmentation scheme.Table.1 shows the quantitative validation of segmentation scheme using Dice similarity coefficient (DSC) and Tanimotto coefficient (TC).

5 Conclusion

Experimental results of various image processing techniques adapted in the proposed work clearly delineates fetal heart structures from the raw clinical ultrasound images. Before applying these techniques, the ultrasound image was more unfavorable to draw unambiguous diagnostic results. Thus the proposed work would be much useful from the position of secondary observer in assisting the radiologists. Moreover, the speckle suppression technique effectively enhances the image structures making the clinical ultrasound images more suitable for application of automatic image analysis techniques for providing automatic decision about the disease diagnosis. The future work is to automatically segment the fetal heart structures.

References

1. Hoffman, J.I., Kaplan, S.: The incidence of congenital heart disease. J. American College of Cardiology Foundation 147, 1890–1900 (2002)
2. Rychik, J., Ayres, N., Cuneo, B., Spevak, P.J., Van Der Veld, M.: American Society of Echocardiography Guidelines and Standards for performance of the fetal Echocardiogram. Journal of the American Society of Echocardiography, 803–810 (July 2004)
3. Goodman, J.: Some fundamental properties of speckle. Journal of Optical Society, 1145–1150 (1976)
4. Deledalle, C.-A.: Loic Denis and Florence Tupin: Iterative Weighted Maximum Likelihood Denoising with Probabilistic Patch based weights. IEEE Trans. in Image Processing, 2661–2672 (December 2009)
5. Alvares, L., Mazorra, L.: Signal and Image Restoration using Shock Filters and anisotropic diffusion. SIAM Journal of Numerical Analysis, 590–695 (1994)
6. Meghoufel, A., Cloutier, G., Crevier-Denoix, N., de Guise, J.A.: Tissue Characterization of Equine Tendons with clinical B-Scan Images using a Shock filter thinning algorithm. IEEE Trans. on Medical Imaging, 596–605 (March 2011)
7. Udupa, J.K., Samarasekera, S.: Fuzzy connectedness and object definition: Theory, Algorithms and applications in image segmentation, Graphical models and image processing. Graphical Models and Image Processing 59(3), 246–261 (1996)
8. Lassige, T.A., Benkeser, P.J., Fyfe, D., Sharma, S.: Comparison of septal defects in 2D and 3D echocardiography using active contour models. Computerized Medical Imaging and Graphics 6, 377–388 (2000)
9. Dindoyal, I., Lambrou, T., Deng, J., Todd-Pokropek, A.: Level set snake algorithms on the fetal heart. In: 4th IEEE International Symposium on Biomedical Imaging, pp. 864–867 (April 2007)
10. Wagner, R.F., Smith, S.W., Sandrik, J.M., Lopez, H.: Statistics of speckle in ultrasound B-Scans. IEEE Transactions on Sonics and Ultrasonics 30(3), 156–163 (1983)

An Improved Firefly Fuzzy C-Means (FAFCM) Algorithm for Clustering Real World Data Sets

Janmenjoy Nayak, Matrupallab Nanda, Kamlesh Nayak, Bighnaraj Naik, and Himansu Sekhar Behera

Department of Computer Science & Engineering
Veer Surendra Sai University of Technology (VSSUT), Burla, Odisha, India
{mailforjnayak,linku92,knayak32,mailtobnaik}@gmail.com, hsbehera_india@yahoo.com

Abstract. Fuzzy c-means has been widely used in clustering many real world datasets used for decision making process. But sometimes Fuzzy c-means (FCM) algorithm generally gets trapped in the local optima and is highly sensitive to initialization. Firefly algorithm (FA) is a well known, popular metaheuristic algorithm that simulates through the flashing characteristics of fireflies and can be used to resolve the shortcomings of Fuzzy c-means algorithm. In this paper, first a firefly based fuzzy c-means clustering and then an improved firefly based fuzzy c-means algorithm (FAFCM) has been proposed and their performance are being compared with fuzzy c-means and PSO algorithm. The experimental results divulge that the proposed improved FAFCM method performs better and quite effective for clustering real world datasets than FAFCM, FCM and PSO, as it avoids to stuck in local optima and leads to faster convergence.

Keywords: Clustering, Fuzzy c-means, Firefly Optimization, PSO.

1 Introduction

Clustering is an unsupervised pattern recognition method based on objective function, which is used to partition the objects into different clusters (groups). In the same cluster the objects have high degree of similarity while the objects in different clusters have high degree of dissimilarity. Clustering algorithms can be broadly classified [1] into Hard, Fuzzy, and Probabilistic. K-means is one of the most admired hard clustering algorithms which takes the input parameter *k*, and partitions a set of *n* objects into *k* clusters so that the resulting intra-cluster similarity is high but the inter-cluster likeness is low. Similarity of the cluster is measured in regard to the mean value of the objects in a cluster, which can be viewed as the cluster's centroid or center of gravity.

Fuzzy c-means (FCM) algorithm has been widely used clustering algorithm and was introduced by Bezdek [2] in 1974.But, the result of FCM technique is dependent on the initialization of center because their search is based on the hill climbing methods and the iterative random selection of centers, causes trapping in the local optima. Also, closeness of data object to the cluster centers determines the degree of

M.K. Kundu et al. (eds.), *Advanced Computing, Networking and Informatics - Volume 1*, Smart Innovation, Systems and Technologies 27,
DOI: 10.1007/978-3-319-07353-8_40, © Springer International Publishing Switzerland 2014

membership of a data object to a cluster. To overcome these shortcomings of the fuzzy c-means algorithm, many researchers have integrated various optimization strategies to it. A hybrid Fuzzy c-means and Fuzzy PSO clustering algorithm introduced by Izakian and Abraham in order to use the merits of both the algorithms. A PSO based fuzzy clustering algorithm [3] is discussed in Li, Liu and Xu. Wang *et.al.* [4], presented a new Fuzzy c-means clustering algorithm based on particle swarm optimization. Runkler and Katz [7], introduced two new methods for minimizing the reformulated version of FCM objective function by PSO. These algorithms are less effective to get the optimal solution due to their stochastic nature.

Firefly algorithm introduced by X.S. Yang, [5] is a new nature stimulated meta-heuristics optimization algorithm, which can be effectively used to find globally optimal solutions for various nonlinear optimization problems. The firefly algorithm is based on the flashing behavior of social insects (fireflies) [6] which attracts its neighboring fireflies. Nature based algorithms have shown effectiveness and efficiency to solve difficult optimization problems. In this paper, first a firefly based fuzzy c-means and subsequently an improved firefly (FA) based fuzzy c-means algorithm has been proposed to cluster the real world datasets. The rest of the paper is outlined in the following manner. Section 2, introduces the fuzzy c-means clustering. In Section 3, the firefly algorithm (FA) and in Section 4, the particle swarm optimization algorithm (PSO) is described. The Section 5 presents the proposed clustering algorithm and Section 6 reports the experimental results. Finally, Section 7 concludes the proposed work.

2 Fuzzy C-Means Algorithm

Let us consider a set of n vectors ($X = (x_{1,}x_2, \dots, x_n)\ 2 \le c \le n$) for clustering into c groups. Each vector $x_i \in R^s$ is described by s real valued measurements which represent the features of the object x_i. A membership matrix known as Fuzzy partition matrix is used to describe the fuzzy membership matrix. The set of fuzzy partition matrices$(c \times n)$ is denoted by M_{fc} and is defined in Eq. 1.

$$M_{fc} = \{W \in R^{cn} |\ w_{ik} \in [0,1], \forall i,k; \sum_{i=1}^{c} w_{ik} = 1, \forall k; \\ 0 < \textstyle\sum_{k=1}^{n} w_{ik} < n, \forall\ i\} \tag{1}$$

where $1 \le i \le c,\ 1 \le k \le n$

From the above defined definitions, it can be found that the elements can fit into more than one cluster with different degrees of membership. The total "membership" of an element is normalized to 1 and a single cluster cannot contain all data points. The objective function (Eq. 2) of the fuzzy c-means algorithm is computed by using membership value and Euclidian distance (Eq. 3).

$$J_m(W,P) = \sum_{\substack{1 \le k \le n \\ 0 \le i \le c}} (w_{ik})^m (d_{ik})^2 \tag{2}$$

where

$$(d_{ik}) = ||x_k - p_i|| \tag{3}$$

where $m \in (1, +\infty)$ is the parameter which defines the fuzziness of the resulting clusters and d_{ik} is the Euclidian distance from object x_k to the cluster center p_i.

The minimization [7]of the objective function J_m through FCM algorithm is being performed by the iterative updation of the partition matrix using the Eq. 4 and Eq. 5.

$$p_i = \sum_{k=1}^{n} (w_{ik})^m x_k / \sum_{k=1}^{n} (w_{ik})^m \tag{4}$$

$$w_{ik}^{(b)} = \sum_{j=1}^{c} 1/[(d_{ik}^{(b)}/d_{jk}^{(b)})^{2/m-1}] \tag{5}$$

The steps of FCM algorithmare as follows:

1. Initialize the number of clusters c.
2. Select an inner product metric Euclidean norm and the weighting metric m (fuzziness)
3. Initialize the cluster prototype $P^{(0)}$, iterative counter b=0.
4. Then calculate the partition matrix $W^{(b)}$ using (5).
5. Update the c fuzzy cluster centers $P^{(b+1)}$ using (4).
6. If $\left\|P^{(b)} - P^{(b+1)}\right\| < \varepsilon$ then stop, otherwise repeat step 2 through 4.

3 Firefly Algorithms (FA)

Fireflies are glowworms that glow through bioluminescence. The firefly algorithm is based on the idealized behavior of the flashing characteristics of fireflies. Three rules has been described [8] for the basic principle of firefly algorithm .

1. All fireflies are unisexual in nature so that one firefly will be fascinated to other fireflies despite of their sex.
2. Attractiveness is proportional to their brightness, thus for any two flashing fireflies, the less bright one will move towards the brighter one and if their distance increases, they both decreases. If there is no brighter one than a particular firefly, it will move randomly.
3. The brightness of a firefly is affected or determined by the landscape of the objective function.

The firefly algorithm is a population based algorithm to locate the global optima of objective functions. In the firefly algorithm fireflies are randomly distributed in the search space. A firefly attracts other neighboring fireflies by its tight intensity. The attractiveness of a firefly is dependent on the brightness of the firefly. The brightness is dependent on the intensity of light emitted by the firefly. The intensity is calculated using the objective function. The intensity is inversely proportional to the distance r between two fireflies, $I \propto 1/r^2$. Each firefly represents a candidate solution such that the fireflies move towards the best solution [9] after each iteration as the firefly with best solution has the brightest glow (intensity).

The firefly algorithm has two important issues: the variation of light intensity and formulation of the attractiveness. The attractiveness of the firefly can be determined by its brightness or light intensity [10] which in turn is associated with the encoded objective function. The objective function in this case is described by (Eq.2).Based on this objective function initially, all the fireflies are randomly dispersed across the search space. The two phases of firefly algorithm are as follows :

(1) **Variation of light intensity**: The objective function values are used to find the light intensity. Suppose there exist a swarm of *n* fireflies and x_i represents a solution for firefly *i*, whereas $f(x_i)$ denotes the fitness value in Eq. 6.

$$I_i = f(x_i),\ 1 \leq i \leq n. \tag{6}$$

(2) **Movement towards attractive firefly**: The attractiveness of a firefly is proportional to the light intensity [11] seen by adjacent flies. Each firefly has its distinctive attractiveness β which describes how strong a firefly attracts other members of the swarm. But the attractiveness β is relative, it will vary with the distance r_{ij} (Eq. 7) between two fireflies *i* and *j* located at x_i and x_j, respectively.

$$r_{ij} = ||x_i - x_j|| \tag{7}$$

The attractiveness function $\beta(r)$ of the firefly is determined by Eq. 8.

$$\beta(r) = \beta_0 e^{-\gamma r^2} \tag{8}$$

Where β_0 is the attractiveness at r=0 and γ is the light absorption coefficient.

The movement of a firefly *i* at location x_i attracted to a more attractive firefly *j* at location x_j is given by Eq. 9.

$$x_i(t+1) = x_i(t) + \beta_0 e^{-\gamma r^2}(x_i - x_j) + \alpha\,(\text{rand} - 0.5) \tag{9}$$

The pseudo code for this algorithm is given as follows.

Objective function f(x), $x = (x_1, ..., x_d)^T$
Generate an initial population of fireflies x_{ik}, i = 1, 2, . . . , n and k = 1, 2, . . . , d
where d=number of dimensions
Maxgen: Maximum no of generations
Evaluate the light intensity of the population I_{ik} which is directly proportional to f (x_{ik})
Initialize algorithm's parameters
While(t <Maxgen)
  For i= 1 :n
    For j= 1: n
      If (I_j<I_i)
        Move firefly i toward j in d- dimension using Eq. (9)
      End if
        Attractiveness varies with distance r viaexp[−r2]
      Evaluate new solutions and update light intensity using Eq. (6)
    End for j
  End for i
Rank the fireflies and find the current best
End while
Post process results and visualization

---

4 Particle Swarm Optimization

Particle swarm optimization (PSO) is a stochastic optimization technique inspired by bird flocking and fish schooling originally designed and introduced by Kennedy and Eberhart 1995 [12][13]. In this algorithm each particle (bird or fish) is initialized with a population of random solutions. The particles fly through the searching space with a velocity[14] that is dynamically adjusted. $X_i = (x_{i1}, x_{i2},\dots, x_{iD})$ represents the ith particle, the best solution is $P_i = (p_{i1}, p_{i2}, \dots, p_{iD})$ also known as p_{besti}. The best solution of all particles [15] is p_g, also called g_{best}. $V_i = (v_{i1}, v_{i2}, \dots, v_{iD})$ is the velocity of particle i. For every generation, the velocity and position changes [16] according to the Eq. 10 and Eq. 11 respectively.

$$v_{id} = wv_{id} + c_1 rand_1(p_{id} - x_{id}) + c_2 rand_2(p_{gd} - x_{id}) \quad \mathbf{(10)}$$

$$x_{id} = x_{id} + v_{id} \quad \mathbf{(11)}$$

w is the inertia weight, which often changes from 0.9 to 0.2. c_1 and c_2 are accelerative constants. $rand_1$ and $rand_2$ are random functions which change between 0 and 1.

5 Proposed Clustering Approach

5.1 Firefly Based Fuzzy C-Means Clustering (FAFCM)

In this work the firefly algorithm applied on the FCM to overcome the shortcomings of the FCM algorithm. The firefly algorithm has been applied on the objective function of FCM algorithm given in Eq. (2).Here the goal is to minimize the objective function of FCM clustering. In the proposed method the cluster centers are used as the decision parameters to minimize the objective function. So, in the context of clustering, a single firefly represents the cluster centers vectors. Each firefly is represented by using Eq. 12.

$$x_i = (p_{i1}, \dots, p_{ij}, \dots, p_{ic}), \quad 2 \le j \le c, \quad \mathbf{(12)}$$

wherep_{ij} represents the jth clustering center vector. Hence, a swarm of n fireflies represents n candidate solutions.As this is a minimization problem the intensity of each firefly is equal to the value of the objective function of FCM . The pseudo code for the FAFCM algorithm is being illustrated as follows.

```
Initialize the population of n fireflies with C random cluster centers of d-dimensions each
Initialize algorithm's parameters
Repeat
  For i=1: n
   For j=1: n
          Calculate light intensity (objective function value) of each firefly by eq.(2),
If (Ij<Ii)
          Move firefly i toward j based on eq.(9) to update the position of fireflies
End if
    End for j
  End for i
Rank the fireflies and find the current best
Do Until stop condition true
Rank the fireflies, obtain the global best and the position of the global best( cluster centers)
```

5.2 Improved Firefly Based Fuzzy C-Means Clustering

After the experimental analysis, it is found the cluster centers can be further refined using the improved FAFCM algorithm, which helps in further minimization of the objective function. The experimental results show that for smaller datasets FAFCM provides better results as compared to FCM but with increase in the size of datasets FCM surpass FAFCM. In the proposed improved algorithm, the clustering is performed in two stages. In the first stage the firefly has been initialized with random values and after a fixed number of iterations the cluster centers is obtained. Then in second stage the obtained cluster centers are used as the initial cluster centers for the FCM algorithm to get the refined cluster centers which gives a more minimized objective function value. The pseudo code of the proposed improved FAFCM clustering algorithm has been described as follows.

Initialize the population of n fireflies with C random cluster centers of d-dimensions each
Initialize algorithm's parameters
Repeat
  For i=1: n
    For j=1: n
      Calculate light intensity (objective function value) of each firefly by eq.(2),
      If (I_j<I_i)
        Move firefly i toward j based on eq.(9) to update position of fireflies (cluster centers)
      End if
    End for j
  End for i
  Ranks the fireflies and find the current best to update current best to next iteration
Do until stop condition true
Rank the fireflies, obtain global best and the position of global best(better cluster centers)
Initialize the FCM center with position of global best,
Then using this center iterate the FCM algorithm,
Repeat
Update the membership matrix by eq.(5)
Refine the cluster centers by eq.(4),
Do until it meets the convergence criteria

---

6 Experimental Analysis

6.1 Parameter Set Up

For optimizing the performance of FAFCM and improved FAFCM, the best values for their parameters are chosen. As far as the experimental results are concerned, these algorithms perform best under the following parameters:

Fuzziness coefficient (m) = 2
Number of Fireflies (n) = 20
Attractiveness (β_0) = 1
Light absorption coefficient (γ) = 1
Randomization Criteria (ε) = 0.00001
For PSO, inertia (w) = 0.9
Accelerating constants (c1, c2) = (0.49, 0.49)

The terminating condition in FCM algorithm is when there is no scope for further improvement in the objective function value. The FAFCM stopping condition is 500 generations (maximum no. of iterations) or no changes in current best in 5 consecutive iterations. In the improved FAFCM algorithm the terminating condition for FAFCM is the maximum number of generations i.e. 50 or no changes in current best in 3 consecutive iterations. The terminating condition for FCM in improved FAFCM is same as the FCM algorithm mentioned above.

6.2 Experimental Results

The main aim of our experimental study is to compare and contrast the proposed algorithm with the other algorithms like FCM, PSO and FAFCM. In order to evaluate the proposed method the experiment is being conducted by considering following three real world datasets and one artificial dataset.

- **Glass Dataset:** The Glass data set is defined in terms of their oxide content as glass type. Nine inputs are based on 9 chemical measurements with one of 6 types of glass. The data set contains 214 patterns.
- **Iris Dataset:** The Iris data set consists of three varieties of flowers setosa, virginica and versicolor. There are 150 instances and 4 attributes that make up the 3 classes.
- **Lung Cancer Dataset:** The Lung cancer data set has 32 points in which all predictive attributes are nominal and takes integer value 0, 1, 2 or 3.
- **Single Outlier Dataset:** The single outlier data set is an artificial data set given as $X_1 = \{-1.2, 0.5, 0.6, 0.7, 1.5, 1.6, 1.7, 1.8\}$.This data set contains 2 clusters.

Table 1. Comparison of objective function value and number of iterations for the iris dataset for various clustering algorithms

Algorithm	Objective Function Value		Average Number of iterations
	Best	Mean	
FCM	60.58	62.63	24
PSO	60.58	61.07	56.25
FAFCM	60.57	60.58	48.15
Improved FAFCM	60.57	60.57	15.75

Table 2. Comparison of objective function value and number of iterations for the glass dataset for various clustering algorithms

Algorithm	ObjectiveFunction Value		Average Number of iterations
	Best	Mean	
FCM	156.14	158.63	42
PSO	178.34	183.24	105.65
FAFCM	176.14	180.02	98.33
Improved FAFCM	154.12	154.76	34

Table 3. Comparison of objective function value and number of iterations for the Lung Cancer dataset for various clustering algorithms

Algorithm	Objective Function Value		Average Number of Iterations
	Best	Mean	
FCM	223.05	227.15	14.25
PSO	260.37	273.49	52
FAFCM	227.16	234.25	43
Improved FAFCM	219.05	219.45	9.23

Table 4. Comparison of objective function and number of iterations for the Single Outlier dataset for various clustering algorithms

Algorithm	Objective Function Value		Average Number of Iterations
	Best	Mean	
FCM	2.04	2.13	7.75
PSO	1.79	1.79	10.35
FAFCM	1.78	1.79	6
Improved FAFCM	1.78	1.78	5

The algorithms are implemented in MATLAB 9 and the experimental results of eight runs of each algorithm are recorded in the tables (Table 1, Table2, Table3, Table4). The tables contain the best and mean objective function values as well as the average number of iterations till convergence for each algorithm on the four datasets taken. The changes in objective function value in each iteration for different datasets

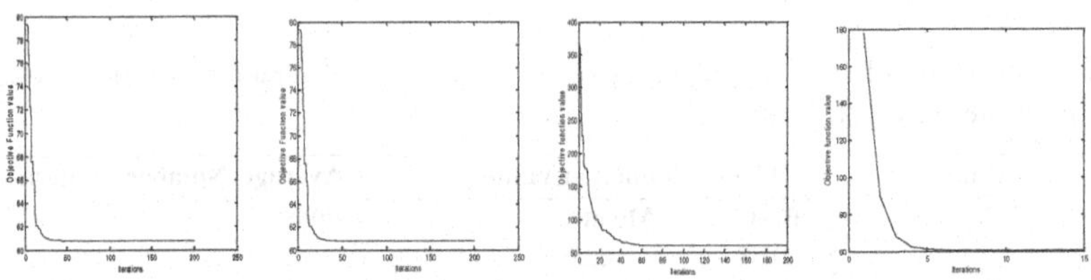

1(a) FCM 1(b)PSO 1(c) FAFCM 1(d)Improved FAFCM

Fig. 1. Typical curves of the FCM objective function for Iris data set using various algorithms (a) FCM, (b) PSO, (c) FAFCM, (d) Improved FAFCM.

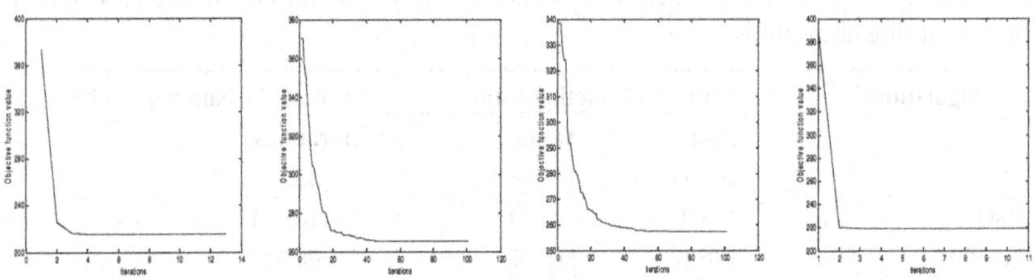

2(a) FCM 2(b)PSO 2(c) FAFCM 2(d) Improved FAFCM

Fig. 2. Typical curves of the FCM objective function for Lung Cancer data set using various algorithms 2(a) FCM, 2(b) PSO, 2(c) FAFCM,2(d) Improved FAFCM.

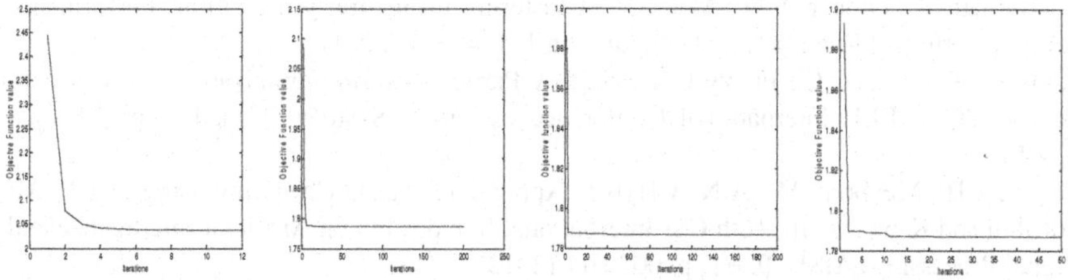

3(a) FCM 3(b)PSO 3(c) FAFCM3 (d) Improved FAFCM

Fig. 3. Typical curves of the FCM objective function for Single Outlier dataset using various algorithms (a) FCM, (b) PSO, (c) FAFCM, (d) Improved FAFCM

have been illustrated in Fig. 1, Fig. 2, and Fig. 3. From the above results, it is apparent that the improved FAFCM gives better and steady results as compared other clustering algorithms in case of all the data sets considered.

7 Conclusion and Future Work

This paper first proposed a firefly based fuzzy c-means algorithm and then an improved FAFCM clustering and the experimental results shows that the improved FAFCM method performs better than other three algorithms. Fuzzy c-means clustering is a very popular clustering algorithm with a wide variety of real world applications, but FCM uses the hill climbing method of search which traps it in the local optima and is also sensitive to initialization. Hence, we have used Firefly; a nature inspired meta-heuristic optimization technique has been used to avoid these problems. After the experimental analysis, it is found that the improved FAFCM shows steady and best results for various data sets considered as compared to the FCM, PSO and FAFCM. The improved FAFCM algorithm leads to faster convergence and minimized objective function value. In future the proposed improved FAFCM can be reformed into hybridization of the current algorithm with different optimization methods, for better performance.

References

1. Izakian, H., Abraham, A.: Fuzzy C-means and fuzzy swarm for fuzzy clustering problem. Expert Systems with Applications 38, 1835–1838 (2011)
2. Bezdek, J.C.: Pattern recognition with fuzzy objective function algorithms, pp. 95–107. Plenum Press, New York (1981)
3. Li, L., Liu, X., Xu, M.: A Novel Fuzzy Clustering Based on Particle Swarm Optimization. In: First IEEE International Symposium on Information Technologies and Applications in Education, pp. 88–90 (2007)
4. Wang, L., et al.: Particle Swarm Optimization for Fuzzy c-Means Clustering. In: Proceedings of the 6th World Congress on Intelligent Control and Automation, Dalian, China (2006)
5. Yang, X.S.: Nature-Inspired Metaheuristic Algorithms. Luniver Press (2008)

6. Senthilnath, J., Omkar, S.N., Mani, V.: Clustering using firefly algorithm: Performance study. Swarm and Evolutionary Computation 1, 164–171 (2011)
7. Runkler, T.A., Katz, C.: Fuzzy Clustering by Particle Swarm Optimization. In: Proceedings of 2006 IEEE International Conference on Fuzzy Systems, Canada, pp. 601–608 (2006)
8. Zadeh, T.H., Meybodi, M.: A New Hybrid Approach for Data Clustering using Firefly Algorithm and K-means. In: 16th CSI International Symposium on Artificial Intelligence and Signal Processing (AISI), Fars, pp. 007-011 (2012)
9. Abshouri, A.A., Bakhtiary, A.: A New Clustering Method Based on Firefly and KHM. Journal of Communication and Computer 9, 387–391 (2012)
10. Yang, X.-S.: Firefly Algorithms for Multimodal Optimization. In: Watanabe, O., Zeugmann, T. (eds.) SAGA 2009. LNCS, vol. 5792, pp. 169–178. Springer, Heidelberg (2009)
11. Yang, X.S.: Firefly Algorithm, Stochastic Test Functions and Design optimization. International Journal of Bio-Inspired Computation 2, 78–84 (2010)
12. Yang, F., Sun, T., Zhang, C.: An efficient hybrid data clustering method based on K-harmonic means and Particle Swarm Optimization. Expert Systems with Applications 36, 9847–9852 (2009)
13. Niknam, T., Amiri, B.: An efficient hybrid approach based on PSO, ACO and k-means for cluster analysis. Applied Soft Computing 10, 183–197 (2010)
14. Huang, K.Y.: A hybrid particle swarm optimization approach for clustering and classification of datasets. Knowledge-Based Systems 24, 420–426 (2011)
15. Chakravarty, S., Dash, P.K.: A PSO based integrated functional link net and interval type-2 fuzzy logic system for predicting stock market indices. Applied Soft Computing 12(2), 931–941 (2012)
16. Shayeghi, H., Jalili, A., Shayanfar, H.A.: Multi-stage fuzzy load frequency control using PSO. Energy Conversion and Management 49(10), 2570–2580 (2008)

On Kernel Based Rough Intuitionistic Fuzzy C-means Algorithm and a Comparative Analysis

B.K. Tripathy[1], Anurag Tripathy[1], K. Govindarajulu[2], and Rohan Bhargav[3]

[1] SCSE, VIT University, Vellore-632014, TN, India
[2] Vignan Institute of Technology and Management, Odisha, India
[3] C-1104, SMR, Vinay Galaxy, Hoodi Circle, Bangalore, 560048, India
tripathybk@vit.ac.in, anurag6742@gmail.com,
govinda_rajulu@rediffmail.com, rb.bhargav@gmail.com

Abstract. Clustering of real life data for analysis has gained popularity and imprecise methods or their hybrid approaches has attracted many researchers of late. Recently, rough intuitionistic fuzzy c-means algorithm was introduced and studied by Tripathy et al [3] and it was found to be superior to all other algorithms in this family. Kernel based counter part of these algorithms have been found to behave better than their corresponding Euclidean distance based algorithms. Very recently kernel based rough fuzzy algorithm was put forth by Bhargav et al [4]. A comparative analysis over standard datasets and images has established the superiority of this algorithm over its corresponding standard algorithm. In this paper we introduce the kernel based rough intuitionistic fuzzy c-means algorithm and show that it is superior to all the algorithms in the sequel; i.e. both normal and the kernel based algorithms. We establish it through experimental analysis by taking different type of inputs and using standard accuracy measures.

Keywords: clustering, fuzzy sets, rough sets, intuitionistic fuzzy sets, rough fuzzy sets, rough intuitionistic fuzzy sets, DB index, D index.

1 Introduction

Clustering is the process of putting similar objects into groups called clusters and putting dissimilar objects into different clusters. In contrast to classification where labeling of a large set of training tuples patterns is necessary to model the groups, clustering starts with partitioning the objects into groups first and then labeling the small number of groups. The large amount of data collected across multiple sources makes it practically impossible to manually analyze them and select the data that is required to perform a particular task. Hence, a mechanism that can classify the data according to some criteria in which only the classes of interest are selected and rests are rejected is essential. Clustering techniques are applied in the analysis of statistical data used in fields such as machine learning, pattern recognition, image analysis, information retrieval, and bioinformatics and is a major task in exploratory data mining. A wide number of clustering algorithms have been proposed to suit the requirements in each field of its application.

M.K. Kundu et al. (eds.), *Advanced Computing, Networking and Informatics - Volume 1*, Smart Innovation, Systems and Technologies 27,
DOI: 10.1007/978-3-319-07353-8_41, © Springer International Publishing Switzerland 2014

There are several clustering methods in literature starting with the Hard C-Means (HCM) [10]. In order to handle uncertainty in data, several algorithms like the fuzzy c-means (FCM) [14] based on the notion of fuzzy sets introduced by Zadeh [19], rough c-means (RCM) [9] based on the concept of rough sets introduced by Pawlak [13], the intuitionistic fuzzy c-means (IFCM) [5] based upon the concept of intuitionistic fuzzy sets introduced by Atanassov [1] have been introduced. Also, several hybrid c-means algorithms like the rough fuzzy c-means (RFCM)[11,12] based upon the rough fuzzy sets introduced by Dubois and Prade [7] and rough fuzzy intuitionistic fuzzy c-means (RIFCM) [3] based upon the rough intuitionistic fuzzy sets introduced by Saleha et al [15] have been introduced. In [3] it has been experimentally established through comparative analysis that RIFCM performs better than the other individual or hybrid algorithms.

Distance between objects can be calculated in many ways, the Euclidean distance based clustering is easy to implement and hence most commonly used. It has two drawbacks; the final results are dependent on the initial centres and it can only find linearly separable cluster. Nonlinear mapping functions used transforms the nonlinear separation problem in the image plane into a linear separation problem in kernel space facilitating clustering in feature space. Kernel based clustering helps in rectifying the second problem as it produces nonlinear separating surfaces among clusters [20, 21]. Replacing the Euclidean distance used in the above algorithms for computing the distance by kernel functions some algorithms have been put forth like the kernel based fuzzy c-means (KFCM) [20] and the kernel based rough c-means (KRCM) [17, 18, 21] were introduced Tripathy and Ghosh. Very recently, the kernel based rough fuzzy c-means [4] was introduced and studied by Bhargav and Tripathy. In this paper we introduce the kernel based rough intuitionistic fuzzy c-means algorithm (KRIFCM) algorithm and provide a comparative analysis of these kernel based as well as the standard c-means algorithm by focusing on RFCM, RIFCM, KRFCM and KRIFCM. Also, we use the Davies-Bouldin (DB) [6] and Dunn (D) indexes [8] to compare the accuracy of performance of these algorithms. We use different types of images and datasets for the purpose of experimental analysis. It has been observed that KRIFCM has the best performance among all these methods.

The paper is divided into seven sections. In section 2 we present some definitions and notations to be used throughout the paper. In Section 3, we deal with c-means clustering algorithms with emphasis on the IFCM algorithm. The proposed kernel based rough intuitionistic fuzzy c-means (KRIFCM) clustering algorithm along is explained in section 5. The complete evaluation is shown in section 6. Evaluation has been performed on synthetic dataset, real dataset and on image dataset. Finally, section 7 concludes the paper.

2 Definitions and Notations

In this section we introduce some definitions and notations to be used throughout this paper.

Definition 2.1: (Zadeh [19]) Let U be a universal set. Then a fuzzy set X on U is defined through its membership function μ_X defined as $\mu_X : U \to [0,1]$ such that

$\mu_X(x)$ for any $x \in U$ is a real number in [0, 1], called the membership value of x in X. The non-membership value $\nu_X(x)$ is defined as $\nu_X(x) = 1 - \mu_X(x)$.

Definition 2.2: (Atanassov [1]) An IFS X on U is given as $X = \{(x, \mu_X(x), \nu_X(x)) \mid x \in U\}$, where $\mu_X(x)$ and $\nu_X(x)$ have values in [0, 1]. They also satisfy the condition $0 \le \mu_X(x) + \nu_X(x) \le 1$.

Every intuitionistic fuzzy set X is associated with intuitionistic degree $\pi_X(x)$, which is known as hesitation degree for each element *x* in X and is defined as $\pi_X(x) = 1 - \mu_X(x) - \nu_X(x)$.

Definition 2.3: (Pawlak [13]) Let U be a universal set and R be an equivalence relation defined over U. Then for any X $\subseteq$ U, the rough set associated with X is a pair of crisp sets called the lower and upper approximations of X with respect to R and are defined as $\underline{R}X = \{x \in U \mid [x]_R \subseteq X\}$ and $\overline{R}X = \{x \in U \mid [x]_R \cap X \neq \phi\}$ respectively.

X is said to be rough with respect to R iff $\underline{R}X \neq \overline{R}X$ and R-definable otherwise. The uncertainty portion associated with a rough set X is called the boundary of X with respect to R. It is denoted by $BN_R(X)$ and is defined as $BN_R(X) = \overline{R}X - \underline{R}X$.

2.1 Performance Indexes

The Davis-Bouldin (DB) [6] and Dunn (D) indexes [8] are two of the most basic performance analysis indexes. As mentioned in ([6], abstract) the DB measure does not depend on neither the number of clusters analysed nor the method of partitioning of the data and can be used to guide a cluster seeking algorithm. Also, it is mentioned that it can be used to compare partitions with either similar or different numbers of clusters. However, the D index is mostly applicable to optimize the value of the number of clusters for efficient clustering. But for a fixed value of the number of clusters the d index is likely to provide an idea of the better algorithm in the sense that the greater the index value the algorithm is better. So, the algorithm having highest value of D index is supposed to be the best algorithm. So, in some sense or other the two indices help in evaluating the efficiency of clustering algorithms.

2.1.1 Davis-Bouldin (DB) Index

The DB index is defined as the ratio of sum of within-cluster distance to between-cluster distance. It is formulated as given below.

$$DB = \frac{1}{c}\sum_{i=1}^{c} \max_{k \neq i}\left\{\frac{S(v_i) + S(v_k)}{d(v_i, v_k)}\right\}, \quad for 1 < i, k < c \tag{1}$$

The within-cluster distance $S(v_i)$ can be formulated independently for different algorithms.

The aim of this index is to minimize the within cluster distance and maximize the between cluster separation. Therefore a good clustering procedure should give value of DB index as low as possible.

2.1.2 Dunn (D) Index

Similar to the DB index the D index is used for the identification of clusters that are compact and separated. It is computed by using the following formula.

$$Dunn = \min_i \left\{ \min_{k \neq i} \left\{ \frac{d(v_i, v_k)}{\max_i S(v_i)} \right\} \right\}, \quad for 1 < k, i, l < c \tag{2}$$

3 Kernel Methods

Distance between objects can be calculated in many ways, the Euclidean distance based clustering is easy to implement and hence most commonly used. It has two drawbacks, firstly the final results are dependent on the initial centers and secondly it can only find linearly separable cluster. Kernel based clustering helps in rectifying the second problem as it produces nonlinear separating hyper surfaces among clusters. Kernel functions are used to transform the data in the image plane into a feature plane of higher dimension known as kernel space.

Nonlinear mapping functions used transforms the nonlinear separation problem in the image plane into a linear separation problem in kernel space facilitating clustering in feature space. Mercer's theorem can be used to calculate the distance between the pixel feature values in kernel space without knowing the transformation function.

3.1 Types of Distance Functions

Definition 3.1.1: (Euclidean Distance) The Euclidean distance d(x, y) between any two objects x and y in any n-dimensional plane can be found using

$$d(x, y) = \sqrt{(x_1 - y_1)^2 + (x_2 - y_2)^2 + \ldots + (x_n - y_n)^2} \tag{3}$$

where, $x_1, x_2, \ldots x_n$ and $y_1, y_2, \ldots y_n$ are attributes of x and y respectively.

Definition 3.1.2: (Kernel Distance) If x is an object then $\varphi(x)$ is the transformation of x in high dimensional feature space where the inner product space is defined by K(x, y) = $\langle \varphi(x), \varphi(y) \rangle$. In this paper we use the Gaussian kernel function.

$$K(x, y) = \exp\left(-\frac{\|x - y\|^2}{\sigma^2} \right) \tag{4}$$

where $\sigma^2 = \frac{1}{N}\sum_{k=1}^{N} \left\| x_k - \bar{x} \right\|^2$ with $\bar{x} = \frac{1}{N}\sum_{k=1}^{N} x_k$.

Here, N is total number of data objects [17]. According to [20, 21] kernel distance function D(x, y) in the generalized form is D(x, y) = K(x, x) + K(y, y) − 2K(x, y) and on applying the property of similarity (i.e., K(x, x) = 1) it can be further reduced to (5).

$$D(x, y) = 2(1 - K(x, y)). \tag{5}$$

4 The c-Means Clustering Algorithms

In this section we shall present the intuitionistic fuzzy algorithm, whose concepts are used in describing the kernel based rough fuzzy intuitionistic fuzzy algorithm.

4.1 Intuitionistic Fuzzy C-Means

As mentioned in section 2, intuitionistic fuzzy sets are generalizations of the fuzzy sets. Here, instead of a specific non-membership value being the one's complement of the membership value is a function by itself having values in the interval [0, 1] and such that the sum of the membership and non-membership values for any element lies in [0, 1]. This introduces the notion of the hesitation function, which can have non-zero values. When this hesitation function has value zero for all the elements, an intuitionistic fuzzy set reduces to a fuzzy set. Using the notion of intuitionistic fuzzy sets the FCM was extended by Chaira [5] to introduce IFCM. It has been established in [5] that IFCM gives better results than the FCM and it has been established by taking many MRI images as input for the comparison purpose.

In this paper we have used Sugeno's Intuitionistic fuzzy generator. Non-membership values are calculated using Sugeno type fuzzy complement $\nu_A(x)$.

$$\nu_A(x) = \frac{1 - \mu_A(x)}{1 + \lambda \cdot \mu_A(x)}, \lambda > 0. \tag{6}$$

Using (6) we derive hesitation degree as

$$\pi_A(x) = 1 - \mu_A(x) - \frac{1 - \mu_A(x)}{1 + \lambda\mu_A(x)}, x \in X \tag{7}$$

To use the concept of intuitionistic fuzzy set, we modify the fuzzy membership μ_{ik} in FCM with the normalized form of μ'_{ik}, where,

$$\mu'_{ik} = \mu_{ik} + \pi_{ik}, \tag{8}$$

where π_{ik} is the hesitation degree of x_k in U_i.

The objective of IFCM [4] is to minimize the cost function given as

$$J(U,V) = \sum_{j=1}^{N}\sum_{i=1}^{c}(\mu'_{ij})^m \left\|x_j - v_i\right\|^2 \tag{9}$$

The computation steps of the IFCM algorithm are same as those of FCM except that at every step we compute μ'_{ik} using (8) and use it in the computation of the new centroids instead of using μ_{ik}. However, this change brings about far superior diagnosis in the brain diseases as shown in [4].

5 Kernel Based Rough Intuitionistic Fuzzy C-Means

Hybridization of the fuzzy and rough concepts provide better results in clustering than the individual ones by taking care of vagueness and uncertainty through the boundary region concept of rough sets and the membership function of fuzzy sets. Also, we have described the superiority of intuitionistic fuzzy set based approach over the fuzzy set approach. In fact intuitionistic fuzzy sets are more appropriate than fuzzy sets in modeling real life situations. This is because of the presence of the hesitation function. So, it is natural to combine the rough set and intuitionistic fuzzy techniques to derive an algorithm which shall have all the advantages of hybrid approach as well as the advantage of intuitionistic fuzzy technique over the fuzzy set technique. As a result, the rough intuitionistic fuzzy c-means (RIFCM) was introduced and studied in [3]. This is obtained as a result of hybridization of the IFCM and RCM algorithms. The resulting clusters are having lower approximations as well as boundary regions where the membership and non-membership values of elements are taken into account. It was established through several measures of accuracy that it provides better results than all the other algorithms in the family.

The objective of this algorithm is to minimize the cost function given in (10).

The parameters w_{low} and w_{up} have the standard meanings. Also μ'_{ij} has the same definition as in IFCM.

In RIFCM, each cluster can be identified by three properties, a centroid, a crisp lower approximation and an intuitionistic fuzzy boundary. If an object belongs in the lower approximation of a cluster then its corresponding membership value is 1 and hesitation value is 0. The objects in the lower region have same influence on the corresponding cluster. If an object belongs in the boundary of one cluster then it possibly belongs to that cluster and potentially belongs to another cluster. Hence the objects in the boundary region have different influence on the cluster. Thus we can say that in RIFCM the membership values of objects in lower region are unity ($\mu'_{ij} = 1$) and for those in boundary region behave like IFCM.

$$J_{RIF}(U',V) = \begin{cases} w_{low} \times A_1 + w_{up} \times B_1 & if\ \underline{BU}_i \neq \phi \ and \ BN(U_i) \neq \phi \\ A_1 & if\ \underline{BU}_i \neq \phi \ and \ BN(U_i) = \phi \\ B_1 & if\ \underline{BU}_i = \phi \ and \ BN(U_i) \neq \phi \end{cases}$$

$$A_1 = \frac{1}{|\underline{BU}_i|} \sum_{x_j \in \underline{BU}_i} x_j \quad and \quad B_1 = \frac{1}{n_1} \sum_{x_j \in BN(U_i)} (\mu'_{ij})^m x_j;$$

$$where \quad n_1 = \sum_{x_j \in BN(U_i)} (\mu'_{ij})^m \tag{10}$$

5.1 The KRIFCM Algorithm

The steps that are to be followed in this algorithm are as given below

1. Assign initial means v_i for c clusters by choosing any random c objects as cluster centres
2. Calculate D_{ik} using the formula (5)
3. Compute U matrix

If $D_{ik} = 0$ or $x_j \in \underline{BU}_i$ then $\mu_{ik} = 1$
Else compute μ_{ik} using

$$\mu_{ik} = \frac{1}{\sum_{j=1}^{c}\left(\frac{D_{ik}}{D_{jk}}\right)^{\frac{2}{m-1}}}.$$

4. Compute π_{ik} using (7).
5. Compute μ'_{ik} using equation (8) and normalize.
6. Let μ'_{ik} and μ'_{jk} be the maximum and next to maximum membership values of object x_k to cluster centroids v_i and v_j.
   If $\mu'_{ik} - \mu'_{jk} < \varepsilon$ then
   $x_k \in \overline{B}U_i$ and $x_k \in \overline{B}U_j$ and x_k cannot be a member of any lower approximation.
   Else $x_k \in \underline{B}U_i$
7. Calculate new cluster means by using

$$V_i = \begin{cases} w_{low}A + w_{up}B & if\ |\underline{BU}_i| \neq \phi \ and\ |BN(U_i)| \neq \phi \\ B & if\ |\underline{BU}_i| = \phi \ and\ |BN(U_i)| \neq \phi \\ A & ELSE \end{cases}$$

Where $A = \frac{\sum_{x_k \in \underline{BU}_i} x_k}{|\underline{BU}_i|}$, $B = \frac{\sum_{x_k \in BN(U_i)} (\mu'_{ik})^m x_k}{\sum_{x_k \in BN(U_i)} (\mu'_{ik})^m}$.

8. Repeat from step 2 until termination condition is met or until there are no more assignment of objects.

It aims at maximizing the between-cluster distance and minimizing the within-cluster distance. Hence a greater value for the D index proves to be more efficient.

6 Experimental Analysis

The evaluation of the algorithm has been done in 2 parts. We have implemented the algorithms and used two types of inputs for the purpose. The first type of input is the zoo dataset, which is numeric by character and is taken from the UCI repository. The second type of inputs comprises of three different kinds of images; the cell, the iris and a football player.

We have made a comparison of four algorithms; RFCM, RIFCM, KRIFCM and KRIFCM and used two of the well known accuracy measures; the DB index and the D index to measure their efficiency. In [3] it was established by us that RIFCM is the best among the family of algorithms; HCM, FCM, RCM, IFCM, RFCM and RIFCM. Again, it was shown in [20, 21] and [17, 18] respectively that the kernel versions KFCM and KRCM perform better than their normal counterparts. In [ ] again we established that KRFCM is more efficient than RFCM. So, our comparative analysis in this paper establishes that KRFCM is the best among all these 11 algorithms.

Even though the proposed as well as existing algorithms have been applied on the cell or iris image for the purpose of comparison of their efficiency, they can also be used in various other fields where clustering of data is required or even were pattern recognition is a must. Further improvements by modifying threshold values can be easily done so as to apply these algorithms in industries which work on data analysis.

So far we have observed that the computations of the complexities of the C-Means algorithms are not found in the literature. However, it is obvious that the computational complexities of the hybrid algorithms are definitely more than the individual algorithms. But the computational complexity of the RFCM, RIFCM, KRFCM and KRIFCM algorithms are same. Similarly the computational complexities of the FCM, IFCM, KFCM and KIFCM algorithms are same. But the comparison of the computational complexities of all the algorithms can be done. Also, one can compare the efficiency gained through the hybridization algorithms at the cost of increase in the computational complexities.

6.1 Image Dataset

As discussed above we have processed a number of images using the four algorithms to obtain the resultant images as well as the DB and Dunn index values. Figure 1 represents the images for the cell image. Figure 2 shows the corresponding ones for the iris image and figure 3 shows the same for the football player image. The DB and D index values for these images for the four algorithms are provided in table 1, table 2 and table 3 respectively. Also, we have provided a graphical presentation in figures 4 and 5 of these values along with these values for the zoo dataset.

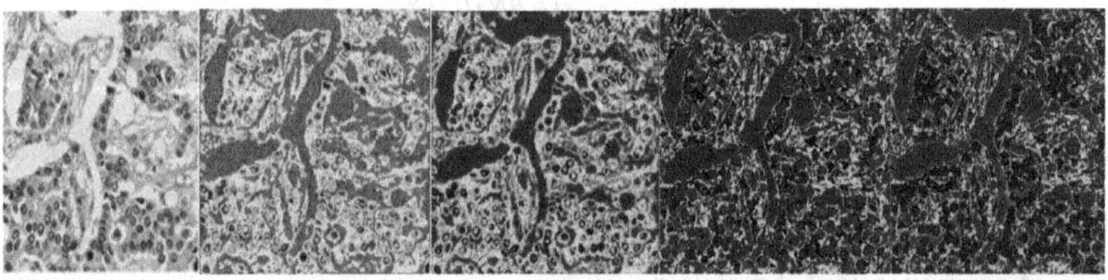

1(a) Oiginal Image 1(b) KRFCM 1(c) KRIFCM 1(d) RFCM 1(e) RIFCM

Fig. 1.

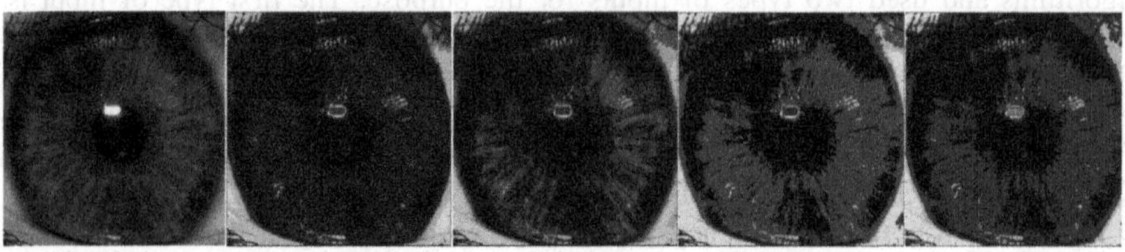

2(a) Oiginal Image 2(b) KRFCM 2(c) KRIFCM 2(d) RFCM 2(e) RIFCM

Fig. 2.

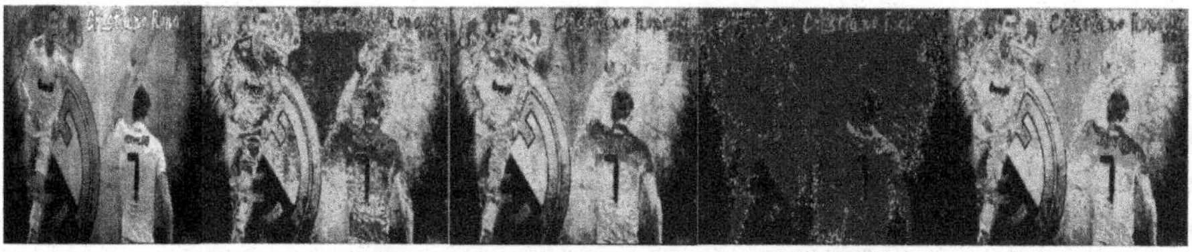

3(a) Oiginal Image 3(b) KRFCM 3(c) KRIFCM 3(d) RFCM 3(e) RIFCM

Fig. 3.

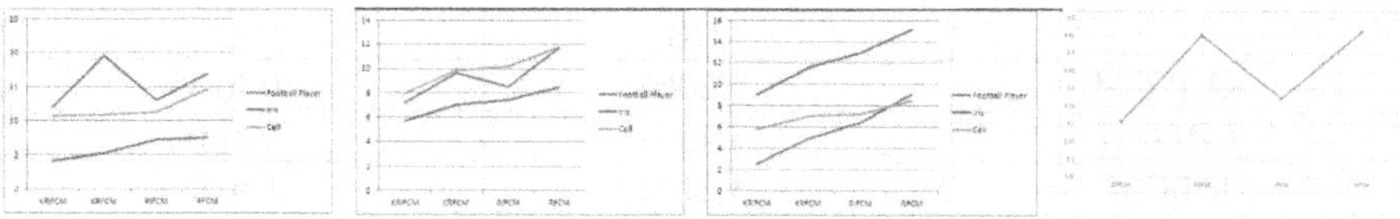

Fig. 4. DB Index Graph

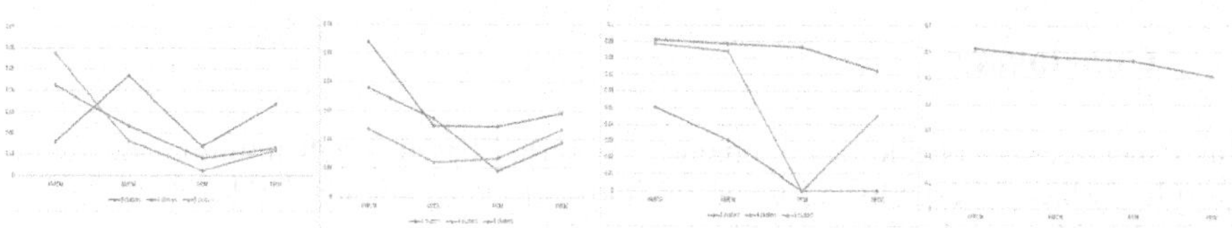

Fig. 5. D Index Graph

Table 1. Football player

	3		4		5	
	DB	D	DB	D	DB	D
KRIFCM	12.014168	0.21750	7.16949	0.03845	2.5738	0.151256
KRFCM	19.47022	0.0470983	9.620288	0.02371	4.90621	0.0164716
RIFCM	13.05654	0.097858	8.5031241	0.124946	6.43955	0.144344
RFCM	16.874518	0.074668	11.707188	0.0828045	9.0494	0.094005

Table 2. Iris Image

	3		4		5	
	DB	D	DB	D	DB	D
KRIFCM	4.1559	0.0886	5.69435	0.0415	9.0432	0.0912
KRFCM	5.232	0.0841	6.9893	0.0405	11.5647	0.0845
RIFCM	7.221926	0.018191	7.4151	0.0411324	12.962991	0.038557
RFCM	7.47037	0.018163	8.444334	0.0396752	15.15012	0.01265

Table 3. Cell Image

	3		4		5	
	DB	D	DB	D	DB	D
KRIFCM	10.7453	0.084332	7.9553	0.10035	5.7763	0.2000796
KRFCM	10.8655	0.06775	9.8764	0.08963	7.04552	0.140076
RIFCM	11.2420896	0.05389655	10.154453	0.088511	7.1997494	0.137886
RFCM	14.5945	0.0528044	11.713193	0.084505	8.461676	0.127872

6.2 Numeric Dataset

In Table 4 below we present the values of the DB and D indices obtained by taking the zoo dataset for the four algorithms; RFCM, IRFCM, KRFCM and KRIFCM. It has been observed that KRIFCM provides the lowest value for DB index and has the highest value for the Dunn index. Also, figures 4 and 5 provide the graphical comparison of these performances.

Table 4. Zoo dataset

	DB	D
KRIFCM	3.2064	0.6135
KRFCM	3.4514	0.5805
RIFCM	3.2742	0.56523
RFCM	3.462	0.5071

7 Conclusion

This paper focuses on the development of a kernel based rough intuitionistic fuzzy algorithm which is established to be the most efficient among the 11 clustering c-means algorithms including the hard c-means, 5 uncertainty based extensions of it; the FCM, IFCM, RCM, RFCM and RIFCM as well the 5 corresponding kernel based algorithms; the KFCM, KIFCM, KRCM, KRFCM and KRIFCM. All the algorithms have been tested against three different type of images; the Ronaldo image, the cell image and the iris image. Also a standard dataset, the soybean dataset is considered. Two indices of measuring accuracies; the DB index and the D index are used for the comparison purpose. Also, the number of clusters was varied to be 3, 4 and 5. In almost all the cases it is observed that KRIFCM has better accuracies than its counterparts. Also, we have tried to put some comments on the computational complexities of these families of algorithms.

References

1. Atanassov, K.T.: Intuitionistic Fuzzy Sets. Fuzzy Sets and Systems 20(1), 87–96 (1986)
2. Bezdek, J.C.: Pattern Recognition with Fuzzy Objective Function Algorithms. Kluwer Academic Publishers (1981)
3. Bhargav, R., Tripathy, B.K., Tripathy, A., Dhull, R., Verma, E., Swarnalatha, P.: Rough Intuitionistic Fuzzy C-Means Algorithm and a Comparative Analysis. In: ACM Conference, Compute 2013, pp. 978–971 (2013) 978-1-4503-2545-5/13/08
4. Bhargava, R., Tripathy, B.: Kernel Based Rough-Fuzzy C-Means. In: Maji, P., Ghosh, A., Murty, M.N., Ghosh, K., Pal, S.K. (eds.) PReMI 2013. LNCS, vol. 8251, pp. 148–155. Springer, Heidelberg (2013)
5. Chaira, T., Anand, S.: A Novel Intuitionistic Fuzzy Approach for Tumor/Hemorrhage Detection in Medical Images. Journal of Scientific and Industrial Research 70(6) (2011)
6. Davis, D.L., Bouldin, D.W.: A cluster separation measure. IEEE Transactions on Pattern Analysis and Machine Intelligence PAMI-1(2), 224–227 (1979)

7. Dubois, D., Prade, H.: Rough fuzzy sets model. International Journal of General Systems 46(1), 191–208 (1990)
8. Dunn, J.C.: A fuzzy relative of the ISODATA process and its use in detecting compact well-separated clusters, pp. 32–57 (1973)
9. Lingras, P., West, C.: Interval set clustering of web users with rough k-mean. Journal of Intelligent Information Systems 23(1), 5–16 (2004)
10. Macqueen, J.B.: Some Methods for classification and Analysis of Multivariate Observations. In: Proceedings of 5th Berkeley Symposium on Mathematical Statistics and Probability, pp. 281–297. University of California Press (1967)
11. Maji, P., Pal, S.K.: RFCM: A Hybrid Clustering Algorithm using rough and fuzzy set. Fundamenta Informaticae 80(4), 475–496 (2007)
12. Mitra, S., Banka, H., Pedrycz, W.: Rough-Fuzzy Collaborative Clustering. IEEE Transactions on System, Man, and Cybernetics, Part B: Cybernetics 36(4), 795–805 (2006)
13. Pawlak, Z.: Rough sets. Int. Jour. of Computer and Information Sciences 11, 341–356 (1982)
14. Ruspini, E.H.: A new approach to clustering. Information and Control 15(1), 22–32 (1969)
15. Saleha, R., Haider, J.N., Danish, N.: Rough Intuitionistic Fuzzy Set. In: Proc. of 8th Int. Conf. on Fuzzy Theory and Technology (FT & T), Durham, North Carolina (USA), March 9-12 (2002)
16. Sugeno, M.: Fuzzy Measures and Fuzzy integrals-A survey. In: Gupta, M., Sardis, G.N., Gaines, B.R. (eds.) Fuzzy Automata and Decision Processes, pp. 89–102 (1977)
17. Tripathy, B.K., Ghosh, A., Panda, G.K.: Kernel based K-means clustering using rough set. In: 2012 International IEEE Conference on Computer Communication and Informatics, ICCCI (2012)
18. Tripathy, B.K., Ghosh, A., Panda, G.K.: Adaptive K-Means Clustering to Handle Heterogeneous Data Using Basic Rough Set Theory. In: Meghanathan, N., Chaki, N., Nagamalai, D. (eds.) CCSIT 2012, Part I. LNICST, vol. 84, pp. 193–202. Springer, Heidelberg (2012)
19. Zadeh, L.A.: Fuzzy Sets. Information and Control 8(11), 338–353 (1965)
20. Zhang, D., Chen, S.: Fuzzy Clustering Using Kernel Method. In: Proceedings of the International Conference on Control and Automation, Xiamen, China, pp. 123–127 (2002)
21. Zhou, T., Zhang, Y., Lu, H., Deng, F., Wang, F.: Rough Cluster Algorithm Based on Kernel Function. In: Wang, G., Li, T., Grzymala-Busse, J.W., Miao, D., Skowron, A., Yao, Y. (eds.) RSKT 2008. LNCS (LNAI), vol. 5009, pp. 172–179. Springer, Heidelberg (2008)

FuSCa: A New Weighted Membership Driven Fuzzy Supervised Classifier

Pritam Das, S. Sivasathya, and K. Joshil Raj

Department of Computer Science
Pondicherry University, Pondicherry, India
{pritamds6,ssivasathya,joshlion89}@gmail.com

Abstract. The aim of this paper is to introduce a new supervised fuzzy classification methodology (FuSCa) to improve the performance of k-NN (k-Nearest Neighbor) algorithm based on the weighted nearest neighbor membership and global membership derived from the training dataset. In this classification method, the test object is assigned a class label having the maximum membership value for that corresponding class while a weighted membership vector is found after utilizing the Global and Nearest-Neighbor fuzzy membership vectors along with a global weight and a k-close weight respectively. FuSCa is compared with other approaches using the standard benchmark data-sets and found to produce better classification accuracy.

Keywords: Fuzzy membership, Supervised classification, Machine learning, Data-mining, Nearest neighbor, Weighted membership.

1 Introduction

Classification algorithms are designed to learn a function which maps a large vector of attributes into one of several classes. In supervised classification model, some known objects are described by a large set of vectors where each vector is composed of a set of attributes and a class label which specifies the category the objects. Supervised k-Nearest Neighbor (k-NN) classification algorithm has been rigorously studied and modified worldwide in last decades to improve its accuracy and efficiency. Although the idea behind k-NN algorithm is very simple, it is widely applied in many real life applications due to its better accuracy rate.

A new method (FuSCa) has been proposed in this paper to improve the accuracy of k-NN algorithm in classification problems by deriving two membership vectors, Global Membership Vector (GMV) and K-Close Membership Vector (KMV) depending on the whole training dataset and nearest neighbor instances for each test data respectively. A Class Determinant Function (CDF) is imposed on GMV and KMV to get a Weighted Membership Vector (WMV) which represents the degree of belonging a test data record to each class.

M.K. Kundu et al. (eds.), *Advanced Computing, Networking and Informatics - Volume 1*, Smart Innovation, Systems and Technologies 27,
DOI: 10.1007/978-3-319-07353-8_42, © Springer International Publishing Switzerland 2014

2 Preliminary

2.1 Basics of Classification Problem

Supervised classification method builds a concise model of the distribution of class labels in terms of predictor features then it assigns class labels to the testing instances where the values of the predictor features are known, but the value of the class label is unknown. Let that each instance i of a given dataset is described by both a vector of n attribute values $x_i=[x_{i1},x_{i2},\ldots, x_{in}]$ and its corresponding class label y_i, which can take any value from a set of values $Y=\{y_1,y_2, \ldots, y_m\}$. Thus, x_{iq} specifies the value of the q-th attribute of the i-th instance. The training and test datasets are represented by D_{TRAIN} and D_{TEST} respectively. D_{NN} contains k- Nearest Neighbor records based on the distance between $x \in D_{TEST}$ and $x_i \in D_{TRAIN}$. Therefore, $D_{NN} \subseteq D_{TRAIN}$ and

$$D_{TRAIN}=\{(x_i,y_j) \mid i=1,2\ldots,h\}; \qquad D_{NN}=\{(x_i,y_j) \mid i=1,2\ldots,k\}; \qquad D_{TEST}=\{x_i \mid i=1,2\ldots,r\}$$

Where h, k, r represents the total number of records in D_{TRAIN}, D_{NN}, D_{TEST} and k may be any arbitrary value between 1 and h. Each test data record from D_{TEST} will be assigned a class label based on the prediction provided by the classifier using the knowledge of the training set D_{TRAIN}.

2.2 Related Work

Numerous methods such as, neural networks [2], [11], support vector machines [2], [11], rough sets[8], [9], decision trees[2], fuzzy sets[1], [4], Bayesian Networks[12],rule based classifiers[18] etc. have been proposed in Machine Learning literatures for effective classification tasks. Among them, the nearest neighbor based classifier i.e; k-NN is used in real life applications due to its simplicity and standard performance. According to Joaquín Derrac *et al.* [1], the first fuzzy nearest neighbor classifier was introduced by Jów´ik [3] in 1983 where every neighbor uses its fuzzy membership (array) to each class for the voting rule which brings the final classification result. Later in 1985, Keller *et al.* [4] proposed a new fuzzy membership based classification method popularly known as Fuzzy KNN which uses three different methods for computing the class memberships. The best performing method is considered for each instance x (x is in class i) to compute the k- nearest neighbors from the training data. F. Chung-Hoon *et al.* [5] proposed a new type-2 fuzzy k-nearest neighbor method in 2003. In 2005, T.D. Pham[6], [7] introduced the kriging computational scheme for deter-mining optimal weights to be combined with different fuzzy membership grades. S. Hadjitodorov [7] applied intuitionistic fuzzy set theory to develop the fuzzy nearest neighbor classifier with the non-membership concept in 1995. On the other hand, R. Jensen *et al.*[8] in 2011 presented the FRNN-FRS and FRNN-VQRS techniques for fuzzy-rough classification which employ the fuzzy rough sets and vaguely quantified rough sets, respectively. Recently in 2012, M. Sarkar [9] proposed a fuzzy-rough uncertainty based classification method which incorporates the lower and upper approximations of the memberships to the decision rule. Another recent work from Feras Al-Obeidat *et al.*[10-13] used a new methodology named as PSOPRO based on the fuzzy indifference relationship and optimized its parameters using PSO (Particle Swarm Optimization) algorithm.

Fuzzy methods which have been used to produce fuzzy classification rules are also proved to be effective in supervised machine learning. For example, Emel Kızılkaya Aydogan *et al.* [18] introduced a fuzzy rule-based classifier (FRBCS) with hybrid heuristic approach (called hGA) to solve high dimensional classification problems in linguistic fuzzy rule-based classification systems in 2012. All of these methods have shown effective improvements over k-NN algorithm.

2.3 K-Nearest Neighbor Classifier

The k-nearest neighbor [4], [11] algorithm computes the distance or similarity between each test example z = (x', y') and all the training examples $(x_i,y_j) \subseteq D_{TRAIN}$ to determine its nearest neighbor list, D_{NN}. The test example is classified based on the distance weighted voting using equation (1).

$$y' = \overset{argmax}{v} \sum_{(x_i,y_j)\in D_{NN}} w_i \times I(v = y_j) \tag{1}$$

where v is a class label, y_j is the class label for one of the nearest neighbors and I(.) is an indicator function that returns the value 1 if its argument is true and 0 otherwise. Distance between x' and x_i is $d(x',x_i)$ and corresponding weight, $w_i = \frac{1}{d(x',x_i)^2}$

2.4 Fuzzy Membership System

Let a conventional crisp subset A of an universal set of objects (U) is commonly defined by specifying the objects of the universe which are the members of A. The characteristic function of A may specify an equivalent way of defining A as, μ_A: U $\rightarrow\{0,1\}$ where for all x $\in$U. Mathematically, it can be expressed as,

$$\mu_A(x) = \begin{cases} 1 & x \in A \\ 0 & x \notin A \end{cases} \tag{2}$$

In fuzzy membership system[4], characteristic functions are generalized to produce a real value in the interval [0, 1]. i.e.; μ_A: U $\rightarrow[0,1]$.

Let the set of sample vectors is$\{x_1, x_2, \ldots, x_n\}$. The degree of membership of each vector in each of c classes is specified by a fuzzy c partition. So, It is denoted by a cxn matrix U, where the degree of membership of x_k in class i is $\mu_{ik}=\mu_i(x_k)$ for i=1,2…,c; and k=1,2,…,n. The following properties hold for U to be a fuzzy partition.

$$\sum_{i=1}^{c} \mu_{ik} = 1 \;; \qquad 0 < \sum_{k=1}^{m} \mu_{ik} < n \;; \qquad \mu_{ik} \in [0,1] \tag{3}$$

3 Proposed Algorithm and Implementation

Suppose the training data(D_{TRAIN}), test data (D_{TEST}), and nearest neighbor data(D_{NN}) sets is defined as described in section-2.1. The adjusted weight, $\eta_i(x)$, for each distance d_i (Euclidean Distance) between test data x to training data $x_i$is formulated below in equation(4). Here d_i is the Euclidean distance between test data and i^{th} training data. The fuzzy membership vectors (global and k-close) for each test data are found using these adjusted weights.

$$\eta_i(x) = 1/(1 + d_i) \quad (4)$$

Where, $d_i = \sqrt{\sum_{q=1}^{n}(x_q - x_{iq})}$

For each test data we define the Global Membership Vectorand K-Close Membership Vector as GMV=(GM_1, GM_2, ...GM_m) and KMV=(KM_1, KM_2, ...KM_m) respectively. Here, m is the number of classes present in training dataset. Each membership value in GMV and KMV follows the properties listed in equation (3). These fuzzy membership vectors are evaluated using equations (5) and (6) given below.

$$GMj = \frac{\sum_{(x_i,y_j)\epsilon D_{TRAIN}} \eta_i(x)}{\sum_{i=1}^{h} \eta_i(x)} \quad (5)$$

where, $\sum_{(x_i,y_j)\epsilon D_{TRAIN}} \eta_i(x)$= Total adjusted weight of jth class label records in D_{TRAIN}

$\sum_{i=1}^{h} \eta_i(x)$ = Total adjusted weight of all class label records in D_{TRAIN}.

$$KMj = \frac{\sum_{(x_i,y_j)\epsilon D_{NN}} \eta_i(x)}{\sum_{i=1}^{k} \eta_i(x)} \quad (6)$$

Where, $\sum_{(x_i,y_j)\epsilon D_{NN}} \eta_i(x)$ = Total adjusted weight of jth class label records in D_{NN}.

$\sum_{i=1}^{k} \eta_i(x)$ = Total adjusted weight of all class label records in D_{NN}.

and 'h' and 'k' are the total number of records in D_{TRAIN} and D_{NN} respectively. AWeighted Membership Vector (WMV)is found after deriving the GMV and KMV vectors for the test data $x \in D_{TEST}$ using a special Class Determinant Function(CDF) given in equation(7). It indicates a Determinant Function Vector, DFV as (F_1, F_2, ..., F_m).

$$< DFV >= w_g \times < GMV > + w_k \times < KMV > \quad (7)$$

Where, $F_j = w_g \times GM_j + w_k \times KM_j$. Here,'x' represents the scalar multiplication with vectors and '+' denotes the vector addition.$w_g \in (0,1)$ and $w_k \in (0,1)$ are the global weight and k-close weight respectively. These two variables define the weightage given to global membership vector and k-close membership vector respectively. The Weighted Membership Vector (WMV) = (g_1, g_2,....,g_m) may be achieved by normalizing the values in Determinant Function Vector, DFV.

The last step is to assign each test data record, $x \in D_{TEST}$ to the right class $y_j$applying the following rulegiven in equation (8). Where,$j \in \{1,2 \ldots, m\}$.

$$x \in y_j \text{ if and only if } \max(g_1, g_2,\ldots,g_m)=g_j \quad (8)$$

4 Experimental Analysis

The proposed FuSCa classification algorithm is implemented in java using the open source java packages of WEKA3.7.9 [14] machine learning tool developed by Waikato University. Performance of FuSCa is evaluated on 25 benchmark data sets from the UCI repository [15]. In this paper, only the results, obtained from six popular UCI datasets, are displayed.

4.1 Performance Evaluation

To evaluate the performance of the proposed FuSCa classifier, 10-fold cross validation is used for each standard benchmark dataset. The accuracy results are compared with other standard machine leaning classifiers and displayed in Table 1. These are C4.5 J48, 1-NN, 3-NN, 5-NN, SMO, SMO with Polynomial-Kernel(SVM2), SMO with RBF-Kernel (SVMG) and logistic regression (LogR). In this case, we used WEKA [14] open source implementations for this purpose. As observed from these results, FuSCa performs well on most of the benchmark dataset.

Table 1. Comparative classificationaccuracy (%) results for each dataset.

DATA SET	FuSCa	C4.5	1-NN	3-NN	5-NN	SVM	SVM2	SVMG	LogR
Breast	97.42	95.90	97.07	96.93	96.93	97.51	96.05	96.63	96.63
Diabetes	77.86	77.60	78.12	77.86	77.73	77.47	76.56	77.86	78.65
Glass	79.44	75.70	79.44	77.10	73.83	75.70	77.10	78.04	73.83
Heart	84.81	82.96	83.33	82.59	83.70	84.81	78.52	83.70	84.81
Iris	94.67	93.33	94.00	94.67	94.67	94.00	92.67	92.67	92.67
Letter	91.41	77.50	90.92	89.60	89.04	89.00	94.20	94.16	86.10

4.2 Friedman Test

To compare the multiple classifiers, Friedman test [16], [17] is considered as one of the best non-parametric statistical test methods. The following steps determine the Friedman rankings of multiple classifiers.

a. Collect the results observed for each pair of algorithm and data set.
b. For each observed data set i, ranks are defined as an integer value from 1 (best algorithm) to k (worst algorithm). Ranks are denoted as $r_i^j (1 \leq j \leq k)$. Where k is the number of classifiers used in comparison.
c. For each algorithm j, average the ranks obtained for all datasets to obtain the final rank denoted by, $R_j = \frac{1}{n}\sum_i r_i^j$

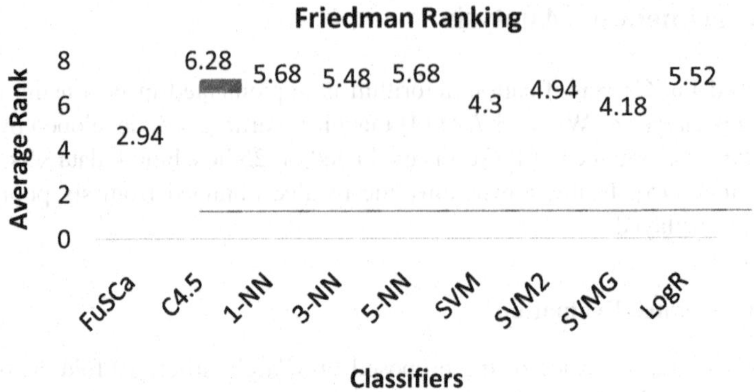

Fig. 1. Friedman accuracy ranking

Fig.1 shows the average rankings of different classifiers based on Friedman's method over 25 benchmark data sets from UCI ML Repository [15]. The height of each column is proportional to the ranking and the lower a column is, the better its associated algorithm is.

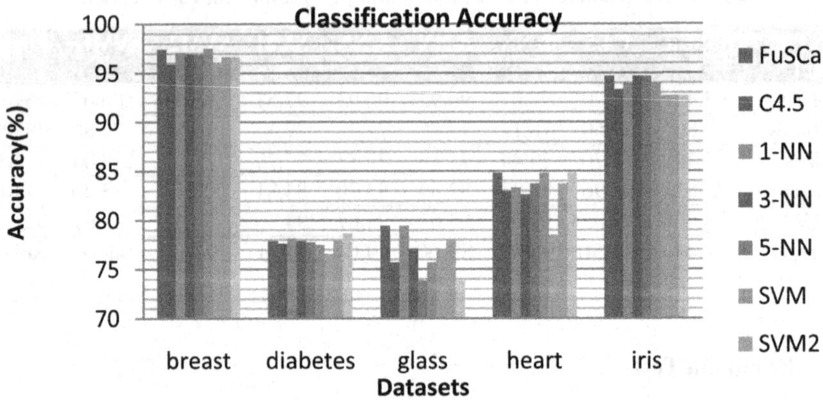

Fig. 2. Comparison of accuracy results of different classifiers

In Fig.2, the graph presents the classification accuracy percentage predicted by different classifiers for six popular datasets available in UCI Machine Learning repository.

5 Conclusion

In this paper, a new methodology is used for classification with fuzzy membership to improve the nearest neighbor classification technique. This supervised learning technique is evaluated using 10-fold cross validation among several well-known UCI Machine Learning datasets and compared with some standard algorithms which are available in open source WEKA data mining tool. Non-parametric Friedman statistical test is applied to compare FuSCa with multiple classifiers. It was also observed that very less additional computational cost is required for this proposed fuzzy mechanism compared to original KNN method. The results indicate that FuSCa has significantly improved the performance of the nearest neighbor based classification methods using the Global and K-close fuzzy membership values.

References

1. Derrac, J., García, S., Herrera, F.: Fuzzy nearest neighbor algorithms: Taxonomy, experimental analysis and prospects. Information Sciences 260, 98–119 (2014)
2. Kotsiantis, S.B.: Supervised Machine Learning: A Review of Classification Techniques. Informatica 31, 249–268 (2007)
3. Jówík, A.: A learning scheme for a fuzzy k-NN rule. Pattern Recognition Letters 1, 287–289 (1983)
4. Keller, J.M., Gray, M.R., Givens, J.A.: A fuzzy k-nearest neighbor algorithm. IEEE Transactions on Systems Man, and Cybernetics 15, 580–585 (1985)
5. Chung-Hoon, F., Hwang, C.: An interval type-2 fuzzy k-nearest neighbor. In: Proceedings of the 12th IEEE International Conference on Fuzzy Systems, pp. 802–807 (2003)
6. Pham, T.D.: An optimally weighted fuzzy *k*-NN algorithm. In: Singh, S., Singh, M., Apte, C., Perner, P. (eds.) ICAPR 2005. LNCS, vol. 3686, pp. 239–247. Springer, Heidelberg (2005)
7. Hadjitodorov, S.: An intuitionistic fuzzy sets application to the k-NN method. Notes on Intuitionistic Fuzzy Sets 1, 66–69 (1995)
8. Jensen, R., Cornelis, C.: Fuzzy-rough nearest neighbour classification. IEEE Transactions on Rough Sets 13, 56–72 (2011)
9. Sarkar, M.: Fuzzy-rough nearest neighbor algorithms in classification. Fuzzy Sets and Systems 158, 2134–2152 (2012)
10. Al-Obeidat, F., Belacel, N., Carretero, J.A., Mahanti, P.: An evolutionary framework using particle swarm optimization for classification method PROAFTN. Applied Soft Computing 11, 4971–4980 (2011)
11. Tan, P.-N., Steinbach, M., Kumar, V.: Introduction to Data Mining. Pearson Education, Inc. (2006)
12. Carvalho, A.M., Roos, T., Oliveira, A.L.: Discriminative Learning of Bayesian Networks via Factorized Conditional Log-Likelihood. Journal of Machine Learning Research 12, 2181–2210 (2011)
13. Witten, H.: Data Mining: Practical Machine Learning Tools and Techniques. Morgan Kaufmann Series in Data Management Systems (2005)
14. WEKA software, Machine Learning, The University of Waikato, Hamilton, New Zealand, `http://www.cs.waikato.ac.nz/ml/weka/`
15. UCI Machine Learning Repository, `http://mlearn.ics.uci.edu/MLRepository.html`
16. Friedman, M.: The use of ranks to avoid the assumption of normality implicit in the analysis of variance. Journal of the American Statistical Association 32, 674–701 (1937)
17. Garcıa, S., Herrera, F.: An Extension on "Statistical Comparisons of Classifiers over Multiple Data Sets" for all Pairwise Comparisons. Journal of Machine Learning Research 9, 2677–2694 (2008)
18. Aydogan, E.K., Karaoglan, I., Pardalos, P.M.: hGA: Hybrid genetic algorithm in fuzzy rule-based classification systems for high-dimensional problems. Applied Soft Computing 12, 800–806 (2012)

Choice of Implication Functions to Reduce Uncertainty in Interval Type-2 Fuzzy Inferences

Sumantra Chakraborty[1], Amit Konar[1], and Ramadoss Janarthanan[2]

[1] Electronics and Tele-Communication Engineering Department, Jadavpur University, Kolkata-32, India
[2] Computer Science and Engineering Department, TJS Engineering College, Chennai, India
srmjana_73@yahoo.com

Abstract. Selection of implication function is a well-known problem in Type-1 fuzzy reasoning. Several comparison of type-1 implications have been reported using set of (nine) standard axioms. This paper attempts to select the most efficient implication function that results in minimum uncertainty in the interval type-2 inference. An analysis confirms that *Lukasiewicz*-1/*Lukasiewicz*-2 membership function is most efficient in the present context.

1 Introduction

Existing approaches to interval type-2 fuzzy reasoning (IT2 FS) [4], [5], [6] usually consider extension of classical Mamdani type reasoning [3], [7], [9], [10]. There exists ample scope of research to identify suitable implication functions from the available list of implications with an aim to reduce the uncertainty in the inferential space in IT2 fuzzy reasoning. Unfortunately, to the best of our knowledge, there hardly exists any literature that compares the role of implication functions in reducing uncertainty in type-2 inferential space. This paper would compare the relative merits of implication functions in the context of uncertainty reduction in the *footprint of uncertainty* in IT2 FS inferences.

Most of the control [11], instrumentation [13], tele-communication [2] and other real-time applications [15], [16], [17] usually consider an observation of the linguistic variable in the antecedent space to derive the inference [12]. The paper also considers that an observation crisp value $x = x'$ in the antecedent space with an attempt to derive the inference. Computer simulations undertaken confirm that *Lukasiewicz*-1 fuzzy implication function outperforms its competitors with respect to uncertainty [8], [14] measure induced by left and right end point centroids of IT2 fuzzy inference.

From this point on, the paper is organized as follows. Selection of the most efficient implication functions that correspond to minimum uncertainty in the interval type-2 inference is given in section 2. The conclusions are listed in section 3.

2 Selection of the Most Efficient Implication Function that Minimizes the Span of Uncertainty

In this section, first the inference generation technique of Type-1 and Type-2 are discussed. Next, one metric (Z) is defined to measure the span of uncertainty (SOU)

M.K. Kundu et al. (eds.), *Advanced Computing, Networking and Informatics - Volume 1*, 369
Smart Innovation, Systems and Technologies 27,
DOI: 10.1007/978-3-319-07353-8_43,

in terms of UMF and LMF. Using this metric Z, the SOU for different implication function has been discussed. Based on this metric Z, the most efficient implication function has been selected.

2.1 Type-1 and Type-2 Inference Technique

Let X = $\{x_1, x_2, \ldots, x_n\}$ and Y = $\{y_1, y_2, \ldots, y_n\}$ be the universe of discourse,

$\mu_A(x)$ be the type-1 membership function of x in A, for $x \in$ X,

$\mu_B(y)$ be the type-1 membership function of y in B, for $y \in$ Y.

Expressing $\mu_A(x)$ for $x \in$ X as A = $[a_i]_{1\times n}$, where $a_i = \mu_A(x)\big|_{x=x_i}$ and $\mu_B(y)$ for $y \in$ Y as B = $[b_j]_{1\times m}$, where $b_j = \mu_B(y)\big|_{y=y_i}$, we have an implication relation R(x, y) for the rule: *if x is A then y is B*, for $x \in$ X, $y \in$ Y.

Let $\mu_{A'}(x) = [a'_i]_{1\times n} = A'$ (say) be an observed membership function for x is A'. We derive the fuzzy forward inference $\mu_{B'}(y) = [b'_j]_{1\times m} = B'$ by

$$B' = A' \circ R, \tag{1}$$

where R = $[r_{ij}]_{n\times m}$ be the implication relation such that $r_{ij} = f(a_i, b_j)$, where f(. , .) is an implication function.

Given a crisp data point $x' \in$ X, we want to derive fuzzy inference by forward reasoning. Here,

$$B' = A' \circ R,$$

$$= (\mu_A(x') \wedge \mu_A(x)) \circ R$$

$$= (\alpha \wedge A) \circ R \tag{2}$$

$$= (\alpha \wedge (A \circ R)), \tag{3}$$

where $\alpha = \mu_A(x') = \mu_A(x)\big|_{x=x'}$ is a scalar membership in [0, 1].

Now, consider an IT2 rule of the form:

if x is $\tilde{A}$ *then y is* $\tilde{B}$, where $\tilde{A} = [\underline{\mu}_{\tilde{A}}(x), \overline{\mu}_{\tilde{A}}(x)]$ and $\tilde{B} = [\underline{\mu}_{\tilde{B}}(y), \overline{\mu}_{\tilde{B}}(y)]$ be two interval type-2 fuzzy sets. Here, the reasoning is undertaken using the UMF and the LMF pairs of the antecedent and consequent. Thus, we have two relational matrices,

$\underline{R}$: For fuzzy rule- *if x is* $\underline{A}$ *then y is* $\underline{B}$,

and $\overline{R}$: For fuzzy rule- *if x is* $\overline{A}$ *then y is* $\overline{B}$,

where $\underline{A} = \underline{\mu}_{\tilde{A}}(x)$, $\overline{A} = \overline{\mu}_{\tilde{A}}(x)$, $\underline{B} = \underline{\mu}_{\tilde{B}}(y)$ and $\overline{B} = \overline{\mu}_{\tilde{B}}(y)$. Now, for a given crisp data point $x' \in$ X, the extension IT2 inference is formalized below.

The IT2 inference is $\tilde{B}' = [\underline{\mu}_{\tilde{B}'}(y), \overline{\mu}_{\tilde{B}'}(y)] = [\underline{B}', \overline{B}']$, where $\underline{B}' = (\underline{\alpha} \wedge (\underline{A} \circ \underline{R}))$, and $\overline{B}' = (\overline{\alpha} \wedge (\overline{A} \circ \overline{R}))$, here $\underline{\alpha} = \underline{\mu}_{\tilde{A}}(x')$ and $\overline{\alpha} = \overline{\mu}_{\tilde{A}}(x')$ (following equation (3)).

2.2 Metric of Span of Uncertainty (SOU) in Terms of UMF and LMF

Here, we define a metric in terms of UMF and LMF, and this metric indirectly measures the span of uncertainty. The span of uncertainty is $(c_r - c_l)$, where c_r and c_l represent the right and left end point centroid of IT2 FS [9]. The equation of c_r and c_l are given as [18]

$$c_l = \left(\int_{-\infty}^{c_l} x\overline{\mu}_{\tilde{A}}(x)dx + \int_{c_l}^{\infty} x\underline{\mu}_{\tilde{A}}(x)dx\right) \Bigg/ \left(\int_{-\infty}^{c_l} \overline{\mu}_{\tilde{A}}(x)dx + \int_{c_l}^{\infty} \underline{\mu}_{\tilde{A}}(x)dx\right) \quad (4)$$

$$c_r = \left(\int_{-\infty}^{c_r} x\underline{\mu}_{\tilde{A}}(x)dx + \int_{c_r}^{\infty} x\overline{\mu}_{\tilde{A}}(x)dx\right) \Bigg/ \left(\int_{-\infty}^{c_r} \underline{\mu}_{\tilde{A}}(x)dx + \int_{c_r}^{\infty} \overline{\mu}_{\tilde{A}}(x)dx\right). \quad (5)$$

Now, the metric, Z, for the FOU = $[\underline{B}', \overline{B}']$ is defined as

$$Z = \sum_{\forall j} (\overline{b}'_j - \underline{b}'_j), \quad (6)$$

here, $\overline{b}'_j$ is the jth element of $\overline{B}'$ and $\underline{b}'_j$ is the jth element of $\underline{B}'$.

Now, we prove the fact that- *if the value of Z is minimized then the SOU would be minimized* by Theorem 1 and 2.

Theorem 1: If $\overline{\mu}'_{\tilde{A}}(x) < \overline{\mu}_{\tilde{A}}(x)$ at least for one $x = x_i$ and $c_r < x_i < c_l$, then the SOU is reduced.

Proof: We have divided the proof in two parts:

i) *If $\overline{\mu}'_{\tilde{A}}(x) < \overline{\mu}_{\tilde{A}}(x)$, at least for one $x = x_i$ and $c_r < x_i < c_l$, then left-end point centroid increases*

Simplifying equation (4), we have

$$\int_{-\infty}^{c_l} \overline{\mu}_{\tilde{A}}(x)(c_l - x)dx = \int_{c_l}^{\infty} \underline{\mu}_{\tilde{A}}(x)(x - c_l)dx$$

$$\Rightarrow \int_{-\infty}^{c_l} \overline{\mu}'_{\tilde{A}}(x)(c_l - x)dx < \int_{c_l}^{\infty} \underline{\mu}_{\tilde{A}}(x)(x - c_l)dx, \text{ (Since, } \overline{\mu}'_{\tilde{A}}(x) < \overline{\mu}_{\tilde{A}}(x)\text{)} \quad (7)$$

$$\Rightarrow \int_{-\infty}^{c'_l} \overline{\mu}'_{\tilde{A}}(x)(c'_l - x)dx = \int_{c'_l}^{\infty} \underline{\mu}_{\tilde{A}}(x)(x - c'_l)dx \quad \text{(Let } c'_l \geq c_l\text{)}$$

$$\Rightarrow c'_l = \left(\int_{-\infty}^{c'_l} x\overline{\mu}'_{\tilde{A}}(x)dx + \int_{c'_l}^{\infty} x\underline{\mu}_{\tilde{A}}(x)dx\right) \Bigg/ \left(\int_{-\infty}^{c'_l} \overline{\mu}'_{\tilde{A}}(x)dx + \int_{c'_l}^{\infty} \underline{\mu}_{\tilde{A}}(x)dx\right) \quad (8)$$

The above equation shows that c_l' is the left-end point of FOU=[$\underline{\mu}_{\tilde{A}}(x)$, $\bar{\mu}'_{\tilde{A}}(x)$], where $c_l' > c_l$.

ii) *If $\bar{\mu}'_{\tilde{A}}(x) < \bar{\mu}_{\tilde{A}}(x)$, at least for one $x = x_i$ and $c_r < x_i < c_l$, then right-end point centroid decreases*

Again, simplifying equation (5), we obtain

$$\int_{-\infty}^{c_r} \underline{\mu}_{\tilde{A}}(x)(c_r - x)dx = \int_{c_r}^{\infty} \bar{\mu}_{\tilde{A}}(x)(x - c_r)dx$$

$$\Rightarrow \int_{-\infty}^{c_r} \underline{\mu}_{\tilde{A}}(x)(c_r - x)dx > \int_{c_r}^{\infty} \bar{\mu}'_{\tilde{A}}(x)(x - c_r)dx, \text{ (Since, } \bar{\mu}'_{\tilde{A}}(x) < \bar{\mu}_{\tilde{A}}(x)\text{)} \tag{9}$$

$$\Rightarrow \int_{-\infty}^{c_r'} \underline{\mu}_{\tilde{A}}(x)(c_r' - x)dx = \int_{c_r'}^{\infty} \bar{\mu}'_{\tilde{A}}(x)(x - c_r')dx \quad \text{(Let } c_r' < c_r\text{)}$$

$$\Rightarrow c_r' = \left(\int_{-\infty}^{c_l'} x\underline{\mu}_{\tilde{A}}(x)dx + \int_{c_l'}^{\infty} x\bar{\mu}'_{\tilde{A}}(x)dx \right) \Big/ \int_{-\infty}^{c_l'} \underline{\mu}_{\tilde{A}}(x)dx + \int_{c_l'}^{\infty} \bar{\mu}'_{\tilde{A}}(x)dx \tag{10}$$

The above equation shows that c_r' is the right-end point of FOU=[$\underline{\mu}_{\tilde{A}}(x)$, $\bar{\mu}'_{\tilde{A}}(x)$], where $c_r' < c_r$.

Thus, for $\bar{\mu}'_{\tilde{A}}(x) \leq \bar{\mu}_{\tilde{A}}(x)$, left-end point increases and right-end point decreases, i.e., SOU is reduced. □

Theorem 2: If $\underline{\mu}'_{\tilde{A}}(x) > \underline{\mu}_{\tilde{A}}(x)$, at least for one $x = x_i$ and $c_r > x_i > c_l$, then the SOU is reduced.

Proof: Similar to the proof of Theorem 1. □

2.3 Measure of SOU for Various Fuzzy Implication Function

Here, we evaluate Z for different fuzzy implication function.

***Mamdani* Implication:** By Mamdani's implication [1] the $(i, j)^{\text{th}}$ element in relational matrix is written as

$r_{ij} = \text{Min}(a_i, b_j)$

Now, following the equation (3), the jth element of inferred matrix B', i.e., b_j', is written as

$$b_j' = \alpha \wedge ([a_i] \text{ o } [r_{ij}]_{\forall i})$$
$$= \alpha \wedge ([a_i] \text{ o } [\underset{\forall i}{Min}(a_i, b_j)])$$
$$= \alpha \wedge [\bigvee_{\forall i} (a_i \wedge a_i \wedge b_j))] = \alpha \wedge [\bigvee_{\forall i} (a_i)] \wedge b_j = (\alpha \wedge M_A) \wedge b_j, \tag{11}$$

where $\alpha = \mu_A(x)\big|_{x=x'}$ and the maximum value of A, $M_A = \bigvee_{\forall i} (a_i)$.

Now, following equation (6), we obtain the Z for IT2 inference using Mamdani's implication as

$$Z_L = \sum_{\forall j}[(\bar{\alpha} \wedge M_{\bar{A}} \wedge \bar{b}_j) - (\underline{\alpha} \wedge M_{\underline{A}} \wedge \underline{b}_j)] \tag{12}$$

Kleen-Dienes: By *Kleen-Dienes* implication the $(i, j)^{th}$ element in relational matrix is written as

$r_{ij} = \text{Max}((1- a_i), b_j)$

Now, following the equation (3), the jth element of inferred matrix B', i.e., b'_j, is written as

$$\begin{aligned} b'_j &= \alpha \wedge ([a_i] \circ [r_{ij}]_{\forall i}) \\ &= \alpha \wedge ([a_i] \circ [\underset{\forall i}{Max}((1\text{-}a_i), b_j)]) \\ &= \alpha \wedge [\bigvee_{\forall i} ((a_i \wedge (1-a_i)) \vee (a_i \wedge b_j))] \\ &= \alpha \wedge [\{\bigvee_{\forall i} ((a_i \wedge (1-a_i))\} \vee \{\bigvee_{\forall i} (a_i \wedge b_j))\}] \\ &= \alpha \wedge [(A \circ \neg A) \vee (\bigvee_{\forall i} (a_i) \wedge b_j)] \\ &= \alpha \wedge [(A \circ \neg A) \vee (M_A \wedge b_j)] = (\alpha \wedge M_A) \wedge ((A \circ \neg A) \vee b_j), \end{aligned} \tag{13}$$

where $\alpha = \mu_A(x)|_{x=x'}$, $A = [a_i]$ and $\neg A = [(1\text{-}a_i)]$.

Now, following equation (6), we obtain the Z for IT2 inference using *Kleen-Dienes* implication as

$$Z_K = \sum_{\forall j}[\{\bar{\alpha} \wedge M_{\bar{A}} \wedge ((\bar{A} \circ \neg\bar{A}) \vee \bar{b}_j))\} - \{\underline{\alpha} \wedge M_{\underline{A}} \wedge ((\underline{A} \circ \neg\underline{A}) \vee \underline{b}_j\}] \tag{14}$$

Lukasiewicz: By *Lukasiewicz* implication the $(i, j)^{th}$ element in relational matrix is written as

$r_{ij} = \text{Min}\{1, (1- a_i + b_j)\}$

Now, following the equation (3), the jth element of inferred matrix B', i.e., b'_j, is written as

$$\begin{aligned} b'_j &= \alpha \wedge ([a_i] \circ [r_{ij}]_{\forall i}) \\ &= \alpha \wedge ([a_i] \circ [\underset{\forall i}{Min}\{1, (1\text{-}a_i + b_j)\}]) = \alpha \wedge [\bigvee_{\forall i} (a_i \wedge (1-a_i + b_j))] \end{aligned} \tag{15}$$

Now, following equation (6), we obtain the Z for IT2 inference using *Lukasiewicz* implication as

$$Z_L = \sum_{\forall j}[\{\bar{\alpha} \wedge (\bigvee_{\forall i} (\bar{a}_i \wedge (1\text{-}\bar{a}_i + \bar{b}_j))\} - \{\underline{\alpha} \wedge (\bigvee_{\forall i} (\underline{a}_i \wedge (1\text{-}\underline{a}_i + \underline{b}_j))\}] \tag{16}$$

***Lukasiewicz-1*:** Following *Lukasiewicz* implication we obtain the Z as

$$Z_{L1} = \sum_{\forall j}[\{\bar{\alpha} \wedge (\bigvee_{\forall i}(\bar{a}_i \wedge ((1-\bar{a}_i+(1+\lambda)\bar{b}_j)/(1+\lambda\bar{a}_i)))\}$$

$$-\{\underline{\alpha} \wedge (\bigvee_{\forall i}(\underline{a}_i \wedge ((1-\underline{a}_i+(1+\lambda)\underline{b}_j)/(1+\lambda\underline{a}_i)))\}], \text{ where } \lambda > -1. \quad (17)$$

***Lukasiewicz-2*:** Following *Lukasiewicz* implication we obtain the Z as

$$Z_{L2} = \sum_{\forall j}[\{\bar{\alpha} \wedge (\bigvee_{\forall i}(\bar{a}_i \wedge (1-(\bar{a}_i)^w+(\bar{b}_j)^w)^{1/w}))\}$$

$$-\{\underline{\alpha} \wedge (\bigvee_{\forall i}(\underline{a}_i \wedge (1-(\underline{a}_i)^w+(\underline{b}_j)^w)^{1/w}))\}], \text{ where } w > 0. \quad (18)$$

2.4 Comparative Study

Table 1, the expression of Z for all the implication function has been summarized. To get the reduced SOU in IT2 FS, we need to select minimum value of Z. Explanation of Table 1 is briefly given below.

i) Mamdani implication: $Z_M \geq 0$ holds as $\bar{b}_j \geq \underline{b}_j$ for all *j*, $\bar{\alpha} \geq \underline{\alpha}$ and $M_{\bar{A}} \geq M_{\underline{A}}$ always hold. Thus, Mamdani implication based inference produces valid FOU.
ii) Kleen-Dienes: Here, $Z_K \geq 0$ when

$$(\bar{A} \circ \neg\bar{A}) > (\underline{A} \circ \neg\underline{A})$$

which may not occur always. Thus, Kleen-Dienes sometimes results in invalid FOU. However, above condition minimizes metric Z.

Table 1. Validity and SOU of various implication functions

Implication Function	Validity of FOU	SOU
Mamdani	ALWAYS VALID	USED AS REFERENCE
Kleen-Dienes	VALID IF $(\bar{A} \circ \neg\bar{A}) > (\underline{A} \circ \neg\underline{A})$	$Z_K < Z_M$
Lukasiewicz	VALID IF $\bigvee_{\forall i}(\bar{a}_i \wedge (1-\bar{a}_i+\bar{b}_j)) > \bigvee_{\forall i}(\underline{a}_i \wedge (1-\underline{a}_i+\underline{b}_j))$	$Z_L < Z_M$
Lukasiewicz-1	VALID FOR SELECTIVE λ	$Z_{L1} < Z_I$, WHERE I = M, K, L
Lukasiewicz-2	VALID FOR SELECTIVE w	$Z_{L2} < Z_I$, WHERE I = M, K, L

iii) *Lukasiewicz:* Here, $Z_L \geq 0$ when $\bigvee_{\forall i}(\bar{a}_i \wedge (1-\bar{a}_i+\bar{b}_j)) > \bigvee_{\forall i}(\underline{a}_i \wedge (1-\underline{a}_i+\underline{b}_j))$, which may not hold always. Thus, sometimes it may result in invalid FOU.

iv) Lukasiewicz-1: In *Lukasiewicz-1*, the value of λ >-1. However, it is observed that for only selective range of λ, $Z_{L1} \geq 0$. So, the FOU is valid for selective range of λ. The choosing of λ also reduces the span of uncertainty.
v) Lukasiewicz-2: In *Lukasiewicz-2*, the value of *w*> 0. However, it is observed that for only selective range of *w*, $Z_{L2} \geq 0$. So, the FOU is valid for selective range of *w*. The choosing of *w* also reduces the span of uncertainty.

To avoid the drawback and get the benefit of *Lukasiewicz* implication we switch to *Lukasiewicz-1* or, -2 implications. In *Lukasiewicz-1* or, -2 implications, there are a tuning factors (λ and *w*) that helps in to select a valid FOU as well as reduced Z (i.e., reduced SOU). Using computer simulation we can search a proper value of λ or *w* that will reduce SOU as well as prevent the crossing of LMF to UMF. An example is given below.

Example 1: Given antecedent UMF, $\overline{A}$ = [0.2 0.9 0.9 0.9 0.1] and LMF, $\underline{A}$= [0.1 0.3 0.5 0.3 0.1]. Consequent UMF $\overline{B}$ = [0.2 0.9 0.9 0.9 0.1] and LMF, $\underline{B}$ =[0.1 0.3 0.5 0.5 0.1], and crisp input value $x = x_3$.
$\overline{\alpha} = \overline{A}(x_3) = 0.9$ and $\underline{\alpha} = \underline{A}(x_3) = 0.5$.
Mamdani inference: $\overline{B}'$ = [0.2 0.9 0.9 0.9 0.1] and $\underline{B}'$ = [0.1 0.3 0.5 0.5 0.1].
Conclusion: Valid FOU, Z = 1.5.
Kleen-Dienes inference: $\overline{B}'$ = [0.2 0.9 0.9 0.9 0.2] and $\underline{B}'$ = [0.5 0.5 0.5 0.5 0.5].
Conclusion: Invalid FOU as $\underline{B}' > \overline{B}'$.
Lukasiewicz inference: $\overline{B}'$ = [0.3 0.9 0.9 0.9 0.2] and $\underline{B}'$ = [0.5 0.5 0.5 0.5 0.5].
Conclusion: Invalid FOU as $\underline{B}' > \overline{B}'$.
Lukasiewicz-1 inference: $\overline{B}'$ = [0.5637 0.9 0.9 0.9 0.5013] and $\underline{B}'$ = [0.5 0.5 0.5 0.5 0.5]. Conclusion: Valid FOU and reduced SOU at λ = -0.858, Z = 1.265.
Lukasiewicz-2 inference: $\overline{B}'$ = [0.5282 0.9 0.9 0.9 0.5068] and $\underline{B}'$ = [0.5 0.5 0.5 0.5 0.5]. Conclusion: Valid FOU and reduced SOU at w = 2.25, Z = 1.235.
Here, we search the suitable value of λ for greater than -1 and *w* for greater than 0 that can produce minimum Z satisfying $\overline{B}' \supseteq \underline{B}'$. This searching technique is done by simple Matlab programming.

3 Conclusion

Existing analysis with traditional implication functions reveal that *Lukasiewicz* implication satisfy nine standard axioms [19], and thus is widely used for type-1 fuzzy reasoning. While extending type-1 reasoning to IT2 FS, it is observed that *Lukasiewicz* implication poses a restriction in IT2 inference generation because of the possibility of having the LMF crossing the UMF in the inferential space. While comparing suitability of implication functions to reduce uncertainty in IT2 inference, it is observed that *Lukasiewicz*-1/*Lukasiewicz*-2 implication are good choices for their flexibility in controlling the relational surface by a parameter λ and w. Experiments reveal that for a suitable λ/w in the positive real axis, it yields the reduced Z, resulting in a reduction in the span of uncertainty.

References

1. Zadeh, L.A.: Fuzzy sets. Information and Control 8(3), 338–353 (1965)
2. Jammeh, E.A., Fleury, M., Wagner, C., Hagras, H., Ghanbari, M.: Interval type-2 fuzzy logic congestion control for video streaming across IP networks. IEEE Trans. on Fuzzy Systems 17(5), 1123–1142 (2009)
3. Mendel, J.M., John, R.I.: Type-2 fuzzy sets made simple. IEEE Trans. on Fuzzy Systems 10, 117–127 (2002)
4. Mendel, J.M.: Advances in type-2 fuzzy sets and systems. Information Sciences 177(1), 84–110 (2007)
5. Mendel, J.M.: Type-2 fuzzy sets and systems: An overview. IEEE Computational Intelligence (2007)
6. Mendel, J.M., John, R.I., Liu, F.: Interval type-2 fuzzy logic systems made simple. IEEE Trans. on Fuzzy Systems 14, 808–821 (2006)
7. Wu, D., Mendel, J.M.: Uncertainty measures for interval type-2 fuzzy sets. Information Sciences 177(23), 5378–5393 (2007)
8. Mendel, J.M., Wu, H.: Type-2 fuzzistics for nonsymmetric interval type-2 fuzzy sets: forward problems. IEEE Trans. on Fuzzy Systems 15(5), 916–930 (2007)
9. Mendel, J.M., Wu, D.: Perceptual computing: Aiding people in making subjective judgments, vol. 13. Wiley-IEEE Press (2010)
10. Wu, D., Mendel, J.M.: Linguistic summarization using IF–THEN rules and interval type-2 fuzzy sets. IEEE Trans. on Fuzzy Systems 19(1), 136–151 (2011)
11. Wu, D., Tan, W.: Type-2 fuzzy logic controller for liquid level process. In: Proc. IEEE Int. Conf. on Fuzzy Systems, pp. 242–247 (2005)
12. Lascio, L.D., Gisolfi, A., Nappi, A.: Medical differential diagnosis through type-2 fuzzy sets. In: Proc. IEEE Int. Conf. on Fuzzy Systems, pp. 371–376 (2005)
13. Hagras, H.: A hierarchical type-2 fuzzy logic control architecture for autonomous mobile robots. IEEE Trans. on Fuzzy Systems 12, 524–539 (2004)
14. Coupland, S., John, R.: Geometric type-1 and type-2 fuzzy logic systems. IEEE Tran. on Fuzzy Systems 15(1), 3–15 (2007)
15. Karnik, N.N., Mendel, J.M.: Centroid of type-2 fuzzy set. Information Sciences 132, 195–220 (2001)
16. Aguero, J.R., Vargas, A.: Calculating functions of interval type-2 fuzzy numbers for fault current analysis. IEEE Trans. Fuzzy Syst. 15(1), 31–40 (2007)
17. Juang, C.-F., Huang, R.-B., Lin, Y.-Y.: A recurrent self-evolving interval type-2 fuzzy neural network for dynamic system processing. IEEE Trans. Fuzzy Syst. 17(5), 1092–1105 (2009)
18. Mendel, J.M., Wu, H.: Type-2 fuzzistics for nonsymmetric interval type-2 fuzzy sets: forward problem. IEEE Trans. Fuzzy Systems 15(5), 916–930 (2007)
19. Klir, G.J., Yuan, B.: Approximate reasoning: Fuzzy sets and fuzzy Logic (2002)

Detection of Downy Mildew Disease Present in the Grape Leaves Based on Fuzzy Set Theory

Dipak Kumar Kole[1], Arya Ghosh[2], and Soumya Mitra[1]

[1] Department. of Computer Science and Engineering
St. Thomas' College of Engineering & Technolgy, Kolkata, India
[2] Department. of Computer Science and Engineering
ABACUS Institute of Engineering & Management, India
{dipak.kole,mitra92.soumya}@gmail.com, arya_mca_05@yahoo.com

Abstract. Agriculture has a significant role in economy of the most of the developing countries. A significant amount of crops are damaged in every year due to fungi, fungus, bacteria, Phytoplasmas, bad weather etc. Grapes are one of the most widely grown fruit crops in the world with significant plantings in India. Grapes are used in the production of wine, brandy, or non-fermented drinks and are eaten fresh or dried as raisins. Sometimes grape plants are affected by downy mildew, a serious fungal disease. Therefore, farmers try to detect the stage of the disease in plant at an early stage so that they can take necessary steps in order to prevent the disease from spreading to others parts of the fields. This article presents a novel technique for detection of downy mildew disease present in the grape leaves based on fuzzy importance factor. The proposed technique uses some digital image processing operations and fuzzy set theory concept. We experimented on thirty one diseased and non-diseased images and got 87.09% success. The experimental results reveal that the proposed technique can effectively detect the present of downy mildew disease in the grape leaves.

Keywords: fuzzy value, downy mildew disease, energy.

1 Introduction

Grapes are one of the most widely grown fruit crops in the world with significant plantings in India. Grapes are used in the production of wine, brandy, or non-fermented drinks and are eaten fresh or dried as raisins. It is a very important cash crop in India. But sometimes grape plants are affected by downy mildew, a serious fungal disease caused by Plasmoparaviticola. It is always difficult to be identified by using naked-eye observation method directly. So, it is difficult to diagnose it accurately and effectively by using traditional plant disease diagnosis method that is mainly dependent on naked-eye observation [5]. Downy mildew disease of grape leaves is among the oldest plant disease known to man. It is a highly destructive disease of grapevines in all grape-growing areas of the world. Early literature on grape cultivation mentions this devastating disease and its ability to destroy entire

M.K. Kundu et al. (eds.), *Advanced Computing, Networking and Informatics - Volume 1*,
Smart Innovation, Systems and Technologies 27,
DOI: 10.1007/978-3-319-07353-8_44, © Springer International Publishing Switzerland 2014

grape plantation [4]. Since downy mildew discovery, numerous studies have been conducted on the life cycles of downy mildew pathogen and its management. The information gained from these studies has enabled to develop best management practices that reduce the impact of the diseases. Today, worldwide epidemic losses are rare, though the diseases can occur at significant levels in particular fields or throughout a particular growing region. If it is not controlled the powdery mildew fungus not only affects fruit yield and quality, but it also reduces vine growth and winter hardiness.

Significant progress has been made in the use of image processing approaches to detect various diseases in other crops. The work by Boso *et al.* confirmed that image processing can provide a means of rapid, reliable and quantitative early detection of these diseases [1].Weizhen *et al.* has emphasized that both quality and quantity of agricultural product are highly reduced by the various plant diseases [10].

A semi-automated image processing method developed by Peresotti *et al.* (2011) to measure the development of the downy mildew pathogen by quantifying the sporulation of the fungi on the grapevine [6]. In order to quantify the sporulation the image was converted to 8-bit format using median-cut colour quantization [3] and then the contrast was adjusted to enhance the white sporulated area in the leaf. An image recognition method reported by Li *et al.* (2012) for the diagnosis of vines with downy mildew and powdery mildew disease [5]. Images were pre-processed using nearest neighbour interpolation to compress the image prior to removal by a median filter and finally diseased regions were segmented via a K-means [11], [12] clustering algorithm. Then a SVM classifier was evaluated for performance in disease recognition.

In this paper, we present efficient technique to classify the downy mildew diseased and non-diseased grape leaves based on K-means and fuzzy set theory. This proposed technique consists of two stages. First one is feature reduction stage in which dominating features are obtained from a list of twenty features. Second stage is the detection of downy mildew disease present in grape leaves. In the first stage, we create a feature matrix after calculating all twenty features of the given diseased images. After that a normalized feature matrix is generated in which K-means algorithm is applied on individual feature (column). Each cluster is represented by a linguistic symbol and the fuzzy value of each linguistic symbol (cluster) for each feature is calculated with respect to total number of diseased images. A fuzzy matrix is generated from normalized feature matrix. Then fuzzy value of each cluster for each feature is calculated. Those features are selected whose maximum fuzzy value is greater than predefined threshold value. In the second stage, we first generate the values of selected features of all test images. Then each normalized feature value in the normalized feature matrix is replaced by a fuzzy value of the closest cluster of the respective feature. The average fuzzy value is calculated for each image. Those images are detected as downy mildew disease present in grape leaves, whose average fuzzy value are greater than or equal to the experimental threshold value.

The rest of the paper is organized as follows. Preliminaries are given in Section 2. Proposed method is described in Section 3. In Section 4, we present experimental result. Concluding remarks appear in Section 5.

2 Preliminaries

2.1 Diseases in Grape Leaves

The six most common diseases of grapes are black rot, downy mildew, powdery mildew, deadarm, crown gall, and gray mold. Downy Mildew Fig. 1 is a serious fungal disease caused by the *Plasmoporaviticola*. Itattacks the leaves, shoots, fruit, and their tendrils during their immature stage. Spores formed in the infected part of plants are blown to leaves, shoots, and blossom clusters during cool, moist weather in the spring and early summer. The fungus attacks shoots, tendrils, petioles, leaf veins, and fruit. This disease appears first on fruits as dark red spots. Then these spots are gradually are circular, sunken, ashy-gray and in late stages these are surrounded by a dark margin. The spots vary in size from 1/4 inch in diameter to about half the fruit.

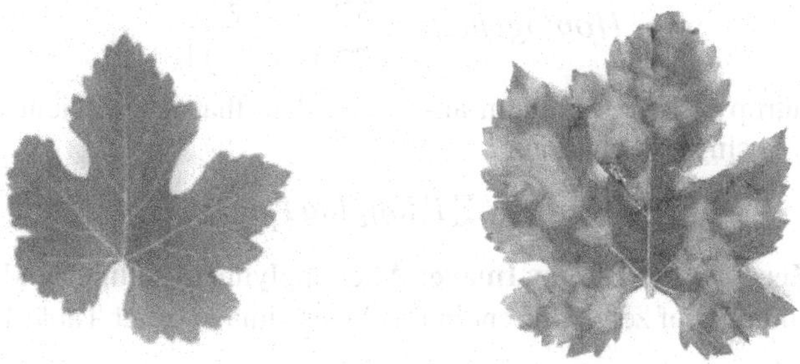

a. Disease Free Grape Leaf b. Downy Mildew on Grape Leaf

Fig. 1. Disease Free and Diseased Images

2.2 Featured used in the Proposed Techniques

Mean: The local mean is a measure of average gray level in neighborhood S(x; y).

Inertia: Also known as Variance and contrast, Inertia measures the local variations in the matrix.

$$Inertia = \sum_i \sum_j (i - j)^2 c(i, j) \tag{1}$$

Correlation: It measures the sum of joint probabilities of occurrence of every possible gray-level pair.

$$Correlation = \sum_i \sum_j \frac{(i-\mu_i)(j-\mu_j)c(i,j)}{\sigma_i \sigma_j} \tag{2}$$

$$\mu_i = \sum_i i \sum_j c(i,j), \mu_j = \sum_j j \sum_i c(i,j), \quad \sigma_i = \sum_i (i-\mu_i)^2 \sum_j c(i,j) \quad (3)$$

$$\sigma_j = \sum_i (j-\mu_j)^2 \sum_j c(i,j) \quad (4)$$

Energy: Also known as uniformity or the angular second moment, It provides the sum of squared elements in the matrix.

$$Energy = \sum_i \sum_j (c(i,j))^2 \quad (5)$$

Homogeneity: It measures the closeness of the distribution of the elements in the matrix to their diagonal elements respectively.

$$Homogeneity = \sum_{i,j} \frac{c(i,j)}{1+|i-j|} \quad (6)$$

Entropy: Entropy is the minimum amount of data that is sufficient to describe an image without losing information.

$$\text{Entropy}=-\sum P_i log_2 \, log \, P_j \quad (7)$$

Number of Zeros in the Binary Image: After applying Otsu thresholding algorithm, we count the number of zeros present in this binary image (refer Table 1).

Table 1. Feature number (F_NO) with respective Feature name (F_NAME)

F_NO	F_Name	F_NO	F_Name	F_NO	F_Name	F_NO	F_Name
1	Hue Mean	2	Hue Standard Deviation	3	Hue Entropy	4	Saturation Mean
5	Saturation Standard Deviation	6	Saturation Entropy	7	Value Mean	8	Standard Deviation
9	Entropy Value	10	Enteria	11	Correlation	12	Energy
13	Homogeneity	14	Image Mean	15	Image Standard Deviation	16	Image Entropy
17	Image Mode	18	Image Peaks	19	Number of Zeroes	20	Number of connected Components

2.3 Fuzzy Value

A fuzzy concept is a concept of which the meaningful content, value, or boundaries of application can vary considerably according to context or conditions, instead of being fixed once and for all [2], [8], [9]. This generally means the concept is vague, lacking a fixed, precise meaning, without however being meaningless altogether [7].

3 Proposed Technique

The proposed technique consists of two stages: first one is the feature selection stage based on fuzzy set theory with respect to diseased images from a list of twenty features and second is the verification stage for disease or non-disease image.

3.1 Feature Selection Based on Fuzzy Value

In this phase, we first determine twenty feature values on all diseased images (say, m number of diseased images). After doing this, a two-dimension matrix having size m x 20 is generated. Since the range set for different feature are not same, we make the range set of all features to same range in (0, 1) interval. Thus a normalized featured matrix is generated to compare any feature with other features. K-means algorithm are applied on individual column of the normalized featured matrix considering the value of K=3. After clustering on each column, linguistic symbols are assigned to each cluster of a particular column where each linguistic symbol represents a cluster of the corresponding column. Then the frequency of individual linguistic symbol is calculated. After that the fuzzy value of that linguistic symbol is determined with respect to total number of diseased images. Store the cluster information (cluster centers and fuzzy values) for each feature and choose the linguistic symbol corresponding to each feature having maximum fuzzy value. Those features are selected whose maximum fuzzy value is greater than predefined threshold value.

The following algorithm describes the feature selection based on fuzzy value.

Algorithm: Feature Selection based on Fuzzy Value

Input: Input all diseased images (say m).

Output: The list of features whose maximum fuzzy value are greater than predefined threshold value.

Step1: Generate a Feature Matrix of dimension $F = (f_{ij})_{m \times n}$
Where f_{ij} represent the jth feature of the ith image, n = number of features and m = total number of diseased images.

Step2: Generate Normalized feature matrix $NF = (q_{ij})_{mxn}$ where $q_{ij}=f_{ij} / \text{Max}\{f_{ik}\}$ k=1 to m

Step3: Recreate the normalized feature matrix with linguistic values using K-means algorithm on individual normalized feature (column), where each linguistic symbol represents a cluster among 'K' clusters (say, $L_1, L_2, \ldots, L_k$).

Step4: For each feature (column) calculate the fuzzy value for each linguistic symbol using the following formula: z_x = frequency(L_x)/m, where m = number of images and x=1 to K

Step5: Store the cluster information (cluster centers and fuzzy values) for each feature.

Step6: Select the linguistic symbol corresponding to each feature having maximum fuzzy value.

Step7: After calculating ***n*** maximum fuzzy values, each of which correspond to a feature, select those features whose maximum fuzzy value is greater than predefined threshold value. Let us assume ***p*** number of features have been selected.

3.2 Detection of Diseased Image

In the second phase, we took a list of images which will be verified as diseased or non-diseased images. First we calculate the value of the selected features generated from the previous algorithm on the list of images. We transform that feature matrix into normalized feature matrix. Then we convert the matrix into a matrix of fuzzy values where each feature value of an image will be replaced by a fuzzy value for that, we first find out the cluster in which the distance between this feature value and centroid of each cluster is minimum. We replace the feature value in the normalized feature matrix by the fuzzy value of the selected cluster. After getting this fuzzy matrix we find out the average value of all fuzzy values for each image (row wise).

We got a threshold value experimentally which gives us maximum success rate. Depending on this experimental threshold value, we conclude the list images whose average fuzzy value is less than the threshold are considered diseased free images and the list of images whose average fuzzy value are greater than or equal to the threshold are considered as diseased images.

The following algorithm describes the detection of diseased images.

Algorithm: Detection of Downy Mildew disease on grape leaves.

Input: Set of m number of images to detect downy mildew on grape leaves
Output: Identification of diseased images

Step1: Extract those features $\{f_1, f_2, \ldots, f_p\}$ from the images which have been determined by theFeature Selection Algorithm. $F = (f_{ij})_{m \times p}$, where f_{ij} represent the jth feature of ith image, p = number of features.

Step2: Load Cluster information corresponding to each feature (generated by Feature Selection Algorithm)

Step3: For each ith image (i = 1 to m)

3.1 Determine the cluster (corresponding to each feature) to which each feature value ($f_1, f_2, \ldots, f_p$) of the ith image must belong to. The feature value belongs to the cluster from whose center its distance is the smallest.

3.2 fz_{ij} = fuzzy value of the cluster to which the jth feature of the ith image belongs to.

3.3 Calculate fuzzy value of ith image, $fv_i = (\sum fz_{ij})/p$ for j= 1 to p.

Step4: Depending upon a predefined threshold for fuzzy value (*fv*), images are marked as diseased or disease free.

4 Experimental Results

We have applied feature selection algorithm on a list of diseased images on fuzzy set theory to get the most dominating features from a list of twenty features. In this stage, we have considered the value of k in k means is 3and fuzzy threshold value is 0.5. This algorithm reduces from twenty features to nine features. In the second stage a list of images are considered for detection for downy mildew diseases present in grape leaves. We experimentally got a threshold value 0.47 for which the success rate for detection of downy mildew disease is 87.09%.We have experimented on 31 images whose result is shown in the following Table 2. Column 3 of Table 2 represents number of features with list of feature number selected from feature selection algorithm.

Table 2. Experimental result on thirty one images

No. of Cluster	Fuzzy Threshold for Feature Reduction	No. of Feature used after Reduction	Threshold	No. of false positive	No. of false negative	Truly Detected	Success Rate
3	0.5	9,{2,3,5,10,12.14.18,19,20}	0.47	2	2	27	87.09%

5 Conclusion

This paper presents a novel technique for detection of downy mildew disease present in the grape leaves. The proposed technique uses a proposed feature reduction algorithm which finds out the most dominating features from a list of features in this domain based. For that it uses a K-means algorithm and fuzzy set theory concept. The dominating features are obtained for only diseased images. Those features are calculated from a list of test images and latter classified by the proposed detection of diseased images algorithm into diseased and non-diseased images. The above experimental result shows that our technique yields good results.

Acknowledgement. The authors would like to thank Dr. Amitava Ghosh, ex-Economic Botanist IX, Agriculture dept., Govt. of West Bengal, presently attached as Guest Faculty, Dept. of Genetics and Plant Breeding, Institute of Agriculture, Calcutta University, for providing grape leaves images with scientist's comments.

References

1. Boso, S., Santiago, J.L., Martínez, M.C.: Resistance of Eight Different Clones of the Grape Cultivar Albariño to Plasmoparaviticola. Plant Disease 88(7), 741–744 (2004)
2. Maurus, V.B., James, N.M., Ronald, W.M., Patrick, F.: Inheritance of Downy Mildew Resistance in Table Grapes. J. Amer. Soc. Hort. Sci. 124(3), 262–267 (1999)
3. Heckbert, P.: Color image quantization for frame buffer display. In: Proceedings of 9th Annual Conference on Computer Graphics and Interactive Techniques, SIGGRAPH 1982, pp. 297–307 (1982)
4. Hofmann, U.: Plant protection strategies against downy mildew in organic viticulture. In: Proceedings of International Congress of Organic Viticulture, pp. 167–174 (2000)
5. Li, G., Ma, Z., Wang, H.: Image recognition of grape downy mildew and grape powdery mildew based on support vector machine. In: Li, D., Chen, Y. (eds.) CCTA 2011, Part III. IFIP AICT, vol. 370, pp. 151–162. Springer, Heidelberg (2012)
6. Peressotti, E., Duchêne, E., Merdinoglu, D., Mestre, P.: A semi-automatic non-destructive method to quantify downy mildew sporulation. Journal of Microbiological Methods 84, 265–271 (2011)
7. Dietz, R., Moruzzi, S.: Cuts and clouds. Vagueness, Its Nature, and Its Logic. Oxford University Press (2009)
8. Ridler, T.W., Calvard, S.: Picture thresholding using an iterative selection method. IEEE Transactions on Systems, Man and Cybernetics 8, 630–632 (1978)
9. Haack, S.: Deviant logic, fuzzy logic: beyond the formalism. University of Chicago Press, Chicago (1996)
10. Weizheng, S., Yachun, W., Zhanliang, C., Hongda, W.: Grading Method of Leaf Spot Disease Based on Image Processing. In: Proceedings of the 2008 International Conference on Computer Science and Software Engineering, vol. 6, pp. 491–494 (2008)
11. Samma, A.S.B., Salam, R.A.: Adaptation of K-Means Algorithm for Image Segmentation. International Journal of Information and Communication Engineering 5(4), 270–274 (2009)
12. Redmond, S.J., Heneghan, C.: A method for initialising the K-means clustering algorithm using kd-trees. Pattern Recognition Letters 28, 965–973 (2007)

Facial Expression Synthesis for a Desired Degree of Emotion Using Fuzzy Abduction

Sumantra Chakraborty[1], Sudipta Ghosh[1], Amit Konar[1], Saswata Das[1], and Ramadoss Janarthanan[2]

[1] Electronics and Tele-Communication Engineering Department, Jadavpur University, Kolkata-32, India
[2] Computer Science and Engineering Department, TJS Engineering College, Chennai, India
srmjana_73@yahoo.com

Abstract. Modulating facial expression of a subject to exhibit emotional content is an interesting subject of current research in Human-Computer Interactions. The ultimate aim of this research is to exhibit the emotion-changes of the computer on the monitor as a reaction to subjective input. If recognizing emotion from a given facial expression (of a subject) is referred to as forward (deductive) reasoning, the present problem may be considered as abduction. The logic of fuzzy sets has widely been used in the literature to reason under uncertainty. The present problem of abduction includes different sources of uncertainty, including inexact appearance in facial expression to describe a given degree of a specific emotion, noisy ambience and lack of specificity in features. The logic of fuzzy sets, which has proved itself successful to handle uncertainty in abduction, thus can be directly employed to handle the present problem. Experiments undertaken reveal that the proposed approach is capable of producing emotion-carrying facial expressions of desired degrees. The visual examination by subjective experts confirms that the produced emotional expressions lie within given degrees of emotion-carrying expressions.

1 Introduction

Emotion recognition by facial expression analysis and synthesizing emotion in facial expressions are often regarded as complementary pair of problems. Although a lot many research papers on emotion recognition are available, there hardly exist fewer papers on emotion synthesis. This article provides a novel approach to synthesizing emotion on facial expression of subjects by a process of fuzzy abduction.

The sense of abduction [5], [7], [10] appears here for the inherent nature of the problem, where the user is concerned about detecting necessary changes in facial patterns to satisfy a given requirement in the degree of emotion-content appearance. The abduction problem in turn is formulated in the settings of determining optimal solution of a fuzzy implication relation, which is covered in detail in [7], [12]. The emotion synthesis problem undertaken here thus is an offshoot of the work undertaken in [12] with emphasis on the scope of application of the underlying theory in the synthesis problem.

M.K. Kundu et al. (eds.), *Advanced Computing, Networking and Informatics - Volume 1*, Smart Innovation, Systems and Technologies 27,
DOI: 10.1007/978-3-319-07353-8_45, © Springer International Publishing Switzerland 2014

Experiments undertaken reveal that the proposed technique is capable of correctly selecting the nearest emotion-contained facial expression from a video of the subject when the input degree of emotion is supplied.

The paper is divided into the following three sections. Section 2 provides the determination technique of desired facial features for a given degree of emotion by abductive reasoning. The experiment to have desired facial features from a given degree of emotion is undertaken in Section 3. Conclusions are listed in section 4.

2 Determination of Desired Facial Features for a Given Degree of Emotions by Fuzzy Abduction

This section aims at determining the desired facial features of a given subject to ensure a desired degree of specific emotions. We presume that the facial expressions of the person under consideration for varied degrees of emotion are available. Now, for an unknown degree of emotion, we need to determine the facial features that correctly represent the desired facial expression containing the required degree of emotion. The present problem has been formulated here as an abductive reasoning problem. There exists several research works on fuzzy abduction. Some of the well-known techniques in this regard are listed here [12], [7], and [16]. Among these, the work reported in [12] is the simplest. We here attempt to solve the problem by this method.

2.1 Principles of Fuzzy Abduction

We first present a few definitions that would be useful to explain the principles of fuzzy abduction.

Definition 1: A matrix **Q** is the **pre-inverse** of a fuzzy matrix **R** , if **Q o R =I'→I**, where **I** is the identity matrix and **I' → I** means **I'** is close enough to **I** and o denotes max-min composition operation [17]

Definition 2: $\mathbf{Q}_{\mathbf{best}}$ is called the **best pre-inverse** of **R** if $\|(\mathbf{Q}_{\mathbf{best}}\ \mathbf{o}\ \mathbf{R}) - \mathbf{I}\| \le \|(\mathbf{Q}\ \mathbf{o}\ \mathbf{R}) - \mathbf{I}\|$ for all real matrix Q, with elements $0 \le q_{ij} \le 1$, where $\| \boldsymbol{\delta} \|$ denotes the sum of the square of the elements of matrix $\boldsymbol{\delta}$.

To determine **Q**, the pre-inverse matrices of **R** of dimension (n × n) let us consider the k-th row and i-th column of (**Q o R**), given by

$$(\mathbf{QoR})_{k,i} = \bigvee_{j=1}^{n} (q_{kj} \wedge r_{ji})$$

To obtain **Q**, one has to satisfy **Q o R = I'**, sufficiently close to **I**, which requires

$$\bigvee_{j=1}^{n} (q_{kj} \wedge r_{jk}) \text{ to be close to 1 (criterion 1)}$$

and $$\bigvee_{j=1, i \neq k}^{n} (q_{kj} \wedge r_{ji}) \text{ to be close to 0 (criterion 2)}$$

By [12], we write the choice of q_{kj}, $\forall k, j$ from the set $\{r_{j1}, r_{j2}, \ldots., r_{jk}, \ldots., r_{jn}\}$ is governed by the following two criteria

i) $(q_{kj} \wedge r_{jk})$ is to be maximized.

ii) $(q_{kj} \wedge r_{ji})$ is to be minimized for $1 \leq \forall i \leq n$ but $i \neq k$.

The above two criteria can be combined to a single criterion as depicted below

$$(q_{kj} \wedge r_{jk}) - \bigvee_{j=1, i \neq k}^{n} (q_{kj} \wedge r_{ji}) \text{ is to be maximized, where } q_{kj} \varepsilon \{r_{j1}, r_{j2}, \ldots., r_{jk}, \ldots., r_{jn}\}.$$

Procedure of pre-inverse is designed based on the last criterion.

Procedure of Pre-inverse (Q, R);
Begin

```
    For k: =1 to n
     For j: =1 to n
         For w: =1 to n
             compute αw: = (rjw ∧ rjk) -   ∨ (rjw ∧ rji)
                                       j=1,i≠k (to n)
         End For;
         sort (αw ,βw)   || this procedure sorts the elements of the array αw
                            and saves them in βw in descending order ||
         For w:= 1 to n-1
           if β1 = βw+1
             qkj : = rjw ;
             print qkj;
           End For;
         End For;
        End For;
      End.
```

2.2 Determining Features for a Desired Degree of Emotion for Emotion Synthesis Applications

Consider a fuzzy production rule with antecedent describing fuzziness of facial features and consequent describing fuzziness of degree of emotion. One illustrative rule is given below for convenience.

Rule: If *end-to-end-lower-lip* is LARGE then *happiness* is HIGH.

Let R(*end-to-end-lower-lip*, *happiness*) be a implication relation. Given the membership functions (MFs) of *end-to-end-lower-lip* is LARGE and *happiness* is HIGH, we can construct R by standard implication functions [15]. Now suppose, we know the degree of *happiness* to be HIGH, we need to determine the membership of *end-to-end-lower-lip* to be LARGE. This is possible by the following two steps.

Step 1: Compute pre-inverse of R by the procedure mentioned above.
Step 2: Use the provided degree of membership of *happiness* to be HIGH, use the following operation to derive the membership of *end-to-end-lower-lip* to be LARGE. Formally,

$$\mu_{LARGE}\,(\textit{end-to-end-lower-lip}) = \mu_{HIGH}\,(\textit{happiness}) \circ R^{-1},$$

where R^{-1} is the pre-inverse of R.

3 Experiments

In this experiment, we have taken fourteen different faces of a person (see Fig. 1). Now, three steps are given to formulate desired facial features of the parson when degree of emotion is provided.

3.1 Membership Function Construction

The pixel value for *end-to-end-lower-lip* of these fourteen images are collected as [88 89 91 94 95 99 101 103 105 106 108 108 108 108] (see Fig. 2). Maximum (*Mx*), minimum (*Mn*) and median (*Med*) pixel value of *end-to-end-lower-lip* are *Mx* = 108, *Mn* = 88, and *Med* = 99. Now, the *end-to-end-lower-lip* is fuzzified into three fuzzy sets as HIGH, MEDIUM and LOW using the variable *end-to-end-lower-lip*, L_i = [88 90 92 94 96 98 100 102 104 106 108] in the mathematical expression given below.

$$\mu_{\text{HIGH}}(L_i) = Max\left(0.1, \frac{L_i - Med}{Mx - Med}\right) = Max\left(0.1, \frac{L_i - 99}{108 - 99}\right) \tag{1}$$

$$\left.\begin{aligned} \mu_{\text{MEDIUM}}(L_i) &= \left(\frac{L_i - Mn}{Med - Mn}\right) = \left(\frac{L_i - 88}{99 - 88}\right), \textit{for } L_i \le 99 \\ &= \left(\frac{Mx - L_i}{Mx - Med}\right) = \left(\frac{108 - L_i}{108 - 99}\right), \textit{for } L_i > 99 \end{aligned}\right\} \tag{2}$$

$$\mu_{\text{ML-Low}}(L_i) = Max\left(0.1, \frac{Med - L_i}{Med - Mn}\right) = Max\left(0.1, \frac{99 - L_i}{99 - 88}\right). \tag{3}$$

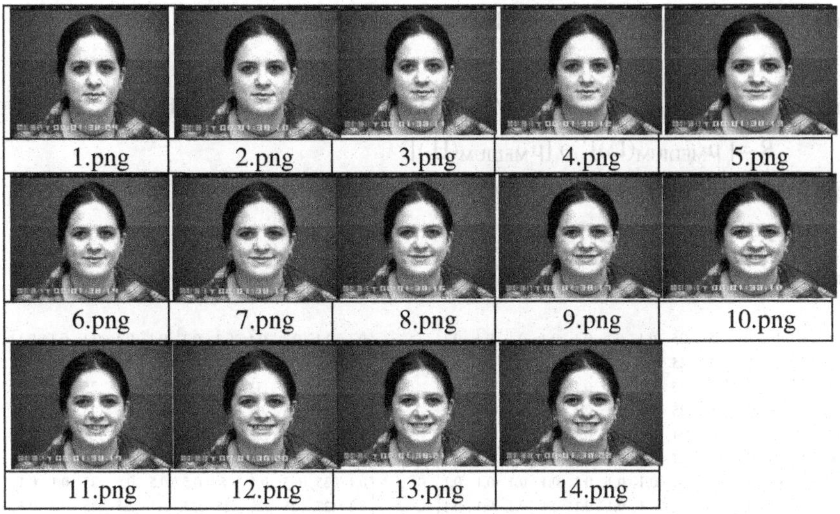

Fig. 1. Fourteen different faces of a person

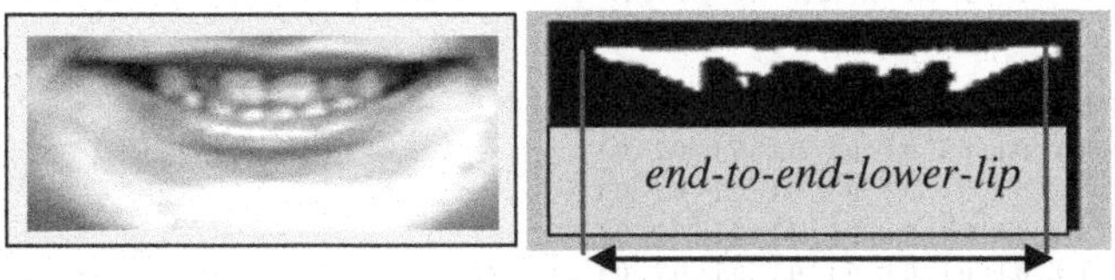

Fig. 2. End-to-end-lower-lip measurement from a cropped mouth portion

The graphical representation of membership function of *end-to-end-lower-lip* (LOW, MEDIUM, HIGH) and the percentage of *happiness* (divided equally) are given in Fig. 3.

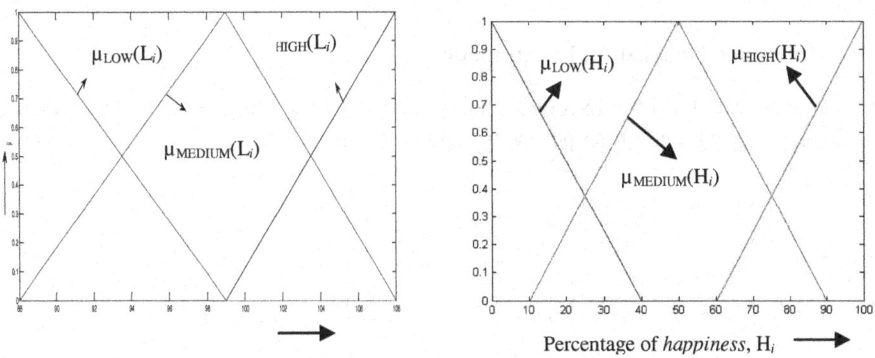

Fig. 3. Membership functions of end-to-en-lower-lip Length Low, Medium and High

3.2 Implication Relation Construction

Using fuzzy rules and membership function we first construct relational matrices (using Mamdani's implication). Fuzzy rules and corresponding relational matrices are

Rule 1: If *end-to-end-lower-lip* is HIGH then *happiness* is HIGH.

$$R_1=[\mu_{HIGH}(L_i)]^T \circ [\mu_{HIGH}(H_i)]$$

Rule 2: If *end-to-end-lower-lip* is MEDIUM then *happiness* is MEDIUM.

$$R_2=[\mu_{MEDIUM}(L_i)]^T \circ [\mu_{MEDIUM}(H_i)]$$

Rule 3: If *end-to-end-lower-lip* is LOW then *happiness* is LOW.

$$R_3=[\mu_{LOW}(L_i)]^T \circ [\mu_{LOW}(H_i)]$$

The relational matrices are given below.

$$R_1=\begin{bmatrix} 1 & .75 & .5 & .25 & 0.1 & 0.1 & 0.1 & 0.1 & 0.1 & 0.1 & 0.1 \\ 1 & .75 & .5 & .25 & 0.1 & 0.1 & 0.1 & 0.1 & 0.1 & 0.1 & 0.1 \\ .66 & .66 & .5 & .25 & 0.1 & 0.1 & 0.1 & 0.1 & 0.1 & 0.1 & 0.1 \\ .33 & .33 & .5 & .25 & 0.1 & 0.1 & 0.1 & 0.1 & 0.1 & 0.1 & 0.1 \\ 0.1 & 0.1 & 0.1 & 0.1 & 0.1 & 0.1 & 0.1 & 0.1 & 0.1 & 0.1 & 0.1 \\ 0.1 & 0.1 & 0.1 & 0.1 & 0.1 & 0.1 & 0.1 & 0.1 & 0.1 & 0.1 & 0.1 \\ 0.1 & 0.1 & 0.1 & 0.1 & 0.1 & 0.1 & 0.1 & 0.1 & 0.1 & 0.1 & 0.1 \\ 0.1 & 0.1 & 0.1 & 0.1 & 0.1 & 0.1 & 0.1 & 0.1 & 0.1 & 0.1 & 0.1 \\ 0.1 & 0.1 & 0.1 & 0.1 & 0.1 & 0.1 & 0.1 & 0.1 & 0.1 & 0.1 & 0.1 \\ 0.1 & 0.1 & 0.1 & 0.1 & 0.1 & 0.1 & 0.1 & 0.1 & 0.1 & 0.1 & 0.1 \\ 0.1 & 0.1 & 0.1 & 0.1 & 0.1 & 0.1 & 0.1 & 0.1 & 0.1 & 0.1 & 0.1 \end{bmatrix} \quad R_2=\begin{bmatrix} 0.1 & 0.1 & 0.1 & 0.1 & 0.1 & 0.1 & 0.1 & 0.1 & 0.1 & 0.1 & 0.1 \\ 0.1 & 0.1 & 0.1 & 0.1 & 0.1 & 0.1 & 0.1 & 0.1 & 0.1 & 0.1 & 0.1 \\ 0.1 & .25 & .33 & .33 & .33 & .33 & .33 & .33 & .25 & 0.1 & 0.1 \\ 0.1 & .25 & 0.5 & .66 & .66 & .66 & .66 & 0.5 & .25 & 0.1 & 0.1 \\ 0.1 & .25 & 0.5 & .75 & 1 & 1 & .75 & 0.5 & .25 & 0.1 & 0.1 \\ 0.1 & .25 & 0.5 & .75 & .75 & .75 & .75 & 0.5 & .25 & 0.1 & 0.1 \\ 0.1 & .25 & 0.5 & 0.5 & .5 & 0.5 & 0.5 & 0.5 & .25 & 0.1 & 0.1 \\ 0.1 & .25 & .25 & .25 & .25 & .25 & .25 & .25 & .25 & 0.1 & 0.1 \\ 0.1 & 0.1 & 0.1 & 0.1 & 0.1 & 0.1 & 0.1 & 0.1 & 0.1 & 0.1 & 0.1 \\ 0.1 & 0.1 & 0.1 & 0.1 & 0.1 & 0.1 & 0.1 & 0.1 & 0.1 & 0.1 & 0.1 \\ 0.1 & 0.1 & 0.1 & 0.1 & 0.1 & 0.1 & 0.1 & 0.1 & 0.1 & 0.1 & 0.1 \end{bmatrix}$$

$$\text{and } R_3=\begin{bmatrix} 0.1 & 0.1 & 0.1 & 0.1 & 0.1 & 0.1 & 0.1 & 0.1 & 0.1 & 0.1 & 0.1 \\ 0.1 & 0.1 & 0.1 & 0.1 & 0.1 & 0.1 & 0.1 & 0.1 & 0.1 & 0.1 & 0.1 \\ 0.1 & 0.1 & 0.1 & 0.1 & 0.1 & 0.1 & 0.1 & 0.1 & 0.1 & 0.1 & 0.1 \\ 0.1 & 0.1 & 0.1 & 0.1 & 0.1 & 0.1 & 0.1 & 0.1 & 0.1 & 0.1 & 0.1 \\ 0.1 & 0.1 & 0.1 & 0.1 & 0.1 & 0.1 & 0.1 & 0.1 & 0.1 & 0.1 & 0.1 \\ 0.1 & 0.1 & 0.1 & 0.1 & 0.1 & 0.1 & 0.1 & 0.1 & 0.1 & 0.1 & 0.1 \\ 0.1 & 0.1 & 0.1 & 0.1 & 0.1 & 0.1 & 0.1 & 0.1 & 0.1 & 0.1 & 0.1 \\ 0.1 & 0.1 & 0.1 & 0.1 & 0.1 & 0.1 & 0.1 & .25 & .25 & .25 & .25 \\ 0.1 & 0.1 & 0.1 & 0.1 & 0.1 & 0.1 & 0.1 & .25 & 0.5 & 0.5 & 0.5 \\ 0.1 & 0.1 & 0.1 & 0.1 & 0.1 & 0.1 & 0.1 & .25 & 0.5 & .75 & .75 \\ 0.1 & 0.1 & 0.1 & 0.1 & 0.1 & 0.1 & 0.1 & .25 & 0.5 & .75 & 1 \end{bmatrix}.$$

3.3 Inference Generation by Abduction

Following above algorithm (Section 2.1), we construct the pre-inverse matrices of R_1, R_2 and R_3 as Q_1, Q_2 and Q_3 respectively. These are given below.

$$Q_1=\begin{bmatrix} 1 & .75 & .1 & .1 & .1 & .1 & .1 & .1 & .1 & .1 & .1 \\ .1 & .1 & .5 & .1 & .1 & .1 & .1 & .1 & .1 & .1 & .1 \\ .1 & .1 & .1 & .25 & .1 & .1 & .1 & .1 & .1 & .1 & .1 \\ ..1 & .1 & .1 & .1 & .1 & .1 & .1 & .1 & .1 & .1 & .1 \\ .1 & .1 & .1 & .1 & .1 & .1 & .1 & .1 & .1 & .1 & .1 \\ .1 & .1 & .1 & .1 & .1 & .1 & .1 & .1 & .1 & .1 & .1 \\ .1 & .1 & .1 & .1 & .1 & .1 & .1 & .1 & .1 & .1 & .1 \\ .1 & .1 & .1 & .1 & .1 & .1 & .1 & .1 & .1 & .1 & .1 \\ .1 & .1 & .1 & .1 & .1 & .1 & .1 & .1 & .1 & .1 & .1 \\ .1 & .1 & .1 & .1 & .1 & .1 & .1 & .1 & .1 & .1 & .1 \\ .1 & .1 & .1 & .1 & .1 & .1 & .1 & .1 & .1 & .1 & .1 \end{bmatrix}, Q_2=\begin{bmatrix} 1 & .1 & .1 & .1 & .1 & .1 & .1 & .1 & .1 & .1 & .1 \\ .1 & .1 & .1 & .1 & .1 & .1 & .1 & .1 & .1 & .1 & .1 \\ .1 & .1 & .25 & .33 & .33 & .1 & .33 & .33 & .33 & .1 & .1 \\ ..1 & .1 & .1 & .1 & .1 & .1 & .1 & .1 & .1 & .1 & .1 \\ .1 & .1 & .1 & .1 & .1 & .1 & .1 & .1 & .1 & .1 & .1 \\ .1 & .1 & .1 & .1 & .1 & .1 & .1 & .1 & .1 & .1 & .1 \\ .1 & .1 & .1 & .1 & .1 & .1 & .1 & .1 & .1 & .1 & .1 \\ .1 & .1 & .1 & .1 & .1 & .1 & .1 & .1 & .1 & .1 & .1 \\ .1 & .1 & .1 & .1 & .1 & .1 & .1 & .1 & .1 & .1 & .1 \\ .1 & .1 & .1 & .1 & .1 & .1 & .1 & .1 & .1 & .1 & .1 \\ .1 & .1 & .1 & .1 & .1 & .1 & .1 & .1 & .1 & .1 & .1 \end{bmatrix}, Q_3=\begin{bmatrix} .1 & .1 & .1 & .1 & .1 & .1 & .1 & .1 & .1 & .1 & .1 \\ .1 & .1 & .1 & .1 & .1 & .1 & .1 & .1 & .1 & .1 & .1 \\ .1 & .1 & .1 & .1 & .1 & .1 & .1 & .1 & .1 & .1 & .1 \\ ..1 & .1 & .1 & .1 & .1 & .1 & .1 & .1 & .1 & .1 & .1 \\ .1 & .1 & .1 & .1 & .1 & .1 & .1 & .1 & .1 & .1 & .1 \\ .1 & .1 & .1 & .1 & .1 & .1 & .1 & .1 & .1 & .1 & .1 \\ .1 & .1 & .1 & .1 & .1 & .1 & .1 & .1 & .1 & .1 & .1 \\ .1 & .1 & .1 & .1 & .1 & .1 & .1 & .25 & .1 & .1 & .1 \\ .1 & .1 & .1 & .1 & .1 & .1 & .1 & .1 & .5 & .1 & .1 \\ .1 & .1 & .1 & .1 & .1 & .1 & .1 & .1 & .1 & .75 & .1 \\ .1 & .1 & .1 & .1 & .1 & .1 & .1 & .1 & .1 & .1 & 1 \end{bmatrix}.$$

Now, let the observed antecedent for 80 % *happiness* (μ_{HIGH}) is given as B′=[0.1 0.1 0.1 0.1 0.1 0.1 0.25 0.5 0.75 1]. Therefore, the fuzzy inference is

$$A' = (B' \circ Q_1) \cup (B' \circ Q_2) \cup (B' \circ Q_3) = [0.1\ 0.1\ 0.1\ 0.1\ 0.1\ 0.1\ 0.1\ 0.25\ 0.5\ 0.75\ 1].$$

3.4 Results

The centroid value of A′

$$c = \frac{0.1\times 88 + 0.1\times 90 + 0.1\times 92 + 0.1\times 94 + 0.1\times 96 + 0.1\times 98 + 0.1\times 100 + 0.25\times 102 + 0.5\times 104 + 0.75\times 106 + 1\times 108}{0.1+0.1+0.1+0.1+0.1+0.1+0.1+0.25+0.5+0.75+1}$$
$$= 103.375$$

Thus, for 80% happiness (degree of emotion) we obtain pixel value of *end-to-end-lower-lip* (facial feature) by abductive reasoning. Now, we can check our database for the image which is similar to the pixel value 103.375 (or very near to this pixel value) of *end-to-end-lower-lip* and show that image as the output image for 80% Happy. The MATLAB code for this application has helped us to execute this program on the above images to give the required facial expression.

9.png

We have done the same experiment on five different faces but for page restriction it is not given here.

4 Conclusions

Synthesis of emotion is an important issue in the next generation human-computer interface design. The paper introduced a novel approach to select the right facial expression representative of the desired degree of a specific emotion. Principle of fuzzy abduction seems to provide an important mechanism to handle the present problem. The selected facial expression describing the facial image truly represents the desired emotion of the subject. Complexity of the algorithms is determined primarily by the computation of the inverse fuzzy relation.

References

1. Arnould, T., et al.: Backward-chaining with fuzzy “if... then...” rules. In: Proc. 2nd IEEE Inter. Conf. Fuzzy Systems, pp. 548–553 (1993)
2. Arnould, T., Tano, S.: Interval-valued fuzzy backward reasoning. IEEE Trans. Fuzzy Systems 3(4), 425–437 (1995)
3. El Ayeb, B., et al.: A New Diagnosis Approach by Deduction and Abduction. In: Proc. Int’l Workshop Expert Systems in Eng. (1990)

4. Bhatnagar, R., Kanal, L.N.: Structural and Probabilistic Knowledge for Abductive Reasoning. IEEE Trans. on Pattern Analysis and machine Intelligence 15(3), 233–245 (1993)
5. Bylander, T., et al.: The computational complexity of abduction. Artificial Intelligence 49, 25–60 (1991)
6. de Campos, L.M., Gámez, J.A., Moral, S.: Partial Abductive Inference in Bayesian Belief Networks—An Evolutionary Computation Approach by Using Problem-Specific Genetic Operators. IEEE Transactions on Evolutionary Computation 6(2) (April 2002)
7. Chakraborty, S., Konar, A., Jain, L.C.: An efficient algorithm to computing Max-Min inverse fuzzy relation for Abductive reasoning. IEEE Trans. on SMC-A, 158–169 (January 2010)
8. Charniak, E., Shimony, S.E.: Probalilistic Semantics for Cost Based Abduction. In: Proc., AAAI 1990, pp. 106–111 (1990)
9. Hobbs, J.R.: An Integrated Abductive Framework for Discourse Interpretation. In: Proceedings of the Spring Symposium on Abduction, Stanford, California (March 1990)
10. Pedrycz, W.: Inverse Problem in Fuzzy Relational Equations. Fuzzy Sets and Systems 36, 277–291 (1990)
11. Peng, Y., Reggia, J.A.: Abductive Inference Models for Diagnostic Problem-Solving. Springer-Verlag New York Inc. (1990)
12. Saha, P., Konar, A.: A heuristic algorithm for computing the max-min inverse fuzzy relation. Int. J. of Approximate Reasoning 30, 131–147 (2002)
13. Yamada, K., Mukaidono, M.: Fuzzy Abduction Based on Lukasiewicz Infinite-valued Logic and Its Approximate Solutions. In: FUZZ IEEE/IFES, pp. 343–350 (March 1995)
14. Petrantonakis, P.C., Hadjileontiadis, L.J.: Emotion Recognition from EEG Using Higher Order Crossings. IEEE Transactions on Information Technology in Biomedicine 14(2) (March 2010)
15. Klir, G.J., Yuan, B.: Approximate reasoning: Fuzzy sets and fuzzy Logic (2002)
16. Chakraborty, A., Konar, A., Pal, N.R., Jain, L.C.: Extending the Contraposition Property of Propositional Logic for Fuzzy Abduction. IEEE Transaction on Fuzzy Systems 21, 719–734 (2013)
17. Konar, A.: Artificial Intelligence and Soft Computing. CRC Press LLC (2000)

A Novel Semantic Similarity Based Technique for Computer Assisted Automatic Evaluation of Textual Answers

Udit Kr. Chakraborty[1], Samir Roy[2], and Sankhayan Choudhury[3]

[1] Department of Computer Science & Engineering
Sikkim Manipal Institute of Technology, Sikkim
[2] Department of Computer Science & Engineering
National Institute of Technical Teachers' Training & Research, Kolkata
[3] Department of Computer Science & Engineering
University of Calcutta, Kolkata
{udit.kc,samir.cst,sankhayan}@gmail.com

Abstract. We propose in this paper a unique approach for the automatic evaluation of free text answers. A question answering module has been developed for the evaluation of free text responses provided by the learner. The module is capable of automatically evaluating the free text response of the learner S_A to a given question Q and its model text based answer M_A on a scale [0, 1] with respect to the M_A. This approach takes into consideration not only the important key-words but also stop words and the positional expressions present in the learners' response. Here positional expression implies the pre-expression and post-expression appearing before and after a keyword in the learners' response. The results obtained on using this approach are promising enough for investing into future efforts.

Keywords: evaluation, learners' response, evaluation, keywords, pre-expression, post-expression.

1 Introduction

Evaluation is an important and critical part of the learning process. The evaluation of learners' response decides not only the amount of knowledge gathered by the learner but also contributes towards refinement of the learning process. The task requires the evaluator to have the required knowledge and also to be impartial, benevolent and intelligent. However, all of these qualities may not always be present in human evaluators, who are also prone to fatigue. Reasons similar to these and the requirement of performing the task of evaluation on a larger scale necessitates the implementation of auto-mated systems for evaluation of the learners' response. Such mechanized processes would not only be free from fatigue and partiality but also be able to evaluate across geographical distances if implemented in e-Learning systems whose importance, popularity and penetration is on the rise.

M.K. Kundu et al. (eds.), *Advanced Computing, Networking and Informatics - Volume 1*, Smart Innovation, Systems and Technologies 27,
DOI: 10.1007/978-3-319-07353-8_46,

However, the task of machine evaluation is easier said than done for reasons of complexity in natural languages and the lack of our abilities in understanding them. These reasons have given rise to the popularity of other types of assessment techniques namely multiple choice questions, order matching, fill in the blanks etc., which in spite of their own roles are not fully reliable for evaluation of fulfillment of learning outcomes. Whether the efficacy of questions requiring free text responses are more than the other types is debatable, but it is beyond contention that free text responses test the learners' ability to explain, deduce and logically derive apart from other parameters which are not brought out by the other types of question answering systems.

The problem with the evaluation of free text responses lie in the variation in answering and evaluation. Since the learners' response is presented in his unique style and words, the same answer can be written in different ways due to the richness in form and structure of natural languages. Another problem is in the score assigned to the answer since the score assigned can vary from one individual to another.

The computational challenge imposed by this task is immense, because, to determine the degree of correctness of the response the meaning of the sentence has to be extracted. The semantic similarity which is a means of finding the relation that exists between the meaning of the words and meaning of sentences also needs to be considered.

The work presented in this paper proposes an automated system that evaluates the free text responses of the learner. The approach is in deviation from the currently existing techniques in a few areas and considers not only important keywords but also the words before and after them. Unlike n-grams technique, the number of words before and after a keyword is not fixed and varies depending on the occurrence of the next keyword. The current work is limited to single sentence responses only.

2 Previous Work

Interest in question answering has shifted from factoid questions to descriptive questions [1], for reasons already discussed. A number of systems using different techniques have been developed for the evaluation of the free text response of learners namely: Intelligent Essay Assessor [2] developed by Landauer, Foltz and Laham based on Latent Semantic Analysis (LSA) [3] is used for the evaluation of learner essays. Apex a web based learning application developed by Dessus *et al.* [4] is also based on LSA technique and is used for the evaluation of learners' responses. Atenea developed by Perez *et al.* [5] is based on the Bi-Lingual Evaluation Understudy (BLUE) [6] Algorithm and is capable of evaluating the free text responses of learners in both English and Spanish irrespective of the language the learner wishes to answer. C-Rater [7] developed by Education Testing Service (ETS) uses Natural Language Processing (NLP) techniques for the evaluation of short responses provided by the learners. Auto-mark developed by Mitchell *et al.* [8] is a software system which uses NLP techniques and is capable of evaluating free text responses provided to descriptive questions.

However, a large scale acceptance of these systems, have not yet taken place and a complete replacement of the human evaluator is still a long distance away.

3 Proposed Methodology

The Answer Evaluation (AE) Module consists of two parts, one for the teacher and other for the learner. The role of the teacher would be to fix the model answer for a given question and fix the parameters of evaluation. This is similar to preparing a solution scheme of evaluation by the teacher which is referred to while actually evaluating the responses written by learners. The learners merely use the AE module to type in there textual responses to the questions presented.

The task chalked out for the AE module can be stated as:

Given a question Q, its model text based answer M_A and a learner response S_A, the AE module should be able to evaluate S_A on a scale of [0, 1] with respect to M_A.

- If the S_A is completely invalid or contradictory to M_A, then it is an incorrect response and a value 0 is returned.
- If the S_A is exactly same as the M_A or is a paraphrase of the M_A, or is a complete semantic match, then it is a correct response and a value 1 is returned.
- If the S_A is non-contradictory and is a partial semantic match for M_A, then the response is partially correct and a value greater than 0 and less than 1 is returned depending upon the match.

This work is built upon the understanding that, an answer to a question is a collection of keywords and their associated pre and post expressions which augment sense to the keywords in the context of the question and also establishes links between them. Unlike the nugget approach, which considers only keywords as the building blocks, the current approach considers the preceding and following sets of words as well. The choice of pre and post-expression is not based on the popular n-gram technique but the occurrence of the next keyword. It is also worth mention that unlike other natural language processing approaches, we do not remove the stop words from a response as we consider these to be important information carriers. Fig.1. presents the idea of how the answers are perceived by the model.

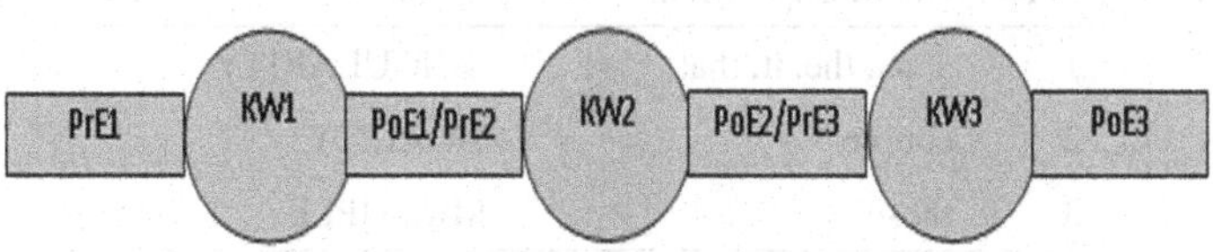

Fig. 1. Schematic diagram of the models perception of an answer

Since the system deals with natural language answers, we do not decide or attach any weight to the order in which the key-words appear while evaluating the responses. Also, it may be possible that a particular part of the response acts as post-expression for a keyword and pre-expression for the next keyword, in which case it will be considered twice depending upon the solution scheme presented by the teacher.

Each pre-expression and post-expression is again broken up into four parts, namely logic, certainty, count and part-of, which are the expected types of senses that these expressions attach to the keywords. There is however, no fixed order in which the words belonging to any of these categories would appear and it is also possible that they do not appear at all. Lists of words have been pre-pared to be belonging to each of these four categories and they are as shown in Table 1, 2, 3 and 4 respectively.

Table 1. Logic Expressions and their associated logic

S.No.	Logic Expressions	Logic
1	and	Conjunction
2	or	Disjunction
3	either-or	Exclusive Disjunction
4	only if	Implication
5	if and only if	Equivalence
6	just in case	Bi-conditional
7	not both	Alternative Denial
8	neither – nor	Joint Denial
9	not it is false that it is not the case that	Negation
10	is	Equality

Table 2. Certainty Expressions and their classes

S.No.	Certainty Expressions	Certainty Classes
1	usually, likely, unlikely	Class A
2	certainly, most certainly, definitely	Class B
3	most probably, most likely	Class C
4	probably	Class D

Table 3. Count Expressions and their meanings

S. No.	Count Expressions	Count
1	a, an, the, it, that	SINGULARITY
2	couple	DUALITY
3	those	MULTIPLE

Table 4. Part-of Expressions

S. No.	Part of Expressions
1	belong
2	into
3	for
4	like

3.1 Preprocessing Tasks

Prior to the evaluation process, the teacher has to perform the following pre-processing tasks to prepare the system to be able to evaluate the learners' response. This comprises of typing in the model response, identifying the model phrase from the complete response, identifying the keywords, the post and pre expressions for each key word and the categorization of the words in post and pre expressions into their rightful sense conveying types.

The model answer is the answer prepared by the human evaluator and presents the benchmark against which the learners' response would be evaluated. This answer consists of a central part, which we call the model phrase, M_P, and represents the core of the answer.

To ensure that the reproduction of your illustrations is of a reasonable quality, we advise against the use of shading. The contrast should be as pronounced as possible.

If screenshots are necessary, please make sure that you are happy with the print quality before you send the files.

The steps are listed below in the order of their occurrence:

Step 1: The model answer M_A is created.
Step 2: A model phrase M_P is identified within the model answer M_A.
Step 3: Keywords are identified and listed.
Step 4: All KW are marked with their associated part of speech.
Step 5: For each KW:
  Step 5.1: Synonyms having same POS usage are listed.
Step 6: Weights are associated to each KW depending on importance and relevance. Sum of all weights to be equal to 1.
Step 7: For every KW the pre-expression and the post-expression are extracted and words/phrases put in their respective sense brackets, i.e., logic, certainty, count or part-of.

3.2 Steps for the Evaluation of the Learner Response

Once the pre-processing is done, the system is ready to read in the learners' response and evaluate it. The aim is to evaluate the response and return a score in the range of 0 to 1. The algorithm for performing the same is as follows:

Algorithm Eval_Response:
Evaluates the learners' text based response.

Variables:
S_A (Learner's Response), M_A(Model Answer), M_P (Model Phrase), KW (Keyword), PrE(Pre-expression), PoE (Post-expression), KW_S(Score from a keyword), KW_Weight (Weight of a keyword), PrE_S(Score of a pre-expression), PoE_S(Score of a post-expression), Marks(Total marks).

Step 1: Set Marks = 0
Step 2: String Compare (S_A,M_A)

Step 3: If (S_A=M_A), Marks = 1

Step 4: ElseIf (Search (S_A, M_P) = Success), Marks = 0.85 //search the learners'response for the model phrase.

Step 5: Else For every KW Search (KW,S_A) //search for the keyword in S_A

Step 5.1: If KW found: Evaluate (PrE, PoE) //Evaluate both pre and post-expressions

KW_S = PrE_S*KW_Weight*PoE_S

//Score of keyword is calculated

Step 5.2: Else Search every synonym of current keyword in S_A.If synonym match found. GOTO Step 5.1, Else KW_S = 0

Step 5.3: Marks = Marks + KW_S

Step6: Stop

While evaluating the pre-expression (and the post-expression), we search whether the words listed in the solution scheme appear in the pre-expression. If they do, then they are put in the sense bracket that they are expected to belong. Otherwise, we check whether a valid substitution of the word appears in the pre-expression of the learners' response. The valid substitutions of logic, certainty, count or part-of word is maintained in a list along with their weights. For example, if the pre-expression in the solution scheme shows that the count expression is 'a' and the learners' response contains no 'a' in the pre-expression but a 'the', then the count expression for that particular keyword is taken as 'the' in the learners' response and is given a score of 0.5 instead of 1 (had it been 'a'), because 'a' signifies 'one in many', while the word 'the' signifies 'the only one'.

Each sub-field, namely logic, certainty, count and part-of, in the solution scheme may not be filled. Under such circumstances, we do not search for those expressions in the learners' response and such non-active fields do not contribute to the score. The exception to this rule is the logic sub-field, which contributes 1 to the score if, either the logic is inactive or the sub-field is active and the word is found in the learners' response. If the logic sub-field is active in the scheme, but the word is not present in the learners' response, the contribution becomes 0. If the field is active but the word is substituted, then the score changes appropriately.

Finally, every pre-expression and post expression score (PrE_S/PoE_S) is evaluated according to the expression given in Eq.1

$$\text{PrE\_S} = \frac{Logic*(Certainty+Count+Part\_of}{No.of\ active\ sub\_fields} \tag{1}$$

4 Experiments and Results

The methodology discussed in the previous sections was employed to test the correctness in comparison to a human evaluator. While performing the tests, we considered single sentences responses only. The human evaluators were kept unaware of the method to be employed by the automated system; however, since the automatic evaluation would return fractional values between 0 and 1, human evaluators were asked to score up to 2 decimal places. Two such tests and their details are presented here along with some findings on the results.

4.1 Set 1

Question: What is an annotated parse tree?
Model Answer: A parse tree showing the attribute value at each node is called an annotated parse tree.
Model Phrase: A parse tree showing attribute value at each node.

The question was presented to 39 learners', and the responses evaluated by the automated system and also by human evaluators, based on the model response specified and shown in Table 5. The correlation co-efficient between the two evaluators was calculated and found to be equal to 0.6324, with 30% cases having difference not more than 10%.

4.2 Set 2

Question: What is the advantage of representing data in AVL search tree than to represent data in binary search tree?
Model Answer: In AVL search tree, the time required for performing operations like searching or traversing is short. e.g. worst case complexity for searching in BST (O(n)) worst case complexity for searching in AVL search tree (log(n)).
Model Phrase: In AVL search tree, the time required for performing operations like searching or traversing is short.

This question was asked to a group of 50 learners' and the responses similarly evaluated with model response as in Table 6. The results found returned a correlation co-efficient of 0.6919, with 68% instances of not more than 10% difference in the marks allotted by the system and the human evaluator.

It is observed that the performance of the system tends to improve on increasing the volume of the response. However, the in-crease in the volume of the response in this case also meant an increase in the number of keywords. As a matter of fact, both tests were conducted on keyword heavy samples. Whether the performance would change on having heavier pre and post expressions is yet to be explored.

Table 5. Scheme for the evaluation of Set 1

Pre-Expression				KW	Wt.	Post-Expression			
L	C	O	P			L	C	O	P
-	-	A	-	parse	0.2	-	-	-	-
-	-	-	-	tree	0.05	-	-	-	-
-	-	-	-	showing	0.2	-	-	the	-
-	-	-	-	attribute	0.3	-	-	-	-
-	-	-	-	value	0.05	-	-	-	-
-	-	each	-	node	0.2	-	-	-	-

Legends: L: Logic; C: Certainty; O: Count; P: Part of

Table 6. Scheme for the evaluation of Set 2

Pre-Expression				KW	Wt. of KW	Post-Expression			
L	C	O	P			L	C	O	P
-	-	the	-	time	0.1	-	-	-	-
-	-	-	-	Required	0.1	-	-	-	-
-	-	-	for	Performing	0.05	-	-	-	-
-	-	-	-	Operation	0.05	-	-	-	like
-	-	-	-	Searching	0.1	or	-	-	-
-	-	-	-	Traversing	0.1	-	-	-	-
is	-	-	-	short	0.2	-	-	-	-
-	-	-	for	e.g.	0.01	-	-	-	-
-	-	-	-	worst	0.01	-	-	-	-
-	-	-	-	case	0.01	-	-	-	-
-	-	-	-	Complexity	0.015	-	-	-	-
-	-	-	for	Searching	0.01	-	-	-	-
-	-	-	in	BST	0.03	-	-	-	-
-	-	-	-	(O(n))	0.07	-	-	-	-
-	-	-	-	worst	0.01	-	-	-	-
-	-	-	-	case	0.01	-	-	-	-
-	-	-	-	Complexity	0.015	-	-	-	-
-	-	-	for	Searching	0.01	-	-	-	-
-	-	-	in	AVL	0.02	-	-	-	-
-	-	-	-	search	0.005	-	-	-	-
-	-	-	-	tree	0.005	-	-	-	-
-	-	-	-	(log(n))	0.7	-	-	-	-

5 Conclusion

The work is aimed at developing a novel system that is capable of evaluating the free text responses of learners'. Unlike the widely followed bag-of-words approach, the work mentioned here takes positional expressions, keyword and even stop words into consideration during evaluation. The proposed method generates a fuzzy score taking into consideration all mentioned criteria. The score generated by the system does not deviate too much from the human evaluator and further tests may produce still better results.

References

1. Lin, J., Fushman, D.D.: Automatically Evaluating Answers to Definition Questions. In: Proceedings of Human Language Technology Conference and Conference on Empirical Methods in Natural Language Processing (HLT/EMNLP), pp. 931–938 (2005)

2. Foltz, P.W., Laham, D., Landauer, T.K.: The Intelligent Essay Assessor: Applications to Educational Technology. Interactive Multimedia Education Journal of Computer Enhanced Learning 1(2) (1991)
3. Landauer, T.K., Foltz, P.W., Laham, D.: An Introduction to Latent Semantic Analysis. Discourse Processes 25(2&3), 259–284 (1998)
4. Dessus, P., Lemaire, B., Vernier, A.: Free-text Assessment in a Virtual Campus. In: Proceedings of the 3rd International Conference on Human-Learning Systems, pp. 2–14 (2000)
5. Perez, D., Alfonseca, E.: Adapting the Automatic Assessment of Free-Text Answers to the Learners. In: Proceedings of the 9th International Computer-Assisted Assessment (CAA) Conference (2005)
6. Papineni, K., Roukos, S., Ward, T., Zhu, W.: BLEU: a Method for Automatic Evaluation of Machine Translation. In: Proceedings of the 40th Annual Meeting on Association for Computational Linguistics, pp. 311–318 (2002)
7. Leacock, C., Chodorow, M.: C-rater: Automatic Content Scoring for Short Constructed Responses. In: Proceedings of the 22nd International FlAIRS Conference, pp. 290–295 (2009)
8. Mitchell, T., Russell, T., Broomhead, P., Aldridge, N.: Towards Robust Computerized Marking of Free-Text Responses. In: Proceedings of 6th International Computer Aided Assessment Conference, Loughborough (2002)

Representative Based Document Clustering

Arko Banerjee[1] and Arun K. Pujari[2]

[1] College of Engineering and Management, Kolaghat, WB, India
arko.banerjee@gmail.com
[2] University of Hyderabad, Andhra Pradesh, India
arun.k.pujari@gmail.com

Abstract. In this paper we propose a novel approach to document clustering by introducing a representative-based document similarity model that treats a document as an ordered sequence of words and partitions it into chunks for gaining valuable proximity information between words. Chunks are subsequences in a document that have low internal entropy and high boundary entropy. A chunk can be a phrase, a word or a part of word. We implement a linear time unsupervised algorithm that segments sequence of words into chunks. Chunks that occur frequently are considered as representatives of the document set. The representative based document similarity model, containing a term-document matrix with respect to the representatives, is a compact representation of the vector space model that improves quality of document clustering over traditional methods.

Keywords: document clustering, sequence segmentation, word segmentation, entropy.

1 Introduction

Document clustering is an unsupervised document organization method that put documents into different groups called clusters, where the documents in each cluster share some common properties according to a defined similarity measure. In most document clustering models similarity between documents is measured on basis of matching with single words rather than matching with phrases. The motivation of this paper is to bring the effectiveness of phrase based matching in document clustering. Work related to phrase based document clustering that has been reported in literature is limited. Zamir *et al.* [3][4] proposed an incremental linear time algorithm called Suffix Tree Clustering (STC), which creates clusters based on phrases shared between documents. They claim to achieve $nlog(n)$ performance and produce high quality clusters. Hammouda et al [5] proposed a document index model which implements a Document Index Graph that allows incremental construction of a phrase-based index of the document set and uses an incremental document clustering algorithm to cluster the document set.

In this paper we introduce a representative based document clustering model. The model involves three main phases: sequence segmentation, document representation, and clustering. Let D be a set of documents containing documents

M.K. Kundu et al. (eds.), *Advanced Computing, Networking and Informatics - Volume 1,* Smart Innovation, Systems and Technologies 27,
DOI: 10.1007/978-3-319-07353-8_47,

$d_1, d_2, ..., d_r$. Sequence segmentation begins with converting each document d_i into an ordered sequence of words say s_i by removing stop-words and spaces between words. Let $S \leftarrow \bigcup_{i=1}^{r} s_i$ and $S =< e_1, e_2, .., e_N >$, where each e_i is an alphabet. Then parsing of the sequence S is performed using the entropy and frequency measures to produce a set of chunks. The first phase is an essential pre-processing method to gain valuable proximity information between words. The document representation phase begins with identifying chunks that occur frequently and are selected as representative chunks.The document set is then represented as a term-document matrix where each document is represented by a feature vector which contains metric scores such as binary score (presence or absence of a term in the document), TF (i.e., within-document term frequency) or TF.IDF with respect to the selected representatives. This phase reduces the high dimensionality of the feature space, which in turn improves the clustering efficiency and performance. In the final phase the target documents $d_1, d_2, ..., d_r$ are grouped into distinct clusters by applying clustering algorithms on the term-document matrix.To demonstrate the effectiveness of the model, we have run our method on several datasets and found promising results.

The rest of this paper is organized as follows. Section 2 describes boundary entropy and frequency as sequence segmentation measures. Section 3 explains our proposed unsupervised sequence segmentation algorithm that we implemented in the first phase of our model. Section 4 introduces the concept of representative based document clustering method with a toy example which is implemented in the second and third phases. Section 5 provides the detailed experimental evaluation of our method with other existing algorithms. Finally, some concluding remarks and directions of future research are provided in Section 6.

2 Boundary Entropy and Frequency as Segmentation Measures

A successful sequence segmentation algorithm seeks to maximize the unpredictability between subsequences or chunks. To do that, alphabet that succeeds and alphabet that precedes a chunk, should have their unpredictability maximized with respect to the chunk. The boundary entropy of a chunk is a measure that expresses the magnitude of unpredictability at its end boundary and hence we measure the boundary entropies of each chunk and its reverse chunk to predict both boundaries.

To find boundary entropy of all the chunks and reverse chunks in S of length less than equal to n, an ngram TRIE of depth $n+1$ is generated by sliding a window along the sequence S. Let $w^1_{i,j} =< e_i, e_{i+1}, ..., e_j >$ represents a chunk of length n starting and ending at i^{th} and j^{th} position in S, respectively, where $1 \leq i < j \leq N$ and $j-i \leq n-1$. Also let $w^2_{i,j} =< e_j, e_{j-1}, ..., e_i >$ represents the corresponding reverse chunk of the said chunk $w^1_{i,j}$. A chunk or a reverse chunk in S is represented by a node of the TRIE consisting of two frequency fields. For example, in Fig. 1 a TRIE with depth 2 is generated using the sequence $\{e\ a\ b\ c\ a\ b\ d\}$. Every chunk and the reverse chunk of length 2 or less in the sequence is represented by a node in the tree. The right and left frequency fields represent frequencies of the chunk and the reverse chunk, respectively. For example, the chunk $\{a\ b\}$ and the reverse chunk $\{b\ a\}$ occurs twice. The reverse chunk $\{a\ c\}$ occurs once but the chunk $\{a\ c\}$ does not occur. Let a node n_k represents the frequencies of a chunk and a reverse chunk in S by f_k^1 and f_k^2, respectively. If n_k has m children, then they would be denoted by $n_{k,1}, ..., n_{k,m}$ having frequencies $f_{k,1}^t, ..., f_{ki,m}^t$, respectively, where $t = 1, 2$.

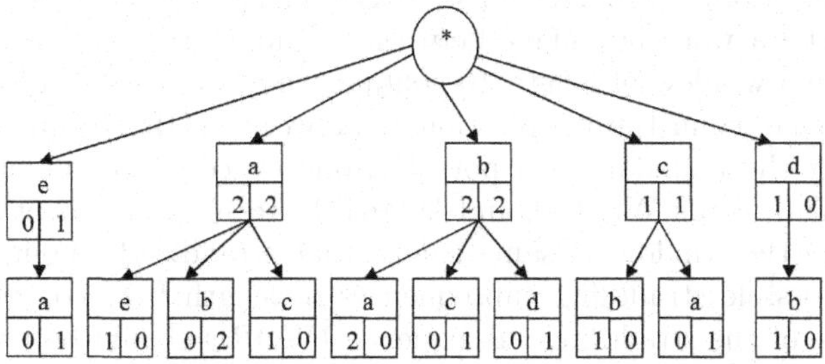

Fig. 1. Two way TRIE structure of the sequence {e a b c a b d}

The conditional probability of an alphabet succeeding/preceding a chunk is measured by the frequency of the alphabet succeeding/preceding the chunk divided by the frequency of the chunk, which we denote by $Pr(n_{kh}^t) = f_{kh}^t / f_k^t, t = 1, 2$. The boundary entropy of the chunk/reverse chunk is then the summation of entropy of the conditional probabilities of all such succeeding/preceding alphabets, which we denote by $H(n_k^t) = -\sum_{h=1}^{m} Pr(n_{kh}^t) \log Pr(n_{kh}^t)$, $t = 1, 2$. For example, the node **a** as a chunk in Fig. 1 has low entropy equal to 0 because it has only one child **b**. Hence there is a chance that $\{a\ b\}$ forms a chunk. Whereas $\{b\}$ as a chunk with high entropy equal to 1.0 has a high chance of being the end of a chunk. Therefore we expect the segmented sequence as $\{\ e\ a\ b*\ c\ a\ b*\ d\ \}$, where * denotes the end of a chunk. Again $\{b\}$ as a reverse chunk has an entropy equal to 0 hence high chance of$\{b\ a\}$ to form a reverse chunk. Whereas $\{a\}$ as a reverse chunk has an entropy equal to 1.0 hence a high chance of being the end of a reverse chunk. Therefore after combining forward and backward segmentation the expected segmented sequence should be $\{e\ \#a\ b{*}c\#a\ b{*}d\}$, where # denotes the end of a reverse segment.

By sliding a window of length n along a sequence, where n varies from 2 (chunks with one alphabet not considered) to a maximum window size, say W(< depth of the TRIE), each position in the sequence receives separate $(W - 1)$ boundary entropy values for chunks and for reverse chunks. We maintain a final boundary score at each position by adding all $(W - 1)$ boundary entropy values separtely for chunks and for reverse chunks and we call them the *forward entropy score* and *backward entropy score*, respectively. Let the *forward entropy score* and *backward entropy score* are denoted by H_j^1 and H_j^2 for a position j in S, respectively. Then $H_j^t \leftarrow \sum_{1 \le i, 0 < j-i \le (W-1)} H(Find\_Trie\_Node(w_{i,j}^t)), t = 1, 2$

and $1 < j \leq N$. Here the $Find\_Trie\_Node$ function returns the node of the TRIE that represents $w^t_{i,j}$. We will utilize the said scores to predict word boundaries in our segmenation algorithm described in section 4. In the following we derive the frequency information of chunks in the sequence to perform segmentation.

We make an effort to determine the boundary between two consecutive chunks by comparing frequencies of the chunks with subsequences that straddle the common boundary. We explain the situation using the following example. Fig. 2 shows a sequence of ten alphabets W O R D 1 W O R D 2 that represents two consecutive words WORD1 and WORD2. The goal is to determine whether there should be a word boundary between "1" and "W". Let us assume that at an instance a window of length 10 contains the words and it considers a hypothetical boundary in the middle that is between WORD1 and WORD2. To check whether the boundary is a potential one we count the number of times the two chunks (words WORD1 and WORD2) are more frequent than subsequences inside the window of same length that straddle the boundary. Fig. 2 shows four possible straddling subsequences of length five. For example, frequency of one of the straddling subsequence 1WORD should be low compared to WORD1 or WORD2, since WORD2 has less chance of occurring just after WORD1. We maintain a score that is incremented by one only if WORD1 or WORD2 are more frequent than that of 1WORD. So, if both possible chunks are higher in frequency than a straddling subsequence, the score to the hypothetical boundary location is incremented twice. In our algorithm we consider a window of length $2n$, where n varies from 2 to the maximum length W. A window of length $2n$ has its hypothetical boundary separating alphabets at n^{th} and $(n+1)^{th}$ positions, hence number of straddling subsequences across the boundary would be $(n-1)$. Therefore, by comparing all $(n-1)$ straddling subsequences with both chunks inside the right and left windows the boundary location receives total $2(n-1)$ scores.To normalize the scores of different window length we average the scores by dividing it with $2(n-1)$. The final score at each location is then the sum of all $(n-1)$ average scores contributed by chunks inside the window having length of 2 to W. The final score at each location of the sequence, which we call the *frequency score*, measures the potentiality of the location to be a word boundary and is utilized by segmentation algorithm in section 4. Let the *frequency score* at j^{th} position in S using a window of length $2n(2 \leq n \leq W)$ is denoted by F_j. If a straddling sequence starts inside the window from the kth position in S, then F_j can be written as $F_j \leftarrow \sum_{n=2}^{W}(\sum_{k=j-n+2}^{j}(\delta_{Freq(Find\_Trie\_Node(w^1_{k,k+n-1}))<Freq(Find\_Trie\_Node(w^1_{j-n+1,j}))}+$

$$\delta_{Freq(Find\_Trie\_Node(w^1_{k,k+n-1}))<Freq(Find\_Trie\_Node(w^1_{j+1,j+n}))}))/2(n-1),$$

where $\delta_{TRUE} = 1$ and $\delta_{FALSE} = 0$. Here $Freq$ function returns the frequency of a node in the TRIE.

In most iterations the window may partially contain two consecutive actual words of different length or it may contain one of the two actual words, where in both cases the words or part of words contained by the window contribute properly to the boundary score. For a long window the left and right windows

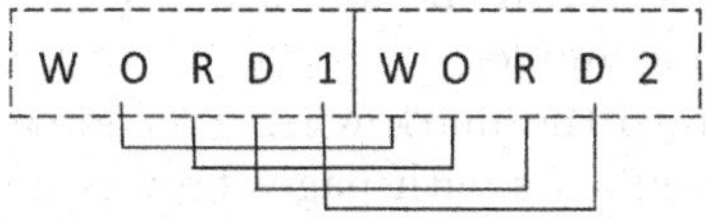

Fig. 2. Detection of word boundary by comparing frequencies of WORD1 and WORD2 with all four possible straddling subsequences of length five

would contain many consecutive actual words. In that case, like all straddling subsequences across the boundary, the consecutive words in the left and right windows would occur mostly once in the sequence. So they would contribute almost nothing to the boundary score. As window length is increased after a certain length, the contribution becomes mostly zero, which is a kind of getting rid of window-size parameter. In the following section we explain the segmentation algorithm that segments an input sequence using boundary entropy and frequency scores.

3 The Sequence Segmentation Algorithm

The segmentation method implements three segmentation scores at jth position in S namely, forward entropy score H_j^1, backward entropy score H_j^2 and frequency score F_j (derived in section 2) to perform three separate segmentations of S say, SS_{ent}^1, SS_{ent}^2 and SS_{freq}, respectively . Instead of taking threshold values from the user the method cuts the text at locations that have locally maximum segmentation scores. We implemented entropy scores and frequency scores separately to perform segmentation of sequences, and discovered that chunking is more robust if both entropy and frequency works together.Therefore the segmentation method combines all the three segmented sequences into a final consensus sequence. Cuts that are common in all three segmentations are considered to be very accurate but less in number. To get moderate number of accurate cuts the segmentation method considers cuts that are common in at least two segmentations. We will call the segmentation method the *Consensus Segmentation using Entropy and Frequency* (CSEF) algorithm. If SS is denoted as the final segmentation of the sequence S derived by CSEF, then $SS \leftarrow (SS_{ent}^1 \cap SS_{ent}^2) \cup (SS_{ent}^1 \cap SS_{freq}) \cup (SS_{ent}^2 \cap SS_{freq})$. In the following we compare CSEF with an existing sequence segmentation algorithm using a Toy example.

The CSEF algorithm is a linear time sequence segmentation algorithm that requires the maximum window length as the only input parameter. The CSEF algorithm is a robust unsupervised algorithm that works without the help of an existing lexicon and performs good even for small input document set. We compare CSEF with the Voting Experts (VE) algorithm of Cohen et al [1]. The VE algorithm is a linear time one way sequence segmentation algorithm that detects chunks by giving voting scores for each location in the sequence by maintaining a TRIE structure. In the following, we compare the performances of CSEF and VE with an example.

Input: Miller was close to the mark when he compared bits with segments. But when he compared segments with pages he was not close to the mark. his being close to the mark means that he is very close to the goal. segments may be identified by an information theoretic signatures. page may be identified by storage properties. bit may be identified by its image.
Output of VE: Mi*lle*rwas*close*tot*he*ma*rk*whe*nhe*co*mpa*re*dbi*ts* wi* thse*gmen*ts*But*whe*nhe*co*mpa*re*dse*gmen*ts*wi*thpa*ge*she*was *not*close *tot*he*ma*rkhi*sbe*ingc*lose*tot*he*ma*rkme*ans*that*he*isve* ryclose *tot*he*goa*lse*gmen*ts*ma*ybe*ide*nti*fie*dbya*ni*nfo*rmati*ont* he*ore*ti* csi *gnat*ure*spa*ge*ma*ybe*ide*nti*fie*dbys*tora*ge*pro*pe *rtie*sbi*tma*y be*ide *nti*fie*dbyi*tsi*mage
Output of CSEF: Mill*er*was*close*to*the*mark*when*he*compar*ed* bitswit h* segments*But* when*he*compa*red*segments*with*page*she*was*not*clo se* to*the*mark*his*bei*ng* close*to*the*mark*me*an*stha*the*isvery*close * to*the*goal*segments*may*beid*ent*ifi*ed by*an*inf*or*ma*tion*the*oretic *si*gn*atures*page*may*beid*ent*ifi*edby*st*or*age*prop*er* ti*es*bit*may *beid*ent*ifi*edby*itsimage

Overall the output of CSEF shows better performance than that of VE. In experimental section we show that though VE performs better when the input document is large in size, the CSEF outperforms VE over most of the benchmarks. We have also compared (but not reported) results of CSEF with another improved version of VE called Bootstrap Voting Algorithm (BVE) by Cohen et al [2]. We found that BVE does not perform better than CSEF for most of the datasets, whereas BVE suffers from higher space and time complexity due to maintaining a knowledge tree as another voting expert. In the next section we introduce the concept of *Representative based Document Clustering*, where we implement CSEF to perform document clustering and explain it with a toy example.

4 Representative Based Document Clustering

We implement CSEF algorithm to find word segmentation to perform document clustering. The frequency fields in TRIE are updated with normalized frequencies of chunks. Chunks occurring more than mean frequency are marked as representatives of the document set. Representative chunks may represent stems of important terms(words or phrases) derived unsupervisely. By sliding the representatives along each document we compute the corresponding feature vector and gradually form the term-document matrix. A clustering algorithm is implemented on the term-document matrix to group the documents $\{d_1, d_2, ..., d_r\}$ into clusters. We explain the document clustering method with a toy example. Here input documents and their outputs are separated with semicolons.

A Toy Example

Input(set of documents formed from the input of section 3): Miller was close to the mark when he; compared bits with segments. But when he; compared segments with pages he was not; close to the mark. his being close to; the mark means that he is very close to; the goal. segments may be identified by an; information theoretic signatures. page may; be identified by storage properties. Bit; may be identified by its image.;

Representatives found after chunking by CSEF:
close; to; the; mark; when; he; segments; page; may; beid;
Given K=3, Output of K-means:

Cluster1: compared bits with segments. But when he; compared segments with pages he was not; *Cluster 2*: the goal. segments may be identified by an; be identified by storage properties. bit; may be identified by its image; information theoretic signatures. page may;*Cluster 3*: Miller was close to the mark when he; close to the mark. his being close to; the mark means that he is very close to;

In the following experimental section we produce the performance of our clustering method on some large datasets.

5 Experiments

In this section, we present an empirical evaluation of our representative based document clustering method in comparison with some other existing algorithms on a number of benchmark data sets[11][10].The results of only four document datasets are recorded as for most of other datasets we got almost similar results. The first five columns of Table 1 summarize the basic properties of the data sets. To compare the performances of CSEF with VE, F_1 score[13] is used to determine quality of segmentation from both methods. The F_1 score is the harmonic mean of precision and recall and reaches its best value at 1 and worst score at 0. In the 6th and 7th column of Table 1 F_1 scores of VE and CSEF are recorded, respectively. The results show that CSEF achieves better performance on all the four data sets.

Table 1. Detailed description of datasets along with comparison of VE and CSEF

Data	Source	points	words	classes	F_1 score of VE	F_1 score of CSEF
Tr31	TREC	927	10128	7	0.58	0.71
Tr41	TREC	878	7454	10	0.55	0.68
Tr45	TREC	690	8261	10	0.56	0.68
re0	Reuters-21578	1504	2886	13	0.52	0.61

The datasets considered in the experiment have their class labels known to the evaluation process. To measure the accuracy of class structure recovery by a clustering algorithm we use a popular external validity measure called Normalized Mutual Information (NMI) [12]. The value of NMI equals 1 if two clusterings are identical and is close to 0 if one is random with respect to the other. Thus larger values of NMI indicate better clustering performance. We have chosen K-means and Cluto, by Ying Zhao *et al.* [9], as our document clustering algorithms. In table 2 CSEF-Cluto and CSEF-Kmeans denote our representative based clustering model with Cluto and Kmeans, respectively. In most of the cases number of representatives chosen by CSEF is half the number of the words in the document set. We compare our results with four other document clustering algorithms namely, Clustering via local Regression (CLOR)[6], Spectral clustering with normalized cut (NCUT)[7], Local learning based Clustering Algorithm (LLCA1) and its variant (LLCA2)[8]. The NMI performances of the four algorithms are taken from the paper "Clustering Via Local Regression, by Jun Sun *et al.* [6] and are verified. The output of the algorithms depends on a parameter k, which they have mentioned as neighbourhood size. By drawing NMI/k graphs they have provided the outcomes that we recorded numerically in Table 2. *max* and *avg* in Table 2 means maximum and average NMI values attended in the range of k=5 to 120, respectively. For re0 data, only the best result they have recorded which happened for k=30.

Table 2. Comparison of performances of clustering algorithms

	tr41 (max/avg)	tr45(max/avg)	tr31(max/avg)	re0 (for k=30)
LLCA1	0.63/0.62	0.61/0.55	0.53/0.5	0.3905
LLCA2	0.63/0.6	0.61/0.53	0.53/0.47	0.3847
NCUT	0.64/0.6	0.57/0.55	0.53/0.45	0.4030
CLOR	0.66/0.64	0.65/0.61	0.57/0.5	0.4302
Cluto	0.6751	0.6188	0.6410	0.3753
CSEF-Cluto	0.7390	0.7338	0.6512	0.4134
Kmeans	0.33	0.31	0.28	0.15
CSEF-Kmeans	0.38	0.38	0.28	0.22

Table 2 shows that K-means does not give satisfactory results when applied alone but with CSEF quality of results improve. Generally in practice, Cluto produces very good results compared to other algorithms. But it performs even better when associated with CSEF, which demonstrates the effectiveness of representative based similarity approach to document clustering. Here we see that CSEF-Cluto outperforms other algorithms for most of the datasets.

6 Conclusions and Future Work

In this paper, we proposed a new approach to document clustering which is based on a representative based similarity concept that uses the idea of consensus sequence segmentation. Experimental results on many data sets show that our method together with a good clustering algorithm improves the quality of the result and outperforms other well known clustering algorithms. In the future, we want to do a deeper analysis on the underlying reason for good performance of our algorithm and also to understand when it fails to give good results. We are also looking for comparing our method with other phrase-based algorithms. A more sound concept we need to develop that would resist in generating unnecessary small segments. We will also try to automatically detect optimum window size in CSEF to make it fully automatic.

References

1. Cohen, P., Adams, N., Heeringa, B.: Voting experts: An unsupervised algorithm for segmenting sequences. Journal of Intelligent Data Analysis (2006)
2. Hewlett, D., Cohen, P.: Bootstrap Voting Experts. In: IJCAI, pp. 1071–1076 (2009)
3. Zamir, O., Etzioni, O.: Web Document Clustering: A Feasibility Demonstration. In: Proc. 21st Ann. Int'l ACM SIGIR Conf., pp. 45–54 (1998)
4. Zamir, O., Etzioni, O.: Grouper: A Dynamic Clustering Interface to Web Search Results. Computer Networks 31(11-16), 1361–1374 (1999)
5. Hammouda, K., Kamel, M.: Efficient Phrase-Based Document Indexing for Web Document Clustering. IEEE Trans. Knowl. Data Eng. 16(10), 1279–1296 (2004)
6. Sun, J., Shen, Z., Li, H., Shen, Y.: Clustering Via Local Regression. In: Daelemans, W., Goethals, B., Morik, K. (eds.) ECML PKDD 2008, Part II. LNCS (LNAI), vol. 5212, pp. 456–471. Springer, Heidelberg (2008)
7. Shi, J., Malik, J.: Normalized Cuts and Image Segmentation. IEEE Transactions on Pattern Analysis and Machine Intelligence 22(8), 888–905 (2000)
8. Wu, M., Scholkopf, B.: A local learning Approach for Clustering. In: Advances in Neural Information Processing Systems, vol. 19 (2006)
9. Zhao, Y., Karypis, G.: Empirical and Theoretical Comparisons of Selected Criterion Functions for Document Clustering. Machine Learning 55, 311–331 (2004)
10. Lewis, D.D.: Reuters-21578 text categorization test collection, `http://www.daviddlewis.com/resources/testcollections/reuters21578`
11. TREC: Text REtrieval Conference, `http://trec.nist.gov`
12. Strehl, A., Ghosh, J.: Cluster Ensembles - A Knowledge Reuse Framework for Combining Multiple Partitions. Journal of Machine Learning Research 3, 583–617 (2002)
13. Van Rijsbergen, C.J.: Information Retrieval, 2nd edn. Dept. of Computer Science, University of Glasgow (1979)

A New Parallel Thinning Algorithm with Stroke Correction for *Odia* Characters

Arun K. Pujari[1], Chandana Mitra[2], and Sagarika Mishra[2]

[1] School of Computer & Info. Sciences
University of Hyderabad
Arun.k.pujari@gmail.com
[2] SUIIT, Sambalpur University
Sambalpur, Odisha, 768019
chandanamitra@ymail.com

Abstract. There are several thinning algorithms reported in literature in last few decades. Odia has structurally different script than that of other Indian languages. In this paper, some major thinning algorithms are examined to study their suitability to skeletonize Odia character set. It is shown that these algorithms exhibit some deficiencies and vital features of the character are not retained in the process. A new parallel thinning technique is proposed that preserves important features of the script. Interestingly, the new algorithm exhibits stroke-preservation which is a higher level requirement of thinning algorithms. Present work also discusses a concept of stroke correction where a basic stroke is learnt from the original image and embedded on the skeleton.

1 Introduction

Automatic character recognition is becoming extremely important in digital preservation of cultural heritage. The importance of OCR for Indian languages cannot be underestimated as massive volumes of print-media resources representing the cultural and historical heritage of India are available in several Indian scripts. Several recognition techniques for printed characters of different Indian scripts have been reported in literature [11-13]. Like Devanagari and Bangla scripts, *Odia* script derives its origin from Brahmi script. But *Odia* scripts are mostly circular in nature and do not have accompanying top horizontal line. *Many character recognition methods use a thinning stage to facilitate shape analysis and stroke identification.* Thinning is a process of skeletonization of a binary image to extract the skeleton of the shape of interest. The essential requirement of thinning algorithm are topology preservation and shape preservation. Mask-based iterative thinning algorithms delete successive layers of pixels on the boundary of the pattern until only a skeleton remains. These masks aim at identifying boundary pixels that are likely to be crucial in preserving geometric and topological properties. According to the way they examine pixels, these algorithms can be classified as sequential or parallel. Many thinning algorithms have been developed in the past few years for general application as well as for OCR [1-4, 6-11, 14-18].

M.K. Kundu et al. (eds.), *Advanced Computing, Networking and Informatics - Volume 1*,
Smart Innovation, Systems and Technologies 27,
DOI: 10.1007/978-3-319-07353-8_48, © Springer International Publishing Switzerland 2014

In this paper, we examine working of 10 major algorithms in terms of shape preservation, stroke preservation, over-erosion, preservation of junction points, 3- or 4-way branch preservation which may play important roles in subsequent recognition process of *Odia* characters. The main contributions are the following. Experimental analysis of major thinning algorithms on *Odia* characters is reported. A new algorithm is proposed by blending judiciously the steps of earlier methods. The new algorithm is shown empirically to perform better than any of the earlier algorithms. The characteristics of the language is taken into account and in addition to preservation of topology, strokes are also preserved.

2 Existing Thinning Algorithms

In this section, ten major mask-based thinning algorithms are examined. The objective is to analyze efficacy of these algorithms to capture the inherent characteristics of *Odia* script. It is worth mentioning here that some characters of this script are of circular shape joining a vertical stroke to the right (e.g., ଖ, ସ, ମ, ଅ). Most thinning algorithms face no difficulty in retaining a vertical line in isolation. But when a curvilinear strokes joins the vertical line (e.g., ଗ, ଘ, ଖ), the line is over-eroded. As a result, a critical portion of the stroke is not retained and hence, ambiguity in recognition arises (as in ଗ and ର). Another crucial stroke in *Odia* script is short slanting line from which a curvilinear shape originates (as in ଧ). It is observed that most thinning algorithms do not preserve this stroke. A representative set of character images are taken (Figure 1) to illustrate the efficiency of the existing algorithms. The observations made here are not restricted to this representative set but are found in most instances during our experimentation.

One of the most-cited thinning algorithms, ZS algorithm [17] works in two sub-iterations. There are some major shortcomings of ZS algorithm [15], namely, excessive erosion at junction points; non-unit width skeleton; distortion of short slanting stroke; and complete deletion of patterns which can be reduced to 2×2 square. When ZS algorithm is applied to *Odia* characters, the strokes are distorted and misrepresented on many instances. For example, in Figure 5d, short slanting stroke of *dhYa* (ଧ୍ୟ) is distorted. Similarly, for the character *E*(ଏ), in Figure 3d, ZS algorithm returns slant line of two-pixel width. ZS algorithm is modified in [11] as LW algorithm with an aim to handle slanting lines. We notice that over-erosion at junction point persists in LW (Figure 6e). The vertical stroke of *Sa* (ସ) is distorted and can be confused with tail-stroke of R (ର). Kwon et al. [9] propose an enhancement of ZS algorithm (KGK Algorithm) with additional post processing conditions. Contrary to the claims in [9], diagonal line is non-unit width for *E* (ଏ) (Figure 3c) and there are also spurious strokes at the boundary. The skeleton of the compound character *Nka* (ଙ୍କ) is not proper (Figure 4c). ZW algorithm [18] uses a 4×4 neighbourhood and it is shown [4] that ZW algorithm does not preserve topology. Our experiment with ZW algorithm reveals that due to overerosion, even vertical strokes of *Odia* characters are not preserved. ZW algorithm results in over erosion as can be seen for *dhYa* (ଧ୍ୟ) in Figure 5e and spurious error around the contour (Figure 3e). ZW algorithm's performance is better than that of ZS algorithm for diagonal stroke at top left portion of *dhYa,* (Figure 5e). CWSI87[2] algorithm proposed by Chen et al [2] uses 4×4 window and it has one restoring mask in addition to the thinning mask. A pixel is deletable if it matches with any thinning mask and does not match the restoring mask. This algorithm retains the circular shapes but cannot remove the contour noise point for *Odia* characters. For the character, *Nka* (ଙ୍କ), we notice (Figure 8a) that the small circular shape at top-right corner is preserved. Readers may compare the output with KGK algorithm for the same character where the circular shape is more of a rectangular shape (Figure 5a). But on the other hand CWSI algorithm is unable to retain straight line strokes. CWSI algorithm suffers from many other shortcomings such as contour noise, stroke-end bifurcation, retention of vertical lines and joint distortion (Figure 8b). H89 [6] algorithm preserves the horizontal, vertical lines properly. But it fails to remove noise points and to retain curvilinear strokes. Circular strokes, predominant in *Odia* script, are not retained. In GH92 [6] curves and lines are maintained, but closeby junction points are merged. It does not guarantee unit width skeleton (Figure 6d). CCS95 [3] algorithm uses 5 different thinning masks and 10 restoring masks. In this method, the skeleton is not guaranteed to be connected for *DhYa* (ଢ୍ୟ) and *NTha* (ଣ୍ଠ) (Figure 5a and 4a). BM99 [1] uses two different types of thinning masks and one restoring mask. It is unable to find skeleton of unit width and cannot retain vertical strokes with joints (Figure 6a). Recently, a robust algorithm [10] is proposed which performs better than ZS and KGK algorithm. However, we notice that its performance is not satisfactory for *Odia* character. For example, small line segments in the character get deleted (Figure 6a). A small diagonal stroke at bottom left corner gets smoothen. There is extra noise along contour as seen in Figure 13b. For the character *ma* and *sa,* small top portion and bottom portion of the straight line are either deleted or curve.

3 Stroke Preserving Thinning Algorithm

In this section, a new thinning technique is proposed which blends the strengths of aforementioned thinning algorithms. It is observed that some of the existing algorithms emphasize on topology preservation, some on connectivity and some on single pixel width skeleton. Many of the existing algorithms do have advantages that can be blended together to generate a new method of thinning. But when considered in isolation, none of these is able to yield satisfactory results for *Odia* script. The steps are sequenced carefully to ensure that algorithms that are able to delete contour noise are used first. Similarly, those which can restore slant-lines are used later. The proposed method has the following major components. In its first k iterations, it uses two subiterations of ZS Algorithm. Two subiterations are used only for fixed number of steps and then to switch over to apply thinning and restoration masks of CCS95 Algorithm. This step terminates when there is no deletable pixel. After completion of this step, the algorithm applies the postprocessing step of KGK Algorithm. The pseudo code of the algorithm is given below.

Algorithm-Stroke preserving thinning with stroke correction

```
INPUT: Binary image of isolated character with black pixel as 1
     do for k iterations
         do for each pixel p
              use masks of ZS algorithm to mark pixels to be deletable
         end do
     delete all deletable pixels
end do
do while no deletable pixel found
     do for all pixels p
         if p satisfies thinning mask of CCS95 then mark the pixel deletable
     end do
     do for every deletable pixel p
         if p satisfies any of the restoring masks of CCS95 then unmark p
     end do
  delete all pixels that are marked deletable
 end while
do while no deletable pixel found
     do for all pixels p
         if p satisfies postprocessing masks of KGK, then mark p as deletable
     end do
  delet all deletable pixels
 end while
call stroke correction step
```

4 Stroke Corrections

Our experimental study show that the new algorithm preserves topolgy, connectedness, shape, strokes and other desirable features of *Odia* character. The algorithm is able to skeletonize some important strokes to near-perfect accuracy. For instance, vertical lines can be preserved but over-erosion at the joints cannot be fully avoided. We propose a novel idea of stroke correction stage. If the thinning algorithm retains the shape of the stroke to a unambiguous state, then the stroke can be recognized by a separate process and can be replaced by a perfect stroke. For instance, if the thinning algorithm returns a vertical line eroded at junction points, we introduce a separate method to recognize the straight line and then replace the eroded line with a perfect line.

5 Experimental Results

Experiments to evaluate the performance of the proposed algorithm in comparison to earlier algorithms. Our dataset consists of isolated characters of 2000 characters of several fonts and sizes. Documents are scanned at 300dpi with high-resolution scanner. Characters are isolated by connected component method. These are size-normalized to 48×48. In the present experiments, only the alphabets and compound characters are taken leaving aside numerals and modifiers. k is 2% of image size and hence k =1. In Figure 3-7, comparison of the proposed method with CCS, CWSI, KGK, ZS and ZW algorithms for 5 typical characters, namely sa (ସ), E(ଏ), NTha(ଣ୍ଠ), Nka(ଙ୍କ), and dhYa(ଧ୍ୟ) are reported in Figures 3-5. Most of the characters chosen for illustration contain typical strokes and are compound characters. These algorithms perform better than others for other scripts. For the sake of illustration, Figure 6 gives the output of the proposed algorithm and of BM99, H89, Robust, GH92 and LW for the character ma(ମ) and aw(ଅ). It is evident from Figures 4-5 that the end points and junction points are either shrunk or deleted by CWSI, KGK and ZW. CCS algorithm does not preserve connectivity and ZS algorithm yields 2-pixel width skeleton for slanting lines.

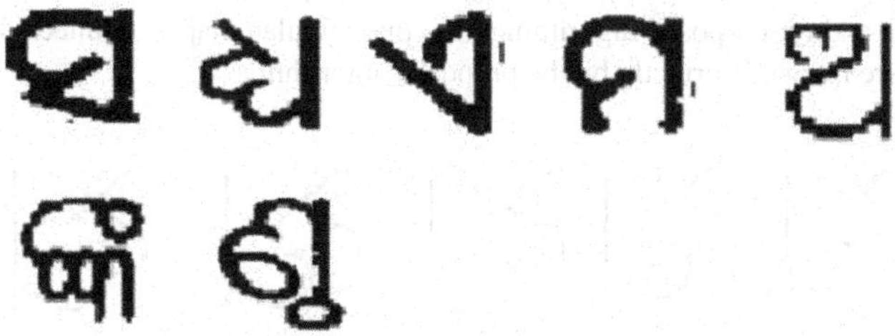

Fig. 1. Representative set of charater images for illustration

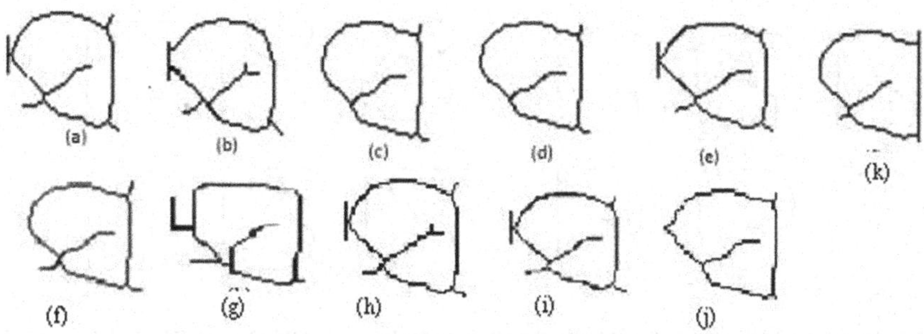

Fig. 2. Skeleton of sa (ସ) by thinning algorithms (a) CCS (b) CWSI (c) KGK (d) ZS (e) ZW (f) LW (g) H89 (h) GH92 (i) BM99 (j) robust and (k) proposed algorithm. The diagonal line, vertical line and junction points have been restored distinctly by the proposed method.

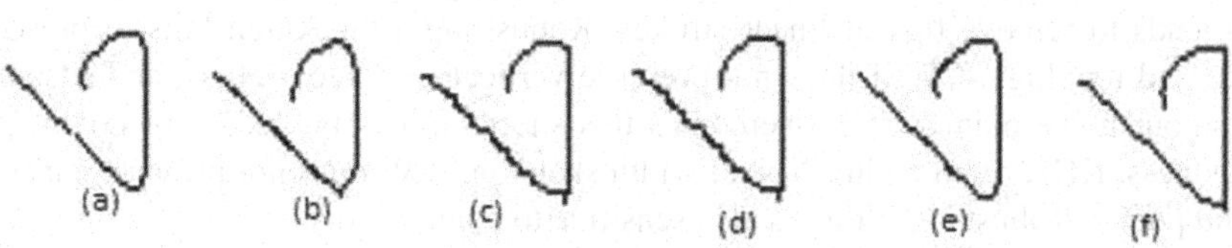

Fig. 3. Skeleton of e (ଏ) by algorithms (a) CCS (b) CWSI (c) KGK (d) ZS (e) ZW and (f) proposed algorithm

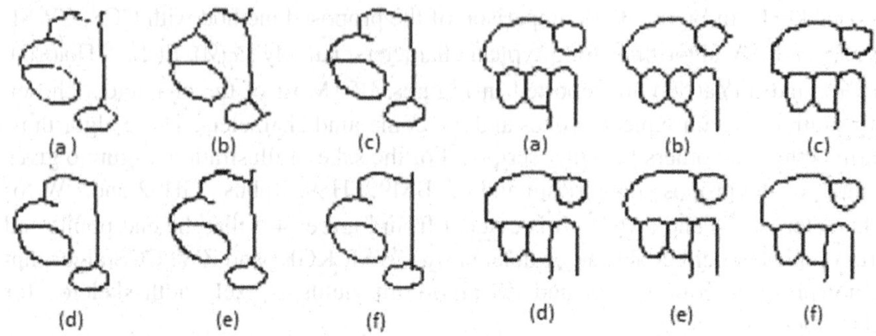

Fig. 4-5. Skeleton of NTha (ଣ୍ଠ) and Nka (ଙ୍କ) by thinning algorithms (a) CCS (b) CWSI (c) KGK (d) ZS (e) ZW and (f) proposed algorithm. Note that circular shapes, connectedness and short linear strokes are retained correctly by the proposed algorithm

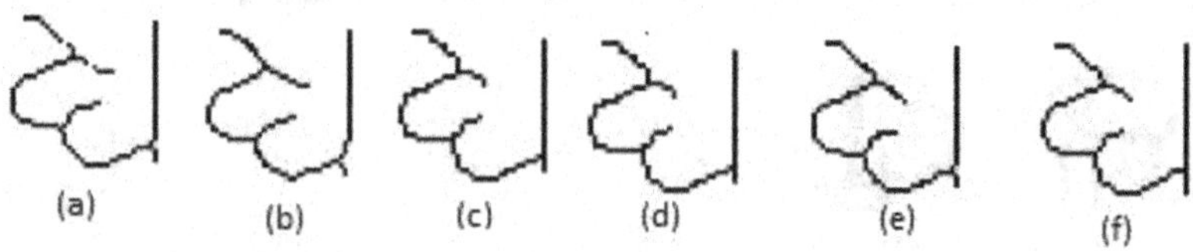

Fig. 6. Result of algorithms (a) CCS (b) CWSI (c) KGK (d) ZS (e) ZW (f) proposed algorithm

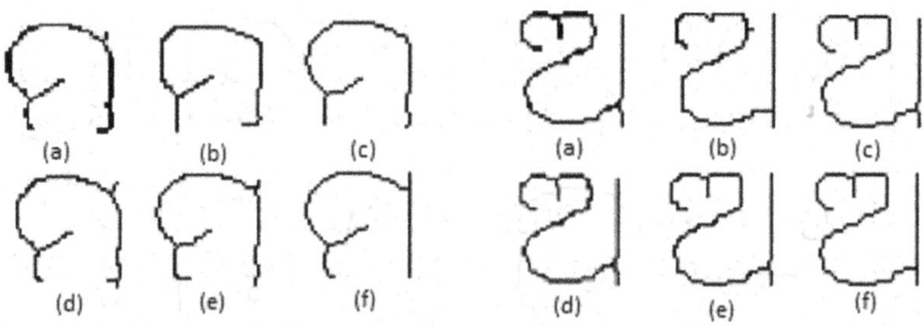

Fig. 7-8. Result of algorithms (a) BM99 (b) H89 (c) Robust (d) GH92 (e) LW and (f) proposed algorithm. Note that many outputs are 2-pixel width skeletons.

From our extensive experimentation, we observe that CCS, CWSI and ZW are not able to remove contour noise. BM99, LW and GH92 yield skeletons of 2-pixel width. H89 tends to remove tips of linear strokes. Robust algorithm often fails to preserve shape and topology. CWSI does not preserve vertical or diagonal lines and sensitive to contour noise points. CCS overcomes this shortcomings but does not retain connectedness. KGK often yields 2-pixel width skeleton. LW does not preserve junction or end points. Robust algorithm is also sensitive to contour noise.

6 Conclusions

The present work is concerned with investigating existing thinning algorithms for their suitability to a specific script, namely *Odia* script. It is observed that none of the existing algorithms is very suitable and has some shortcoming or other. By combining some steps of different algorithms, a new algorithm is proposed that not only preserves the desirable features like shape, topology, connectivity etc but also retains basic strokes. This is very useful for subsequent recognition process. A new concept of stroke correction is also introduced here. We propose to integrate this with our recognizer in future and extend the concept of stroke correction to many other basic strokes.

References

1. Bernard, T.M., Manzanera, A.: Improved low complexity fully parallel thinning algorithm. In: ICIAP (1999)
2. Chin, R.T., Wan, H.K., Stover, D.L., Iversion, R.D.: A one-pass thinning algorithm and its parallel implementation. CVGIP 40, 30–40 (1987)
3. Choy, S.S.O., Choy, C.S.T., Siu, W.C.: New single-pass algorithm for parallel thinning. CVIU 62, 69–77 (1995)
4. Couprie, M.: Note on fifteen 2-D parallel thinning algorithm. In: IGM 2001, pp. 1–21 (2006)
5. Datta, A., Parui, S.K.: A robust parallel thinning algorithm for binary images. Pattern Recognition 27, 1181–1192 (1994)
6. Guo, Z., Hall, R.W.: Fast fully parallel thinning algorithm. CVGIP 55, 317–328 (1992)
7. Hall, R.W.: Fast parallel thinning algorithm: parallel speed and connectivity preservation. CACM 32, 124–131 (1989)
8. Kong, T.Y., Rosenfeld, A.: Digital topology: introduction and survey. CVGIP 48, 357–393 (1989)
9. Kwon, J.S., Gi, J.W., Kang, E.K.: An enhanced thinning algorithm using parallel processing. In: ICIP, vol. 3, pp. 752–755 (2001)
10. Lam, L., Suen, C.Y.: An evaluation of parallel thinning algorithm for character recognition. IEEE Tr. PAMI 17, 914–919 (1995)
11. Negi, A., Bhagvati, C., Krishna, B.: An OCR system for Telugu. In: ICDAR 2001, pp. 1110–1114 (2001)
12. Pal, U., Chaudhuri, B.B.: Indian script character recognition-A survey. Pattern Recognition 37, 1887–1899 (2004)
13. Pujari, A.K., Naidu, C.D., Jinga, B.C.: An adaptive and intelligent character recognizer for Telugu scripts using multiresolution analysis and associative measures. In: ICVGIP (2002)
14. Lu, H.E., Wang, P.S.P.: An improved fast parallel algorithm for thinning digital pattern. In: IEEE CVPR 1985, pp. 364–367 (1985)
15. Tarabek, P.: A robust parallel thinning algorithm for pattern recognition. In: SACI, pp. 75–79 (2012)
16. Wang, P.S.P., Hui, L.W., Fleming, T.: Further improved fast parallel thinning algorithm for digital patterns. In: Computer Vision, Image Processing and Communications Systems and Appln., pp. 37–40 (1986)
17. Zhang, T.Y., Suen, C.Y.: A fast parallel algorithm for thinning digital patterns. CACM 27, 236–239 (1984)
18. Zhang, Y.Y., Wang, P.S.P.: A modified parallel thinning algorithm. In: CPR, pp. 1023–1025 (1988)

Evaluation of Collaborative Filtering Based on Tagging with Diffusion Similarity Using Gradual Decay Approach

Latha Banda and Kamal Kanth Bharadwaj

School of Computer and System Sciences
Jawaharlal Nehru University, India
{latha.banda,kbharadwaj}@gmail.com

Abstract. The growth of the Internet has made difficult to extract useful information from all the available online information. The great amount of data necessitates mechanisms for efficient information filtering. One of the techniques used for dealing with this problem is called collaborative filtering. However, enormous success of CF with tagging accuracy, cold start user and sparsity are still major challenges with increasing number of users in CF. Frequently user's interest and preferences drift with time. In this paper, we address a problem collaborative filtering based on tagging, which tracks user interests over time in order to make timely recommendations with diffusion similarity using gradual decay approach.

Keywords: Collaborative filtering, collaborative tagging, interestingness drift, gradual decay approach.

1 Introduction

With the growth of information in internet it is difficult to manage the data in websites. To manage data in websites collaborative filtering has been used. There are three types of CF: 1) Memory based, uses user rating data to compute similarity between users or items. 2) Model based CF, Models are developed using data mining, machine learning algorithms to find patterns based on training data and 3) Hybrid collaborative filtering, number of applications combines the memory-based and the model-based CF algorithms. These overcome the limitations of native CF approaches. However CF suffers from Sparsity, the user-item matrix used for collaborative filtering is extremely large and sparse which uses large datasets brings about the challenges in the performances of the recommendation. Scalability problems as the numbers of users and items grow traditional CF algorithms will suffer serious scalability problems. Cold start user; there is relatively little information about each user, which results in an inability to draw inferences to recommend items to users. To solve these limitations tagging was used in CF.

Tag is a freely chosen word or sentence by a user. These tags categorize content by using simple keywords. Tag cloud is a visual representation of data and there are more clouds are available in websites (i) Data cloud is a data which uses font size or

M.K. Kundu et al. (eds.), *Advanced Computing, Networking and Informatics - Volume 1*, Smart Innovation, Systems and Technologies 27,
DOI: 10.1007/978-3-319-07353-8_49, © Springer International Publishing Switzerland 2014

colour to indicate numerical values. (ii) Text clouds are representation of word frequency in a given text. (iii) Collocate cloud examines the usage of a particular word instead summarizing an entire document. Our work in this paper is evaluation of Collaborative Filtering based on tagging using Gradual decay approach. Firstly we have generated the results of data sets using gradual decay approach with recent time stamp and then the results of CF and CF with tagging were evaluated which gives accurate and efficient results.

2 Related Work

In this section, we briefly explain several major approaches we used in our research such as Collaborative Filtering, Collaborative Filtering based on Tagging, diffusion similarity and Gradual decay approach.

2.1 Collaborative Filtering

Collaborative filtering is the process of filtering for information using techniques involving collaboration among multiple users, agents, viewpoints, data sources, etc. Applications of collaborative filtering typically involve very large data sets. CF can be categorized into memory based CF and item based CF. in user based, predictions are calculated based on who share the same rating patterns. In item based CF, predictions are based on relationship between pair of items.

2.2 Collaborative Filtering Based on Tagging

Recently tagging has grown in popularity on the web, on sites that allow users to tag bookmarks, photographs and other content [usage patterns]. Tag is freely created or chosen by a user in social tagging systems which may not have a structure.

Fig. 1. Collaborative filtering based on tagging and Tag cloud

Tagging can be categorized into two types: (i) Static tags or like/dislike tags in which a user post thumbs up for like and thumbs down for dislike. (ii) Dynamic tags in which a user give his opinion or comment on item. These are also called as popular tags. Collaborative filtering based on tagging is shown in Fig.1.

2.3 Neighborhood Formation Using Diffusion Similarity

Neighbourhood set is a group [1], which is having similar taste and preferences. In the process of neighbourhood formation the size must be large to get the accurate results. For this the various similarity methods can be used such as, Pearson correlation [2] coefficient is given by Eq. 1.

$$corr(x,y) = \frac{\sum_{s \in S_{xy}} (r_{x,s} - m_x)\ (r_{y,s} - m_y)}{\sqrt{\sum_{s \in S_{xy}} (r_{x,s} - m_x)^2}\sqrt{\sum_{s \in S_{xy}} (r_{y,s} - m_y)^2}} \quad (1)$$

The Eq. (1) is not appropriate because it select only common items for both users. To compute similarity of multiple features Euclidian distance can be used

$$d(x,y) = \frac{1}{z}\sum_{i=1}^{z} \sqrt{\sum_{j=1}^{N} (x_{i,j} - y_{i,j})^2} \quad (2)$$

The other similarity measure is cosine similarity technique in which inverse user frequency is applied, the similarity between users, u and v is measured by the following equation:

$$sim(u,v) = cos(\vec{u},\vec{v}) = \frac{\sum_{t \in T} (A_{u,t} .iuf_t)\ (A_{v,t} .iuf_t)}{\sqrt{\sum_{t \in T} (A_{u,t} .iuf_t)^2}\sqrt{\sum_{t \in T} (A_{v,t} .iuf_t)^2}}, \quad (3)$$

Here 'l' is the total number of users in the system and nt, the number of users tagging with tag t. iuft is the inverse user frequency for a tag t: iuft = (log(l/nt) [3]. Here to get accuracy [4] in neighbourhood formation we have used diffusion based similarity [5]. In this if the user has rated any item then the value is set to 1, otherwise set to 0. If the user has used any tag then the value is set to 1 otherwise set to 0. The resource that an item ∝ gets from V reads:

$$r_{\propto v} = \frac{a_{v\propto}}{k(v)} \quad (4)$$

Here $a_{u\alpha} = 1$ if the item is collected otherwise it is 0. Here U has collected objectα. V is the target user on α, k(v) is the degree of target user.

The item based [6] similarity is defined by using S_{uv}

$$S_{uv} = \frac{1}{K(v)}\sum_{\alpha \in 0} \frac{a_{u\alpha} a_{v\alpha}}{K(\alpha)} \quad (5)$$

The tag based similarity is defined by using S'_{uv}

$$S'_{uv} = \frac{1}{K(v)}\sum_{t \in T} \frac{a_{ut} a_{vt}}{K(t)} \quad (6)$$

2.4 Prediction and Recommendation

Once the neighbourhood formation is done according to the similarity of users the prediction can be done for the new user who has not rated the items and these items can be recommended to the new user. In this paper we have used diffusion similarity to get accuracy of recommendation in collaborative tagging

3 Proposed Work CF Framework Based on Tagging with Diffusion Similarity Using Gradual Decay Approach

Fig.2 illustrates our approach in three phases: (i) Timestamp and gradual decay approach applied on collection of data. (ii) Collaborative tagging has been implemented using diffusion similarity. (iii) Get results by using Prediction and recommendation.

3.1 Collaborative Filtering Based on Tagging Using Gradual Decay Approach

The timestamp have been applied on collection of datasets, it gives the data in a sequence order. Here we use timestamp to get recent data so that more recommendation can be made for the user. As time increase the importance of data is going to be diminishes. So the recent data is more important rather than the old data. Here we have used gradual decay approach to get recent data more. In this we have given least weight to old data and more weight to recent data. Here the data we have chosen from year 2002 to year 2013 (shown in Table.1). The details are given in Fig.2.

4 Experiments

We have conducted several experiments on different datasetsamazon.com, Grouplens, Movielens, Flickr.com and Youtube.

To examine the effectiveness of our new scheme for collaborative filtering based on tagging with diffusion similarity using gradual decay approach. In this we address the following issues.

(i) The datasets have been rearranged using timestamp method and gradual decay approach.
(ii) We have conducted experiments on collaborative filtering based on tagging.
(iii) We have compared these approaches in terms of mean absolute error.

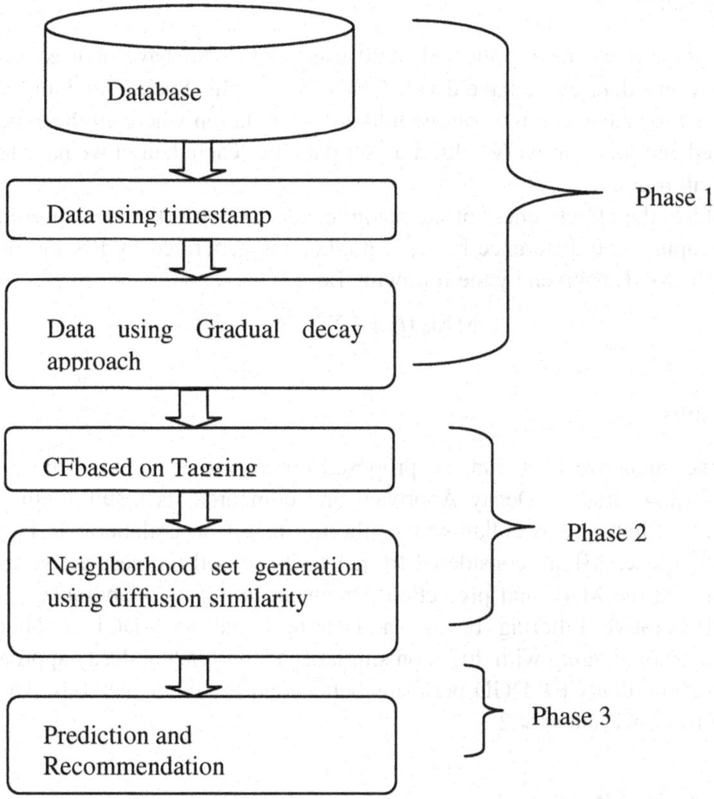

Fig. 2. Architecture of the proposed approach for item recommendation (CFT-DGD)

Table 1. Gradual decay approach applied on collaborative filtering based on tagging

Year	%of data
2002-2003	10%
2003-2004	20%
2004-2005	30%
2005-2006	40%
2006-2007	50%
2007-2008	60%
2009-2010	70%
2010-2011	80%
2011-2012	90%
2012-2013	100%

4.1 Datasets

For every dataset we have collected 5000 users data who have at least visited 40 items. For every dataset we have divided 1000 users split. Such a random separation was intended for the executions of one fold cross validation where all the experiments are repeated one time for every split of a user data. For each dataset we have tested set of 30% of all users.

To evaluate the effectiveness of the recommender systems, the mean absolute error (MAE) computes the difference between predictions generated by RS and ratings of the user. The MAE is given by the following Eq. (7)

$$\text{MAE}\ (i) = \frac{1}{n_i}\sum_{j=1}^{n_i}\left|pr_{i,j}-r_{i,j}\right|. \tag{7}$$

4.2 Results

In this experiment we have run the proposed collaborative tagging with Diffusion Similarity using Gradual Decay Approach and compared its results with classical Collaborative filtering and collaborative filtering based on collaborative tagging. In this the active users[7] are considered from 1 to 20. Based on these active users, we have computed the MAE and prediction percentage of CF (Collaborative Filtering), CFT (Collaborative Filtering based on Tagging), and CFT-DGD (Collaborative filtering based on tagging with diffusion similarity using gradual decay approach) and the results show that CFT-DGD performs better than the other methods. The results are shown in split.1 and split.2

4.3 Analysis of Results

For CFT-DGD, out of 10 runs for each active user, the run with the best weights was chosen and plotted the results from CF and CFT as shown in Fig. 3 and Fig. 4 for split.1 and split.2 respectively. The results summarized in Table 2 and Table 3 shows total average of MAE and correct predictions for all the algorithms that collaborative tagging with diffusion similarity using gradual decay approach outperforms rather than CF and CFT for all 5 splits. MAE for CFT-DGD was always smaller than the corresponding values for CF and CFT and this is described in Fig. 3 and Fig. 4.

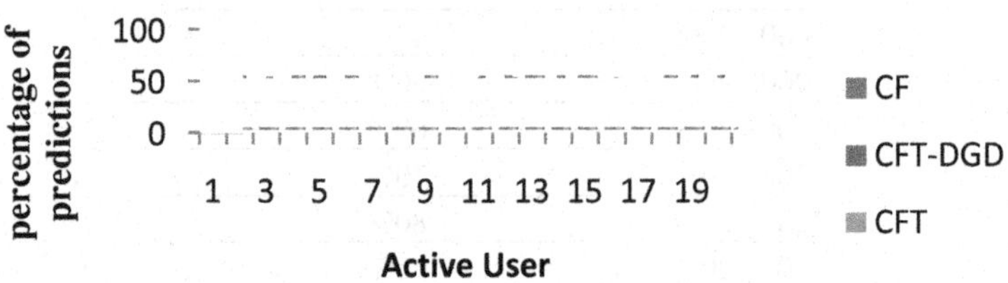

Fig. 3. Correct predictions percentage for active user of split1

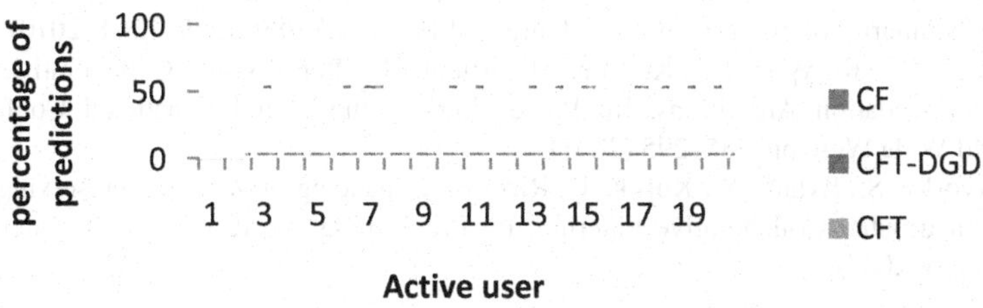

Fig. 4. Correct predictions percentage for active user of split2

Table 2. Total MAE for CF,CFT and CFT-DGD

Split	MAE (CF)	MAE (CFT)	MAE (CFT-DGD)
1	0.890	0.802	0.762
2	0.902	0.792	0.742
3	0.801	0.770	0.682

Table 3. Total correct predictions for CF,CFT and CFT-DGD

Algo	CorrectPred's (%)	MAE (AVG)
CF	41.54	0.864
CFT	47.82	0.778
CFT-GIM	54.28	0.728

5 Conclusion

We have proposed a framework collaborative filtering. Our approach targets the user interests over time in order to make timely recommendations with diffusion similarity using gradual decay approach. Experimental results show that our proposed scheme can significantly improve the accuracy of predictions.

References

1. Omahony, M.P., Hurley, N.J., Silvestre, G.C.M.: An Evolution of Neighbourhood Formation on the Performance of Collaborative Filtering. Journal of Artificial Intelligence Review 21(3-4), 215–228 (2004)
2. Adomavicius, G., Tuzhilin, A.: Toward the next generation of recommender systems A survey of the state-of-the-art and possible extensions. Journal of IEEE Transactions on Knowledge and Data Engineering 6(17), 734–749 (2005)
3. Nam, K.H., Ji, A.T., Ha, I., Jo, G.S.: Collaborative filtering based on collaborative tagging for enhancing the quality of recommendation. Electronic Commerce Research and Applications 9(1), 73–83 (2010)
4. Anand, D., Bharadwaj, K.K.: Enhancing accuracy of recommender system through adaptive similarity measures based on hybrid features. In: Nguyen, N.T., Le, M.T., Świątek, J. (eds.) ACIIDS 2010, Part II. LNCS (LNAI), vol. 5991, pp. 1–10. Springer, Heidelberg (2010)

5. Shang, M.S., Zhang, Z.K., Zhou, T., Zhang, Y.C.: Collaborative filtering with diffusion-based similarity on tripartite graphs. Journal of Physica A 389, 1259–1264 (2010)
6. Sarwar, B., Karypis, G., Konstan, J., Riedl, J.: Item based Collaborative Filtering Recommendation Algorithms. In: Proceedings of the 10th International Conference on World Wide Web, pp. 285–295 (2001)
7. Berkvosky, S., Eytani, Y., Kuflik, T., Ricc, F.: Enhancing privacy and preserving accuracy of a distributed Collaborative Filtering. In: Proceedings of ACM Recommender Systems, pp. 9–16 (2007)

Rule Based Schwa Deletion Algorithm for Text to Speech Synthesis in Hindi

Shikha Kabra, Ritika Agarwal, and Neha Yadav

Department of Computer Science and Engineering,
Banasthali University, Rajasthan
{Kabra.shikha1990,Ritika.90agr,Yadavneha2511}@gmail.com

Abstract. This paper provides the solution of Schwa deletion while converting grapheme into phoneme for Hindi language. Schwa is short neutral vowel and its sound depends onto adjacent consonant. Schwa deletion is a significant problem because while writing in Hindi every Schwa is followed by a consonant but during pronunciation, not every Schwa followed by a consonant is pronounced. In order to obtain good quality TTS system, it is necessary to identify which Schwa should be deleted and which should be retained. In this paper we provide various rules for Schwa deletion which are presently applicable to a limited length of word and will try to further extend it in future.

Keywords: Schwa, block, phoneme, vowel, consonant.

1 Introduction

Speech is the general mean of the human communication except for the people who are not able to speak. Person who is visually impaired is even possible to communicate with other person with the help of Speech but is not able to communicate with the computer because the interaction between person and computer is based on written [2] text and images. So to overcome this problem speech synthesis is needed in natural language. Speech synthesis is a process of converting text into natural speech by using NLP and DSP, the two main component [1], [2] of TTS system. The NLP part consists of tokenization, normalization; G2P conversion and DSP part consist of waveform generation. Finding correct and natural pronunciation for a word is an important task and G2P is responsible for it. G2P accept word as an input and convert it into its phonetic transcription later this transcription is used for waveform generation. Hence it is necessary to generate correct phonetic transcription which is done by deleting Schwa from the word, when it is not pronounced. Our paper present an algorithm which resolves problem of schwa and layout of the paper is as follows. Here, we first introduce the Hindi writing system and Schwa problem then discussed the terminology we have used during algorithm. In next sections, we have elaborated algorithm and its result.

M.K. Kundu et al. (eds.), *Advanced Computing, Networking and Informatics - Volume 1*,
Smart Innovation, Systems and Technologies 27,
DOI: 10.1007/978-3-319-07353-8_50, © Springer International Publishing Switzerland 2014

1.1 Hindi Writing System and Schwa

The Hindi script (writing script) is called Devnagari [3] script which is most common writing system having 33 consonant and 11 vowels. Each vowel letter has 2 forms. One is [5] dependent form which indicate that vowel (other than Schwa) is attached to consonant like सा = स+ ा and another one is [5] independent form which shows the vowel occurs alone in whole word like अ in अपना. Schwa problem – Every [4] consonant letter by itself automatically include a short "a" vowel sound eg- सकता (sakawA). In linguistics ,this sound has a special name "Schwa" which is sometime pronounced but the significant problem is to identify [5]which Schwa should be pronounced and which should not, for example while pronouncing सकता (sakawA) ,Schwa(a) followed by 's' is pronounced but Schwa (a)followed by 'k' is not pronounced.

2 Terminology Used

This algorithm is performed on individual word as input. Here we divided a word into different blocks on the basis of consonant(C) and vowel (V). Before discussing the rules we discuss some terminology related to our approach.

- **Block** – All consonant followed by a vowel [6] binds in one block .Here, consonant ∈ (belongs to) any consonant in English or ' ' (NULL).We have made following algorithm for up to 4 block length only (refer Table. 1).

Table 1. Division of words into blocks of different sizes

Word	W-X Notation	Block Size
क्या	kyA	1
स्नान	snAna	2
अपना	a pa nA	3
परंपरा	pa ramparA	4

We divide the input words into two categories.

Simple – For every block of word, if every vowel followed by a consonant or vowel appears alone then word belong to simple category.

Mathematically, ∀ block ∈ CV or V
For example:आपकाA pa kA (V CV CV)

Complex – In any block of word, if a consonant follows a consonant then word belong to complex category.

Mathematically,∃ block ∈ CCV or CCCV
For example- स्वतंत्रsvawanwra (CCV CV CCCV)

3 Algorithm for Schwa Deletion

This section describe an algorithm which resolves the problem of Schwa deletion with the help of Sound effects(vowel and consonant).Rules we have used for Schwa deletion are recursive in nature. We have implemented rules up to 4 block length words for both simple and complex categories. Output of each block is further concatenated which leads to resultant phoneme.

3.1 For 1 Block Length Word

If block length = = 1 ..(1)

3.1.1 Word ∈ Simple (V or CV)

If

block contain V || CV then Schwa is retained or no change

Example: व (va), के (ke), या (yA)

3.1.2 Word ∈ Complex (CCV or CCCV or CCCCV)

If

Rightmost CV end with ya or va or ra (य orव or र) || rightmost V ∈ any vowel except Schwa 'a' then Schwa is retained or no change.

Example: श्यSya(CCV) ->Sya, क्याkyA(CCV) ->kyA

Else

Delete Schwa'a' or last V

Example: ब्दbxa ->bx

3.2 For 2 Block Length Word

If block length == 2 ..(2)

3.2.1 Word ∈ Simple (V or CV)

If

Last block's V ∈any vowel except Schwa (a) then no change

Example : रानीrAnI (CV CV) ->rAnI

Else

. Delete Schwa 'a' from last block and concatenate with rest of blocks

Example :रसrasa (CV CV) –>ras

3.2.2 Word ∈ Complex (CCV or CCCV or CCCCV)

If

Rightmost CV end with(ya or va or ra) || (in rightmost CV , If C ∈ any consonant && V ∈ vowel except Schwa 'a') than Schwa is retained or no change.

Example : अन्यa nya (V CCV) ->anya ; अंशु a nSu(V CCV) ->anSu

Else

Delete Schwa 'a' from last block and concatenate with rest of blocks

Example- सन्तsanwa(CV CCV) –>sanw ; श्याम SyA ma (CCV CV)->SyAm

3.3 For 3 Block Length Word

If block length == 3 ..(3)

3.3.1 Word ∈ Simple (V or CV)

If

Second block's V contain Schwa 'a' && third block 's V ∈ any vowel except Schwa 'a' than divide whole word (all three block)into ratio of 2:1 block.

First part of size 2 blocks is passed to equation (2) and second part of size 1 block is passed to equation (1) after this concatenate the results of both equations.

Example : चलताca la wA (CV CV CV) ->ca la wA(2:1) ->cal + wA (concatenate) ->calwA

Else

Divide whole word into 1:2 blocks. Here, first part of size 1 block will pass to equation (1) and second part of size 2 block will pass to equation (2) and then concatenate the results of both equations.

Example: चमकca ma ka (CV CV CV) ->ca ma ka(1:2) ->ca + mak (concatenate) ->camak

3.3.2 Word ∈ Complex (CCV or CCCV or CCCCV)

If

Consonant followed by consonant (CC) occurs in (third block) || (second && third block) || (first block), divide word (all three blocks) in ratio of 1:2 block.

First part of length 1 block is passed to equation (1) and second parts of length 2 blocks are passed to equation (2) and then concatenate the result of both equations.

Example:

कलंकसंबंधव्यापित

Ka la nkasambanXavyA pi wa

cv cv ccv cv ccv ccvccv cv cv

ka la nka (1:2) sambanXa(1:2) vyA pi wa(1:2)

ka + lank sa + mbanXvyA + piw

kalanksambanXvyApiw

Else

If

Second block's V is Schwa 'a' && consonant followed by consonant(CC) occurs in second block && third block's V ∈ any vowel except Schwa 'a', then divide whole word(all three block) in ratio 2:1.

First part of size 2 blocks passed to equation (2) and second part of size 1 block is passed to equation (1) and then concatenates the res sult of both equations.

Example : जंगलीjangalI (CV CCV CV) ->jangalI (2:1) ->jang + lI(concatenate) ->JanglI

Else

Divide the whole word (all three block) in ratio of 1:2 blocks.

First part of size 1 block passed to equation (1) and second part of size 2 block passed to equation (2) then concatenate the result of both equations.

Example : जंगलjanga la (CV CCV CV) ->janga la(1:2) ->ja + ngal (concatenate) ->jangal

3.4 For 4 Block Length Word

If block length = = 4 .. (4)

3.4.1 Word ∈ Simple (V or CV)

If

Third block's vowel ends with 'a' && fourth block end with any vowel except Schwa 'a' than divide whole word in ratio of 1:3 block.

First part of size 1 block is passed to equation (1) and second part of size 3 block is passed to equation (3) then concatenate the results of both equations.

Example- चमकताca ma kawA (CV CVCV CV) ->ca ma kawA(1:3)->
ca + makwA(concatenate) ->camakwA

Else

Divide whole word (all 4 blocks) in ratio of 2:2 .

First part of block size 2 passed to equation (2) and second part of block size 2 also passed to equation (2) than concatenate results of both equations.

Example : आसपासA sapAsa (V CV CV CV) –> A sapAsa(2:2) ->
A s + pA s (concatenate) ->AspAs

3.4.2 Word ∈ Complex (CCV or CCCV or CCCCV)

If

Rightmost CV is (ya or va or ra)|| last block's V ∈any vowel except Schwa 'a' than Schwa is retained or no change.

Example : राजस्थानी rAjasWAnI (CV CV CCV CV) ->rAjasWAnI
अनिवार्य a nivArya (CV CVCV CCV) ->anivArya

Else

Delete Schwa 'a' from last(delete V from rightmost CV)

Example : हस्ताक्षरha swAkSara (CV CCV CCV CV) ->haswAkSar

4 Results

In this section we have presented the accuracy of above algorithm with the help of graph. Above algorithm will not provide correct solution for compound words like जनसंख्या(janasanKyA). Morphological analyzer[7-9] could be used to segregate compound words into individual base word and then concatenate the results after applying algorithm on individual base word. For example:

Compound Word :janasanKyA (grapheme)

Using Morphological analyzer to segregate compound word into base words and apply algorithm to individual base words.

Baseword 1 : jana (grapheme) →jan (phoneme)

Baseword2 : sanKyA(grapheme) →sanKyA (phoneme)

Concatenate the base word phoneme to get resultant phoneme of compound word as jan + sanKyA = jansanKyA (phoneme)

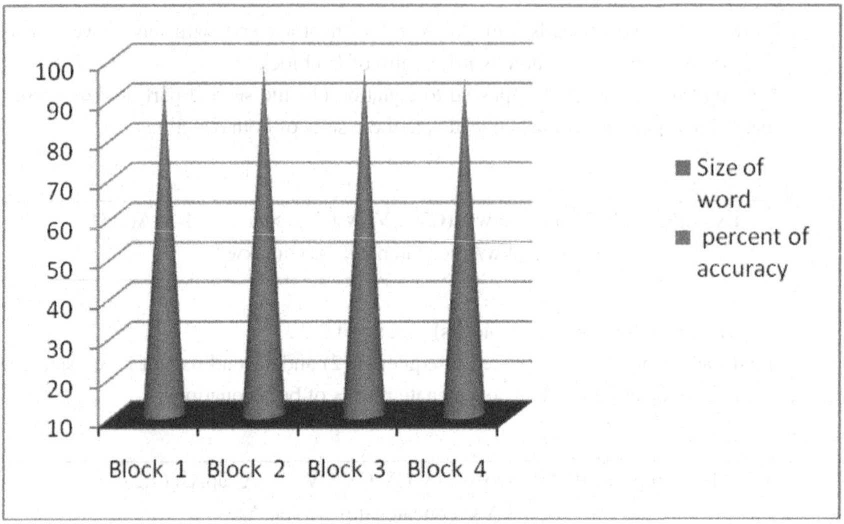

Fig. 1. Result Analysis

Fig.1.Shows the results of Rule based algorithm for Schwa deletion applied on up to 4 block length word as stated above. Results are manually tested on1700 base word shaving length up to 4 blocks only and got the overall accuracy of 96.47%. Further, number of correct result and accuracy for each block size words is shown in Table2.

Table 2. Test Results

Bloc	No. of words	Correct Result	Accuracy (in %)
1	2	18	93.5
2	6	58	98.1
3	5	48	97.2
4	4	37	94.5

5 Conclusion and Future Work

In this paper, we have presented a rule based algorithm for solving the problem of Schwa deletion in Hindi. We categorized words in blocks of length up to 4 and then applied rules, based on presence of vowel and consonant. We have manually tested algorithm on 1700 base words having up to 4 block length and got 96.47 % accuracy. To increase accuracy Morphological analyzer is used. In future, we will try to extend algorithm for greater block length along with handling of exceptions arised.

References

1. Wasala, A., Weerasinghe, R., Gamage, K.: Sinhala Grapheme-to-Phoneme Conversion and Rules for Schwa Epenthesis. In: Proceedings of the COLING/ACL 2006 Main Conference Poster Sessions, pp. 890–897 (2006)
2. Choudhury, M.: Rule-Based Grapheme to Phoneme Mapping for Hindi Speech Synthesis In: 90th Indian Science Congress of the International Speech Communication Association
3. Data sheet for W-X notation, `http://caltslab.uohyd.ernet.in/wx-notation-pdf/Gujarati-wx-notation.pdf`
4. Narasimhan, B., Sproat, R., Kiraz, G.: Schwa-Deletion in Hindi Text-to-Speech Synthesis. International Journal Speech Technology 7(4), 319–333 (2004)
5. Singh, P., Lehal, G.S.: A Rule Based Schwa Deletion Algorithm for Punjabi TTS System. In: Singh, C., Singh Lehal, G., Sengupta, J., Sharma, D.V., Goyal, V. (eds.) ICISIL 2011. CCIS, vol. 139, pp. 98–103. Springer, Heidelberg (2011)
6. Choudhury, M., Basu, A.: A rule based algorithm for Schwa deletion in Hindi. In: International Conf. on Knowledge-Based Computer Systems, pp. 343–353 (2002)
7. Jurafsky, D., Martin, J.H.: Speech and Language Processing: An Introduction to Natural Language Processing, Computational Linguistics, and Speech Recognition. Pearson Education (2000)
8. Dabre, R., Amberkar, A., Bhattachrya, P.: A way to them all:A compound word analyzer for Marathi. In: ICON (2013)
9. Deepa, S.R., Bali, K., Ramakrishnan, A.G., Talukdar, P.P.: Automatic Generation of Compound Word Lexicon for Hindi Speech Synthesis

Unsupervised Word Sense Disambiguation for Automatic Essay Scoring

Prema Nedungadi and Harsha Raj

Amrita Vishwa Vidyapeetham
Amritanagar, Tamil Nadu - 641112, India
prema@amrita.edu

Abstract. The reliability of automated essay scoring (AES) has been the subject of debate among educators. Most systems treat essays as a bag of words and evaluate them based on LSA, LDA or other means. Many also incorporate syntactic information about essays such as the number of spelling mistakes, number of words and so on. Towards this goal, a challenging problem is to correctly understand the semantics of the essay to be evaluated so as to differentiate the intended meaning of terms used in the context of a sentence. We incorporate an unsupervised word sense disambiguation (WSD) algorithm which measures similarity between sentences as a preprocessing step to our existing AES system. We evaluate the enhanced AES model with the Kaggle AES dataset of 1400 pre-scored text answers that were manually scored by two human raters. Based on kappa scores, while both models had weighted kappa scores comparable to the human raters, the model with the WSD outperformed the model without the WSD.

Keywords: Latent Semantic Analysis (LSA), SVD, AES, Word Sense Disambiguation.

1 Introduction

With large classrooms, teachers often find it difficult to provide timely and high quality feedback to students with text answers. With the advent of MOOC, providing consistent evaluations and reporting results is an even greater challenge. Automated essay scoring (AES) systems have been shown to be consistent with human scorers and have the potential to provide consistent evaluations and immediate feedback to students [4].

A teacher evaluates student essays based on students' understanding of the topic, the writing style, grammatical and other syntactic errors. The scoring models may vary based on the question, for example, in a science answer the concepts may carry more weight and the grammatical errors may be less important while in a language essay grammar, spelling and syntactical errors may be as important as the content. Hence, for each essay, AES learns the concepts from learning materials and the teachers scoring model from previously scored essays. Most AES systems today that use LSA consider the important terms as

M.K. Kundu et al. (eds.), *Advanced Computing, Networking and Informatics - Volume 1*, Smart Innovation, Systems and Technologies 27,
DOI: 10.1007/978-3-319-07353-8_51, © Springer International Publishing Switzerland 2014

a bag of words and cannot differentiate the meaning of the terms in the context of the sentence. In order to improve the accuracy of AES, we incorporate unsupervised word sense disambiguation as part of the pre-processing. In contrast with conventional WSD approaches, we did not just take the senses of the target word for scoring; but measure the similarity between gloss vectors of the target word, and a context vector comprising the remaining words in the text fragment containing the words to be disambiguated resulting in better WSD [3].

The rest of the paper is organized as follows: we first present existing approaches to AES. Then we discuss the system architecture of proposed AES system that incorporates WSD. Next we train and test the system and compare the grading accuracy of the AES system that incorporates WSD with the base AES system. Finally, we show using weighted kappa scores that the proposed model has a higher inter-rater agreement than the previous model.

2 Existing Systems

Automated essay scoring systems are very important research area in educational system and may use NLP, Machine Learning and Statistical Methods to evaluate text answers. In this section, we discuss existing essay scoring systems and word sense disambiguation methods.

2.1 Project Essay Grader (PEG)

Project Essay Grade (PEG) is the first research based on scoring essays by computer. It uses measures to evaluate the intrinsic quality of the essay for grading. Proxes denote the estimation of the intrinsic variables such as fluency, diction, grammar, punctuation, etc., Apart from the content, PEG grading is done based on the writing quality.Using multiple regression technique, training in PEG needs to be done for each essay set used. Page's latest experiments achieved results reaching a multiple regression correlation as high as 0.87 with human graders.

2.2 E-Rater

The basic method of E-Rater is similar to PEG. In addition, E-rater measures semantic content by using a vector-space model. Document vector for the essay to be graded are constructed and its cosine similarity is computed with all the pre-graded essay vectors. The essay takes the score of the essay that it closely matches. E-rater cannot detect humour, spelling errors or grammar. It evaluates the content by comparing the essays under same score.

2.3 Intelligent Essay Assessor (IEA)

IEA is an essay grading technique, where a matrix is built from the essay documents, and dimension reduced by the SVD technique. Column vectors are created for each ungraded essay, with cell values based on the terms (rows) from the original matrix. The average similarity scores from a predetermined number of sources that are most similar to this is used to score the essay.

2.4 Corpus-Based Word Sense Disambiguation (WSD)

A corpus-based WSD method uses supervised learning techniques to generate a classifier from training data that are labelled with the sense of the term. The classifier is then used to predict the sense of the target word in novel sentences.

2.5 Graph-Based Word Sense Disambiguation Using Measures of Word Semantic Similarity

This word sense disambiguation in this work is an unsupervised method that uses weighted graph representation for word sense dependencies in text [2]. It uses multiple semantic similarity measures for centrality based algorithm on the weighted graph to address the problem of word sense disambiguation.

3 Automated Essay Scoring with WSD

In our previous work, we had described an automatic text evaluation and scoring tool A-TEST that checks for surface features such as spelling errors and word count and also uses LSA to find the latent meaning of text [5]. In this paper, we discuss the enhancements to the existing system that incorporates word sense disambiguation. Though LSA systems need a large number of sample documents and have no explanatory power, they work well with AES systems [1].

Our AES system is designed to learn and grade essays automatically and first learns important terms from the golden essays or the course materials. Next it uses a set of pre-scored essays that have been given a grade by human raters manually as the training set to create the scoring model. These essays are pre-processed as a list of words or terms with stop-word removal, stemming, lemmatizing, and tokenized.

Algorithm 1. Learn course material with WSD

Input: Golden Essays or Course Material
Output: The Reduced Matrix with the correct sense of the terms.

Step1: Preprocess the training essay set (spelling correction, stop word, lemmatizing)
Step2: Extract the sense of each word in every sentence of the essay (WSD).
Step3: Generate the Term-by-document matrix using the output from WSD
Step4: Decompose A into U, V and (singular value decomposition)

3.1 WSD Process

Next we enhance the AES to include a modified version of a WSD algorithm [3] that identifies the sense or meaning of a word in the context of the sentence by considering the sense of every other word in the sentence instead of '?'.The most likely sense is extracted using the word senses from WordNet, and selecting the sense which has highest similarity with the remaining words in the sentence. Once the sense is determined, this sense is fixed and used for determining the sense of the remaining words thus reducing the time complexity of the WSD.

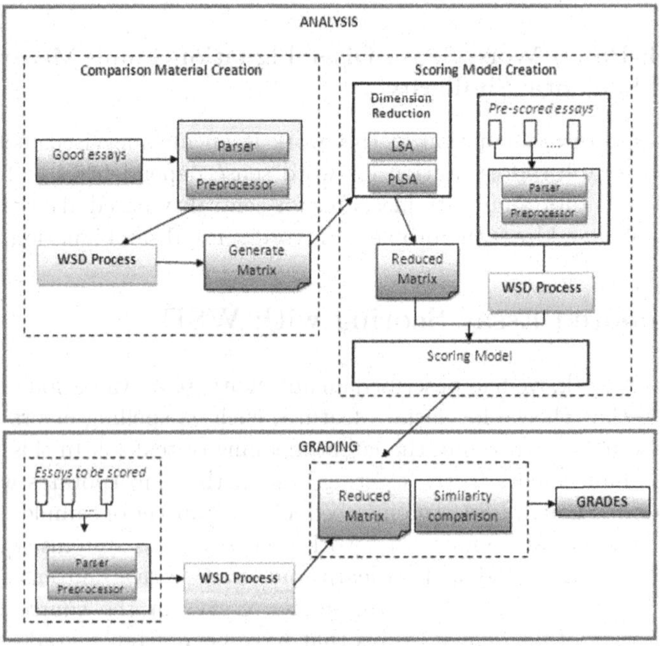

Fig. 1. WSD enhanced AES

The output of the WSD process (Refer Fig. 1) is used by the next phase, LSA to determine the correct of the word used in each sentence of essays [1]. A sense-by-document matrix is created which represents the golden essays, followed by LSA for dimensionality reduction.The document vectors of the pre-scored training essays are created after the pre-processing and WSD process and then compared with the reduced matrix to find the similarity score of the best match. A scoring model is derived using the similarity score, the spelling error and the word count of the essay.

3.2 Scoring Model Using Multiple Regression Analysis

The scoring model was determined using system R, with the similarity score, grade corresponding to the best match, the number of spelling errors and word count.

Scoring Model with WSD

$$Score = diff.in\_spell\_errors * 0.053306 + (Spell - error)^1/4 * 1.008228 + Word\_count * 0.002787 + Similarity\_Measure * 0.277596$$

Scoring Model without WSD

$$Score = diff.in\_spell\_errors * 0.043408 + (Spell - error)^1/4 * 0.793543 + Word\_count * 0.004804 + Similarity\_Measure * 0.174913$$

The last step is the grading phase. Document vectors from the essay to be graded go through the same preprocessing and WSD process and derive the similarity measure of the best match. The scoring model is used to determine the grades for the new essays.

4 Results and Performance Evaluation

We used the Kaggle dataset to test the new AES with WSD and compared the results to the AES without the WSD processing. We tested our model using 304 essays from dataset of 1400 pre-scored essays while the remaining were used to train the AES. There were two human rater scores for each essay that ranged from 1 to 6, where 6 was the highest grade.

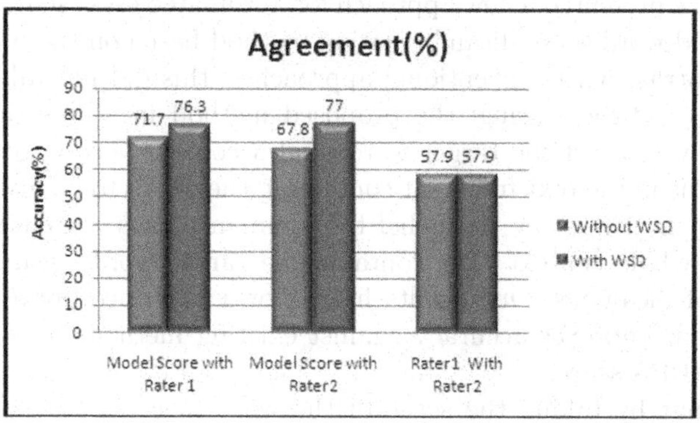

Fig. 2. Agreement between model scores with Rater1 and Rater2

The scores from the first human rater were used to learn the scoring model. The agreement between the model-scores with human raters is shown in Fig. 2. AES with WSD could correctly classify 232 essays from the 304 essays in the testing phase while the AES without WSD could classify 218 out of the 304 essays. The percentage of this is illustrated in the figure. It is interesting to note that the human raters only agreed 57.9% of the time.

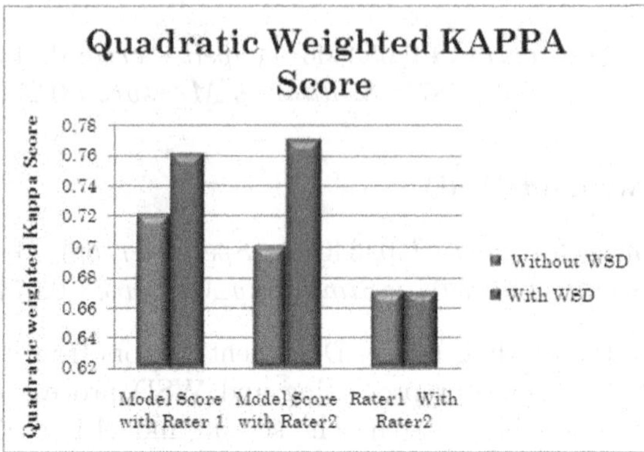

Fig. 3. Quadratic weighted Kappa score

4.1 Kappa Score

While both the models showed good inter-rater agreement, the quadratic weighted Kappa's score show a significant improvement in inter-rater reliability using the AES system with WSD to the base AES system (Refer Fig. 3).

5 Conclusion

This project has presented a new approach for automated essay scoring by taking similarity-based word sense disambiguation method based on the use of context vectors. In contrast with conventional approaches, this did not take the senses of the target word for scoring; the proposed method measures the similarity between gloss vectors of the target word, and a context vector comprising the remaining words in the text fragment containing the words to be disambiguated. This has been motivated by the belief that human beings disambiguate words based on the whole context that contains the target words, usually under a coherent set of meanings. Our results have shown that incorporating WSD to the AES system improves accuracy against existing methods, as evaluated by taking the KAPPA score.

We show that by taking the sense of the word in context to all the other words in the sentence, we improved the inter-rater agreement of our model with the human raters. Though our prediction model worked as well as the manually evaluated teacher model, additional enhancements such as n-grams, grammar specific errors, and other dimensionality methods such as PLSA can further improve the inter-rater accuracy and are planned as further work.

The proposed system is applicable for essay with raw text. In future the proposed work will be extended towards the grading of essays containing text, tables and mathematical equations.

Acknowledgment. This work derives direction and inspiration from the Chancellor of Amrita University, Sri Mata Amritanandamayi Devi. We thank Dr.M

Ramachandra Kaimal, head of Computer Science Department, Amrita University for his valuable feedback.

References

1. Valenti, S., Neri, F., Cucchiarelli, A.: An overview of current research on automated essay grading. Journal of Information Technology Education 2, 319–330 (2003)
2. Sinha, R., Mihalcea, R.: Unsupervised Graph-based Word Sense Disambiguation Using Measures of Word Semantic Similarity. In: IEEE International Conference on Semantic Computing (2007)
3. Abdalgader, K., Skabar, A.: Unsupervised similarity-based word sense disambiguation using context vectors and sentential word importance. ACM Trans. Speech Lang. Process. 9(1) (2012)
4. Kakkonen, T., Myller, N., Sutinen, E., Timonen, J.: Comparison of Dimension Reduction Methods for Automated Essay Grading. Educational Technology and Society 11(3), 275–288 (2008)
5. Nedungadi, P., Jyothi, L., Raman: Considering Misconceptions in Automatic Essay Scoring with A-TEST - Amrita Test Evaluation & Scoring Tool. In: Fifth International Conference on e-Infrastructure and e-Services for Developing Countries (2013)

Disjoint Tree Based Clustering and Merging for Brain Tumor Extraction

Ankit Vidyarthi and Namita Mittal

Malaviya National Institute of Technology Jaipur, India
{2012rcp9514,nmittal.cse}@mnit.ac.in

Abstract. Several application areas like medical, geospatial and forensic science use variety of clustering approaches for better analysis of the subject matter. In this paper a new hierarchical clustering algorithm called Disjoint Tree Based Clustering and Merging, is proposed which clusters the given dataset on the basis of initially generating maximum possible disjoint trees followed by tree merging. Proposed algorithm is not domain specific and be used for both data points and image. For the result analysis, the algorithm is tested on medical images for clustering of the abnormality region (tumor) from brain MR Images. Proposed algorithm was also compared with the standard K-Means algorithm and the implementation results shows that the proposed algorithm gives significant results for tumor extraction.

Keywords: Clustering; Disjoint Trees, Tree Merging, Tumor.

1 Introduction

Clustering helps to bring items as closely as possible on the basis of similarity and forming a group. The formed group, which was called cluster, has low variance within the cluster and high variance between the clusters which gives the intension that within the group the items are more similar in certain sense as compared to other items in different clusters. Clustering can also be used in many application fields, including machine learning, pattern recognition, image analysis, information retrieval, and bioinformatics. Since there is no prescribed notation of cluster [1] that's why there are numerous clustering algorithms exists but every algorithm have a single motive of grouping the data points. Some of the mainly used clustering models [2] are Connectivity model, Density Model, Centroid based model and Graph based Models.

The literature helps to find that existing clustering algorithms for images had cluster out the region based on either of some predefined threshold value like Otsu threshold based clustering [4] which results in generation of cluster which was dependent on specific threshold value or clustering was dependent on randomization. Some of the randomization algorithms like K-means and K-centroid [3] give significant results with certain limitations like initial selection of the value of k and the selection of the random centroid point for each k clusters. These clustering approaches were known as exclusive clustering as the pixel which belongs to a definite cluster doesn't include in any other cluster.

M.K. Kundu et al. (eds.), *Advanced Computing, Networking and Informatics - Volume 1*, Smart Innovation, Systems and Technologies 27,
DOI: 10.1007/978-3-319-07353-8_52, © Springer International Publishing Switzerland 2014

Medical Image Computing is one such field where the interaction of various domain experts of medicine, computer science, data science, and mathematics field contribute collectively [6]. The main focus of the experts is to extract the relevant information which was present in the medical image for better analysis of the subject matter. Various imaging modalities are present for the studying of these abnormalities like Computed Tomography (CT), Magnetic Resonance (MR), Positron Emission Tomography (PET) and Magnetic Resonance Spectroscopy (MRS) [9]. Among all imaging modalities MR imaging is widely used for the diagnosis of the tumor patients as it provides the good contrast over the soft tissues of the body without taking the use of radiations while other technique like CT scans and PET uses radiations to identify the problems which were quite harmful to the body. These MR images are being analyzed by the radiologist to manually segment the abnormality regions.

The rest of this paper is organized as follows: Section 2 discusses about the previous literature survey. Section 3, describes the proposed clustering approach for tumor extraction. Section 4, gives the results followed by conclusion in Section 5.

2 Related Work

Segmentation of the abnormality from the MR imaging modalities is a huge concern for the radiologist. In the past various distinguished approaches and methods have been proposed based on different approaches like edge detection [10], Otsu threshold [7] and fuzzy logics [5, 8] for fast and effective segmentation. The literature suggests that all the previously applied methods and approaches basically dependent on initial selection of some threshold value which divides the image into sub regions based on that threshold value.

An edge based approaches like modified Ant Colony Optimization (ACO) algorithm using probabilistic approach [11] and a canny edge based approach with watershed [10] algorithm are used for the extraction of the tumor region from brain MR image gives satisfactory results but due to weak gradient magnitude the algorithm works effectively well only for high contrast images and degrades performance with low contrast images. The various clustering based segmentation approaches like k-means and k-centroid [3], works well for the extraction of tumors but as the algorithm was based on initial selection of the value of K and random selection of k-centroids so every time when the algorithm runs shows variation. These methods have the limitation that they work well only for convex shape problems.

In [7], authors used hybrid technique for tumor segmentation from MR image using Otsu threshold with fuzzy c-means algorithm. Otsu threshold was used to find out the homogenous regions in an image followed by the segmentation of these homogenous regions using fuzzy clustering approach. Another hybrid mechanism based upon Otsu and Morphological operation [12] was also used for clustering out the tumor region from MR image. Morphological operation was used to cluster the tumor region based on the selection of seed point which was identified using the Otsu threshold. These algorithms gave significant results but the initial selection of the threshold value for Otsu algorithm is still a concern. Hierarchical clustering algorithm like HSOM with FCM [18] was used for abnormality (tumor) extraction from medical images. HSOM basically used for mapping the higher dimensional data to the lower dimensional space.

The key idea of such is to place the closely weighted nodes of the vector graph together and using fuzzy c-means to form the cluster using adaptive threshold value.

Some of the automatic tumor segmentation algorithms [13, 14] cluster the tumor region based on the automatic selection of the seed point which was used by the fuzzy connectedness approach. Such an approach resolves problem of random selection of seed point but for the extraction of the tumor, algorithm was still dependent on certain threshold value. Some other clustering approaches based on symmetry analysis [15], area constrained based clustering [16] and probabilistic approach using Hidden Markov Model [17] was also tested for tumor extraction but all the algorithms somehow used threshold value to segment the tumor region from MR image.

3 Proposed Approach

For the extraction of the tumor region in MR image, a new hierarchical clustering approach is introduced which is based upon the concept of graph theory. Proposed approach focuses on generating disconnected trees followed by merging of the trees for clustering for the tumor region inside an image. The algorithm is free from any initial selection of the threshold value. The proposed clustering algorithm is given in Algorithm 1.

Algorithm 1. Disjoint Tree based clustering and Merging

Input: *Brain MR Image* (I)
Output: *Clustered Image* (out)
Step 1: *Preprocessing*
f← FilterImage(I)
gr←Gray(f)
Step 2: *Disjoint Tree Generation*
V ←{p_i ; p_i ∈ gr}
for each v_i∈ V
for each v_k∈ V && v_i != v_k
E ←Findedge (v_i ,v_k)
V′ ← UpdateSet(V)
T ← FindTree(V,E)
end for
end for
return (T)
Step 3: *Tree Merging*
k← no of clusters
while (k)
for each t_i∈ T
d← FindDistance (t_i , t_j)
m← minimum(d)
c← MinimumDistanceCluster(t_i , t_j)
T′ ← UpdateSet (T)
end for
out← clusterset(c)

3.1 Disjoint Tree Generation

After preprocessing of the input brain MR image, disjoint trees be generated based on the concept of graph theory. Intensity values of the image are considered as vertices of the graph. As per the definition of the clustering, the main focus is to bring closely the items of similar type, thus key idea of making disjoint trees is to connect the intensity values which are same at one tree. The key concept is shown by an example in Fig. 1. A sample matrix is shown in Fig.1(a) and its corresponding disjoint trees are shown in Fig. 1(b) (for simplicity only two trees are shown).

$$\begin{pmatrix} 0 & 1 & 2 & 0 & 4 \\ 2 & 4 & 1 & 3 & 0 \\ 3 & 0 & 2 & 4 & 1 \\ 1 & 0 & 3 & 2 & 4 \end{pmatrix}$$

Fig. 1. (a). sample Matrix

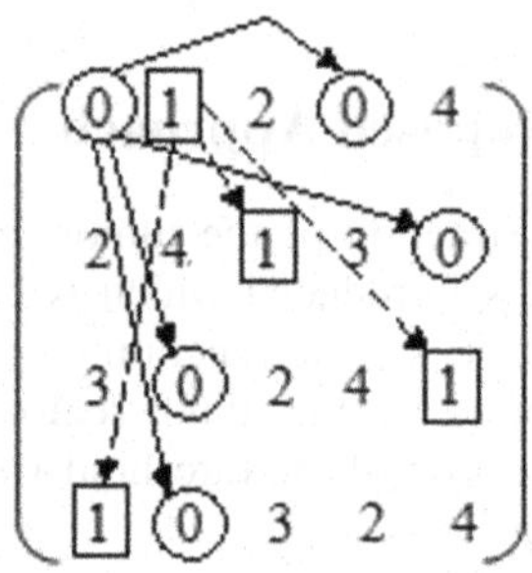

Fig. 1. (b). Corresponding Disjoint Trees

Each tree consists of items of the items in form of intensity values of similar type. When the node is initialized for a tree then the algorithm will find all such nodes in an image having same intensity value as root node and join to the tree which results in generation of the disjoint trees having variation in root values. Initially algorithm assumes that all these disjoint trees are forming clusters having same intensity values as shown in Fig. 2.

3.2 Tree Merging

In this phase, the corresponding generated clusters from above step were merged on the basis of minimum distance criteria. Since for an 8 bit image, the pixel range for gray image lies between in a range of 0-255. Thus for an image of 8 bit the maximum possible generated cluster will be lay in the range of:

$$1 \leq C \leq 256 \tag{1}$$

As the size of the image increases, the corresponding number of clusters was also increased. In a worst case (for our research work we are using 8bit image only) the maximum number of clusters be 256, while in average case the range of cluster lies between as shown in equation (1), which itself is a huge number and also increases complexity. Now these clusters were merged on the basis of finding minimum distance between the trees. Before applying the distance criteria, all the generated clusters as shown in Fig. 2 were replaced by the mean value of the cluster. These values are used for finding the minimum distance from each cluster of tree to every other cluster as shown in Fig. 3.

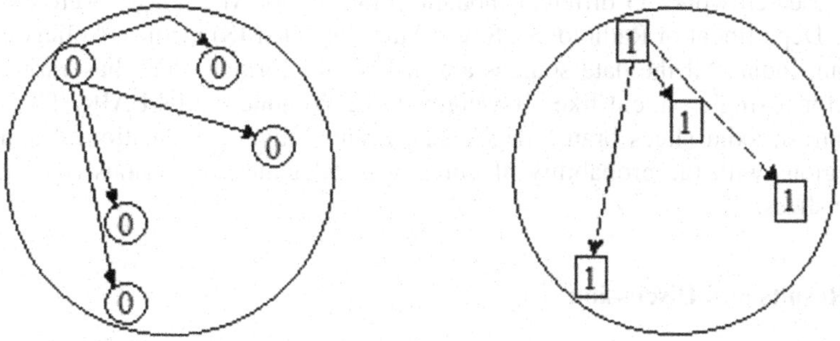

Fig. 2. Generated Clusters corresponding to Disjoint Trees

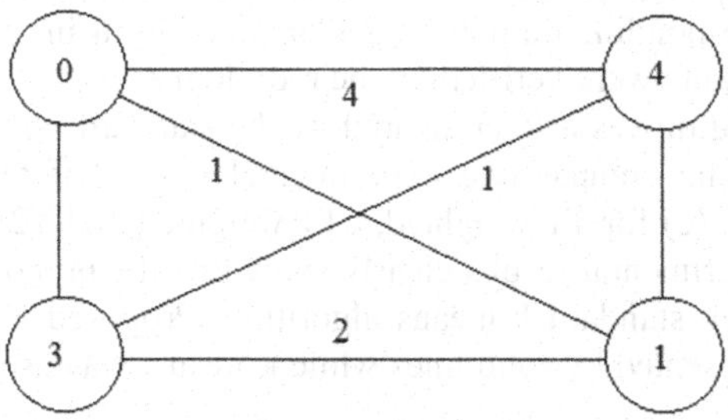

Fig. 3. Cluster Distance

On the basis of minimum distance between the clusters, the corresponding clusters were merged to form a new cluster whose value is being replaced by the maximum of the value of the corresponding clusters to be merged as shown in Fig. 4.

Fig. 4. Corresponding Merged Clusters

The key idea for replacing the value of new cluster with maximum of the value joining clusters is for maximizing the inter cluster distance as much as possible and minimizing intra cluster distance. Thus the formed cluster is differentiated to each other as much as possible. Since proposed methodology uses intensity values for tree formation and merging, thus higher intensity value gives brightness and lower intensity value gives dullness to an image. On the basis of this key concept it was found thatmaximum the difference formed between two clusters better be the visibilitybetween the clustered regions of an image seen.

4 Experimental Results

4.1 Dataset Used

For the research work, 25 different modality brain tumor MR images were collected from the Department of Radiologist, Sawai Man Singh (SMS) medical College Jaipur, Rajasthan, India. All the data sources are in DICOM format which has various modalities for a single patient like T1-weighted, T2-Weighted, T1-FLAIR, T2-FLAIR, post contrast, axial slices brain images. In current study, segmentation of abnormal brain regions with the probability of tumor was calculated on various types of MR Imaging slices.

4.2 Results and Discussion

Proposed algorithm was tested on various MR images having tumor of variable shape and size. Fig. 5(a), (b), and (c) shows the input MR image of T1-weighted, eT1-weighted and T2-Flair post contrast images which was used for study. Fig. 5(d), (e), and (f) shows the corresponding output images obtained by using the proposed algorithm. Experimental results were verified by the radiologist.

The proposed algorithm was also compared by the standard k-means algorithm on various values of k. The comparative experimental results of the algorithm were shown in Fig. 6(a), (b), (c) for T1 weighted, eT1 weighted and T2 flair post contrast brain MR images. Experimental results clearly show that the proposed algorithm will less execution time then standard k-means algorithm. Proposed algorithm computational values were represented by solid lines while k-means via dashed line.

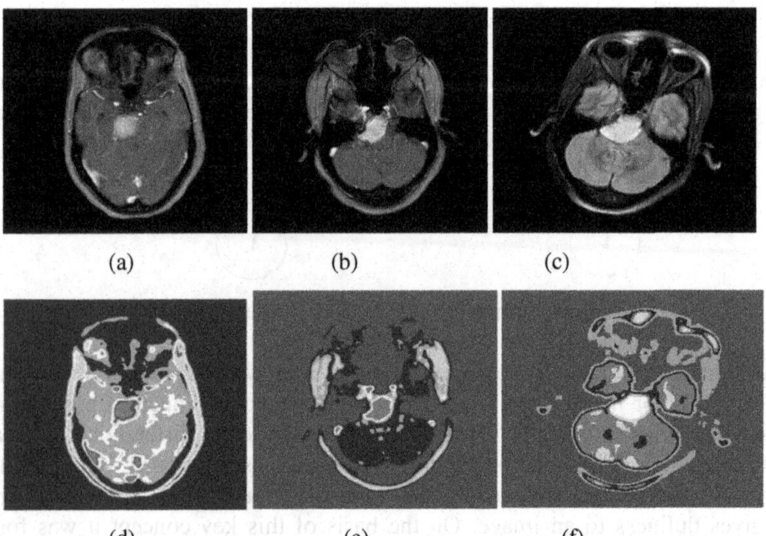

Fig. 5. Illustration of the input MR post contrast images (a) T1 weighted (b) eT1 weighted (c) T2 flair proposed algorithm output at k =4: (d) T1 weighted (e) eT1 weighted (f) T2 flair

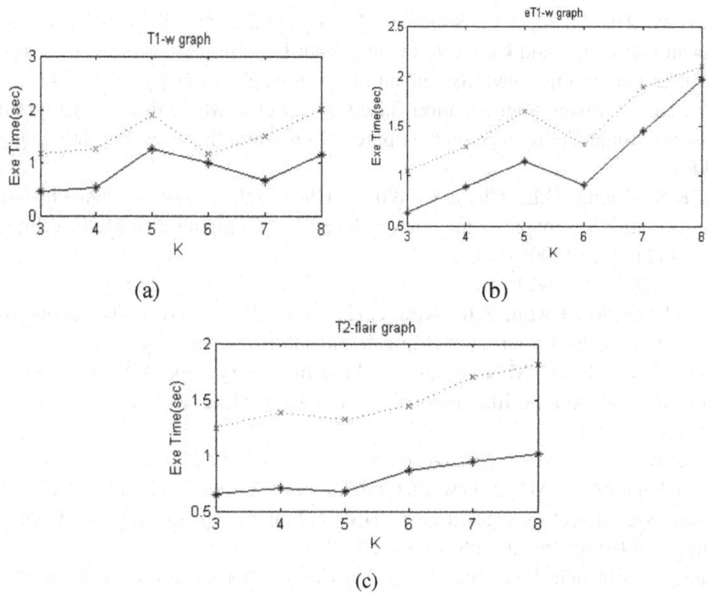

Fig. 6. Comparative simulation results of k-means algorithm with proposed algorithm on the basis of execution time for (a) T1 weighted (b) eT1 weighted (c) T2 flair post contrast brain MR images

5 Conclusion

In this paper a new hierarchical clustering algorithm is proposed for the extraction of the tumor region from brain MR Image. Proposed algorithm is based on the concept of graph theory where the image is being clustered on the basis of finding the maximum possible disjoint trees in an image followed by the merging of the trees and generating k clusters using minimum distance metric concept. The proposed algorithm was tested on various values of k and the experimental results show that the proposed algorithm gives significant results by giving maximum possible abnormality region in one cluster on k = 4. Proposed algorithm was also compared by the standard K-means algorithm for various runs.

Acknowledgement. We are thankful to the Sawai Man Singh (SMS) Medical College Jaipur for providing us the original brain tumor images. We would also like to thanks Dr. Sunil Jakhar, MD, Radio Diagnosis, Department of Radiology, SMS Medical College Jaipur for helping us in verifying the results.

References

1. Estivill-Castro, V.: Why so many clustering algorithms - A Position Paper. ACM SIGKDD Explorations Newsletter 4(1), 65–75 (2000)
2. Han, J., Kamber, M.: Data Mining: concepts and techniques, 2nd edn. Morgan Kaufmann Publishers (2006)

3. Moftah, H.M., Hassanien, A.E., Shoman, M.: 3D Brain Tumor Segmentation Scheme using K-mean Clustering and Connected Component Labeling Algorithms. In: 10th International Conference on Intelligent System and Design Application, pp. 320–324 (2010)
4. Zhang, J., Hu, J.: Image Segmentation Based on 2D Otsu Method with Histogram Analysis. In: International Conference on Computer Science and Software Engineering, pp. 105–108 (2008)
5. Chuang, K.S., Tzeng, H.L., Chen, S., Wu, J., Chen, T.J.: Fuzzy c-means clustering with spatial information for image segmentation. Journal of Computerized Medical Imaging and Graphics 30(1), 9–15 (2006)
6. [Online] `http://en.wikipedia.org/wiki/Medical_image_computing`
7. Nyma, A., Kang, M., Kwon, Y.K., Kim, C.H., Kim, J.M.: A Hybrid Technique for Medical Image Segmentation. Journal of Biomedicine and Biotechnology (2012)
8. Mohamed, N.A., Ahmed, M.N., Farag, A.: Modified fuzzy c-mean in medical image segmentation. In: 20th Annual International Conference on Engineering in Medicine and Biology Society (1998)
9. [Online] `http://en.wikipedia.org/wiki/Medical_imaging`
10. Maiti, I., Chakraborty, M.: A new method for brain tumor segmentation based on watershed and edge detection algorithms in HSV colour model. In: National Conference on Computing and Communication Systems (2012)
11. Soleimani, V., Vincheh, F.H.: Improving Ant Colony Optimization for Brain MRI Image Segmentation and Brain Tumor Diagnosis. In: First Iranian Conference on Pattern Recognition and Image Analysis (2013)
12. Vidyarthi, A., Mittal, N.: A Hybrid Model for the Extraction of Brain Tumor in MR Image. In: International Conference on Signal Processing and Communication (2013)
13. Weizman, L., Sira, L.B., Joskowicz, L., Constantini, S., Precel, R., Shofty, B., Bashat, D.B.: Automatic segmentation, internal classification, and follow-up of opticpathway gliomas in MRI. Journal of Medical Image Analysis 16, 177–188 (2012)
14. Harati, V., Khayati, R., Farzan, A.: Fully Automated Tumor Segmentation based on Improved Fuzzy Connectedness Algorithm in Brain MR Images. Journal of Computers in Biology and Medicine 41, 483–492 (2011)
15. Khotanlou, H., Colliot, O., Atif, J., Bloch, A.: 3D brain tumor segmentation in MRI using fuzzy classification, symmetry analysis and spatially constrained deformable models. Journal of Fuzzy Sets and System 160, 1457–1473 (2009)
16. Niethammer, M., Zach, C.: Segmentation with area constraints. Journal of Medical Image Analysis 17, 101–112 (2013)
17. Solomon, J., Butman, J.A., Sood, A.: Segmentation of Brain Tumors in 4D MR Images using Hidden Markov Model. Journal of Computer Methods and Programs in Biomedicine 84, 76–85 (2006)
18. Logeswari, T., Karnan, M.: An Improved Implementation of Brain TumorDetection Using Segmentation based onHierarchical Self Organizing Map. International Journal of Computer Theory and Engineering 2(4), 591–595 (2010)

Segmentation of Acute Brain Stroke from MRI of Brain Image Using Power Law Transformation with Accuracy Estimation

Sudipta Roy[1], Kingshuk Chatterjee[2], and Samir Kumar Bandyopadhyay[3]

[1] Department of Computer Science and Engineering
Academy of Technology, Adisaptagram, West Bengal, India
[2] E.C.S.U, Indian Statistical Institute, Kolkata-700108
Department of Computer Science and Engineering
[3] University of Calcutta, 92 A.P.C. Road, Kolkata-700009, India
sudiptaroy01@yahoo.com, kingshukchaterjee@gmail.com,
skb1@vsnl.com

Abstract. Segmentation of acute brain stroke and its position is very important task in medical community. Accurate segmentation of brain's abnormal region by computer aided design (CAD) system is very difficult and challenging task due to its irregular shape, size, high degree of intensity and textural similarity between normal areas and abnormal regions areas. We developed a new method using power law transformation which gives very fine result visually and from quantifiable point of view. Our methods gives very accurate segmented tumor output with very low error rate and very high accuracy.

Keywords: Brain Stroke, MRI of Brain, Power Law Transformation, Segmentation, Accuracy Estimation, CAD system, Abnormal Region.

1 Introduction

Stroke cannot be considered a diagnosis in itself. Stroke refers to any damage to the brain or spinal cord caused by a vascular abnormality, the term generally being reserved for when symptoms begin abruptly. Stroke is anything but a homogeneous entity, encompassing disorders as different as rupture of a large blood vessel that causes flooding of the subarachnoid space with blood, the occlusion of a tiny artery supplying a small but strategic brain site and thrombosis of a venous conduit obstructing outflow of blood from the brain. When physicians speak of stroke, they generally mean there has been a disturbance in brain function, often permanent, caused by either a blockage or a rupture in a vessel supplying blood to the brain. In order to function properly, nerve cells within the brain must have a continuous supply of blood, oxygen, and glucose (blood sugar). If this supply is impaired, parts of the brain may stop functioning temporarily. If the impairment is severe, or lasts long enough, brain cells die and permanent damage follows. Because the movement and functioning of various parts of the body are controlled by these cells, they are affected

M.K. Kundu et al. (eds.), *Advanced Computing, Networking and Informatics - Volume 1*, Smart Innovation, Systems and Technologies 27,
DOI: 10.1007/978-3-319-07353-8_53, © Springer International Publishing Switzerland 2014

also. The symptoms experienced by the patient will depend on which part of the brain is affected.

Symptoms of stroke include numbness, weakness or paralysis, slurred speech, blurred vision, confusion and severe headache. R K. Kosioret *et al.*[1] proposed automated MR topographical score using digital atlas to develop an objective tool for large-scale analysis and possibly reduce interrater variability and slice orientation differences. The high apparent diffusion coefficient lesion contrast allows for easier lesion segmentation that makes automation of MR topographical score easier. Neuro imaging plays a crucial role in the evaluation of patients presenting acute stroke symptoms. While patient symptoms and clinical examinations may suggest the diagnosis, only brain imaging studies can confirm the diagnosis and differentiate hemorrhage from ischemia with high accuracy. C S. Kidwell *et al.*[2] compared the accuracy of MRI and CT for detection of acute intra cerebral hemorrhage in patients presenting with acute focal stroke symptoms and when it became apparent at the time of an unplanned interim analysis that MRI was detecting cases of hemorrhagic transformation not detected by CT. Thus MRI image used is better than CT image for brain stroke detection. Chawla *et al.*[3] detected and classifed an abnormality into acute infarct, chronic infarct and hemorrhage at the slice level of non-contrast CT images. A two-level classification scheme is used to detect abnormalities using features derived in the intensity and the wavelet domain. Neuroimaging in acute stroke [4], [5], [6] is essential for establishment of an accurate diagnosis, characterization of disease progression, and monitoring of the response to interventions. MRI has demonstrably higher accuracy, carries fewer safety risks and provides a greater range of information than CT.

We use the data set of "Whole Brain Atlas" image data base [7] which consist of T1 weighted, T2 weighted, proton density (PD) MRI image and use the part of the dataset slice wise. The paper is organized as follows: In section 2 we describes our proposed methodology ; In section 3 we discussed about our results; In section 4 we do quantification and accuracy estimate of the method with mathematical metric ; and finally in section 5 we conclude our paper.

2 Proposed Methodology

A RGB image has been taken as an input and converted into gray image first. Binarization is very effective preprocessing methods for most of the segmentation for the MRI imaging technique. Due to large variation of background and foreground of MRI of brain images maximum number of binarization technique fails, and here a binarization method developed by Roy *et al.*[9] has been selected with a global threshold value which has been selected by standard deviation of the image. Global thresholding using standard deviation of image pixels gives very good results and binarize each interesting part of the MRI image. A image I[i, j] and h is the intensity of each pixel of the gray image. Thus the total intensity of the image is defined by:

$$T = \sum_{I} h[I]$$

The mean intensity of the image is defined as the mean of the pixel intensity within that image and the mean intensity is defined as Imean by:

$$I_{mean} = \frac{1}{T} \sum_{(i,j)\epsilon I} I[i,j]$$

The standard deviation Sd of the intensity within a image is the threshold value of the total image is defined by:

$$S_d = \sqrt{\frac{1}{T-1} \sum_{i,j\epsilon I} (I[i,j] - I_{mean})^2}$$

or

$$S_d = \sqrt{\frac{1}{T-1} \sum_{i,j\epsilon I} I^2[i,j] - T{I_{mean}}^2}$$

Here the threshold intensity as global value i.e. the threshold intensity of the entire image is unique. The binarized image using standard deviation intensity Sd of the image pixel of a image I[i, j] or matrix element for I[i, j] is given by :

$$I[i,j] = 1 \qquad if I[i,j] \geq S_d$$

$$I[i,j] = 0 \qquad if I[i,j] < S_d$$

After complimenting the binary image the two dimensional wavelet decompositions [10] have been done using 'db1' wavelet up to second level and re-composition of the image has been done using the approximate coefficient. The objectives of these two steps are to remove the detailed information from the complementary image which helps to remove skull from the brain. The previous process have done as separation of skull to the brain portion as skull and brain may or may not connected, thus previous process gives the surety that skull and brain portion are not connected and results in decrease in size of the complementary image to half of the original image, moreover due to reduction of size and removal of detailed information the white pixel of the complementary image come closure and form a complete ring. Then interpolation method is used to resize the image of the previous step to the original size then re-complement of the image is done and these results produce the complete separation between brain and skull. Then labeling of the image has been done using union find method. Except maximum area all other component are removed, and in this process skull and other artefact are removed. Actually removal of artefact and skull improve the detection quality and reduce the error rate of false detection.

As the image contains one pixels structure are discrete in nature that's why quick-hull [11], [12], [13] algorithm for Convex Hulls is used here to generate original image without artefact and skull. It is computed for these one pixel and the entire pixels inside the convexhull are set to one and outside it are set to zero. Here if some of the brain part where set to zero during the binarization stages due to wrong evaluation this step ensure that the error is corrected. Then obtained binarized image is multiplied with the original image pixel wise and produce the desired results i.e. MRI of brain image without artifacts and border and is stored 'r'. Power-law transformations [14] have the basic form

$$s = cr^y$$

Where c and y are positive constants and sometimes power law equation can be written as high to account for an offset (that is, a measurable output when the input is zero)

$$s = c(r + \epsilon)^y$$

However, offsets typically are an issue of display calibration and as a result they are normally ignored, using power law a family of possible transformation can obtained simply by varying the value of y and curves generated with values of y>1 have exactly the opposite effect as those generated with values of y<1. Using this power law transformation we increase the visibility of the abnormal region(stroke) by taking y as 4 and we can segment easily the abnormal region (like acute stroke speaks nonsense words, acute stroke speech arrest caused by the portion of brain) by summing average intensity of s with standard deviation of s of non black region.

$$\text{Total} = \frac{1}{T} \sum_{(i,j)\epsilon I} s[i,j] + Sd(s)$$

Depending on this total threshold intensity we can find the abnormal region of brain stroke from MRI of brain. Acute stroke region are the more intense part of the brain and also the different intensity form the normal brain intensity, so intensity of tumor is more than that of other region. Then first select average intensity value of the non black region and calculate the intensity value by standard deviation. As the intensity level of acute stroke and other abnormalities are greater than that of other region of the brain that's why using the total intensity thresholds value that can easily detect the abnormality and other abnormalities of brain from MRI. For abnormality position detection we need to calculate centroid of the brain stroke region, this is done by weighted mean of the pixels and mathematical formulation of calculation of centroid is given below:

$$X_{cood} = \sum\nolimits_{n=1}^{p} x_n I_n / \sum\nolimits_{n=1}^{p} I_n$$

$$Y_{cood} = \sum\nolimits_{n=1}^{p} y_n I_n / \sum\nolimits_{n=1}^{p} I_n$$

Where p is the total number of pixels; in is the individual pixel weigh. After calculating the centroid we can easily find the abnormal regions or brain stroke position from the center position of the brain. Here distance between brain centers to centroid of the abnormal regions has been calculated as center to centroid distance using X coordinate and Y coordinate named as X_{cood} and Y_{cood}. Distance between brain top positions to stroke regions top position named as TT, distance between brain left position to stroke regions left position named as LL, distance between brain right position to stroke regions right position named as RR, and brain bottom position to stroke regions bottom are named as BB.

3 Results and Discussion

Conventional T2-weighted images and MR PD images have converted to high signal in the lesion and showed a large area of abnormal signal in the region clinically suspected: the portion of left hemisphere supplied by the middle cerebral artery. Abnormally bright signal is seen here because of the presence of excess water which has a prolonged relaxation time. As tissue has become infracted and edematous, the sulcus is no longer identifiable and compares the infracted side with the normal right side. Thus we can easily find the position of the infected region by our algorithms and also hemisphere which are shown below very distinctly. This automated accurate size and positioning of centroid detection help for diagnosis purpose.

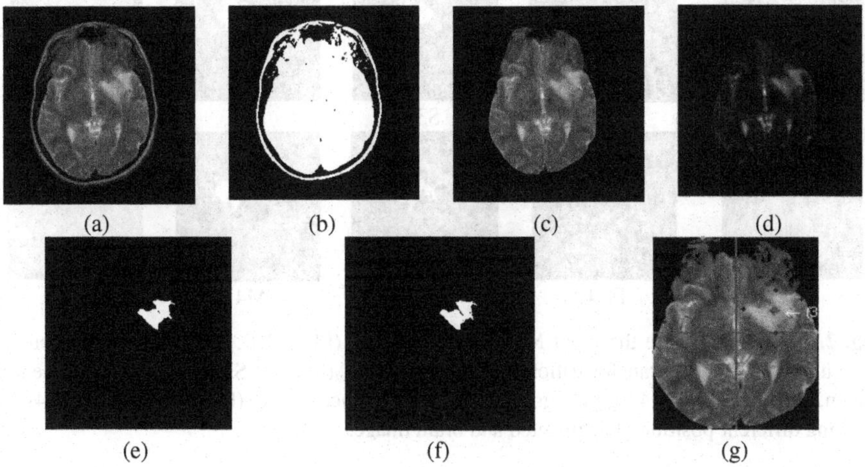

Fig. 1. (a) original input image, (b) binarized image, (c) without artefact and skull of brain, (d) after using power law transformation, (e) segmented region, (f) ground truth image, (g) position of abnormality(stroke)

Fig. 1(b) is the binarized output for an input Fig. 1(a) by the standard deviation method and this binarization is very helpful for removing the skull of the brain. Fig. 1(c) is the output of the brain without skull and artifact which helps to detect the brain stroke region and its helps accurate (without false detection) detection. Fig. 1(d) is output of brain image using power law transformation and from this image abnormal region is truly visible and finally in Fig. 1(e) is the segmented part. Finally in Fig. 1(g) shows the position of abnormal part according to the hemisphere. Some other results of brain MRI images from standard database (slice wise) with ground truth image are shown in below.

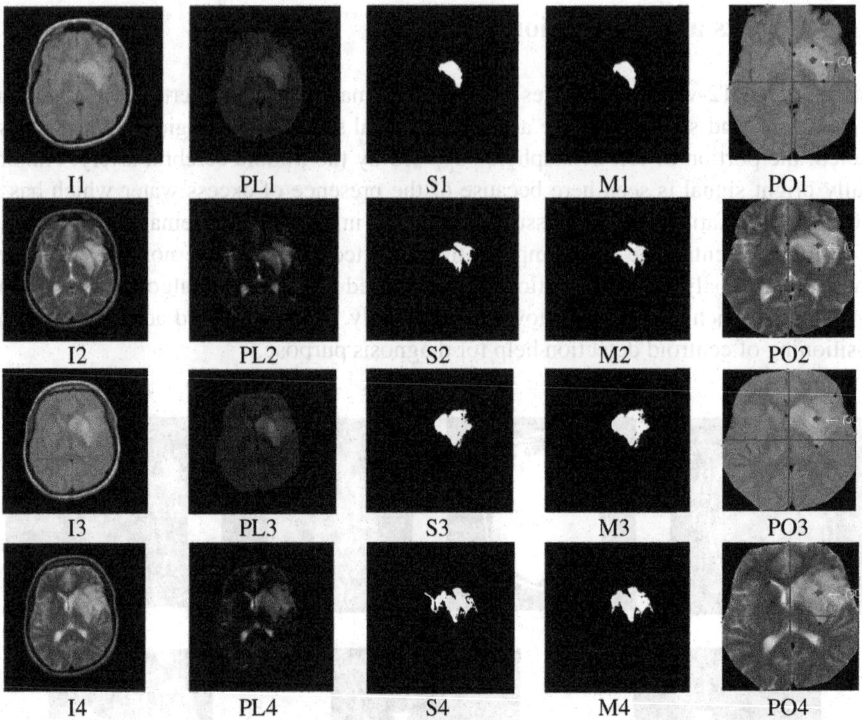

Fig. 2. (I1, I2, I3, I4) are the input MRI of brain image, (PL1, PL2, PL3, PL4) are the output after using power law transformation and without skull, (S1, S2, S3, S4) are the segmented region, (M1, M2, M3, M4) are the ground truth or reference image, (PO1, PO2, PO3, PO4) are showing different position of segmented and brain image.

Table 1 shows the qualifications of different type of brain MR images that are used in thestudy to measure centroid, segmented area and their locations. The centroid position are defined by 'X' and 'Y' coordinate (here X_{cood}, Y_{cood}). Positive X coordinate means centroid located right part of the brain from viewer side and negative X means left part of the brain from viewer side and TT,BB, LL,RR are described in the methodology section.

4 Quantification and Accuracy Estimation

Characterizing the performance of image segmentation methods is a challenge in image analysis. An important difficulty we have to face in developing segmentation methods is the lack of a gold standard for their evaluation. Accuracy of a segmentation technique refers to the degree to which the segmentation results agree with the true segmentation. Although physical or digital phantoms can provide a level of known "ground truth", they are still unable to reproduce the full range of imaging characteristics and normal and abnormal anatomical variability observed in clinical data.

Table 1. Distances between different positions of segmented to the brain portion and area of segmented area in pixels

Image name	Xcood	Ycood	TT	BB	LL	RR	Segmented area
12_T2	34.0889	21.6207	83.0963	111.3059	89.0674	35.4683	972
13_PD	24.1981	24.2943	70.0928	102.2155	83.6481	34.1760	952
13_T2	33.7338	28.8135	66.3702	100.7025	80.5295	30.3645	1394
14_PD	30.2732	26.4251	61.0574	99.8098	78.7718	20.2485	2582
14_T2	30.5932	27.4891	37.1214	99.6393	68.5930	00.0000	2168
16_PD	27.6816	23.3254	50.9902	91.7061	84.0536	30.4631	1994
16_T2	25.8965	22.8745	63.8643	97.8981	87.8766	32.7527	1903
17_PD	17.8586	10.6768	57.8705	77.4726	85.000	39.3192	1245
17_T2	23.3428	12.5841	55.6606	80.6475	86.7888	42.6549	872
18_PD	21.5468	13.1462	54.7814	89.0056	89.1852	39.8121	673
18_T2	20.4809	15.8391	44.8412	92.5758	82.7893	40.6584	669

First let AV and MV denote [15] the area of the automatically and manually segmented objects and |x| represents the cardinality of the set of voxels x. In the following equations Tp = MV ∩ AV, Fp = AV – Tp and Fn = MV – Tp denote to the "true positive", "false positive" and "false negative" respectively. The Kappa index between two area is calculated by the following equation:

$$K_i(AV, MV) = \frac{(2|AV \cap MV|)}{(|AV| + |MV|)} * 100\%$$

The similarity index is sensitive to both differences in size and location. The Jaccard index between two areas is represented as follow:

$$J_i(AV, MV) = \frac{|AV \cap AM|}{|Tp + Fn + Fp|} * 100\%$$

This metric is more sensitive to differences since both denominator and numerator change with increasing or decreasing overlap. Correct detection ratio or sensitivity is defined by the following equation:

$$C_d = \frac{|AV \cap MV|}{MV} * 100\%$$

False detection ra tio (F_d) is same except 'AV-Tp' in place of 'AV MV'. The Relative Error [8] (RE) for stroke region can be calculated as "AV" stroke area using automated segmentation , "MV" is stroke area using manual segmentation using expert.

$$RE = \frac{(AV - MV)}{MV} * 100\%$$

Table 2 shows the specifications of different type of brain MR images that are used in the study to measure the accuracy level and different type of image metrics. The results that are obtained in this study from Table2 are described here after.

Table 2. Value of different metric for accuracy estimation from ground truth image

Image name	**AV**	**MV**	**RE in %**	**Tp**	**Fp**	**Fn**	**Kappa index**	**Jacard index**	**% of Cd**	**% of Fd**
12_T2	972	927	4.8543	927	45	00	0.9763	0.9537	100.00	4.8544
13_PD	952	968	1.6528	932	20	36	0.9708	0.9433	96.280	2.0700
13_T2	1394	1297	7.4787	1297	97	00	0.9639	0.9304	100.00	7.48
14_PD	2582	2584	0.0773	2582	00	02	0.9996	0.9992	99.920	0.00
14_T2	2168	2222	2.4302	2000	168	222	0.9112	0.8368	90.010	7.560
16_PD	1994	1821	8.6760	1821	173	00	0.9547	0.9132	100.00	9.5003
16_T2	1903	1872	1.6559	1850	53	32	0.9801	0.9560	98.824	2.8648
17_PD	1245	1291	3.5631	1245	00	46	0.9819	0.9644	96.440	0.00
17_T2	872	815	6.9938	801	71	14	0.9496	0.9040	98.282	8.7116
18_PD	673	691	2.6049	666	07	25	0.9765	0.9542	96.380	1.010
18_T2	669	622	7.5562	609	60	13	0.9419	0.8841	97.909	9.6946

The similarity index or similar is sensitive to both differences in size and location. For the similarity index, differences in location are more strongly reflected than differences in size and Si > 70% indicates a good agreement. In our experiment Kappa index reaches above 90% maximum times thus according to the Kappa index our methods produce very good results. Jacard index is more sensitive to differences since both denominator and numerator change with increasing or decreasing overlap and in our methodology maximum times it gives greater than 90% which indicates our experiment promising. Correct detection ratio indicates the correct detection area normalized by the reference area and is not sensitive to size. Therefore correct detection ratio solely cannot indicate the similarity and should be used with false detection ratio or other volume metrics. False detection ratio shows the error of the segmentation and indicates the volume that is not located in the true segmentation. Using this metric with the correct detection ratio can give a good evaluation of the segmentation. Maximum times our methodology gives above 95% correct detection ratio and below 5% false detection ratio with very low relative area error (maximum times below 5%). However, the overlap measure depends on the size and the shape complexity of the object and is related to the image sampling. Thus our methodology gives very good results visually as well as quantifiably.

5 Conclusion

We have argued a completely automatic segmentation system of brain stroke from MRI Brain for PD as well as T2 type of images and the system can simply adapted to different attainment sequences as presented in Section 2 and Section 3, which are the main advantages of our method. Numerous variables, such the age of the stroke, systemic blood pressure, collateral flow, and the treatment very much depend on the size and location of the area of infarction and from that point of view it successfully execute visually as well as mathematically. However, since we have more data coming in the future, to make the system understanding with sequence different MRI, we may need to capture new knowledge and tune the corresponding rules or parameters for more accurate CAD system.

References

1. Kosior, R.K., Lauzon, M.L., Steffenhagen, N., Kosior, J.C., Demchuk, A., Frayne, R.: Atlas-Based Topographical Scoring for Magnetic Resonance Imaging of Acute Stroke. American Heart Association 41(3), 455–460 (2010)
2. Kidwell, C.S., Chalela, J.A., Saver, J.L., Starkman, S., Hill, M.D., Demchuk, A.M., Butman, J.A., Patronas, N., Alger, J.R., Latour, L.L., Luby, M.L., Baird, A.E., Leary, M.C., Tremwel, M., Ovbiagele, B., Fredieu, A., Suzuki, S., Villablanca, J.P., Davis, S., Dunn, B., Todd, J.W., Ezzeddine, M.A., Haymore, J., Lynch, J.K., Davis, L., Warach, S.: Comparison of MRI and CT for Detection of Acute Intracerebral Hemorrhage. The Journal of the American Medical Association 292(15) (2004)
3. Chawla, M., Sharma, S., Sivaswamy, J., Kishore, L.: A method for automatic detection and classification of stroke from brain CT images. In: Proceedings of IEEE Engineering in Medicine and Biology Society, pp. 3581–3584 (2009)
4. Perez, N., Valdes, J., Guevara, M., Silva, A.: Spontaneous intracerebral hemorrhage image analysis methods: A survey. Advances in Computational Vision and Medical Image Processing Computational Methods in Applied Sciences 13, 235–251 (2009)
5. Smith, E.E., Rosand, J., Greenberg, S.M.: Imaging of Hemorrhagic Stroke. Magnetic Resonance Imaging Clinics of North America 14(2), 127–140 (2006)
6. Merino, J.G., Warach, S.: Imaging of acute stroke. Nature reviews. Neurology 6, 560–571 (2010)
7. Online, http://www.med.harvard.edu/AANLiB/cases/
8. Roy, S., Nag, S., Maitra, I.K., Bandyopadhyay, S.K.: A Review on Automated Brain Tumor Detection and Segmentation from MRI of Brain. International Journal of Advanced Research in Computer Science and Software Engineerin 3(6), 1706–1746 (2013)
9. Roy, S., Dey, A., Chatterjee, K., Bandyopadhyay, S.K.: An Efficient Binarization Method for MRI of Brain Image. Signal & Image Processing: An International Journal (SIPIJ) 3(6), 35–51 (2012)
10. Barber, C.B., Dobkin, D.P., Huhdanpaa, H.: The Quickhull Algorithm for Convex Hulls. ACM Transactions on Mathematical Softwar 22(4), 469–483 (1996)
11. Daubechies, I.: Ten lectures on wavelets. CBMS-NSF conference series in applied mathematics. SIAM Ed. (1992)
12. Mallat, S.: A theory for multiresolution signal decomposition: the wavelet representation. IEEE Transactions on Pattern Analysis and Machine Intelligence 11(7), 674–693 (1989)
13. Meyer, Y.: Ondelettesetopérateurs, Tome 1, Hermann Ed. (1990); English translation: Wavelets and operators. Cambridge Univ. Press (1993)
14. Gonzalez, R.C., Woods, R.E.: Digital Image Processing, 2nd edn. (2002)
15. Khotanlou, H.: 3D brain tumors and internal brain structures segmentation in MR images. Thesis (2008)

A Hough Transform Based Feature Extraction Algorithm for Finger Knuckle Biometric Recognition System

Usha Kazhagamani and Ezhilarasan Murugasen

Department of Computer Science and Engineering,
Pondicherry Engineeing College, India
ushavaratharajan@gmail.com, mrezhil@pec.edu

Abstract. Finger Knuckle Print is an emerging biometric trait to recognize one's identity. In this paper, we have developed a novel method for finger knuckle feature extraction and its representation using Hough transform. Hough transform plays a significant role in locating features like lines, curves etc., present in the digital images by quantifying its collinear points. This paper formulates the Elliptical Hough Transform for feature extraction from the captured digital image of finger knuckle print (FKP). The primary pixel points present in the texture patterns of FKP are transformed into a five dimensional parametric space defined by the parametric representation in order to describe ellipse. The discrete coordinate of the five dimensional coordinate spaces along with its rotational angle are determined and characterized as the parameters of elliptical representation. These parametric representations are the unique feature information obtained from the captured FKP images. Further, this feature information can be used for matching various FKP images in order to identify the individuals. Extensive experimental analysis was carried out to evaluate the performance of the proposed system in terms of accuracy. The obtained results shows the lowest error rate of EER = 0.78%, which is found to be remarkable when compared to the results of existing systems presented in the literature.

1 Introduction

Finger knuckle print of a person has inherent structure patterns present in the outer surface of the finger back region which is found to be highly unique and stable birth feature (invariant throughout the lifetime). These patterns have greater potentiality towards the unique identification of individuals which in turn contributes a high precision and rapid personal authentication system [1].

The various other hand based biometric modalities are fingerprints, palmprints, finger geometry, hand geometry and hand vein structures etc., Fingerprints have greater vulnerability towards tapping of finger patterns when it is left on the surface of the image acquiring device. On the other hand, the feature information extracted from finger geometry and hand geometry are found to have less discriminatory power when size of the data sets grows exponentially [2]. In palm prints, the region of interest captured for feature extraction is large in size which may result in computational overhead. The hand vein structures biometric modality has the main drawback of complex capturing system to conquer the vein patterns of the hand dorsum surface.

M.K. Kundu et al. (eds.), *Advanced Computing, Networking and Informatics - Volume 1*, Smart Innovation, Systems and Technologies 27,
DOI: 10.1007/978-3-319-07353-8_54, © Springer International Publishing Switzerland 2014

Unlike fingerprints, finger knuckle prints are very difficult to scrap because it is captured by means of contact less capturing devices and patterns captured are present in the inner surface of the finger knuckle. Moreover, the presence of phalangeal joint in the finger knuckle surface creates flexion shrinks which has rich texture patterns like lines, creases and wrinkles and size of the captured finger knuckle print is very small when compared to palm prints, which reduces the computational overhead.

In general, the recognition methods for finger knuckle print matching are categorized into two broad categories viz., Geometric based methods and Texture based methods [3]. In texture analysis methods, the feature information is extracted by means of analyzing the spatial variations present in the image. In this analysis, the mathematical models are used to characterize spatial variation in terms of feature information. The spatial quantifiers of an image can be derived by analyzing the different spectral values that is regularly repeated in a region of large spectral scale. Texture analysis on the digital image results in a quantification of the image the texture properties.

Kumar and Ravikanth [4] were the first to explore texture analysis method for finger knuckle biometrics. In their work, the feature information of captured finger knuckle print is extracted by means of statistical algorithms viz, principal component analysis, linear discriminant analysis and Independent component analysis. Further, Zhang *et al.* [5] derives a band limited phase only correlation method (BPLOC) for matching finger knuckle prints based on texture analysis. Linlin shen *et al.* [6] used two dimensional Gabor filters to extract feature information from the finger knuckle print image texture and uses hamming distance metric to identify the similarities and the differences between the reference and input images of finger knuckle print. Furthermore, Chao lan *et al.* [7] proposes a new model based texture analysis methods known as complex locality preserving projection approach for discriminating the finger knuckle surface images. Lin Zhang *et al.* [8] used ridgelet transform method to extract the feature information from the captured FKP images. A.Merounica *et al.*, [9] have shown the implementation of Fourier transform functions for deriving the feature information from the finger knuckle region and palm region.

From the study conducted, it has been found that texture methods provides high accuracy rate when the captured image is of high quality images. But, in the case of any missing information in the input image and also in the case of low quality images like noisy images, these methods fail to work. This paper addresses this problem by proposing new methodology for feature extraction from finger knuckle print based on Elliptical Hough Transform. This feature extraction process is robust towards missing information and well tolerant to noise when the FKP images are captured partially and by using low quality sensors respectively.

2 The Proposed System Design

The following Fig.1 illustrates the design of the proposed biometric recognition system based on finger knuckle print. The proposed biometric system gets the images of

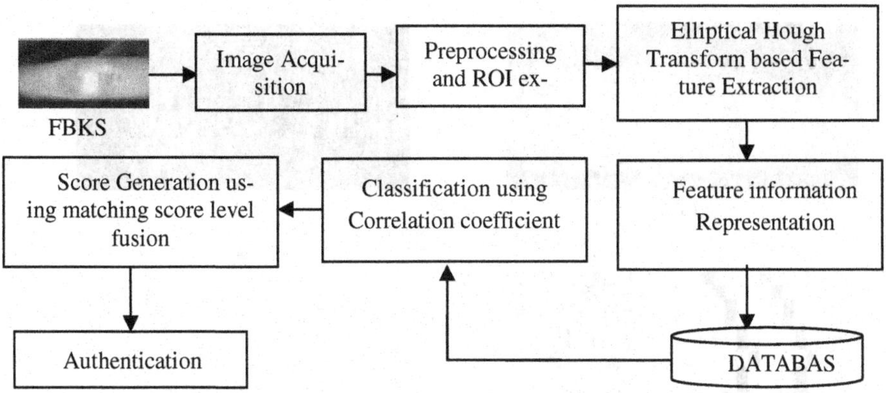

Fig. 1. Block diagram of proposed biometric authentication system based on finger knuckle print

right index finger knuckle, left index finger knuckle, right middle finger knuckle and left middle finger knuckle are given as input to the proposed model.

Initially, preprocessing and extraction of region of interest of the finger knuckle print is done based on edge detection method. Secondly, Elliptical Hough Transform based feature extraction process is done to derive the feature information from the finger knuckle print. Thirdly, the extracted feature information is represented to form the feature vector. This vector information is passed to the matching module which implemented using correlation coefficient. Finally, matching scores generated from different fingers are fused based on matching score level fusion to make the final decision on identification.

2.1 Preprocessing and ROI Extraction

The preprocessing and ROI extraction is done on the captured FKP in order to extract a portion of the image which is rich in texture patterns. These processes enable to extract highly discriminative feature information from the captured FKP image even though it varies according to their scaling and rotational characteristics. The preprocessing of the captured FKP image is done by incorporating the coordinate system [10]. This is achieved by defining the x- axis and y-axis for the captured image. The base line of the finger is taken as x-axis. The y-axis for the knuckle patterns in the finger is defined by means of convex curves determined from the edge record of the canny edge detection algorithm [11]. The curvature convexities of the obtained convex curves are determined for finger knuckle patterns. The Y axis of the finger knuckle print is derived by means of the curvature complexity which is nearly equal to zero at the center point. By this method, ROI is extracted from the FKP which of 110x180 pixel size.

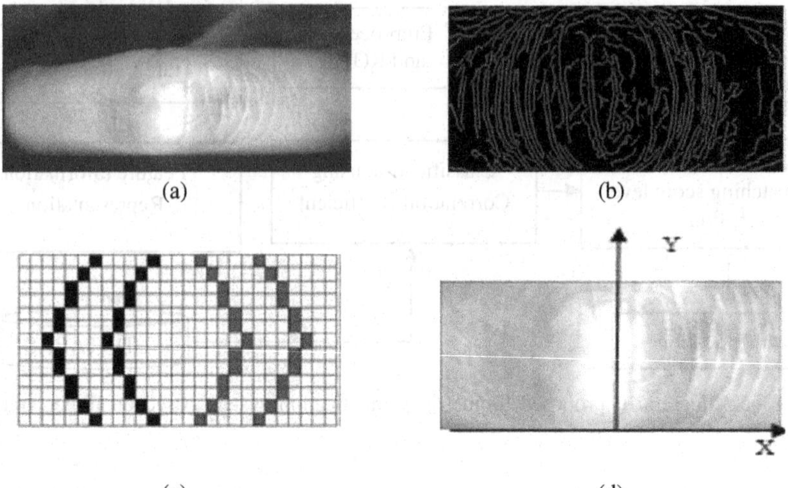

Fig. 2. (a) Captured finger knuckle print,(**b**) FKP image subjected to canny edge detection algorithm,(**c**) Convex Curves of FKP image,(d) Coordinated system for the FKP image

The above Fig. 2 (a), (b),(c), and (d) illustrates the captured finger knuckle print, edge image of the FKP detected using canny edge detection algorithm, convex curves detected from FKP image and coordinated system constructed for FKP image.

2.2 Elliptical Hough Transform Based Feature Extraction Method

Elliptical Hough Transform (EHT) [12] is used to isolate an elliptical structure from the knuckle part based on the primary pixel points obtained from the FKP image patterns. An elliptical structure posses a unique property which states that, the summation of distance obtained from the every point on the curved line to the two specific points (termed as foci points) located horizontally inside the curved line remains constant. From the FKP image, the foci points and a point on a curved line can be derived from the convex curves obtained through the preprocessing of the captured image. The following figure 3 shows the convex curves representing the foci points and base points on the curve. The convex curve obtained through preprocessing of the FKP, curves both left wards and rightwards from the phalangeal joint point. While preprocessing any number of curves are obtained according to the patterns of the captured finger knuckle print. The obtained curves can be classified as innermost curve and outer most curves of finger knuckle print. From this, the edge points of the innermost curve on either side of the phalangeal joint is considered as foci points and the base point of the outer most right side convex curve is considered as the point on the curved line. Here, the foci points are $(f_{x1}, f_{y1}), (f_{x2}, f_{y2})$ and the base points on the curves termed as(b_x, b_y).

In Cartesian coordinates the elliptical structure can be using the foci points and curve base points using the following Eq. 1,

$$S_d = \sqrt{(b_x - f_{x1})^2 + (b_y - f_{y1})^2} + \sqrt{(b_x - f_{x2})^2 + (b_y - f_{y2})^2} \quad (1)$$

Where S_d is the sum of the distances from the two foci points to the base point of the curve.

The set of curve points can be defined by the relation shown in Eq. 2.

$$f(f_{x1}, f_{y1}, f_{x2}, f_{y2}, S_d), (b_x, b_y) = \sqrt{(b_x - f_{x1})^2 + (b_y - f_{y1})^2} + \sqrt{(b_x - f_{x2})^2 + (b_y - f_{y2})^2} \quad (2)$$

The set of curve points with the forms the ellipse using the curve tracing algorithm [15] , with following Eq. 3.

$$f(f_{x1}, f_{y1}, f_{x2}, f_{y2}, S_d), (b_x, b_y) = \sqrt{(b_x - f_{x1})^2 + (b_y - f_{y1})^2} + \sqrt{(b_x - f_{x2})^2 + (b_y - f_{y2})^2} - S_d = 0 \quad (3)$$

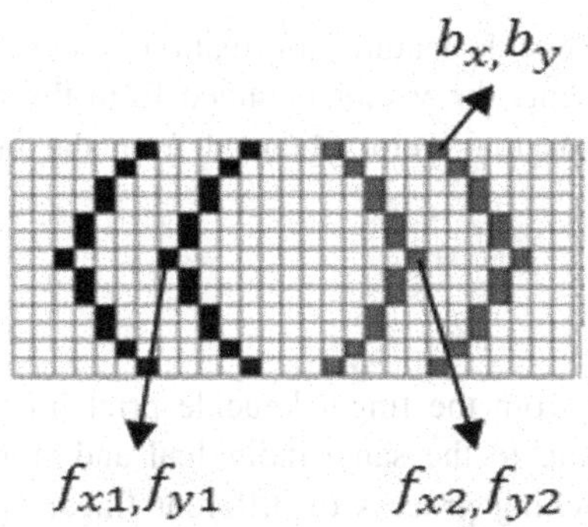

Fig. 3. Representation of knuckle foci points and base points

Hence the knuckle pixel points are identified from the pattern generated from the captured finger knuckle prints, which are transformed in to a parametric representation to derive in the form of elliptical structure. The knuckle feature information can be obtained by analyzing the elliptical structure.

2.3 Feature Information Representation

The knuckle feature information is obtained by examining the elliptical structure derived from the finger knuckle patterns. The feature vector represents the parameters of the knuckle elliptical structure which can be derived from the following equations.

The distance from the center to the focus point defined as knuckle foci distance K_f and given by Eq. 4.

$$K_f = \frac{\sqrt{(b_x - f_{x1})^2 + (b_y - f_{y1})^2}}{2} \quad (4)$$

The primary and secondary axes of the knuckle elliptical structure, which can be defined as primary knuckle axis (K_p) and secondary knuckle axis (K_s),are given by Eq. 5 and Eq. 6.

$$K_p = \frac{S_d}{2} \quad (5)$$

$$K_s = \sqrt{\left(\frac{S_d}{2}\right) - {K_f}^2} \quad (6)$$

Further, knuckle base point angle (K_a), can be derived from the knuckle elliptical structure as given by Eq. 7.

$$K_a = \tan^{-1}\left(\frac{f_{y2} - f_{y1}}{f_{x2} - f_{x1}}\right) \quad (7)$$

The obtained finger knuckle feature information is stored in the vector named as $V_{ref} and V_{inp}, V_{ref}$ represents the vector obtained from the registered finger knuckle image and V_{inp} represents the vector obtained from the input image of the finger knuckle image.

2.4 Classification

The feature information from the finger knuckle print image were derived for four different fingers belonging to the same individual and stored in the corresponding feature vector. The matching process of different finger knuckle prints is done by means of finding the correlation between reference and input vectors. The derivation of correlation coefficient [13] for the classification process is done by means of the following Eq.8.

$$\rho = \frac{\sigma_{xy}}{\sigma_x \sigma_y} \quad (8)$$

where x *and* y are values taken from reference and input vectors respectively.

If the value of ρ is close to 1, then high degree of similarity is identified, when the value of ρ diminishes to 0 indicates the dissimilarity.

2.5 Matching Score Level Fusion

Matching Score level fusion scheme is adopted to consolidate the matching scores produced by the knuckle surfaces of the four different fingers [14]. In the Matching score level, different rules can be used to combine scores obtained by from each of the finger knuckle. In this paper, sum rule is used. In the sum rule, say that $S_1, S_2, S_3, and\ S_4$ represent score values obtained from finger knuckle surface of left index finger, right index finger, left middle finger and right middle finger respectively. The final score S_f is computed using the following Eq. 9.

$$S_f = S_1 + S_2 + S_3 + S_4 \tag{9}$$

From the obtained final score the authentication decision is taken.

3 Experimental Analysis and Results Discussion

The evaluation of the finger knuckle biometric recognition system based on Hough transform is done by using PolyU database [15]. In this, finger knuckle print is captured using automated low cost contactless method using low resolution camera and in peg free environment. The knuckle images were collected from 165 persons. Extensive experiments were conducted and performance analysis was done based on the parameters viz., genuine acceptance rate and equal error rate.

Table 1. Performance of the Proposed System based on Genuine Acceptance Rate

Finger Knuckle Print	**Genuine Acceptance Rate %**		
	FAR = 0.5%	**FAR=1%**	**FAR=2%**
Left Index (LI)	78.34	84.78	88.26
Left Middle (LM)	79.89	85.9	90.27
Right Index (RI)	77.69	84.31	89.52
Right Middle (RM)	79.8	85.79	92.98
LI +LM	81.56	87.62	90.75
LI+RI	82.9	88.78	93.25
LI+RM	80.67	89.45	91.56
LM+RI	81.45	88.6	91.98
LM+RM	82.49	89.98	92.01
RM+RI	80.69	87.89	92.78
LI + RI+RM	83.89	92.56	93.56
LI+LM+RI	84.54	93.78	94.79
LI+LM+RM	83.76	94.89	94.68
RI+RM+LM	85.78	95.34	95.85
All the four fingers	90.65	95.98	97.62

The genuine acceptance rate is computed by manipulating the number of genuine matches and imposter matches corresponding to the total number of matches done with the system. Equal error rate is a point at which false acceptance rate and false rejection rate becomes equal. The above shown Table 1 illustrates the values of the genuine acceptance for the different combination of fusion.

From the above tabulated results, it is obvious that the finger knuckle print can be considered as one of the reliable biometric identifier. The following graphical illustration in Fig.4 depicts the some of the tabulated results. From the results obtained, it is clear that the accuracy of the system is considerably better when the single sample of finger knuckle surface is used. Further, the accuracy gets increases as when fusion of two, three finger knuckle regions was done. The best part of the accuracy can be accomplished by fusing the entire four fingers knuckle regions.

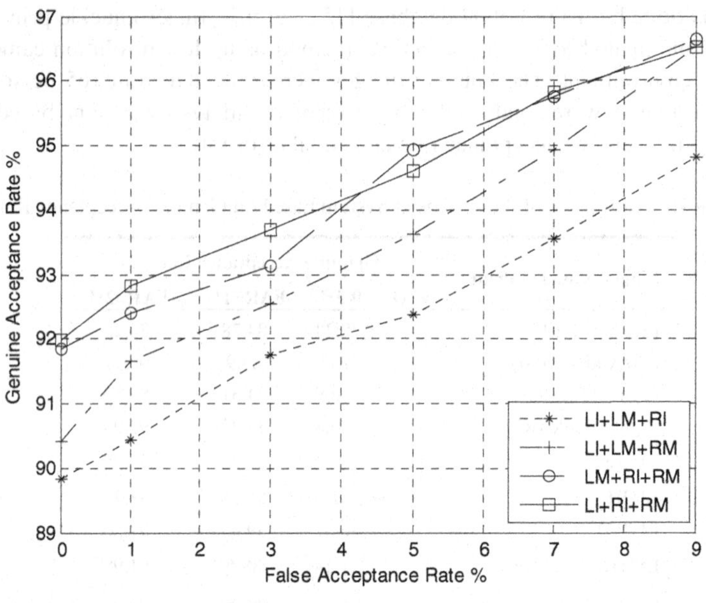

Fig. 4. Recognition rates obtained using EHT based feature extraction

The proposed Hough transform based feature extraction method for finger knuckle biometric recognition system is compared with the other geometric and texture based methods of feature extraction. The following Table 2 illustrates comparative analysis of the existing methods with the proposed methodology.

From the comparative analysis chart shown, it is obvious that the proposed Elliptical Hough transform based feature extraction method outperforms the existing methodology with high accuracy of Recognition rate = 98.26% and lowest error rate of EER = 0.78%.

Table 2. Comparative analysis of the Performance of the Proposed System

References	Dataset	Recognition Methods	Results
A. Kumar *et al.* [4]	Newly created database – 105 users with 630 Images	Principal Component Analysis, Linear Discriminant Analysis, Independent Component Analysis	EER – 1.39%
Zhang *et al.* [5]	PolyU database for Knuckle database.	Band Limited Phase only correlation method	EER - 1.68%
Chao lan *et al.* [7]	PolyU database for Knuckle database.	Complex locality preserving approach	EER - 4.28%
Lin Zhang *et al.* [8]	PolyU database for Knuckle database.	Ridgelet Transform	EER - 1.26%
A.Merounica *et al.*[9]	PolyU database for Knuckle database.	Fourier Transform	EER-1.72%
This paper	PolyU database for Knuckle database	Elliptical Hough Transform based Feature Extraction	EER = 0.78%

4 Conclusion

This paper presents a novel method for feature extraction based on Elliptical Hough Transform for the implementation of personal recognition system using finger knuckle prints. In this, EHT is formulated to isolate the elliptical structures present in the finger knuckle print and feature information obtained from those structures was represented to identify the individuals. The rigorous experiments were conducted and well promising results were achieved.

References

1. Hand-based Biometrics. Biometric Technology Today 11(7), 9–11 (2000)
2. Kumar, A., Zhang, D.: Combining fingerprint, palm print and hand shape for user authentication. In: Proceedings of International Conference on Pattern Recognition, pp. 549–552 (2006)

3. Rowe, R.K., Uludag, U., Demirkus, M., Parthasaradhi, S., Jain, A.K.: A multispectral whole-hand biometric authentication system. In: Biometric Symposium (2007)
4. Kumar, A., Ravikanth, C.: Personal Authentication Using Finger Knuckle Surface. IEEE Transactions on Information Forensics and Security 4(1), 98–110 (2009)
5. Zhang, L., Zhang, L., Zhang, D.: Finger-Knuckle-Print Verification Based on Band-Limited Phase-Only Correlation. In: Jiang, X., Petkov, N. (eds.) CAIP 2009. LNCS, vol. 5702, pp. 141–148. Springer, Heidelberg (2009)
6. Shen, L., Bai, L., Zhen, J.: Hand-based biometrics fusing palm print and finger-knuckle-print. In: 2010 International Workshop on Emerging Techniques and Challenges for Hand-Based Biometrics (2010)
7. Jing, X., Li, W., Lan, C., Yao, Y., Cheng, X., Han, L.: Orthogonal Complex Locality Preserving Projections Based on Image Space Metric for Finger-Knuckle-Print Recognition. In: 2011 International Conference on Hand-based Biometrics (ICHB), pp. 1–6 (2011)
8. Zhang, L., Li, H., Shen, Y.: A Novel Riesz Transforms based Coding Scheme for Finger-Knuckle-Print Recognition. In: 2011 International Conference on Hand-based Biometrics (2011)
9. Meraoumia, A., Chitroub, S., Bouridane, A.: Fusion of Finger-Knuckle-Print and Palmprint for an Efficient Multi-biometric System of Person Recognition. In: 2011 IEEE International Conference on Communications (2011)
10. Bao, P., Zhang, L., Wu, X.: Canny Edge Detection Enhancement by Scale Multiplication. IEEE Transactions on Pattern Analysis and Machine Intelligence 27(9), 1485–1490 (2005)
11. Zhanga, L., Zhanga, L., Zhanga, D., Zhub, H.: Online Finger-Knuckle-Print Verification for Personal Authentication, vol. 1, pp. 67–78 (2009)
12. Olson, C.F.: Constrained Hough Transforms for Curve Detection. Computer Vision and Image Understanding 73(3), 329–345 (1999)
13. Shen, D., Lu, Z.: Computation of Correlation Coefficient and Its Confidence Interval in SAS, vol. 31, pp. 170–131 (2005)
14. Hanmandlu, M., Grover, J., Krishanaadsu, V., Vasirkala, S.: Score level fusion of hand based Biometrics using T-Norms. In: 2010 IEEE International Conference on Technologies for Homeland Security, pp. 70–76 (2010)
15. Alen, J.V., Novak, M.: Curve Drawing Algorithms for Raster displays. ACM Transactions on Graphic 4(2), 147–169 (1985)

An Efficient Multiple Classifier Based on Fast RBFN for Biometric Identification

Sumana Kundu and Goutam Sarker

Computer Science and Engineering Department, NIT Durgapur, INDIA
sumana.kundu@yahoo.co.in, g_sarker@ieee.org

Abstract. In this present paper a modified Radial Basis Function Network (RBFN) based multiple classifiers for person identification has been designed and developed. This multiple classification system comprises of three individual classifiers based on a modified RBFN using Optimal Clustering Algorithm (OCA). These three individual classifiers perform Fingerprint, Iris and Face identification respectively and the super classifier performs the final identification based on voting logic. The technique of using the modified RBFN and OCA to design each classifier for fingerprint, iris and face is efficient, effective and fast. Also the accuracies of the classifiers are substantially moderate and the recognition times are quite low.

Keywords: Multiple Classifier, Fingerprint Identification , Iris Identification, Face Identification, OCA, RBFN, BP Learning, Holdout method, Accuracy.

1 Introduction

Biometric technologies have immense importance in various security, access control and monitoring applications. Single modal biometric systems repeatedly face significant restrictions due to noise in sensed data, spoof attacks, data quality, non universality and other factors. Multimodal biometric authentication or multimodal biometrics is the approach of using multiple biometric traits from a single user in an effort to improve the results of the authentication process.

Various multimodal biometric systems can be found which propose traditional fusion technique and use two biometric features for identification.

In paper [10] a fingerprint-iris fusion based identification system was proposed. This proposed framework developed a fingerprint and iris fusion system which utilized a single Hamming Distance based matcher for identification. Here iris feature extraction was based on Daugman's approach and was implemented by Libor Masek. Fingerprint feature extraction was based on Chain code to detect minutiae. A simple accumulator based fusion approach was employed here. The fingerprint, iris and fused accuracy are 60%, 65% and 72% respectively.

Paper [11] proposed an iris and fingerprint fusion based technique based on Euclidean distance matching algorithm. Here preprocessed and normalized data is given to the Gabor filters and then extracted features were used for matching. The accuracy of the multimodal system is 99.5% for threshold 1 compare to 99.1% and 99.3% for threshold 0.1 and 0.5 respectively.

M.K. Kundu et al. (eds.), *Advanced Computing, Networking and Informatics - Volume 1*, Smart Innovation, Systems and Technologies 27,
DOI: 10.1007/978-3-319-07353-8_55, © Springer International Publishing Switzerland 2014

In paper [13] a fingerprint and iris fusion based recognition technique was proposed using conventional RBF neural network. The features of fingerprint were extracted by Haar wavelet based method and iris features were extracted by Block sum method. The testing time for fingerprint, iris and fusion modality are 0.35, 0.19 and 0.12 seconds respectively.

In our previous works [8], [9],[12], we have developed unimodal systems based on RBFN with optimal clustering algorithm (OCA) and BP network for rotation invariant, clear as well as occluded fingerprint and face identification and location invariant localization.

In this paper, we propose a multiple classifier for person identification where each classifier operates on different aspects of the input. There are three individual classifiers for Fingerprint, Iris and Face identification and also a Super classifier which give the proper identification of person based on voting logic considering the result of three individual classifiers. The ANN model which is adopted here is the Radial Basis Function (RBF) network with Optimal Clustering Algorithm (OCA) for training units and Back Propagation (BP) learning for classification.

1.1 Classifier Performance Evaluation

Holdout Method

In order to obtain a good measure of the performance of a classifier, it is necessary that the test dataset has approximately the same class distribution as the training dataset. The training dataset is that portion of the available labeled examples that is used to build the classifier. The test dataset is that portion of the available labeled examples that is used to test the performance of the classifier. In holdout method [3] the reason for keeping aside some of the labeled examples for a test dataset is to test the performance of the constructed classifier.

Accuracy

The accuracy [3] of a classifier is the probability of its correctly classifying records in the test dataset. In practice, accuracy is measured as the percentage of records in the test dataset that are correctly classified by a classifier. If there are only two classes (say C and not C), then the accuracy is computed as:

$$Accuracy = \frac{a+d}{a+b+c+d} \tag{1}$$

where 'a', 'b', 'c' and 'd' are defined in the matrix in Fig.1.

		Actual class	
		C	Not C
Predicted Class	C	a	b
	Not C	c	d

Fig. 1. Confusion Matrix (2 class)

In evaluation of a classifier with accuracy the overall performance of the classifier is reflected irrespective of the individual performance evaluation for each and every class or category. This is more appropriate to find out the system evaluation through a particular numeric value.

2 Overview of the System and Approach

2.1 Preprocessing

Fingerprint Images

The finger print image has to be preprocessed before learning as well as recognition. There are several steps in preprocessing.

- **Conversion of RGB fingerprint images to grayscale images:** The first step of preprocessing is to convert training database or test database images into gray scale images. Those images may / may not be blurred and /or noisy.
- **Removal of noise from the images:** The fingerprint images may be noisy. To remove those noises from fingerprint patterns, we use 2-D Median Filter.
- **De blurring the images:** In this process we use blind deconvolution algorithm to deblur and get sharp images. The blind deconvolution algorithm can be used effectively when no information about the distortion is known.
- **Background elimination:** In this step the background of fingerprint patterns are removed.
- **Conversion of grayscale images into binary images:** This process converts the grayscale image into binary image (2D matrix file).
- **Image Normalization:** This process normalizes all fingerprint image patterns into lower dimensions and of same size.
- **Conversion of binary images into 1D matrix:** In the last step of preprocessing, 2D matrix fingerprint files are converted into 1D matrix files. This set is the input to the Optimal Clustering Algorithm (OCA).

Iris Images

Preprocessing of the Iris image is also required before feature extraction. Iris preprocessing comprises of several steps.

- **Conversion of RGB iris images to grayscale images:** The first step of preprocessing is to convert training database or test database images into gray scale images.
- **Iris Boundary Localization:** The radial-suppression edge detection algorithm [1] is used to detect the boundary of iris which is similar to Canny edge detection technique. In the radial-suppression edge detection, a non-separable wavelet transform is used to extract the wavelet transform modulus of the iris image and then radial non maxima suppression is used to retain the annular edges and simultaneously remove the radial edges. Then an edge thresholding is utilized to remove the isolated edges and determine the final binary edge map.

  Now circular Hough Transformation [2] is used to detect final iris boundaries and deduce their radius and center. The Hough transform [2] is defined, as in (2), for the circular boundary and a set of recovered edge points x_j, y_j (with $j=1,...,n$).

$$H(x_c, y_c, r) = \sum_{j=1}^{n} h(x_j, y_j, x_c, y_c, r) \tag{2}$$

where, $$h(x_j, y_j, x_c, y_c, r) = \begin{cases} 1, & if\ g(x_j, y_j, x_c, y_c, r) = 0 \\ 0, & otherwise \end{cases} \tag{3}$$

with, $$g(x_j, y_j, x_c, y_c, r) = (x_j - x_c)^2 + (y_j - y_c)^2 - r^2 \tag{4}$$

For each edge point (xj, yj), g(xi, yj, xc, yc, r) = 0 for every parameter (xc, yc, r) that represents a circle through that point. The triplet maximizing H corresponds to the largest number of edge points that represents the contour of interest.

- **Extract the iris:** In this step, we remove the other parts of the eyes images such as eyelids, eyelashes and eyebrows and extract the iris.
- **Conversion into Binary images:** This process converts the image into binary image (2D matrix file).
- **Image Normalization:** This process normalizes all patterns into lower dimensions and of same size.
- **Conversion of binary images into 1D matrix:** In the last step of preprocessing, 2D matrix iris files are converted into 1D matrix files. This set is the input to the Optimal Clustering Algorithm (OCA).

Face Images

The face images also have to be preprocessed before learning as well as recognition. The preprocessing steps for face images are almost same as fingerprint preprocessing i.e. removal of noise, de blurring, background removal, conversion of RGB images to grayscale images, conversion of gray scale images to binary images, image normalization and finally conversion of binary images into 1D matrix files. The difference is at background elimination step. Here, in this step the background of face patterns are removed using Gaussian Model. Here a sum-of-Gaussians model (2 Gaussians) is used for the color at each pixel results in a simple way of separating the background.

2.2 Theory of Operation

Our proposed system comprises of three individual classifiers. Each classifier consists of a modified Radial Basis Function Network (RBFN) [8], [9], [12]. In this approach, Optimal Clustering Algorithm (OCA) [8], [9], [12] is used to form groups of the input data set. The elements of the same group are similar with some similarity measure and those in different groups are dissimilar with the identical measure. The mean "μ" and standard deviation "σ" of each cluster formed by OCA with approximated normal distribution output function are used for each basis unit of Radial Basis Function Network (RBFN). The RBFN with Back Propagation network classifier [8], [9], [12] is used for pattern identification.

Optimal Clustering Algorithm (OCA)

OCA [9, 12] is a partitioned clustering method. In this type of clustering the user has to specify a threshold "T" (maximum allowable distance of each point from the cluster mean), which is a measure of the degree of similarity among the members of the same clusters and that of dissimilarity among members of the different clusters.

Back Propagation (BP) Network

The training of a network by Back Propagation (BP) [4], [5], [6], [8], [9], [12] involves three stages. The feed forward of the input training pattern, the calculation and back propagation of the associated error and the adjustment of the weights. After training, application of the net involves only the computations of activation of the feed forward phase. The output is calculated using the current setting of the weights in all the layers. The weight update is given by,

$$w_{ij}(t+1) = w_{ij}(t) + \Delta w_{ij}(t) \tag{5}$$

where,
$$\Delta w_{ij}(t) = \eta . \text{error} \tag{6}$$

Here "t" is the index to indicate the presentation step for the training pattern at step t (iteration number) and "η" is the rate of learning.

In the BP learning algorithm [9],[12] of our developed system initialization of η are different for three individual classifiers.

The optimum weight is obtained when:

$$w_{ij}(t+1) \approx w_{ij}(t) \tag{7}$$

Radial Basis Function Network (RBFN)

The RBFN [6], [7], [8], [9], [12] consists of three layers namely an input layer for pattern presentation, a hidden (clustering) layer containing 'basis units' and an output (classification) layer. The clustering outputs (mean μ, standard deviation σ and corresponding approximated normal distribution output functions) are used in 'basis units' of RBFN. Thus, OCA is the first phase learning and using Back Propagation (BP) learning we get the optimal weights in the BP network, which is the second phase of learning. The hidden layer use neurons with RBF activation functions describing local receptors. Then output node is used to combine linearly the outputs of the hidden neurons. Here "m" denotes the number of inputs while "Q" denotes the number of outputs. In fig.2 it is shown that there is one hidden layer with RBF activation function,

$\emptyset_1 \ldots\ldots\ldots\ldots \emptyset_{m1}$, and output layer with linear activation function.

$$\mathrm{R} = \mathrm{w}_1 \emptyset_1 (||\mathrm{x} - \mu_1||) + \cdots + \mathrm{w}_{\mathrm{m1}} \emptyset_{\mathrm{m1}} (||\mathrm{x} - \mu_{\mathrm{m1}}||) \tag{8}$$

where, $||\mathrm{x} - \mu||$ distance of $x = (x_1, \ldots\ldots, x_m)$ from vector μ.

and $\mathrm{w}_1, \ldots\ldots, \mathrm{w}_{\mathrm{m1}}$ are the weights learned through classification.

The hidden units use Radial Basis Functions (RBF). The output depends on the distance of the input x from the center μ.

$\emptyset_\sigma(||x - \mu||)$, where μ is called center and σ is called standard deviation or spread.

A hidden neuron is more sensitive to data points near its center. For Gaussian RBF this sensitivity may be tuned by adjusting the spread σ, where a larger spread implies less sensitivity. So, the output of the hidden units,

$$y = e^{\frac{-(x-\mu)^2}{\sigma^2}} \tag{9}$$

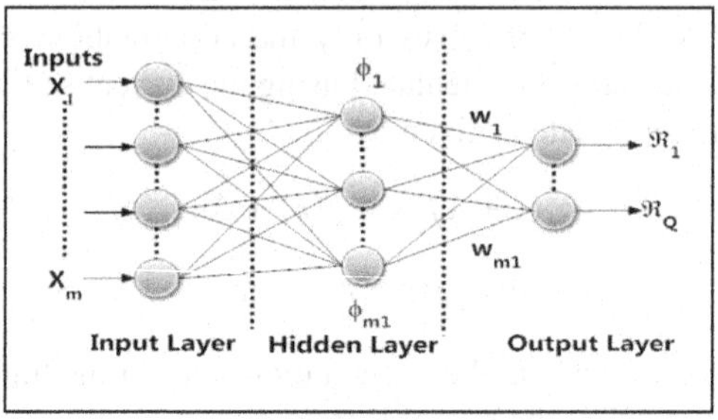

Fig. 2. RBFN Architecture

Identification Learning

We use three training databases for three individual classifiers. Each database consists of different image patterns i.e. fingerprint, iris and face. After preprocessing, all the patterns are fed individually as input to the RBFN of individual classifiers. When the networks have learned all the different patterns (fingerprint, iris, face) of different qualities or expressions and of different angles or views for all different people, the network is ready for identification of learned image patterns Fig.3.

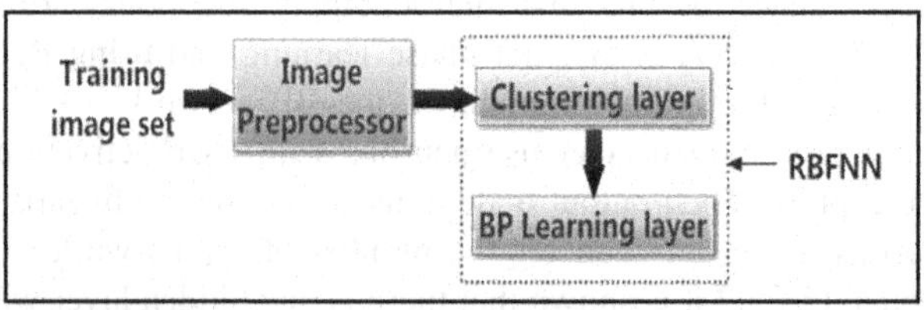

Fig. 3. Block diagram for learning identification

Identification with Test Images

The test images from the test image sets are fed as input to the preprocessor. Our proposed system consists of three different preprocessor, each for fingerprint, iris and face images. The preprocessed patterns are fed as input to the previously trained BP networks of three individual classifiers. If any of the 3 outputs of the network is active high, then the corresponding pattern is identified or recognized. If the majority of the classifiers are concluding a particular person, then the super classifier or integrator has to conclude that person as the decision of the overall identification system (refer Fig. 4).

Algorithm for person identification for super classifier
Input: *Test folder containing known pattern sets .Each pattern set consists of one fingerprint, one iris and one face pattern.*
Output: *Identification of the given test pattern set.*
Steps:

1. *Get the pattern set to test.*
2. *Preprocess each pattern (fingerprint, iris, face) individually with the preprocessor of individual three classifiers.*
3. *Feed these as inputs to the trained BP networks of individual classifiers for pattern recognition.*
4. *If the* n^{th} *outputs* $(1 \leq n \leq 3)$ *of the BP networks are approximately 1, then conclude the respective pattern as n.*
5. *Get the results of three classifiers.*
6. *Input this result to super classifier.*
7. *If all classifiers outputs as* $n^{th}(1 \leq n \leq 3)$ *person to the pattern set, then super classifier concludes that the pattern set is of* n^{th} *person.*
8. *If any two classifiers outputs as* $n^{th}(1 \leq n \leq 3)$ *person to the pattern set and one classifier output as different, then super classifier concludes that the pattern set is of* n^{th} *person.*
9. *If all three classifier outputs as different, then the super classifier conclude that the pattern set as 'unidentifiable person'.*
10. *If any more testing is required, go to step 1.*
11. *Stop.*

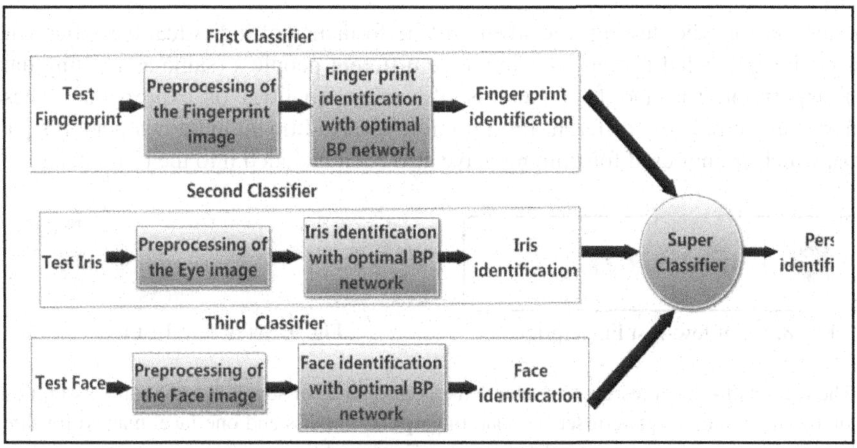

Fig. 4. Block diagram for testing identification

3 Result and Performance Analysis

We have used the training and test dataset for fingerprints samples from FVC database (http://www.advancedsourcecode.com/fingerprintdatabase.asp), Iris samples from MMU2 database (http://pesona.mmu.edu.my/~ccteo/) and Face samples from FEI database (http://fei.edu.br/~cet/facedatabase.html).

3.1 Training Set

We have used three different training databases for three different classifiers which contains three different people's fingerprint, iris and face images. In fingerprint database for each person's finger print there are three different qualities of fingerprints and also three different angular (0, 90, 180 degrees) finger prints. In iris database left and right eye images of three different qualities for each person has been taken. Finally in face database for each person's face, three different expressions and also three different angular views of face images i.e. frontal view, 90 degree left side view and 90 degree right side view are taken (refer Fig.5, Fig. 6, Fig,7).

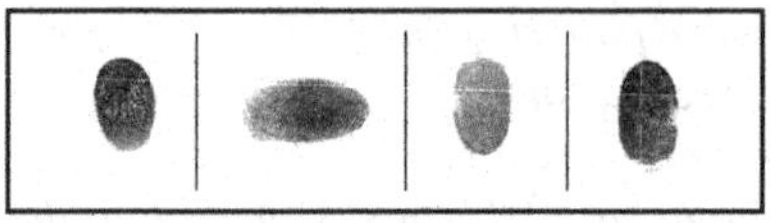

Fig. 5. Samples of few training Finger print images

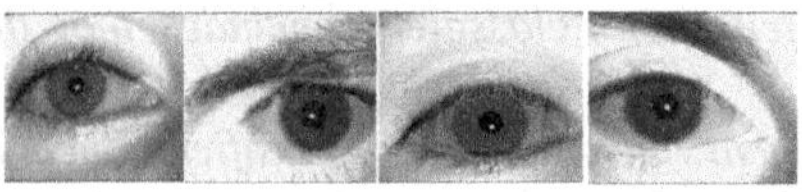

Fig. 6. Samples of few training Iris images

Fig. 7. Samples of few training Face images

3.2 Test Set

The test set for label testing to evaluate the performance of individual classifiers and test set for unlabeled testing contains three different people's (same as training data set) fingerprint, iris and face patterns of various qualities or expressions. These patterns are completely different from training set. Additional unknown person's patterns, which are not used for training, have also been included into the test database.

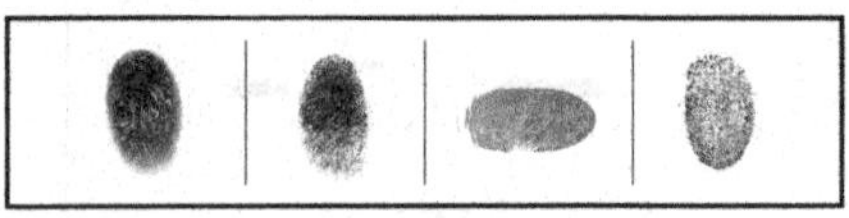

Fig. 8. Set of few Test Fingerprint

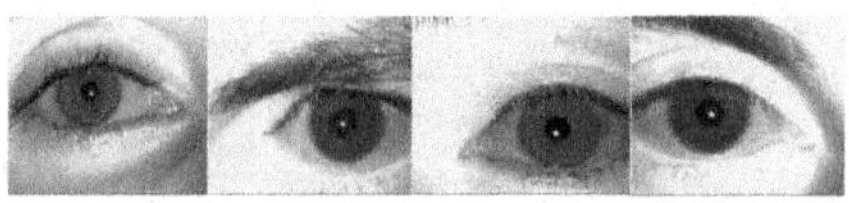

Fig. 9. Set of few Test Iris

The test sets for label testing to evaluate the performance of super classifier and also for unlabeled testing contains pattern set (i.e. one fingerprint, one iris and one face images) for three different people (same as training data set) of various qualities or expressions. These patterns and also some additional unknown people's pattern set are completely different from training set (refer Fig. 8, Fig. 9, Fig. 10, Fig. 11).

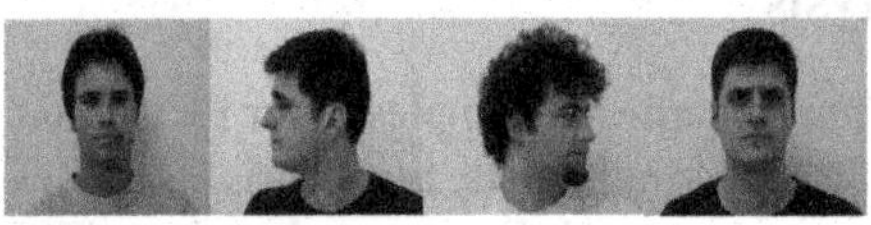

Fig. 10. Set of few Test Face

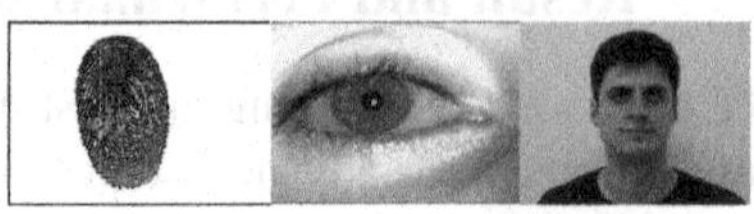

Fig. 11. A test set for Super classification

3.3 Experimental Results

The proposed system was made to learn on a computer with Intel Xeon E5506,2.13 GHz Quad core processor with 12 GB RAM and Windows 7 64-bit Operating System. We used MATLAB 2013a to develop the multiple classifiers.

Table 1. Accuracy of the Classifiers (Holdout Method)

Classifiers	Accuracy
First Classifier (Fingerprint)	91.67%
Second Classifier(Iris)	94.44%
Third Classifier(Face)	91.67%
Super Classifier	97.22 %

Table 2. Learning time of the Classifiers

Classifiers	Training Time(seconds)	Performance Evaluation Time * (seconds)	Total Learning Time (seconds)
First Classifier (Fingerprint)	91.06	0.69	91.75
Second Classifier (Iris)	17.37	0.63	18.00
Third Classifier (Face)	21.91	0.69	22.60
Super Classifier	130.34	0.66	131.00

* We take one third of the labeled test samples (9 samples) to find out performance evaluation time for each classifier

From Table 1 we find that the accuracy of the individual classifiers is 91.67%, 94.44%, 91.67% and the accuracy of the super classifier is 97.22%.Thus, it is evident that the super classifier is more efficient for biometric authentication than considering single classifiers individually. Similarly in Table2, the proposed approach shows overall low performance evaluation time (< 1 second) for the standard test data set. Hence, the proposed approach shows improvement both in terms of accuracy and performance evaluation time as compared to techniques mentioned in Section 1.

4 Conclusion

A multiple classification system has been designed and developed using a modified Radial Basis Function Network (RBFN) with Optimal Clustering Algorithm (OCA) and BP learning. This multiple classifier consists of three individual classifiers for Fingerprint, Iris and Face identification. The Super classifier concludes the decision for identification considering these classifier's result based on voting logic. This system can identify the authenticated person. Due to the application of our OCA based modified RBFN, the individual learning systems are fast. The performance in terms of accuracy with Holdout method in each classifier is moderately high. Also the training and testing time are moderately low for different types of fingerprints, iris, and face patterns. Proposed RBFN multiple classification is efficient, effective and faster compared to other conventional multimodal identification technique.

References

1. Huang, J., You, X., Tang, Y.Y., Du, L., Yuan, Y.: A novel iris segmentation using radial-suppression edge detection. Signal Processing 89, 2630–2643 (2009)
2. Conti, Y., Militello, C., Sorbello, F.: A Frequency-based Approach for Features Fusion in Fingerprint and Iris Multimodal Biometric Identification Systems. IEEE Transactions on Systems, Man, and Cybernetics—Part C: Applications and Reviews 40(4), 384–395 (2010)
3. Pudi, V.: Data Mining. Oxford University Press, India (2009)
4. Sarker, G.: A Multilayer Network for Face Detection and Localization. IJCITAE 5(2), 35–39 (2011)
5. Revathy, N., Guhan, T.: Face Recognition System Using Back propagation Artificial Neural Networks. IJAET III(I), 321–324 (2012)
6. Sarker, G.: A Back propagation Network for Face Identification and Localization. Accepted for publication in ACTA Press Journals 202(2) (2013)
7. Aziz, K.A.A., Ramlee, R.A., Abdullah, S.S., Jahari, A.N.: Face Detection using Radial Basis Function Neural Networks with Variance Spread Value. In: International Conference of Soft Computing and Pattern Recognition, pp. 399–403 (2009)
8. Sarker, G., Kundu, S.: A Modified Radial Basis Function Network for Fingerprint Identification and Localization. In: International Conference on Advanced Engineering and Technology, pp. 26–31 (2013)
9. Kundu, S., Sarker, G.: A Modified Radial Basis Function Network for Occluded Fingerprint Identification and Localization. IJCITAE 7(2), 103–109 (2013)
10. Baig, A., Bouridane, A., Kurugollu, F., Qu, G.: Fingerprint – Iris Fusion based Identification System using a Single Hamming Distance Matcher. In: Symposium on Bio-Inspired Learning and Intelligent Systems for Security, pp. 9–12 (2009)
11. Lahane, P.U., Ganorkar, S.R.: Fusion of Iris & Fingerprint Biometric for Security Purpose. International Journal of Scientific & Engineering Research 3(8), 1–5 (2012)
12. Bhakta, D., Sarker, G.: A Rotation and Location Invariant Face Identification and Localization with or Without Occlusion using Modified RBFN. In: Proceedings of the 2013 IEEE International Conference on Image Information Processing (ICIIP 2013), pp. 533–538 (2013)
13. Gawande, U., Zaveri, M., Kapur, A.: Fingerprint and Iris Fusion Based Recognition using RBF Neural Network. Journal of Signal and Image Processing 4(1), 142–148 (2013)

Automatic Tortuosity Detection and Measurement of Retinal Blood Vessel Network

Sk. Latib[1], Madhumita Mukherjee[2], Dipak Kumar Kole[1], and Chandan Giri[3]

[1] St. Thomas' College of Engineering & Technology, Kolkata, West Bengal, India
[2] Purabi Das School of Information Technology, BESUS, Shibpur, West Bengal, India
[3] Department of Information Technology, IIEST, Shibpur, West Bengal, India
{sklatib,madhumita07,dipak.kole,chandangiri}@gmail.com

Abstract. Increased dilation and tortuosity of the retinal blood vessels causes the infant disease, retinopathy of prematurity (ROP). Automatic tortuosity evaluation is a very useful technique to prevent childhood blindness. It helps an ophthalmologist in the ROP screening. This work describes a method for automatically detection of a retinal image into low, medium or highly tortuous. The proposed method first extracts a skeleton retinal image from the original retinal image to get the overall structure of all the terminal and branching nodes of the blood vessel based on morphological operations. Then, it separates each branch and rotates it so that partitioning process is easier which follows a recursive way. Finally, from the partitioned vessel segments, the tortuosity is calculated and the tortuous symptom of the image is detected. The results have been compared with the eye specialist's analysis on twenty five images which gives good result.

1 Introduction

Normal retinal blood vessels are straight or gently curved. In some diseases, the blood vessels become tortuous, i.e. they become dilated and take on a serpentine path. The dilation is caused by radial stretching of the blood vessel and the serpentine path occurs because of longitudinal stretching. The tortuosity may be local; occurring only in a small region of retinal blood vessels or it may involve the entire retinal vascular tree. Many diseases such as high blood flow, angiogenesis and blood vessel congestion produce tortuosity. In Retinopathy of Prematurity (ROP), the growth of retinal vessels does not reach the periphery of the retina. This vessel is not fully vascularized. So, preterm birth carries many complications. The early indicator of ROP is a whitish gray demarcation line between normal retina and anteriorly undeveloped vascularized retina. Hence the infants should be screened for ROP to receive appropriate treatment for preventing permanent loss of vision.

Therefore, tortuosity measurement is a problem related to clinical correlation. There are many methods to measure tortuosity. It is measured by based on the ratio between arc length and the chord length [4]. They measured tortuosity using relative length variation. Seven integral estimates of tortuosity based on the curvature of the vessels are presented by Hart et al. [6]. For a better accuracy of the tortuosity measurement, Buitt et al. [8] generalized Hart's estimates to 3D images obtained by

M.K. Kundu et al. (eds.), *Advanced Computing, Networking and Informatics - Volume 1*, Smart Innovation, Systems and Technologies 27,
DOI: 10.1007/978-3-319-07353-8_56, © Springer International Publishing Switzerland 2014

MRA. Grisan et al. [3] proposed another method satisfying all the tortuosity properties. Sukkaew et al. [5] present a new method based on partitioning and used the tortuosity calculation formulae proposed by Grisan et al. [3].

But, all the above methods do not separate each branch of the retinal image and calculate tortuosity of branches in skeleton image. So, these methods are complicated and they couldn't separate a single branch of a vessel. The method proposed by Sukkaew et al. [5], the maximum allowable interpolation point calculation is not fast and takes large time. Again, these methods may not always give correct result. Because, a circular arc with larger radius is non tortuous. Although, the ratio between arc length and the chord length could be very large.

In this work, we have proposed a method to measure tortuosity of the retinal blood vessels by breaking the images into branches and calculate tortuosity for every branch curve separately. It rotates every curve to calculate maximum distance between curves and chord easily and then produce a new formula for tortuosity measurement which satisfies all the properties of tortuosity. The proposed method is also better than the existing work as those methods could not separate each branch curve.

Rest of the paper is organized as follows. Section 2 explains some properties of blood vessels that are used in the proposed method.Section3deals with the proposed method for tortuosity detection. Section 4 presents results and explanation of such results by comparing with eye specialist's analysis. Section5 describes and overall conclusion of the total process.

2 Background

Blood Vessels have many branches and furcation. It is necessary to find this furcation and branches.

2.1 Determination of Furcation and Branches

The blood vessel branches and furcation are determined by the number of neighboring pixels [2]. A point with two neighboring pixels is defined as one of the elements of a branch. A point with three neighboring pixels is defined as a bifurcation and point with four neighboring pixels is defined as a furcation. Figure 1 shows determination of furcation and branches.

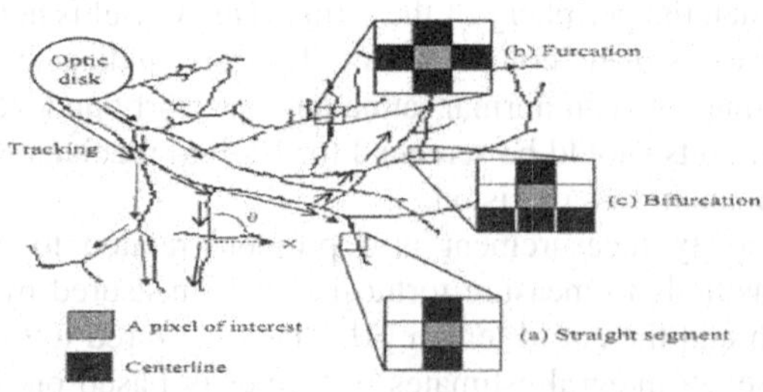

Fig. 1.(a) Straight segment (b) Furcation point (c) Bifurcation

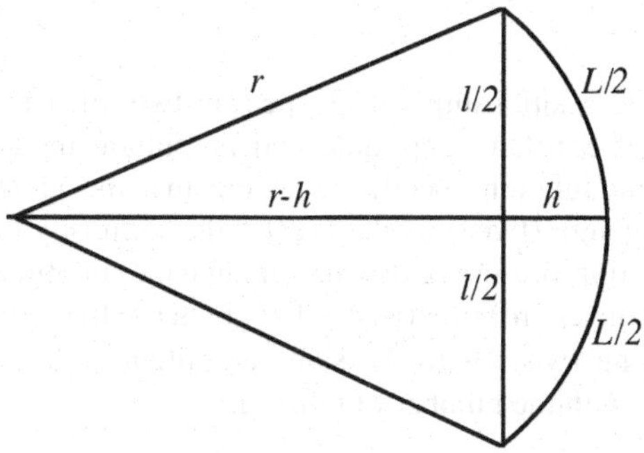

Fig. 2. Relation between arc and chord

2.2 Relation Between Arc and Chord:

The simple measure of a vessel tortuosity is the ratio between length difference of arc and chord and length of the chord. This length variation is measured by using the arrow heights and chord lengths. The tortuous vessel consists of a series of circular arcs. It can be assumed that each individual curvature is characterized by its chord length l_i and arrow height h_i.

Now, from Figure 2, tortuosity is then the relative length variation. i.e., $\frac{\Delta l}{l} \cong \frac{8}{3}\left(\frac{h}{l}\right)^2$ (1)

2.3 Tortuosity Features:

Blood Vessel tortuosity measures have some well-defined properties.

2.3.1. Affine Transformations

Affine transformations of a vessel are translation, rotations and scaling. These transformations are related to the geographical position and orientation of vessels in the retina, and do not alter the clinical perception of tortuosity.

2.3.2. Composition

Composition properties are dealt when two vessel curves are merged into a single one, or when various segments of the same vessel with different tortuosity measures, build up to give the total vessel tortuosity.

For two adjacent continuous curves s_1 and s_2, the combination of the two is defined as

$$s_3 = s_1 \oplus s_2 \quad (2)$$

Since the two composing curves are belong to the same vessel, without loss of generality it can be assumed the continuity of s_3. Again a new composition property is proposed, such that a vessel s, combination of various segments s_i, will have tortuosity measure greater than or equal to any of its composing parts:

$$T(s_i) \leq T(s_1 \oplus s_2 \oplus .... \oplus s_n). \quad \forall\ i=1....n,\ s_i \subseteq s \quad (3)$$

2.3.3. Modulation

There exists a monotonic relationship with respect to two other properties. They are frequency modulation at constant amplitude and amplitude modulation at constant frequency. Frequency modulation specifies the rule that the greater the number of changes in the curvature sign (twist) is, the greater the tortuosity of the vessel. Similarly, amplitude modulation preserves that the greater the amplitude (maximum distance of the curve from the chord) of a twist is the greater is the tortuosity.

For two vessels having twists with the same amplitude, the tortuosity difference changes simultaneously with the number of twists Ω:

$$\Omega\,(s_1) \leq \Omega\,(s_2) \;\Rightarrow\; T\,(s_1) \leq T\,(s_2) \tag{4}$$

Conversely, for two vessels with the same number of twists (with the same frequency), the difference in tortuosity changes simultaneously with the difference in amplitude $\alpha$of the twists:

$$\alpha\,(s_1) \leq \alpha\,(s_2) \Rightarrow T\,(s_1) \leq T\,(s_2) \tag{5}$$

3 Proposed Method

The proposed method is described in five sections as follows. The proposed method deals with binary images because it takes less time and space for processing and morphological image processing can be used easily which are efficient and fast in extracting image features whose shape is known priori like vascular structure. Morphological processing is also resistant of noise. The captured image is binarised to get the blood vessel structure clearly. Binary image is used as the input to algorithm that performs useful tasks. This algorithm can handle tasks ranging from very simple tasks to much more complex recognition, localization, and inspection tasks.

3.1 Vessel Skeleton Extraction

This section consists of some processes. At first, Original retinal RGB image is converted to gray image. Then, Log filter [5] is used for low-contrast image edge detection. After that, the image is converted to binary image. However, the image is skeletonized [1] to get single pixel wide by morphology technique. By using morphology, bridge operation and pruning process is applied on it simultaneously. After that, the spur pixels are removed depending on requirements and thin, clean operations are applied on it by using morphological image processing. Figure3 illustrates this algorithm.

3.2 Breaking Into Branches

First of all endpoints of branches are extracted by using morphology. To get the branch, it traverses from first endpoint to bifurcation and after getting the branch morphological bridge and thin operations are applied on remaining image. Again, the resultant image is processed by traversing from first endpoint to bifurcation for getting the branch and this process runs recursively until all endpoints are used. On the other side, if extracted branch length is greater than a threshold value, that branch is stored in a matrix and in this recursive way all branches are stored respectively. In Figure 4, some original images and extracted vessel skeletons from those images have shown. In Figure 5, extracted branches from skeleton images have shown. In Figure 6, shows an example after execution of steps for one branch.

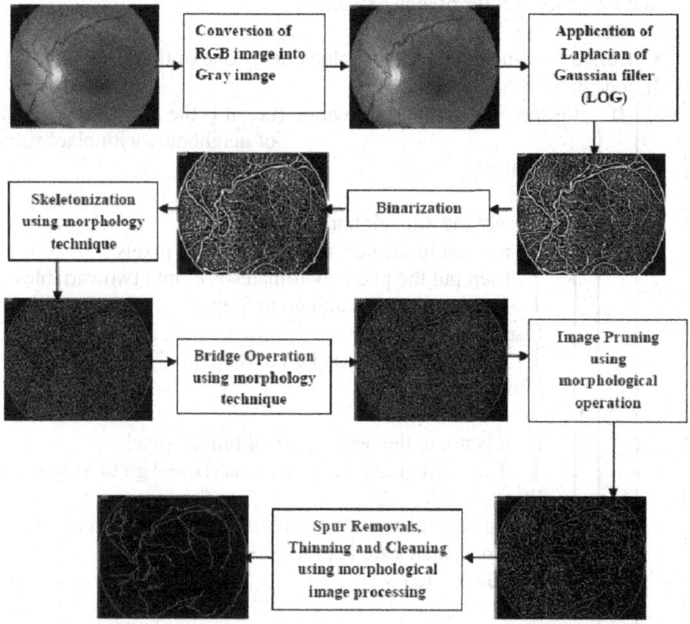

Fig. 3. Process flow diagram of the proposed series of process to extract blood vessels' skeleton

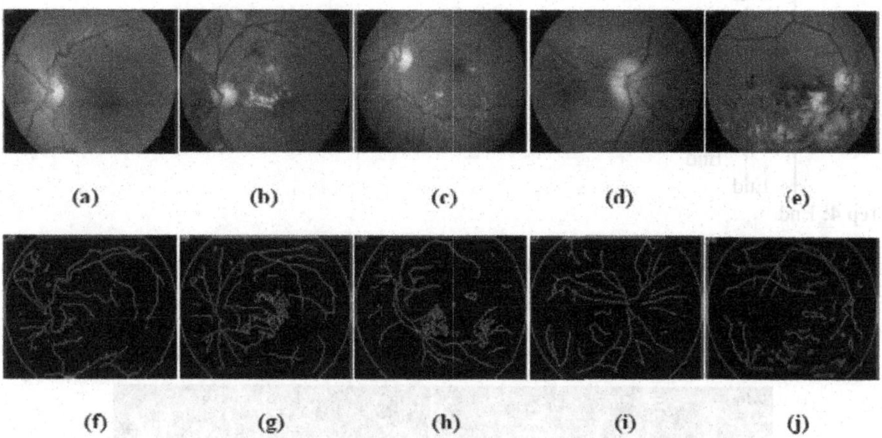

Fig. 4.(a),(b),(c),(d) and (e) the original images. (f),(g),(h),(i) and (j) are respectively detected vessels' skeleton

3.2.1. Algorithm of Extraction of a Branch:

Input: Branch pixel endpoint with coordinate(r, c)
Output: Total separate vessel branch

Step 1: Endpoints are stored in a matrix and create new empty matrix to store branch pixels
Step 2: Take the first endpoint for branch extraction
Step 3: For variable k=1 to 8

```
Calculate the 8-neighbour pixel value one by one of that pixel point
If  one pixel value is 1
    If  it is the condition of bifurcation (i.e., it is the 3rd in between the 3 number
                                            of neighbours with pixel value 1)
        Then stop
    Else
        If  it is not the 8th neighbor
            If  it is not in the new matrix of branch pixels
                Then put the pixel coordinates(r, c) into two variables
                Increment k by 1 and go to Step 3
            Else
                Increment k by 1 and go to Step 3
            End
        Else
            If  it is not in the new matrix of branch pixels
                Then store the pixel in that matrix and go to Step 3
            Else
                In new matrix of branch pixels store the previous pixel whose
                coordinates are kept in variables
                Go to Step 3
            End
        End
    End
Else
    If  it is not the 8th neighbor
        Increment k by 1 and go to Step 3
    Else
        In new matrix of branch pixels store the coordinate variables
        Go to Step 3
    End
End
```

Step 4: End

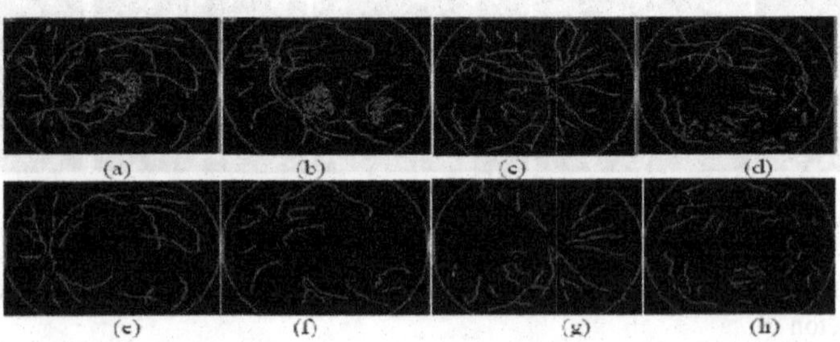

Fig. 5.(a),(b),(c),(d) the skeleton images. (e),(f),(g),(h)are extracted branches respectively from skeleton image

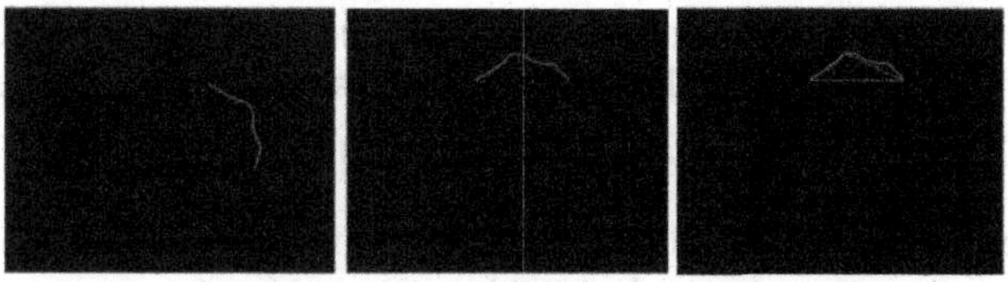

(a) (b) (c)

Fig. 6.(a) One extracted branch,(b) After rotation,(c) vessel partitioning into segments and finding maximum distance

3.3 Rotating Vessel Branches and Partitioning into Segments by Finding Maximum Distance from Chord

This step deals with finding a set of chords. At first, a vessel branch B is rotated at such an angle that the joining of two end points is a straight line parallel to x-axis. The angle is obtained from the straight line equation joining two end points of the branch. After rotation, the boundary points are joined by a chord C. At last the maximum distance between the chord and the branch can be measured easily. The following steps are follows for finding a set of chords.

Step 1: Find both end points of a branch, suppose *(r1,c1)* and *(r2,c2)*.

Step 2: Construct a straight line joining these two ends. The equation is of the form, $r = mc+k$ where $m=tan\theta$

Step 3: Apply rotation of the branch at that particular angle θ, around its center point.

Step 4: Get the chord *C* by joining both end points after rotation.

Step 5: Find maximum distance from chord and the vessel, which cuts vessel branch. i.e., $\rho_0 = \text{argmax}|C - B|$

Step 6: This divides vessel branch into two curves B_1 and B_2.

Step 7: Then repeat step 1 to step 4 for every branch curve B_1 and B_2.

Step 8: Then find the next maximum distance,$\rho_1 = \text{argmax}\{\max|C_1 - B_1|, \max|C_2 - B_2|\}$

Step 9: This procedure runs recursively until one maximum value is less than a threshold *T*. i.e., $\max|C_k - B_k| < T, \forall_k$ where, *T*=5 or 6 or 10.

Step 10: Apply the same procedure for every branch curve. i.e., go to Step 1.

3.4 Tortuosity Measurement

Calculation of tortuosity is done by the following equation.

$$\tau = \frac{1}{length(C)} \sum_{k=1}^{n} \frac{length(B_k)}{length(C_k)}, \quad (6)$$

where$C$ is chord of the main vessel branch and k is the segment number and $n$the total number of segments.$B_k$is kth curve and C_k is kth chord of the segment. The tortuosity measure has a dimension of 1/length that is interpreted as tortuosity density. Tortuosity is measured by relative length variations of vessel curve and chord, where vessel curve is partitioned in many small segments so that there is a small length variation of each individual chord and segment. The equation 13 formula is applied for every branch curve. The length(B_k) is the length of a segment of branch curve and

$\text{length}(C_k)$ is the length of a segment of chord. $\text{length}(B_k)$is measured by counting no of pixels of B_k , whereas the $\text{length}(C_k)$ is measured by D8 distance.

$$\text{i.e. D8 (p,q)= max(|x − s|, |y − t|)} \quad (7)$$

In this case, the pixel(s,t) with D8 distance from pixel(x,y) less than or equal to some value r from a square centered at (x,y).

3.5 Tortuosity Detection

Here, several most tortuous branches for detection purpose have used. In order to detect low, medium or highly tortuous images the following condition is applied. High tortuosity occurs if m > Th1 otherwise, it seems to be a medium tortuosity. Where, m is average tortuosity of the most tortuous branches and Th1 is one threshold. If m > Th2, it means medium tortuosity otherwise low tortuosity has identified. Where, Th2 denotes another threshold.

In our proposed method, images with 384 X 512 pixels are used where Th1 is .045 and Th2 is .035. (These thresholds are obtained as per analysis of eye specialist.)

4 Experimental Results

The above proposed method has tested on images obtained from an Eye Department of Ramakrishna Mission Seva Pratisthan, Kolkata, West Bengal, India. The data base contains 25 color retinal images with 384 X 512 pixels. Our result matches with the visual detection. Also eye specialist of that hospital deliver comments after analyzing those images and our result matches with their analysis.Figure9 shows the images which are used with their tortuosity value. Here tortuosity for every branch is measured by the formula and maximum five tortuous branches taken into account to detect how much tortuous the image is. From these, five branches mean(*m*) is derived.

Observations done tortuosity measurement by different methods like normal visualization and Eye specialist's analysis suggests that images which are highly tortuous have tortuosity value greater than .045 and images which are low tortuous have tortuosity value less than .035. Images which are medium tortuous have tortuosity value between 0.035 and 0.045. For this reason, two thresholds .035 and .045 have been chosen for tortuosity detection.

So, mean(*m*) is checked by the following conditions,

m<.035 ------> low tortuosity
.045 >m>.035 ------> medium tortuosity
m>.045 ------>high tortuosity

The experimental result of tortuosity measurement and detection is presented in Table1.Throught the simulation threshold value, T=6 have been considered for every image.

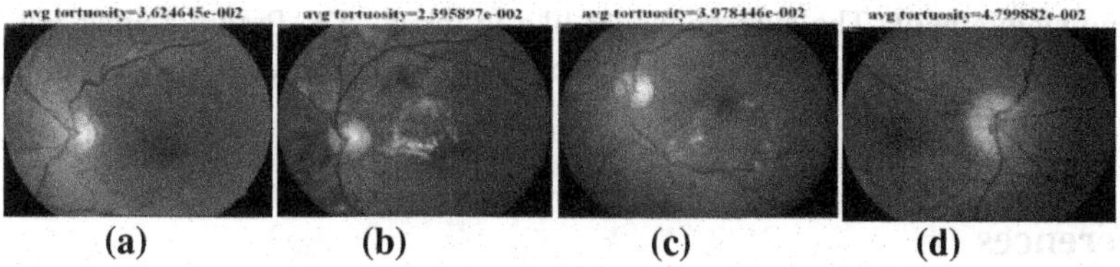

Fig. 7.(a), (b), (c), (d) images with tortuosity value

Table 1. Tortuosity measurement for different branches and detection

Image No.	BRANCH_1	BRANCH_2	BRANCH_3	BRANCH_4	BRANCH_5	MEAN(m)	DETECTION
Figure. 9(a)	0.053846154	0.04237109	0.036690034	0.025498891	0.022826087	0.036246451	MEDIUM
Figure. 9(b)	0.037254902	0.025833333	0.025721721	0.016260163	0.014724712	0.023958966	LOW
Figure. 9(c)	0.075892857	0.034744644	0.031058991	0.030392157	0.026833631	0.039784456	MEDIUM
Figure. 9(d)	0.091817043	0.054013644	0.039408867	0.033082707	0.021671827	0.047998817	HIGH

Table 1 represents the tortuous symptom of some images by high, medium or low range. According to eye specialists' view high tortuosity generally occurs due to obstruction in the blood flow through the veins, which might again be caused due to various diseases. Tortuosity, being merely a symptom can only be cured if the causing disease is treated. Failure to do so ultimately leads to rupture of veins leading to hemorrhage. Tortuosity may also occur in some people only as a normal variant, i.e., an anomaly without any particular reason, may also be congenital. In Table 1, those tortuous images are mentioned as medium or low range.

5 Conclusion

In this work, a method to measure tortuosity and produce a formula for tortuosity measurement is proposed. The proposed formula meets all the features of tortuosity i.e., the composition property is satisfied by summation, the amplitude modulation is satisfied by the ratio of the arc length over the chord length for every turn curve, and by means of the curve splitting, frequency modulation has satisfied. In this work, the rotation operation helps to rotate each branch and measures maximum distance according to threshold to divide the curve into segments. Again this method is threshold based and partitioning process gives best result. The detection process depends on the mean of the most tortuous segments of the blood vessel network. The main advantage of this method is, it breaks the image into branches so that every branch curve may use easily. So, from this aspect, this method is better than other existing methods. In future, the tortuosity detection process will be further improved. Then more clinical cases will be included in this study for early detection of disease. The techniques employed in this study will help in diagnostic accuracy as well as in reducing the workload of eye specialists.

Acknowledgement. The authors would like to thank Dr. Pradip Kumar Pal, Senior visiting surgeon, Eye Department, Ramakrishna Mission Seva Pratisthan, Kolkata for providing images and clinical advice.

References

1. Jerald Jeba Kumar, S., Madheswaran, M.: Automated Thickness Measurement of Retinal Blood Vessels for Implementation of Clinical Decision Support Systems in Diagnostic Diabetic Retinopathy. World Academy of Science, Engineering and Technology 64 (2010)
2. Hatanaka, Y., Nakagawa, T., Aoyama, A., Zhou, X., Hara, T., Fujita, H., Kakogawa, M., Hayashi, Y., Mizukusa, Y., Fujita, A.: Automated detection algorithm for arteriolar narrowing on fundus images. In: Proc. 27th Annual Conference of the IEEE Engineering in Medicine and Biology, Shanghai, China, September 1-4 (2005)
3. Grisan, E., Foracchia, M., Ruggeri, A.: A novel method for the automatic evaluation of retinal vessel tortuosity. In: Proc. 25th Annual International Conference of the IEEE Engineering in Medicine and Biology, Cancun, Mexico (September 2003)
4. Lotmar, W., Freiburghaus, A., Bracher, D.: Measurement of vessel tortuosity on fundus photographs. Graefe's Archive for Clinical and Experimental Ophthalmology 211, 49–57 (1979)
5. Sukkaew, L., Uyyanonvara, B., Makhanov, S.S., Barman, S., Pangputhipong, P.: Automatic Tortuosity-Based Retinopathy of Prematurity Screening System. IEICE Trans. Inf. & Syst. E91-D(12) (December 2008)
6. Hart, W., Golbaum, M., Cote, B., Kube, P., Nelson, M.: Measurement and classification of retinal vascular tortuosity. Int. J. Medical Informatics 53, 239–252 (1999)
7. Sukkaew, L., Uyyanonvara, B., Barman, S., Fielder, A., Cocker, K.: Automatic extraction of the structure of the retinal blood vessel network of premature infants. J. Medical Association of Thailand 90(9), 1780–1792 (2007)
8. Bullitt, E., Gerig, G., Pizer, S., Lin, W., Aylward, S.: Measuring tortuosity of the intracerebral vasculature from MRA images. IEEE Trans. Med. Imaging 22(9), 1163–1171 (2003)

A New Indexing Method for Biometric Databases Using Match Scores and Decision Level Fusion

Ilaiah Kavati[1,2], Munaga V.N.K. Prasad[2], and Chakravarthy Bhagvati[1]

[1] University of Hyderabad, Hyderabad-46, India
[2] Institute for Development and Research in Banking Technology, Hyderabad-57, India
kavati089@gmail.com, mvnkprasad@idrbt.ac.in,
chakcs@uohyd.ernet.in

Abstract. This paper proposes a new clustering-based indexing technique for large biometric databases. We compute a fixed length index code for each biometric image in the database by computing its similarity against a preselected set of sample images. An efficient clustering algorithm is applied on the database and the representative of each cluster is selected for the sample set. Further, the indices of all individuals are stored in an index table. During retrieval, we calculate the similarity between query image and each of the cluster representative (*i.e.*, query index code) and select the clusters that have similarities to the query image as candidate identities. Further, the candidate identities are also retrieved based on the similarity between index of query image and those of the identities in the index table using voting scheme. Finally, we fuse the candidate identities from clusters as well as index table using decision level fusion. The technique has been tested on benchmark PolyU palm print database consist of 7,752 images and the results show a better performance in terms of response time and search speed compared to the state of art indexing methods.

Keywords: Palm print, Indexing, Clustering, Sample images, Match scores, Decision level fusion.

1 Introduction

Biometric identification refers to automated method of identifying individuals based on their physiological and/or behavioral characteristics. However, in biometric identification systems, the identity corresponding to a query image is typically determined by comparing it against all images in the database [1]. This exhaustive matching process increases the response time and the number of false positives of the system. Therefore, an efficient retrieval technique is required to increase the search speed and reduce the response time of the system. The retrieval technique should be such that, instead of comparing the query image with every image in the database it has to retrieve a small set of images from the database to which the actual matching process is applied. This retrieval can be done by, a) partitioning the database into groups of similar images in order to facilitate and accelerate the search process, b) indexing methods.

M.K. Kundu et al. (eds.), *Advanced Computing, Networking and Informatics - Volume 1*,
Smart Innovation, Systems and Technologies 27,
DOI: 10.1007/978-3-319-07353-8_57, © Springer International Publishing Switzerland 2014

The partitioning of the database into groups can be done by either classification [3,4] or clustering [9]. Authors in [3], [4] partitions the database into predefined groups or classes. The class of the query identity is first calculated and compared only with the entries present in the respective class during the search process. However, the classification methods suffers with, uneven distribution of images in the predefined classes and image rejection [8]. On the other hand, clustering method organizes the database into natural groups (clusters) in the feature space such that images in the same cluster are similar to each other and share certain properties; whereas images in different clusters are dissimilar [9]. Clustering method does not use category labels that tag objects with prior identifiers, *i.e.,* class labels.

The other approach to reduce number of matching for identification is indexing. In traditional databases, records are indexed in an alphabetical or numeric order for efficient retrieval. But biometric data do not have any natural sorting order to arrange the records [9]. Hence, traditional indexing approaches are not suitable for biometric databases. Biometric indexing techniques are broadly categorized into, a) point [5], [6], b) triplet of points [11], [12], [13], and c) match score [14], [15], [16] based approaches. In [5], [6], authors extracted the key feature points of the biometric images and mapped them into a hash table using geometric hashing. Authors in [11], [12], [13], computed the triplets of the key feature points and mapped them into a hash table using some additional information. However, the limitation of these indexing methods is that all of them deal with variable length feature sets which make the identification system statistically unreliable.

In recent years, indexing techniques based on fixed length match scores also investigated for biometric identification. Maeda et al [14], computes a match score vector for each image by comparing it against all the database images and stored these vectors permanently as a matrix. Though, the approach achieves quicker response time, it takes linear time in worst case and also storing of match score matrix leads to increase in the space complexity. Gyaourova et al. [15] improved the work on match scores by choosing a small set of sample images from the database. For every image in the database a match score vector (index code) was computed by matching it against the sample set using a matcher and stored this match score vector as a row in an index table. However, a sequential search is done in the index space for identification of best matches which takes linear time and is prohibitive for a database containing millions of images. Further, authors in [16],used Vector Approximation (VA+) file to store the match score vectors and *k-NN* search, palm print texture to retrieve best matches. However, the performance of VA+ file method generally degrades as dimensionality increases [17]. To address these problems, this paper proposes an efficient clustering-based indexing technique using match scores. We compute a fixed-length index code for each input image based on match scores. Further, we propose an efficient storage and retrieval mechanism using these indices.

Rest of the paper is organized as follows. The proposed indexing technique has been discussed in Section 2. Section 3 describes the proposed retrieval technique. Section 4 presents the experimental results and performance of the proposed system against other indexing methods in the literature. Conclusions are given in Section 5.

2 Proposed Indexing Methodology

This section discusses our proposed methodology for indexing the biometric databases. Let $S = \{s_1, s_2,..., s_k\}$ be the sample image set, and $M_x = \{m(x,s_1), m(x,s_2), ... m(x,s_k)\}$ be the set of match scores obtained for an input image x against each sample image in S. We describe M_x as the index code of image x i.e., the index code of an image is the set of its match scores against the sample set. The match score between two images is computed by comparing their key features in Euclidean space. The match scores obtained are usually in the range 0-100.

Further, we store the index code of each individual in a 2D Index table A Fig. 1. Each column of the table corresponds to one sample image in the sample set. If image x has a match score value $m(x,s_i)$ with sample image s_i its identity (say x) is put in location $A(m(x,s_i), s_i)$. It can be seen from Fig. 1 that, each entry of the table $A(m(x,s_i), s_i)$ contains a list of image identities (*i.e.*, *IidList*) from the database whose match score is $m(x,s_i)$ against sample image s_i.

The motivation behind this concept is that, images that belong to a same user will have approximately similar match scores against a third image (say sample image s_i). Let q be a query image, we can thus determine all similar images from this index table by computing its match score against the sample image and selecting all images (*i.e.*,*IidList*) that have approximately similar match score against the sample image.

Match score \ Sample image	s_1	....	s_i	....	s_k
0	*IidList*		*IidList*		*IidList*
1	*IidList*		*IidList*		*IidList*
. .	. .	....	. .	....	. .
100	*IidList*		*IidList*		*IidList*

Fig. 1. Index table organization

2.1 Selection of Sample Image Set

The selection of sample images from the database plays a crucial role in the performance of the system. Images which are more generic (*i.e.*, very different from one another) and represent the qualities of the entire database should be selected for sample set. In this paper, we use an efficient dynamic clustering algorithm (*i.e.*,leader algorithm) for the selection of sample image set.

Leader clustering algorithm [7] makes only a single pass through the database and finds a set of leaders as the cluster representatives (which we call sample images). In this work, we use the match score between the images to determine the cluster similarity. The motivation of using match score as similarity measure is that, usually similar images will have almost same features and so their match score is high *i.e.*, images in the same cluster will have high match score between them. Leader clustering algorithm uses a user specified similarity threshold and one of the image as the starting leader. At any step, the algorithm assigns the current image to the most similar cluster (leader) or the image itself may get added as a leader if its match score similarity with the current set of leaders does not qualify based on the user specified threshold. Finally, each cluster will have a listing of similar biometric identities and is represented with an image called leader. The found set of leaders acts as sample set of the database. The major advantage of dynamic clustering (such as leader algorithm) is that, new enrollments can be done with a single database scan and without affecting the existing clusters which is useful for clustering and indexing large databases.

3 Retrieval of Best Matches (Identification)

This section proposes an efficient retrieval system to identify a query image. Fig. 2 shows the proposed method of identification. When a query image is presented to the identification system, the technique retrieves the candidate identities from the clusters as well as from the 2D Index table which are similar to the query. Finally, the proposed system fuses the candidate identities (evidences) of both strategies to achieve better performance.

Although there are other strategies like multi-biometrics [18], [19] (such as multi-sensor, multi-algorithm, multi-sample, etc.) to retrieve multiple evidences for personal identification, we want to make a full use of the intermediate results in the process of computing index code in order to reduce the computational cost. It is easy to see from the Fig. 2 that, when computing the index code for a query image to identify the possible matches (Candidate list2) from the index table, we can get set of match scores against cluster representatives. Using them, we can also retrieve the candidate identities (Candidate list1) as additional evidence from the selected clusters whose representative match score greater than a threshold.

Let $G = \{g_1, g_2,..., g_k\}$ be the set of clusters, and $M_q = \{m(q,s_1), m(q,s_2), \ldots m(q,s_k)\}$ be the index code of a query image q where $m(q,s_i)$ be the match score value of q against sample image s_i. The retrieval algorithm use $(m(q,s_i), s_i)$ as index to the Index table and retrieve all the images (*IidList*) found in that location as similar images to the query into a Temporary list. In other words we retrieve all the images from the index table whose match score value against the sample image is equal to the query image. We also retrieve images from the predefined neighborhood of the selected location into the Temporary list. Finally, we give a vote to each retrieved image. Further, we also retrieve images from cluster g_i as similar images to the query, if $m(q,s_i)$ $\geq$ similarity threshold *i.e.*, clusters are selected whose representative is similar to the query image. We store the retrieved cluster images into Candidate list 1. We repeat this process for each match score value of the query index code. In our next step, we accumulate and count the number of votes of each name in Temporary list. Finally, we sort the all the individuals in descending order based on the number of votes received and select the individuals whose vote score greater than a predefined threshold into Candidate list 2.

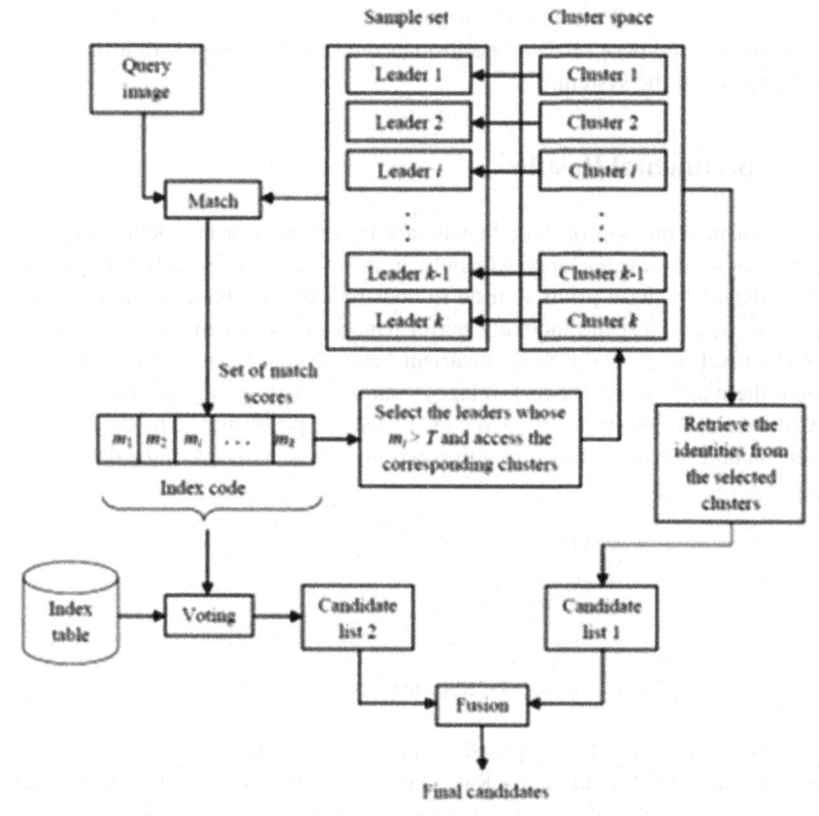

Fig. 2. Block diagram of the proposed identification system

3.1 Fusion of Decisions Output

The performance of uni-modal biometric systems may suffer due to issues such as limited population coverage, less accuracy, noisy data and matcher limitations [18]. To overcome the limitations of uni-modal biometrics and improve the performance, fusion of multiple pieces of biometric information has been proposed. Fusion can be performed at different levels such as data, feature, match score and decision level. In this paper, we use decision level fusion method. With this method, the decisions output (candidate identities) obtained from the cluster space and Index table are combined using, a) union of candidate lists, b) intersection of candidate lists.

The union fusion scheme combines the candidate list of the individual techniques. This fusion scheme has the potential to increase the chance of finding correct identity even if the correct identity is not retrieved by some of the techniques *i.e.*, the poor retrieval performance of one technique will not affect the overall performance. However, this scheme often increases the search space of the database. With intersection fusion scheme, the final decision output is the intersection of the candidate lists of the individual techniques. This type of fusion can further reduce the size of the search space. However, the poor retrieval performance of one technique will affect the overall performance of the system.

4 Experimental Results

We experimented our approach on benchmark PolyU palm print database [10] consist of 7,752 gray-scale images, approximately 20 prints each of 386 different palms. The first ten images of these prints is used to construct the database, while the other ten images are used to test the indexing performance. We segment the palm images to 151 x 151 pixels and use the Scale Invariant Feature Transform (SIFT features [2]) to compute the match score between images. Samples of some segmented images from PolyU database is shown in Fig. 3.We evaluated the performance of the system with different sample set sizes and chosen the optimum value as 1/3rd of the database.

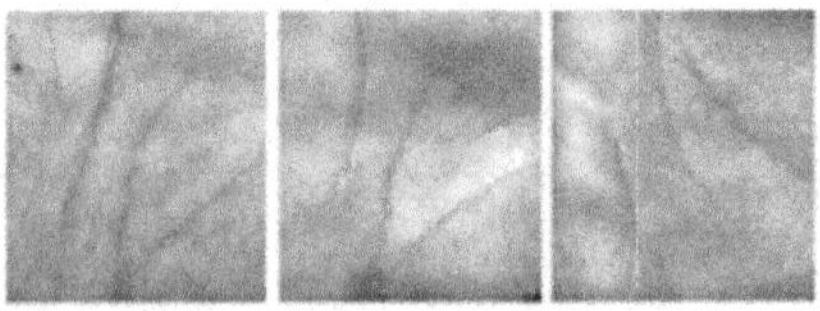

Fig. 3. Samples of segmented palm print images of PolyU database

The performance of the proposed technique is evaluated using two measures, namely Hit Rate (HR), Penetration Rate (PR); where HR is defined as the percentage of test images for which the corresponding genuine match is present in the candidate list and PR is the average search space in the database to identify a test image (*i.e.*, average candidate list size). The performance curves plotting the HR against PR at various thresholds are shown in Fig. 4(a). It is observed that the proposed fusion techniques performs well (as their PR is very less) compared to individual techniques. Further it can be seen that, the union fusion performs well to the intersection fusion. Table 1 shows the performance of the proposed techniques for PolyU database.

Table 1. PR (where HR=100%) of our proposed techniques for PolyU database

Clustering	Indexing	Intersection fusion	Union fusion
48.7%	18.8%	12.4%	10.2%

4.1 Retrieval Time

We analyze the retrieval time of our algorithm with big-O notation. Let q be the query image, k be the number of sample images chosen, and N be the number of enrolled user in the database. To retrieve the best matches for a query image, our algorithm computes the match score of query against the each sample image and retrieves the images, a. from the index table whose match scores against that sample image are nearer to query, b. as well as from respective cluster of the sample image if its match score is greater than similarity threshold. This process takes O(1) time. However, there are *k* sample images, so the time complexity our algorithm is O(*k*). On the hand linear search methods requires O(*N*). Thus our approach takes less time than the linear search approach as *k*<<*N*.

4.2 Comparison with Other Indexing Techniques

The proposed technique has been compared with existing match score base indexing techniques [15] [16]. The performance of proposed technique against technique in [15] can be seen from the Fig. 4(b). It is seen that our algorithm performs with less PR compared to the [15]. Further, authors in [15] performed linear search over the index space to retrieve the best matches. This process takes considerable amount of time i.e. O(N). Finally, the system in [16] achieves only a maximum of 98.28% HR for PolyU database. It can be inferred that the proposed system enhanced the indexing performance.

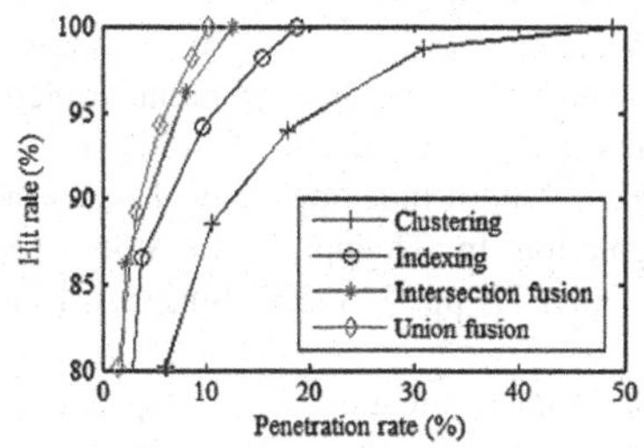

(a)

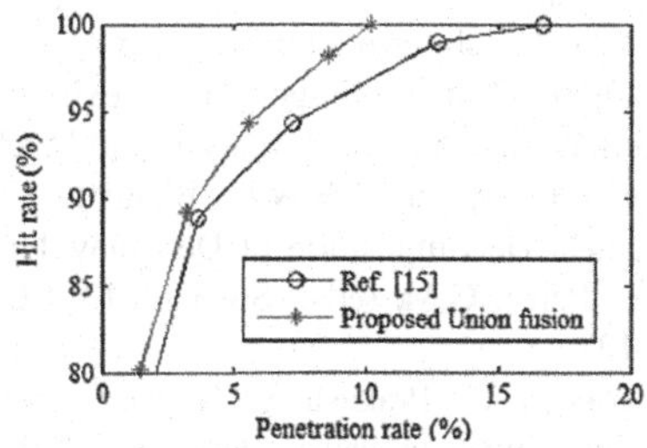

(b)

Fig. 4. HRVs PR of different techniques on PolyU database, (a) Our proposed methods, (b) Comparison with Ref [15]

5 Conclusions

In this paper, we propose a new clustering based indexing technique for identification in large biometric databases. We compute a fixed length index code for each biometric image using the sample images. Further, we propose an efficient storing and searching method for the biometric database using these index codes. We efficiently used the intermediate results in the process of computing index code that retrieve multiple evidences which improves the identification performance without increasing computational cost. Finally, the results shows the efficacy of our approach against state of art indexing methods. Our technique is easy to implement and can be applied to any large biometric database.

References

1. Jain, A.K., Pankanti, S.: Automated fingerprint identification and imaging systems. In: Advances in Fingerprint Technology, 2nd edn. Elsevier Science (2001)
2. Lowe, D.G.: Distinctive image features from scale-invariant keypoints. International Journal on Computer Vision 60, 91–110 (2004)
3. Henry Classification System International Biometric Group (2003), http://www.biometricgroup.com/HenryFingerprintClassification.pdf
4. Wu, X., Zhang, D., Wang, K., Huang, B.: Palmprint classification using principal lines. Pattern Recognition 37, 1987–1998 (2004)
5. Boro, R., Roy, S.D.: Fast and Robust Projective Matching for Finger prints using Geometric Hashing. In: Indian Conference on Computer Vision, Graphics and Image Processing, pp. 681–686 (2004)
6. Mehrotra, H., Majhi, B., Gupta, P.: Robust iris indexing scheme using geometric hashing of SIFT keypoints. Journal of Network and Computer Applications 33, 300–313 (2010)
7. Jain, A.K., Murty, M.N., Flynn, P.J.: Data Clustering: A Review. ACM Computing Surveys 31(3) (1999)
8. Maltoni, D., Maio, D., Jain, A.K., Prabhakar, S.: Handbook of Fingerprint Recognition. Springer (2003)
9. Mhatre, A., Palla, S., Chikkerur, S., Govindaraju, V.: Efficient search and retrieval in biometric databases. Biometric Technology for Human Identification II 5779, 265–273 (2005)
10. The PolyU palmprint database, http://www.comp.polyu.edu.hk/biometrics
11. Bhanu, B., Tan, X.: Fingerprint indexing based on novel features of minutiae triplets. IEEE Transactions on Pattern Analysis and Machine Intelligence 25, 616–622 (2003)
12. Kavati, I., Prasad, M.V.N.K., Bhagvati, C.: Vein Pattern Indexing Using Texture and Hierarchical Decomposition of Delaunay Triangulation. In: Thampi, S.M., Atrey, P.K., Fan, C.-I., Perez, G.M. (eds.) SSCC 2013. CCIS, vol. 377, pp. 213–222. Springer, Heidelberg (2013)
13. Jayaraman, U., Prakash, S., Gupta, P.: Use of geometric features of principal components for indexing a biometric database. Mathematical and Computer Modelling 58, 147–164 (2013)
14. Maeda, T., Matsushita, M., Sasakawa, K.: Identification algorithm using a matching score matrix. IEICE Transactions in Information and Systems 84, 819–824 (2001)
15. Gyaourova, A., Ross, A.: Index Codes for Multi biometric Pattern retrieval. IEEE Transactions on Information Forensics and Security 7, 518–529 (2012)
16. Paliwal, A., Jayaraman, U., Gupta, P.: A score based indexing scheme for palmprint databases. In: International Conference on Image Processing, pp. 2377–2380 (2010)
17. Weber, R., Schek, H., Blott, S.: A quantative analysis and performance study for similarity search methods in high-dimensional spaces. In: Proceedings of the 24th Very Large Database Conference, pp. 194–205 (1998)
18. Ross, A., Nandakumar, K., Jain, A.K.: Handbook of Multibiometrics. Springer, New York (2006)
19. Kumar, A., Shekhar, S.: Personal identification using multibiometrics rank-level fusion. IEEE Transactions on Systems, Man and Cybernetics 41, 743–752 (2011)

Split-Encoding: The Next Frontier Tool for Big Data

Bharat Rawal[1], Songjie Liang[2], Anthony Tsetse[3], and Harold Ramcharan[1]

[1] Department of Computer and Information Sciences Shaw University
Raleigh, NC, USA
[2] SoroTek Consulting, Inc.
[3] Department of Computer and Information Sciences State University of New York
Fredonia, NY, USA
{brawal,hramcharan}@shawu.edu, jeffliang1@gmail.com,
Anthony.Tsetse@fredonia.edu

Abstract. This paper proposes Split-encoding mechanism, an alternative to available encoding and compression techniques for transmitting large medical images. In the Split-protocol, a client establishes a connection with one connection server, then transfers data concurrently from the multiple data servers located on subnets, while providing client connection anywhere along the network. The separation of data transfer from connection establishment is completely transparent to the client. Split-encoding technique reduces network impact, avoids redundancy of data, and improves data transmission time; offers better reliability and Security. The Split-protocol was successfully engaged in defending against the DoS/DDoS attack. This paper describes the design and initial implementation of Split-encoding, serving as a basis for application implementation. It also reports the results of quantitative studies regarding the execution of the initial execution.

1 Motivation

Immense data is generated from our day to day operations. As billions of devices are connected to networks, possibility of congestion increases substantially, thus requiring new ways to reduce network traffic and enhance data storage capacity. Gaming and medical images are generating major Peer to Peer traffic on today's Internet. In North America, 53.3 % of network traffic is contributed by P2P communication [13, 14]. Today, even though there are uncountable techniques available to compress and decompress data before transmitting on the Internet. Implementing various security protocols and techniques adds additional traffic on the network, which affects the overall functioning of the network. In this paper, we propose a simple data encoding technique that resides on top of existing data compression techniques and security protocols implemented with a Split - protocol.

Critical medical conditions like cardiology and neurological traumas need complete information in the minimum amount of time, so that a patient's life can be spared [6]. As shown in Fig. 1, any medical event will trigger transmission of medical data. The data servers (DSs) send relevant data and images simultaneously to the respective team member. The cardiologist will have all necessary information related

M.K. Kundu et al. (eds.), *Advanced Computing, Networking and Informatics - Volume 1*,
Smart Innovation, Systems and Technologies 27,
DOI: 10.1007/978-3-319-07353-8_58, © Springer International Publishing Switzerland 2014

to cardiology, while the radiologist will receive all data such as X-ray images, MRI, CT-Scan etc.

In this paper, we propose the implementation of the Split - encoding approach on top of exiting data compression techniques for faster transmission of medical images and securing database servers in the health information system. The physicians using web browsers with no direct access to the PACS information system will encounter slower network speeds [15].

2 Introduction

To demonstrate the split protocol concept, we describe web server and client applications that are designed and implemented using a bare machine computing (BMC) paradigm [18], also known as dispersed operating system computing (DOSC). Bare PC applications are easier and friendlier to use in the demonstration. In the BMC model, operating system (OS) or kernel is completely eliminated and application objects (AO) [3] carry their own network stack and drivers to move independently on a bare PC. All BMC applications are self-contained, self-managed and self-executable, therefore granting the application programmer to solely control the applications and their execution environments. [1 - 7], and [12, 17, 19] have been presented on the BMC concept.

The rest of the paper is organized as follows. Section 3 discusses related work. Section 4 describes design and implementation. Section 5 outlines the Split-encoding schemes. The Section 6 presents experimental results and Section 7 contains the conclusion.

3 Related Work

As per NVIDIA, device to device transfer rate on GeForce 9600 GT is approximately 57.6 GB/Sec. In contrast to the transfer rate from host to device which is around 2 GB/Sec; it is much faster for device to device transfer [9].

Image compression may either be lossy or lossless. Lossless compression is used primarily for archival purposes and medical imaging instead of lossy, which is more suitable for natural images such as pictures where minor loss of pixels can be acceptable to achieve a substantial reduction in bit rate.

In the grid paradigm, we still need to address issues such as how to reduce differences in finish times between elected replica servers, to avoid traffic congestion resulting from transmitting the same blocks in different links among hosts and clients [8]. The medical images of diverse modalities, such as computerized tomography (CT), magnetic resonance image (MRI), positron emission tomography (PET), and single photon emission computed tomography (SPECT) and histology images [10]. According to Manimurugan, a novel technique referred to as adaptive threshold-based block classification, is for medical image compression, and can be applied to all variations of medical images. Also, this method introduces a computational algorithm to classify the blocks based on the adaptive threshold value of the variable [11].

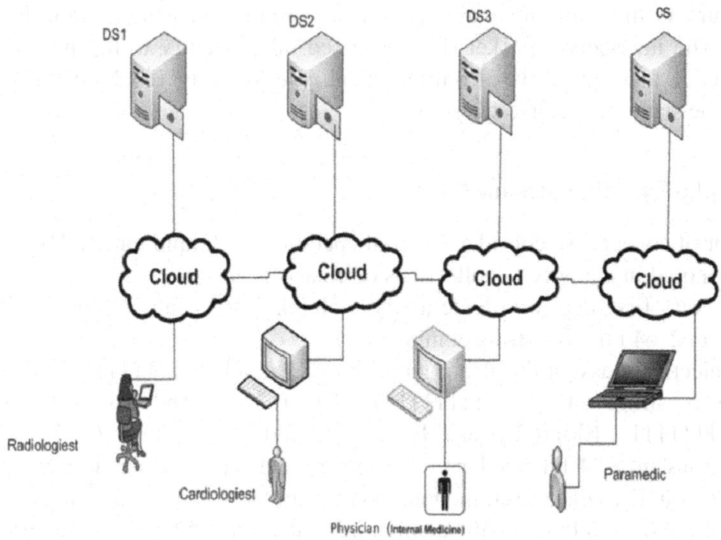

Fig. 1. Multi-Client / Multi-Server Split Architecture

A wavelet based compression scheme uses the correlation analysis of wavelet coefficients like "Wavelet-based Medical Image Compression with Adaptive Prediction", but makes it simpler and more accurate by excluding the requirement of selection of basis function and through the quantization of prediction error in coarse bands [20]. Splitting protocol at a client server architecture level is different from migrating TCP connections, processes or Web sessions; splicing TCP connections; or masking failures in TCP-based servers. For example, in migratory TCP (M-TCP), a TCP connection is migrated [5]. The Split-protocol was successfully employed in defending against the DoS / DDoS attack [12].

4 Design and Implementation

Split-protocol client server architecture design and implementation reproduced here differ from traditional client and server designs. As the traditional client server architecture is modified in this approach, we have designed and implemented a client and a server based on a bare PC, where there is no traditional OS or kernel running on the machine. This made our design simpler and easier to make modifications to conventional protocol implementations. Fig. 1 depicts a high level design structure of a node and server in a bare PC design. The large data file is circulated on various DSs. Each client and a server consist of a TCP state table (TCB), which consists of the state of each request. Each TCB entry is made unique by using a hash table with key values of IP address and a port number. The CS and DS TCB table entries are referred by IP3 and Port#. The Port# in each case is the port number of the request initiated by a client. Similarly, the TCB entry in the client is referenced by IP1 and Port# [4].

The TCB tables form the key system component in the client and server designs. A given entry in this table maintains complete state and data information for a given request. The inter-server packet (ISP) is based on this entry to be shipped to a DS when a GET message arrives from the client. We use only 168 Byte ISP packets to transfer the state of a given request [4].

4.1 Split-Encoding Scheme # 1

For proof of concept we consider a partial encoding technique for a 4.71GB JPG file, which is stored in memory as collections of binary 0s and 1s.

In the very first step, we divide a whole chunk of data into various 64 bit strings resulting in 2^64 possible string combinations.

We selectively assign the string number to S1 to 1111111111111…11 all 64 1s, S2 to a string of (1111111111111…10) (63 1s and last one zero), S3 (111111111111…100) (62 1s and last two 0s) and so on up to # 16777216 (possible combinations in a 24 bit number). Considering the fact that our data are generated using 128 ASCII character set, the total combinations of the 8 byte string are 128^8 = 72057594037927936.It is worth mentioning that these strings can be formed with repeated characters (e.g. aaaaaaaa). However, if non-printable characters and rarely used characters are reasonably removed from the 128 ASCII character set, the total possible combinations of 8 byte strings can reasonably be reduced to 96^8. We are limiting our number assigned to any string to be up to maximum 16777216. The total combinations of any 64 bit data will approximately be 96^8 = 7213895789838336

If any 64 bit combination is different from our selected 16777216 string combinations, we will transmit those 64 bits as it is without encoding. Considering the fact that in real life communication, we use certain fixed patterns based on language and numbering system, the possibility of non-occurrence reduces and the possibility of occurrence increases for a particular pattern. Whether the information is encoded or ordinary is determined by a special bit in DM1 packet (DM1 packet is 168 Bytes large) [4]. The regular data will be supplied by CS with the information sequence# of a particular piece of 64 bit data. However, this type of situation occurs very rarely.

The Ethernet MTU is 1500 bytes, and can contain up to 1460 byte payload. To transfer 4.71GB file will require 3463921 Ethernet frames (packets). Each frame# will correspond to a chronological order of string numbers. We assign the unique string number of 24 bits to 64bit data string made of 1s and 0s (keeping in mind that we are only encoding partial data)

For every individual Ethernet frame, we can send up to 486 unique string (pattern) numbers (1460 bytes). We can make DS1 to send odd numbered frames and DS2 send even numbered frames (i.e. If a DS1 sends frame #1, 3, 5… and DS2 will send frame#2, 4, 6, … and so on.

Based on the frame sequence number, receiving end will arrange all Ethernet frames. Since each frame contains 486 strings of 8 Bytes, when receivers decode it will convert into 486×8 =3888 Bytes data. Normally one frame sends 1460 Bytes of data (excluding 20 byte IP header and 20 byte TCP header). So the encoded data transaction is 2.66 times more than a normal data transaction. In other words, we are sending only 38% frames, that mean it will reduces 62% of network traffic.

Typically in our experiment we encountered very few non-encoded data strings. However, when we increase the data size in the Terabyte the probability of occurrence of non-encoded string is higher.

Let us compute the probabilities of occurrence of two strings #1 and #2.

Assume probabilities of occurrence of two strings #1 and #2 are P (1) and P (2) respectively.

Neither the probability of string #1 nor string #2 is affected by the occurrence (or an occurrence) of the other event, so the two events should be independent.

Therefore, the conditional probability of occurrence for strings #1 and #2 is as follows:

$P(1|2) = P(1)$, which is the same as
$P(2|1) = P(2)$

Similarly, the probability of occurrence of two strings #1 and #2 is as follows:

$$P(1 \cap 2) = P(1)\, P(2) \tag{1}$$

Based on the above information, the occurrence of any string does not affect the occurrence other string. The total available numbers to be assigned to any 64 bit piece of data are 96^8 = 7213895789838336. So the probability of any one piece of 64bit data to be assigned a number is

$$P = (2^{24}) / (96^{8}) = 2.32568e{-9}. \tag{2}$$

So the probability of any one piece of 64 bit data NOT to be assigned a number is

$$NP = 1 - P = 1 - 2.32568e{-9} = 0.999999998. \tag{3}$$

For given set of strings S1, S2… Sr (r=16777216) and at length k= 64, r strings occur independently of each other in a random fashion. This 64bit string is made of m substring length k (m= 8, k= 8). Substrings also, occur in independently of each other and in random fashion.

The probability that there is a string T of length *k* that is a substring of each of S1, S2… Sr (r=16777216).

According to Double Independence Model.

$$P(\varsigma) = 1 - \Pi_{\gamma \in C_k} (1_{-} P^{r}_{\gamma})^{\Omega(\gamma)}) \tag{4}$$

C_k - Set of all compositions for strings of length k
P_γ– Probability of occurrence for strings with composition (γ)
$\boldsymbol{\Omega}(\gamma)$ – Number of strings with the composition γ
$P(\varsigma)$ - the probability that at least one string of length k occurs as a common substring to all the random strings.

The results will offer good approximation when the number of random strings is low, but poor approximations when there are a large number of strings.

For proof of concept we took 4.71GB data JIF file, there are 632165499 strings (64 bits per string). The probability of occurrence of any string is 2.32568e-9.

For the majority of the time the data we are use falls within those 16777216 string numbers and in our data block many strings appear repeatedly. To transfer 4.71 GB file requires 3463921 Ethernet frames, if a frame carries 1460 byte data.

We estimate the following 2 numbers (encoded and non-encoded) in the way below:

3888*x + 1460*y = 4.71*1024*1024*1024
x/y = 124820/138575

So # of encoded frames x =918029; the non-encoded # y=1019195.

In our typical experiment we found 918370 frames were encoded and 1019400 were non encoded frames. We have noticed that there are multiple frames for encoded as well as non encode frame received in the data. Transferring entire file it took only 875 seconds as compared to 1400 second with our previous MC/MS results it reduces almost 35.6% transmission time and 60% network traffic.

```
//convert from binary to decimal
public static String convertBinToDec( String s )
{
//check
if ( isNumber(s,2) == false ) return "-1";
BigInteger bi = new BigInteger(s, 2);
return bi.toString();
}    //convert from decimal to binary
public static String convertDecToBin( String s )
{
//check
if ( isNumber(s, 10) == false ) return "-1";
BigInteger dec = new BigInteger(s, 10);
return dec.toString(2);. BigDec2Bin conversion
method B [20].
```

When we scan the 64 bit string, there can be only two possibilities. Either it does exist in our database with the corresponding string number or doesn't. These events are independent, and if one of the two outcomes is defined as a success, then the probability of exactly x successes out of N trials (events) are given by.

$$P(x) = N!/(x!(N-x)!)\ \ (\pi^{x}\,(1-\pi)^{(N-x)}\,) \tag{5}$$

where the probability of success =0. 5

N = 3463921

x = 124820

For the binomial approximation we found following Mean =1731960. 5, Variance = 865980.25 and Standard Deviation 930.58. The high standard deviation shows that the data is widely spread.

5 Experimental Results

5.1 Experimental Setup

The experimental setup involved a prototype server cluster consisting of the Dell OptiPlex 790 PCs with Intel R-core (TM) i5-2400 CPU @3.10GHz Processor, 8 GB RAM, and an Intel 1G NIC on the motherboard. All systems were connected to a Linksys 16 port 1 Gbps Ethernet switch. Bare PC clients were used to stress test the servers. The bare PC Web clients capable of generating 5700 requests/Sec were used to create workload.

5.2 Performance Measurement

For scanning binary strings, we used a modified program provided by creativecommons.org [18], and then placed the timer on it. For binary 330 digits, the program (SuffixTree1.java) took < 1minute to build the suffix tree. For number of binary digits =330, it took 0.146 seconds to create the suffix tree for this number. However, when we tried a 3000+ digit long number, it was out of memory. For a number of binary digits =3403 it shows Exception error telling the overhead limit exceeded at "SuffixTreeNode. Insert". In addition, we tried another program, which is found on code.google.com. It is much faster to create suffix tree than the first one, but it could only handle binary 220 digit number or so. It took 0.009 seconds to create the suffix tree for this number.

Snappy is a compression/decompression library. It aims for very high speeds and reasonable compression. For instance, compared to the fastest mode of zlib, Snappy has an order of magnitude faster for most inputs, but the resulting compressed files are anywhere from 20% to 100% bigger. On a single core of a Core i7 processor in 64-bit mode, Snappy compresses at about 250 MB/sec or more and decompresses at about 500 MB/sec or more [16].

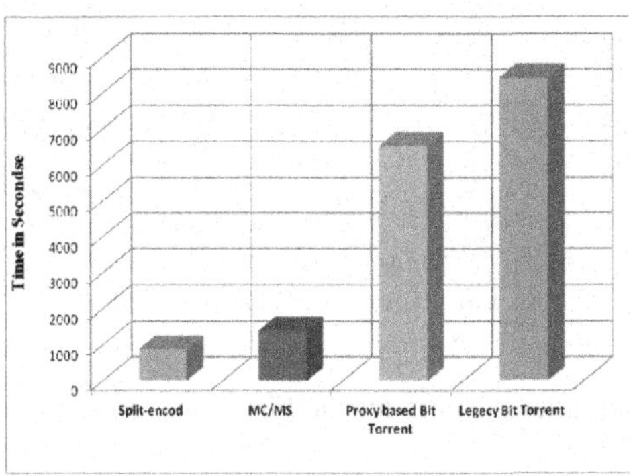

Fig. 2. Data Transmission Time for (4.71GB) [6]

Fig. 2 represents the comparison between legacy BitTorent and regular server systems. It can be seen that this configuration, TCP/Http transaction time for the Split - encode system of five DS servers offers factor of six times faster data transmission compared to legacy Bit Torrent. This is a significant reduction in data transmission time.

Fig. 3 illustrates the Image Retrieval Time for CR Chest 7.1 MB files for System of 3 Servers. We can notice that there is little or no significant improvement over MC/MS System. That means for smaller file sizes Split-encoding do not offer any advantages over other techniques.

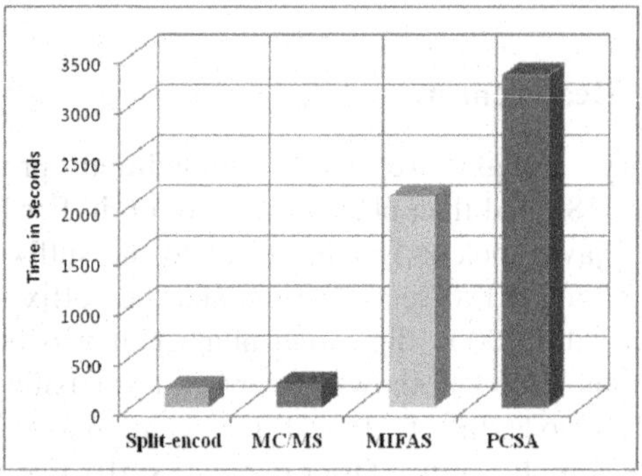

Fig. 3. Image Retrieval Times for CR Chest 7.1 MB file for System 3 Servers [6]

As shown in Fig. 4. MC/MS or Split-encode dose not add any advantages over other not suitable for smaller data sizes.

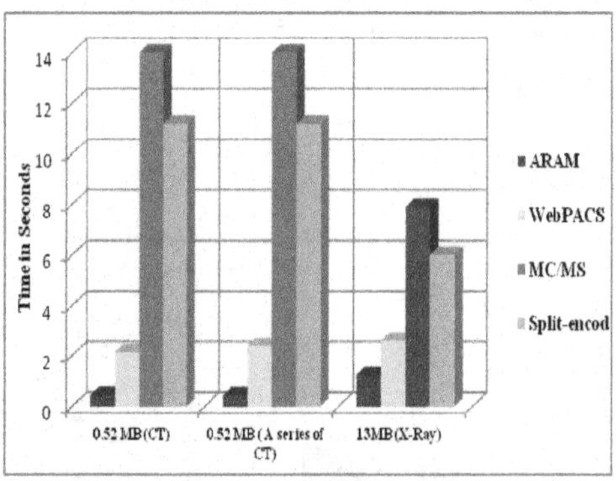

Fig. 4. Compare Query/Retrieval Times for the First Image from Local Grid Node and Web PACS using ARAM [20]

6 Conclusion

This paper proposes an augmented 'mission-critical' application as the motivation for this work. In our research, we extended the novel concept of split protocols and their applications to facilitate encoding on top of existing data compression techniques. Various encoding and data compression techniques were presented.

The novel, Split-encoding technique avoids connection bottleneck and single point failure of the server as well as client. It provides a better network solution for transfer of large files from an individual data source, reduces network traffic, and improves system performance at all levels of a network. Split-encoding can be used to transfer and store Big Data. Equations 1-5 describe mathematical expression which confirms empirical results. The Split-encoding technique possesses inherently failover ability in connection servers and data servers and also provides reliability, simplicity and scalability for building high-performance and client server systems.

References

1. Rawal, B., Karne, R., Wijesinha, A.L.: Splitting HTTP Requests on Two Servers. In: Third International Conference on Communication Systems and Networks (2011)
2. Rawal, B., Karne, R., Wijesinha, A.L.: Insight into a Bare PC Web Server. In: 23nd International Conference on Computer Applications in Industry and Engineering (2010)
3. Karne, R.K.: Application-oriented Object Architecture: A Revolutionary Approach. In: 6th International Conference, HPC Asia (2002)
4. Rawal, B., Karne, R., Wijesinha, A.L.: Split Protocol Client Server Architecture. In: 17th IEEE Symposium on Computers and Communication (2012)
5. Rawal, B., Karne, R., Wijesinha, A.L.: A Split Protocol Technique for Web Server Migration. In: 2012 International Workshop on Core Network Architecture and Protocols for Internet (2012)
6. Rawal, B.S., Berman, L.I., Ramcharan, H.: Multi-Client/Multi-Server Split Architecture. In: The International Conference on Information Networking (2013)
7. Rawal, B.S., Phoemphun, O., Ramcharn, H., Williams, L.: Architectural Reliability of Split-protocol. In: IEEE International Conference on Computer Applications Technology (2013)
8. Yang, C.-T., Lo, Y.-H., Chen, L.-T.: A Web-Based Parallel File Transferring System on Grid and Cloud Environments. In: International Symposium on Parallel and Distributed Processing with Applications (2010)
9. NVIDIA. Nvidia Cuda Compute Unified Device Architecture. Programming Guide v. 2.0 (2008)
10. Bhadoria, S.N., Aggarwal, P., Dethe, C.G., Vig, R.: Comparison of Segmentation Tools for Multiple Modalities in Medical Imaging. Journal of Advances in Information Technology 3(4), 197–205 (2012)
11. Alagendran, B., Manimurugan, S.: A Survey on Various Medical Image Compression Techniques. International Journal of Soft Computing and Engineering 2(1) (2012)
12. Rawal, B., Ramcharan, H., Tsetse, A.: Emergent of DDoS Resistant Augmented Split Architecture. In: IEEE 10th International Conference HONET- CNS (2013)

13. Lua, E.K., Crowcroft, J., Pias, M., Sharma, R., Lim, S.: A Survey and Comparison of Peer-to-Peer Overlay Network Schemes. IEEE Journal of Communications Surveys & Tutorials 7(2), 72–93 (2005)
14. http://torrentfreak.com/bittorrent-still-dominates-global-internet-traffic-101026/
15. Koutelakis, G.V., Lymperopoulos, D.K.: A grid PACS architecture: Providing data-centric applications through a grid infrastructure. In: Proceedings of the 29th Annual International Conference of the IEEE EMBS (2007)
16. https://code.google.com/p/snappy/
17. Karne, R.K.: Application-oriented Object Architecture: A Revolutionary Approach. In: 6th International Conference, HPC Asia (2002)
18. [Online] http://vassarstats.net/binomialX.html
19. Ford, G.H., Karne, R.K., Wijesinha, A.L., Appiah-Kubi, P.: The Performance of a Bare Machine Email Server. In: Proceedings of 21st International Symposium on Computer Architecture and High Performance Computing, pp. 143–150 (2009)
20. Ramesh, S.M., Shanmugam, A.: Medical image compression using wavelet decomposition for prediction method. International Journal of Computer Science and Information Security 7(1) (2010)

Identification of Lost or Deserted Written Texts Using Zipf's Law with NLTK

Devanshi Gupta, Priyank Singh Hada, Deepankar Mitra, and Niket Sharma

Manipal University Jaipur, ComputerScience Department, Jaipur, India
{devanshigupta06,pshada1,deepankar.mitra4ever,
niket.sharma1408}@gmail.com

Abstract. Sometimes it becomes very difficult to identify the valuable text written by some great personalities; especially when the text is not having a signature or the author is anonymous. Deserted manuscripts or documents without a title or heading can be an additional pain. It might happen that the work of dignitaries are lost or only some part of their valuable piece of work is found available in the libraries or with other storage media's. By deploying Zipf's law with the NLTK module available in python, this problem can be solved to a great extent, helping save the originality of the valuable texts and not leaving them unidentified. This can also be helpful in some real time data analysis where frequency plays an important role; plagiarism detection in written texts is one such example. NLTK is a strong toolkit which helps in extracting, segmenting, parsing, tagging and searching etc. of many natural languages with the help of python modules. In this paper it has been tried to combine Zipf's Law with NLTK to come up with a tool to identify the anonymous or deserted valuable texts.

Keywords: Keywords: NLTK, Zipf's law, NLP.

1 Introduction

Dealing with Natural Language Processing (NLP) is a fascinating and active research field. With the help of Natural Language processing Toolkit (NLTK) module available in Python, graphical analysis of natural languages has also become possible. The graphical user interface present in NLTK helps to plot and study graphs to understand the results of natural language processing in a much better an efficient way. NLTK suits both linguists and researchers as it has enough theory as well as practical practice examples in the NLTK book itself [1], [2]. NLTK is an extensive collection of documentation, corpora and few hundreds exercises making it a huge framework providing a better understanding of the natural language's and their processing. NLTK is entirely self-contained providing raw as well as annotated text versions and also simple functions to access these [3], [4].

Moving to Zipf's law; it is based on power law distribution which emphasizes that if given a corpus containing utterances of a natural language, then the frequency of any word in the given corpus is inversely proportional to the rank in frequency table so formed. It implies that the word which is occurring most frequently will occur

M.K. Kundu et al. (eds.), *Advanced Computing, Networking and Informatics - Volume 1*, Smart Innovation, Systems and Technologies 27,
DOI: 10.1007/978-3-319-07353-8_59,

approximately twice as the frequency of the second most frequently occurring word in the same corpus and so on. The general Zipf's law studies the influence relation that frequency is suffered by rank where rank, is equivalent to the independent variable and frequency can be looked upon as the dependent variable. It is observed that the randomly generated distribution for frequency of words in texts is quite similar to Zipf's law. Occurrence frequency of any word is similar to the inverse power law taking into consideration its word rank and the exponent of the inverse power law is found almost very close to 1, because the transformation from the word's length to its corresponding rank stretches or expands an exponential behavior to power law function behavior [5].

In this paper Zipf's law has been deployed along with the available corpus in NLTK module in python in order to generate plots. Further, analysis of these plots have been done to show how a particular style of writing hampers a writer in his writing and usage of words and grammar in the related context. The frequency of words in various texts has been calculated and the results have been obtained in the form of graphical plots which has helped to show the existence of similarity between various texts by the same writer/ author.

2 Background Theory

Many independent works have already been done using either NLTK or Zipf's law. But, a combination of two has made no considerable contribution in the present influenced world of natural language processing.

Several works has been done in NLP wherein two parsers have been fused together [6]. In this paper the author has described how in present scenario where natural language processing applications and its implementations are equally existent for the programmers having null linguistic knowledge were being able to build some specific NLP linguistic systems. Two parsers were fused together to achieve more accuracy and generality were tested on a corpus showing a higher level of accuracy in the results.

Other previous works in NLTK include a focus on computational semantics software and how it is easy to do computational semantics with NLTK [7]. This paper shows how Python gives an advantage to those students who are not exposed to any programming language in the merits of Prolog and Python when compared relatively. Zipf's law was introduced by George Kingsley Zipf and proposed it in Zipf 1935, 1949. It is an empirical law which is formed using the mathematical statistics, and it is referred that various kinds of data types have been studied in the fields of physics as well as social sciences which can be aptly approximated in accordance with Zipfian distribution, which is very much related to discrete mathematics probability distributions of power law family.

On the other hand Zipf's law like distribution have been implemented in various real time and internet analysis, where an analysis has been done to check the revenue of 500 top firms' of China and its rank and frequency following Zipf's like distribution with inverse power law where the slope is found to be close to 1 [8].

3 Motivation

Significant work has been done independently using NLTK or Zipf's Law. This work has tried to combine the above two where NLTK has been used to get some of the texts of the Project Gutenberg to do the analysis and Zipf's law has helped to plot and study those analyses.

Project Gutenberg is an electronic text archive which preserves cultural text, which is freely available to all and a small section of texts have been included in NLTK creating a Gutenberg Corpus, which is a large body of text [9]. To see the distribution of words and texts and the frequency of words used Zipf's law has been implemented on some of the texts of the Gutenberg Corpus in NLTK. The Gutenberg Corpus of NLTK contains three texts by Chesterton, namely; Chesterton-ball, Chesterton-brown and Chesterton-Thursday and three by Shakespeare, namely; Shakespeare-hamlet, Shakespeare-Macbeth and Shakespeare-Caesar, amongst other texts in the corpus. These are the various texts written and composed by the same author/ writer. The word type and its frequency of occurring in the text singly and comparatively has been checked using NLTK module and using Zipf's law, comparative graph has been plot to analyze the results of the research.

Here it is tried to see how much the writing is hampered, influenced and varies according to the speech and types of words used by a particular author/writer. Writing, usage of words, frequency of using words, adjectives, presenting those phrases and everything differs and varies person to person. Even a common man has a different way of writing essays and texts than another common man.

Often it is found that the problem of text kept or preserved anonymously, sometimes it is done deliberately on author's request but sometimes it's not. This valuable text or cultural texts' identification becomes important so that its master's originality can be kept intact. This can be done by employing Zipf's law on the text and counting the words and its corresponding frequency in the respective text. Here after plotting a comparative graph of the same with some identified text which is thought to be written by the same author as this unidentified one.

The results can then show that whether the text has been identified is according to one which is assumed or not.

A number of corpora from NLTK have been studied and various comparison results have been plotted to study the Zipf's law and its scope and slope over natural languages and words from daily usage of various language but mainly English. As a first step the frequency distribution is calculated and stored in a file for some nine corpuses from a suite of corpuses in NLTK namely Reuters Corpus, Project Gutenberg, Movie Reviews Corpus, State Union Corpus, Treebank, and Inaugural Corpus from USA, Webtext Corpus, Nps Chat Corpus, and Cess Corpus. Frequency occurrence of words, word, and its rank are also calculated in a file and finally a comparison plot is generated among various corpuses and texts of NLTK.

4 Getting Started

Zipf's law is explained as let f (w) be the frequency of a word w in any free text. It is supposed that the most frequently occurring word is ranked as one, i.e. all words in a text are ranked according to the frequency of occurrence in the text. According to Zipf's law; frequency of occurrence of a word of any type is inversely proportional to its rank.

$$f \propto 1/r$$
$$f = k/r$$
$$f * r = k$$

where k is a constant.

Power law is when the frequency of any event varies as the power of some attributes of that event the frequency is said to follow a power law. Significance of power law here is taken in sense of logarithmic scale because values here are very large. So to denote these values on scale, a power law is used in form of log scale to demonstrate these values on x-y axes.

At first it is started with counting words in the text and calculating how many times each word is appearing in the given text. Each word in each line is read by stripping of the front and back whitespaces around the word, converting each word into lowercase to make the things a little more manageable. A dictionary is maintained to store all the words with their corresponding frequency.

Next step is assigning the rank to each word. Rank of a value is one plus the number of higher values. This implies that if an item x is taken wherein it is said that x is the fourth highest value in the list means its rank is four then it is said that there are three other items whose values are higher than x. Therefore, to assign a rank to each word by its corresponding frequency, it should be known that how many words have higher frequency than that word. So grouping is to be done first i.e. group the words according to their frequency. Certain boxes are taken and each box is labeled according to the frequency of the word i.e. putting all words appearing three times in a corpus, in a box labeled as three or 3 and so on. To check how many words have higher frequency than a given word, the label of the box is checked to which a selected word belongs to and then adding up the number of words in the boxes with larger labels.

Now putting up all the words having the same frequency into a box representing that box with the corresponding frequency, then to determine the rank, see which box it belongs to, identify all boxes that keep words with higher frequency then add number of words that belong to the box identified in the previous step and lastly add one to the resulting sum. To put all words with the same frequency into a box with that frequency, a dictionary is used and the dictionary created earlier is used to lookup for the frequency of the word and a rank function is created with three parameters as word, frequency of word and group of words in box dictionary.

It is also observed that many real systems don't show true power law behavior because they are either incomplete or possibly inconsistent with the undertaken conditions under which it is expected for power laws to emerge. In general Zipf's law doesn't hold for subsets of objects or events or a union or combination of Zipfian sets. Some missing elements produce deviations from a pure Zipfian or Zipf's law in the subset. The line is sometimes not well fit, for highest rank words and lowest rank words which are overestimated and underestimated respectively.

5 Result Analysis

At first Zipf's law is implemented on text 2: Sense and Sensibility by Jane Austen 1811 and text 4: Inaugural Address Corpus of NLTK and found that Zipf's law is obeyed with some deviations at the extremes and it is also seen that there are quite some similarities in these two texts (Fig. 1(a)), i.e. the way in which the two texts are written and the usage of words and its frequencies is quite similar and in fact the graph intersects at a place near the lower mid signifying that some of the words usage is same.

Similarly implementing Zipf's law on text 2: Sense and Sensibility by Jane Austen 1811 and text 8: Personals Corpus of NLTK, it is found that there are no similarities between words of these two corpuses (Fig. 1(b)) i.e. the frequencies of using words is not similar and a considerable gap is maintained among the two texts.

With the result analyses and comparison on these three texts of NLTK it is found that how one text i.e. text 2 is quite similar to text 4 but the same text i.e. text 2 has absolutely no similarities with text 8; while all the texts deviate at the extremities due to underestimated or overestimated values at the two extremes respectively.

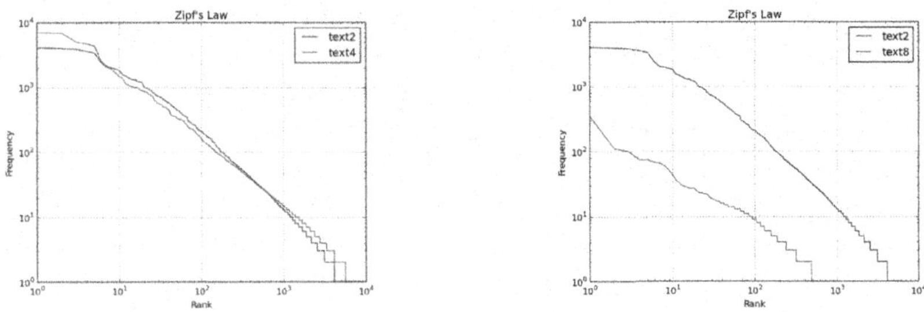

(a) (b)

Fig. 1. (a) Comparision results of text 2: Sense and Sensibility by Jane Austen 1811 and text 4: Inaugural Address Corpus of NLTK, (b) Comaprision results of text 2: Sense and Sensibility by Jane Austen 1811 and text 8: Personals Corpus of NLTK

Now analyzing the results of the main objective i.e. to see how well it works for the texts (taken as sample) by Shakespeare and Chesterton, two great authors and writers and how important it is too see a similarity in their existing texts so that the analyses of these results can be used to test some valuable unidentified or anonymous text; where these two tests and these two authors are just taken to check the results.

When simply the words of the two texts i.e. Chesterton Ball and Chesterton Brown, composed by Chesterton are compared (no frequency count is considered), a graph with almost similar and overlapping pattern is generated till the lower mid and a little deviating in the later part (Fig. 2(a)), which shows that the words used are quite similar.

And when the comparison of these two texts according to the frequency of occurrence of words is done it is found that the graph is completely overlapping except for a few words at the lower extremity. This means that the frequency of using words is very similar in these two texts by the same writer/ author (Fig. 2(b)).

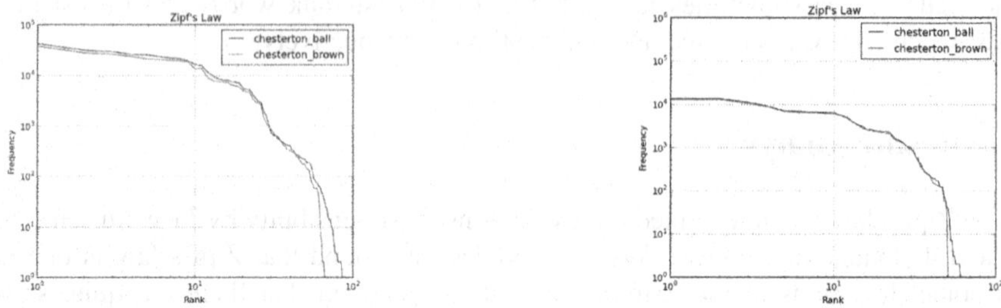

(a) (b)

Fig. 2. (a) Comparision of two different texts written by Chesterton (Brown and Ball), **(b)** Comparison according to the frequency of words in the text Ball and Brown by Chesterton

When the comparison for all the three texts available by Chesterton in Project Gutenberg namely; Ball, Brown and Thursday is done, it is seen that in the composition Thursday the frequency of words used is a little less though the pattern followed is similar but at lower extremity, the frequency of words of text Thursday matches with that of Brown. Here also the frequency of occurrence of words is compared (Fig. 3).

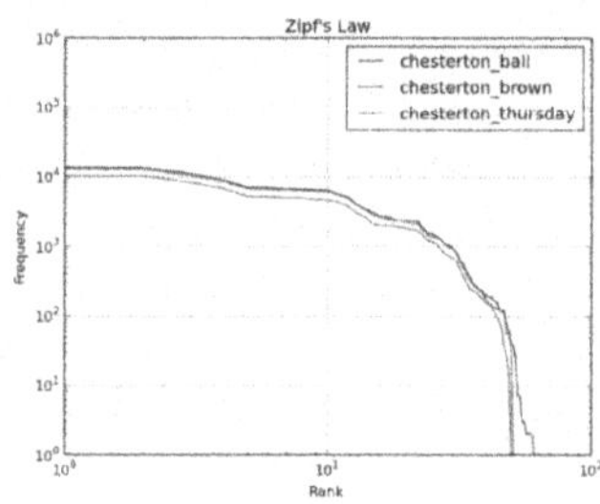

Fig. 3. Comparison according to the frequency of words in the text Ball, Brown and Thursday by Chesterton

In the figure below, the results say that when the frequency of occurrence of words in both the compositions by Shakespeare are compared i.e. Shakespeare_Macbeth and Shakespeare_caesar, it is noticed that the words usage is almost similar in both the texts and also the graph does not touch the axes at the lower end, which means that there are no words in the text that are used almost negligibly or with very less frequency and hence the usage of words is distributed in a certain pattern (Fig. 4(a)), the words and their frequency count is used for the analyses.

When all the three plays by Shakespeare namely; Caesar, Macbeth and Hamlet, available in NLTK's Project Gutenberg, when included to test the results on the basis of the frequency of occurrence of words, it was noticed that the graph for text Hamlet touches the axis at the lower end signifying that words usage in play Hamlet is different and is not used very frequently making it touch the lower axis. It is also noticed that the graph for Hamlet follows the similar pattern as that of the other two but the words used in Hamlet are with a higher frequency (Fig. 4(b)), this means the pattern followed by a particular writer is similar though the frequency of using words may vary.

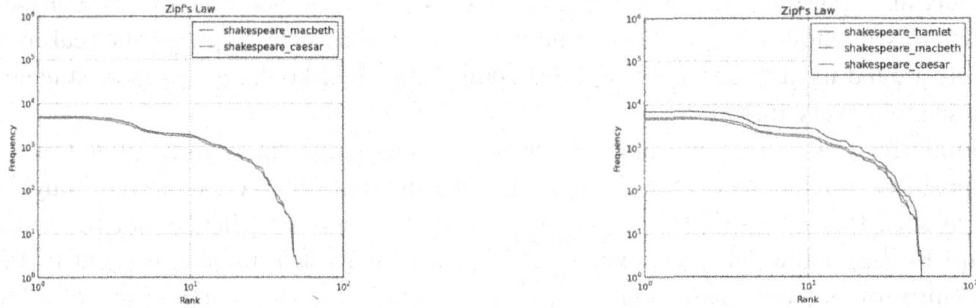

(a) (b)

Fig. 4. (a) Comparison according to the frequency ofoccurence of words with their corresponding ranks in the text Caesar and Macbeth by Shakespeare, (b) Comparison according to the frequency of occurrence words with their corresponding ranks in the text Caesar, Hamlet and Macbeth by Shakespeare

When the texts by Chesterton are compared, here considering Ball and Brown and the compositions by Shakespeare considering here Caesar and Macbeth are compared with each other, it is found that a considerable gap in the usage of words and the frequency with which they appear and are used by the respective authors and writer (Fig. 5), it is seen that a constituent gap is maintained between the writing patterns of various writers and the frequency with which they tend to use words in their texts and compositions follows a certain kind of pattern.

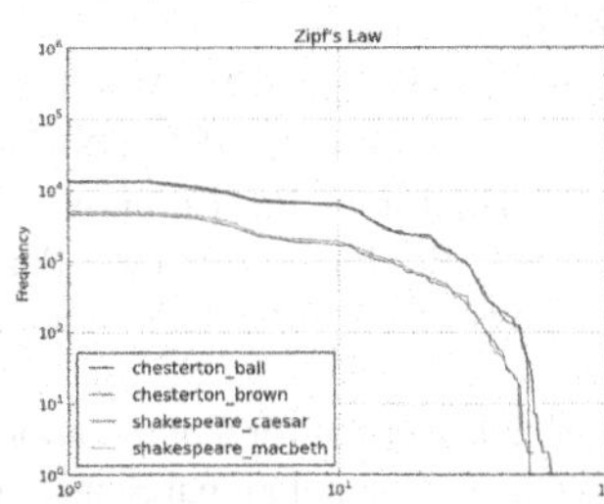

Fig. 5. Comparison according to the frequency of occurrence of words with their corresponding ranks in the text Caesar and Macbeth by Shakespeare and Ball and Brown by Chesterton

6 Conclusion and Future Work

After studying the above results it is concluded that the writing pattern and the usage of words differ from person to person and can help in some major result findings and research. After seeing the result in Fig. 5, it is seen that how the compositions by two people vary and the frequency of usage of words differs producing a considerable gap in the graph.

These results can also help in understanding this kind of pattern which is being followed in many cases and places like the population in largest cities in World, or increase in revenue of an organization over a period of time, to check the progress of students in a year etc., with the help of Zipf's law results generation and its analyses is done to see whether the result is achieved or not. It may also help in some real time aspects around us and their results play a significant role like the progress of students for a school is very important.

Similarly it is very important sometimes to judge the anonymity of a text or composition which can be done through this measure and find the existence. though it is also seen that exact Zipf's law is not followed in here and hence can be said a failure to Zipf's law but the concept can be used to to determine and identify the anonymity of not only texts but THIS can be used in various other aspects as in Intrusion Detection as a future application, where a set pattern of Intrusion can be detected and saved and if any deviation from the existing pattern is found, it can be said that there might be an Intrusion in the organization.

References

1. Bird, S., Klein, E., Loper, E.: Natural Language Processing with Python: Analyzing Text with the Natural Language Toolkit. O'Reilly Media (2009)
2. Lobur, M., Romanyuk, A., Romanyshyn, M.: Using NLTK for Educational and Scientific Purposes. In: 11th International Conference on The Experience of Designing and Application of CAD Systems in Microelectronics, pp. 426–428 (2011)
3. Abney, S., Bird, S.: The Human Language Project: Building a Universal Corpus of the World's Languages. In: Proceedings of the 48th Annual Meeting of the Association for Computational Linguistics, pp. 88–97 (2010)
4. Li, W.: Random Texts Exhibit Zipf's-Law-Like Word Frequency Distribution. IEEE Transactions on Information Theory 38(6), 1842–1845 (1992)
5. Rahman, A., Alam, H., Cheng, H., Llido, P., Tarnikova, Y., Kumar, A., Tjahjadi, T., Wilcox, C., Nakatsu, C., Hartono, R.: Fusion of Two Parsers for a Natural Language Processing Toolkit. In: Proceedings of the Fifth International Conference on Information Fusion, pp. 228–234 (2002)
6. Garrette, D., Klein, E.: An Extensible Toolkit for Computational Semantics. In: Proceedings of the 8th International Conference on Computational Semantics, pp. 116–127 (2009)
7. Chen, Q., Zhang, J., Wang, Y.: The Zipf's Law in the Revenue of Top 500 Chinese Companies. In: 4th International Conference on Wireless Communications, Networking and Mobile Computing, pp. 1–4 (2008)
8. Shan, G., Hui-xia, W., Jun, W.: Research and application of Web caching workload characteristics model. In: 2nd IEEE International Conference on Information Management and Engineering (ICIME), pp. 105–109 (2010)
9. Project Gutenberg Archive, `https://archive.org/details/gutenberg`

An Efficient Approach for Discovering Closed Frequent Patterns in High Dimensional Data Sets

Bharat Singh, Raghvendra Singh, Nidhi Kushwaha, and O.P. Vyas

Indian Institute of Information Technology, Allahabad
India
{bharatbbd1,raghvendra.the.invisible,kushwaha.nidhi12,dropvya}@gmail.com
www.iiita.ac.in

Abstract. The growth in the new technology in the field of e-commerce and bioinformatics has resulted in production of large data sets with few new uniqueness. Microarray datasets consist of a very large number of features (nearly thousands of features) but very less number of rows because of its application type. ARM can be used to analyze such data and find the characteristics hidden in these data. However, most state-of-the-art ARM methods are not able to tackle a datasets containing large number of attributes effectively. In this paper, we have proposed and implemented a modified Carpenter algorithm with different consideration of data structure, which in result give us the better time complexity in compare to simple implementation of Carpenter.

Keywords: High Dimensional Data, Association Rule Mining (ARM), Closed Frequent Pattern, Frequent Pattern, Microarray Data.

1 Introduction

Progress in the bar-code as well as in the retail market and online shopping technology made it very simple for the retailer to collect and store massive amounts of sales data. Such kind of transaction record typically consists of the date and the items bought by the customers. In the recent year, many MNC retail organizations view such databases as an important pieces for marketing strategies. The researcher are keen to develop a information-driven marketing strategies with the help of these databases, which helps marketers to develop and customized marketing strategies [10]. In the latest example, Electronic health records (EHRs) have effective and efficient power to develop the delivery of healthcare services, importantly if it is implemented and used in smooth manner. It can be handled by maintaining an exact problem definition and up-to-date data [1]. As the technology grew advanced in the field of bioinformatics, e-commerce in various gene expression data sets, transactional data bases have been produced. Data sets of extremely high dimensionality pose challenges on efficient calculation to various state-of-the-art algorithms for data mining process. Examples of such

M.K. Kundu et al. (eds.), *Advanced Computing, Networking and Informatics - Volume 1*,
Smart Innovation, Systems and Technologies 27,
DOI: 10.1007/978-3-319-07353-8_60, © Springer International Publishing Switzerland 2014

high dimensional data are microarray gene data, network traffic data, and shopping transaction data. As this data have very large numbers of attributes (genes) in compare to number of instances (rows). Such type of extremely high dimensional data requires special and innovative data mining techniques to discover interesting hidden patterns from it. For example, Classification and Clustering [15] techniques have been applied on such kind of data, with slower processing or lower accuracy or they are fail to handle data. The association rule mining was proposed by P. Cheeseman *et al.* [9]. An example of association rule might be 97% of patient that are suffering from dyspepsia and epigastric pain are commonly followed by heartburn [1], [3].

Data mining is the science of finding hidden patterns and useful data from the raw data, have been gathered from different sources. It is a combination of different streams including machine learning, data warehousing, data collection etc. It helps in extraction of interesting patterns or knowledge from huge amount of data. There are many techniques used for this purpose: Association rules discovery, Sequential Pattern Discovery, Cluster analysis, Outlier Detection, Classifier Building, Data Cube/Data Warehouse Construction, and Visualization.

A set of items that appears in most of the transaction is said to be frequent. If I=i_1, i_2,......, i_k be a set of items, then support for I is the total number of transaction in which item I exists as a subset. The Mining of frequent items is totally different concept to the similarity search. In similarity search, items that have maximal number of matching in common, although the total number of transaction may be very less. The large oddness leads to build up new type of techniques for retrieving the highly frequent item-sets.

Most suitable areas for frequent pattern mining in high dimensional data is Customer Relationship Management (CRM)[11], Web pages Searches and Analysis, Geographical Data Analysis, Drug Synthesis and disease prediction [1]. In this paper, attributes, feature and column are indicating the same meaning. In general, they are characteristics which are required to represent a particular instance of the data sets.

2 Related Work

ARM worked on items 'Ii' that belong to one and more transactions 'Ti' [11]. In the e-commerce place the items might be produced and sold online from a store, and the transactions may be data of buying items made in that store. In this characterization, each transaction is a set of items bought together. The standard framework for association rule mining (Apriori) uses two measure that is support and confidence as a constraint which reduces the searching problem [1], [11], [15]. Using this conception, an association rule, A and B is a pair of items (or sets of items) that are allied to each other based on their frequency of co-occurrence in transactional databases. Let us consider an example, a hypothetical association rule might be insulin diabetes with support 450 and confidence 85%. The rule indicates that 450 patients were suffered from insulin and had diabetes on their problem list, and that 85% of patients with insulin on their list also had diabetes on their problem list [1]. This method can be used to discover disease-disease, disease-finding and disease-drug co-occurrences in medical databases [7], [8].

Since last twenty years, Closed Frequent pattern mining [1], [2], [4], [5], [11] emerged as a vast research area that concentrates a momentous quantity of interest by the researchers. Due to the huge number of features the existing techniques carried out various redundant frequent patterns as we have discussed above. To condense the frequent patterns to a condensed size, mining frequent closed patterns has been proposed. In literature, we have found the followings advanced method to mining closed frequent patterns. The previous data mining techniques could not tackle such data efficiently because of its high dimensionality. The column enumeration (feature) and row enumeration (instance) are basically the type of frequent pattern mining algorithm [1], [4], [5], [6]. Close [14] and Pascal [12] are the techniques which find out closed patterns by performing breadth-first, column enumeration. An Apriori based technique Close [14] was proposed to determine a closed item sets. The problem and limitation of these two techniques is due to the level-wise approach, the number of feature sets enumerated will be extremely large when they are run on a datasets having large number of feature, like biological datasets. CLOSET [3] was known for depth first search column enumeration technique to find closed frequent patterns. For compressed representation it uses FP tree data structure. But it was unable to handle long micro array datasets because of FP's inability of effective compression of long rows and secondly too many possibilities of combinations in column enumeration. Recently CLOSET+[6] was proposed as an enchantment in the former one. Another algorithm CHARM worked similarly as CLOSET but it stored data into vertical format.

Let us consider the bioinformatics data, which is having long Feature sets and very few instances, the evaluation criteria deteriorate due to the long feature permutations. To tackle such kind of difficulties, a new approach based on row enumeration, Carpenter [2], has been developed to handle and perform on microarray data sets as an alternative. Carpenter techniques have their own limitation, it consider a single enumeration only, which is unable to reduce the data if we have a data set with high number of features and high number of rows. To handle such problems new algorithm Cobbler [4] has been implemented which can proficiently mine high number of features and high number of rows in the datasets. Based on the nature of data properties Cobbler is designed to transform data automatically between feature enumeration and row enumeration[4]. A new algorithm based on row enumeration tree method, TD-Close [13] is designed and implemented for mining a complete set of frequent closed item sets from high dimensional data.

The problem of mining highly frequent patterns has been the interest of researcher in these recent days, as it produces a large number of redundant pattern.s To reduce these redundant patterns, we will prune certain frequent patterns that are less associated with each other. Therefore, we will manage the whole frequent pattern sets with some reduced number of highly frequent pattern sets and eliminate the redundancy.

The previous data mining techniques could not tackle such data efficiently because of its high dimensionality. The column enumeration (feature) and row enumeration (instance) are basically the type of frequent pattern mining algorithm [1], [4], [5], [6]. In column enumeration techniques like [5], [6], various combination of features are tested to discover the highly frequent patterns. Microarray data have very small instances; these methods have proved more effective in comparison to column enumeration-based algorithms.

In this paper, we have proposed a modified Carpenter algorithm with different consideration of data structure, which in result gives us the better time complexity in compare to simple implementation of Carpenter. Carpenter method was designed to handle high dimensional data which have very large number of features and less number of rows, i.e. n¡¡m. In the next section we will discuss some of the basic, preliminaries and define our problem. After that we will give our proposed approached for highly frequent pattern mining for high dimensional data.

3 Preliminaries

We denote our data set as D=n×m. Let the set of all feature or column be F $= \{f_1, f_2, f_3, ..., f_m\}$. Our dataset also have n number of rows R=$\{r_1, r_2, r_3, ..., r_n\}$, $n \ll m$, where every row r_i is a set of features from F, i.e., $ri \subseteq F$ which can be derived from our notation. For consideration, let see the Table 1, the dataset has 8 features represented by set $\{f_1, f_2, f_3, f_4, f_5, f_6, f_7, f_8\}$ and there are 4 rows,$\{r_1, r_2, r_3, r_4\}$. A row have an feature f_j if f_i have a values 1 in the corresponding row r_i. For example row r_1 contains feature set $\{f_1, f_4, f_5, f_7, f_8\}$ and row r_2 contains feature $\{f_3, f_5, f_6\}$ [2], [4]. We have been used some already defined terms. *Feature support set*, as the maximal set of rows that contains F' while it is given $F' \subseteq F$, denoted by $R(F') \subseteq R$. Similarly, *Row Support Set*, as the maximal set of features common to all the row in R', while it is given that $R' \subset R$, denoted by $F(R') \subset F$. As an example consider Table 1. Let F'=$\{f_1, f_5, f_8\}$ then R(F')=$\{r_1, r_4, r_5\}$ since these are the number of rows that have F'. Consequently, let R'=$\{r_1, r_5\}$ then F(R')= $\{f_1, f_5, f_8\}$ since it is the maximal set of features common to both r_1 and r_5 [2], [4]. We have a record of datasets D that are part of a features set. Our problem is to find all high frequent patterns with respect to the user based support constraints. For this example, we are assuming that the datasets follows the condition $|R| \ll |F|$.

Carpenter method was designed to handle high dimensional data which is having very large number of features and less number of rows, i.e. n¡¡m. But, we have found by analysis that when the number of attributes become too large it does not handle data effectively.

It depends on the use of data structure. Therefore, several versions of Carpenter have been implemented and tested on benchmark datasets. The result shows that, by using a different data structure we can make an efficient Carpenter that performs better than simple implementation of Carpenter.

Table 1. A sample of data set

-	f_1	f_2	f_3	f_4	f_5	f_6	f_7	f_8
r_1	1	0	0	1	1	0	1	1
r_2	0	0	1	0	1	1	0	0
r_3	0	0	0	0	0	1	1	0
r_4	1	0	0	0	1	0	0	1
r_5	1	0	0	0	1	0	0	1

4 Proposed Approach for Discovering Frequent Pattern

The previous discussions clearly show that since the data has too many numbers of rows and too many numbers of columns so the speed of finding frequent patterns will be slow with normal algorithms and without efficient pruning techniques. For Example, enumeration on even mere 25 features will result in a set of nearly 2^{25} patterns, where each pattern will have to be checked to find their number of occurrences to classify them as frequent or infrequent, so we can imagine what will happen if the number of features are largeer than (10000, in the range of thousands). These algorithms thus fail to satisfy our current speed needs. Thus, our basic aim is to solve the problems coming in the way of high dimensional highly frequent pattern [1] mining algorithms and try to give a new algorithm for the same. For making and testing our implementation of high dimensional association rule mining the following approach was used.

First select the dataset from HD domain for analysis and apply preprocessing for the discretization and normalization such that we can remove noise. For this, we used the tool mentioned on the next section. After applying, attributes are increased, we determine appropriate parameters in coding for generating algorithm. The data structures and their functionalities are explained in the next section. Then checked it for the scope of better implementation then the previous one. We have developed six versions for the improvement of our own algorithms. After the implementation we checked the entire versions and came to know the last version has less errors as well as better performance compare with others.

5 Results and Discussion

For testing our implementation we choose pima_indians_diabetes (UCI [1]) dataset and also Ovarian Cancer Dataset[2]. After the preprocessing[3] step like normalization and discretization, we tested the dataset. We used GCC java libraries 4.7.1 and CodeBlock IDE in Linux platform (4GB RAM core i5 3230) for developing the code. Six versions of the algorithms have been developed each are different

[1] `http://archive.ics.uci.edu/ml/datasets/Pima+Indians+Diabetes`

[2] `http://www.biogps.org/dataset/tag/ovarian-cancer`

[3] UCS-KDD CARM Discretisation/ Normalisation (DN) software (Version 2 — 18/1/2005) From `http://cgi.csc.liv.ac.uk/~frans/KDD/Software/`

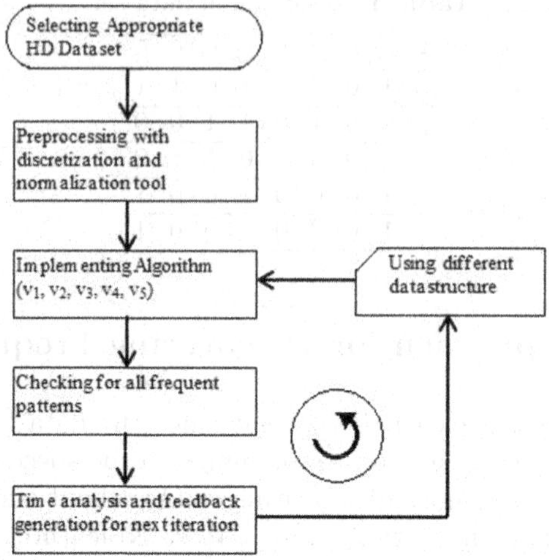

Fig. 1. An approach for Closed Frequent Pattern Mining with different version

in the sense of complexities and time consumption. Result of all the algorithms are shown below for pima_indians_diabetes dataset. Version 1 correspond to Fig. 2 which shows the running time with respect to number of row using different minimum support parameter (MSP). In Fig. 2, Fig. 3, Fig. 4, Fig. 5, Fig. 6, Fig. 7, the x-axis corresponds to number of rows, and time consumption (sec.) is represented by y-axis. Fig. 3. Corresponds to version 2 in which we uses only arrays to provide direct address based memory access instead of data value based access as done in version 1 using STL maps. This version also improves on bugs that are found in previous version. Extra arrays are used to minimize time spent in copying and maintaining proper function calls. Transposed table and FCPs stored in 2D global arrays. As our result shows it did not improve the overall performance of the algorithm so we move towards next version of it. Version 3 uses only arrays to provide direct address based memory access and removes the functions for finding feature set and FCPs. Now we are using Map to store FCPs. For minimizing the memory allocation and de-allocation time we define global arrays and again removing some bugs from the previous version. Fig. 4. Shows that there is improvement in runtime from the previous version 1 and with the increase in number of rows the runtime is changing rapidly. Arrays that are used for checking rows present in every column in TT (Transposed Table) have been removed in version 4. The graph shows slightly improved performance than the previous version with the increased time. The next version Fig. 5 uses depth based 2D array approach to minimize time wasted in memory allocation and de-allocation. Conditional Pointer Table will be made global to avoid continuous memory allocation and copying between iterations and function calls. It will surely increase the runtime with the number of rows in the datasets. Our last version uses a helper table which sacrificed space complexities for speeding up the Pattern Mining by helping in the quick formation of FCPs. This version also uses depth based 2D array for storing CPT also row positions corresponding to given column in transposed table were stored in a 2D array (-1 if row not present else count of position for particular column was used) which helped in direct lookup during new CPT creation. Version 6 graph shows the same trends as shown in the previous graphs but on increasing the number of columns in data this easily defeats the runtime of previous versions. This is the one which has been thoroughly checked for bugs since no further versions were to be maid and thus we accept this as our final implementation. Runtime of this version also has been evaluated and graph displayed in Fig. 8. To do the tests some rows were taken from the 253 rows and 265493 columns of the Ovarian Cancer dataset. All columns were taken for a given rows no column length criteria was taken. The graph shows that on increasing the number of columns the runtime do not increase exponentially instead just an almost linear relationship will be seen comparing with previous results with less number of columns. The total number of columns used was much larger than the previous data sets but the increase in time remains almost only linearly dependent on the number of columns. Fig. 9 shows the time comparison graph of all the versions. Version 1's performance is very low and can be easily defeated by others. Time complexity has been reduced by version 4 & 5. However the versions have some bugs that later on removed in the version 6.

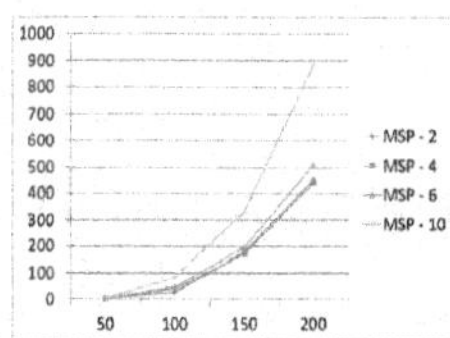

Fig. 2. CA Version 1

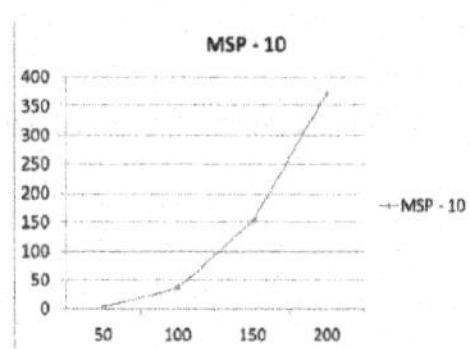

Fig. 3. CA Version 2

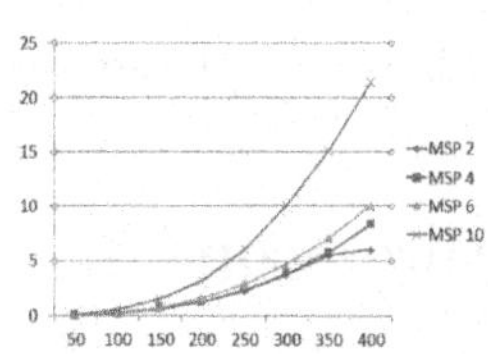

Fig. 4. CA Version 3

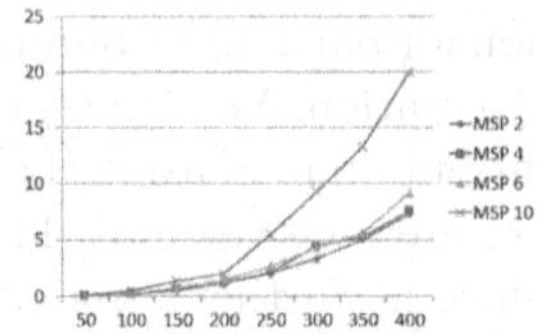

Fig. 5. CA Version 4

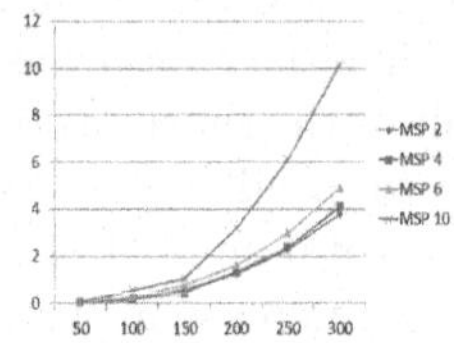

Fig. 6. CA Version 5

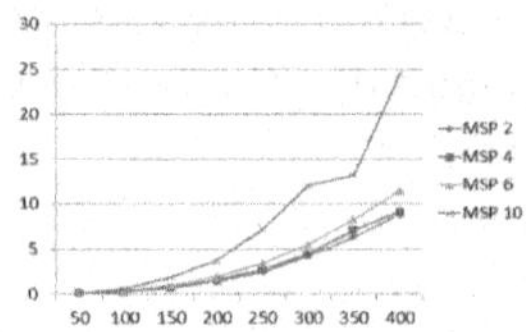

Fig. 7. CA Version 6

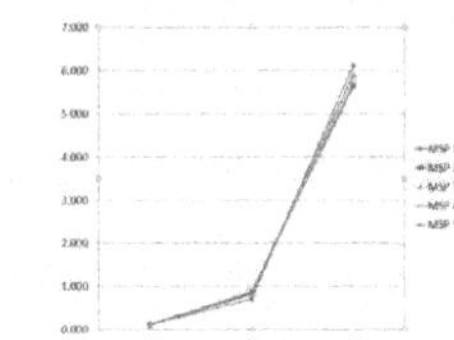

Fig. 8. Runtime of version 6

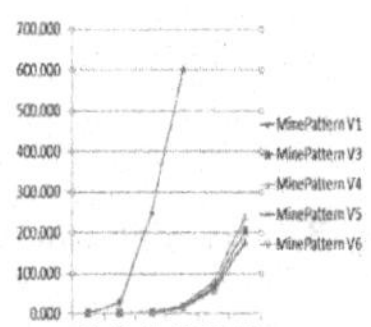

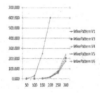

Fig. 9. Time Comparison of all the six versions (MSP=10%)

6 Conclusion and Future Scope

The importance of High Dimensional Data mining Algorithms are very large in the coming days. Thus very efficient implementations of algorithms are needed. There is always scope for huge improvements in the algorithms being currently researched and their current implementations by researchers and student groups.

Testing should be also very rigorous to find the bugs on different kind of datasets. This algorithm can also contribute in finding frequent closed patterns generation in the case of High Dimensional Mining where number of rows is large. What we can do is we can divide the dataset and in same proportion the MSP and find the frequent patterns using these new criteria and then try to search among these true patterns for entire data or guess the patterns based on some other criteria. Actually we think this division method has the potential to solve the problem of generation of large number of highly frequent patterns in the microarray datasets usually in order of 106 for approximately 10000 attributes and the problems related with memory requirements in case of Carpenter algorithm. Many of these features although frequent are not interesting. Thus we need not find all the frequent patterns instead remove them early in the pattern generation period only. Some iterations to increase accuracy of patterns generated can be done as in cross validation at the cost of time to yield better results in case of baskets of data but we will talk about that some other time after our extensive tests we are currently doing. The current implementation would be beneficial for fast processing in the Biomedical Science, Biotechnology for producing results in minimum number of time. Also Business Intelligence needs the quick generation of frequent item sets in huge data.

References

1. Wright, A., McCoy, A., Henkin, S., Flaherty, M., Sittig, D.: Validation of an Association Rule Mining-Based Method to Infer Associations Between Medications and Problems. Ppl. Clin. Inf. 4, 100–109 (2013)
2. Pan, F., Cong, G., Tung, A.K.H.: Carpenter: Finding closed patterns in long biological datasets. In: Proceedings of ACM-SIGKDD International Conference on Knowledge Discovery and Data Mining, pp. 637–642 (2003)
3. Pei, J., Han, J., Mao, R.: CLOSET: An efficient algorithm for mining frequent closed item sets. In: Proceedings of ACM-SIGMOD International Workshop Data Mining and Knowledge Discovery, pp. 11–20 (2000)
4. Pan, F., Cong, G., Xin, X., Tung, A.K.H.: COBBLER: Combining Column and Row Enu-meration for Closed Pattern Discovery. In: International Conference on Scientific and Statistical Database Management, pp. 21–30 (2004)
5. Zaki, M., Hsiao, C.: Charm: An efficient algorithm for closed association rule mining. In: Proceedings of SDM, pp. 457–473 (2002)
6. Wang, J., Han, J., Pei, J.: Closet+: Searching for the best strategies for mining frequent closed item sets. In: Proceedings of 2003 ACM SIGKDD International Conference on Knowledge Discovery and Data Mining (2003)
7. Chen, E.S., Hripcsak, G., Xu, H., Markatou, M., Friedma, C.: Automated Acquisition of Disease: Drug Knowledge from Biomedical and Clinical Documents: An Initial Study. J. Am. Med. Inform. Assoc., 87–98 (2008)
8. Sim, S., Gopalkrishnan, V., Zimek, A., Cong, G.: A survey on enhanced subspace clustering. Data Mining Knowl. Disc. 26, 332–397 (2013)
9. Cheeseman, P.: Auto class: A Bayesian classification system. In: 5th International Conference on Machine Learning. Morgan Kaufmann (1988)
10. Associates, D.S.: The new direct marketing. Business One Irwin, Illinois (1990)

11. Agrawal, R., Srikant, R.: Fast algorithms for mining association rules. In: Proceedings of 1994 International Conference on Very Large Data Bases (VLDB 1994), pp. 487–499 (1994)
12. Bastide, Y., Taouil, R., Pasquier, N., Stumme, G., Lakhal, L.: Mining frequent closed itemsets with counting inference. SIGKDD Explorations 2(2), 71–80 (2000)
13. Hongyan, L., Han, J.W.: Mining frequent Patterns from Very High Dimensional Data: A Top-down Row Enumeration Approach. In: Proceedings of the Sixth SIAM International Conference on Data Mining, pp. 20–22 (2006)
14. Pasquier, N., Bastide, Y., Taouil, R., Lakhal, L.: Discovering frequent closed itemsets for association rules. In: Proceedings of 7th International Conference on Database Theory, pp. 398–416 (1999)
15. Kriegel, H.P., Kröger, P., Zimek, A.: Clustering high-dimensional data: A survey on sub-space clustering, pattern-based clustering, and correlation clustering. ACM Transactions on Knowledge Discovery from Data 3(1), 1–58 (2009)

Time-Fading Based High Utility Pattern Mining from Uncertain Data Streams

Chiranjeevi Manike and Hari Om

Department of Computer Science & Engineering
Indian School of Mines,
Dhanbad - 826 004, Jharkhand, India
chiru.research@gmail.com, hariom.cse@ismdhanbad.ac.in

Abstract. Recently, high utility pattern mining from data streams has become a great challenge to the data mining researchers due to rapid changes in technology. Data streams are continuous flow of data with rapid rate and huge volumes. There are mainly three widow models namely: Landmark window, sliding window and time-fading window used over the data streams in different applications. In many applications knowledge discovery from the data which is available in current window is required to respond quickly. Next the Landmark window keeps the information from the specific time point to the present time. Where as in time-fading model information is also captured from the landmark time to current time but it assigns the different weights to the different batches or transactions. Time-fading model is mainly suitable for mining the uncertain data which is generated by many sources like sensor data streams and so on. In this paper, we have proposed an approach using time-fading window model to mine high utility patterns from uncertain data streams.

Keywords: Data streams, time-fading window, uncertain data, high utility patterns.

1 Introduction

Due to the advancement of information technology, the automation of measurements and data collection is producing unbounded stream of data with rapid rate [1], [2]. These streams contain tremendously huge volumes of data which is not possible to keep in the main memory. There are many reallife applications which are generating streams of data such as surveillance systems, web clicks, computer network traffic, stock tickers etc. [2]. Since, many years several approaches have been developed for mining these data streams for frequent itemsets, however, frequency of an itemset is not a significant measure to discover the interesting patterns. Because in frequent pattern mining the support measure is defined over the binary (0/1) database. Hence, utility mining was introduced [3], in which all measures of an itemsets other than the occurring frequency are considered as utilities (like purchased quantity, price, profit etc.).

M.K. Kundu et al. (eds.), *Advanced Computing, Networking and Informatics - Volume 1,* Smart Innovation, Systems and Technologies 27,
DOI: 10.1007/978-3-319-07353-8_61,

Utility mining is the process of discovering patterns with utility more than the user specified minimum utility threshold. Utility mining may also be considered as an extension of frequent pattern mining [4]; however, it is more complex due to the absence of anti-monotone [5] property and many utility calculations.

Several approaches have been proposed for mining high utility patterns from static as well as dynamic databases to overcome the above barriers. Few algorithms were also proposed and implemented using sliding window and landmark window models [6], [4], [7]. Apart from these models time-fading window model also has significant importance in some real-life applications. The Landmark window keeps the information from landmark time to the current time point where in sliding window only recent information will be maintained and contains fixed number of batches or transactions. Time-fading window also keeps the information from the landmark time to current time point, but it gives the more importance to the data which is arrived recently, that is old batch has got less importance or weight than the recent batch [1], [2]. The main objective of assigning different weights to batches is to avoid early pruning of low utility itemsets, which may become high utility itemsets in future.

Besides data streams with precise data in which user has guarantee that an item present in a transaction or not, but in other cases where data stream has uncertain data [8], [9], in which user has no guarantee that an item presence or absence in a transaction. For example, let us consider the retail market transaction data most of the products has high frequency on the period of the year that is based on the season. But some other products may be sold at any time period which cannot be predicted. Therefore knowledge discovery from the data streams which contain the precise data may not be a big challenge than the discovery from uncertain data. In this paper we have proposed an algorithm to mine high utility patterns from uncertain data streams using time-fading window model. In our approach two data structures those are already used in our previous works, one for mining landmark window (LHUI-Tree,) and another for sliding window (SHUI-Tree).

Remaining part of this paper is organized as follows. In Section 2, preliminaries are given and related work on high utility pattern mining over streams discussed in Section 3. In Section, 4 our proposed algorithm is discussed briefly. Experimental results are presented in Section 5, to show the performance of proposed approach. Finally, conclusions and future enhancements are given in Section 6.

2 Preliminaries

Let $I = \{i_1, i_2, i_3, \ldots i_n\}$ be a finite set of items and DS be a transaction data stream $\{T_1, T_2, T_3, \ldots T_m\}$ where each transaction $T \in DS$ is a subset of I. Each item $i_j \in T_k$ associated with a value called internal utility of an item, denoted as $iu(i_j, T_k)$. For example, internal utility of item a in transaction T_1 and T_3 is 2 and 0 respectively (Fig. 1(a)). Each item $i_j \in I$ associated with a value outside the transaction data stream called external utility, denoted as $eu(i_j)$.

For example, external utility of items a and b is 5, 3 respectively (Fig. 1(b)). Utility of an item i_j in transaction T_k is obtained by multiplying item internal utility and external utility, denoted as $iu(i_j, T_k) \times eu(i_j)$. For example, utility of an item a in transaction T_1 is 10 (Fig. 1).

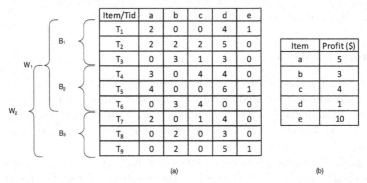

Item/Tid	a	b	c	d	e
T_1	2	0	0	4	1
T_2	2	2	2	5	0
T_3	0	3	1	3	0
T_4	3	0	4	4	0
T_5	4	0	0	6	1
T_6	0	3	4	0	0
T_7	2	0	1	4	0
T_8	0	2	0	3	0
T_9	0	2	0	5	1

(a)

Item	Profit ($)
a	5
b	3
c	4
d	1
e	10

(b)

Fig. 1. Example transaction data stream, utility table

3 Related Work

Utility mining was introduced while discovering patterns and association rules based on user interest [10]. The basic definitions and theoretical model of utility mining was given in [3]. The first property which has similar strategy in reducing search space like Apriori property was observed in [11]. Based on this anti-monotone property several algorithms have been implemented to mine high utility patterns from traditional databases. However, mining of these patterns from evolving data streams becomes more significant, because the utility of a pattern may be increased or decremented as the data arriving continuously in the stream [12]. First approach called, THUI-Mine [13] which is proposed for mining high utility patterns from data streams based on sliding window model. This approach processes the transactions in a batch by batch fashion; first it calculates each item transaction weighted utility. Items those are having the transaction weighted utility [11] more the threshold are considered for further process. This algorithm did not meet the essential requirements of data stream because it requires more than one database scan and cannot be applied for landmark and time fading window models.

To overcome the drawbacks in the first approach another two algorithms MHUI-BIT, MHUI-TID were proposed [14] by representing data in the form of Bitvecot(BIT) and Tidlist(TID). However, MHUI-TID and MHUI-BIT achieved efficiency in terms of memory consumption and execution time over THUI-Mine still need more than one database scan because of level-wise approach applied for itmesets of length more than 2. Afterwards, few approaches were proposed using different window models, except GUIDE algorithm [4], none of them was proposed any methods based on the time-fading window model. As we already discussed earlier, number of sources those are generating stream of data and their generation rate is going on increases. On the other hand mining frequent patterns from the uncertain data streams is another important issue [1], [2]. In this paper we have proposed an approach to mine high utility patterns from uncertain data streams using time-fading window model.

3.1 Proposed Work

In this section, our proposed approach to mine high utility patterns from uncertain data streams is presented. We have implemented our approach using tree structures of landmark window and sliding window, so both of these tree structures are discussed and their time and space complexities are also analyzed in this section.

Our approach comprises of three steps, for each tree structure these three steps are clearly described. First consider the updating of tree structure of landmark window model (LHUI-Tree) and for overall discussion consider the example data stream (Fig. 1) and utility table given in earlier sections. Let us consider the data stream which is dived in to three batches (B_1, B_2 and B_3) of each three transactions. In landmark window model, window keeps the information of each batch from the landmark time. Let us consider landmark time 0 sec, first batch has appeared with in 1 sec, second batch with in another second and so on. In landmark window, window size increases with increasing batches or with increasing transactions in a batch. When the first batch arrives algorithm finds all possible itemsets over the items present first in transaction T_1. In first transaction T_1 items a, d and e are present so possible itemsets over these items $\{a\}, \{d\}, \{e\}, \{a, d\}, \{a, e\}, \{d, e\}$ and $\{a, d, e\}$ and their utilities are 10, 4, 10, 14, 20, 14 and 24 are generated. Next these itemsets are updated in landmark window tree structure (LHUI-Tree), each node in this LHUI-Tree contains item and utility information except root node. Fig. 2, show the information of itemsets in LHUI-Tree after processing T_1.

To decrease the importance of old itemsets with increasing time stamp values, we used common exponential decay function. Lifetime of each itemset depends on the decay factor which is used to assign importance to the itemset. Let us assume that the decay factor set to 0.9, after processing each transaction or batch of transactions each utility of itemset is multiplied with decay factor. In Fig. 2(b) itemset a has time fading utility value (tfu) 19(i.e. $(10 \times 0.9) + 10$), this is calculated using below equation (1) in the case of landmark window tree structure. If we process batch of transactions at a time then algorithm generate the set of all possible values and calculates utilities. Here only the difference is all patterns in a batch will get the same weight. After processing all transactions with in the landmark window results of high utility patterns are generated based on the user specified utility threshold value.

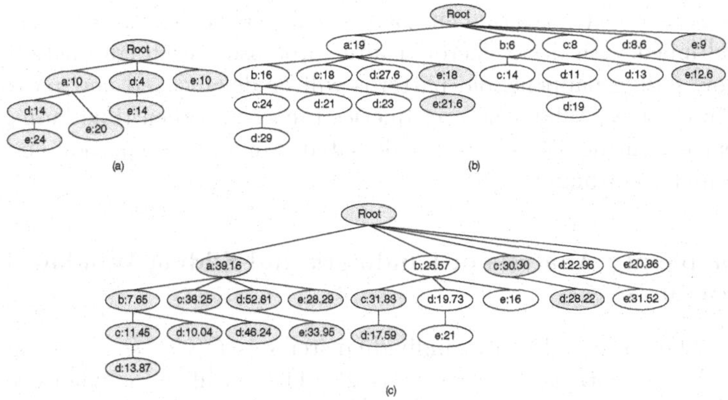

Fig. 2. Landmark window tree(LHUI-Tree)

$$tfu(X, t_n) = \{\Sigma_{i=1}^{n-1} tfu(X, T_i) \times \alpha\} + u(X, t_n) \tag{1}$$

$$tfu(X, t_n) = \Sigma_{i=1}^{n} (tfu(X, T_i) \times \alpha^{(t_n - t_i)}) \tag{2}$$

Next consider sliding window tree structure (SHUI-Tree) to update all transactions in time-fading window model. When a transaction or a batch of transactions arrives in the data stream algorithm generates all possible itemset corresponding utilities and update in tree structure. Unlike in landmark window data structure, it keeps the information of time of an itemset and its utility, here time fading utility of an itemset is calculated using above equation only when the user query encounters.

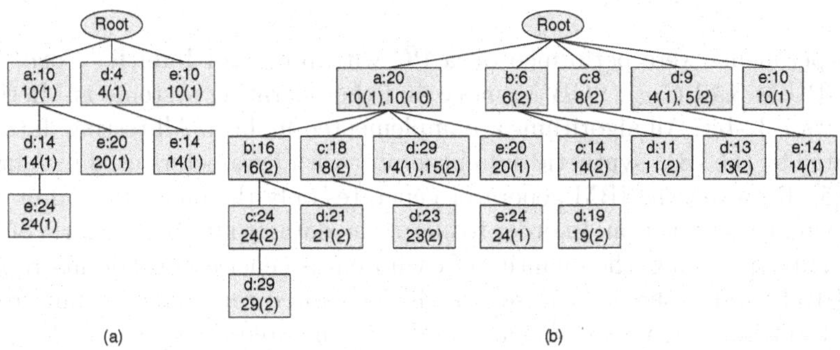

Fig. 3. Sliding window tree(SHUI-Tree)

Fig. 3(a), show how the information of a transaction T_1 is represented in sliding window tree structure. Consider Fig. 3(b) which represents the contents of tree after transaction T_2 is processed by algorithm. Itemset a contains total utility and also include the information of utilities in transactions T_1 and T_2 i.e.

10(1), 10(2), assume transaction number as time of occurrence. Here utility value is not multiplied with decay factor α after processing each transaction. When the window sliding happens, old patterns will be deleted and new patterns are updated in the tree. When the user queries the system then time fading utility of each pattern in the SHUI-Tree is calculated using above equation and results are generated accordingly.

3.2 Complexity Analysis of Landmark and Sliding Window Tree Structures

Total possible worlds [15] of n individual items is $O(2^n)$, so total number of nodes needed in both tree structure is 2^n. There will be a balance of space requirement among these two tree structures or sliding window tree structure may occupy more memory than the other because in each node it keeps the extra information of each pattern that is batch wise or transaction wise information. If we consider the time complexity of landmark window tree structure to update, it will be more than the time complexity of sliding window tree structure. Reason is in landmark window tree structure for each transaction we need to update all nodes to multiply patterns time fading utility by decay factor. In sliding window tree structure no need of updating all patterns. If we consider the response time of a user query, time-fading model which is used landmark window tree structure obviously faster than the response of sliding window tree structure because we need to calculate each node time-fading utility in real time. From the above analysis we can say that, there could be a balance among the time-fading models which are used above two tree structures in terms of execution time and memory consumption.

4 Experimental Results

The experiments were performed on a PC with processor Intel(R) $Core^{TM}$ i7 2600 CPU @ 3.40 G_{HZ}, 2GB Memory and the operating systems is Microsoft Windows 7 32-bit. All algorithms are implemented in Java. All testing data was generated by the IBM synthetic data generator with parameters used in Agrawal *et al.* [5]. However, the IBM generator generates only the quantity of 0 or 1 for each item in a transaction. In order to adapt the dataset into the scenario of high utility pattern mining, the quantity of each item is generated randomly ranging from 1 to 5, and profit of each item is also generated randomly ranging from 1 to 20. To fit in the real time scenario we have generated the profits by following lognormal distribution.

We have compared the performance of our time-fading approach over two tree structures LHUI-Tree and SHUI-Tree. Execution time need and memory consumed by these two tree structures is analysed to get clear insights. Experiments are done over two data sets, one is sparse and dense. Fig. 4 show the performance of approach with two tree structures on dense dataset in terms of execution time and memory. From this we can observe that there is no significant

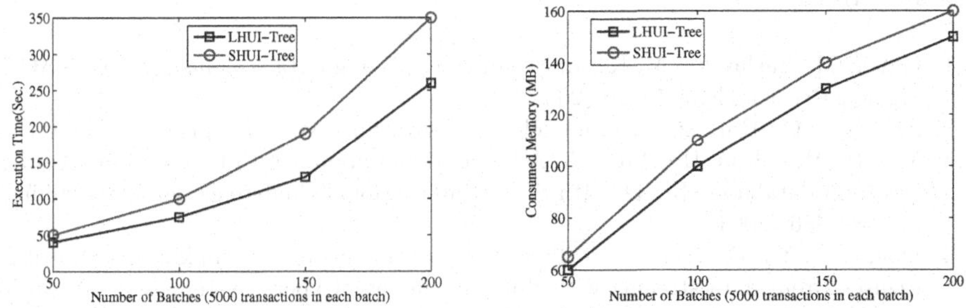

Fig. 4. Performance with varying number of batches over dense dataset

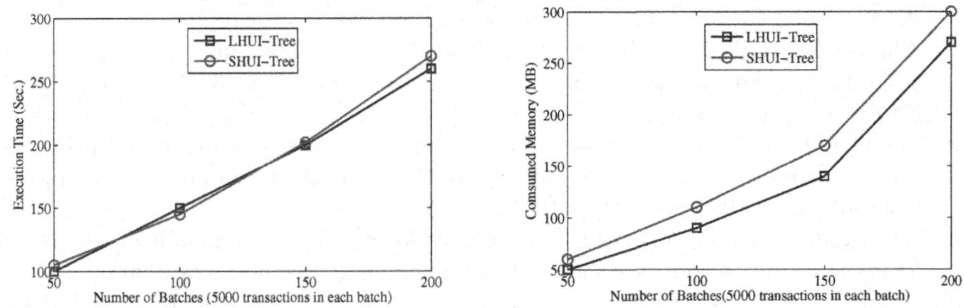

Fig. 5. Performance with varying number of batches over sparse dataset

difference among execution times. Reason is already discussed in above section that landmark need much time to update all patterns after each batch is processed and sliding window tree need more time to generate high utility patterns. From Fig. 4, we can observe that there is slight variation among the amount of memory consumed by both trees. In Fig. 5 performance of our approach with two tree structures on sparse dataset is shown. From this, we can observe that there is a slight variation among two LHUI-Tree and SHUI-Tree consumed memory and execution time respectively.

5 Conclusions

In this paper, we have proposed an approach based on time-fading window model for mining high utility patterns from uncertain data streams. In our approach we used two tree structures to keep the information of patterns in order to give clear insights. Theoretical analysis and experimental results showed that two tree structures provided a trade-off between execution time and memory consumption. In our future, we extend work we will focus on the performance efficiency by applying our method over sparse and dense datasets and analyze the situation where any one of these structures gives better performance in terms of execution time and memory consumption.

References

1. Gaber, M., Zaslavsky, A., Krishnaswamy, S.: Mining data streams: a review. ACM Sigmod Record 34(2), 18–26 (2005)
2. Aggarwal, C.: Data streams: models and algorithms, vol. 31. Springer (2007)
3. Yao, H., Hamilton, H., Butz, C.: A foundational approach to mining itemset utilities from databases. In: The 4th SIAM International Conference on Data Mining, pp. 482–486 (2004)
4. Shie, B.E., Yu, P., Tseng, V.: Efficient algorithms for mining maximal high utility itemsets from data streams with different models. Expert Systems with Applications 39(17), 12947–12960 (2012)
5. Agrawal, R., Srikant, R.: Fast algorithms for mining association rules. In: Proc. 20th Int. Conf. Very Large Data Bases, VLDB, vol. 1215, pp. 487–499 (1994)
6. Ahmed, C., Tanbeer, S., Jeong, B.S.: Efficient mining of high utility patterns over data streams with a sliding window method, pp. 99–113 (2010)
7. Ahmed, C., Tanbeer, S., Jeong, B.S., Choi, H.J.: Interactive mining of high utility patterns over data streams. Expert Systems with Applications 39(15), 11979–11991 (2012)
8. Leung, C.S., Jiang, F.: Frequent itemset mining of uncertain data streams using the damped window model. In: Proceedings of the 2011 ACM Symposium on Applied Computing, pp. 950–955. ACM (2011)
9. Aggarwal, C., Yu, P.: A survey of uncertain data algorithms and applications. IEEE Transactions on Knowledge and Data Engineering 21(5), 609–623 (2009)
10. Chan, R., Yang, Q., Shen, Y.D.: Mining high utility itemsets. In: Third IEEE International Conference on Data Mining, ICDM 2003, pp. 19–26. IEEE (2003)
11. Liu, Y., Liao, W.K., Choudhary, A.: A two-phase algorithm for fast discovery of high utility itemsets, pp. 689–695 (2005)
12. Giannella, C., Han, J., Pei, J., Yan, X., Yu, P.: Mining frequent patterns in data streams at multiple time granularities. Next Generation Data Mining 212, 191–212 (2003)
13. Tseng, V., Chu, C.J., Liang, T.: Efficient mining of temporal high utility itemsets from data streams. In: Second International Workshop on Utility-Based Data Mining, p. 18. Citeseer (2006)
14. Li, H.F., Huang, H.Y., Chen, Y.C., Liu, Y.J., Lee, S.Y.: Fast and memory efficient mining of high utility itemsets in data streams. In: Eighth IEEE International Conference on (ICDM), pp. 881–886. IEEE (2008)
15. Leung, C.K.-S., Jiang, F.: Frequent pattern mining from time-fading streams of uncertain data. In: Cuzzocrea, A., Dayal, U. (eds.) DaWaK 2011. LNCS, vol. 6862, pp. 252–264. Springer, Heidelberg (2011)

Classification for Multi-Relational Data Mining Using Bayesian Belief Network

Nileshkumar D. Bharwad and Mukesh M. Goswami

Department of Information Technology, Dharmsinh Desai University, Nadiad
{nbharwad4588,mukesh.goswami}@gmail.com

Abstract. Multi-Relational Data Mining is an active area of research from last decade. Relational database is an important source of structured data, hence richest source of knowledge. Most of the commercial and application oriented data uses a relational database scheme in which multiple relations are linked through primary key, foreign key relationship. Multi-Relational Data Mining (MRDM) deals with extraction of information from a relational database containing multiple tables related to each other. In order to extract important information or knowledge, it is required to apply Data Mining algorithms on this relational database but most of these algorithms work only on a single table. Generating a single table from multiple tables may result in loss of important information, like the relation between tuples, also it is a not efficient in terms of time and space. In this paper, we proposed a Probabilistic Graphical Model, Bayesian Belief Network (BBN), based approach that considers not only attributes of the table but also the relation between tables. The conditional dependencies between tables are derived from Semantic Relationship Graph (SRG) of the relational database, whereas Tuple Id propagation helps to derive the conditional probability of tables. Our model not only predicts class label of unknown samples, but also gives the value of sample if class label is known.

Keywords: Multi-Relational Data Mining, Relational database, Bayesian Belief Network, Data Mining, Probabilistic graphical model, Semantic Relationship Graph, Tuple Id propagation.

1 Introduction

Most of the real world data are well structured and stored in Relational database that is making it a very important source of information. Many data mining algorithms for classifications like Naïve Bayesian, Support Vector Machine, works on a single table. Traditional algorithms cannot be applied until relational database transformed into a single table.

Further, conversion of relational tables into a single table, semantic information is missed. Deprivation of this data leads to inaccurate classification. The operation of conversion is tedious in terms of times and space. It creates huge universal dataset when we convert them to single flat table (relation). This paper discusses a Bayesian Belief Network based approach for multi relational data mining that builds a probabilistic model directly from multiple tables and also exploits relation between tables as well.

M.K. Kundu et al. (eds.), *Advanced Computing, Networking and Informatics - Volume 1*, Smart Innovation, Systems and Technologies 27,
DOI: 10.1007/978-3-319-07353-8_62, © Springer International Publishing Switzerland 2014

The residual of the Research paper is formed as follows: Next section gives a brief inspection of the surviving work, section III gives a brief discussion of background theory, section IV discusses proposed method with examples, and section V gives conclusions and future works of research study.

2 Related Work

2.1 Inductive Logic Programming (ILP)

For multi-relational classification, Inductive logic programming has attracted many researchers in early days of MRDM [1], [2]. Multi-relational classifier becomes more precise if we find relevant features in non-target relation that differentiate target tuples. Target and non-target relations are connected via multiple join paths. Complex schema needs to search a large number of join paths. It is very time consuming to identify good features and repeatedly explore and link up the relations along different join paths and need to evaluate it. Although, Inductive logic programming approaches are efficient and provides good classification accuracy, but they are not scalable

Many researchers are figuring out on how to build scalable and efficient ILP algorithms. Building a decision tree from stored data is one of the approaches. Efficiency can be improved by evaluating a bunch of quarries that have common prefaces [3]. The main drawback is that the query should be known. To overcome this problem, CrossMine [4] was developed. Also, author proposed an approach to virtual join, known as a Tuple Id Propagation, instead of physical join, which will be discussed in the next section.

The MR-Radix algorithm provides the best efficiency when we compared with the traditional algorithms. Suppose the data is spread over many tables' causes many problems in the practice of data mining. To obviate the above problem Krogel, and Knobbed defined that how the aggregate functions are used in the Multi Relational data mining [5], [6]. In summation to that when it dispense with the logic–based MRDM, which is called as Inductive Logic Programming [7-10].

2.2 Probabilistic Approaches

To get advantage of both logical and probabilistic approaches for knowledge representation and reasoning, probabilistic relational model is used and which is an extension of Bayesian networks for handling relational data. The Naïve Bayesian approach is one of the most widely used, as it is easy to train and give us good classification results. Some work can be found in [8], [10], [12] on a similar concept.

2.3 Graph-Based Approach

Graph-based approaches are mainly used for its underlying representation, which is totally different then logic based approaches. Examples, background knowledge, hypotheses and target concepts are represented as a graph. Graph theory based approaches like frequent substructure graph [13], and greedy based approaches [11] for example, subdue.

3 Background Theory

3.1 Bayesian Belief Network

Bayesian Belief network [8] is a probabilistic graphical model. It shows relationship between a set of random variables and their conditional dependencies using directed acyclic graph. Every node in the graph represents a random variable. Every random variable has its own set of values; Conditional dependencies are represented by edges. Nodes are conditionally independent if there are no incoming edges.

3.2 Semantic Relationship Graph (SRG)

Semantic Relationship Graph (SRG) is a directed acyclic graph G (V, E, W), Vis set of tables, E is a set of directed edges, and an edge (i, j) means table i can be linked to table j by directly joining these two tables. W is a set of link attributes, either Primary Key or Foreign key, which links two tables, which is shown in below Fig. 1.

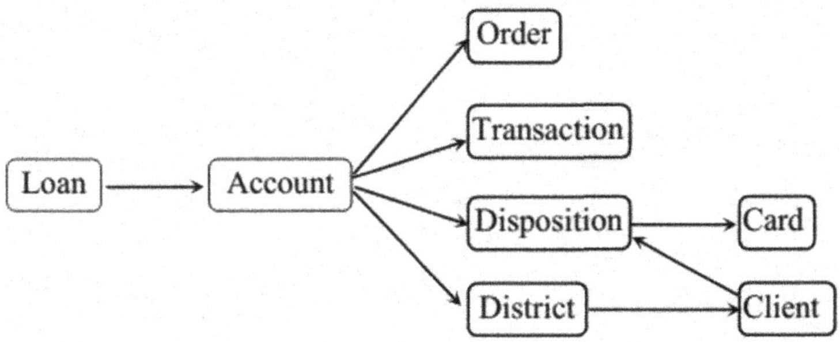

Fig. 1. Semantic Relationship graph for financial data

3.3 Tuple-ID Propagation

Tuple ID propagation is a method for virtually joining non-target relations with the target relation. It is a flexible and effective method and it avoids the high price of physical join [14]. Example of Tuple ID propagation is shown in below Table 1 and Fig. 2.

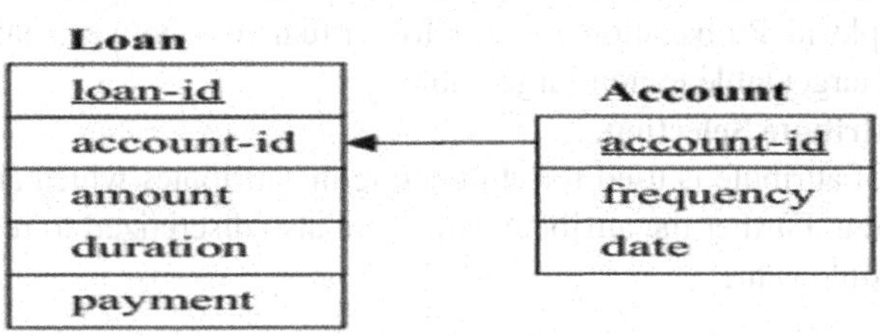

Fig. 2. Loan table and Account table relationship in PKDD cup99 dataset

Table 1. Account table in PKDD cup99 dataset

Loan					
loan-id	account-id	amount	duration	payment	
1	124	1000	12	120	+
2	124	4000	12	350	+
3	108	10000	24	500	-
4	45	12000	36	400	-
5	45	2000	24	90	+

4 Proposed Approach

In this proposed approach weshow the architecture of Classification for Multi-Relational Data Mining using Bayesian Belief Network in (Refer Fig. 3).

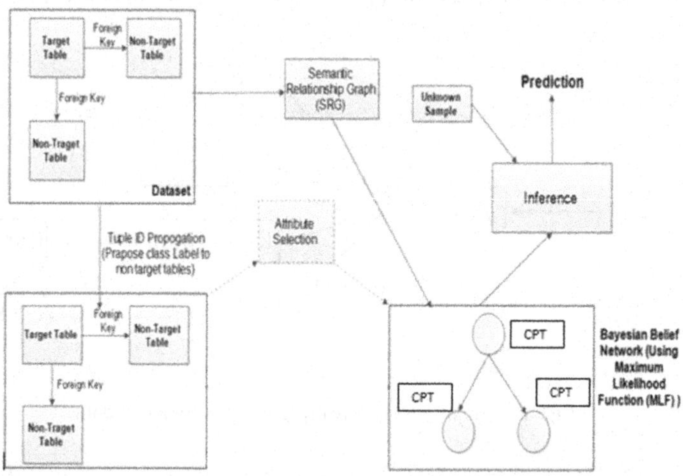

Fig. 3. Architecture of Classification for Multi-Relational Data Mining using Bayesian Belief Network

Step 1: Pre-Processing.
As mentioned in background theory Semantic relationship graph is generated from tables. Tuple id Propagation is used for virtual join between tables and propagate class from target table to non-target tables.

Step 2: Attribute Selection.
Selection of attribute is used for eliminating the attributes which are not related to the classification. Further the attribute values are also discretized to reduce the total number of possible values.

Step 3: Generation of Bayesian Belief Network.
By the use of Semantic relationship graph and Tuple id propagation, Bayesian Belief network is generated. Tables are represented in the form of nodes. Dependency between tables can be inferred from the semantic relationship graph, which is shown as edges. Tables will reducible to compound variables which are formed of the attributes.

The number of attributes and values of each attribute may be large. This may result into complex variables. To avoid this situation, attribute selection used, as mentioned in Step 2.

Step 4: Estimation of conditional probability of tables (CPT) and inference.

Conditional Probability for every table is calculated by considering different values of each and every table variable and their relative probabilities.For example in Fig. 4, there are two tables called Ta and TB.

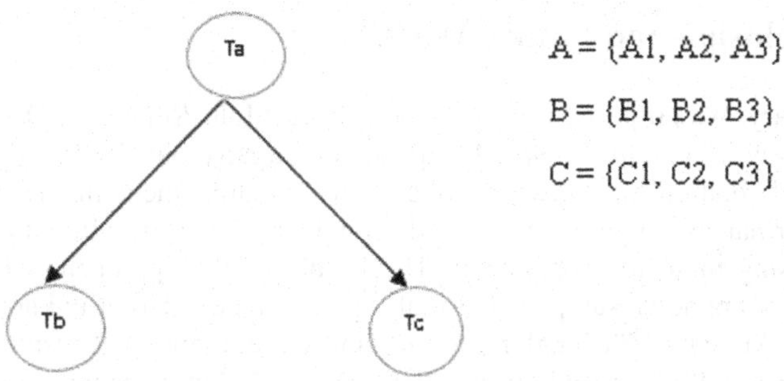

Fig. 4. Relationship between Ta,Tb and Tc

Although, the CPT estimation is a time consuming process, but it is only a one time process.

Ta is an independent (refer Table 2). So P (Ta) is given as,

Table 2. Ta Table with different values

A1	A2	A3	C1	C2
a10	a20	a30	Y	N
a10	a20	a31	N	Y
a10	a21	a30	N	N

Tb is dependent on Ta. For P (B/A) is as follows in Table 3,

Table 3. Tb table with different values

B1	B2	A1	A2	A2	A1	A2	A3
b10	b20	a10	a20	a30	a11	a12	a13
b10	b21	a10	a20	a31	a21	a22	a23

V7, …,V18 are different values and C1 and C2 are class labels. C1 and C2 are class labels in Table 2 and Table 3.

Inference: P (A, B, C) = P (A). P (B/A).P (C/A)

For unknown sample A3,

$$argmax\ A3 = P(A).P\left(\frac{B}{A}\right).P\left(\frac{C}{A}\right)$$

For unknown sample A3, it belongs to class c1 if c1>c2 or it relates class c2.

5 Conclusion and Future Work

In the summary, we proposed a novel approach for Multi-Relational Data Mining, in which accuracy of classifier should improve by finding out dependencies between attributes and relationship between tables and including them in the classification model. Performance can be improved by keeping tables as it is, without joining tables physically using tuple Id propagation. The beauty of the approach is not only that predict class of unknown samples, but it also predicts the value of unknown sample if class label is known. The implementation and the experimental results of the proposed approach will be carried out as future work. The implementation will work on kddcup99 dataset.

References

1. Sašo, D., Lavrač, N.: An introduction to inductive logic programming. In: Relational Data Mining, pp. 48–73 (2001)
2. Lavrac, N., Dzeroski, S.: Inductive Logic Programming: Techniques and Applications. Ellis Horwood (1994)
3. Blockeel, H., Dehaspe, L., Demoen, B., Janssens, G., Ramon, J., Vandecasteele, H.: Improving the Efficiency of Inductive Logic Programming through the Use of Query Packs. J. Artificial Intelligence Research 16, 135–166 (2002)
4. Yin, X., Han, J., Yang, J., Yu, P.S.: CrossMine: Efficient classification across multiple database relations. In: Boulicaut, J.-F., De Raedt, L., Mannila, H. (eds.) Constraint-Based Mining. LNCS (LNAI), vol. 3848, pp. 172–195. Springer, Heidelberg (2006)
5. Muggleton, S.H.: Inverse Entailment and Progol. New Generation Computing 13(3-4), 245–286 (1995)
6. Muggleton, S., Feng, C.: Efficient Induction of Logic Programs. In: Proceedings of Conference on Algorithmic Learning Theory (1990)
7. Pompe, U., Kononenko, I.: Naive Bayesian classifier within ILP-R. In: Proceedings of the 5th International Workshop on Inductive Logic Programming, pp. 417–436 (1995)

8. Heckerman, D.: Bayesian networks for data mining. Data Mining and Knowledge Discovery 1(1), 79–119 (1997)
9. Ceci, M., Appice, A., Malerba, D.: Mr-SBC: A Multi-relational Naïve Bayes Classifier. In: Lavrač, N., Gamberger, D., Todorovski, L., Blockeel, H. (eds.) PKDD 2003. LNCS (LNAI), vol. 2838, pp. 95–106. Springer, Heidelberg (2003)
10. Flach, P., Lachiche, N.: 1BC: A first-order Bayesian classifier. In: Džeroski, S., Flach, P.A. (eds.) ILP 1999. LNCS (LNAI), vol. 1634, pp. 92–103. Springer, Heidelberg (1999)
11. Neville, J., Jensen, D., Gallagher, B., Fairgrieve, R.: Simple Estimators for Relational Bayesian Classifiers. In: International Conference on Data Mining (2003)
12. Manjunath, G., Murty, M.N., Sitaram, D.: Combining heterogeneous classifiers for relational databases. Pattern Recognition 46(1), 317–324 (2013)
13. Quinlan, J.R., Cameron-Jones, R.M.: FOIL: A Midterm Report. In: Proceedings of 1993 European Conference on Machine Learning (1993)
14. Yin, X., Han, J., Yang, J.: Efficient Multi-relational Classification by Tuple ID Propagation. In: Proceedings of the KDD-2003 Workshop on Multi-Relational Data Mining (2003)

Noise Elimination from Web Page Based on Regular Expressions for Web Content Mining

Amit Dutta[1], Sudipta Paria[2], Tanmoy Golui[2], and Dipak Kumar Kole[2]

[1] Department of IT, St. Thomas' College of Engineering & Technology, Kolkata
to.dutta@gmail.com

[2] Department of CSE, St. Thomas' College of Engineering & Technology, Kolkata
{pariasudipta,tanmoy.stcet,dipak.kole}@gmail.com

Abstract. Web content mining is used for discovering useful knowledge or information from the web page. So, noisy data in web document significantly affect the performance of web content mining. In this paper, a noise elimination method has been proposedbased on regular expression followed by Site Style Tree (SST). The proposed technique consists of two phases. In the first phase, filtering method based on regular expression is used on web pages to remove noisy HTML tags The filtered document then undergoes to second phase where an entropy based measured is used for removing further noise. The page size is reduced considerably by eliminate a number of lines of code preceded by some predefined noisy HTML tags. The con-sized web document is then used to form Document Object Model (DOM) tree and consequently the Site Style Tree is formed by crawling the pages from the same URL path as of the website. The experiment conducted on some most popular websites like www.amazon.com, www.yahoo.com and www.abcnews.com. The experimental result reveals that the filtering method eliminates a significant amount of noise before introduction of SST, so the overall space and time complexity is reduced compared to other SST based approach.

Keywords: Noise, Web Mining, Web Content Mining, Regular Expression, DOM Tree, Site Style Tree (SST), Node Importance, Composite Importance.

1 Introduction

Nowadays the rapid expansion of the Internet has made the World Wide Web to be a popular place for broadcasting and collecting information. Thus web mining becomes an important task for discovering useful knowledge or information from the Web [1]. Web mining can be classified as: web structure mining, web content mining and web usage mining [2]. It is very common that useful information on the web is often accompanied by a large amount of noise such as banner advertisements, navigation bars, copyright notices, etc. Although such information items are functionally useful for human viewers and necessary for the web site owners, they often hamper automated information gathering and web data mining, e.g., web page clustering, classification, information retrieval and information extraction.

M.K. Kundu et al. (eds.), *Advanced Computing, Networking and Informatics - Volume 1*, Smart Innovation, Systems and Technologies 27,
DOI: 10.1007/978-3-319-07353-8_63,

In the field of web content mining some significant work had been done in the past years. In [3], a GA based model is proposed for mining frequent pattern in web content. In [4][5], web usage mining method proposed for personalization and business intelligence for e-commerce websites. In [6], a method is proposed for coherent keyphrase extraction using web content mining.Noise removal from web pages is an important task which helps to extract the actual content for web content mining in later phase. In [7], a method is proposed to detect informative blocks in news Web pages. In [8], Web page cleaning is defined as a frequent template detection problem in which partitioning is done based on number of hyperlinks that an HTML element has. Web page cleaning is also related to feature selection in traditional machine learning[9]. In feature selection, features are individual words or attributes. A method is proposed for learning mechanisms to recognize banner ads, redundant and irrelevant links of web pages [10]. The HITS algorithm[11]is enhanced by using the entropy of anchor text to evaluate the importance of links. It focuses on improving HITS algorithm to find more informative structures in Web sites.

In this work, noisy contents are removed partially using tag based filtering method based on regular expressions[12] and then an entropy based measure is incorporated to determine remaining noisy contents in the web page and eliminate them.In this paper, Section 2,3,4 and 5 represent preliminaries, proposed method along with the necessary algorithms, experimental results and conclusion respectively.

2 Preliminaries

Noise: Information blocks except actual or main content blocks are referred as noise. Noises can be classified into two types:*Local Noise or Intra-page Noise* (noisy information blocks within a single web page, Ex: Banner advertisements, navigational guides, decoration pictures etc.) and *Global Noise or Inter-page Noise* (noises on the web which are usually no smaller than individual pages, Ex: Mirror sites, legal/illegal duplicated Web pages, old versioned Web pages etc.)

Regular Expression: A regular expression (abbreviated regex) is a sequence of characters that forms a search pattern, mainly for use in pattern matching with strings, or string matching, i.e. "find and replace"-like operations.Example of a regular expression for matching a string within double quotes might be: " [^"] *"

DOM Tree: DOM tree is a data structure which is used to represent the layout of a HTML web page. HTML tags are the internal nodes and the contents e.g. texts, images or hyperlinks enclosed by the tags are the leaf nodes of DOM tree. In the DOM tree, each solid rectangle represents an internal node. The shaded box represents the actual content of the node e.g., for the tag IMG, the actual content is src=myimage.jpg. Fig. 1 shows the DOM tree corresponding to the segment of HTML code.

```
<BODY bgcolor=RED>
<p> ….. </p>
<div>
<TABLE width=400
height=400>
  ….
</TABLE>
<imgsrc=”myimage.jpg”
width=100 height=100 >
<TABLE width=200 height=200>
  ….
</TABLE>
</div>
<p> ….. </p>
</BODY>
```

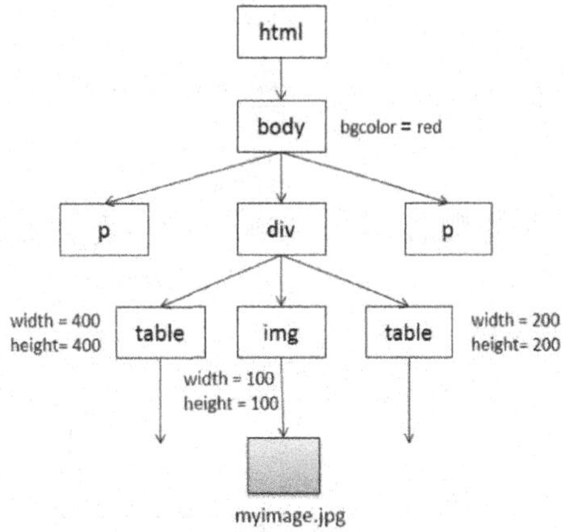

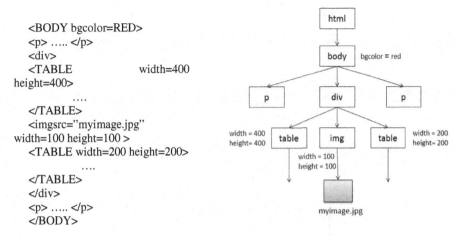

Fig. 1. A DOM Tree corresponding to the HTML Code

Site Style Tree (SST)[13]**:** Site Style Tree (SST) is a data structure which is created by mapping the nodes of individual DOM trees of each web page of the website. SST consists of two types of nodes: (i) Style Node and (ii) Element Node

Style Node- A style node S in the Style tree represents a layout or presentation style, which has two components, denoted by (*Es, n*), where,*Es* is a sequence of element nodes and *n* is the number of pages that has this particular style at this node level.

Element Node - An element node E has three components, denoted by (*TAG, Attr, Ss*), where, *TAG* is the tag name, *Attr*is the set of display attributes of *TAG* and *Ss* is a set of style nodes below E.

Example of an SST is given in Fig. 2 as a combination of DOM trees d1 and d2. A count is used to indicate how many pages have a particular style at a particular level of the SST. Both d1 and d2 start with BODY and the next level below BODY is P-DIV-P. Thus both BODY and P-DIV-P has a count of 2. Below the DIV tag, d1 and d2 diverge i.e. two different presentation styles are present. The two style nodes (represented in dashed rectangle) are DIV-TABLE-TABLE and DIV-A-IMG havingpage count equal to 1.

So, SST is a compressed representation of the two DOM trees. It enables us to see which parts of the DOM trees are common and which parts are different.

To determine an element node in SST to be noisy an entropy based measure is usedwhich evaluates the combined importance i.e. both presentation and content importance. Higher the combined importance of an element node is, more likely it is the main content of the web pages.

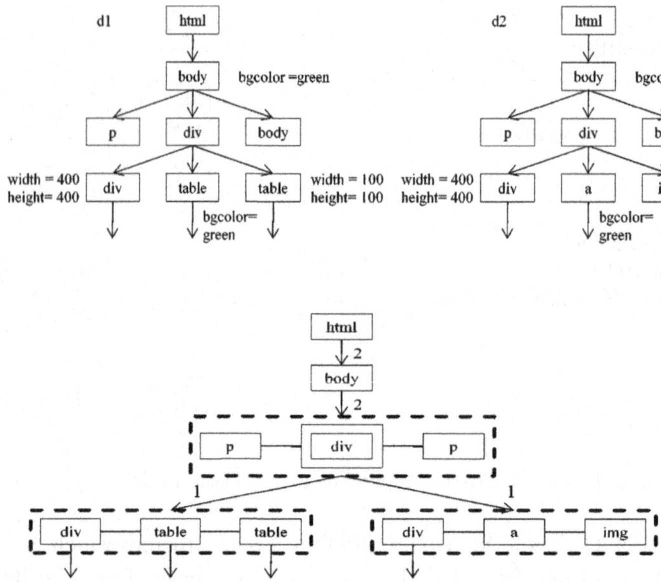

Fig. 2. DOM Trees and the Site Style Tree

Node Importance

Definition: For an element node E in the SST, let $m$be the number of pages containing E and l be the number of child style nodes of E (i.e., $l = |E.Ss|$), the *node importance* of E, denoted by *NodeImp*(E), is defined by,

$$NodeImp(E) = \begin{cases} -\sum_{i=1}^{l} p_i \log_m p_i & ,if\ m > 1 \\ 1 & ,if\ m = 1 \end{cases} \quad (1)$$

where $p_i$is the probability that a Web page uses the i^{th}style node in $E.Ss$.

Considering only the node importance, an element cannot be saidto be a noisy item because it does not consider the importance of its descendants. Hence, composite importance is considered to measure the importance of an element node and its descendants.

Composite Importance:

Definition (internal node): For an internal element node E in the SST, let $l = |E.Ss|$. The *composite importance* of E, denoted by *CompImp(E)*, is defined by,

$$CompImp(E) = (1 - \gamma^l)NodeImp(E) + \gamma^l \textstyle\sum_{i=1}^{l}(p_i\,(CompImp(S_i)) \quad (2)$$

where, $p_i$is the probability that *E* has the *ith*child style node in *E.Ss*. In the above equation, *CompImp(*S_i*)* is the *composite importance* of a style node $S_i(\in E.Ss)$, which is defined by,

$$CompImp(S_i) = \frac{\sum_{j=1}^{k} CompImp(E_j)}{k} \quad (3)$$

where,$E_j$is an element node in $S_i.E$, and $k = |S_i Es|$, which is the number of element nodes in S_i.

In (2), $\gamma$is the attenuating factor, which is set to 0.9.$\gamma$is directly proportional to the weight of NodeImp(E) i.e. it increases the weight of *NodeImp(E)* when *l*is large and it decreases the weight of *NodeImp(E)*when *l* is small.

Definition (leaf node): For a leaf element node *E* in the SST, let *l* be the number of features (i.e., words, image files, link references etc.) appeared in *E* and let *m*be the number of pages containing *E*, the composite importance of *E* is defined by,

$$CompImp(E) = \begin{cases} 1 & ,if m = 1 \\ 1 - \frac{\sum_{i=1}^{l} H(a_i)}{l} & ,if m > 1 \end{cases} \quad (4)$$

where, $a_i$is an actual feature of the content in *E*. *H(*a_i*)* is the information entropy of $a_i$within the context of *E*,

$$H(a_i) = -\sum_{j=1}^{m} p_{ij} \log_m p_{ij} \quad (5)$$

where, $p_{ij}$is the probability that $a_i$appears in *E* of page *j*.

Definition (noisy): For an element node *E* in the SST, if all of its descendants and itself have composite importance less than a specified threshold *t*, then the*E* is said to be *noisy*.

Definition (maximal noisy element node): If a noisy element node *E* in the SST is not a descendent of any other noisy element node, then *E* is called *maximal noisy element node*.

Definition (meaningful): If an element node *E* in the SST does not contain any noisy descendent, then *E* is said to be *meaningful*.

Definition (maximal meaningful element node): If a meaningful element node *E* is not a descendent of any other meaningful element node, then*E* is said to be a *maximal meaningful element node*.

3 Proposed Method

Initiallya filtering method is appliedbased on regex to eliminate contents enclosed by negative tags. But filtering does not ensure removing all the noisy information present. So both the layouts and the contents of web pages in a given web site need to be analyzed. Site Style Tree concept is used for this purpose which is created by combining DOM Trees of the web pages in the website. After creating the SST,an entropybased measure is incorporated for evaluating the node importance in the SST for noise detection.The steps involved in our proposed technique can be represented by the following block diagram (refer Fig.3.).

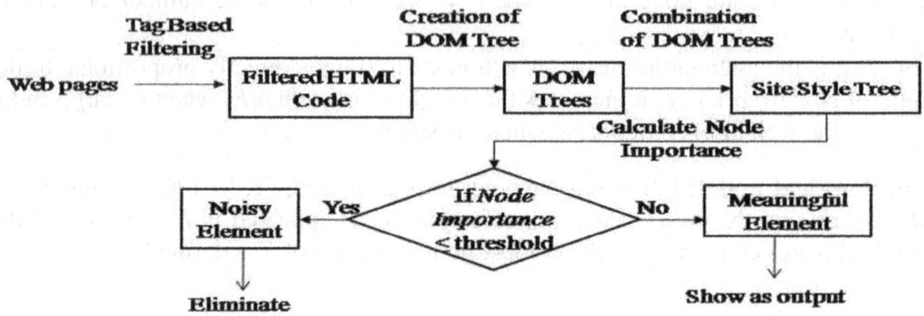

Fig. 3. Block Diagram for Noise Elimination

Assumptions: In this work, typical commercial websites have been considered in which an element with more presentation styles and more diversity is more important. We are focusing on detecting & eliminating local or intra-page noises of a webpage.

3.1 Filtering

Depending on the content ofHTML tags in a web page the tags can be divided into two types: a) positive tag and b) negative tag[14]. Positive tag contains useful information. All the tags except positive tags are referred to as Negative tags. In this work some tags have been defined as negative tags to remove noisy data from a web page. The following tags are considered as negative tag:Anchor tag (<a>), Style tag (<style>), Link tag (<link>), Script tag (<script>), Comment tag (<!-- … -->) and No Script tag (<noscript>)

3.2 Regular Expression

The regex <!--.*?--> is used to eliminate HTML comment line present in a particular web page. To eliminate contents enclosed by script tag, the regex is defined as<script\b[^>]*>.*?</script>where, \b indicates a word boundary and [^>]* matches all characters except ‘>’. Similarly, regex for style and link tag will be <style\b[^>]*>.*?</style>and <link\b[^>]*> respectively. Regex for fetching URL from anchor tag (<a>) is given by, <a\b[^>]* href="[^"]*"|'[^']*'|[^'">]*>.*?</a>. After fetching URL of the website, Naïve String Matching Algorithm [15]is applied to determine whether the URL is noisy or meaningful.In this technique the URL of the web site is used as the pattern which will be checked with the fetched URL of the anchor tag and if any valid shift is present then it can be concluded that the hyperlink reference is related to the website and regarded as useful data. If no valid shift is obtained then the hyperlink reference contains noisy elements like banner advertisements of other website (local noise) or mirror sites, duplicate web page etc. (global noise).

3.3 Noise Detection in SST [13]

The creation of DOM Trees and SST has been discussed already in the earlier section. Some existing algorithmsare used in our techniquelike *MarkNoise(EN)* and *MarkMeaningful(EN)*which marks all maximal noisy and maximal meaningful Element Nodes(EN) respectively. The descendants of maximal noisy nodes will be removed from the SST. Some element nodes in the SST may be neither noisy nor meaningful. In this case, we simply go down to the lower level nodes and check whether the descendants of that node are meaningful or noisy. For easy presentation of the algorithm, it is assumed that the DOM tree of the page is converted to a style tree with only one page (called a Page Style Tree or PST) by *BuildPST(P)* algorithm.The $MapSST(E,E_{pst})$ algorithm maps PST of the page to the SST.

3.4 Proposed Algorithm

Given a website, the system first randomly crawls a number of web pages from the website (Step 1). Next it remove the contents enclosed by negative tags using regular expressions (Step 2-5).Then the SST is built (Step 6-10). By calculating the composite importance of each element node in the SST, the maximal noisy nodesand maximal meaningful nodes can be found out (Step 11-13). To eliminate noise from a new page *P*, its PST will be mapped to the SST to eliminate noises (Step 14-17).

Step 1:Randomly crawl *k* pages from the given Web site *S*
Step 2:Apply the regular expressions to remove contents enclosed by <script/>, <style/>, <link/>, <!--> and<noscript/> tags
Step 3:Apply regular expression to extract URL address from <a/> tag
Step 4:4.1T ← extracted URL from <a/> by regular expression
4.2P ← URL address of the web page
4.3NAÏVE_STRING_MATCHER (T, P)
Step 5:Remove all the contents enclosed by <a/> for which valid shift is not present
Step 6:Set null SST with virtual root *E* (representing the root);
Step 7:**for** each page *W* in the *k* pages **do**
7.1 BuildPST(W);
7.2 BuildSST(E, E_w)/*E and $E_w$are the element node of SST & PST respectively */***Step 8:*** **end for**
Step 9:CalcCompImp(*E*);
Step 10:MarkNoise(*E*);
Step 11:MarkMeaningful(*E*);
Step 12:**for** each target Web page *P* **do**
12.1E_{pst}= BuildPST(*P*) /* representing the root */

12.2 MapSST(E, E_{pst})/*E and $E_{pst}$are the element node of SST & PST respectively */

***Step 13:*end for**

4 Experimental Results

The above algorithm has been implemented using Java, in 32-bit Windows XP Professional with Service Pack 2. The web browser used for the purpose is Google Chrome Version 32. The processor used is Core 2 Duo, 3.06GHz and 2.00GB RAM.

Table 1. Line based statistics

Websites	Lines	<!-->	<script />	<style />	<link />	<a />	<noscript />	Total
Yahoo	1460	5.70%	19.30%	30.30%	0.06%	0.41%	6.60%	62.37%
Amazon	4251	0.68%	30.50%	32.31%	0.26%	0.60%	1.15%	65.50%
MakeMyTrip	2102	10.00%	18.71%	10.00%	0.43%	0.23%	0.14%	38.78%
Quikr	1185	2.46%	44.93%	3.55%	0.26%	0.05%	1.01%	52.26%
Flipkart	2200	0.60%	51.04%	0.70%	0.04%	0.06%	0.22%	52.66%
abcnews	1819	14.02%	14.52%	0.05%	0.33%	0.20%	0.27%	29.00%
Myntra	1678	1.86%	12.07%	0.60%	0.48%	0.88%	0.11%	16.00%
Yatra	1952	3.37%	15.55%	1.24%	0.35%	0.34%	0.00%	20.85%
Snapdeal	989	1.19%	32.90%	0.00%	0.68%	2.04%	0.00%	36.81%
Rediff	1116	0.98%	65.67%	0.26%	0.17%	0.17%	0.18%	67.43%

Table 2. Overall statistics

Website	Filtering	Site Style Tree	Total Noise	Content
Yahoo	62.37%	8.28%	70.65%	29.35%
Amazon	65.50%	5.78%	71.28%	28.72%
MakeMyTrip	38.78%	30.35%	69.13%	30.87%
Quikr	52.26%	13.92%	66.18%	33.82%
Flipkart	52.66%	21.64%	74.30%	25.70%
abcnews	29.00%	16.49%	45.49%	54.51%
Myntra	16.00%	47.68%	63.68%	36.32%
Yatra	20.85%	31.30%	52.15%	47.85%
Snapdeal	36.81%	37.61%	74.42%	25.58%
Rediff	67.43%	2.24%	69.67%	30.33%

For the purpose of experiment ten popular commercial websites have been taken. As these websites are dynamic in nature so its contents are varied from time to time. The snapshots of these websites are taken on 05.01.2014. In the initial phase a comparative analysis is done (Table 1) on how much percentage of noisy element (number of lines enclosed by negative tag) are there in those site.

In the second phase, elimination of further noise (that is not removed by filtering) is done by SST. In our experiment the threshold value (t) is initialized to 0.2. The following table shows the total percentage of noise that can be removed by the combined method of filtering and SST. The last column of the Table 2 shows the amount of content present on those websites, which shows a reality of the experiment.

5 Conclusion

In this paper the technique to detect and remove local noisy elements from web pages has been discussed. The proposed technique incorporates two phases: filtering based on regular expressions and an entropy based measure in SST. From the experimental result it is evident that the filtering method eliminates a considerable amount of noisy (9^{th}column of Table 1) element before introduction of SST. So number of element nodes in SST will be less compared to the technique where only SST methodis involved for noise removal [13]. As the number of nodes in SST gets reduced, the space requirement for storing nodes becomes less as well as the traversal time for tree operations like insertion, deletion, searching etc. is reduced. Thus, overall spaceand time complexity will be reduced. Furthermore the performance of web content mining will also be improved because actual contents of the web pages can be extracted easily after noisy elements removal. Experimental resultsshow that our proposed method is highly effective. The proposed method ensures partial removal of the global noises present in web pages. Hence, the proposed method can be extended further to detect and remove all the global noises from web pages.

References

1. Han, J., Chang, K.C.-C.: Data Mining for Web Intelligence. IEEE Computer 35(11), 64–70 (2002)
2. Srivastava, J., Desikan, P., Kumar, V.: Web Mining - Concepts, Applications, and Research Directions. In: Chu, W., Lin, T.Y. (eds.) Foundations and Advances in Data Mining. STUDFUZZ, vol. 180, pp. 275–307. Springer, Heidelberg (2005)
3. Sabnis, V., Thakur, R.S.: Department of Computer Applications, MANIT, Bhopal, India, GA Based Model for Web Content Mining. IJCSI International Journal of Computer Science Issues 10(2), 3 (2013)
4. Eirinaki, M., Vazirgiannis, M.: Web mining for web personalization. ACM Transactions on Internet Technology 3(1), 1–27 (2003)
5. Abraham, A.: Business Intelligence from Web Usage Mining. Journal of Information & Knowledge Management 2(4), 375–390 (2003)
6. Turney, P.: Coherent Keyphrase Extraction via Web Mining. In: Proceedings of the Eighteenth International Joint Conference on Artificial Intelligence, pp. 434–439 (2003)
7. Lin, S.-H., Ho, J.-M.: Discovering Informative Content Blocks from Web Documents. In: Proceedings of Eighth ACM SIGKDD International Conference on Knowledge Discovery and Data Mining, pp. 588–593 (2002)
8. Bar-Yossef, Z., Rajagopalan, S.: Template Detection via Data Mining and its Applications. In: Proceedings of the 11th International Conference on World Wide Web (2002)

9. Yang, Y., Pedersen, J.O.: A comparative study on feature selection in text categorization. In: International Conference on Machine Learning (1997)
10. Kushmerick, N.: Learning to remove Internet advertisements. In: Proceedings of Third Annual Conference on Autonomous Agents, pp. 175–181 (1999)
11. Kao, J.Y., Lin, S.H., Ho, J.M., Chen, M.S.: Entropy-based link analysis for mining web informative structures. In: Proceedings of Eleventh International Conference on Information and Knowledge Management, pp. 574–581 (2002)
12. Fried, J.: Mastering regular expressions. O'Reilly Media Inc. (2006)
13. Lan, Y., Bing, L., Xiaoli, L.: Eliminating Noisy Information in Web Pages for Data Mining. In: Proceedings of Ninth ACM SIGKDD International Conference on Knowledge Discovery and Data Mining, KDD 2003, pp. 296–305 (2003)
14. Kang, B.H., Kim, Y.S.: Noise Elimination from the Web Documents by using URL paths and Information Redundancy (2006)
15. Cormen, T.H., Leiserson, C.E., Ronald, R.L., Clifford, S.: Introduction to Algorithm. The MIT Press (2009)

Modified Literature Based Approach to Identify Learning Styles in Adaptive E-Learning

Sucheta V. Kolekar, Radhika M. Pai, and M.M. Manohara Pai

Department of Information and Communication Technology
Manipal Institute of Technology, Manipal University, Karnataka, India
{sucheta.kolekar,radhika.pai,mmm.pai}@manipal.edu

Abstract. To effectively understand the adaptation approaches in content delivery on E-learning, learner's learning styles need to be identified first. There are two main approaches that detect the learning styles: Questionnaire based and Literature based. The main challenge of Adaptive E-learning is to capture the learner's learning styles while using E-learning portal and provide adaptive user interface which includes adaptive contents and recommendations in learning environment to improve the efficiency and adaptability of E-learning. To address this challenge the literature based approach requires to be modified according to learner's usage of e-learning portal and should generate learner's profile according to standardized learning style model. The study focuses on engineering students and the learning style model considered is Felder-Silverman Learning Style Model. The paper presents the analysis of log data which is captured in log files and database. Analysis of obtained results show that the captured usage data is useful to identify the learning styles of the learners and the types of contents is proved important factor in literature based approach.

Keywords: Adaptive E-learning, Felder-Silverman Learning Style Model, Web Logs, Behavioral Model.

1 Introduction

The evolution of E-learning is focusing mainly on adaptation in education. Each learner has his/her own learning style which indicates how he/she learns most effectively from courses. Adaptive E-learning system allows students to learn by themselves so that it would improve learning effect and overcome the problems of traditional learning.

Effective Adaptive E-learning has different adaptation approaches which mainly focus on content delivery and adaptive user interface on E-learning portal. Main input factor to adaptation approach is learner's learning preferences which should be captured according to standard learning style model. There are two main approaches that detect the learning styles: Questionnaire based and Literature based [1], [2].

In questionnaire based, learner's learning preferences will be captured by analyzing the responses made to the questions. The problem with this approach is that sometimes it will produce inaccurate result and is not capable of tracking the changes in learning styles. Literature-based approach is a new methodology which is depends on learner's interaction towards learning portal. The dynamic nature of learner's learning styles can

M.K. Kundu et al. (eds.), *Advanced Computing, Networking and Informatics - Volume 1*,
Smart Innovation, Systems and Technologies 27,
DOI: 10.1007/978-3-319-07353-8_64,

be captured using this approach. Support of learning style model is making this approach efficient and accurate.

Several learning style models have been proposed by Myers-Briggs, Kolb and Felder-Silverman out of which Felder-Silverman Learning Style Model (FSLSM) focuses specifically on aspects of the learning styles of engineering students and it is successfully been proved in questionnaire approach and as a behavioral model in literature based approach. Learning styles represent a combination of cognitive, effective and psychological characteristics that indicate the learner's way of processing, grasping, understanding and perception capabilities towards the course contents.

The paper discusses about the Prototype version of e-learning portal which has been designed and developed to test the proposed framework of e-learning for identifying learning styles of learners. It is based on the different types of contents and nature of page visits in order to modify the literature based approach.

2 Related Works

Grapf *et al.* [2] discussed about Literature based approach where authors have analyzed the behaviors of 127 learners for object oriented modeling course in Learning Management System (LMS) moodle using Felder-Silverman Learning Style Model. Behavior patterns associated with thresholds are determined according to learner's activities. In this paper, authors have done the experimentation on moodle which is a static web portal that supports only static user interface for all learners.

Literature based approach of detecting learning styles in e-learning portal is mentioned by P.Q.Dung *et al.* [3]. Literature based approach is modified with multi-agent architecture where learning styles of the learners have been identified by conducting tests and storing it in a database. Later different types of contents are provided to learners and monitored as to identify whether the learning styles are changing as per the behavior or not. The mentioned approach is time saving but static in nature where the students learning styles are changing depending on structure of the course and learner's interest. Also the learning material is not divided into content types based on FSLSM.

Salehi *et al.* [4] have addressed the recommendation of course material for e-learning system. Authors have mentioned the content based recommendation and similarity based recommendation. However the learner's profiles have not been classified based on learning styles which is the main factor of recommendation with respect to learner's individual preferences.

The Web-based Education System with Learning Style Adaptation (WELSA) described by Popescu *et al.* [5] which is an intelligent e-learning system. It adapts the course contents to the learning preferences of each learner. Authors have considered different learning objects as a Resource Type (e.g. Exercise, Simulation, diagrams and experiment etc.) and not focused on different types of learning contents where learners are trying to get knowledge by understanding and processing the contents with supportive resources.

Liu *et al.* [6] have described combined mining approach of web server logs and web contents. The approach is useful to facilitate better adaptive presentation and navigation as a part of Adaptive E-learning system. The mentioned novel approach of combination is useful to find out the user navigation patterns and page visiting sequences.

Evolution and new trends of learning systems are reviewed in the context of learning paradigms and methodologies by Simi *et al.* in [7]. Authors have given overview of personalization as a learning process which is closely related to recommendation and leads to the adaptation in contents on e-learning portal based learner's preferences.

3 Methodology

The main objective of proposed work is to identify the learning styles using web usage mining in literature based approach. A work is carried out by capturing the access patterns of the learners in W3C extended log formats and in database. The W3C log files give the usage of different pages accessed as per the learners login and page visit sequence. The database log gives the usage of different files accessed as per the course contents and time spent on that page.

3.1 Felder-Silverman Learning Style Model (FSLSM)

The model adopted is the one suggested by Felder and Silverman (1988) for engineering education, which classifies students according to their position in several scales that evaluate how they perceive and process information [8], [9]. The initial levels of parameters considered according to the FSLSM are shown in Table 1. The parameters for specific learner can be captured in web logs of E-learning Portal. The captured logs can be further analyzed to modify literature based approach. Based on analysis the learners can be classified into FSLSM categories.

Table 1. Parameters for FSLSM

Contents	Act	Ref	Sen	Int	Vis	Ver	Seq	Glo
Text		X		X		X	X	
Videos		X	X	X	X	X		
Demo/PPT		X	X			X		X
Exercise	X			X				
Forum	X			X				
Index of Topics	X		X					X

3.2 E-learning Framework

Adaptive E-learning is dealing with educational systems that adapt the learning content as well as the user interface with respect to pedagogical and didactical aspects. The framework shown in Fig. 1 explains about the approach of Adaptive E-learning portal where the first step is to recognize the learning styles of individual student according to the actions or navigation that he or she has performed on portal using Web Usage Mining(WUM). Web log mining is substantially the important part of WUM algorithm. It involves transformation and interpretation of the log information to predict the patterns as per different learning styles using optimized clustering technique. Ultimately these patterns are useful to classify various defined profiles depending on content types.

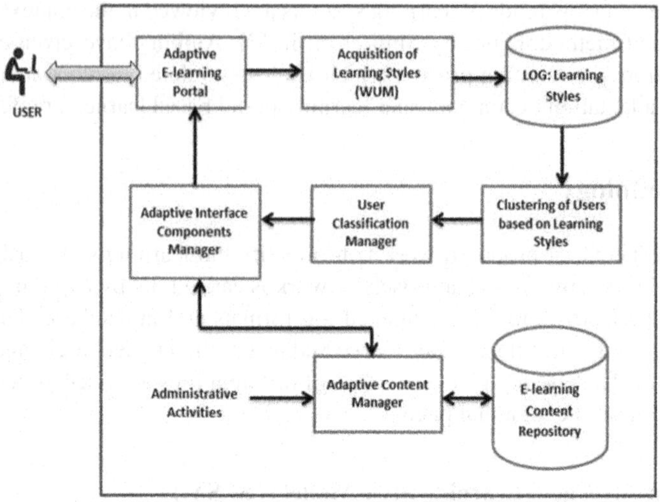

Fig. 1. Framework of E-learning

The clustered learner profiles with learning patterns are the input for user classification manager. Here classification algorithm can be used to identify different kind of learners based on the learning style of Felder and Silverman Model. After identifying the categories of learners the Interface component manager is changing the graphical representation of user interface based on small applications of e-learning portal. The contents of e-learning should be adaptive enough to change the interface components according to the learning style. This adaptive interface components can be generated using adaptive contents based on user classification with the help of administrative activities and e-learning content repository. This content is an important input factor for Interface Component Manager (ICM).

The paper describes about the approach of learning styles acquisition through usage log data of e-learning portal and the detailed analysis of log data.

4 Experimentation Details

According to the framework mentioned in Fig. 1, implementation of e-learning portal has been started using Microsoft Visual Studio 2008 and Microsoft SQL server 2005. The portal is deployed on IIS server to provide access to all learners in the Intranet. The log file option of IIS server is set to W3C extended log file format for the portal which will capture usage details of learners who are accessing the portal.

The prototype version of portal is made accessible for First year engineering students for Problem Solving Using Computers subject with the following topics: Arrays, Strings and Pointers.Each topic is made available for study in different file formats such as text (doc/pdf), video(mpeg/mp4) and demo (ppt/pptx). Also an exercise on 1D array, 2D array and Strings for learners is generated. Around 75 learners have registered in portal out of which 30 learners have accessed the topics which are mentioned above. The log files of learners who are accessing portal are maintained in W3C log file along with the time spent and access details of files are captured in database against the specific session of the learner.

5 Experimentation Design

In this work, we have designed an experimental algorithms to analyze the records of usage which are captured in log records and database. In the following subsections the algorithms of the system are described in detail [10], [11].

1. **Session Identification of Learner:** Session can be described as the time spent on portal by a learner from the moment he/she logged in to the moment he/she logged out. So session identification of learner is the process of collecting individual access sessions of each learner from the log records and accumulate it to get total session time. Algorithm 1 is used to get total session time of each and every learner on portal.

Algorithm 1. Total time spent on the portal by user

INPUT: A finite set of Learners L = $L_1, L_2, ..., L_N$ and Sessions S = $S_1, S_2, ..., S_Q$

OUTPUT: $Session_{Time_j}$ = Time spent on portal in one session and $TotalSession_{Time_i}$ = Total time spent on portal by one Learner in Different Sessions

initialize $Session_{Time_j} \leftarrow 0, LogIn_{Time} \leftarrow 0, LogOut_{Time} \leftarrow 0, TotalSession_{Time_i} \leftarrow 0$
for *each Learner* L_i where $i \leftarrow 1$ **to** N **do**
  for *each Session* S_j where $j \leftarrow 1$ **to** Q **do**
    if L_i *is logged in* **then**
      $LogIn_{Time} \leftarrow t$
    end if
    if L_i *is logged out* || L_i *idle for threshold time* **then**
      $LogOut_{Time} \leftarrow t'$
    end if
    $Session_{Time_j} \leftarrow LogOut_{Time} - LogIn_{Time}$
  end for
  $TotalSession_{Time_i} \leftarrow TotalSession_{Time_i} + Session_{Time_j}$
end for

2. **Content Identification:** Content identification is analyzed by calculating the time spent on specific file in one all the sessions of that learner. Since sessions are segmented between many sub sessions, time spent of specific file in one session of the learner is need to be considered first. The detailed steps are given in Algorithm 2.

3. **Content Type Identification:** Content type identification is useful to understand what type of contents learners are more interested in. It is analyzed by calculating the frequency of accessing specific type of file by all learners. There are three types of contents are available on portal i.e. pdf, ppt and mp4. The Algorithm 3 describes the steps to identify frequency of each file type which is accessed by a learner in all sub sessions.

Algorithm 2. Total time spent on specific file by user in all the sessions

INPUT: A finite set of Learners L = $L_1, L_2, ..., L_N$ and Sessions S = $S_1, S_2, ..., S_Q$
OUTPUT: $TFile_j$ = Time spent on specific file in one session and Total $Duration_i$ = Total time spent on a specific file in all sessions by Learner L_i

```
initialize End_Time ← 0, Start_Time ← 0, TotalDuration_i ← 0, TotalSession_Time_i ← 0
for each Learner L_i where i ← 1 to N do
    for each Session S_j where j ← 1 to Q do
        if "File" is accessed then
            Start_Time ← t
            {t is the system time at Learner clicked}
        end if
        if L_i is clicked back button || L_i clicked other link || L_i idle for threshold time then
            End_Time ← t'
            {t' is the system time at Learner unclicked}
        end if
        TFile_j ← End_Time − Start_Time
    end for
    TotalDuration_i ← TotalDuration_i + TFile_j
end for
```

Algorithm 3. Number of times a specific type of file accessed by learners

INPUT: A finite set of Learners L = $L_1, L_2, ..., L_N$ and Sessions S = $S_1, S_2, ..., S_Q$
OUTPUT: Frequency of accessing specific type of file

```
initialize FileType ← NULL, Freq_PDF ← 0, Freq_PPT ← 0, Freq_MP4 ← 0
for each Learner L_i where i ← 1 to N do
    for each Session S_j where j ← 1 to Q do
        if "File" is accessed then
            get FileType
            switch FileType
            case "pdf":
            Freq_PDF ← Freq_PDF + 1
            break
            case "ppt":
            Freq_PPT ← Freq_PPT + 1
            break
            case "mp4":
            Freq_MP4 ← Freq_MP4 + 1
            break
            end switch
        end if
    end for
end for
```

4. **Topic Identification:** Topic identification portal can be analyzed by counting the number of times the specific topic is accessed by all learners in all their respective sessions. Algorithm 4 is used to calculate the count of specific topic.
5. **Page Identification:** Algorithm 5 is used to calculate the count of specific page accessed on portal. There are many different pages available on portal which are related to learner's requirements. Page identification will give count of specific page accessed by learners. These details are useful to get the usage patterns of the learners with respect to portal.

The analysis has been done on IIS log records and Database and result has been discussed through different graphs in next section.

Algorithm 4. Number of times learners accessed specific topic

INPUT: A finite set of Learners L = $L_1, L_2, \ldots, L_N$ Topics T = $T_1, T_2, \ldots, T_M$ and Sessions S = $S_1, S_2, \ldots, S_Q$
OUTPUT: T_{ik} = Number of times one Learner accessed specific $Topic_k$ in all sessions $TopicCount_k$ = Number of times all the Learners accessed specific $Topic_k$

```
initialize T_ik ← 0, T_ijk ← 0, TopicCount_k ← 0
for each Topic T_k where k ← 1 to M do
    for each Learner L_i where i ← 1 to N do
        for each Session S_j where j ← 1 to Q do
            if Topic T_k is accessed then
                T_ijk ← T_ijk + 1
            end if
        end for
        T_ik ← T_ik + T_ijk
    end for
end for
for each Learner L_i where i ← 1 to N do
    TopicCount_k ← TopicCount_k + T_ik
end for
```

Algorithm 5. Number of times learners accessed specific pages on portal

INPUT: A finite set of Learners L = $L_1, L_2, \ldots, L_N$ Pages P = $P_1, P_2, \ldots, P_R$ and Sessions S = $S_1, S_2, \ldots, S_Q$
OUTPUT: P_{ik} = Number of times one Learner accessed specific $Page_k$ in all sessions
$PageCount_k$ = Number of times all the Learners accessed specific $Page_k$

```
initialize P_ik ← 0, P_ijk ← 0, PageCount_k ← 0
for each Page P_k where k ← 1 to R do
    for each Learner L_i where i ← 1 to N do
        for each Session S_j where j ← 1 to Q do
            if Page P_k is accessed then
                P_ijk ← P_ijk + 1
            end if
        end for
        P_ik ← P_ik + P_ijk
    end for
end for
for each Learner L_i where i ← 1 to N do
    PageCount_k ← PageCount_k + P_ik
end for
```

6 Results and Discussion

1. **Analysis done from database: Total time spent on specific file by learners in all their sessions -** Fig. 2 shows that the total time spent on specific file by learners in their all sessions. Learners who have been accessed portal are spent time on files in different sessions. The graph is showing the result of most frequently accessed files by learners in specified duration.
2. **Snapshot of report at instructor side: Number of times a specific type of file accessed by learners -** Fig. 3 shows the report which an instructor can generate to get learner wise count of accessing different types of files. As per implementation, the portal is supporting only for PDF, PPT and Video files. Depending on frequency of accessing specific types of file, learners interest in specific material can be identified.
3. **Analysis done from database: Number of times learners accessed different topics -** Fig. 4 shows the number of times learners accessed the topics with the material. This analysis will give the interest in specific topic and requirement of providing good material on different topics. The analysis can be further captured

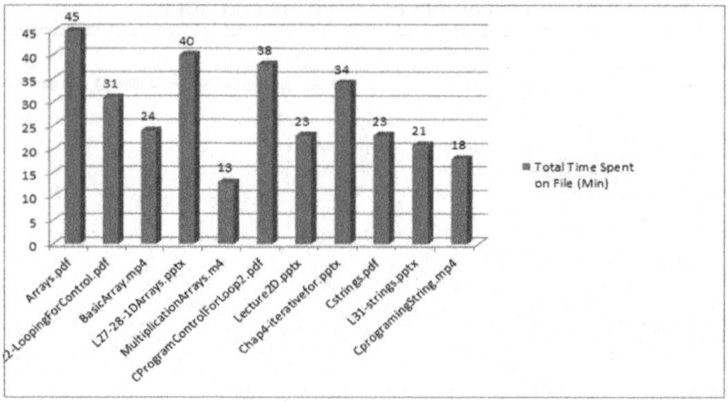

Fig. 2. Time spent on specific file by learners

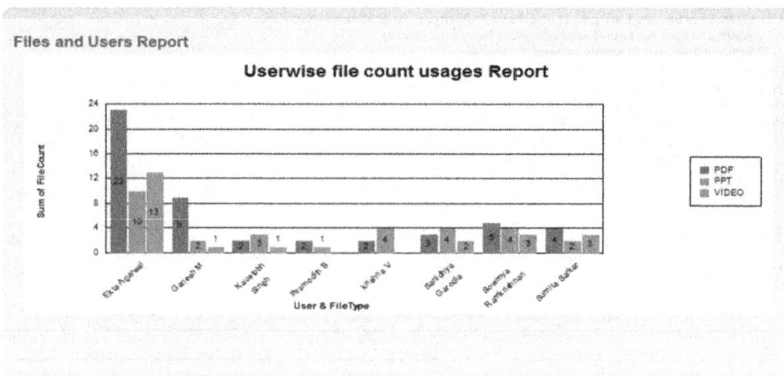

Fig. 3. Number of times Learners accessed specific type of file

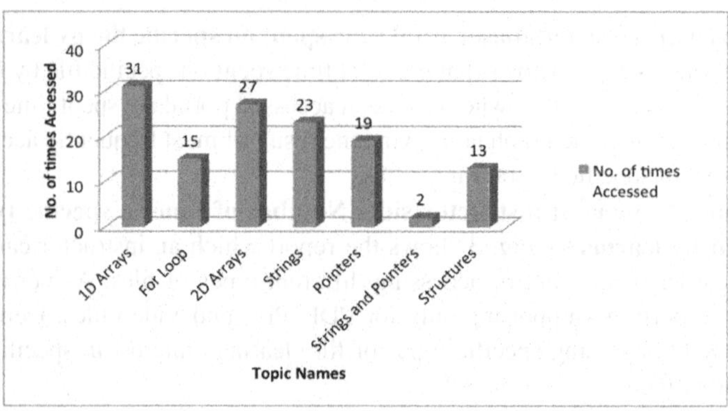

Fig. 4. Analysis of Number of times Learners Accessed Different Topics

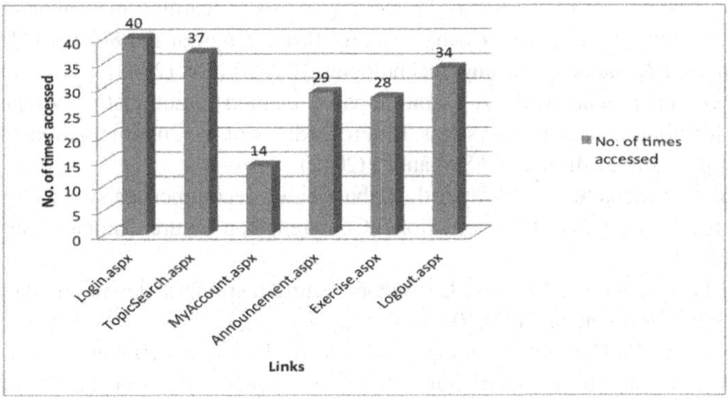

Fig. 5. Analysis of number of times learners accessed different pages

into accessing the topics as per learner's requirement, e.g., how many learners accessed previous concept or advanced concept topic after accessing main topic.

4. **Analysis done from W3C log files: Number of times learners accessed different pages on portal -** Fig. 5 shows that learners have not only accessed TopicSearch to access different files but also accessed other pages such as exercise and announcement pages. This analysis is important to understand the behavior of learner's on portal.

7 Conclusion

Online courses learning cannot be complete without incorporating the learning styles of the learners in e-learning portal which leads to adaptive e-learning approach. In this context, one challenge is to develop a portal which will be able to the identify learning styles of learners and change the user interface components as per learner's requirement.

In the proposed work, the prototype version of e-learning portal has been developed to capture the usage data of learners through log files and database. The initial level of analysis with respect to time spent on different types of files and number of times files accessed is done.

Our method of capturing the learning styles comprises not only IIS log files but also database entries where important factors of learning styles are captured. The usage data is useful to identify the learning styles of the learners and classify the according to FSLSM four dimensions such as Preprocessing, Perception, Input and Understanding and eight categories of mentioned dimensions like Active/Reflective, Sensing/Intuitive, Visual/Verbal and Sequential/Global.

References

1. Popescu, E., Trigano, P., Badica, C.: Relations between learning style and learner behaviour in as educational hypermedia system: An exploratory study. In: Proceedings of 8th IEEE International Conference on Advanced Learning Technologies, ICALT 2008 (2008)

2. Grapf, S., Kishnuk, L.T.C.: Identifying learning styles in learning management system by using indications from student's behaviour. In: Proceedings of the 8th IEEE International Conference on Advanced Learning Technologies, ICALT 2008 (2008)
3. Dung, P.Q. and Florea, A.M.: A literature-based method to automatically detect learning styles in learning management systems. In: Proceedings of the 2nd International Conference on Web Intelligence, Mining and Semantics (2012)
4. Salehi, M., Kmalabadi, I.N.: A hybrid attribute-based recommender system for e-learning material recommendation. In: International Conference on Future Computer Supported Education (2012)
5. Popescu, E., Badina, C., Moraret, L.: Accomodating learning styles in an adaptive educational system. Informatica 34 (2010)
6. Liu, H., Keselj, V.: Combined mining of web server logs and web contents for classifying user navigation patterns and predicting user's future requests. In: Data and Knowledge Engineering, Elsevier (2007)
7. Simi, V. and Vojinovi, O. and Milentijevia, I.: E-learning: Let's look around. In: Scientific Publications of the State University of Novi Pazar Series A: Applied Mathematics, Informatics and Mechanics (2011)
8. Felder, R.M., Silverman, L.K.: Learning and teaching styles in engineering education 78(7), 674–681 (2011)
9. Felder, R.M., Spurlin, J.: Applications, reliability and validity of the index of learning styles. International Journal of Engineering 21, 103–112 (2005)
10. Son, W.M., Kwek, S.W., Liu, F.: Implicit detection of learning styles. In: Proceedings of the 9th International CDIO Conference (2013)
11. Abraham, G., Balasubramanian, V., Saravanaguru, R.K.: Adaptive e-learning environment using learning style recognition. International Journal of Evaluation and Research in Education, IJERE (2013)

A Concept Map Approach to Supporting Diagnostic and Remedial Learning Activities

Anal Acharya[1] and Devadatta Sinha[2]

[1] Computer Science Deparment, St Xavier's College., Kolkata, India
[2] Computer Science and Engg. Deparment, Calcutta University, Kolkata, India
anal_acharya@yahoo.com

Abstract. Due to rapid advancement in the field of computer communication there has been a lot of research in development of Intelligent Tutoring System (ITS). However ITS fails to pinpoint the exact concept the student is deficient in. We propose the development of an Intelligent Diagnostic and Remedial learning system which aims to diagnose the exact concept the student is deficient in. The proposed system composed of three modules is derived from David Ausubel's theory of meaningful learning which consists of three learning elements. The system is implemented in mobile environment using Android Emulator. Finally an experiment was conducted with a set of 60 students majoring in computer science. Experimental results clearly show that the system improves the performance of the learners for whom they are intended.

Keywords: Theory of Meaningful Learning, Remedial Learning, M-Learning, Concept Mapping, Android Emulator, t-test.

1 Introduction

In the recent years there has been tremendous advancement in the field of computer networks and communication technology. This has led to a lot of progress in the field of e-Learning. One of the particular areas of e-learning that has attracted a lot of researchers is ITS. An ITS is a complete system that aims to provide immediate and customized instruction and feedback to the learners without the intervention of human tutor [12]. A significant contribution in this area has been the work of Johnson [10] in which he proposed a authoring environment for building ITS for technical training for IBM-AT class of computers. Vasandani [11] in his work built an ITS to organize system knowledge and operational information to enhance operator performance. There has been a lot of work in ITS in Mobile Learning (M-Learning) environment as well. In 2005, Virvou*et al.*[7] implemented a mobile authoring tool which he called Mobile Author. Once the tutoring system is created it can be used by the learners to access learning objects as well as the tests. Around the same time Kazi [8] proposed Voca Test which is an ITS for vocabulary learning using M-Learning approach. However as identified by Chen-Hsiun Lee *et al.* [6], evaluations conducted via online tests in a ITS do not provide a complete picture of student's learning as they show

M.K. Kundu et al. (eds.), *Advanced Computing, Networking and Informatics - Volume 1*, Smart Innovation, Systems and Technologies 27,
DOI: 10.1007/978-3-319-07353-8_65, © Springer International Publishing Switzerland 2014

only test scores. They do not help the learner identify the exact area where he is deficient. Thus in this work we propose an Intelligent Diagnostic and Remedial Learning System (IDRLS) which will help the learner identify the concepts he is deficient in and what are the related concepts he should revise. Notable examples in this area are the work of Hwang [3] where he has used a fuzzy output to provide learning guidance. [6] has used the Apriori algorithm to generate concept maps which has been used to generate learning guidance.

The theoretical framework for this study refers to the theory of Meaningful Learning proposed by David Ausubel [2] in 1963. If a person attempts to learn by creating a connection to something he already knows then he experiences meaningful learning. On the other hand, if a person attempts to learn by memorizing some information, he attempts rote learning. Thus in meaningful learning, he is able to relate the new knowledge to the relevant concept he already knows. Ausubel thus advocates the concept of reception learning where new concepts are obtained by adding questions and clarifications to the old concepts. For this, he advocates two methods: signaling which indicates important information and advanced organizers which indicates the sequence between these. These psychological foundation led to the development of Concept Maps in Cornell University by Joseph D Novak in 1972 [1]. In brief, let C1 and C2 be two concepts. If learning the concept C1 is a prerequisite to learning concept C2 then the relationship C1 $\rightarrow$ C2 exists [4]. Rounding up, we construct a mapping between Ausubel's learning elements [7,9] to the modules that may be used for implementation of IDRLS. This mapping is displayed in Table 1.

Table 1. Relationship between Ausubel's Learning elements and our proposed implementation

Ansubel's Learning Elements	Ubiquitous Learning Solution	Proposed Modules for implementation
Lesson Organization	Construct Concept Maps	Concept Map Generator Module
Presentation of Learning Task or Material	Design materials that could be used by learners for learning purpose	Learning Module
Strengthening Cognitive Organization	Conduct examination and send feedback to learners in the form of SMS	Evaluation Module

The organization of the paper is as follows. The next section discusses the architecture of proposed IDRLS in details. We then implement this architecture in M-Learning environment using Android Emulator [5]. Finally, we conclude by discussing a set of experiments which validate that Diagnostic and Remedial learning has indeed been useful to the students.

2 System Overview

Fig. 1shows the architecture of the learning system. The learning system resides on My SQL database which is used as a web server. It supports activities like storing learning objects, examination quizzes, test questions etc. The learning system uses Oracle database management system for storing the student model as shown in Fig. 2. The learning system may be accessed from a Computer or a mobile/hand held device. This could be an Android based device.

As indicated in Table 1, the architecture consist of three components: (i) A Hash based algorithm module to generate concept maps (ii) A module to store and access the learning objects and all data relating to learning (iii) A module to conduct tests and store marks.

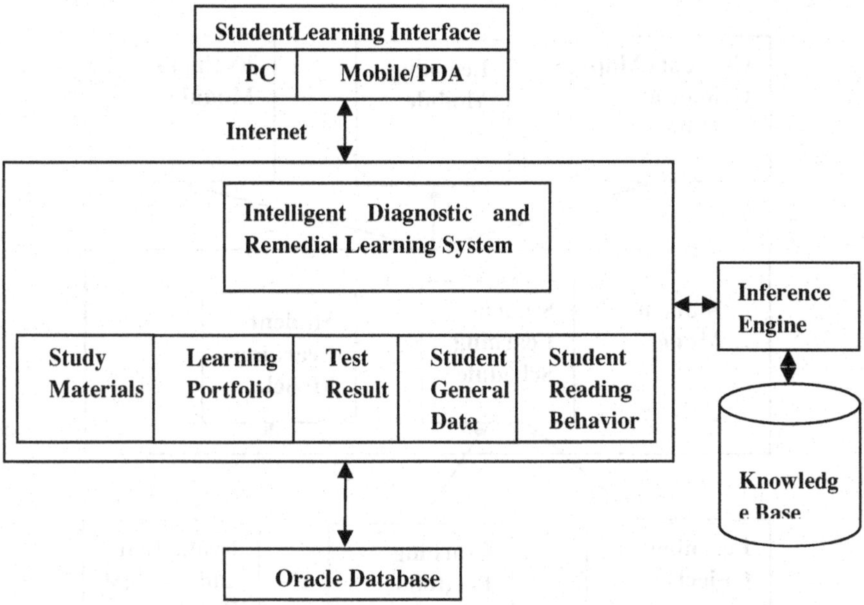

Fig. 1. Architecture of IDRLS

For the first module the inputs required are Test Item Relation Table (TIRT) and Answer Sheet Summary table(ASST) as indicated by [4]. TIRT stores the degree of association between test item Q_i and concept C_j. ASST stores the student's answers for a test. Direct Hashing and Pruning (DHP) algorithm may be applied on these data sets to generate a set of association rules between the concepts along with their weights. Removing the redundant association rules the final concept map is generated. ASST and TIRT is stored in the Knowledge base as shown in Fig. 1. The inference engine generates the Concept Map and stores these in the Knowledge base. These are called Learning Portfolio in totality. The main functionality of the second module is to store the study materials corresponding to each concept. It will also store the student reading behavior. The reading behavior gives an indication of the documents accessed corresponding to each concept and the period of study. It will also store the general student data. This contains basic information relating to the students like enrollment number, their previous academic record, the marks secured in the pretest etc. Classification algorithms may be applied on these data to predict the subset of these who may fail in post test [13]. IDRLS is intended specifically for them. The aim of the third module is to conduct tests and store marks. Thus a question bank is necessary which will store questions, answer choices, degree of difficulty, priority of selection, solution of the question. Test results will store the enrollment number along with the concept name and the marks secured in the examination of the concept. The modules along with data sores are shown in the schematic diagram in Fig. 2.

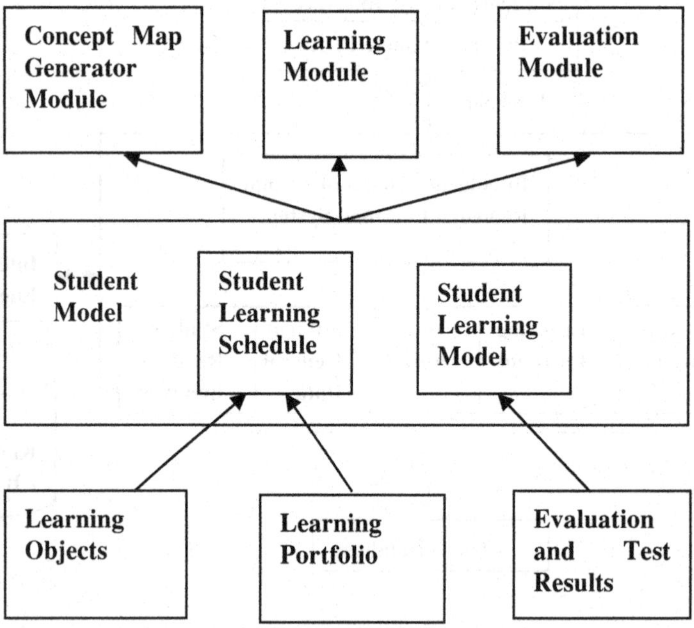

Fig. 2. Functional components of IDRLS

3 Implementation in M-Learning Environment

In this section we develop a simplified implementation of IDRLS in a mobile environment using Android Emulator. 'Introduction to Java Programming' is a course that is taught to most students who major in computer science. The course is divided into a set of concepts and the relationships between these concepts are shown using the concept mapping in Fig. 3.

Table 2. Concept Map Table

Concept id	Concept Name
C1	Variable and Data Type
C2	Operators
C3	Library Functions
C4	Loops

The concept names may be simplified as shown in the Table 2. Due to paucity of space we show some of the records only. The relationship between the concepts are shown in Table 3.

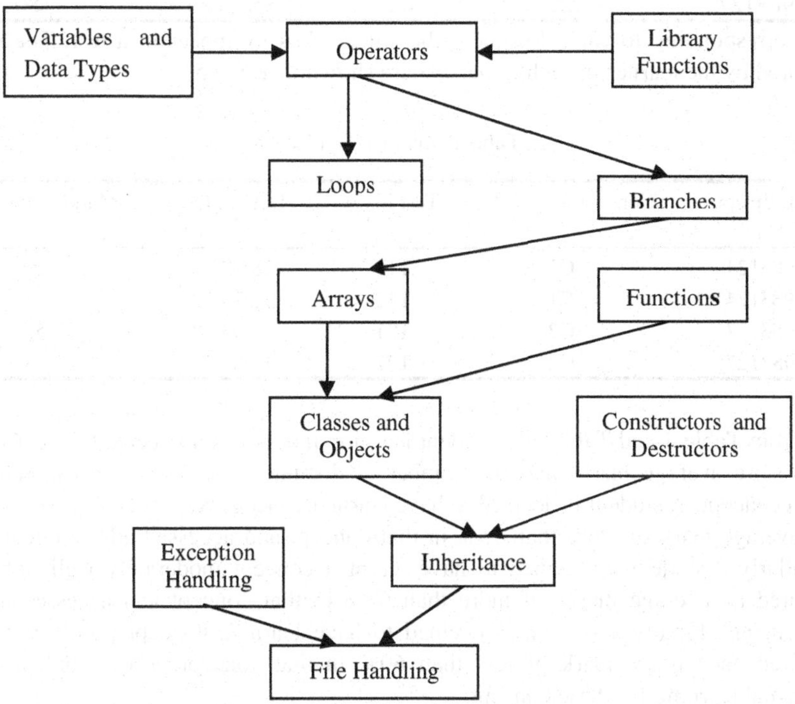

Fig. 3. Concept Map for the course 'Introduction to Java Programming'

Table 3. Concept Relationship table

Parent Concept	Child Concept	Relationship
C1	C2	R1
C2	C4	R2
C2	C5	R3
C3	C5	R4

We next store the learner reading behavior. We assume that each concept can be studied from two documents for the sake of simplicity. The learner is deemed to have studied a concept if he has accessed the two documents. The date of access of each document and duration of study of each is also stored (Table 4).

Table 4. Student Reading Behavior

Enrollment No.	Concept id	Doc id	Date of access	Duration of study(Min)
983124	C1	C11	28/07/13	127
983124	C1	C12	13/08/13	145
983127	C2	C21	23/07/13	176
983127	C2	C22	15/06/13	180

Corresponding to each document the student has to appear in a test. The marks secured by the learner in each of the test are shown in Table 5.

Table 5. Learner Test Result

Enrollment No.	Concept id	Test id	Date of test	Marks Secured
983124	C1	T11	28/08/13	75
983124	C1	T12	13/09/13	65
983127	C2	T21	23/08/13	52
983127	C2	T22	15/07/13	58

From Table 4 and Table5 the student learning model is constructed (Table 6). This will store average marks and the number of documents accessed corresponding to each concept. A student is deemed to have learnt a concept very well if he has secured an average mark of more than 75% in that concept and accessed all the documents. Similarly a student is deemed to have learnt a concept moderately well if he has secured an average marks of more than 60% in that concept and accessed all the documents. Finally a student is deemed to have learnt a concept poorly if he has secured an average mark of less than 60% in that concept. These student needs remedial learning for these concept.

Table 6. Student Learning Model

Enrollment No.	Concept id	Average Marks	Documents Accessed
983124	C1	70	2
983124	C2	55	2

As an example, suppose that a learner has performed poorly in the concept C4. From Table 2 Table 3 it can be immediately deduced that the prerequisite to this concept are concepts C2 and C3. Thus learner is sent an SMS advising him to revise these concepts. A typical table storing these diagnostic SMS is shown in Table 7.

Table 7. Learning Diagnosis

Enrollment No.	Concept Examined	Result	Diagnosis	Mobile No.	Date	Time
983124	C1	Moderately Well learnt	NULL	+919830635250	13/09/13	12:39
983124	C2	Poorly Learnt	Revise concepts C1	+919830635250	23/08/13	17:09

The proposed work has been implemented in a mobile environment using Android Emulator. At the start the learner log into the system using his Email-id and password

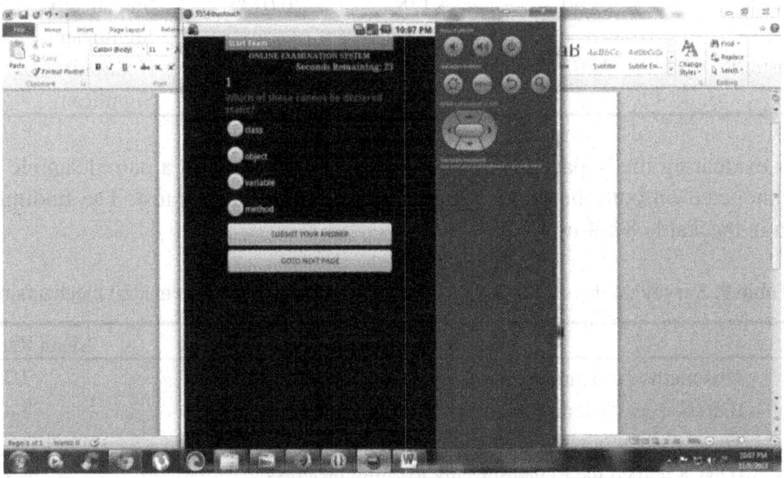

Fig. 4. A typical test question for the concept 'Classes and Objects'

The student then learns the concepts in the pre defined sequence. After learning the concepts document wise the students are to appear in a test. The tests are multiple choice types. A question for a test corresponding to module C9 is shown in Fig.4. These tests are evaluated and the corresponding diagnostic message is SMS to the students.

4 Experiments

An experiment and a survey were conducted to evaluate the effects IDRLS had on the learners. The course 'Introduction to Java Programming' was offered to 60 undergraduate students majoring in computer science under University of Calcutta. The students were given to study in a conventional manner. A pretest was conducted at the end of this learning process. It was found that 18 students failed to pass in this exam. We call them weak students. These students then used IDRLS for remedial learning. At the end of this study, a post test was conducted on both cluster of students (strong and weak) to evaluate the effect IDRLS had on weak students. A survey was also conducted to determine the impact of remediation mechanism offered by IDRLS on these students. The entire scheme is shown in Fig. 5.

Table 8. t-test results of independent samples of two clusters of students.

	Difference in mean	t-value	Two tailed p value	Significance
Pretest	22.60	4.178	0.0139	Statistically Significant
Posttest	7.50	1.8	0.1552	Not Statistically Significant

For evaluating the impact IDRLS had on the weak students, a paired sample t-test was conducted on both these clusters after the pretest and post test. The findings are shown in the Table 8 below:.

Table 9. Survey on the usefulness of IDRLS as a diagnostic and remedial mechanism

Serial	Survey Question	Mean Rating
1.	This method of learning is interesting and challenging.	3.7
2.	IDRLS is user friendly and easy to use.	3.4
3.	The learning contents of IDRLS are useful and up to date.	3.8
4.	IDRLS helped me to measure my learning progress.	4.1
5.	The feedback provided by the diagnostic SMS of IDRLS helped me pinpoint the exact concept I was deficient.	3.9
6.	I would like to use IDRLS for other courses in future.	4.2

The pass mark is taken at 50%.. It is seen that after the pretest the difference in marks between the two clusters is statistically significant whereas after posttest this difference is not statistically significant. This indicates that IDRLS has been successful for remedial learning

Our next objective is to conduct a survey to find out the usefulness of IDRLS on the failed students. Students were given a set of questions and then asked to give their feedback using a 5 point Likert scale where 1 denotes strongly disagree, 2 denotes disagree, 3 denotes neutral, 4 denotes agree and 5 denotes strongly. The survey questions and their results are shown in Table 9.The above results show that IDRLS has had an impact on the failed students

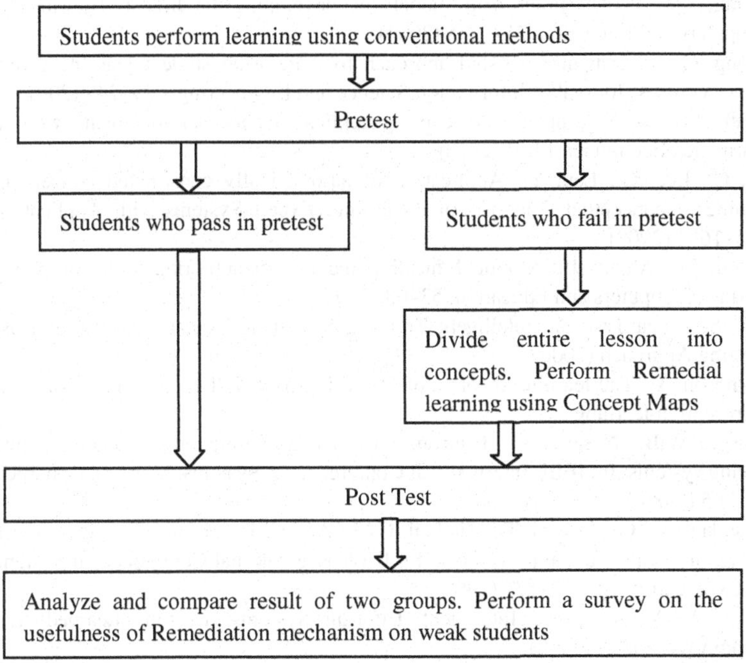

Fig. 5. Experimental Procedure

5 Conclusion

In this work we propose an Intelligent System which can identify the concepts the students are weak in and suggest remedial lesson plan for these. The developed system consists of three modules which are derived from Ausubel's theory of meaningful learning. A simplified version of this is implemented in M-Learning environment using Android Emulator. The students who use this system provided a good feedback of it. Thus the major research objective of this work has been twofold: Firstly, to develop a diagnostic and remedial system based on the psychological foundations of learning and implement it on a certain platform and more significantly, to verify and validate that this form of remediation has indeed been useful for the students. We plan to extend the M-Learning architecture so that this mechanism may be accessed from PC s' as well.

References

1. Novak, J.D., Alberto, C.J.: Theoretical Origins of Concept Map, How to construct them and their used in Education. Reflecting Education 3(1), 29–42 (2007)
2. Pendidican, F.: Learning Theories. Mathematika dan Ilnam Alam, Universitas Pendidikan Indonesia

3. Hwang, G.: A conceptual map model for developing Intelligent Tutoring Systems. Computers and Education, 217–235 (2003)
4. Hwang, G.: A computer assisted approach to diagnosing student learning problems in science courses. Journal of Information Science and Engineering, 229–248 (2007)
5. Pocatilu, P.: Developing mobile learning applications for Android using web services. Informatica Economica 14(3) (2010)
6. Lee, C., Lee, G., Leu, Y.: Application of automatically constructed concept maps of learning to conceptual diagnosis of e-learning. Expert Systems with Application 36(2), 1675–1684 (2009)
7. Virvou, M., Alepis, E.: Mobile Education features in authoring tools for personalized tutoring. Computers and Education, 53–68 (2005)
8. Kazi, S.: Voca Test: An intelligent Tutoring System for vocabulary learning using M-Learning Approach (2006)
9. Thompson, T.: The learning theories of David P Ausubel. The importance of meaningful and reception learning
10. Johnson, W.B., Neste, L.O., Duncan, P.C.: An authoring environment for intelligent tutoring systems. In: IEEE International Conference on Systems, Man and Cybernetics, pp. 761–765 (1989)
11. Vasandani, V., Govindaraj, T., Mitchell, C.M.: An intelligent tutor for diagnostic problem solving in complex dynamic systems. In: IEEE International Conference on Systems, Man and Cybernetics, pp. 772–777 (1989)
12. Psotka, J., Mutter, S.A.: Intelligent Tutoring Systems: Lessons Learned. Lawrence Erlbaum Associates (1998)
13. Sai, C.: Data Mining Techniques for Identifying students at risk of failing a computer proficiency test required for graduation. Australasian Journal of Educational Technology 27(3), 481–498 (2012)

Data Prediction Based on User Preference

Deepak Kumar[1], Gopalji Varshney[2], and Manoj Thakur[1]

[1] School of Basic Science
Indian Institute of TechnologyMandi, India
[2] Hindustan College of Science and Technology, Mathura, India
{deepaktyagi12,manojpma,gopalji.varshneya}@gmail.com

Abstract. Recommender systems are admittedly the widely used applications over the internet E commerce sites and thus the success of a recommender system depends on the time and accuracy of the results returned in response to information supplied by the users.Now-a-days, many big E commerce systems and even social networking portals provide the facility of recommendation on their sites, thus underscoring the demand for effective and accurate recommendation system. But still most of the recommender systems suffer from the problem of cold start, sparsity and popularity bias of the provided recommendation. Also, these recommender systems are unable to provide the recommendation to someone with unique taste. This paper describes a preference based recommender system and collaborative filtering approach which is used to solve prediction and recommendation problem. The collaborative filtering aims at learning predictive model of user interests, behavior from community data or user preferences. It describes a family of model-based approaches designed for this task. Probabilistic latent semantic analysis (PLSA) was presented in the context of data retrieval or text analysis area.This work can be used to predict user ratings in the recommendation system context. It is based on a statistical modeling technique which introduces a latent class of variables in a mixture model setting to discover prototypical interest profiles and user communities. The main advantages of PLSA over memory-based methods are an explicit, constant time prediction and more accurate compact model. It can be used to mine for the user community. The experimental results show substantial improvement in prediction accuracy over existing methods.

Keywords: Recommender systems, Information Search and Retrieval, Information filtering, Probabilistic Latent Semantic Analysis, Collaborative Filtering, mixture models, latent semantic analysis, machine learning.

1 Introduction

Collaborative filtering (CF) [5] uses the notions of other users that are similar to an active user, as a filter.Collaborative filtering (CF) are corresponding to content based filtering and the main aim of learning models of user interests or preferences, activities from community data, that is, a data of available user preferences.The user input or interaction beyond the profile created from previous interactions and annotations is

M.K. Kundu et al. (eds.), *Advanced Computing, Networking and Informatics - Volume 1*,
Smart Innovation, Systems and Technologies 27,
DOI: 10.1007/978-3-319-07353-8_66, © Springer International Publishing Switzerland 2014

not mandatory.Until now, the dominant model for performing collaborative filtering in recommendation systems (RS) [6] has been based on memory-based techniques or nearest neighbor regression.All first generation RS uses the same fundamental approach first identifying users that are alike to some active user for which recommendations have to be fixed, and then compute predictions and recommendations based on judgments and the predilections of these like-minded or similar users.

Content-based filtering (CBF) [7], [12] and retrieval builds on the essential postulation that users are able to create queries that express their information or interests needs in term of the essential characteristics of the items needed.It is difficult to identify appropriate descriptors such as themes, keywords, genres, etc., which can be used to define interests. In some cases, for example E- commerce, users may be at least inattentive or unacquainted of their interest. In both cases, it would like to recommend items and predict user likings without requiring the users to explicitly formulate a query.

The technique proposed is a generalization of a statistical technique is known as probabilistic Latent Semantic Analysis (PLSA) [10] that was originally investigated for information retrieval [11]. PLSA has some similarity with clustering methods in which latent variables for user community are introduced, yet the community would be overlapped and users are not partitioned. It is closely related to matrix decomposition techniques and dimension reduction technique such as Singular Value Decomposition (SVD), which have been used in the context of recommendation systems [12], [13].The main differences with regard to dependency networks and Bayesian network is the fact that the latter learning structures directly on the observable, while PLSA is based on a latent basis model that introduces the notion of user communities or group of items.The advantage over SVD and PCA based dimension reduction methods is this approach offers can build onstatistical techniques for inference, a probabilistic semantics and model selection. However, this approach shares with all the above techniques an assumption that predictions are calculated in the "user-centric" perspective.

The paper also describes the implementation of probabilistic latent semantic indexing (PLSI) [2] and their application on a collation of movies data set. The recommender system is to make available the user with recommendations that reflect the user's personal interest and to induce the user to expectation and explore the given recommendations. Recommender system that creates personal recommendations attains their goal by maintaining profiles forthe users that have their own preferences. The user profiles are used as filters; items that match the user's preferences will slide through and be presented as recommendations.The systems that produce individual recommendations as output and an effective guide tousers in the personalized way to fascinating or suitable objects in a large space of alternatives.

2 Model for Collaborative Filtering

In conventional recommender systems, there is a collection of item and information about user interest in the form of rating giving by the user to the particular item. After doing some preprocessing (rating out of five, user profile, etc.) over the item in the data set, we create a user / item matrix [1], [4] which act as a representation of demand of items and it is considered to be the most appropriate form of input for most of a recommender systems. The anatomy of a conventional recommender system is depicted in Fig. 1.

The recommender system returns the list of items desired by the user against the user interest based on a Similarity Function [8], which evaluates whether a particular item is relevantly or similar to the item liked by the user. If the item is found to be relevant or highly similar with respect to the user's items, the items are added to the final list of recommended items to be returned as results.

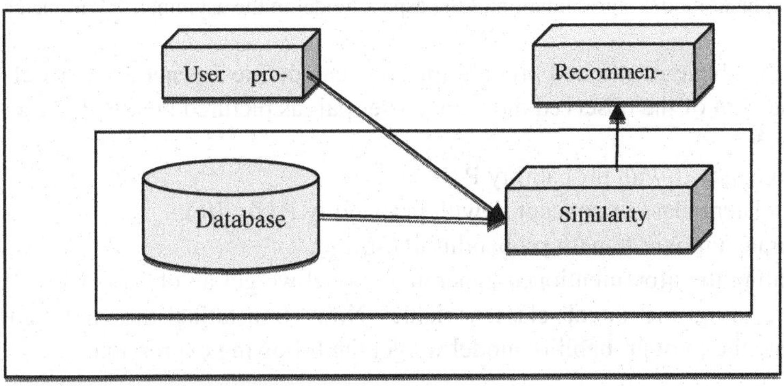

Fig. 1. Anatomy of a conventional recommender system

To improve the accuracy of the recommender system, researchers have tried to cluster similar items to improve the quality of the prediction results. In this case, however the item is not matched individually with each item but it is matched with cluster cancroids. The clusters that are extremely similar to a given item is returned as the result.

We are trying to do similar things, where we are attempting to create clusters of items contained in the items collected by drawing out the relationship between different items. Once create the clusters, it can be used for improving the result of recommender system. Hence we are also able to solve the sparsity and popularity bias problems.

3 The Aspect Model

The kernel of PLSA is a statistical model designated as the aspect model. The aspect model has been formerly proposed in the domain of linguistic modeling, where it was referred as an aggregate Markov model. Assuming an observation being the appearance of a specific word in a particular user the aspect model associates a hidden or latent class variable (denoting the concepts) $Z_k \in \{Z_k, \ldots\ldots, Z_k\}$With each observation [2], [3]. We also acquaint ourselves with the following probabilities:

- $P(U_i)$The probability that the occurrence of a user will be observed in a specific user U_i.
- P $(M_i|Z_k)$The conditional probability that a specific movie $M_i$will be conditioned on the latent class variable Z_k.
- P $(Z_k\ |U_k)$ User specific probability distribution over the inherent concept space.

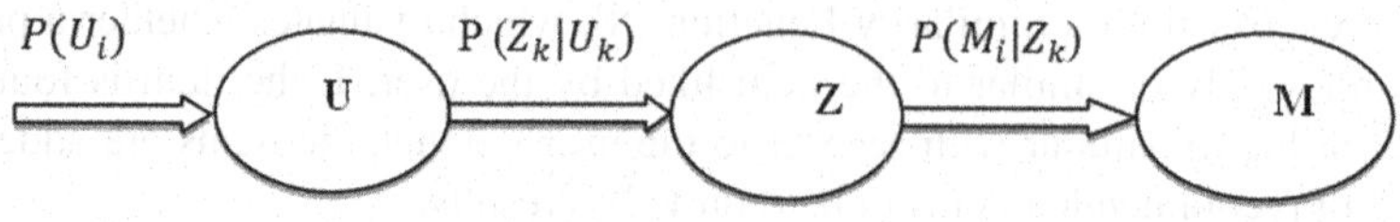

Fig. 2. Graphical model representation of the aspect model in the asymmetric Parameterization

Using the above mentioned probabilities one can define a generative model for the accompaniment of the observed movie and user pair as pictured below along with Fig. 2 and Fig. 3.

- Select a user $U_i$with probability P (U_i).
- Pick a latent class or concept $Z_k$with probability P $(Z_k\ |U_k)$.
- Generate a movie $M_i$with probability P $(M_i|Z_k)$.

Translating the aforementioned generative model we get an observed pair (U_i,M_i), while discarding the latent class variable. Now, translating the aforementioned process into the joint probability model we get the following expression.

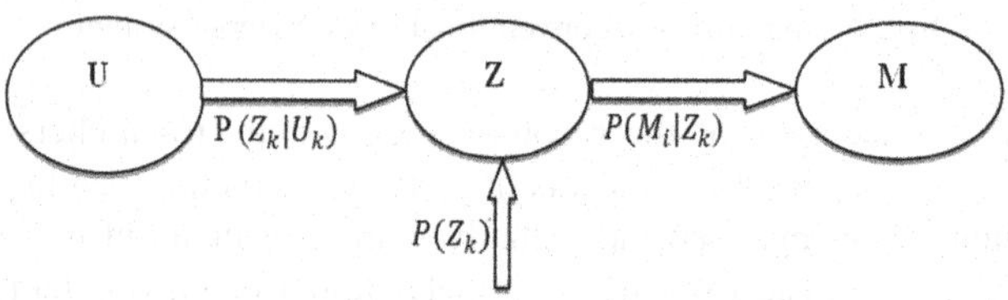

Fig. 3. Graphical model representation of the aspect model in the symmetric parameterization

$$\mathrm{P}\ (U_i,M_i) = \mathrm{P}\ (U_i)\ \mathrm{P}\ (M_i|U_i) \tag{1}$$

where, $\mathrm{P}\ (M_i|U_i) = \sum_{z=1}^{k} \mathrm{P}\ (M_i|Z_k)\mathrm{P}\ (Z_k\ |U_k)$,

To determine the joint probability of the observed pair as in Eq. 1, we perform a summation over all possible choices Z_k. Aspect model embodies a conditional independence assumption similar to most of the latent class models, that the movie and use are generated independently, but conditioned on the topic variable or the latent class variable [12]. Also, the conditional probability distribution$P(M_i|U_i)$ is visualized as the convex combination of the K aspects $P(M_i|Z_k)$. Now applying the likelihood principle, we could obtain P (U_i), P $(M_i|Z_k)$ and P $(Z_k\ |U_k)$ by the maximization of the following log-likelihood function [4].

$$£=\sum_{i=0}^{X}\sum_{i=0}^{Y} n(U_i, M_i) \log P(U_i, M_i) \quad (2)$$

where$n(U_i, M_i)$ denotes the count or term frequency of jth word in the ith document. The symmetric version of the aspect model can also be represented by using Bayes' Theorem to reverse the conditional probability $P(Z_k|U_k)$

$$P(U_i, M_i)=\sum_{k=0}^{K} P(Z_k)P(M_i|Z_k)P(U_i|Z_k) \quad (3)$$

Another simplifying assumption incorporated in the aspect model concerns the cardinality of the latent class variable, which must be a comparatively smaller quantity than both terms and documents, i.e. K << min (X, Y) and the principle behind such an assumption is that when satisfied the latent class variable K acts as a bottleneck variable towards a better prediction of terms [9]. Fig. 4 shows the plate notation model of PLSA where the shaded circles denote the observed variable and the non-shaded circle denotes the latent variable.

The standard course of action followed for maximum likelihood estimation when latent variables are involved is Expectation Maximization (EM) algorithm [4] which iterates two steps until the model converges. The two steps involved are:

- *E-Step* or expectation step computes the posterior probability estimates of the latent variables provided the current best estimates of the involved parameters.
- *M-Step* or maximization step modifies the parameters on the basis of expected completion data log-likelihood function which depends on the posterior probability computed in the E-Step.

In PLSA, the posterior probability of the latent variable computed in the E-Step is as follows:

$$P(Z_k|U_i, M_i) = \frac{P(Z_k)P(M_j|Z_k)P(U_i|Z_k)}{\sum_{a=1}^{k} P(Z_a)P(M_a|Z_a)P(U_a|Z_a)} \quad (4)$$

In M-Step we finally get the following updated values of the parameters:

$$P(M_j|Z_k) = \frac{\sum_{i=1}^{x} n(u_i,M_j)P(Z_k|U_k,M_j)}{\sum_{b=1}^{y}\sum_{i=1}^{x} n(u_i,M_b)P(Z_k|U_i,M_b)} \quad (5)$$

$$P(U_i|Z_k) = \frac{\sum_{j=1}^{y} n(u_i,M_j)P(Z_k|U_k,M_j)}{\sum_{c=1}^{x}\sum_{j=1}^{y} n(u_c,M_j)P(Z_k|U_c,M_j)} \quad (6)$$

$$P(Z_k) = \frac{1}{\sum_{i=1}^{x}\sum_{j=1}^{y} n(u_i,M_j)}\sum_{i=0}^{X}\sum_{i=0}^{Y} n(u_i, M_j)P(Z_k|U_k, M_j) \quad (7)$$

The aforementioned E and M-Step is alternated until some stopping criterion is met, which could be a convergence condition such as a threshold. Early stopping could also be an alternative for stopping criterion where one stops when updating the parameters of training data is not affecting the performance. Approaching a local maximum of the log-likelihood function in aspect model is guaranteed but approaching the global maximum is not sure, but depends extensively on the data in use and the initial values of the parameters [3].

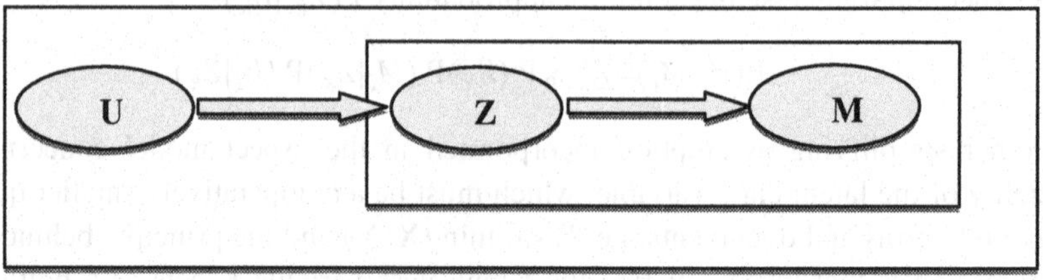

Fig. 4. Graphical model representation of the aspect model in the symmetric

4 Implementation Details

The paper aims to present an explicit comparative study of different successful approaches being deployed in diverse Recommender systems, and to make it viable there is a need for a suitable software framework, where these approaches can be tested on various platforms and it must be deployed in a way to support experimentation. Theapproach discussed in the paper is implemented in Java andon a standard PC with 2.5GHZ CPU and @GB RAM. The data set used in implementation taken from The Movie Len. Before proceeding further with the results, the next Section gives a very concise outline of the data set used to test and experiment.

4.1 Dataset

The Movie data set being employed for the experimentation is a collection of movie dataset. This data set consists of:

- Demographic information for the users (Name, sex (M/F), age, occupation, zip).
- 100,000 from 943 users on 1682 movie ratings (1-5).
- A useris considered as an active user if he/she has rated at least 20 movies.

The data collected through the MovieLens web site during the seven-month period from September 19th, 1997 through April 22nd, 1998. This data have been represented according to the requirement– users who did not complete profile information or had less than 20 ratings were not considered.

Table 1. From K = 15, The clusters obtained from PLSA

Interested Group [IG]	Movie Name	Movie Name	Movie Name	Movie Name	Movie Name	Movie Name
IG[1]	Posino	Bad Boys	Antnio's Line	Desperdo	Quiz Show	Blade Runner
IG[2]	Pulp Fiction	Exotica	Net	Forrest Gump	Flipper	Twister
IG[3]	Love Bug,	Lone Star	Ghost and the Darness	Swingers	Kansas City	Basic Instinct
IG[4]	GoldnEye	Bravheart	Batman Forever	Nadja	Dolores Claiborne	Lion King,
IG[5]	The Mask	Natural Born Killers	Dolores Claiborne	Copycat	Stargate	Faster Pussycat! Kill! Kill
IG[6]	Heavy Metal	Toy Story	Die Hard	Jude	Richard III	Strange Days
IG[7]	Fargo	Striptease	Kansas City	The Ab-yss	Cinema Paradiso	Top Gun
IG[8]	Mary Popins	Cinderella	The Sound of Music	Dumbo	Psycho	Resevoir Dogs
IG[9]	Aristocats, The	Nikita	Dead Poets Society	Field of Dreams	Diabolique	Chasing Amy
IG[10]	Moll Flanders	Ridicule	Home Alone	Fifth Element, The	Unforgiven	Men in Black
IG[11]	Godfather, The	Akira	Blues Brothers	Platoon	Cinema Paradiso	Cold Comfort Farm
IG[13]	Secrets & Lies	Fly Away Home	Paradise Lost	Devil's Advocate, The	Murder at	Dances with Wolves
IG[14]	The Re-mains of the Day	The Piano	Titanic	The Sound of Music	Dumbo	Bio-Dome
IG[15]	Murder	Schindler's List	Showgirls	Lost Highway	3 Ninjas:	Donnie Brasco

5 Results and Analysis

The decomposition of user ratings may lead to the discovery of interesting patterns and regularities that describe user interests as well as disinterest. To that extent, we have to find a mapping of a quantitative PLSA model into a more qualitative description suitable for visualization. We propose to summarize and visualize each user community, corresponding to one of the k possible states of the latent variable Z, in the following way. We sort items within each community or interest group according to their popularity within the community as measured by the probability $P(y|z)$. The highly popular items are used to characterize a community and we anticipate these items to be descriptive of the types of items that are relevant to the user community.It can also be understood that the accuracy of the model is quite miserable without the user-specific, which is a crucial component of the proposed model.It is also quite significant that this effect is obtained with an example involving a relatively small number of communities. Table 1 displays the interest groups (Intested Group= GI) extracted by a multinomial PLSA model with $k = 15$. We conclude from this that the assumption of the user-specific rating scales encodes useful prior knowledge.

Overall, we believe that patterns and regularities extracted with PLSA models can be helpful in understanding shared interests of users and correlations among ratings for different items. The ability to automatically discover communities as part of the collaborative filtering process is a trait which pLSA shares with only few other methods such as clustering approaches, but which is absent in all memory-based techniques.

6 Conclusion

Prediction is one the greatest challenges faced by researchers in Recommender system. Since the paper concerns itself with the effective extraction of the statically structure of a collection of items and hence focuses on the pattern of interest. This motivated us to experiment with the dataset which contains movies, talking in different type and here we show the result of the application of these approaches to avoid the problem of cold start, sparsity and popularity bias of the product recommendation.

This paper presents the results obtained by the application of previous approachesbeing used in recommender systems and that of Probabilistic Latent Semantic Indexing along with the results it also presented the analysis of the results.A descriptive, comparative analysis of the recommender filtering techniques like collaborative filtering (LSA), content based filtering and PLSA was done and the results elaborated and studied on various platforms such as clustering, Log-Likelihood values and convergence.This also presents the results and analysis of these advances. A close examination of the displayed results suggests that PLSA is a better approach than LSA and far better than other content base filtering in various aspects. We get faster and good results with a high degree of optimization with PLSA, whereas LSA produces good results, but is a very slow process and suffers from the over fitting.

Acknowledgements. All the companies/products/service names are used for identification purposes and may be the trademarks of their respective owners. I also want to give my thanks to MovieLens web to providing movies dataset for my research experiment.

References

1. Burke, R.: Hybrid recommender systems: Survey and experiments. User Modeling and User-Adapted Interaction 12(4), 331–370 (2002)
2. Hofmann, T.: Probabilistic Latent Semantic Indexing. In: Annual ACM Conference on Research & Development in Information Retrieval, pp. 50–57 (1999)
3. Landauer, K., Foltz, P.W., Laham, D.: An Introduction To Latent Semantic Analysis. Discourse Processes 25, 259–284 (1998)
4. Bilmes, J.A.: A Gentle Tutorial of the EM Algorithm and its Application to Parameter Estimation for Gaussian Mixture and Hidden Markov Models. International Computer Science Institute (1998)
5. Adomavicius, G., Sankaranarayanan, R., Sen, S., Tuzhilin, A.: Incorporating Contextual Information in Recommender Systems Using a Multidimensional Approach. ACM Transactions in Information Systems 23(1) (2005)
6. Ansari, A., Essegaier, S., Kohli, R.: Internet Recommendations Systems. Journal of Marketing Research, 363–375 (2000)
7. Baeza-Yates, R., Ribeiro-Neto, B.: Modern Information Retrieval. Addison-Wesley (1999)
8. Deshpande, M., Karypis, G.: Item-Based Top-N Recommendation Algorithms. ACM Transactions on Information Systems 22(1), 143–177 (2004)
9. Billsus, D., Brunk, C.A., Evans, C., Gladish, B., Pazzani, M.: Adaptive Interfaces for Ubiquitous Web Access. Communications of the ACM 45(5), 34–38 (2002)
10. Buhmann, M.D.: Approximation and Interpolation with Radial Functions. In: Dyn, N., Leviatan, D., Levin, D., Pinkus, A. (eds.) Multivariate Approximation and Applications, Cambridge University Press (2001)
11. Burke, R.: Knowledge-Based Recommender Systems: Encyclopedia of Library and Information Systems. In: Kent, A. (ed.), vol. 69 (supple. 32). Marcel Dekker (2000)
12. Adomavicius, G., Tuzhilin, A.: Toward the next generationof recommender systems: A survey of the state-of-the-art and possible extensions. IEEE Transactions on Knowledge and Data Engineering 17(6), 734–749 (2005)
13. Caglayan, A., Snorrason, M., Jacoby, J., Mazzu, J., Jones, R., Kumar, K.: Learn Sesame - A Learning Agent Engine. Applied Artificial Intelligence 11, 393–412 (1997)

Automatic Resolution of Semantic Heterogeneity in GIS: An Ontology Based Approach

Shrutilipi Bhattacharjee and Soumya K. Ghosh

School of Information Technology
Indian Institute of Technology Kharagpur, India
shrutilipi.2007@gmail.com, skg@iitkgp.ac.in

Abstract. To facilitate the access of geographic information, spatial data infrastructures (SDI) are being set up within regions, countries and internationally. Consequently, the interoperable access of the information across disparate geospatial databases has become essential. *Semantic heterogeneity* is one of the crucial issues to get resolved for efficient and accurate retrieval of spatial data. This work has adopted an ontology based approach to overcome the *semantic heterogeneities*, usually exists in different heterogeneous data repositories. The proposed method is used for *semantic* matching between user query concepts and the target concepts in the spatial databases. It overcomes the problem of manual intervention during matching process; hence making it automated.

Keywords: Geographic Information System, Semantic heterogeneity, Ontology, WordNet.

1 Introduction

Geospatial information is the key for any decision support system in the geographic information system (GIS) domain [1]. The discovery of suitable data sources by keyword based search in the catalogue becomes inaccurate due to the existing heterogeneity. Heterogeneity in the database can be categorized as *syntactical heterogeneity*, *structural heterogeneity* and *semantic heterogeneity*. The *structural* and *syntactical* interoperability in the database can be ensured by standardizing data model (metadata) or meta-information regarding the data format and structure. *Open Geospatial Consortium* (OGC) has standardized *geographic markup language* (GML) based application schema as the basis for sharing and integrating spatial data at service level. Although the specified standards ensure the *syntactic* interoperability in the data integration process, but unable to address the *semantic heterogeneity* issue. In this regard, an ontology based approach can be introduced to resolve *semantic heterogeneity* in GIS [2][3]. It is a key factor for enabling interoperability in the *semantic* web by allowing the GIS applications to agree on the spatial concepts semantically when communicating with each other. Ontology can be regarded as "*a specification of a conceptualization. A conceptualization is an abstract, simplified view of the world that we wish to represent for some purpose.*"

M.K. Kundu et al. (eds.), *Advanced Computing, Networking and Informatics - Volume 1*, Smart Innovation, Systems and Technologies 27,
DOI: 10.1007/978-3-319-07353-8_67, © Springer International Publishing Switzerland 2014

Many researches have been attempted in this field, however, there exists an enormous scope for further work. The major hindrance in sharing the data across the organization in the GIS application is their own proprietary standards for data formats. Some initial attempts were made to obtain GIS interoperability which involves the manual conversion of geographic data from one proprietary format into another. Otherwise some centralized standard file format can be used for this practice. However, these formats can also lead to some information loss [4]. Buccella *et al.* [5] have studied different geographic information exchange formats for the integration of data sources with the help of ontologies for representing the *semantics.* Same is done through service oriented architecture in [6]. Yue *et al.* [7] have addressed the issue of geospatial service chaining through *semantic* web services for improving the automation. The tools like OntoMorph, PROMPT [8] etc. have also tried to automate the process of *semantic* query resolution. However, the component that automatically identifies the *semantically* similar concepts in the database, is also missing here. In this regard, the main objectives of this work can be stated as follows,

- Building the spatial feature ontology with all possible features in the available data repositories.
- Three level semantic matching (database, relational, and attribute based) between the user query concept and the concepts in the ontology.

This work focuses on the requirement of dynamic mapping between *semantically* similar concepts of the user query and the spatial databases. It facilitates the *semantic* search using some thesaurus based word relators like *WordNet* [9], Nexus [10] etc. An algorithm has been proposed which iteratively searches for the similar concepts from any database in the pool of repositories, by consulting the ontology, by facilitating the semantic search. Three levels of checking is introduced here. They are in database, relation and attribute level. *Semantic* matching is done for all the three levels in succession until a proper match is found. This approach is more efficient than the traditional matching with spatial features, represented as the relational schema in the databases.

This paper is organized in four sections. Section 1 gives an overview about the *semantic heterogeneity* related to GIS. For automated ontology based *semantic* matching between spatial features in the data repositories and the query concept, a suitable framework and along with a methodology is presented in the Section 2. Corresponding implementation procedure and the results are shown in Section 3. Finally, the conclusion is drawn in Section 4.

2 Proposed Semantic Matching Framework

The proposed framework is composed of two main components and their sub components; they are the *geospatial data repositories* and the *semantic query resolution engine.* The framework is depicted in Fig. 1.

In this framework, the *geospatial data repositories* consist of a pool of spatial data sources, provided by different stakeholders which are heterogeneous in na-

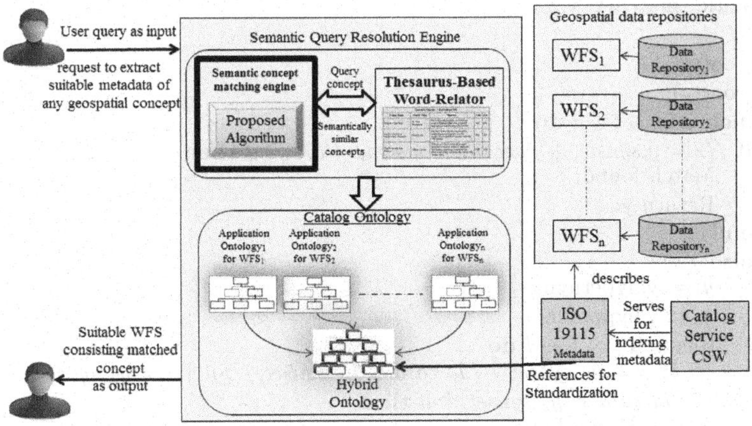

Fig. 1. Semantic matching framework

ture. Each repository publish its own *featureDescription* through *Web Feature Service* (WFS)

which ensures *structural* and *syntactic* interoperability (as GML is the standard) but *semantic heterogeneity* issues still exist. All these features are populated in the catalogue for suitable indexing and the retrieval of data which can be extracted through *catalog service for the web* (CSW)[2]. *Semantic query resolution engine* consists of two components, a *semantic concept matching engine* and a *thesaurus based word relator* which provide spatial vocabularies and their *semantic* associations. *Semantic concept matching engine* is where the proposed algorithm resides. When a user requests to retrieve the suitable metadata for any geospatial feature, it will be processed via a *semantic query resolution engine.* Here, the proposed algorithm consults the thesaurus based word relator to retrieve all similar concepts from the databases. In this study, *WordNet* is used which can be interpreted as a lexical ontology. It then identifies the similar concept in one or more than application ontology that corresponds to a specific WFS of a particular data source. Once the semantically similar concept is found, the actual data can be fetched that corresponds to the user query. The details of the *semantic* matching method is described in Algorithm 1. Some useful terminologies related to this algorithm are *synonym*[4], *hyponym*[4], *hypernym*[4], *meronym.*

First, the direct matching of the query concept with the database based on keyword based searching is done. If any matches found, it directly returns that concept as the output. Else, it calls Algorithm 1 with three arguments (x, y, z), where x is the input concept, y is the corresponding level (e.g., database, relation, attribute, etc.) of database where the *semantic* matching is supposed to be carried out and z is the whole *WordNet* dictionary. In *semantic* matching process, the concept is matched with the *synset* and *gloss* of each of y in *WordNet.* The *jaccard coefficient* is used to calculate the similarity between x and y. If no match found, a threshold value l is chosen. The *hypernym* or *hyponym* of y

```
Semantic_match(x, y, z)
{
x = root(x);
y = root(y);
foreach y.synset ∈ z do
    if ((x = y.synset) || Jaccard(x, y.semantic)) ≠ 0 then
        Match found;
        Return y;
    end
    else
        y1 = y.hypernym;
        y2 = y.hyponym;
        foreach i = 1 → l do
            if x = y1.synset || (Jaccard(x, y1.synset)) >0 || x = y2.synset ||
            Jaccard(x, y2.synset) >0 then
                Match found;
                Return y;
                Break;
            end
            else
                y1 = y1.hypernym;
                y2 = y2.hyponym;
            end
        end
        if i = l + 1 then
            Continue;
        end
        else
            Break;
        end
    end
end
if all y.synset is processed & no match found then
    Return NULL;
end
}
```

Algorithm 1. Semantic matching using word relator

upto l level is matched with x. If this level also could not find any match, it tries to match x in its subsequent level, otherwise y's corresponding concept is returned as the output. If the query concept is not found to be matching with any attributes of the concepts in the repository pool also, it is assumed that no database matches with query concept and the algorithm returns the output as NULL.

3 Implementation and Results

This section identifies some implementation details for the evaluation of the proposed algorithm. *Protégé 3.3.1* [11] is used as the ontology editor. OntoLing is a *Protégé* plug-in that allows for linguistic enrichment of the ontology concepts. The ontology used for this case study is built from the spatial features related to the *Bankura* district of West Bengal, India. It consists of the spatial features like *road, land coverage, water-body, facility* etc. These features are considered as the concepts in the ontology hierarchy and divided into sub-concepts. The hierarchy along with its instances is shown in Fig. 2.

The *synonyms, hypernym* and *homonyms* of the query string are populated in *rdf:label* of each of the concept in the ontology. This matching process is same as the relational level matching. *rdf:comment* represents the meaning of the database concepts. The *jaccard similarity* measure corresponds to the matching with the attribute values. Ontology is updated with all *synonyms, hypernyms, homonyms* and *meronyms* using OntoLing plug-in. For this study, the threshold value l is taken as 2; i.e., upto 2nd level *hypernyms* and *hyponyms* of the query string in the *WordNet* tree, are considered for matching, however, this threshold can be varied as per the application requirement. Table 1 shows some example cases of *semantic* matching for this given case study. Fig. 3. depicts the *semantic* matching process in *Protégé* for the query string "Highway". Proposed method returns the spatial feature "Road" as the matched concept in the database, as "Highway" is the *hyponym* of "Road".

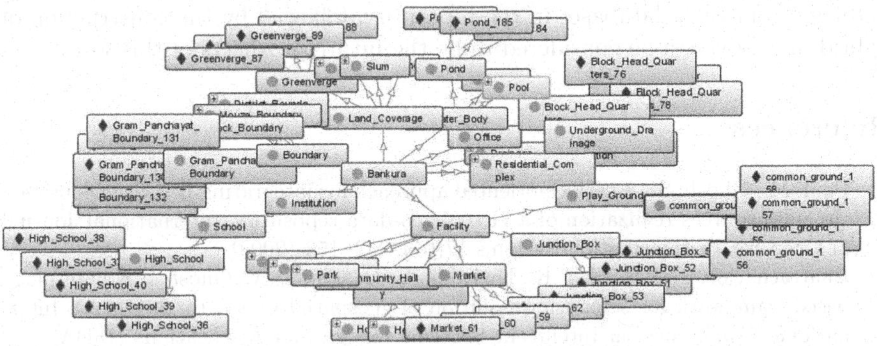

Fig. 2. Ontology of spatial features of Bankura district, India

Table 1. Example cases of semantic matching

Query concept	Matched concept in the ontology	Semantic relation
Highway	***Road***	*hyponym*
Medicine	***Medical Facilities***	*meronym*
Nursing home	***Hospital***	*synonym*
Financial organization	***Bank***	*hypernym*

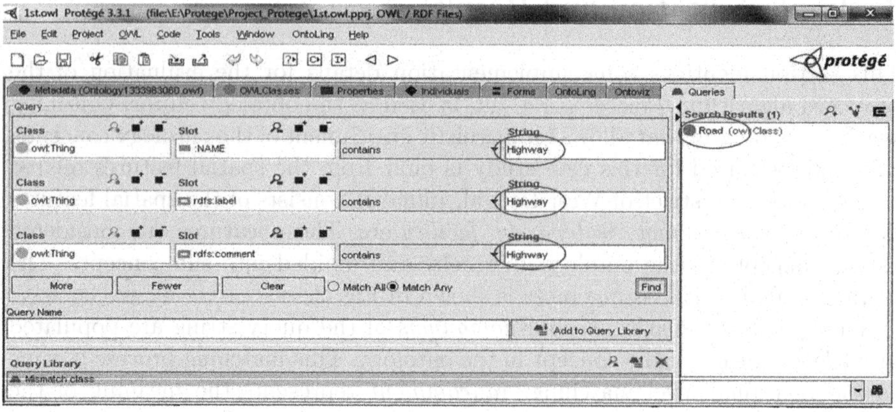

Fig. 3. "Highway" is the *hyponym* of "Road"

4 Conclusion

This paper proposes a *semantic* matching algorithm between user queries and spatial databases using ontology. The whole matching process is divided into three levels, database, relation and attribute, respectively. It is a multi-strategy method based on direct and the *semantic* matching by linguistic-similarity between the user query concepts and the database features. This matching is automated and does not need any manual intervention. Building the application ontology from each databases in the repository, followed by the construction of hybrid ontology, can be considered to be the future prospects of this work.

References

1. Paul, M., Ghosh, S.: A service-oriented approach for integrating heterogeneous spatial data sources realization of a virtual geo-data repository. International Journal of Cooperative Information Systems 17(01), 111–153 (2008)
2. Bhattacharjee, S., Prasad, R.R., Dwivedi, A., Dasgupta, A., Ghosh, S.K.: Ontology based framework for semantic resolution of geospatial query. In: 2012 12th International Conference on Intelligent Systems Design and Applications (ISDA), pp. 437–442. IEEE (2012)
3. Bhattacharjee, S., Mitra, P., Ghosh, S.: Spatial interpolation to predict missing attributes in gis using semantic kriging. IEEE Transactions on Geoscience and Remote Sensing 52(8), 4771–4780 (2014)
4. Fonseca, F.T., Egenhofer, M.J.: Ontology-driven geographic information systems. In: Proceedings of the 7th ACM International Symposium on Advances in Geographic Information Systems, GIS 1999, pp. 14–19. ACM (1999)
5. Buccella, A., Cechich, A., Fillottrani, P.: Ontology-driven geographic information integration: A survey of current approaches. Computers & Geosciences 35(4), 710–723 (2009)
6. Alameh, N.: Service chaining of interoperable geographic information web services. Internet Computing 7(1), 22–29 (2002)

7. Yue, P., Di, L., Yang, W., Yu, G., Zhao, P., Gong, J.: Semantic web services-based process planning for earth science applications. International Journal of Geographical Information Science 23(9), 1139–1163 (2009)
8. Klein, M.: Combining and relating ontologies: an analysis of problems and solutions. In: IJCAI-2001 Workshop on Ontologies and Information Sharing, pp. 53–62 (2001)
9. Miller, G.A.: Wordnet: a lexical database for english. Communications of the ACM 38(11), 39–41 (1995)
10. Jannink, J.F.: A word nexus for systematic interoperation of semantically heterogeneous data sources. PhD thesis, stanford university (2001)
11. Gennari, J.H., Musen, M.A., Fergerson, R.W., Grosso, W.E., Crubézy, M., Eriksson, H., Noy, N.F., Tu, S.W.: The evolution of protégé: an environment for knowledge-based systems development. International Journal of Human-Computer Studies 58(1), 89–123 (2003)

Web-Page Indexing Based on the Prioritized Ontology Terms

Sukanta Sinha[1,2], Rana Dattagupta[2], and Debajyoti Mukhopadhyay[1,3]

[1] WIDiCoReL Research Lab, Green Tower, C-9/1, Golf Green, Kolkata 700095, India
[2] Computer Science Dept.,Jadavpur University, Kolkata 700032, India
[3] Information Technology Dept., Maharashtra Institute of Technology, Pune 411038, India
{sukantasinha2003,debajyoti.mukhopadhyay}@gmail.com,
ranadattagupta@yahoo.com

Abstract. In this world, globalization has become a basic and most popular human trend. To globalize information, people are going to publish the documents in the internet. As a result, information volume of internet has become huge. Tohandle that huge volume of information, Web searcher uses search engines. The Web-page indexing mechanism of a search engine plays a big role to retrieve Web search results in a faster way from the huge volume of Web resources. Web researchers have introduced various types of Web-page indexing mechanism to retrieve Web-pages from Web-page repository. In this paper, we have illustrated a new approach of design and development of Web-page indexing. The proposed Web-page indexing mechanism has been applied on domain specific Web-pages and we have identified the Web-page domain based on an Ontology. In our approach, first we prioritize the Ontology terms that exist in the Web-page content then apply our own indexing mechanism to index that Web-page. The main advantage of storing an index is to optimize the speed and performance while finding relevant documents from the domain specific search engine storage area for a user given search query.

Keywords: Domain Specific Search, Ontology, Ontology Based Search, Relevance Value, Search engine, Web-page Indexing.

1 Introduction

In recent years, the growth of the World Wide Web (WWW) has been rising at an alarming rate and contains a huge amount of multi-domain data [1]. As a result, there is an explosion in information and web searcher uses search engines to handle that information. There are various parameters used by the search engines to produce better search engine performance, Web-page indexing is one of them. Nowadays, Web researchers have already introduced some efficient Web-page indexing mechanism like Back-of-the-book-style Web-page indexes formally called "Web site A-Z indexes", "Human-produced Web-page index", "Meta search Web-page indexing", "Cache based Web-page indexing", etc [2].

M.K. Kundu et al. (eds.), *Advanced Computing, Networking and Informatics - Volume 1*, Smart Innovation, Systems and Technologies 27,
DOI: 10.1007/978-3-319-07353-8_68, © Springer International Publishing Switzerland 2014

In our approach, we have introduced a new mechanism for Web-page indexing. This is fully domain specific Ontological approach, where each Ontology term is treated as a base index. Ontology index number assigned based on their weight value [3-4]. In our proposed mechanism, first we retrieve dominating and sub-dominating Ontology terms for a considered Web-page from the domain specific Web-page repository, then apply primary and secondary attachment rule according to our proposed mechanism.

The paper is organized in the following way. In Section 2, we have discussed the related work on Web-page indexing. The proposed architecture for domain-specific Web-page indexing is given in Section 3. All the component of our architecture is also discussed in the same section. Experimental analyses and conclusion of our paper is given in Section 4 and 5 respectively.

Definition 1.1: Dominating Ontology Term- Ontology term which holds maximum Ontology term relevance value in the considered Web-page.

Definition 1.2: Sub-dominating Ontology Terms- Ontology terms which hold successive maximum Ontology term relevance values other than dominating Ontology term in the considered Web-page.

Rule 1.1: Primary Attachment (P1, P2 ...) – All the dominating Ontology terms for all Web-pages are indexed with the primary attachment of their respective Ontology term.

Rule 1.2: Secondary Attachment (S1, S2 ...) - All the sub-dominating Ontology terms for all Web-pages are indexed with the secondary attachment of their respective Ontology term.

2 Related Works

The main advantage of storing an index is to optimize the speed and performance while finding relevant documents from the search engine storage area for a user given search criteria. In this section, we are going to discuss the existing Web-page indexing mechanism and their drawbacks.

Definition 2.1: Ontology –It is a set of domain related key information, which is kept in an organized way based on their importance.

Definition 2.2: Relevance Value –It is a numeric value for each Web-page, which is generated on the basis of the term Weight value, term Synonyms, number of occurrences of Ontology terms which are existing in that Web-page.

Definition 2.3: Seed URL –It is a set of base URL from where the crawler starts to crawl down the Web pages from the Internet.

Definition 2.4: Weight Table – This table has two columns, first column denotes Ontology terms and second column denotes weight value of that Ontology term. Ontology term weight value lies between '0' and '1'.

Definition 2.5: Syntable - This table has two columns, first column denotes Ontology terms and second column denotes synonym of that ontology term. For a particular ontology term, if more than one synonym exists, those are kept using comma (,) separator.

Definition 2.6: Relevance Limit –It is a predefined static relevance cut-off value to recognize whether a Web-page is domain specific or not.

Definition 2.7: Term Relevance Value – It is a numeric value for each Ontology term, which is generated on the basis of the term Weight value, term Synonyms, number of occurrences of that Ontology term in the considered Web-page.

Back-of-the-book-style Web-page indexes formally called "Web site A-Z indexes". Web site A-Z indexes have several advantages. But search engines language is full of homographs and synonyms and not all the references found will be relevant. For example, a computer-produced index of the 9/11 report showed many references to George Bush, but did not distinguish between "George H. W. Bush" and "George W. Bush" [5].

Human-produced index has someone check each and every part of the text to find everything relevant to the search term, while a search engine leaves the responsibility for finding the information with the enquirer. It will increase miss and hit ratio. This approach is not suitable for the huge volume of Web data [6].

Metadata Web indexing involves assigning keywords or phrases to Web-pages or websites within a meta-tag field, so that the Web-page or website can be retrieved by a search engine that is customized to search the keywords field. This may be involved using keywords restricted to a controlled vocabulary list [7].

Cache based webpage indexing has produced search result quickly because the result information stored into cache memory. On the other hand while an irregular search string encountered, the search engine cannot produce faster search result due to information not available in the cache memory. Irregular search strings always come because of the huge volume of internet information and user [8-9].

3 Proposed Approach

In our approach, we have proposed a new mechanism for indexing domain specific Web-pages. Before going forward with the new indexing mechanism, we need to make sure all the inputs are available in our hands. Those inputs are domain specific Web-page repository, set of Ontology terms, Weight table and Syntable [10]. One of our earlier work, we have created the domain specific Web-page repository [11]. We have used that repository as an input of our proposed approach.

3.1 Extraction of Dominating and Sub-Dominating Ontology Terms

In this section, we will discuss how to extract dominating and sub-dominating Ontology terms. We will illustrate this by using one example Fig. 1.

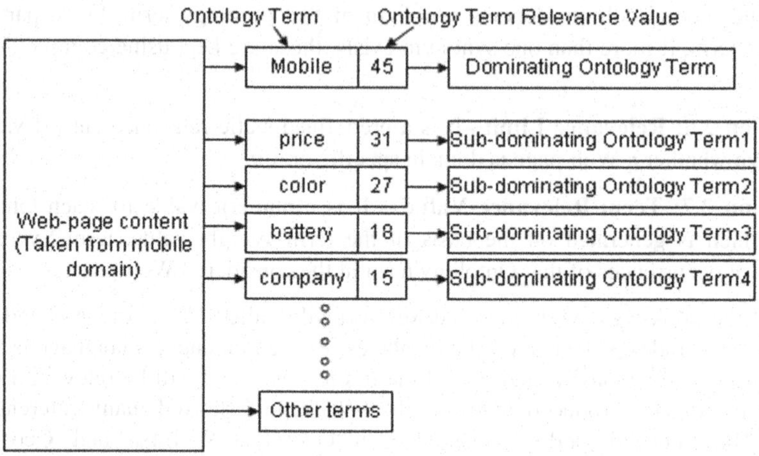

Fig. 1. Example of Extracting Dominating and Sub-dominating Ontology Terms

Consider a 'Mobile' domain Web-page. First extract the Web-page content then apply definition 1.1 and 1.2. We have found that Ontology term 'Mobile' holds term relevance value 45, which is maximum and according to our definition 1.1 Ontology term 'Mobile' becomes dominating Ontology term. Ontology term 'price', 'color', 'battery' and 'company' holds term relevance value 31, 27, 18 and 15 respectively, which are greater than all other Ontology terms excluding 'Mobile' Ontology term. Now according to our definition 1.2, Ontology term 'price', 'color', 'battery' and 'company' become sub-dominating Ontology term 1, sub-dominating Ontology term 2, sub-dominating Ontology term 3 and sub-dominating Ontology term 4 respectively. If number of sub-dominating Ontology term increased then secondary attachments also increases proportionally to store them (refer Rule 1.2), which increases indexing memory size. For that reason, we have used four sub-dominating Ontology terms as a threshold value. Some rare cases, we found multiple Ontology term holds same term relevance value that time we will prioritize dominating and sub-dominating Ontology terms according to their lower term weight value, i.e., consider the higher value of the number of occurrences of that Ontology term in the considered Web-page content.

3.2 Proposed Algorithm of Web-Page Indexing

Proposed algorithm briefly describes the mechanism of Web-page indexing based on the prioritized Ontology terms for a set of domain specific Web-pages.

```
Input    : Domain specific Web-pages
```

```
Output  : Indexed all the Web-pages
1. Select a Web-page (P) from domain specific Web-page re-
   pository
2. Extract Dominating Ontology Term (D)
3. Extract Sub-Dominating Ontology Terms (SDi where 0<i≤4
   and i is an integer)
4. Add Web-page identifier (P_ID) of P with Primary at-
   tachment of D
5. Add Web-page identifier (P_ID) of P with Secondary at-
   tachment of SDi where 0<i≤4 and i is an integer
6. Repeat step 1-5 until all the Web-pages get indexed
7. End
```

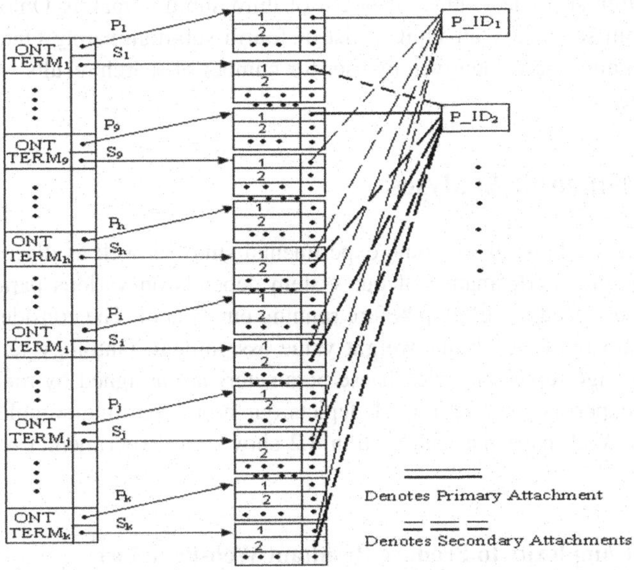

Fig. 2. Web-page structures after applying our indexing mechanism

A pictorial diagram of Web-page structures after applying our indexing mechanism is shown in Fig. 2. Each Ontology term maintains two tables. One table used for storing primary attachments and other one used for storing secondary attachments (refer Rule 1.1 and 1.2). In Fig. 2, (P_1,…, P_9,…,P_h, …, P_k) and (S_1,…, S_9,…,S_h, … , S_k) all are pointing primary and secondary attachment table of their corresponding Ontology terms respectively. Each Web-page has only one primary attachment and four secondary attachments. In the Fig. 2, (P_ID_1, P_ID_2,….) representing Web-page identifier of each considered domain specific Web-pages. Solid lines are denoting primary attachment, which pointing primary attachment of dominating Ontology term. Dotted lines are denoting secondary attachments, which pointing secondary attachment of sub-dominating Ontology terms.

3.3 Web-page Retrieval Mechanism

Web-page retrieval from Web search engine resources are an important role of a Web search engine. We are retrieving a resultant Web-page list from our data store based on the user given dominating and sub-dominating Ontology terms, relevance range, etc. According to our prototype, we are giving a flexibility to the user does not use the search string, directly select the search tokens from the drop down lists. As a result, it reduces the search string parsing time and miss hit ratio due to user's inadequate domain knowledge. Our prototype uses below formula to produce a resultant Web-page list based on the user given relevance range.

(50% of 'x' from the primary attachment list of dominating Ontology term +
20% of 'x' from secondary attachment list of first sub-dominating Ontology term +
15% of 'x' from secondary attachment list of second sub-dominating Ontology term +
10% of 'x' from secondary attachment list of third sub-dominating Ontology term +
5% of 'x' from secondary attachment list of fourth sub-dominating Ontology term),
where 'x' is a numeric value given by user for number of search results want to see in the result page.

4 Experimental Analyses

In this section, we have given some experimental study as well as discussed how to set up our system. Performance of our system depends on various parameters, and those parameters need to be setup before running our system. The considered parameters are domain relevance limit, weight value assignment, Ontology terms, domain specific Web-page repository, etc. These parameters are assigned by tuning our system through experiments. Section 4.1 depicts our prototype time complexity to produce resultant Web-page list and Section 4.2 shows the experimental results of our system.

4.1 Time Complexity to Produce Resultant Web-Page List

We have considered 'k' number of Ontology terms. We have kept them in a sorted order according to their weight value. While finding user given dominating Ontology term primary attachment link, our prototype required maximum $O(\log_2 k)$ time using binary search mechanism (refer Fig. 2). On the other hand while finding other four user given sub-dominating Ontology term secondary attachment links, our prototype required $4O(\log_2 k)$ times. In the second level, our prototype reached from primary and secondary attachment to the Web-pages just spending constant time because there is no iteration required. Finally, our prototype time complexity becomes $[5O(\log_2 k) +5c] \approx O(\log_2 k)$ to the retrieve resultant Web-page list, where 'c' is a constant time required to reach the primary and secondary attachment to Web-pages.

4.2 Experimental Result

It is very difficult to compare our search results with the existing search engines. Most of the cases, existing search engines do not hold domain specific concepts. It is very important that while comparing two systems both are on the same page, i.e., contains same resources, environment, system platforms, search query all are same. Few existing cases, where search engine gives an advanced search option to the Web searchers, but not match with our domains. Anyhow we have produced few data to measure our proposed prototype performance. To produce the experimental results, we have compared the two systems (before and after applying Web-page indexing mechanism) performances. In table 1, we have given a performance report of our system. To measure accuracy, we have applied our set of search query multiple times, which has shown in table 2.

Table 1. Performance report of our system

Number of Search Results	Time Taken (in Seconds)		Total Number of Web-pages in the Repository
	Before applying Web-page indexing	After applying Web-page indexing	
10	0.530973451	0.392156863	5000
20	1.085972851	0.860215054	5000
30	1.753246753	1.409921671	5000
40	2.394014963	2.008368201	5000
50	3.018108652	2.683363148	5000

Table 2. Accuracy of Our System

Number of Search Results	Avg. No. of Relevant Results	Avg. No. of Non-Relevant Results	Total Number of Web-pages in the Repository
10	8.7	1.3	5000
20	17.2	2.8	5000
30	26.4	3.6	5000
40	34.6	5.4	5000
50	43.6	6.4	5000

5 Conclusions

In this paper, we have proposed a prototype of a domain specific Web search engine. This prototype has used one dominating and four sub-dominating Ontology terms to produce Web search results. All the Web-pages are indexed according to their dominating and sub-dominating Ontology terms. According to our experimental results, Web-page indexing mechanism produced faster result for the user selected dominating and sub-dominating Ontology terms. According to our prototype, we are giving a flexibility to the user does not use the search string, directly select the search tokens from the drop down lists. As a result, it reduces the search string parsing time and miss hit ratio due to user's inadequate domain knowledge.

This prototype is highly scalable. Suppose, we need to increase the supporting domains for our prototype, then we need to include the new domain Ontology and other details like weight table, syntable, etc. of that Ontology. In a single domain there does not exist huge number of ontology terms. Hence, the number of indexes should be lesser than a general search engine. As a result, we can reach the web-pages quickly as well as reducing index storage cost.

References

1. Willinger, W., Govindan, R., Jamin, S., Paxson, V., Shenker, S.: Scaling phenomena in the Internet. Proceedings of the National Academy of Sciences, 2573–2580 (1999)
2. Diodato, V.: User preferences for features in back of book indexes. Journal of the American Society for Information Science 45(7), 529–536 (1994)
3. Spyns, P., Meersman, R., Jarrar, M.: Data modelling versus ontology engineering. SIGMOD Record Special Issue 31(4), 12–17 (2002)
4. Spyns, P., Tang, Y., Meersman, R.: An ontology engineering methodology for DOGMA. Journal of Applied Ontology 5 (2008)
5. Diodato, V., Gandt, G.: Back of book indexes and the characteristics of author and non-author indexing: Report of an exploratory study. Journal of the American Society for In-formation Science 42(5), 341–350 (1991)
6. Anderson, J.D.: Guidelines for Indexes and Related Information Retrieval Devices. NISO Technical Report 2, NISO-TR02-1997 (1997)
7. Manoj, M., Elizabeth, J.: Information retrieval on Internet using meta-search engines: A review. CSIR, 739–746 (2008)
8. Brodnik, A., Carlsson, S., Degermark, M., Pink, S.: Small forwarding tables for fast routing lookups. In: Proceedings of ACM SIGCOMM 1997 (1997)
9. Chao, H.J.: Next Generation Routers. IEEE Proceeding 90(9), 1518–1558 (2002)
10. Gangemi, A., Navigli, R., Velardi, P.: The OntoWordNet Project: Extension and Axiomatization of Conceptual Relations in WordNet. In: Meersman, R., Schmidt, D.C. (eds.) CoopIS 2003, DOA 2003, and ODBASE 2003. LNCS, vol. 2888, pp. 820–838. Springer, Heidelberg (2003)
11. Mukhopadhyay, D., Biswas, A., Sinha, S.: A New Approach to Design Domain Specific Ontology Based Web Crawler. In: 10th International Conference on Information Technology, pp. 289–291 (2007)

A Hybrid Approach Using Ontology Similarity and Fuzzy Logic for Semantic Question Answering

Monika Rani, Maybin K. Muyeba, and O.P. Vyas

Indian Institute of Information Technology, Allahabad, India
School of Computing, Mathematics and Digital Technology Manchester Metropolitan University, U.K.
Indian Institute of Information Technology, Allahabad, India
{monikarani1988,dropvyas}@gmail.com, m.muyeba@mmu.ac.uk

Abstract. One of the challenges in information retrieval is providing accurate answers to a user's question often expressed as uncertainty words. Most answers are based on a Syntactic approach rather than a Semantic analysis of the query. In this paper our objective is to present a hybrid approach for a Semantic question answering retrieval system using Ontology Similarity and Fuzzy logic. We use a Fuzzy Co-clustering algorithm to retrieve collection of documents based on Ontology Similarity. Fuzzy scale uses Fuzzy type-1 for documents and Fuzzy type-2 for words to prioritize answers. The objective of this work is to provide retrieval systems with more accurate answers than non-fuzzy Semantic Ontology approach.

Keywords: Question and Answering, Fuzzy Ontology, Fuzzy type-1, Fuzzy type-2, Semantic Web.

1 Introduction

The Educational Semantic web aims to discover knowledge using educational learning areas such as personal learning, education administration and knowledge construction [1]. Semantic web (web 3.0) is providing data integrity capabilities by not only machine readability but also machine analysis. Education is improving by using Semantic web approaches like large number of Online student sharing data semantically also student portal help student to be connected everywhere to the update of class. Electronic text-book [2] provides open context from source like openstax, ck12.org, crowd sourcing, NCERT etc. Massive open online courses (Moocs) example coursera, udacity, khan, Edx, TED-Ed also small virtual classes are found on Internet easily.

The web is naturally Fuzzy in nature, so text document and building Ontogy requires Fuzzy approach. To improve the education Semantic web, the first step is a Semantic question answering systems where uncertain words are questions. To implement such as a system, a Fuzzy Ontology approach can be utilized by use Fuzzy logic (Fuzzy type-1, Fuzzy type-2) [3] levels for text retrieval. A Fuzzy Scale is proposed for two levels, first for membership of document (Fuzzy type-1) and membership of the word (the words having uncertainty (synonyms) as Fuzzy type-2). A Fuzzy Co-clustering algorithm is used to simultaneously cluster documents and words and hence handle

M.K. Kundu et al. (eds.), *Advanced Computing, Networking and Informatics - Volume 1*,
Smart Innovation, Systems and Technologies 27,
DOI: 10.1007/978-3-319-07353-8_69, © Springer International Publishing Switzerland 2014

the overlapping nature of documents in terms of membership functions. To get even more meaningful results Semantic Fuzzy Ontology is proposed as a hybird approach for question answering. The question answering system is based on Semantic approaches as well as Ontology driven representation of knowledge. The rest of the paper is organized as follows: Section 2 gives a background, Section 3 presents a methodology including comparisons and use of Fuzzy type-2, and a conclusion in Section 4.

2 Background

2.1 Question and Answering System

In information retrieval, the challenge is to find accurate answers to questions asked by the user. Questionnaire Mining helps to give accurate answers by handling complex words for which Fuzzy type-2 and linguistic variables can be considered. Thus a Fuzzy Ontology information retrieval system (FOIRS) [4] can play a vital role in understanding Semantic relationships. FOIRS provide basis to find corelationship between user query terms with the document terms.The user query, analysis can be done in much the same way as syntax analysis and also as Semantic analysis for question answering. If the user wants to search any information which is already present in the database for example, if our digital library stores information about painting created by Ravi on subject Irises, nature, soil etc. and the user queries the database with conjunction between the keywords like "Painting" and "Ravi" and "Irises" then no accurate result will be return as keyword are not enough basis to reach accurate answer. While in Semantic system considers the structure of sentence as set of objects, set of functions and various relationships between them. So if user input query example "Painting" by "Ravi" with subject "Blossom" retrieve accurate result, even the query terms can vary like "Painting" by "Ravi" with subject "Irises" will also give accurate result because Ontology define "Blossom" as subclass-hierarchy of Irises in Knowledge graph representation. The Ontology plays a vital role in understanding such ambiguous user questions and helps retrieve appropriate answers. Ontology is way toward Semantic analysis for the question answering search engines like Google, Yahoo etc. For these purposes, Ontology indexing ensemble with Semantic relation among terms is useful.

The question answering (QA) systems main challenge is to retrieve accurate answers to questions [5] asked by users not only based on keywords, but also on Semantic bases, summarized as:

1. Syntax query based retrieval
2. FAQ (question templates) based retrieval
3. Semantic query based retrieval
4. Ontology based retrieval approach
5. Transparent query based approach

Word Net and link grammar approaches toward scaling QA [6] for the web can prove helpful tools in recommender systems and feedback analysis.

2.2 Ontology

Ontology plays an important role in development of knowledge based systems to describe Semantic relationship among entities. Ontology basically describes a formal conceptualization of a domain of interest. Fuzzy Ontology [7], can help in understanding Semantic relationships by applying Fuzzy logic to deal with vagueness of data. Fuzzy type-1 can deal with Crisp membership, whereas Fuzzy type-2 deal with Fuzzy membership as described by Table 1. Scientifically Fuzzy type-1 set as model for words is incorrect as it is unable to deal with uncertainty. As Words means different things to different people so they are uncertainty [8] in nature. Uncertainty about words and be further classified into two types:

1. Intra uncertainty: This is uncertainty that a person has about the word.

2. Inter uncertainty: This is uncertainty that a group of people have about the word.

Table 1. Comparison of Fuzzy type 1 and Fuzzy type 2

Fuzzy type-1	Fuzzy type-2
Level 1	Level 2
Membership Document	Membership of words(synonyms)
Uncertainty is in range [0,1]	Uncertainty is measured by an additional dimension
Two dimension	Three dimension
Notation use A	Notation use tilde A

The proposed methodology uses Fuzzy concepts like linguistic variable and Fuzzy type-2 for information retrieval. In Fuzzy type-2 Model can deal with the uncertainty of words. Fuzzy type-2 reduces to Fuzzy type-1 in case where there is no uncertainty exit in scenario. Ontologies play important role in Information extraction.

Ontology represents knowledge in a graph conceptual diagram using Semantic approach rather than Syntactic approach where each node show either document or word. Various Ontology match a user query and finally retrieves the Ontology for the query Knowledge based (short - path), corpus based (Co-occurrence), Information content and probability of encountering an instance. Then Ontology matching is used as a solution to the Semantic heterogeneity problem. Applying reasoning from an Ontology to text data play an important role in question answering system.

Ontology Similarity:

An edge count method can be used for calculating similarity [9] between a keyword question and hierarchical ontology tree to obtain Semantic relations. For two similar words, the return value is 1 whereas for two dissimilar words return 0 represented as an equation:

$S_t(t1,t2) = (e^{xd} - 1)/(e^{xd} + e^{ys} - 2)$

Where d = depth of tree, S= shorted path length, x and y are smoothing factors and $S_t(t1,t2)$ = similarity value ranging from 0 to 1.

Protégé OWL plug-in [10] shows a major change in describing information of various Ontology by adding new facilities. OWL Ontology can be categorized as OWL Lite, OWL Full and OWL DL [10]. OWL DL can be considered as the extension of OWL Lite. Similarly OWL full is an extension to OWL DL. Semantic web use RDB2onto, DB2OWL and check d2rq etc. to match between Ontology and database. Ontologies do not only represent lexical knowledge, but complex world knowledge about events. Ontologies can be created by Protégé tools, software and after that, use Protégé Java API or translate the Ontology into a rule base using Fuxi [11].

2.3 Data Clustering

In hard clustering data elements are partitioned in such a way that any single data element can belong to only one cluster rather than to many clusters. Fuzzy clustering [12] represents data elements are partitions data in such a way that data can belong to two or more clusters with the degree of belongingness, between overlapping between the cluster can be seen.

3 Methodology Description

The user enters a source string as a question. The first objective of the machine is to Syntactically analyze the text from the source. Only after that the Lexical Analysis can be done for each term in a question and then they are tokenized by removing stop words present in the user's question. The next step is linguistic preprocessing; POS (part of speech) are tagged in such a way that Syntactic analysis can be done easily as shown in Fig. 1, as flow diagram. In POS, a tree is created to differentiate between each question term and label; each term is labeled as a noun, a verb or adjective. The Structural sequence is identified by POS. Then questions can be interpreted for its Semantic meaning. WordNet tool shows the results for all available synonyms of such word which are nouns and verbs. This tool represents knowledge which is also useful for creating a lexical Ontology for the domain knowledge. A word can be processed Semantically by WordNet tools. The groups of words describing the same intension are called synsets. The edge-count method is used to match for question Similarity with the existing Ontology. Fuzzy Co-cluster is used to present collection of answers and Fuzzy scale (Fuzzy type-1 for document and Fuzzy type-2 for words) in order to score the collection obtained by Fuzzy Co-clustering. The final result is the matrix where x-axis represents "Ontology Similarity" and y -axis represents "keywords".

Our proposed Algorithm is as follows:

- Input text in search engine (Question).
- Parse the question for structural analysis.
- Remove stop words for keyword extraction.
- Use WordNet tool to get synonyms of a word in the keyword. Generate all possible combinations of synonyms.
- Retrieval is based on the Semantic Ontology Similarity (edge-count method) match for question; where question is matched with the answer on the basis of existing Ontologies.

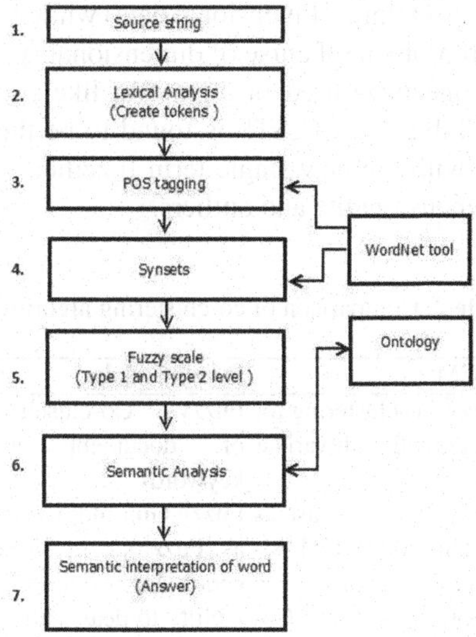

Fig. 1. Flow diagram of semantic question answer

- Result is obtained from the matrix where the x-axis represents "Ontology similarity" and y-axis represents "keywords".
- Use Fuzzy Co-cluster to retrieve answers by using Semantic Ontology Similarity.
- Retrieve the final answer from matrix by prioritizing answers obtained by Fuzzy Co- clustering using Fuzzy scale.

Fuzzy Co-cluster manages data and features into two or more clusters at the same point of time. Here it can be observed overlapping structure of web documents is represented in the cluster with the degree of belongingness for each web document. Reasons to choose the proposed Fuzzy Co-clustering in our case are:

a. Fuzzy Co-clustering is a technique to manage cluster data (Document) and features (Words) [12] into two or more clusters at the same point of time. Here bi-clustering (Co-clustering) has the ability to capture overlap between web documents and words mentioned in the documents. The degree of belongingness for each document and word are mentioned in Co-clustering.

b. The Fuzzy Co-clustering has the following advantages over the traditional clustering:

1. Dimensionality Reduction as the feature is stored in the overlapping form for various clusters.
2. Fuzzy Co-clustering provides efficient results in situations which are vague and uncertain.
3. Interpretability of document clusters becomes easy.
4. Improvement in accuracy due to local model of clustering.
5. Fuzzy membership functions improve representation of overlapping clusters in answers by using Semantic Ontology Similarity.

c. Fuzzy type-2 deals with 3-D (three dimensional data) while FCC_STF [12] algorithm has the ability to deal with problem of curse of dimensionality and outliers.
d. Fuzzy Co-clustering concept is used in algorithm like FCCM, Fuzzy codok and FCC_STF as describe in Table 2. FCC_STF is found to be the best in comparison to FCCM and Fuzzy codok with the new single term fuzzifier approach. FCC_STF is a solution to the curse of dimensionality and outliers.

Table 2. Comparison of co-clustering algorithm

Categories	FCCM	Fuzzy Codok	FCC_STF
Algorithm for Co-clustering	Fuzzy Co-clustering for categorical multivariate	Fuzzy Co-clustering of document and keywords	Fuzzy Co-clustering with Single Term Fuzzifier
Fuzzifier	Fuzzy entropy is use as Fuzzifier in FCCM Algorithm	Fuzzy Gini index is use as Fuzzifier in Fuzzy codok Algorithm	Single Term Fuzzifier is use in FCC_STF Algorithm
Advantage	Algorithm for Co-clustering	Ability to deal with the exponential problem	Clipping for negative value and renormalization take place
Disadvantage	Overflow (exponential) problem	Negative membership	-

To retrieve accurate answers Semantic processing plays an important role. Fuzzy scale (Fig. 2) is an approach towards the Semantic analysis of the question at level 1 and level 2.

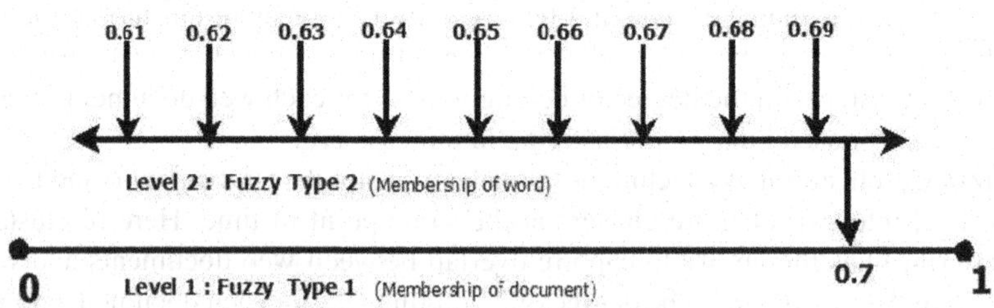

Fig. 2. Fuzzy Scale

Level 1 represents the membership of document in a cluster ($\mu = 0.7$).Here Fuzzy type-1 is used for the document as it unable to deal with uncertainty. Whereas Level 2 represents the membership of word in a cluster ($\mu = 0.61 - 0.69$) using Fuzzy type-2. As one word can have different meanings to different users, so uncertainty come into play. Fuzzy type-2 [8] has ability to deal with uncertainty which can be helpful to deal with the synonyms present in the user question, while Fuzzy type-1 considers no uncertainty.

Calculating Score:

Score = (Membership of document (A) + Membership of Word ($\tilde{A}$))/Number of document (N) = (A+$\tilde{A}/N$). Upper membership function for word ($\mu = 0.69$). Lower membership function for word ($\mu = 0.61$).

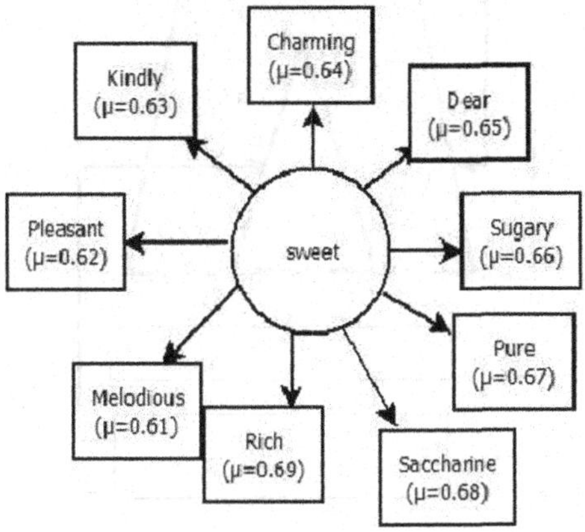

Fig. 3. Membership of word (Fuzzy type)

Fuzzy type-2 is used for computation of the Word as it has the ability to deal with linguistic uncertainty. Whereas Fuzzy type-1 has the crisp membership like for document ($\mu = 0.7$). Fuzzy type-2 has a Fuzzy membership for synonymous words ($\mu = 0.61 - 0.69$), it can be called as Fuzzy-Fuzzy set. Here the computation of word is applied to find appropriate synonym for each question. An Exact synonym helps in obtaining the meaning of the question. So to retrieve appropriate answer Semantic analysis of each query term along with synonyms is a must.

"sweet" is a vague term which we use in our common life every day in common language. The term sweet depends on perception based assessment. The Same word "sweet" has different meanings. When a user types the term "sweet" in the search engine as question this term is treated as a vague term. But uncertainty arises in associating the word "sweet" particularly to sugar. Here uncertainty can arises because the term "sweet" can be associated to describe behaviors like kind, melodious, musical not only to the sugar. In Fig. 3, various memberships of word "sweet" are described. Let us consider the following statements where the term "sweet" needs to be checked for a similar context with respect to its meaning, for which Fuzzy linguistic rules can be applied. Then according to the context of the word membership of word can be applied. For example:

- "Sarah is such a sweet little girl. she's always looking after her brother." - Kindly (μ= 0.63).
- "This tea is too sweet for me to drink, how much sugar is in it?" - sugary (μ= 0.66).

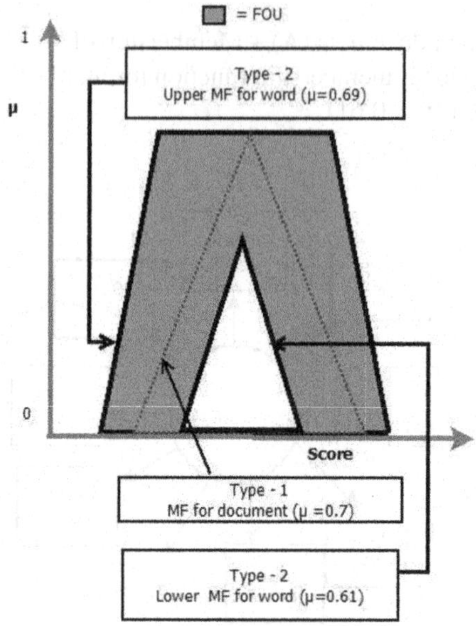

Fig. 4. Foundation of uncertainty (FOU)

Fuzzy type-2 can be visualized by plotting footprints of uncertainty (FOU) in a 2-D domain representation form as shown in Fig. 4. Fuzzy type-2 represents three dimensions data whereas Fuzzy type-1 represents two dimensional data. The uniform color represents the uniformity of possibilities; due to this uniformity Fuzzy type-2 is called Interval type-2 represented by IT2. Till now there is not much progress in IT2 as it's unable to choose best secondary member functions, but computation of words can be an emerging field to use it.

4 Conclusion and Future Work

We have proposed a hybrid approach for Semantic question answering based on Semantic Fuzzy ontology for retrieval systems. Fuzzy Co-clustering is used to retrieve the answers by matching user's question with the existing hierarchical Ontology. Fuzzy scale is use to prioritize the answers retrieved by matrix using Fuzzy Co-clustering. For Fuzzy Co-clustering FCC_STF algorithm is preferred to FCCM and Fuzzy codok. Future work will implement this hybrid approach based on the proposed Semantic Fuzzy Ontology with various applications including e-learning and intelligent web search systems. Users not only get syntactic answers, but also Semantic answers based on the question terms. The proposed question answering system provides a gateway for deep web search along with surface web search.

References

1. Ohler, J.: The semantic web in education. Educause Quarterly 31(4), 7–9 (2008)
2. Agrawal, R.: Computational education: The next frontier for digital libraries (2013)
3. Mendel, J.M.: Type-2 fuzzy sets and systems: an overview. IEEE Computational Intelligence Magazine 2(1), 20–29 (2007)
4. Gallova, S.: Fuzzy ontology and information access on the web. IAENG International Journal of Computer Science 34(2) (2007)
5. Kwok, C., Etzioni, O., Weld, D.S.: Scaling question answering to the web. ACM Transactions on Information Systems 19(3), 242–262 (2001)
6. Guo, Q., Zhang, M.: Question answering system based on ontology and semantic web. In: Wang, G., Li, T., Grzymala-Busse, J.W., Miao, D., Skowron, A., Yao, Y. (eds.) RSKT 2008. LNCS (LNAI), vol. 5009, pp. 652–659. Springer, Heidelberg (2008)
7. Kaladevi, A.C., Kangaiammal, A., Padmavathy, S., Theetchenya, S.: Ontology extraction for e-learning: A fuzzy based approach. In: 2013 International Conference on Computer Communication and Informatics (ICCCI), pp. 1–6 (2013)
8. Mendel, J.: Fuzzy sets for words: why type-2 fuzzy sets should be used and how they can be used. presented as two-hour tutorial at IEEE FUZZ, Budapest, Hongrie (2004)
9. Benamara, F., Saint-Dizier, P.: Advanced relaxation for cooperative question answering. In: New Directions in Question Answering. MIT Press, Massachusetts (2004)
10. Bobilloa, F., Stracciab, U.: Aggregation operators for fuzzy ontologies. Applied Soft Computing 13(9), 3816–3830 (2013)
11. Lord, P.: The semantic web takes wing: Programming ontologies with tawny-owl. arXiv preprint arXiv:1303.0213 (2013)
12. Rani, M., Kumar, S., Yadav, V.K.: Optimize space search using fcc_stf algorithm in fuzzy co-clustering through search engine. International Journal of Advanced Research in Computer Engineering & Technology 1, 123–127 (2012)

Ontology Based Object-Attribute-Value Information Extraction from Web Pages in Search Engine Result Retrieval

V. Vijayarajan[1], M. Dinakaran[2], and Mayank Lohani[1]

[1] School of Computing Science and Engineering, VIT University, Vellore, Tamil Nadu, India
[2] School of Information Technology and Engineering, VIT University, Vellore, Tamil Nadu, India
{vijayarajan.v,dinakaran.m}@vit.ac.in, mayanklohani@outlook.com

Abstract. In this era, search engines are acting as a vital tool for users to retrieve the necessary information in web searches. The retrieval of web page results is based on page ranking algorithms working in the search engines. It also uses the statistical based search techniques or content based information extraction from each web pages. But from the analysis of web retrieval results of Google like search engines, it is still difficult for the user to understand the inner details of each retrieved web page contents unless otherwise the user opens it separately to view the web content. This key point motivated us to propose and display an ontology based O-A-V (Object-Attribute-Value) information extraction for each web pages retrieved which will impart knowledge for the user to take the correct decision. The proposed system parses the users' natural language sentence given as a search key into O-A-V triplets and converts it as a semantically analyzed O-A-V using the inferred ontology. This conversion procedure involves various proposed algorithms and each algorithm aims to help in building the taxonomy. The ontology graph has also been displayed to the user to know the dependencies of each axiom given in his search key. The information retrieval based on this proposed method is evaluated using the precision and recall rates.

Keywords: O-A-V (Object-Attribute-Value), NLP (Natural Language Processing), Knowledge Representation, Taxonomy, Light-Weight Ontology, WordNet, RDF, RDFS

1 Introduction

Web is huge but it is not intelligent enough to understand the queries made by the user and relate them real or abstract entities in the world. It is a collection of text documents to other resources, linked by hyperlinks and URLs (Uniform Resource Locator).

M.K. Kundu et al. (eds.), *Advanced Computing, Networking and Informatics - Volume 1*, Smart Innovation, Systems and Technologies 27,
DOI: 10.1007/978-3-319-07353-8_70, © Springer International Publishing Switzerland 2014

1.1 Semantic Web

Semantic web is the next level of web which treats it as a knowledge graph rather than a collection of web resources interconnected with hyperlinks and URLs. It also aims at adding semantic content to web pages and providing machine processable semantics. With these, web agents will be able to perform complex operations on behalf of the user.

"Semantic Web is about common formats for integration and combination of data drawn from diverse sources and how the data relates to real world objects. It provides a common framework that allows data to be shared and reused across applications, enterprise and community boundaries" [1].

Linked data describes a method of publishing structured data so that it can be interlinked and become more useful. Rather than using Web Technologies to serve web pages for human readers, it uses these technologies to share information in a way that can be read automatically by computers enabling data from different sources to be connected and queried [2].

Reasoning is the capacity for consciously making sense of things, applying logic, for establishing and verifying facts, and changing or justifying practices, institutions, and beliefs based on new or existing information [3]. With Semantic web intelligence, the web agents will be able to reason the content on web and draw inferences based on the relations between various web resources.

1.2 Ontology

Ontology is the philosophical study of the nature of being, becoming, existence, or reality, as well as the basic categories of being and their relations. Any entity whether real or abstract have certain characteristics which relate to certain entities in the real world and interact with them. Ontologies deal with the existence of entities, organizing them into groups based on their similarity, developing hierarchies and studying relations among them which allows us to draw inferences based on their classification. We can also study how they interact with other discrete entities in the world and finally develop our own ontologies based on our domain of interest. In computer science and information science, an ontology formally represents knowledge as a set of concepts within a domain using a shared vocabulary to denote the types, properties and interrelationships of those concepts [4].

Ontologies act as the building blocks for the infrastructure of semantic web. They will allow to transform the existing web of data into web of knowledge. They will allow for knowledge sharing among the various web applications and enable intelligent web services.

1.3 Knowledge Representation

It is the application of logic and ontology to the task of constructing computable models for some domain [5]. Knowledge representation and reasoning are the backbones of semantic web. There is no absolute knowledge representation methodology, it solely depends on the type application and how it uses the acquired knowledge.

- Object-Attribute-Value Triplets used to represent facts about objects and their attributes.
- Uncertain Facts represent uncertainty about facts in O-A-V Triplets using numerical values.
- Fuzzy Facts represent uncertainty about facts using terms from natural language.
- Semantic Networks represent semantic relations between concepts.They are designed after the psychological model of the human associative memory and attempt to reflect cognition.

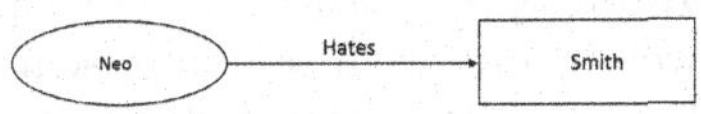

Fig. 1. An O-A-V triplet

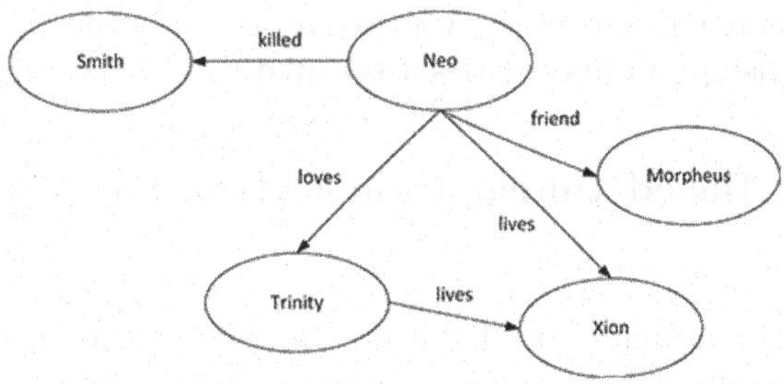

Fig. 2. A simple semantic network

1.4 WordNet

The WordNet [6] is a large lexical database of English Language. It groups closely related words into an unordered set called Synsets which are interlinked by means of conceptual-semantic and lexical relations. Although it is considered as an upper ontology by some, WordNet is not strictly an ontology. However, it has been used as a linguistic tool for learning domain ontologies.

1.5 RDF and RDFS

The Resource Description Framework [7] is an official W3C Recommendation for Semantic Web data models. RDF and RDFS (Resource Description Framework Schema) can be used to design an efficient framework to describe various resources on the web in such a way that they are machine understandable. A resource description in RDF is a list of statements (triplets), each expressed in terms of a Web resource (an object), one of its properties (attributes), and the value of the property. RDF is used to describe instances of ontologies, whereas RDF Schema encodes ontologies providing the semantics, vocabulary and various relationships in the domain.

2 Related Work

- Google's Knowledge Graph [8] is a knowledge base used by Google to enhance its search engine's search results with semantic-search information gathered from a wide variety of sources. It works at the outer level, drawing semantic relations among various resources providing us with the best web results. On the other hand our model works at the inner level, drawing semantic relations inside each web document providing meaningful insight to content available with each web link improving the user's web search experience.
- DBpedia [9] is a project aiming to extract structured content from the information created as part of the Wikipedia project. DBpedia allows users to query relationships and properties associated with Wikipedia resources.

3 Ontology Based Information Extraction from Web Pages

Since most of the information available on the web is in natural language and not machine understandable there is no way to understand the data and draw out semantic inferences. Ontologies can be used to model the information in a way that could be easily interpreted by machines.

Sentence Structure. A typical clause consists of a subject and a predicate, where the predicate is typically a verb phrase - a verb together with any objects and other modifiers (Refer Fig. 3).

3.1 Proposed Architecture for O-A-V Evaluation

On passing the text through the proposed model, it is broken down into clauses which are then tokenized and passed through the WordNet analyzer. The WordNet analyzer provides characteristic properties for each lemma like POS(part of speech), synonyms, hypernyms, hyponyms, etc. Later on, an object is created for each of these individuals and is added to the ontology. On passing the clause

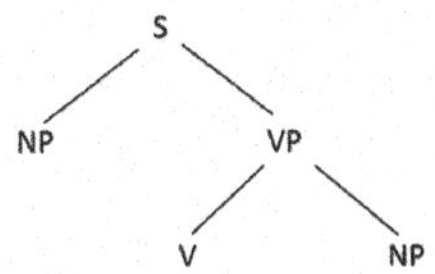

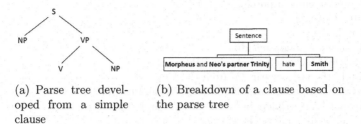

(a) Parse tree developed from a simple clause

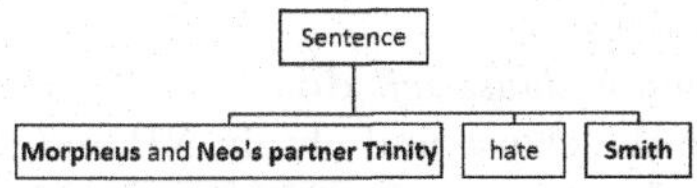

(b) Breakdown of a clause based on the parse tree

Fig. 3. Using parse tree structure for extracting clause

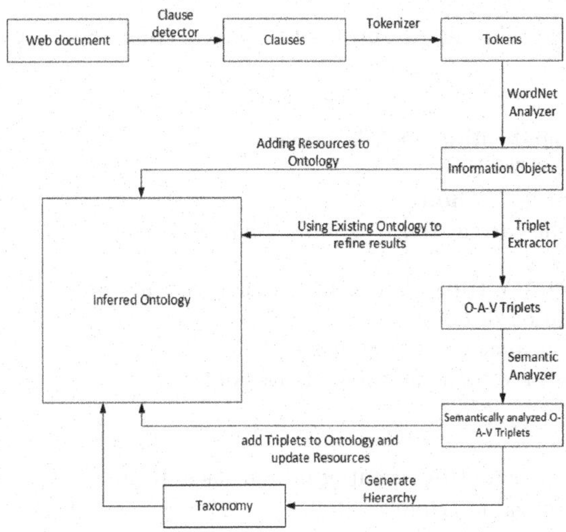

Fig. 4. Proposed architecture for ontology based information extraction

through the triplet extractor it continuously searches for nested and direct relations using the existing ontology. the extracted O-A-V triplets are then passed through a semantic analyzer which determines the true form of the various objects in the O-A-V triplet based on the context in which it has been used. These triplets and updated individuals are added to the ontology along with the generation of a taxonomy. At the end of all these processes a well defined semantic network is developed which can then be used to enhanced search engine web results providing the user with a completely enhanced search experience. Refer Section 3 for more details on 'Enhancing User Search Experience'.

3.2 Algorithms

For extracting nested relations like X's Y's Z, the triplet extractor continuously checks for relations creating an empty individual which can be later on updated based on their future occurance. The individuals are then classified based on the

Data: web document,ontology
Result: Ontology
extract clauses;
while *no more clause left* **do**
  analyze the clause and obtain NP and the VP ;
  obtain the last occuring V from the VP;
  extract compound entities from the NP and the VP;
  create O-A-V triplets between subjects and objects;
  semantically analyze the extracted O-A-V triplets;
  create a semantic network by adding the triplets and individuals to the ontology;
  develope a taxonomy;
end

Algorithm 1. Developing an ontology from the content in a web document

Data: NP
Result: compound entities,O-A-V Triplets
while *not end of NP* **do**
  if *next token* $\notin$ *N* **then**
    create the current token as individual in ontology;
  else
    create O-A-V triplet between current token and next token with V as a combination of both;
    update current token with value of V;
    set class of V with the value of class of A;
  end
end

Algorithm 2. Extracting compound entities from NP, O-A-V represents Object-Attribute-Value triplet

context in which they are used, like"Tommy" will represent a "dog" based on the relation "Sam's dog Tommy" but not on the convention that we have always used a name like Tommy to refer to a dog.

Data: O-A-V Triplet
Result: semantically analyzed O-A-V triplets
if $O \in$ *Ontology and* $O \notin$ *class of V* **then**
  V represents a property or a characteristic of O rather than him;
else
  set class of O with the value of class of V;
end

Algorithm 3. Semantic analysis of direct relations

For analyzing direct relations like X is Y, the semantic analyzer determines the group to which both the individuals belong to, compares them, and accordingly updates the O-A-V triplet based on previous occurance of both the objects and their attribute value, refer Fig. 6 and Fig. 7.

Data: unordered set of Individuals
Result: taxonomy
while *no individual left* **do**
  extract hypernyms for each individuals;
  arrange the individuals in order of appearance in their hierarchies;
  find common ancestors between the individuals up their hierarchies;
  add individuals to a common class having common ancestors;
  remove these individuals and add the ancestor as another individual in the given set;
end

Algorithm 4. Developing a taxonomy

In order to develop a hierarchy among the various identified groups, hypernyms of all the groups are acquired using WordNet (based on their usage) and common ancestors are determined for each entity going up the hierarchy level. This process is continued until we reach the top level entity (Thing). With all

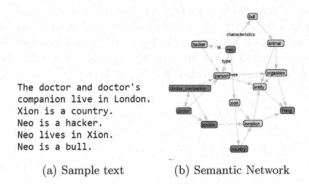

```
The doctor and doctor's
companion live in London.
Xion is a country.
Neo is a hacker.
Neo lives in Xion.
Neo is a bull.
```

(a) Sample text

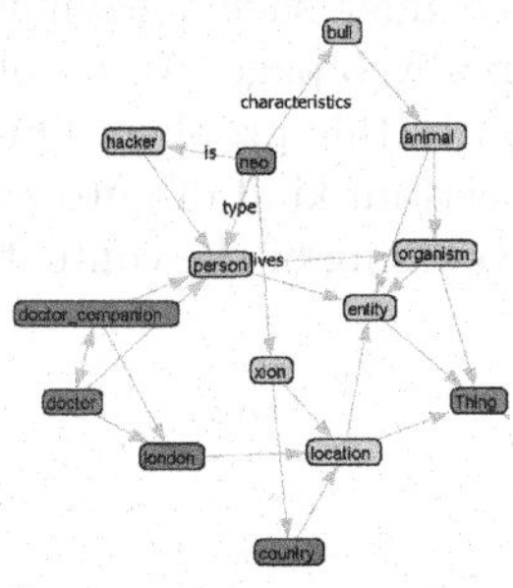

(b) Semantic Network

Fig. 5. Semantic network developed for the given sample text

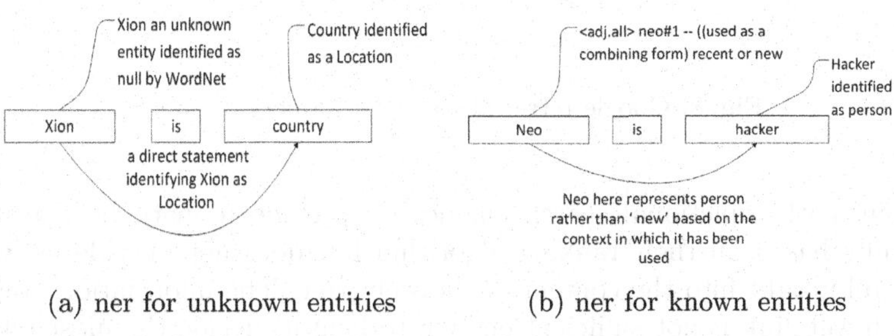

(a) ner for unknown entities (b) ner for known entities

Fig. 6. Named entity recognition(ner) for O-A-V Triplets

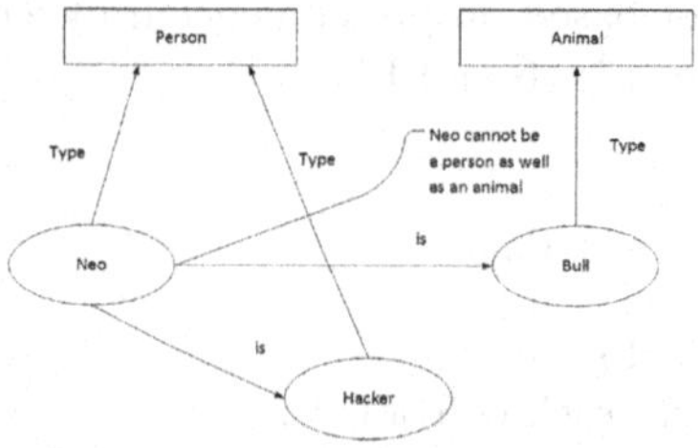

Fig. 7. Semantic analysis of direct relations

the individuals classified into groups along with their relations and a hierarchy, a taxonomy is developed.

4 A Light Weight Ontology Based Search Engine

A browser can recognize the type of content in a web page using the meta-data provided but without understanding the semantics. A sentence like "Karen is a cow" is just another piece of text it has to render but actually it might be expressing Karen's behavior or simply implying that Karen is a cow. A browser has no means to infer such meanings by just reading the plain unstructured text available in a web page. An ontological representation of the web page is a possible solution to this problem. Ontologies act as computational models and enable us with certain kind of automated reasoning. They will enable semantic analysis and processing of the content of the web page.

Fig. 8. Google retrieval result for a query "Neo"

The currently functional search engines do provide us with best available web results based on their ranking algorithm but does not provide us with a meaningful insight into the content of the web page. The information available with each web link is not sufficient enough to help us decide the most relevant web page, refer Fig.8. To get information we always go to Wikipedia without even checking the other web results provided to us by the search engine. In a way we are bound to various websites based on their reputation and neglect valuable information that might we be available with other web pages.

We need a way so that the user can actually get to know the type of content available in a web page without having to go through the complete list of web pages provided by the search engine and also widen his knowledge about the web links he should visit.

4.1 Proposed Architecture for O-A-V Representation

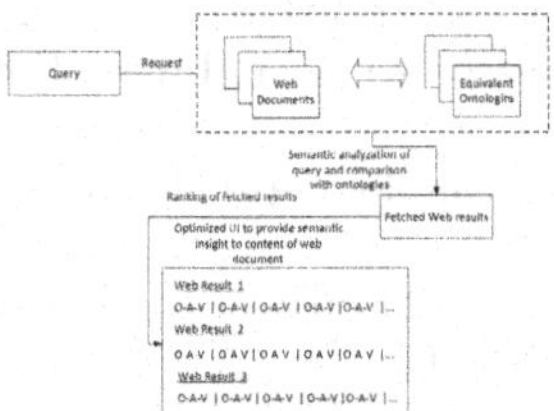

(a) Proposed architecture

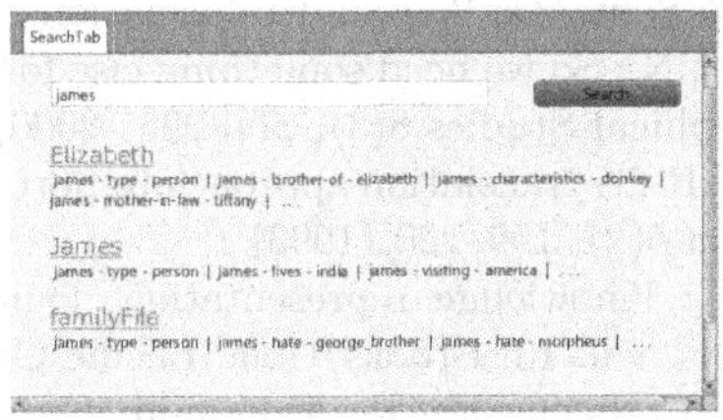

(b) Web search results along with their corresponding O-A-V triplets

(c) A graphical display showing results on search query for individual document (on pressing *i Fig. 10(b)*)

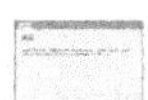

(d) Content within a web result (Elizabeth)

Fig. 9. Proposed Architecture for Ontology Based Search Engine providing the user with a meaningful insight to the content within each Web Result by providing Semantically extracted Triplets rather than a random sentence from the document containing the queried keywords

5 Conclusions and Future Work

The amount of information on the web has increased exponentially in recent years. Going through every web result is time consuming for an impatient user. Providing the top web results does not complete the task if the user still has to browse through them. Providing semantically extracted O-A-V triplets with each web link will the provide user with valuable insight and also helps save time. Later on, we would try providing semantic links within the web documents so that they could be used for integration as well as validation of the relations between various web resources.

References

1. Berners-Lee, T., Hendler, J., Lassila, O.: The semantic web. Scientific American 284(5), 28–37 (2001)
2. Bizer, C., Heath, T., Berners-Lee, T.: Linked data-the story so far. International Journal on Semantic Web and Information Systems (IJSWIS) 5(3), 1–22 (2009)
3. Kompridis, N.: So we need something else for reason to mean. International Journal of Philosophical Studies 8(3), 271–295 (2000)
4. Gruber, T.R.: A translation approach to portable ontology specifications. Knowledge Acquisition 5(2), 199–220 (1993)
5. Sowa, J.F.: Knowledge representation: logical, philosophical, and computational foundations, vol. 13. Brooks/Cole, Pacific Grove (2000)
6. Miller, G.A., Beckwith, R., Fellbaum, C., Gross, D., Miller, K.: Introduction to wordnet: An on-line lexical database. International Journal of Lexicography 3(4), 235–244 (1990)
7. McBride, B.: The resource description framework (RDF) and its vocabulary description language RDFS. In: Handbook on Ontologies, pp. 51–65 (2004)
8. Singhal, A.: Introducing the knowledge graph: things, not strings. Official Google Blog (2012)
9. Auer, S., Bizer, C., Kobilarov, G., Lehmann, J., Cyganiak, R., Ives, Z.: Dbpedia: A nucleus for a web of open data. In: 6th International Semantic Web Conference, 2nd Asian Semantic Web Conference, pp. 722–735 (2007)

Effects of Robotic Blinking Behavior for Making Eye Contact with Humans

Mohammed Moshiul Hoque, Quazi Delwar Hossian, and Kaushik Deb

Faculty of Electrical & Computer Engineering
Chittagong University of Engineering and Technology
Chittagong-4349, Bangladesh
moshiulh@yahoo.com

Abstract. Establishing eye contact with the target human plays a central role both in human-human and human-robot communications. A person seems that s/he meet eye contact with the other when they are looking at each other eyes. However, such a looking behavior alone is not enough, displaying gaze awareness behavior also necessary for making successful eye contact. In this paper, we proposed a robotic framework that can establish eye contact with the human in terms two phases: gaze crossing and gaze awareness. We evaluated a primitive way of displaying robot'sgaze behaviors to its partner and confirmed that the robot withsuch gaze behaviors could give stronger feeling of being lookedat and better feeling of making eye contact than ones with absent of such behaviors.A preliminary experiment with a robotic head shows the effectiveness of using blinking behavior as a gaze awareness modality for responding partner.

Keywords: Human-robot interaction, Eye contact, Gaze awareness, Eye blinks.

1 Introduction

Human-robot interaction (HRI) is an interdisciplinary research field aimed at improving the interaction between human beings and robots and to develop robots that are capable of functioning effectively in real-world domains, working and collaborating with humans in their daily activities. For robots to be accepted into the real world, they must be capable to behave in such a way that humans do with other humans. Although a number of significant challenges remained unsolved related to the social capabilities of robots, the robot that can meets eye contact with human is also an important research issue in the realm of natural HRI. Eye contact is a phenomenon that occurs when two people cross their gaze (i.e. looking at each other). Meeting eye contact is considered as an important prerequisite to initiate a natural interaction process between two partners [1]. Moreover, it plays several important roles in face-to-face communication, such as, regulating turn taking [2], and joint visual attention [3].

It seems that we can make eye contact if we establish gaze crossing. Psychological studies show however, that this gaze crossing action alone may not be enough to make a successful eye contact event. In addition, each party must be aware of being

M.K. Kundu et al. (eds.), *Advanced Computing, Networking and Informatics - Volume 1*, Smart Innovation, Systems and Technologies 27,
DOI: 10.1007/978-3-319-07353-8_71,

looked at by the other [4]. Yoshika *et al.* [5] also mentioned that simply staring is not always sufficient for a robot to make someone feel that they are being looked at. Thus, after crossing the gaze with the intended recipient the robot should interpret the human looking response and display gaze-awareness which is an important behavior for humans to feel that the robot understands his/her attentional response. In order to display awareness explicitly the robot should use some verbal ore non-verbal actions. Therefore, in order to set up an eye contact event with the human, robots should satisfy two important conditions: (i) gaze crossing, and (ii) gaze awareness. In this work, we develop a robotic head that can detect, and track the human head and body. By using the information of head and body of the human, the robotic head adjust its head position to meet the face-to-face orientation with him/her. This will achieve the first condition. After meeting in face-to-face, the robot displays its eye blinking actions which satisfy the second condition.

2 Related Work

Several robotic systems were developed to establish eye-contact with the humans which are broadly classified in two ways: gaze crossing and gaze crossing with gaze awareness. In gaze crossing based, robots are supposed to make eye contact with humans by turning their eyes (cameras) toward the human faces [6], [7].All of these studies focus only on the gaze crossing function of the robots as making its eye contact capability and gaze awareness functions are absent. Several robotic systems were incorporates gaze awareness functions by facial expression (i.e., smiling) [8]. To produce smiling expression, they used a flat screen monitor as the robot's head and display 3D computer graphics(CG) images. A flat screen is unnatural as a face. Moreover, these models used to produce the robot's gaze behavior are typically not reactive to the human partner's actions. Yoshikawa *et al.* [5] used a communication robot to produce the responsive gaze behaviors of the robot. This robot generates a following response and an averting response against the partner's gaze. They showed that the responsive robotic gazing system increases the feeling of people of being looked at. However, it is unknown how the robot produce gaze awareness behavior while it deals with the people. Moreover, the robotic heads that are used in previous studies were mechanically very complex and as such expensive to design, construct and maintain. A recent work that used a robot Simon to produce the awareness function [9]. The Simon blink its ear when it hearing an utterance. Although they consider the single person interaction scenario, they did not used ear blinks as a gaze awareness purpose rather use to create interaction awareness.

3 System Overview

For the HRI experiments we developed a robotic head (Fig. 1). This figure shows a prototype of the robotic head and the corresponding outputs of its several software modules. The head consists of a spherical 3D mask, an LED projector (3M pocket projector, MPro150), a laser range sensor (URG-04LX by Hokuyo Electric Machinery), three USB cameras (Logicool Inc., Qcam), and a pan-tilt unit (Directed Perception Inc., PTU-D46). The LED projector projects CG-generated eyes on the mask.

In the current implementation, one USB camera is attached to the robot's head (as shown in Fig. 1). The other two cameras and the laser sensor are affixed to the tripods, placed at an appropriate position for observing participants' heads as well as their bodies.

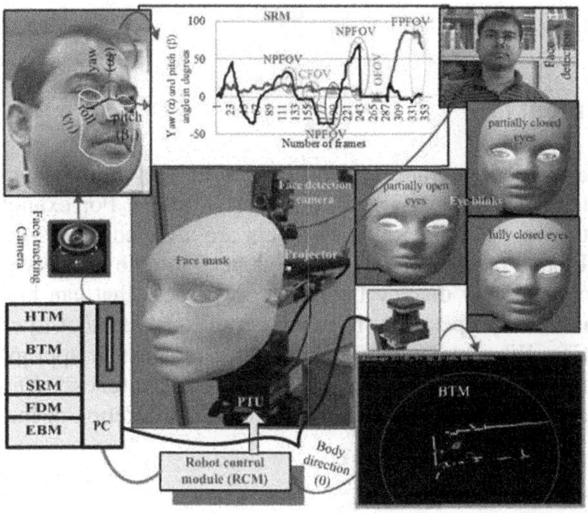

Fig. 1. A prototype robotic head that consists of five modules: HTM (head tracking module), SRM (situation recognition module), BTM (body tracking module), EBM (eye blinks module), and RCM (robot control module)

The system utilizes a general purpose computer (Windows XP). The proposed system has five main software modules: HTM, BTM, SRM, ECM, and RCM. To detect, track, and compute the direction of the participant's head in real time (30 frame/sec), we use FaceAPI by Seeing Machines Inc. It can measure 3D head direction within a $3^0$of error. The human body model is consequently represented with the center coordinates of an ellipse [x, y] and its principle axis direction (θ). These parameters are estimated in each frame by a particle filter framework. Using a laser sensor, the BTM can locate a human body within errors of 6cm for position, and 6^0 for orientation. To assess the current situation (i.e., where the human is currently looking), one observes the head direction estimated by the HDTM. From the results of the HDTM, the SRM recognizes the existing viewing situation (with 99.4% accuracy) and the direction of the relevant object in terms of yaw (α), and pitch (β) movements of the head by using a set of predefined rules. For each rule, we set the values of (α), and (β) by observing several experimental trials. For example, if the current head direction (of the human with respect to the robot) is within $-10^0 \leq \alpha \leq +10^0$ and $-10^0 \leq \beta \leq +10^0$ and remains in the same direction for 30 frames, system recognizes the situation as the central field of view. The results of the SRM are sent to the PTUCM to initiate the eye contact process, and the robot turns toward the human based on the results provided by the BTM. The robot considers that the participant has responded against its actions if s/he looks at the robot within the expected time-frame. If this step is successful, the FDM detect the participant's face. After detecting his/her face, the FDM sends the results to the EBM for exhibiting eye blinks to let the human know that the robot is aware of his/her gaze. The EBM generates the eye blinks to create the feeling of making eye contact. All the robot's head actions are performed by the PTU, with the actual control signal coming from several modules. The details description of the robotic head has described in [10].

4 Experiments

The way one person looks at another and how often they blink seems to have a big impact on what kind of impression they make on others. For example, it has been pointed out that people's impressions of others are affected by the duration they are looked at, and the rate of blinking [11]. Robots need to able not only to detect human gaze but also to accurately display their gaze awareness that can be correctly interpreted by humans. We propose eye blinking action for the robot to create such gaze awareness function. When using blinking, we need to consider appropriate number of blinking or blink duration to effectively convey gaze awareness. We performed two experiments to verify the effect of eye blinks and blink duration of the robot to establish eye contact with the human.

4.1 Experiment 1: Effect of Eye Blinks

A total of 36 participants (28 males, and 8 females) participated in the experiment. The average age of participants was 25.6 years (SD = 3.75). They were all graduate students at Saitama University, Japan. We deliberately concealed the primary purpose of our experiment. There was no remuneration for participants.

Experimental Design: To assign a task for the human, we hung five paintings (P1-P5) on the wall at the same height (a bit above the eye level of participants sitting on the chair). In the experiment, participants were asked to sit down on the chair one at a time and asked to look around the paintings. We prepared two robotic heads. Both were the same in appearance. We programmed each to show different behaviors. One was (i) Gaze crossing robot (GCR). Initially it is static and is looking in a direction not toward the human face. During watching the paintings, it shakes its head to attract the participant attention toward it. If the participant looks at the robot, it will adjust its head position to ensure the face-to-face orientation (i.e., between the robot and the participant). However, it behaved in two ways after crossing the gaze with the participant in two experimental trials. One was, it will produce blinking action after crossing the gaze and another was, it did not produce any blinking action. The second was (ii) gaze averting robot (GAR). We placed GAR to the right of the rightmost painting and GCR to the left of the leftmost painting. An USB camera and a laser sensor were positioned in front of the participant to track his/her head as well as body. Two video cameras were placed in appropriate positions to capture all interactions. Fig. 2 shows the experiment set up. Before the experiment, we explained to the participants that the purpose of experiment was to evaluate the suitable robot's behavior to make them feel that they have made eye contact with the robot.

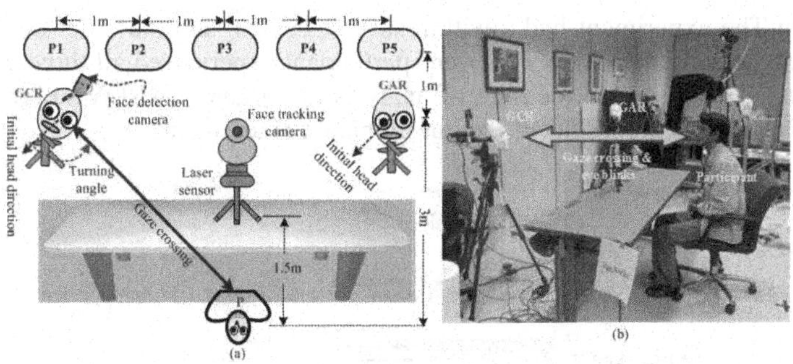

Fig. 2. Experimental scenario: (a) Schematic setting, (b) a scene of the experiment

Conditions: To verify the effect of eye blinks, we prepared the following two conditions: *(i) Gaze averting robot (GAR):*A gaze avoidance phenomenon occurs when a human (H1) avoids looking at another human (H2). If the participant looking at the robot it avoids his/her gaze by rotating its head to another direction by 120^0at a pan speed of 180^0/second.*(ii) Gaze crossing without blinksrobot (GC+WBR):* The robot turns its head toward the participant from its initial position. The robot recognizes his/her face while s/he is looking at it, waits about five seconds, and then turns its head toward another direction.*(iii) Gaze crossing with blinks robot (GC+BR):* The robot turns its head toward the participant. The robot recognizes his/her face while s/he is looking at it. After detecting the face of the participant, the robot starts blinking its eyes about five seconds, and then turns its head toward another direction.

Measurements: We measured the following two items in the experiment:*(A) Impression of robot:* We asked participants to fill out a questionnaire for each condition (after three interactions). The measurement was a simple rating on a Likert scale of 1 to 7, where 1 stands for the lowest and 7 for the highest. The questionnaire had the following two items: *(i) The feeling of looking at:* Did you feel like the robot was looking at you? *(ii) The feeling of making eye contact:* Did you think that behaviors of the robot were created your feeling of making eye contact with it?*(B) Gazing time:*We measure the total time that are spent by the participants to gazing at the robot in each method. This time is measured from the beginning of gaze crossing action of the robot to the end of the participant's looking at it before turning head to another direction by observing the experimental videos.

Predictions: As previous studies suggested, not only gaze crossing but also gaze awareness is an important factor for making eye contact. Simply staring is not always sufficient for a robot to make someone feel being looked at. Although the GC+WBR try to make eye contact using gaze crossing function, it lacks to create gaze awareness function, which is implemented in the GCR. On the other hand, both of these functions are absent in GAR. Moreover, participants' eyes may be coupled with the robot's eye during blinking. Based on this consideration, we predicted the following: *Prediction 1:*Robots with a gaze crossing and eye blinks (GC+BR) functions will give their partners a stronger feeling of being looking at and making eye contact than robots that do not include these functions (i.e., GC+WBR and GAR).*Prediction 2:*Participants will spend more time on gazing at the robot with gaze crossing and blinking function (GC+BR) than that of the robots without gaze crossing and blinking actions (GAR and GC+WBR).

Results: The experiment had a within-subject design, and the order of all experimental trials was counterbalanced. Every participant experienced all three conditions. We conducted the repeated measure of analysis of variance (ANOVA) for all measures. Fig.3(a) shows the results of the questionnaire assessment.

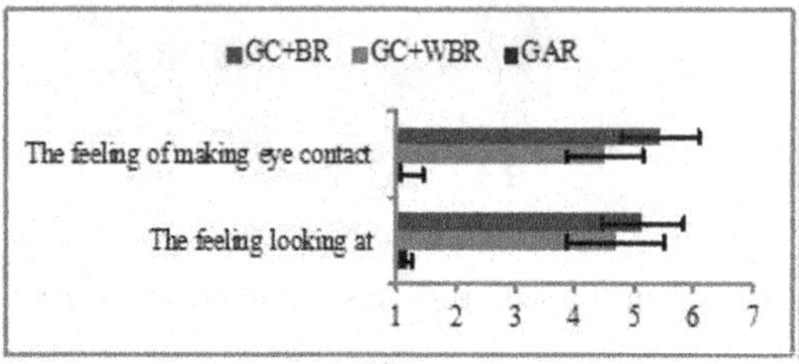

(a)

Robots	Gazing time (seconds)	
	M	SD
GAR	1.75	0.37
GC+WBR	3.75	0.42
GC+BR	4.52	0.43

(b)

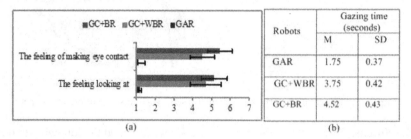

Fig. 3. Evaluation results: (a) participants responses on each question. Error bars indicates the standard deviation, (b) total time spent on gazing

Concerning the feeling of looking at, ANOVA analysis shows that the differences between conditions were statistically significant [$F(2,70)=369.8$, $p<0.01$, $\eta^2=0.88$].We conducted multiple comparisons with the Bonferroni method that showed significant differences between robots GAR and GC+BR ($p<0.001$), and between robots GC+BR and GC+WBR ($p=0.02$). Concerning the feeling of making eye contact, a significant main effect was found [$F(2,70)=288.1$,$p<0.0001$, $\eta^2 = 0.84$]. Multiple comparisons with the Bonferroni method also revealed significant differences between pairs (GC+BR vs. GAR: $p<0.001$, and GC+BR vs. GC+WBR: $p=0.0005$). From these statistical analyses, we confirmed that robots with eye blinks succeeded in giving participants a stronger feeling of being looked at and making eye contact than those without eye blinks. These results verified prediction 1.

In order to evaluate the gazing time, we observed a total of 108(36 × 3) interaction videos for all robots. Fig. 3(b) summarizes the mean values of time that the participants are spent on gazing at the robots. ANOVA analysis showed that there are significant differences that the participants spending to gaze at the robot in each method [$F(2,70)=374.2$, $p<0.0001$, 2=0.88]. Multiple comparisons with the Bonferroni method revealed significant differences between pairs (GC+BR vs. GAR: $p<0.001$, and GC+BR vs. GC+WBR: $p<0.0001$). Results also indicate that the participant looks significantly longer in proposed method (4.52s) than the other methods (1.75sfor GAR and 3.75s for GC+WBR). Thus, the prediction 2 has been supported.

4.2 Experiment 2: Effect of Blink Duration

Ten new participants participated in this experiment. Their average age was 26.3 years (SD=3.68). All participants were graduate students at Saitama University. The purpose of this experiment is to evaluate the appropriate blink duration for creating the feeling of gaze awareness.

Experimental Design: The experimental setting was the same as describe in Section 4.1. We set the robot to perform eye blinking after detecting the face of the human according to the conditions. The experiment had a within-subject design, and the order of all experimental trials was counterbalanced. Gaze awareness impression of the participant created by eye blinks of the robot may depend on the way of blinks (i.e., blinking rate and duration of blinks). Therefore, to design an effective blinking action as a gaze awareness modality, we adopted three blinking rates and five blink duration conditions: *(i) Blinking rate:* 3parameters are prepared for the blinking rate: *1 blink/s*, *2 blink/s*, and *3 blink/s.(ii) Blinks duration:* Since the human's eye blinking appears every 4.8 seconds; we chose the 1-to-5 second duration for observations. Thus, we prepared five parameters for blink duration: *1s, 2s, 3s, 4s,* and *5s.*

The participants were asked to provide evaluations on a 1-to-7 scale in the questionnaire, where 1 stands for the lowest and 7 for the highest. We asked the participants to provide the following evaluation after experiencing all conditions. *Evaluation*: your blinking preference of the robot.

Results: Participants may fail to observe the fast eye blinking rate. They may get bored or lose their attention for large duration of blinks. Thus, we expected that participants will feel that the moderate number of blinks is more suitable as robot's gaze awareness behavior. Fig. 4(a) summarizes the participants' response in each condition. The values into the boxes indicate mean values of subject impressions on different blinking rate and duration. A two-way repeated-measure of ANOVA [analysis of variance] was conducted with two within-subject factors: blinks duration, and blinking rate. A significant main effect were revealed in the blink duration factor ($F(4,132)=25.08$), $p<0.001, \eta^2=0.16$) and blinking rate factor ($F(2,132)=30.23$, $p<0.001$, $\eta 2=0.28$). The interaction effect between the blinks duration and blinking factor was significant ($F(8,132)=8.31$, $p<0.01$, $\eta^2=0.15$). Fig. 4 (b) also illustrates these results. Concerning 1B/s, multiple comparisons with Benferroni method show that there are significant main effect between conditions (such as 3s vs. 1s: $p<0.0001$, 3s vs. 2s: $p<0.001$, 3s vs. 4s: $p=0.003$, and 3s vs. 5s: $p<0.001$). Multiple comparisons with Benferroni method was conducted among blink duration [3s] and revealed significant differences between 3B/s and 2B/s ($p<0.001$), and between 3B/s and 1B/s ($p<0.0001$) respectively. For 1B/s, participants rated more in 3s condition ($M=5.7$) than in all other conditions and blinking rate [Fig. 4 (a)]. Thus, we may use 3s blink duration at the rate of 1B/s as the preferred blinking action to generate the gaze awareness behaviors of the robot.

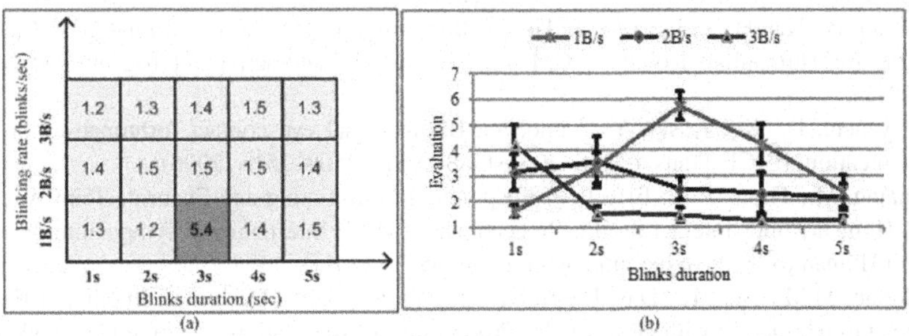

Fig. 4. Mean values of participant's impression on different blinking rate and duration

5 Conclusion

Blinking actions strengthens the feeling of being looked at and it can be used to convey an impression more effectively and colorfully understanding of human social behavior. Experimental results have also confirmed eye blinking actions proved helpful to relay the target participant that the robot was aware of his/her gaze. Participant's eyes couple with the robot's eyes during blinking causes they spent more time on gazing at the robot. This behavior may help the human for identifying the attention shift focus of robots. However, without blinking the robot may fail to create the feeling of eye contact being established, due to the lack of gaze awareness function. Our preliminary studies of eye contact have confirmed that the robot with eye blinking capability (about 3 seconds duration with 1 blinks/s blinking rate) is effective to create such feeling that a human makes eye contact with it. To generalize the finding, we will incorporates this robotic head in a humanoid robot and evaluate the effectiveness of the proposed robotic framework in real scenarios. These are left for future work.

References

1. Argyle, M.: Bodily Communication. Routledge (1988)
2. Kleinke, C.: Gaze and Eye Contact: A Research Review. Psychological Bulletin100(1), 78–100 (1986)
3. Farroni, T., Mansfield, E.M., Lai, C., Johnson, M.H.: Infant Perceiving and Acting on the Eyes: Tests of an Evolutionary Hypothesis. Journal of Exp. Child Psy. 85(3),199– 212 (2003)
4. Cranach, M.: The Role of Orienting Behavior in Human Interaction, Behavior and Environment.The Use of Space by Animals and Men, pp. 217–237. Plenum Press (1971)
5. Yoshikawa, Y., Shinozawa, K., Ishiguro, H., Hagita, N., Miyamoto, T.: The Effects of Responsive Eye Movement and Blinking Behavior in a Communication Robot. In: Proceedings of International Conference on Intelligent Robots and Systems, pp. 4564–4569 (2006)
6. Kanda, T., Ishiguro, H., Imai, M., Ono, T.: Development and Evaluation of Interactive Humanoid Robots. In: Proceedings on Human Interactive Robot for Psychological Enrichmen, vol. 92(11), pp. 1839–1850 (2004)
7. Mutlu, B., Kanda, T., Forlizzi, J., Hodgins, J., Ishiguro, H.: Conversational Gaze Mechanisms for Humanlike Robots. ACM Transactions on Interactive Intelligent Systems1(2) (2012)
8. Miyauchi, D., Kobayashi, Y., Kuno, Y.: Bidirectional eye contact forhuman-robot communication. IEICE Trans. of Info.& Syst. 88-D(11), 2509–2516 (2005)
9. Huang, M., Thomaz, L.: Effects of Responding to, Initiating and Ensuring Joint Attention in Human-Robot Interaction. In: Proceedings of IEEE International Symposium on Robot and Human Interactive Communication, pp. 65–71 (2011)
10. Hoque, M.M., Deb, K., Das, D., Kobayashi, Y., Kuno, Y.: Design an Intelligent Robotic Head to Interacting with Humans. In: Proceedings of International Conference on Computer and Information Technology, pp. 539–545 (2012)
11. Exline, R.V.: Multichannel Transmission of Nonverbal Behavior and the Perception of Powerful Men: The Presidential Debates of 1976. Power, Dominance, and Nonverbal Behavior. Springer-Verlag Inc. (1985)

Improvement and Estimation of Intensity of Facial Expression Recognition for Human-Computer Interaction

Kunal Chanda, Washef Ahmed, Soma Mitra, and Debasis Mazumdar

Centre for Development of Advanced Computing, Kolkata, India
{kunal.chanda,washef.ahmed,soma.mitra,debasis.mazumdar}@cdac.in

Abstract. With the ubiquity of new information technology and media, more effective and friendly methods for human computer interaction (HCI) are being developed. The first step for any intelligent HCI system is face detection and one of the most friendly HCI systems is facial expression recognition. Although Facial Expression Recognition for HCI introduces the frontiers of vision-based interfaces for intelligent human computer interaction, very little has been explored for capturing one or more expressions from mixed expressions which are a mixture of two closely related expressions. This paper presents the idea of improving the recognition accuracy of one or more of the six prototypic expressions namely happiness, surprise, fear, disgust, sadness and anger from the mixture of two facial expressions. For this purpose a motion gradient based optical flow for muscle movement is computed between frames of a given video sequence. The computed optical flow is further used to generate feature vector as the signature of six basic prototypic expressions. Decision Tree generated rule base is used for clustering the feature vectors obtained in the video sequence and the result of clustering is used for recognition of expressions. Manhattan distance metric is used which captures the relative intensity of expressions for a given face present in a frame. Based on the score of intensity of each expression, degree of presence of each of the six basic facial expressions has been determined. With the introduction of Component Based Analysis which is basically computing the feature vectors on the proposed regions of interest on a face, considerable improvement has been noticed regarding recognition of one or more expressions. The results have been validated against human judgement.

Keywords: Human Computer Interaction, Facial Expression Analysis, mixed expressions, Component Based Analysis, gradient based optical flow, decision tree, quantification of score, human psycho-visual judgement

1 Introduction

In the area of Human Computer Interaction (HCI), face is considered to be most important object to model the human behavior precisely. A facial expression is indeed an important signature of human behavior caused due to change in the facial muscles arrangement in response to a person's internal emotional states. For psychologists and

M.K. Kundu et al. (eds.), *Advanced Computing, Networking and Informatics - Volume 1*, Smart Innovation, Systems and Technologies 27,
DOI: 10.1007/978-3-319-07353-8_72, © Springer International Publishing Switzerland 2014

behavioral scientists, facial expression analysis and their quantification has been an active research topic since the seminal work of Darwin [1]. Ekman*et al.* [2] reported their result on categorization of human facial expressions as happiness, surprise, fear, disgust, sadness and anger. They further proposed that facial expressions are results of certain Facial Actions and they introduced the facial action coding system (FACS) [3] designed for human observers to detect subtle changes in facial appearance. However, much research has not been observed in recognizing accurately, one or more of the six basic expressions from two mixed expressions. In real life problems the confusion between two expressions poses an inherent problem of identifying correctly the mixed expressions. Improvement in this aspect based on Component Analysis introduces also the subsequent improvement in facial expression intensity. Recent psychological studies [4] suggests that to understand human emotion only classifying expression into six basic categories is insufficient while the detection of intensity of expression and percentage of each expression in case of mixed mode expression is important in the practical application of human computer interaction.

In the proposed framework we developed a method to extract flow of moving facial muscles or organs based on gradient based optical flow [5] from video sequences and analyze it with some specific parameters to recognize facial expressions. For this we have ensured adequate region of interests based on aspect ratio of face and head tilt within ±15°, to enhance quality of features, most suitable, for estimation of expression intensity. After the generation of optical flow feature vector we used Decision Tree based classifier [6] used to recognize one or more expressions for a given frame in a video sequence. This coupled with the computation of Manhattan distance function gave us the measurement of the intensity of expressions. The Manhattan distance between two vectors is the sum of the differences of their corresponding components. In our work for each frame of a video sequence the components based on projections of optical flow feature vectors for any expression are considered.

2 Optical Flow Based Feature Extraction

In the proposed system, feature computation is done locally within certain regions in which maximum change in frame intensity occurs due to movement of a specific muscle group [3] representing the six basic expressions. To find these regions which are known as the regions of interest, a priori information is gathered based on experiments. The gradient based optical flow is then computed locally on these regions of interest. Optical Flow is the apparent motion of brightness patterns in an image sequence. A gradient based method to compute the components of optical flow (u,v) from a pair of images was proposed by Horn and Schunck in their seminal paper [5]. The algorithm assumes that as points move, their brightness E(x,y) at a point (x,y) does not change significantly within a small time interval. This imposes brightness constraint which generates an ill-posed problem for obtaining two components of optical flow vector. The linear equation obtained from brightness constraint is given by,

$$uEx + vEy + Et = 0 \tag{1}$$

Here, E_x, E_y and E_t are the derivatives of E(x,y) along x, y and time (t) respectively. Horn and Schunck additionally assumed smoothness of the flow vector field, which imposes smoothness constraint on the problem. The equation obtained from smoothness constraint is given by,

$$\nabla^2 u + \nabla^2 v = 0 \tag{2}$$

These two constraint equations are solved iteratively to obtain numerical estimate of optical flow velocity components. In the current work, localized optical flow vector is computed within the 13 windows mentioned as the regions of interest. The direction of orientation of the windows with global horizontal axis is taken as the window symmetry axis. Projection P_{ij} of optical flow vectors $U_{ij}(X)$ is taken on the long symmetry axis of i_{th} window for the j_{th} pixel: $P_{ij=}U_{ij}(X).n_i$ where $U_{ij}(X)=(u_{ij}(x,y),v_{ij}(x,y))$ is the optical flow vector in j_{th} X=(x,y) of i_{th} window computed from two successive significant frames.n_i is the unit vector along the axis of i_{th} window (i= 0,...,12).After computing the optical flow vectors and their projections we compute the mean and standard deviation of the projected flow vectors for consecutive frames. With these we generate the feature vector to represent the basic expressions.These feature vectors are utilized for the recognition of the six expressions by using a rule base generated by a trained decision tree.

3 Rule Based Generation by Training a Decision Tree

A popular classification method called Decision tree [7] is used next to generate a rule base to classify the six basic facial expressions. A Decision tree results in a flow-chart like tree structure where each node denotes a test on an attribute value and each branch represents an outcome of the test. The tree leaves represent the classes. In the current work, the discriminatory power of each attribute is evaluated using Shanon's entropy. The computed 26 dimensional feature vectors are used as attributes to the decision tree since they capture the discriminatory characteristics of optical flow vectors to classify six basic expressions.

The sample space consists of facial expression video database of 50 subjects, divided into training set of 30 subjects and test set of 20 subjects. The CDAC, Kolkata facial expression data is used in our experiment. The videos were taken under constrained illumination and head pose within ±15°.

For training the decision tree, 30 subjects are taken from the database displaying video sequence of 6 basic facial expressions. 10 frames are chosen from each video sequence to compute optical flow. Using the training cases as input a decision tree is constructed.

4 Improvement in Intensity of Expressions Based on Component Based Analysis

In order to investigate the confusion of facial expressions related to mixed expressions we first conducted a comprehensive survey to study the human recognition rate of six basic facial expressions. Afterwards we analyzed the results regarding the confusion of facial expressions and compared the recognition rates of humans with our developed algorithm. We observed that for some group of mixed expressions the results were most confusing while for others the results were less confusing. Ekman and Friesen [3] has suggested particular region of interest for specific expression. From the current literature it has been found that variations of region of interest can reduce the confusion between two expressions. These variations of region of interest have been selected depending on the optical flow vector. Based on our work we have shown that the region of interests, depending on the optical flow vectors, can be modified. The proposed regions of interest are arrived based on the following observations. At first gradient based optical flow is computed globally on the face region comparing the neutral face and the face in the apex of the expression. Based on the spread of the magnitude and direction of optical flow vectors done locally within certain regions located in forehead, two eye brows, two eyes, nose, cheek, mouth and chin we obtain the proposed regions. It may be noted that maximum change in frame intensity occurs due to movement of a specific muscle group for displaying all six basic expressions [8]. The results shown with the help of the confusion matrix explains improved classification for the two expressions creating most confusion.

The survey to study the human recognition of facial expressions was conducted on 200 persons both male and female and age varying from 18 to 60 years. Each of the persons was shown six video sequences created by CDAC, Kolkata which contained one of the six basic expressions. Each person gave a score within a scale of 1 to 10 based on the intensity of expression. The scores are shown in Fig. 1. In this figure a table illustrating the score given by each of 200 persons to whom six video sequences were displayed. An intense gray scale value denotes a high score for a particular facial expression while light gray value denotes low score. From the figure it is obvious that happiness is best distinguished from the other facial expressions. Surprise gets most confused with fear, sometimes with sadness or anger and even less with disgust and happiness. It also known that fear is the hardest to discriminate from the other facial expressions. Fear often gets confused with surprise and sometimes with sadness or anger. Disgust gets also confused with fear and sometimes surprise. Anger also sometimes gets confused with sadness and fear or surprise. Sadness gets little confused with disgust and fear, but gets highly confused with anger.

From the above observations it is found that there are group of two mixed expressions which are most confusing. They are sadness with anger, anger with disgust, disgust with fear, fear with surprise and surprise with fear. Also another set of group of expressions which have less confusion within them are sadness with disgust, sadness with fear, anger with sadness, anger with fear, anger with surprise, disgust with surprise, fear with sadness, fear with anger, surprise with sadness, surprise with anger, surprise with disgust and surprise with happiness. The established literature suggests that there are certain components (region of interests) from the facial image that are responsible for projecting one or more expressions. In our experiment, we categorize the class and intensity of mixed expressions by automatically measuring the optical flow vector generated due to mutual change of intensity pattern for a current frame as compared with the neutral one. While carrying out experiments it has been found that the set of group of two mixed expressions which are most confusing are appropriately justified.

The region of interest responsible for creating confusion between different expressions are first identified using the spread of the optical flow vector in that region.

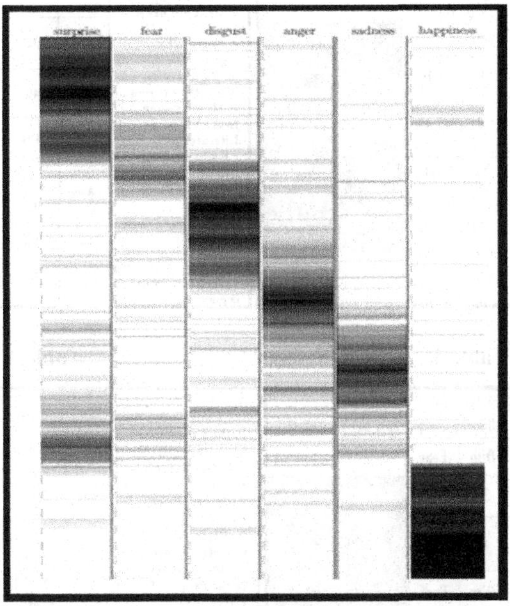

Fig. 1. The table shows result of humans specifying the facial expression visible in each of the video sequences containing one of the six basic expressions of the CDAC Kolkata Facial Expression Database

Based on the observation we have added or eliminated specific region of interest for reducing confusion while recognizing different expressions. The tables below show the comparison between the regions of interest suggested in established literature verses our consideration after conducting experiments.

C-Means clustering [7], [9] aims to partition video sequences with two mixed expressions into two separate clusters in which each of the video sequences belongs to the cluster with the nearest mean of the feature vector based on the proposed regions of interest. A confusion matrix is a specific layout that allows to quickly visualize the results of a clustering algorithm. Here for the first experiment in Table 1, video sequences with sadness or anger as dominant expression have been considered. The confusion matrix generated on the partitioned video sequence with two mixed expressions separate into two clusters based on C-Means clustering depending on the proposed region of interest of sadness are expressed in the first two rows of the table.

Out of 54 video sequences having dominant sadness expression, 28 and 26 video sequences are clustered as sadness and anger respectively. Similarly out of 61 video sequences having dominant anger expressions, 27 and 34 video sequences are clustered as anger and sadness respectively.

Table 1.

Sadness and Anger Considering 'sadness' the following components are suggested			
In the Established Literature:[3][10] i. Between eye inner brow region ii. Chin region iii. Mouth region		Sad	Ang
	Sad	28	26
	Ang	34	27
In the Proposed Literature: i. Between eye inner brow region ii. Chin region		Sad	Ang
	Sad	34	20
	Ang	29	32

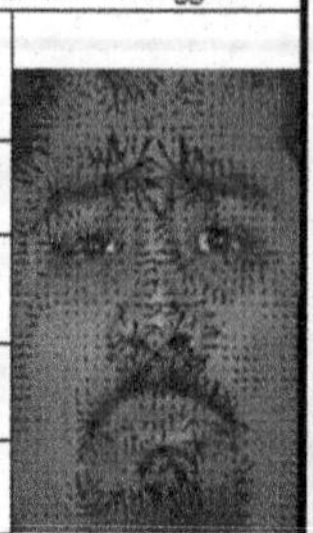

Table 2.

Anger and Disgust Considering 'anger' the following components are suggested			
In the Established Literature:[3][10] i. Right eye brow region ii. Left eye brow region iii. Right eye region iv. Left eye region v. Mouth region		Ang	Disg
	Ang	32	29
	Disg	40	27
In the Proposed Literature: i. Right eye brow region ii. Left eye brow region iii. Right eye region iv. Left eye region v. Right nasalobial region vi. Left nasalobial region		Ang	Disg
	Ang	35	26
	Disg	36	31

Table 3.

Disgust and Fear Considering 'disgust' the following components are suggested			
In the Established Literature:[3][10] i. Root of the nose region ii. Mouth region		Disg	Fear
	Disg	38	29
	Fear	43	14
In the Proposed Literature: i. Middle of the inner brow ii. Root of the nose region iii. Mouth region		Disg	Fear
	Disg	50	17
	Fear	30	27

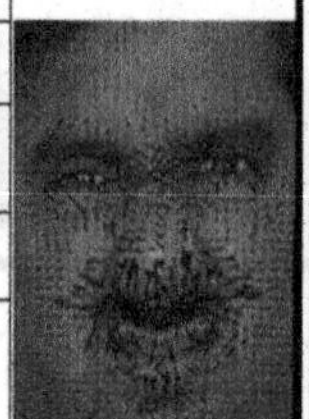

Table 4.

Fear and Surprise Considering 'fear' the following components are suggested			
In the Established Literature:[3][10] i. Right eye brow region ii. Left eye brow region iii. Right eye region iv. Left eye region v. Mouth region		Fear	Surp
	Fear	30	27
	Surp	32	21
In the Proposed Literature: i. Forehead region ii. Right eye brow region iii. Left eye brow region iv. Right eye region v. Left eye region vi. Mouth region		Fear	Surp
	Fear	34	23
	Surp	28	25

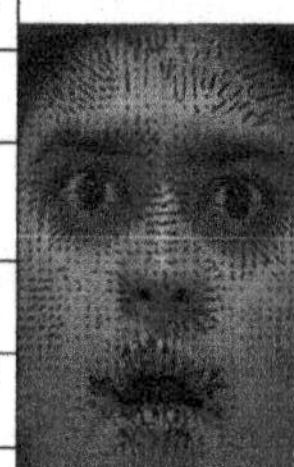

However considering the proposed region of interest for C-means clustering the performance has been increased from 28 to 34 for sadness dominant expression and from 27 to 32 for anger dominant expression. Similar type of confusion matrix for most confusing pair of expressions (as shown from Table 2 to Table 4) reveals that the performance of expression recognition increases by suitable selection of region of interest as proposed in this paper.

5 Estimation of Expression Intensity Based on Manhattan Distance Metric

In the present paper the Manhattan distance function is used to estimate the expression intensity. It computes the distance that gets traveled to get from one node of decision tree to the other for a feature of a particular expression computed between the neutral frame and frames with increasing emotional intensity. The Manhattan distance is also called the L_1 distance. If u = (x_1, y_1) and v = (x_2, y_2) are two points, then the Manhattan Distance between u and v is given by MH (u, v) = $| x_1 - x_2 | + | y_1 - y_2 |$.

Instead of two dimensions, in this paper we have considered features having 26-dimensions, such as a = (x_1, x_2, x_3... x_{26}) and b = (y_1, y_2, y_3... y_{26}) then the above equation can be generalized by defining the Manhattan distance between a and b as MH (a, b) = $| x_1-y_1| + | x_2-y_2| + \ldots + | x_{26}-y_{26}| = \sum | x_i - y_i|$.

In our work the Manhattan distance is computed at each node which denotes a test on an attribute value to give an absolute difference. The sum of the differences through those nodes for which a particular expression is estimated gives the total Manhattan distance which is normalized to give an estimate of intensity of expression for that particular frame with that of a neutral frame.

6 System's Performance Compared to Human Perception

In order to investigate the human recognition rate of facial expressions which are caused by one of the six basic expressions we have conducted a comprehensive survey. This is done so as to analyze the results and compare the performance of our system with manual recognition by human.

For this a joint experiment has been conducted at our institution and Psychology Department of University of Calcutta. The standardization of our system is based on human validation technique. Our survey questions around two hundred people about the facial expressions exhibited in each of the videos. Six sample videos (consists of one clip each of all the six basic facial expressions) were shown to two hundred people and asked to identify the six basic facial expressions in those video clips along with identifying the intensity of emotion by giving a score in the range between 1 and 10 for quantification of the score. Note that this video clips does not provide communication channels and further context information. Therefore, the participants are provided with the same information as current facial expression interpretation algorithms. This makes the comparison of human capabilities and current system algorithms appropriate.

Based on the manual survey's result this section calculates the recognition rate of humans and compares it to the recognition rate of our system's algorithm. For each video sequence, objectively comparing the accuracy requires to know about the correct facial expression. The average score based on the number of persons identifying the correct facial expressions for each of the six video sequences is depicted in Table 5. The next Table6 displays the data based on the performance of our system. Each of the video sequence is made to run through our system and after frame by frame analysis intensity of one of the expression is generated. After completion of analysis on the whole video, average score for that expression is computed based on the number of frames responding for that particular expression. It may be noted that the proposed method considering the suggested regions of interest helps to identify correctly intensity of one of the expression after undergoing frame by frame analysis.

The results show that the scores for each of the expression is within a unit signifying the fact that our system generated score for each of the expression is based on the same line as the average score computed on the number of people on whom survey was conducted.

Table 5.

	Happiness	Surprise	Fear	Sadness	Disgust	Anger
video1		5.4				
video2					5.9	
video3	6.4					
video4			6.0			
video5						6.5
video6				6.7		

Table 6.

	Happiness	Surprise	Fear	Sadness	Disgust	Anger
video1		4.5				
video2					4.8	
video3	7.1					
video4			5.5			
video5						5.1
video6				6.1		

7 Conclusion

In this paper, a Component Based Analysis on the regions of interest of a face is introduced. This work contributes to the recognition of mixed expressions and the confusion involved in the video. The system finds use in the treatment of mental and emotional disorders (patients suffering from autism with low IQ) through the use of psychological techniques designed to encourage communication of conflicts and insight into problems, with the goal being relief of symptoms, changes in behavior leading to improved social and vocational functioning and personality growth. For usage as a psychotherapeutic tool it is essential to capture accurately one or more expressions from mixed expressions and also to keep records of intensity of expression.

References

1. Darwin, C.: The expression of Emotions in Man and Animals. Univ. Chicago Press (1872)
2. Ekman, P.: Facial expressions and emotion. Amer. Psychol. 48, 384–392 (1978); Consulting Psychologists Press(1978)
3. Ekman, P., Friesen, W.V.: Facial Action Coding System (FACS): Manual. Pal Alto, Calf.
4. Ambadar, Z., Schooler, J., Cohn, J.F.: Deciphering the enigmatic face, the importance of facial dynamics in interpreting subtle facial expression. Psychological Science (2005)
5. Horn, B.K.P., Schunck, B.G.: Determing optical flow. Artificial Intelligence 17 (1981)
6. Dunn, J.C.: A Fuzzy Relative of the ISODATA Process and Its Use in Detecting Compact Well-Separated Clusters. Journal of Cybernetics 3, 32–57 (1973)
7. Gupta, G.K.: Introduction to data mining with case studies. Prentice Hall of India Private Limited (2006)
8. Reilly, J., Ghent, J., McDonald, J.F.: Investigating the dynamics of facial expression. In: Bebis, G., Boyle, R., Parvin, B., Koracin, D., Remagnino, P., Nefian, A., Meenakshisundaram, G., Pascucci, V., Zara, J., Molineros, J., Theisel, H., Malzbender, T. (eds.) ISVC 2006. LNCS, vol. 4292, pp. 334–343. Springer, Heidelberg (2006)
9. Bezdak, J.C.: Pattern Recognition with Fuzzy Objective Function Algorithms. Plenum Press, New York (1981)

Cognitive Activity Recognition Based on Electrooculogram Analysis

Anwesha Banerjee[1], Shreyasi Datta[2], Amit Konar[2],
D.N. Tibarewala[1], and Janarthanan Ramadoss[3]

[1] School of Bioscience & Engineering, Jadavpur University, India
[2] Department of Electronics & Telecommunication Engineering, Jadavpur University, India
[3] Department of Computer Science, TJS Engineering College, India
anwesha.banerjee@ymail.com, shreyasidatta@gmail.com,
konaramit@yahoo.co.in, biomed.ju@gmail.com, srmjana_73@yahoo.com

Abstract. This work is aimed at identification of human cognitive activities from the analysis of their eye movements using Electrooculogram signals. These signals are represented through Adaptive Autoregressive Parameters, Wavelet Coefficients, Power Spectral Density and Hjorth Parameters as signal features. To distinctly identify a particular class of cognitive activity, the obtained feature spaces are classified using Support Vector Machine with Radial Basis Function Kernel. An average accuracy of 90.39% for recognition of eight types of cognitive activities has been achieved in a one-versus-all classification approach.

Keywords: Activity Recognition,Electrooculogram,Support Vector Machine.

1 Introduction

Cognitive activity recognition finds applications in the development of ubiquitous computing systems [1]. Eye movements contain a high level of information regarding a person's activities and cognitive context [2]. There are different techniques to measure eye movements [3] that include Infrared Video System (IRVS), Infrared Oculography (IROG), Optical-type Eye Tracking System, Purkinje dual-Purkinje-image (DPI) and Electrooculogram (EOG). EOG has proved to be a very efficient tool for eye movement measurement, being cost effective, easy to acquire, simple to process and work in real time. The EOG system comprises of surface electrodes to be placed on skin surrounding the eye socket, and hence is noninvasive and also portable. The EOG signal [4] measures the potential difference between the retina and the cornea of the human eye. Significant applications of EOG signal analysis include the detection and assessment of many diseases as well as the development of eye movement controlled human computer interfaces and neuro-prosthetic aids [5-7].

The present work aims to recognize human cognitive activities by analysis of eye movements. Such a work can be utilized in the development of a ubiquitous computing environment that is aware of people's cognitive context by only studying eye movements. Eye movement characteristics can model the visual behavior in different

M.K. Kundu et al. (eds.), *Advanced Computing, Networking and Informatics - Volume 1*,
Smart Innovation, Systems and Technologies 27,
DOI: 10.1007/978-3-319-07353-8_73, © Springer International Publishing Switzerland 2014

cognitive contexts. Three major types of eye movements, namely, saccades, fixations and blinks are important in conveying information regarding a person's current activity [2]. The human eyes move constantly in saccades to build a 'mental map' of the visuals that is seen. Fixations are present in between saccades where the eyes focus on a particular location. Blinks are short duration pulses of relatively higher magnitude that are heavily characterized by environmental conditions and mental workload. In [2], features based on these three types of eye movements have been used to recognize cognitive context from EOG signals achieving average precision of 76.1%. In [8] reading activity based on saccades and fixations has been recognized with a recognition rate of 80.2%. In our previous works [9-10] we have classified EOG to recognize directional eye movements from some standard signal features.

In the present work, eye movements are recorded using a two-channel Electrooculogram signal acquisition system developed in the laboratory by surface electrodes from ten subjects while they performed eight different activities. The acquired EOG signals are filtered and four standard signal features, namely, Adaptive Autoregressive Parameters, Hjorth Parameters, Power Spectral Density and Wavelet Coefficients are extracted. These feature spaces are used independently as well as in combinations to classify the EOG signals using a Support Vector Machine with Radial Basis Function kernel, successfully recognizing eight different cognitive activities.

The rest of the paper is structured as follows. Section 2 explains principles and the methodology followed in the course of the work. Section 3 covers the experiments and results. Finally in Section 4 the conclusions are drawn and future scopes are stated.

2 Principles and Methodology

This section describes the process of Electrooculogram signal acquisition and processing to classify different cognitive activities. The methodology followed is illustrated in Fig.1.

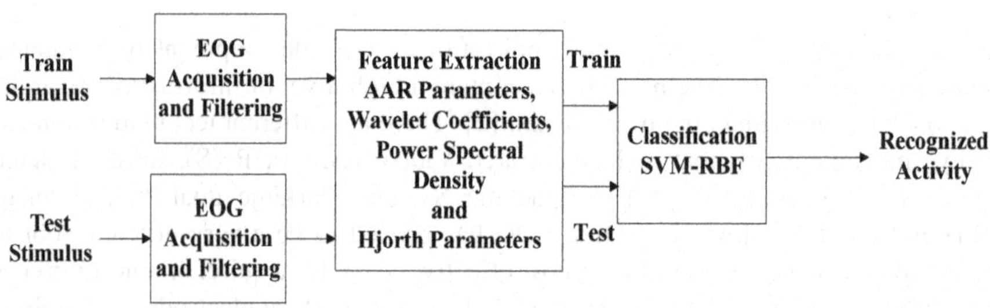

Fig. 1. Flowchart showing EOG Signal Processing and Classification

2.1 Electrooculogram Signal Acquisition and Pre-processing

Electrooculogram (EOG) [4], [11] is the potential between the cornea and the retina which is detected by the skin electrodes placed around the eye socket. EOG signal magnitudes have a range of 5-20 μV per degree of eye ball movement. When the gaze is shifted in the upward direction, the positive cornea becomes closer to the upper electrode, which becomes more positive, with zero potential at the electrode below the eye, and vice versa resulting negative and positive output voltage respectively. The value of EOG amplitude will be positive or negative depending upon the direction in which the eye is moved. Similarly the left and right eye movements can also be explained. Blinks are short duration pulses, having comparatively high amplitude.

The recording of the EOG signal has been done through a two channel system developed in the laboratory using five Ag/AgCl disposable electrodes at a sampling frequency of 256 Hz. Two electrodes are used for acquiring horizontal EOG and two for vertical EOG, while one electrode acts as the reference as shown in Fig. 2(a). The frequency range of the acquired EOG signal is below 20 Hz and the amplitude lies between 100-3500 mV. The circuit designed for signal acquisition has been shown in Fig. 2(b). The signal collected from the electrodes is fed to an instrumentation amplifier, implemented by IC AD620, having a high input impedance and a high CMRR. This output is given to a second order low pass filter with a cut off frequency of 20Hzto eliminate undesirablenoise. Different stages of filters, implemented using IC OP07s, provide various amounts of gain. Amplifier provides a gain of 200 and a gain of 10 is provided by the filter. Thus an overall gain of 2000 is obtained. For any kind of bio-signal acquisition, isolation is necessary for the subject's as well as the instrument's safety. Power isolation is provided by a dual output hybrid DC-DC converter (MAU 108) and signal isolation is achieved by optically coupling the amplifier output signal with the next stage through HCNR 200.For conversion of the signal in digital format, an Analog to Digital Converter is necessary.The Electrooculogram data has been acquired in the LabView 2012 platform for processing in the computer using National Instruments 12-bit ADC.

To eliminate undesirable noise and obtain EOG in the frequency range of 0.1 to 15Hz, the range where maximum information is contained, we implement band pass filtering. A Chebyshev band pass filter of order 6 has been used.

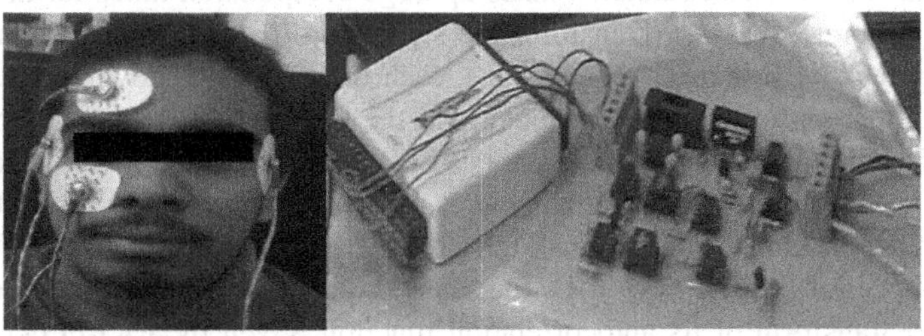

(a) (b)

Fig. 2. Data Acquisition System showing (a) Placement of Electrodes and (b) Acquisition Circuit Snapshot

2.2 Feature Extraction

Two time domain features, namely Adaptive Autoregressive Parameters and Hjorth Parameters, a frequency domain feature, namely, Power Spectral Density and a time-frequency correlated feature, namely, Wavelet Coefficient have been used as signal features in the present work.

An Autoregressive (AR) model is an efficient technique for describing the stochastic behaviour of a time series.EOG signals are time varying or non-stationary in nature, and hence the AR parameters for representing EOG signals should be estimated in a time-varying manner, resulting in Adaptive Autoregressive (AAR) parameters [12] explained by (1) and (2).

$$y_k = a_{1,k} * y_{k-1} + \ldots + a_{p,k} * y_{k-p} + x_k \tag{1}$$

$$x_k = N\{0,\ \sigma^2_{x,k}\} \tag{2}$$

Here $x_k$is a zero-mean-Gaussian-noise processwith time varying variance $\sigma^2_{x,k}$, the index k is an integer to denote discrete, equidistant time points,y_{k-i} with i = 1to p are the p previous sample values, p is the order of the AR model and $a_{i,k}$are the time-varying AR model parameters.In the present work, AAR parameter model was chosen to be of order 6 and AAR estimation was done using Kalman filtering with an update coefficient of 0.0085. The feature space for the EOG data is constructed by taking EOG signals of one second duration as a single instance for each of the two channels.

Wavelet transform [13] provides both frequency and time-domain analysis at multiple resolutions. It evaluates the transient behaviour of a signal by convoluting it with a localized wave called wavelet.

$$X_{WT}(\tau, s) = \frac{1}{\sqrt{|s|}} \int x(t).\psi^*\left(\frac{t-\tau}{s}\right) dt \tag{3}$$

Eq. (3) defines the wavelet transform for a signal$x(t)$ and a mother wavelet$\psi(t)$, shifted by τ amount and scaled by s=1/frequency. Eq. (3) can be sampled to produce the Wavelet series. This evaluation takes up large computation times for higher resolutions.In Discrete Wavelet Transform the signals are passed through high and low pass filters in several stages. At each stage, each filter output is downsampled by two to produce the approximation coefficient and the detail coefficient. The approximation coefficient is then decomposed again, to get the approximation and detail coefficients of the subsequent stages.In the present work,Haar mother wavelet and the fourth level approximate coefficients of discrete wavelet transform has been used.

Power Spectral Density [14] of a wide sense stationary signal $x(t)$is computed from the Fourier transform of its autocorrelation function, given by $S(w)$ in (4), where E denotes the expected value and T denotes the time interval.

$$S(w) = \frac{1}{T} \int_0^T \int_0^T E[x^*(t)x(t')] e^{iw(t-t')} dt dt' \tag{4}$$

For a time varying signal such as EOG PSD should be evaluated by segmenting the complete time series. In the present work PSD has been evaluated using Welch Method [15] that splits the signal into overlapping segments, computes the periodograms of the overlapping segments from their Fourier Transforms, and averages the resulting periodograms to produce the power spectral density estimate. PSD was computed in the frequency range of 1-15Hz using Welch Method with 50% overlap between the signal segments using a Hamming window. The resulting feature vector has dimensions of 15 for each channel of EOG data in the integer frequency points between 1 and 15 Hz.

Activity, Mobility and Complexity, collectively called Hjorth Parameters [16] describe a signal in terms of time domain characteristics.

2.3 Classification

Classification has been carried out using the well-known binary classifier, Support vector machine with Radial Basis Function kernel (SVM-RBF) [17]. For SVM-RBF, the width of the Gaussian kernel is taken as 1 and the one-against-all (OVA) classification accuracies are computed with a particular activity as a class and all the other activities comprising another class.

3 Experiments and Results

EOG data is collected from ten healthy subjects, five male and five female in the age group of 25±5 years using audio visual stimuliaccording to eight activities, namely reading, writing, copying text, web browsing, watching video, playing game, searching for words in a word maze and relaxing.

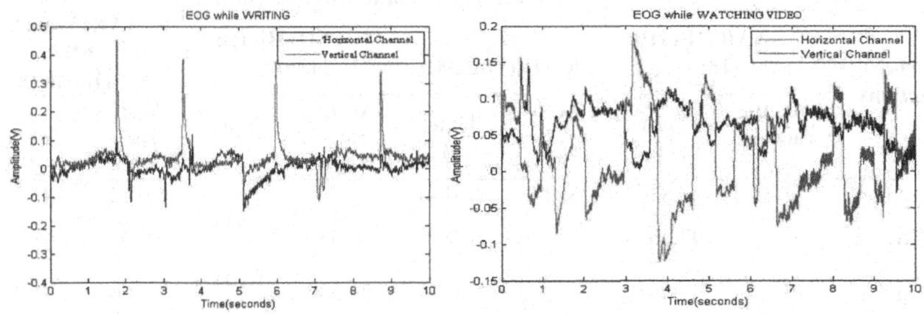

(a) (b)

Fig. 3. Filtered EOG signals from a particular Subject while Subject is (a) Writing and (b) Watching Video

The classifier is trained to recognize these eight activities using features from 60 seconds EOG data of each activity per subject. In the second phase the trained classifier is tested with unknown test stimuli. Classification accuracies are computed from respective confusion matrices. The filtered EOG signal corresponding to 10 seconds of data acquired from Subject 1 for different activities is shown in Fig. 3.

The results of classification in terms of classification accuracy and computation time (including feature extraction and classification) have been tabulated in Table 1 and 2, mentioning each feature vector dimension (in parenthesis), for two channels of EOG and indicating the maximum accuracy in bold for each class.

Table 1. Classification results using SVM-RBF on single feature spaces

Class of Activity	Average Performance over ten subjects							
	AAR (12)		Wavelet (32)		PSD (30)		Hjorth (6)	
	Accuracy (%)	Time (s)	Accuracy (%)	Time (s)	Accuracy (%)	Time (s)	Accuracy (%)	Time (s)
Reading	67.08	47.4801	87.92	0.2358	59.38	1.2815	**90.02**	0.7002
Writing	57.71	42.7067	**87.71**	0.1203	53.75	1.2574	84.58	0.6051
Copying	56.67	42.3419	88.33	0.1102	73.75	1.2538	**93.33**	0.6011
Web Browsing	72.92	42.3738	**90.04**	0.8873	72.75	1.2693	89.38	0.5978
Watching Video	69.37	43.3074	87.29	0.7786	73.75	1.2615	**89.58**	0.6074
Playing Game	50.83	42.5047	90.21	0.7687	52.71	1.2689	**91.04**	0.6384
Word Search	67.08	41.9151	78.54	0.8010	60.62	1.2404	**82.08**	0.6328
Relaxing	77.29	42.6593	87.71	0.7648	74.17	1.2571	**90.83**	0.6026

Table 2. Classification results using SVM-RBF on combined feature spaces

Class of Activity	Average Performance over ten subjects							
	AAR+Hjorth (18)		Wavelet+Hjorth (38)		PSD+Hjorth (36)		AAR+Wavelet+PSD+Hjorth (80)	
	Accuracy(%)	Time (s)	Accuracy (%)	Time(s)	Accuracy (%)	Time(s)	Accuracy (%)	Time(s)
Reading	88.12	53.8342	88.83	0.8146	75	1.4	90.42	41.3734
Writing	87.50	47.5630	85.46	0.8839	86.88	1.3446	90.42	41.6432
Copying	91.04	44.0070	89.33	0.8372	89.38	1.3046	92.08	41.3488
Web Browsing	90.00	45.4924	82.25	0.8706	85.62	1.3116	90.63	41.4337
Watching Video	86.46	46.7490	91.04	0.8258	58.13	1.334	90.21	41.9159
Playing Game	65.83	46.0374	90.71	0.8408	53.54	1.3404	93.13	41.544
Word Search	72.08	45.2229	82.71	0.8552	50.00	1.3446	87.08	41.728
Relaxing	84.17	46.5651	90.63	0.8407	67.50	1.33	89.17	41.8736

It is observed that using single feature spaces Hjorth Parameters provide highest average accuracy of 88.85% over all classes of activities. Thus other features are combined with Hjorth Parameters thereby showing a significant increase in classification accuracy as is evident from Table 2. The combination of all the four features produces the highest mean classification accuracy of 90.39% in 41.60 seconds on an average over all classes of activities.

4 Conclusions and Future Scopes

The present work is concerned with the recognition of eight human cognitive activities from eye movement analysis using Electrooculogram signals. AAR Parameters, Hjorth Parameters, Power Spectral Density and Wavelet Coefficients have been used as signal features and the feature spaces are classified with SVM-RBF classifier achieving a maximum average accuracy of 90.39% over eight cognitive activities.This work applicable in developing cognitive context-aware ubiquitous computing systems based on eye movement study. The work can be extended by incorporating EOG based wearable eye movement trackers that can provide real time analysis. Future works in this direction include the recognition of combinations of activities, the use of other features that can provide better distinction among cognitive activities and using other bio-signals with EOG for the same purpose.

Acknowledgment. This work is supported by Council of Scientific and Industrial Research (CSIR) and Jadavpur University, India.

References

1. Davies, N., Siewiorek, D.P., Sukthankar, R.: Special issue on activity based computing. IEEE Pervasive Computing 7(2) (2008)
2. Bulling, A., Ward, J.A., Gellersen, H., Troster, G.: Eye Movement Analysis for Activity Recognition. In: Proceedingsof 11th International Conference on Ubiquitous Computing, pp. 41–50. ACM Press (2009)
3. Deng, L.Y., Hsu, C.L., Lin, T.C., Tuan, J.S., Chang, S.M.: EOG-based Human–Computer Interface system development. Expert Systems with Application 37(4), 3337–3343 (2010)
4. Arden, G.B., Constable, P.A.: The electro-oculogram. Progress in Retinal and Eye Research 25(2), 207–248 (2006)
5. Stavrou, P., Good, P.A., Broadhurst, E.J., Bundey, S., Fielder, A.R., Crews, S.J.: ERG and EOG abnormalities in carriers of X-linked retinitis pigmentosa. Eye 10(5), 581–589 (1996)
6. Barea, R., Boquete, L., Mazo, M., López, E., Bergasa, L.M.: EOG guidance of a wheelchair using neural networks. In: IEEE International Conference on Pattern Recognition, pp. 668–671 (2000)
7. Banerjee, A., Chakraborty, S., Das, P., Datta, S., Konar, A., Tibarewala, D.N., Janarthanan, R.: Single channel electrooculogram (EOG) based interface for mobility aid. In: 4th IEEE International Conference on Intelligent Human Computer Interaction (IHCI), pp. 1–6 (2012)
8. Bulling, A., Ward, J.A., Gellersen, H., Tröster, G.: Robust Recognition of Reading Activity in Transit Using Wearable Electrooculography. In: Indulska, J., Patterson, D.J., Rodden, T., Ott, M. (eds.) PERVASIVE 2008. LNCS, vol. 5013, pp. 19–37. Springer, Heidelberg (2008)
9. Banerjee, A., Konar, A., Janarthana, R., Tibarewala, D.N.: Electro-oculogram Based Classification of Eye Movement Direction. In: Meghanathan, N., Nagamalai, D., Chaki, N. (eds.) Advances in Computing & Inf. Technology. AISC, vol. 178, pp. 151–159. Springer, Heidelberg (2012)

10. Banerjee, A., Datta, S., Pal, M., Konar, A., Tibarewala, D.N., Janarthanan, R.: Classifying Electrooculogram to Detect Directional Eye Movements. First International Conference on Computational Intelligence: Modeling Techniques and Applications. Procedia Technology 10, 67–75 (2013)
11. Roy Choudhury, S., Venkataramanan, S., Nemade, H.B., Sahambi, J.S.: Design and Development of a Novel EOG Biopotential Amplifier, International Journal of Bioelectromagnetism 7(1), 271–274 (2005)
12. Schlögl, A., Lugger, K., Pfurtscheller, G.: Using adaptive autoregressive parameters for a brain-computer-interface experiment. In:Proceedings of the 19th Annual International Conference of Engineering in Medicine and Biology Society 4, 1533–1535 (1997)
13. Pittner, S., Kamarthi, S.V.: Feature extraction from wavelet coefficients for pattern recognition tasks. IEEE Transactions on Pattern Analysis and Machine Intelligence 21(1), 83–88 (1999)
14. Saa, J.F.D., Gutierrez, M.S.: EEG Signal Classification Using Power Spectral Features and linear Discriminant Analysis: A Brain Computer Interface Application. In: Eighth Latin American and Caribbean Conference for Engineering and Technology (2010)
15. Welch, P.: The use of fast Fourier transform for the estimation of power spectra: a method based on time averaging over short, modified periodograms. IEEE Transactions on Audio and Electroacoustics 15(2), 70–73 (1967)
16. Hjorth, B.: Time domain descriptors and their relation to a particular model for generation of EEG activity. CEAN-Computerized EEG Analysis, 3–8 (1975)
17. Gunn, S.R.: Support Vector Machinesfor Classification and Regression. Technical report, University of Southampton (1998)

Detection of Fast and Slow Hand Movements from Motor Imagery EEG Signals

Saugat Bhattacharyya[1], Munshi Asif Hossain[1], Amit Konar[1], D.N. Tibarewala[2], and Janarthanan Ramadoss[3]

[1] Dept. of Electronics and Telecommunication Engineering, Jadavpur University, Kolkata, India
[2] School of Bioscience and Engineering, Department of Computer Science, Jadavpur University, Kolkata, India
[3] TJS Engineering College, Chennai, India
saugatbhattacharyya@gmail.com, asif_ju@live.com, konaramit@yahoo.co.in, biomed.ju@gmail.com, srmjana_73@yahoo.com

Abstract. Classification of Electroencephalography (EEG) signal is an open area of re-search in Brain-computer interfacing (BCI). The classifiers detect the different mental states generated by a subject to control an external prosthesis. In this study, we aim to differentiate fast and slow execution of left or right hand move-ment using EEG signals. To detect the different mental states pertaining to motor movements, we aim to identify the event related desynchronization/ synchronization (ERD/ERS) waveform from the incoming EEG signals. For this purpose, we have used Welch based power spectral density estimates to create the feature vector and tested it on multiple support vector machines, Nave Bayesian, Linear Discriminant Analysis and k-Nearest Neighbor classifiers. The classification accuracies produced by each of the classifiers are more than 75% with naïve Bayesian yielding the best result of 97.1%.

Keywords: Motor imagery, Brain-computer interfacing, Event related desynchronization/ synchronization, Electroencephalography, Pattern classifiers.

1 Introduction

Brain-computer interfacing (BCI) is a field of study which aims to provide a communication channel between the mental commands generated from the brain of a user and an external device (say, a robot) without any muscular intervention [1]. Earlier studies in BCI aimed at providing rehabilitation to physically challenged patients suffering from Amytrophic Lateral Sclerosis (ALS), cerebral palsy, stroke, paralysis and amputee [2]. But recent advances in BCI has opened newer fields of application in communication and control, gaming, robotics and military [3-5].

Electroencephalography (EEG) is the most frequently used form of brain measure because it is non-invasive, portable, easy to use, widely available, and has a

M.K. Kundu et al. (eds.), *Advanced Computing, Networking and Informatics - Volume 1*, Smart Innovation, Systems and Technologies 27,
DOI: 10.1007/978-3-319-07353-8_74, © Springer International Publishing Switzerland 2014

good temporal resolution [6]. EEG is a non-linear, non-stationary, non-Gaussian signal [1] and for this reason information relating to the mental states of the user cannot be ascertained directly from the raw data. Thus, to extract the relevant features and classify the mental states of the user, an EEG-BCI module is made of the following components: i) Pre-processing, where the incoming signals from the EEG amplifier are filtered to the required frequency bands, ii) Feature Extraction, where the characteristic information from the filtered EEG signals are extracted using signal processing algorithms, and iii) Classification, to decode between the different mental states from the EEG signal [7]. Existing literatures in BCI have widely used time-, frequency-, time-frequency-, and non-linear domain as feature extractors [1, 3, 5, 6, 8] and algorithms like Support Vector Machine (SVM), Linear Discriminant Analysis (LDA), Nave Bayesian and Neural Networks for classification [9].

Different mental states leads to generation of diverse brain signal modalities. A few well known ones are steady-state visually evoked potential (SSVEP), slow cortical potential (SCP), P300, event related desynchronization/synchronization (ERD/ERS) and error related potential (ErRP)[2]. The selection of brain signals (or signal modality), is an important issue in EEG-BCI analysis and it depends on the cognitive task performed by the subject. ERD/ERS originates during motor planning, imagination or execution (also referred to as motor imagery signals)[10] and thus, this signal finds relevance in our study. ERD/S waveforms are pre-dominantly obtained from the alpha (8-12 Hz) and central beta (16-24 Hz) rhythm[2].

Existing literatures [11-14] on motor imagery EEG signal classification have extensively classified between left and right motor imagery signals. In this paper, we aim to classify the speed of execution of the motor imagery tasks given to the subjects. For this purpose, we have designed a two-level hierarchical classification strategy, where the first level classifies between the left and right hand movement and the second level differentiates between fast and slow movement of the respective limb (as decided in the first level). Here, we have employed Welch based Power Spectral Density estimates in the alpha and central-beta band as features and linear discriminant analysis, k-nearest neighbors, naïve Bayesian and Support Vector Machines as classifiers. The results thus obtained, suggests that our proposed classification strategy is successful in classifying the mental states of fast and slow movement for both the left and right hand.

The rest of the paper is organized as follows: Section 2 describes the details of the experiments along with the data processing and classification strategies implemented in this study, the results are discussed in Section 3 followed by the concluding remark in Section 4.

2 Experiments and Methods

In this study, we have separated the fast and slow execution of the left and right hand movement according to the instructions given to the subject. The complete scheme for this study is given in Fig. 1. Following the data acquisition of EEG signals according to different mental states, the first step involves the filtering of the raw signal. In the second step, the features are extracted using Power Spectral Density leading to the formation of the feature vectors. The feature vectors are used as inputs to the classifiers to determine the mental state of the user. Nine right-handed subjects (4 male and 5 female) have participated in this study over 3 separate sessions organized in 3 different days.

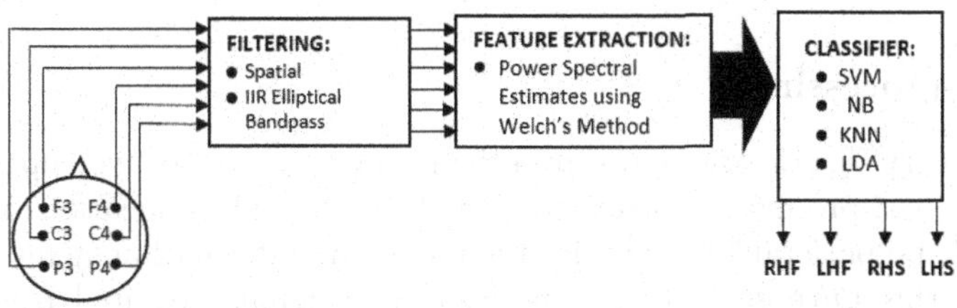

Fig. 1. A block diagram representation of our overall scheme

2.1 Visual Cue

The subjects perform the motor execution tasks based on the instructions given on a visual cue. Each session consists of 100 trials (25 each for right-hand fast move-ment (RHF), left-hand fast movement (LHF), right-hand slow movement (RHS) and left-hand slow movement (LHS)). The timing scheme for each session (as shown in Fig. 2) are as follows: In the first second, a fixation cross '+' is displayed on screen which is an indication to the subject to get ready to perform the task. In the next screen, instruction is given to the subject on the speed of execution, i.e., slow and fast movement. This screen lasts for 1 second following which left or right arrow is displayed according to which the subject moves the respective limb either slowly or briskly (based on the instruction in the previous screen) for 3 seconds. At the end of each trial, a blank screen of 2 seconds is displayed during which the subject can relax.

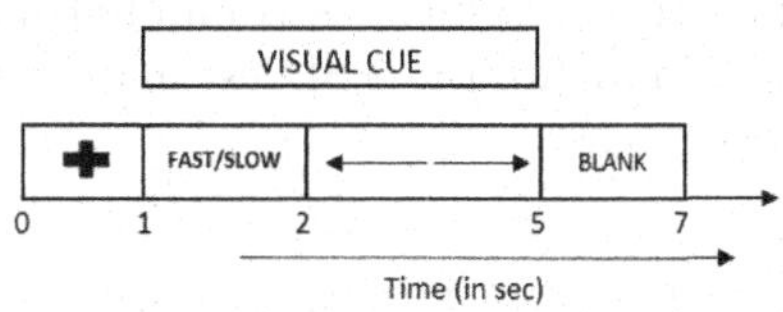

Fig. 2. The timing scheme diagram of a visual cue

2.2 Experimental Setup

In this experiment, a NeuroWin (manufactured by NASAN) EEG machine with 19 electrodes (Fp1, Fp2, F8, F4, Fz, F3, F7, T4, C4, Cz, C3, T5, T6, P4, Pz, P3, T7, O2, O1) is used to record the mental states of the subjects. The left ear is used as a reference and FPz location is grounded. The EEG is recorded using gold plated electrodes and the impedances are kept below 5 kΩ. The EEG signals are amplified, sampled at 250 Hz, and band-pass filtered between 0.5 and 35 Hz.

2.3 Pre-processing

Motor imagery signals originates from the primary motor cortex, supplementary motor area and pre-motor cortex from the brain [2]. Thus, locations in between the frontal and parietal lobe contains the maximum information on motor related tasks. For this purpose, we have analyzed the signals acquired from F3, F4, C3, C4, P3 and P4 electrode locations in this study. Before feature extraction, we spatially filter the signals from the six electrodes using Common Average Referencing (CAR) [2] method to reduce the effect of neighboring locations from these signals. Then, we temporally band pass filter the signals in the bandwidth of 8-25 Hz using an IIR elliptical filter of order 6, pass-band attenuation of 50 dB and stop-band attenuation of 0.5 dB. The merit of selecting elliptical filter lies in its good frequency-domain characteristics of sharp roll-off, and independent control over the pass-band and stop-band ripples [15].

Based on the timing sequence of the visual cue, sample points from the 2nd second to the 5th second (3 seconds in total) are extracted from each trial for data analysis.

2.4 Feature Extraction

Following pre-processing, we apply Welchs' based power spectral density to determine the feature vectors of the EEG signals at each trial. Welchs' method is a parametric method of estimation of power spectral density [16]. Power spectral density (PSD) portrays the distribution of power in the frequency domain. Welch method divides the time series data, $x(n)$ into possibly overlapping segments over a length L, a weighting vector w_k is applied to each segment and a modified periodogram of each segment is computed using a discrete Fourier transform(DFT) thereby averaging the PSD estimates. Thus Welch PSD estimate can thus be summed up as,

$$P(w_k) = \sum_{n=0}^{N-1} |DFT(x_n)|^2 \tag{1}$$

In this paper we have selected a hamming window of size 250 over the complete frequency range and an overlap percentage of 50%. From the whole frequency range of 0 to 125 Hz (since 250 Hz is the sampling frequency), only estimates from the bands of 8-12 Hz and 16-24 Hz are selected to construct the feature vector. The final dimension of the feature vector is 14. Fig. 3 illustrates an example of the power spectral density estimates of fast and slow movement for both limbs based on the EEG obtained from channel location C3.

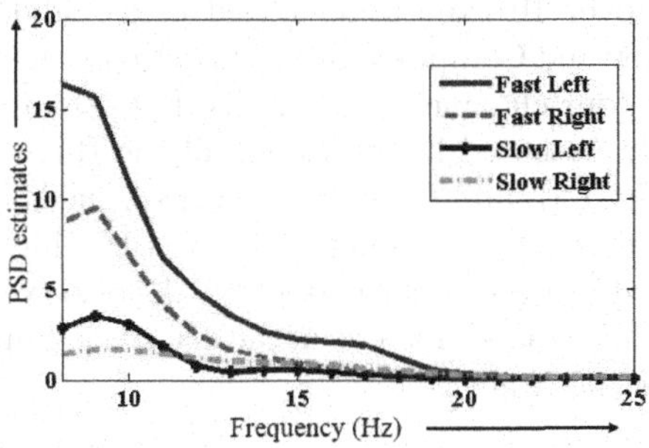

Fig. 3. The power spectral density estimates from electrode C3 for the four different mental tasks performed by a subject

2.5 Classification

The classification scheme implemented in this study is shown in Fig. 4. Two levels of hierarchical classifiers are implemented, where the first level classifies between the left and right hand movement (Classifier 1) and the second level differentiates between fast and slow movement (Classifier 2 and 3). Consequently, the final output is in the form of right-hand fast movement (RHF), left-hand fast movement (LHF), right-hand slow movement (RHS) and left-hand slow movement (LHS).

In this paper, we have used support vector machine (SVM), naïve Bayesian (NB), linear discriminant analysis (LDA) and k-nearest neighbor (kNN) to distinguish between the different levels of classifiers [9].

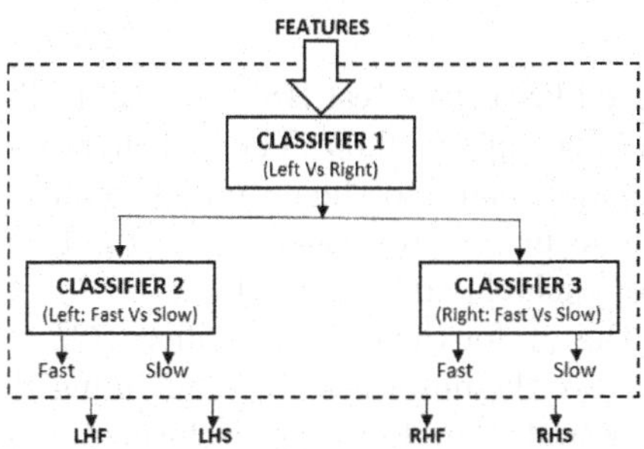

Fig. 4. The hierarchical classification strategy implemented in this study.

3 Results and Discussions

The analysis of the whole experiment is performed in a Matlab 2012b platform on a computer with the following specifications: Processor Intel i7 @ 3.2 GHz, 8 GB RAM.

The total feature vector is partitioned into two separate datasets, training and testing dataset, over 10 runs using k-fold cross validation approach [17]. Classification Accuracy and Computational Time are the metrics used to analyze the dataset using the four classifiers: SVM, NB, LDA, and kNN and the results are shown in Table 1. As noted in the table, all the three classifiers: Classifier 1 (CL1), Classifier 2 (CL2), and Classifier 3 (CL3), yields an accuracy of more than 75%. Also, the three NB classifiers produces the highest accuracy (more than 95%) as compared to the other classifiers. As observed from Table 1, NB classifier maintains a fine trade-off between the accuracies and the computational time taken to process the result.

Table 1. Accuracies of the classifiers obtained during training for 9 subjects

S.ID	SVM			NB			LDA			kNN		
	CL1	CL2	CL3	CL1	CL2	CL3	CL1	CL2	CL3	CL1	CL2	CL3
1	88.2	82.3	76.4	100	100	94.1	82.3	82.3	76.4	88.2	82.3	76.4
2	76.4	76.4	76.4	100	100	94.1	82.3	82.3	76.4	88.2	88.2	82.3
3	88.2	70.5	70.5	100	100	94.1	82.3	76.4	70.5	88.2	76.4	76.4
4	82.3	82.3	76.4	94.1	100	100	88.2	76.4	76.4	82.3	82.3	76.4
5	76.4	70.5	70.5	100	94.1	100	82.3	76.4	70.5	82.3	82.3	76.4
6	88.2	70.5	70.5	100	100	94.1	88.2	88.2	76.4	88.2	82.3	82.3
7	82.3	88.2	82.3	94.1	94.1	100	82.3	88.2	82.3	76.4	76.4	88.2
8	76.4	82.3	82.3	94.1	94.1	94.1	76.4	82.3	82.3	76.4	82.3	88.2
9	82.3	76.4	76.4	94.1	94.1	100	76.4	82.3	82.3	82.3	82.3	82.3
Avg.	82.3	77.7	75.8	97.3	97.3	96.7	82.3	81.7	77.1	83.6	81.7	81.0
C.T.	0.3107			0.0541			0.0330			0.0338		

We have also employed Friedman Test [18] to statistically validate our results. The significance level is set at α =0.05. The null hypothesis here, states that all the algorithms are equivalent, so their ranks should be equal. We consider the mean classification accuracy (from Table 2) as the basis of ranking. Table 2 provides the ranking of each classifier algorithm.

Now, from Table 2, the χ_F^2 for the three classifiers $(CL1, CL2, CL3) = (16.44, 17.76, 20.61) >= 9.488$, So, the null hypothesis, claiming that all the algorithms are equivalent, is wrong and, therefore, the performances of the algorithms are determined by their ranks only. It is clear from the table that the rank of NB is 1, claiming NB outperforms all the algorithms by Friedman Test.

Table 2. Accuracies of the classifiers obtained during training for 9 subjects

S.ID	SVM			NB			LDA			kNN		
	CL1	CL2	CL3	CL1	CL2	CL3	CL1	CL2	CL3	CL1	CL2	CL3
1	2.5	3	3	1	1	1	4	3	3	2.5	3	3
2	4	4	3.5	1	1	1	3	3	3.5	2	2	2
3	2.5	4	3.5	1	1	1	4	2.5	3.5	2.5	2.5	2
4	3.5	2.5	3	1	1	1	2	4	3	3.5	2.5	3
5	4	4	3.5	1	1	1	2.5	2.5	3.5	2.5	2.5	2
6	3	4	4	1	1	1	3	2	3	3	3	2
7	2.5	2.5	3.5	1	1	1	2.5	2.5	3.5	4	4	2
8	3	3	3.5	1	1	1	3	3	3.5	3	3	2
9	2.5	4	4	1	1	1	4	2.5	2.5	2.5	2.5	2.5
Rj	3.06	3.44	3.5	1	1	1	3.11	2.78	3.22	2.83	2.78	2.28

4 Conclusion and Future Direction

The paper proposes a classification strategy towards discriminating between fast and slow movement of the left and right hand. The feature vectors are prepared using a power spectral density estimates using Welch method and are fed as inputs to the SVM, NB, LDA and kNN classifiers. Here, we have incorporated two levels of classification: the first level classifies between left and right movement and the second level classifies between fast and slow movement. The results show that NB classifier yields the best results with 97.1% of average classification accuracy in 0.0541 seconds.

Based on the promising result obtained in this study, we propose to move forward towards real-time control problems in our future study. We intend to control the movement of a robotic arm which can further help in the development of neuro-prosthetics in future.

Acknowledgments. I would like to thank University Grants Commission, India, University of Potential Excellence Programme (Phase II) in Cognitive Science, Jadavpur University and Council of Scientific and Industrial Research, India.

References

1. Daly, J.J., Wolpaw, J.R.: Brain-computer interfaces in neurological rehabilitation. Lancet. Neurol. 7, 1032–1043 (2008)
2. Dornhege, G.: Towards Brain-Computer Interfacing. MIT Press (2007)
3. McFarland, D.J., Wolpaw, J.R.: Brain-computer interface operation of robotic and prosthetic devices. Computer 41(10), 52–56 (2008)

4. Bermudez i Badia, S., Garcia Morgade, A., Samaha, H., Verschure, P.F.M.J.: Using a Hy-brid Brain Computer Interface and Virtual Reality System to Monitor and Promote Cortical Reorganization through Motor Activity and Motor Imagery Training. IEEE Trans. Neural Sys. Rehab. Eng. 21(2), 174–181 (2013)
5. Bordoloi, S., Sharmah, U., Hazarika, S.M.: Motor imagery based BCI for a maze game. In: 4th Int. Conf. Intelligent Human Computer Interaction (IHCI), Kharagpur, India, pp. 1–6 (2012)
6. Millan, J.R., Rupp, R., Muller-Putz, G.R., Murray-Smith, R., Giugliemma, C., Tangermann, M., Vidaurre, C., Cincotti, F., Kubler, A., Leeb, R., Neuper, C., Muller, K.R., Mattia, D.: Combining brain-computer interfaces and assistive technogies: State-of-the-art and challenges. Front. Neurosci. 4, 1–15 (2010)
7. Bhattacharyya, S., Sengupta, A., Chakraborti, T., Konar, A., Tibarewala, D.N.: Automatic feature selection of motor imagery EEG signals using differential evolution and learning automata. Med. & Bio. Eng. & Comp. 52(2), 131–139 (2014)
8. Zhou, W., Zhong, L., Zhao, H.: Feature Attraction and Classification of Mental EEG Using Approximate Entropy. In: 27th Ann. Int. Conf. Eng. Med. & Bio. Soc., pp. 5975–5978 (2005)
9. Theodoridis, S., Koutroumbas, K.: Pattern Recognition, 4th edn., pp. 13–22. Academic Press (2009)
10. Qiang, C., Hu, P., Huanqing, F.: Experiment study of the relation between motion complexity and event-related desynchronization/synchronization. In: 1st Int. Conf. Neural Interface & Cont. 2005, pp. 14–16 (2005)
11. Chai, R., Ling, S.H., Hunter, G.P., Nguyen, H.T.: Mental non-motor imagery tasks classifi-cations of brain computer interface for wheelchair commands using genetic algorithm-based neural network. In: The 2012 Int. Joint Conf. Neural Networks, pp. 1–7 (2012)

A Solution of Degree Constrained Spanning Tree Using Hybrid GA with Directed Mutation

Sounak Sadhukhan, and Samar Sen Sarma

Department of Computer Science and Engineering
University of Calcutta, Kolkata, India
sounaksju@gmail.com, sssarma2001@yahoo.com

Abstract. It is always an urge to reach the goal in minimum effort i.e., it should have a minimum constrained path. The path may be shortest route in practical life, either physical or electronic medium. The scenario can be represented as a graph. Here, we have chosen a degree constrained spanning tree, which can be generated in real time in minimum turn-around time. The problem is obviously NP-complete in nature. The solution approach, in general, is approximate. We have used a heuristic approach, namely hybrid genetic algorithm (GA), with motivated criteria of encoded data structures of graph and also directed mutation is incorporated with it and the result is so encouraging, that we are interested to use it in our future applications.

Keywords: NP-complete, degree constrained spanning tree, graphical edge-set representation, graphical edge-set crossover, and graphical edge-set directed mutation.

1 Introduction

Many real-world situations can be described by a set of points, inter-connected through handshaking theorem (i.e., graphical structure) [1]. Graph plays a vital role in computer science. The *Degree Constrained Spanning Tree* (DCST) is one of the classic combinatorial graph problems. It is a NP-complete problem, so, till now, we do not have an exact algorithm which solves it in polynomial time. Obviously, we now seek guidelines for solving this problem around in a way to compromise. The strategies are usually and roughly divided into two classes: Approximation algorithm and heuristic approaches [2]. Heuristic approaches usually find reasonably good solutions reasonably fast. GA, is one of the famous heuristic search technique, tries to emulate Darwin's Evolutionary process. It deals efficiently, with digital encoded problem, and DCST problem is in perfect unison. Our proposed solutions may not be optimal but acceptable in some sense. After running several times, GA will converge each time, possibly at different optimal chromosomes and the schemata, which promise convergence, are actually indicative of the regions in the search space where good chromosomes may be found. So, therefore, the GA is coupled with a random Depth First Search (DFS) mechanism to find the optimal chromosome in a region. This is a kind of hybridization of our algorithm. The data structure employed here is graphical edge-set representation [3]. In addition, it is also shown, how a problem dependent

M.K. Kundu et al. (eds.), *Advanced Computing, Networking and Informatics - Volume 1*, Smart Innovation, Systems and Technologies 27,
DOI: 10.1007/978-3-319-07353-8_75, © Springer International Publishing Switzerland 2014

hill-climbing approach can be effectively incorporated into the mutation operator, so that, it gives a better solution.

The following section describes the DCST problem. The Solution methodology is explained in Section 3, which includes generation of initial populations, graphical edge-set crossover, graphical edge-set directed mutation; and an experimental comparison to other optimization techniques is given in Section 4. Some concluding remarks are made in Section 5.

2 Degree Constrained Spanning Tree

Problem definition: A simple, symmetric and connected graph G = (*V*, *E*), where *V* is the set of vertices and *E* denotes the set of edges with positive integer *N* • |*V*|. The problem is to find out a Spanning Tree (ST) T in G, which contains no vertices whose degree is larger than *N*, where *N*> 0 [4].

Narula and Ho, first proposed this problem in designing electrical circuits [5]. Hamiltonian path problem is a special edition of this problem, where degree of every vertices of the spanning tree has the upper bound 2. Finding such a path is related to the travelling salesman problem. Hamiltonian path, travelling salesman problem all are the examples of NP-complete problem [4]. In this case, a spanning tree may have degree up to /*V*| - 1. Finding a DCST in a graph is usually a hard task.

3 Solution Methodology

3.1 Representation of Chromosomes in GA in ST Problems

Representation of each individual is very important in GA. Here each individual must representa spanning tree of a graph. In literature, there are several encoding schemes available for representing a spanning tree namely edge encoding, Pr fer encoding, vertex encoding and graphical edge-set representation etc. [3], [6], [7], [8], [9]. These representations have both advantages and disadvantages over another.

There are lots of advantages to use graphical edge-set to represent a spanning tree, over the other encoding schemes. Using this representation, only feasible solutions (spanning tree) are always generated. Each spanning tree can be represented by a unique chromosome and vice-versa. GA operators like crossover and mutation are computationally efficient and incorporated with this representation.

3.2 Generation of Initial Population

In order to create initial population, this initial population generation algorithm always produces the valid ST. We have used two algorithms to generate initial population; one is random algorithm and another is random DFS algorithm. Actually, this mimics classical hybridization of graph optimization algorithms. Random DFS generates 20% individuals of the initial populations. The elitist model of GA is used here.

In the random algorithm, E_T is the subset of *E*, contains the elements, which forms an individual (spanning tree). *V* is the set of vertices in a graph G. Initially set E_T is empty and each vertices of V, formed individual components like, $\{V_1\}$, $\{V_2\},\ldots,\{V_n\}$.Fig. 1shows two individuals (spanning tree) are created from a graph G with seven vertices and twelve edges. The random algorithm for generating initial population is given below. Complexity of this algorithm is O(|E|).

Algorithm Initial_Population(G(V, E))
Step 1: E_T • , S•E.
Step 2: For all edges (i,j) • S in random order, do
Check whether $\{E_T U\ (i,j)\}$ is a circuit
i) To check the circuit we find the end vertices of the edge (i,j).
ii) Assume $V_i$and $V_j$are the end vertices of the edge (i,j). Then check whether $V_i$and $V_j$belong to same or different connected component. If $V_i$and $V_j$are belong to different connected component, then there would be no circuit at all, otherwise a circuit would be present.
Step 3: If no circuit is present then, E_T • $\{E_T U\ (i,j)\}$, S • $S - (i,j)$.
Step 4: If $|V_T| = |V| - 1$, Return E_T.
Step 5: Else if circuit present then ignores the edge and S • $S - (i,j)$.
Step 6: End loop.
Step 7: Stop.

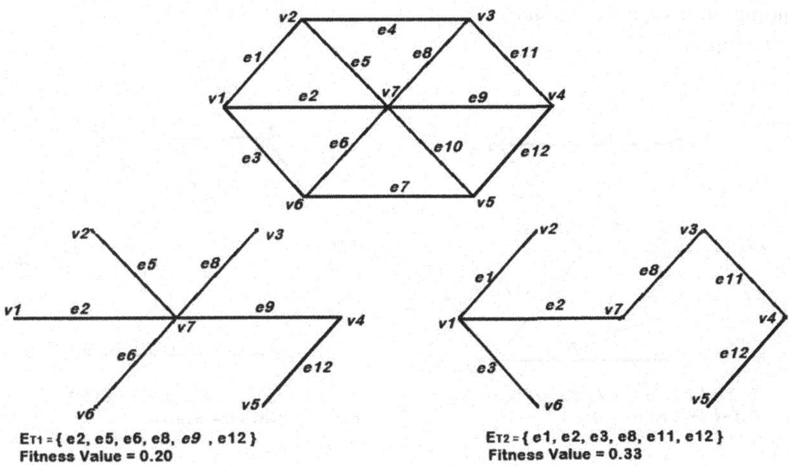

Fig. 1. A graph G and its two individuals are created from the above graph

Fitness function gives a value for each chromosome, which will help us to identify the good individual among the population. Definition of fitness function is varies problem to problem, depending upon the objective function. In this problem, fitness function is a fraction, represented by the reciprocal of maximum degree of each chromosome (degree of tree). ST with lowest degree represents the fittest individual among population.

3.3 Graphical Edge-Set Crossover

After selecting two individuals $T_1$and $T_2$from the population, this algorithm tries to develop two new offsprings C_1 and C_2if possible. Like the initialization procedure, in this algorithm, edges are processed one after the another in a random order and only those edges, that do not form a cycle, are included. Fig. 2 shows that two valid offsprings are genarated from the two individuals shown in Fig. 1.This algorithm takes O(|*E*|) time.

Algorithm Graphical_Edge-set_Crossover(T_1, T_2, |V|, E)
Step 1: Select any two individuals $T_1$and $T_2$from population depending on their fitness values using Roulette Wheel selection.
Step 2: Make some edge sets like, A • E_{T1} U E_{T2}, B • E_{T1} • E_{T2}, C •E – A ($E$is the edge set of graph G), D •A – B, C_1• and C_2• .
Step 3: Take edges from B one by one randomly and add into E_{C1} till all the edges of B are explored and | E_{C1}| • |V| - 1. If any edge creates cycle, discard the edge.
Step 4: If | E_{C1}| < |V| - 1, take edges from C,do the same as Step 3.
Step 5: If | E_{C1}| < |V| - 1, take edges from D,do the same as Step 3.
Step 6: If | E_{C1}| = |V| - 1, then return C_1,if C_1 is valid. Otherwise discard it.
Step 7: Develop $E_{C2,}$ as same way of $E_{C1,}$ from the set D and C respectively.
Step 8: If two offsprings are valid, two individuals from the population are replace depending on their fitness values.
Step 9: Stop.

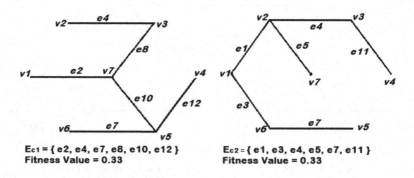

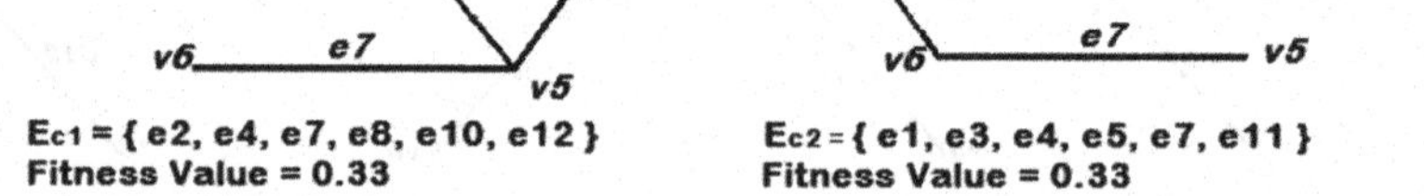

Fig. 2. Two individuals are created through graphical edge-set crossover from the individuals

3.4 Graphical Edge-Set Directed Mutation

The mutation operator brings diversity in the population of the potential solutions and inhibits premature convergence. Mutation alone induces a random walk through the search space. In this problem, we have used directed mutation which uses hill-climbing approach. The directed mutation is based on the principle of addition of an edge from a particular set and deletion an edge from a created cycle in the tree depending on the objective function, which leaves a new valid tree with better or same fitness than the previous one. Fig. 3 gives an idea that how a new individual is created from the old one using graphical edge-set directed mutation algorithm and it has been also shown that how its degree would be better or same as the old individual.The computational effort for the procedure is $O(|E|)$.

Algorithm Graphical_Edge-Set_Directed_Mutation (E_T, $|V|$, E)
Step 1: Identify the vertices with maximum degree and second maximum degree in the tree E_T and put them in the set S (Provided, if second maximum degree is 1, then take only the vertices with maximum degree).
Step 2: Remove all edges adjacent to those vertices of set S and removed edges are put into set R and put the neighbors of S in the set P.
Step 3: The tree E_T becomes E_T• (disconnected). Randomly search an edge from the set (E - E_T) whose end vertices are in P and add it to E_T• (E is the set of all the edges of graph G).
Step 4: Add those removed edges from set R in E_T• one by one in random order, till $|V_T|$ • $|V|$ - 1, in such a way that no circuit has been created.
Step 5: If addition of any edge creates cycle, discard that edge. E_T is replaced with E_T•, if E_T• is a valid tree.
Step 6: Stop.

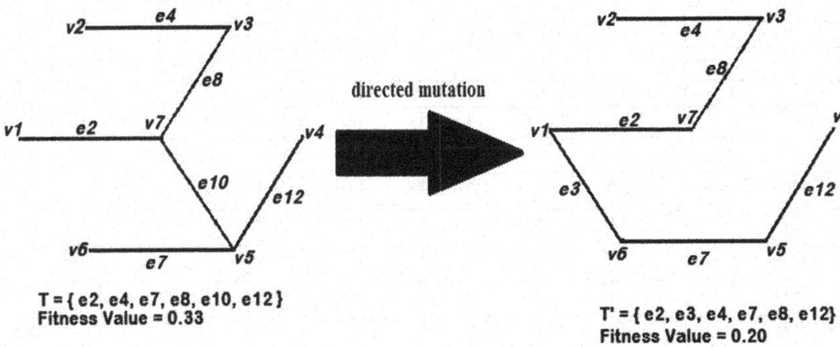

Fig. 3. An individual is created (*right*) through mutation from another individual (*left*)

Lemma 1. The above directed mutation algorithm always creates a feasible solution with the fitness value never greater than the fitness value of the previous.

Proof: Assume the tree contains n vertices. Assume, n_1 and $n_2$are number of vertices with maximum and second maximum degree respectively and number of deleted edges is e_d. So, number of remaining edges will be = $(n - 1- e_d)$. After addition of an edge among the components, the number of edges will be $(n - e_d)$. Then, removed edges are added one after another randomly. If all the removed edges are added, then the number of edges in the tree will be, $(n - e_d+ e_d) = n$. Then, there must exist a cycle which contains the newly added edge. The created cycle is also included the vertices with the maximum degree or second maximum degree. To remove the cycle, this algorithm removes an edge, which is adjacent with a maximum degree vertex or a second maximum degree vertex. Such an edge after being removed leaves a valid tree and degree of the adjacent vertices remains either same or reduced by one. It proves the lemma.

4 Experimental Result and Comparison

We have compared the result of this proposed algorithm with an existing approximation algorithm and genetic algorithm with normal mutation in Table 1. This experiment has been done on random graphs and each of the procedures was executed 10 times on each random graph. Simulation has been done in MATLAB 7.0.1. Then, we have calculated the percentage of errors with respect to the best value we get among the three procedures and plotted it as a graph to compare the percentage of errors in terms of the size of random graphs in Fig. 4. Here we consider, initial population size = 40, p_c= 0.5 and p_m= 0.4, total no. of iterations = 500.

Table 1. The performances in terms of quality of solutions of the three procedures

|V|	|E|	Procedure A				Procedure B				Procedure C			
		B_A	W_A	AVG_A	E_A	B_B	W_B	AVG_B	E_B	B_C	W_C	AVG_C	E_C
20	36	3	3	3.0	0.0	3	3	3.0	0.0	3	4	3.4	13.33
30	150	3	3	3.0	0.0	3	4	3.4	13.33	3	5	3.8	26.60
40	220	3	4	3.5	16.66	3	4	3.7	23.33	4	5	4.1	36.66
50	260	3	4	3.9	30.00	4	4	4.0	33.33	3	5	4.0	33.33
60	369	4	4	4.0	0.0	4	5	4.1	2.50	4	4	4.0	0.0
70	491	4	5	4.1	2.50	4	5	4.2	5.00	4	5	4.3	7.50
80	215	3	4	3.9	30.00	4	4	4.0	33.33	3	6	4.5	50.00
90	186	3	4	3.9	30.00	4	4	4.0	33.33	4	7	4.7	56.66
100	275	3	4	3.9	30.00	4	5	4.5	50.00	4	6	5.2	73.33
120	184	3	4	3.9	30.00	4	5	4.7	56.66	4	6	5.0	66.66
140	2381	4	4	4.0	0.0	5	5	5.0	25.00	4	6	5.3	32.50
160	800	4	4	4.0	0.0	5	5	5.0	25.00	4	6	5.0	25.00
180	1200	4	4	4.0	0.0	4	5	4.9	22.50	4	6	5.1	27.50
200	1600	4	4	4.0	0.0	5	5	5.0	25.00	5	7	5.7	42.50
250	2091	4	5	4.1	2.50	5	6	5.1	27.50	5	6	5.1	27.50

Procedure A, **Procedure B** and **Procedure C** denote Genetic Algorithm with directed mutation, Genetic Algorithm with crossover and normal mutation and Approximation algorithm respectively. $\mathbf{B}_i$, $\mathbf{W}_i$ and $\mathbf{AVG}_i$ denotes the best, worst and average result produced by procedure *i*, $\mathbf{E}_i$= % of error produced by $\boldsymbol{i}$ = [{AVG_i– min(B_A, B_B, B_C)}/ min(B_A, B_B, B_C)] × 100.

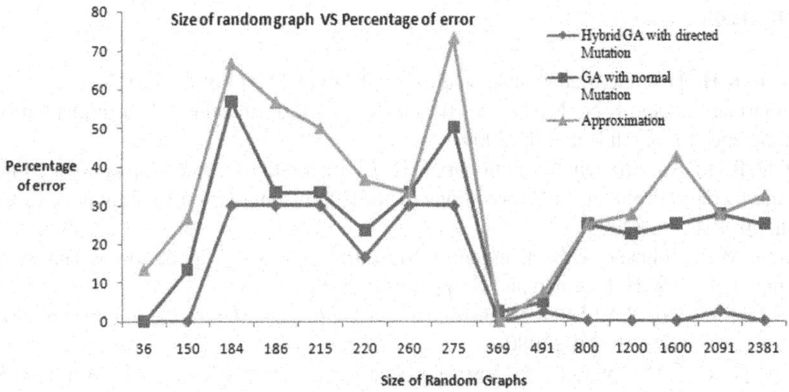

Fig. 4. Percentage of error (*in average*) respect to the size

5 Conclusion

This paper describes an approach to solve degree constrained spanning tree problem using hybrid GA. This method gets the optimal solution, through randomly generating a series of chromosomes, which represents a spanning tree in form of edge-set, and being operated in the way of genetic algorithms.

Our results show that, this approach is competitive with existing traditional approximation algorithm. The research work shows that, it is very important and crucial to choose the encoding and selection strategy in optimizing combinatorial problems and the graphical edge-set representation is really a fine encoding scheme for generating a tree for this problem. Most importantly, our algorithm always produces only feasible candidate solutions, if possible. Initial population, crossover and directed mutation is done in O(|*E*|) time. After each iteration, weaker individuals are obsolete by the more fit individuals, so, the average fitness value of the population is also increased.

This problem has huge importance in wireless ad-hoc network or mobile network at the time of routing. If it is possible to generate a spanning tree dynamically at the time of routing and whose degree is minimum, then noise and congestion will be reduced and routing will be faster. Similarly, it can be used in traffic signaling system, air traffic control, electrical network etc. We are optimistic that our approach will open up a vista of new solutions to a fascicle of practical problems.

Acknowledgements. We acknowledge Prof. Samiran Chattopadhyay of Jadavpur University and Prof. Malay Kumar Kundu of Indian Statistical Institute, Kolkata, for encouragement and help. We also thank Jadavpur University and University of Calcutta for providing necessary support.

References

1. Rosen, K.H.: Discrete Mathematics and its Application.TMH Edition (2000)
2. SenSarma, S., Dey, K.N., Naskar, S., Basuli, K.: A Close Encounter with Intractability. Social Science Research Network (2009)
3. Raidl, R.G.: An Efficient Evolutionary Algorithm for the Degree Constrained Minimum Spanning Tree Problem. In: Proceedings of the 2000 Congress on Evolutionary Computation, pp. 104–111 (2000)
4. Garey, M.R., Johnson, D.S.: Computers and Intractability – A guide to the Theory of NP Completeness. W.H. Freeman and Company (1999)
5. Narula, S.C., Ho, C.A.: Degree-constrained minimum spanning tree. Computers and Operations Research 7, 239–249 (1980)
6. Zhou, G., Gen, M., Wut, T.: A New Approach to the Degree-Constrained Minimum Spanning Tree Problem Using Genetic Algorithm. In: IEEEInternational Conference on Systems, Man and Cybernetics, pp. 2683–2688 (1996)
7. Knowles, J., Corne, D.: A New Evolutionary Approach to the Degree-Constrained Minimum Spanning Tree Problem. IEEE Transaction on Evolutionary Computation 4, 125–134 (2000)
8. Zeng, Y., Wang, Y.: A New Genetic Algorithm with Local Search Method for Degree-Constrained Minimum Spanning Tree problem. In: IEEE International Conference on Computational Intelligence and Multimedia Applications, pp. 218–222 (2003)
9. Zhou, G., Meng, Z., Cao, Z., Cao, J.: A New Tree Encoding for the Degree-Constrained Spanning Tree Problem. In: IEEE International Conference on Computational Intelligence and Security, pp. 85–90 (2007)

Side Lobe Reduction and Beamwidth Control of Amplitude Taper Beam Steered Linear Array Using Tschebyscheff Polynomial and Particle Swarm Optimization

Prarthana Mukherjee[1], Ankita Hajra[1], Sauro Ghosal[1], Soumyo Chatterjee[1], and Sayan Chatterjee[2]

[1] Department of ECE, Heritage Institute Of Technology, Kolkata-700107, India
[2] Department of ETCE, Jadavpur University, Kolkata-700032, India
sayan1234@gmail.com, m.prarthana@rediffmail.com, soumyo33@yahoo.co.in

Abstract. Present paper examines the various aspects of beam steered linear array of isotropic radiators with uniform inter-element spacing. In order to control beam broadening and to achieve side lobe level (SLL) reduction in beam steered array, a novel method has been proposed modifying primarily the search space definition for Particle Swarm Optimization. Tschebyscheff polynomial and PSO has been used to develop the proposed method. Search space for PSO has been defined using Tschebyscheff polynomial for an amplitude taper beam steered linear array. PSO with the information of where to search finds the optimum excitation amplitude of the beam steered linear array to either achieve reduced Side Lobe Level and narrow beamwidth within the beam steering range.

1 Introduction

The synthesis problem of reduced SLL and narrow beamwidth for linear array has been the subject of many investigations till date. Conventional analytical methods such as Tschebyscheff or Taylor Method are widely used [1],[2] for solving synthesis problem related to uniformly space linear array of isotopic elements. For linear array, Dolph had used Tschebyscheff polynomial to obtain a radiation pattern with equal SLL.

Another synthesis problem related to linear array is that of beam steering or beam scanning widely used in Radio Detection and Ranging(RADAR) applications. Beam steering in linear array can be achieved by either varying the phase excitations or varying the frequency of operation [3],[4].

Conventional techniques cannot be used to solve multi objective synthesis problems. This led to the introduction of evolutionary computing techniques [5],[7]. These techniques are used for achieving single objective like SLL reduction or multiple objectives like SLL reduction and null placement. One such evolutionary algorithm that is extensively used for solving different synthesis problems related to linear array is Particle Swarm Optimization (PSO) [6],[7]. PSO has advantages

M.K. Kundu et al. (eds.), *Advanced Computing, Networking and Informatics - Volume 1*, Smart Innovation, Systems and Technologies 27,
DOI: 10.1007/978-3-319-07353-8_76, © Springer International Publishing Switzerland 2014

compared to other optimization technique. Among its advantages are ease of implementation and less sensitivity to optimization parameters [10].

In the present work, by using PSO, SLL reduction and beam broadening problem in beam steered linear array has been addressed. To overcome the search space limitation of PSO a new method using Tschebyscheff polynomial has been proposed. The proposed method has been used in both single and multiple objective optimization problems. Single objective of reduced SLL and multiple objectives of reduced SLL and beam width control have been considered. Efficacy of the proposed method has been illustrated by considering a 12 element array at a desired SLL of -20 dB for different main beam positions.

2 Beam Steered Linear Array and Tschebyscheff Polynomial

In linear array, elements lie along a straight line with equal inter element spacing as shown in Fig.1. Equation (1), represents normalized array factor for even numbered linear array. The inter-element spacing of this even numbered linear array is uniform with constant phase difference [3]. In (1), k=2 / is the wave number, a_n is the excitation amplitude of the nth element, d is the uniform spacing, $_s$is the angle where the main beam has to be steered or placed.

$$AF(\theta) = \sum_{n=1}^{N} a_n \cos\left[\left(\frac{2n-1}{2}\right) kd\left(\cos\theta - \cos\theta_s\right)\right] \tag{1}$$

From (1) it can be said that the array factor of an even numbered linear array is a summation of cosine terms. This form is same as that of the Tschebyscheff polynomials [1]. The unknown coefficients of the array factor can be determined by equating the series representing the cosine terms of the array factor to the appropriate Tschebyscheff polynomial [2] . The order of the polynomial should be one less than the total number of elements of the array.

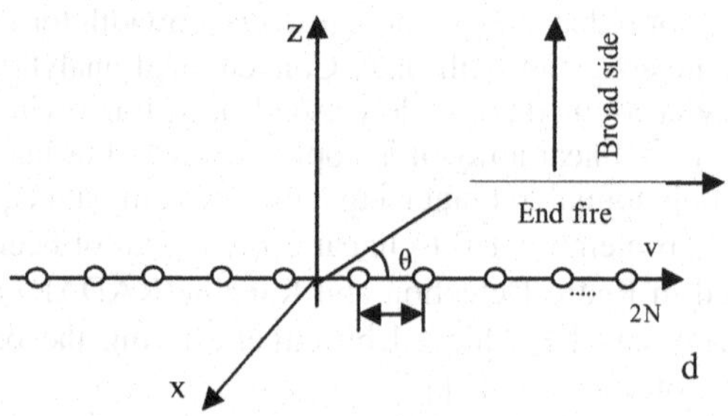

Fig. 1. Geometry of the 2N-isotropic element symmetric linear array placed along the y-axis

3 PSO on Linear Array Design

PSO is a stochastic optimization technique that has been effectively used to solve multidimensional discontinuous optimization problems in a variety of fields [8], [9],[10]. Let the optimization objective is to achieve the desired SLL reduction by varying the excitation amplitude. According to PSO swarm of bees is going to search for the location where desired SLL is present. Swarm of bees here represents possible solution sets. Location of each solution set is being defined by the values of excitation amplitudes. During the search process each solution set updates its location and velocity on two pieces of information. The first is its ability to remember the previous location where it has found the desired SLL (particle best (p_{best})). The second information is all about the location where SLL close to desired SLL is found by all the solution sets (bees) of the swarm (global best (g_{best})) at present instant of time. This process of continuous updating of velocity and position continues until one of the solution set (bee) finds the location of desired SLL within the defined search space. This location gives the value of optimum excitation amplitude.In present application of PSO, particle's velocity is manipulated according to the following equation:

$$v_{n+1} = wv_n + c_1 rand\ ()[p_{bestn} - x_n] + c_2 rand\ ()[g_{bestn} - x_n] \tag{2}$$

In (2), v_n is the velocity of the particle in the nth dimension , x_n is the particle's coordinate in the nth dimension. Parameter w is the inertial weight that specifies the weight by which the particle's current velocity depends on the previous velocity. The starting value w is 0.9 and has been decreased linearly up to 0.4. In this work c_1 is decreased from 2.5 to 0.5 whereas c_2 is increased from 0.5 to 2.5[9]. Function rand ( ), is a random function generating a number in the range [0,1]. After the time step, the new position of the particle is given by:

$$x(t+1) = x(t) + v(t+1) \tag{3}$$

The criterion for termination depends upon either of the two end results stated. The iteration continues until it equals the pre-defined value of number of iterations or till the desired value is acquired.

4 Proposed Method

Conventional PSO algorithm initially does not know where to search for the optimum values. The search space has to be defined randomly in order to begin the optimization process. In the proposed method the excitation amplitude of each array element is calculated using the method developed by Dolph [2]. From the solution upper and lower limit of search space is defined for the optimization process. This facilitates PSO with the knowledge of where to search. In PSO concept of fitness function guides the bees during their search for the optimum position within the search space. If it is required to obtain only the desired SLL (SLL $(dB)_d$) then the following function is used to assess the fitness.

$$f = \left| SLL(dB)_d - \max\left\{ 20\log\left| \frac{AF(\theta)}{AF(\theta)_{max}} \right| \right\} \right| \tag{4}$$

The above fitness function has two parts. First part is the desired SLL and second part is the calculated SLL obtained from PSO algorithm. At each iteration by varying the excitation amplitude PSO tries to reduce the difference value until it reaches 0.To obtain both the desired SLL (SLL $(dB)_d$) and beamwidth following function is used to assess the fitness.

$$f = \left| SLL(dB)_d - \max\left\{ 20\log\left| \frac{AF(\theta)}{AF(\theta)_{max}} \right| \right\} \right| + \beta\left| FNBW_c - FNBW_d \right| \tag{5}$$

In (5), is a positive weighting factor determining the influence of first null beamwidth (FNBW) expression in the fitness function. In (4) and (5) $FNBW_c$ is the calculated FNBW and $FNBW_d$ is the desired FNBW. Desired FNBW is taken at broadside position. Both the fitness functions (4 and 5) are defined in the Side Lobe Region (SLR). The SLR is as defined in (6).

$$SLR = \max\left(\left|AF(\theta)\right|\right)_{\theta \in \theta_{SLL}} ; \theta = \left[0, \theta_{FNL}\right] \cup \left[\theta_{FNR}, 180\right] \tag{6}$$

where $_{FNL}$ and $_{FNR}$ are left and right first nulls around the broadside main beam. For symmetrically specified beamwidth (BW) these values are

$$_{FNR/FNL} = |\text{minimum of}(_{null\_right / null\_left} - s)| \tag{7}$$

Here , $_{null\_left}$ represents the set of null positions (value of 0) in the array factor expression less than s. And $_{null\_right}$ represents the set of null positions in the array factor expression greater than s.

5 Experimental Setup and Results

In this section capability of proposed method has been demonstrated to check beam broadening and to achieve desired SLL in case of beam steered uniformly spaced linear array. If the optimization objective is to achieve only the desired SLL then fitness function as defined in (4) is to be used. Using this fitness function and considering linear array of 12 isotopic elements PSO finds the optimum excitation for different main beam position within the defined scan angle range for a desired SLL of -20 dB. Fig. 2 shows the radiation pattern of an optimized 12 element linear array and conventional DolphTschebyscheff array. The main beam is positioned at 80 degree. Corresponding convergence curve is shown in Fig. 3. Number of iterations in the convergence curves (Fig.3,Fig.5), shows number of iteration cycle in the best run.

Different parameters for this array with different main beam position are summarized in Table I. From Table I it is observed that the desired objective is met for most values of beam steering position. Only for the values of beam steering close to end fire positions (0° or 180°) the desired SLL is not achieved but is close to the desired value.

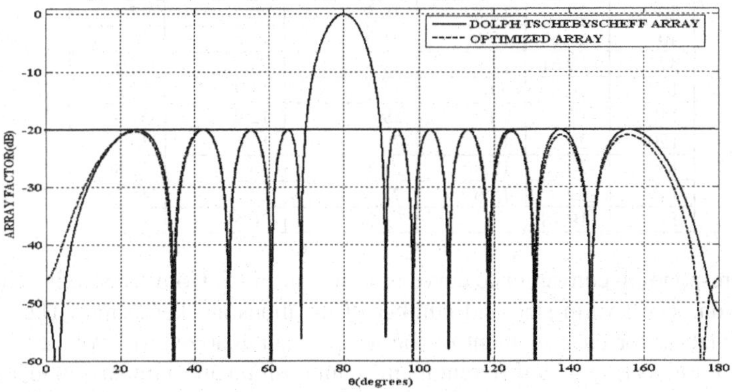

Fig. 2. Radiation pattern of 12 elements DolphTschebyscheff array and optimized linear array with desired SLL of -20 dB and main beam at 80 degree

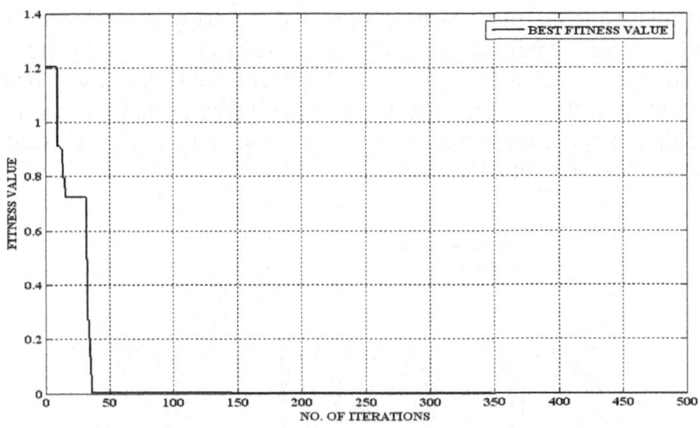

Fig. 3. Convergence curve for 12 elements optimized linear array with main beam placed at 80° and SLL of -20 dB

By using the proposed method both SLL reduction and beamwidth control is to be achieved for beam steered linear array. A 12 element linear array of isotropic elements with uniform spacing of 0.5λ is considered as an example. The desired SLL level is set at -20 dB and with different main beam position broadside FNBW is taken as desired FNBW. The radiation pattern of conventional DolphTschebyscheff array and optimized array is given in Fig.4. Number of elements for both the cases is 12 and main beam position is at 120°.

Table 1. Parameters of uniformly spaced 12 elements optimized array for different main beam position and desired SLL is -20dB

s	SLL (dB)	HPBW (deg.)	First Null position (deg.)		FNBW (deg.)
			Left	*Right*	
80	-20	9.6	11.9	11.3	23.2
70	-20	10	12.5	12	24.5
60	-20	10.6	14.3	13	27.3
50	-20	12.8	16.3	20.7	37
40	-19.66	15.6	21.9	22.5	44.4
37	-19.61	17	29.1	23.5	50.3
100	-20	9.6	11.4	11.4	22.8
110	-20	9.8	19.2	12.2	31.4
120	-20	11	20	13.6	33.6
130	-20	12.8	21.5	16.3	37.8
140	-19.5	15.6	23.5	22	45.5
143	-19.78	17	24.3	26.5	50.8

The corresponding convergence curve is shown in Fig.5. Convergence is achieved at 345th iteration. The swarm size and number of iterations are kept same. The optimum current excitation for each element of the array is summarized in Table 2. Different parameters for this array with different main beam position are summarized in Table 3.

The proposed methods objective was to keep the SLL at -20 dB and FNBW of 22.2 deg. As seen from Table 3 the objective of getting a SLL of -20 dB is achieved up to steering angle of 70° (110°). The FNBW variation stays within 1 degree whereas in the conventional case the variation is up to 1.6 degree. Excitation amplitude of end elements increases as the main beam position of the array is steered in the end fire direction (0° or 180°).When steering angle is varied from 90° to 0° directivity value increases and when it moves close to 0° the directivity value decreases. When steering angle is varied from 90° to 180° directivity value increases and when it moves close to 180° the directivity value decreases. When steering angle is varied from 90° to 0° (or 90° to 180°) HPBW value increases.

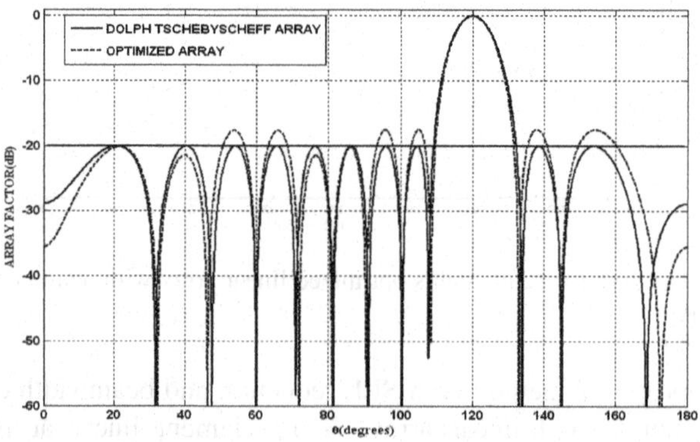

Fig. 4. Radiation pattern of 12 elements DolphTschebyscheff array and optimized linear array at desired SLL of -20 dB with main beam at 120 degree

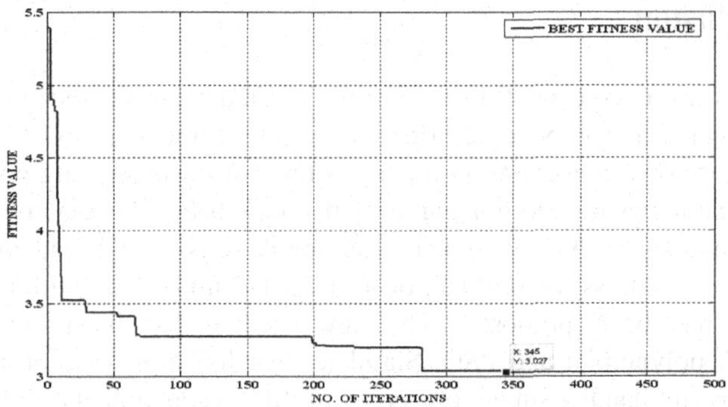

Fig. 5. Convergence curve for 12 elements linear array with main beam placed at 120° and SLL of -20 dB

Table 2. Excitation amplitude and directivity of 12 elements optimized array for different main beam position and SLL of -20dB

$_s$ (deg.)	Directivity (dB)	Optimized array (Excitation amplitude)
80	10.6364	2.0871 1.9724 1.7778 1.4491 1.1959 1.4473
70	10.6302	2.0239 1.9277 1.7185 1.3622 1.2313 1.3079
60	10.6969	1.9053 1.8228 1.6585 1.3679 1.2277 1.6694
50	10.6638	1.7067 1.7487 1.4227 1.4226 1.3871 2.1995
40	10.4833	1.3391 1.1998 1.3032 1.1896 1.8507 2.3195
100	10.6091	2.0547 1.8565 1.7426 1.3091 1.2216 1.2516
110	10.6264	2.0493 1.8972 1.8010 1.3254 1.2642 1.3103
120	10.7001	1.8695 1.8233 1.4441 1.5305 1.1920 1.5459
130	10.6842	1.7738 1.5386 1.3630 1.5058 1.3100 2.0448
140	10.599	1.1952 1.3872 1.2437 1.4313 1.8598 2.0847

Table 3. Parametric comparison between 12 element DolphTschebyscheff array and optimized array with desired SLL of -20dB and FNBW of 22.6 degree

$_s$ (deg.)	**DolphTschebyscheff array (SLL=-20dB)**				**Optimized Array**				
	HPBW (deg.)	**First Null position (deg.)**		**FNBW (deg.)**	**SLL (dB)**	**HPBW (deg.)**	**First Null position (deg.)**		**FNBW (deg.)**
		Left	***Right***				***Left***	***Right***	
80	9.3445	11.5	11.1	22.6	-20	9.4	11.4	11.2	22.6
70	9.798	12.4	11.4	23.8	-20	9.4	11.5	11.2	22.7
60	10.644	13.9	12.1	26	-17.78	10	12.8	11.3	24.1
50	12.0683	16.7	13.3	30	-14.13	11.2	13.6	11.3	24.9
40	14.5053	23.5	15	38.5	-10.01	12.8	15.2	11.4	26.6
100	9.3445	11.1	11.5	22.6	-20	9.4	11.2	11.3	22.5
110	9.798	11.4	12.4	23.8	-20	10	11.2	12	23.2
120	10.644	12.1	13.9	26	-17.48	10	11.1	12.5	23.7
130	12.0683	13.3	16.7	30	-14.25	10.4	11.2	13.4	24.6
140	14.5053	15	23.5	38.5	-9.862	11.4	11.6	15.5	27.1

6 Conclusion

In this paper, various aspects of non-uniformly excited beam steered symmetric linear array with equal SLL has been descried elaborately. Simulated results reveals, for a particular linear array configuration there is a limit of beam steering positions. It has been shown that in beam steered linear array the main beam is not symmetrical around the main beam position. Also with beam steering there is beam broadening in a linear array. In order to address the problem of beam broadening along with reduced SLL a new method has been proposed. The new method has been developed using Tschebyscheff polynomial and PSO. Simulated results shows, the proposed method has the ability to handle single objective of SLL reduction as well as multiple objectives of SLL reduction and beamwidth control for a linear array within the beam steering range. Hence, in multiple objectives scenario proposed method partially achieves the desired objectives and needs further investigation for beam steered linear array with reduced SLL and narrow beamwidth.

References

1. Taylor, T.T.: One Parameter Family of Line Sources Producing Modified Sin(πu)/πu Patterns. Hughes Aircraft Co. Tech. Mem. 324, Culver City, Calif. Contract AF 19(604)-262-F-14 (1953)
2. Dolph, C.: A Current Distribution for Broadside Arrays which Optimizes the Relationship between Beamwidth and Side Lobe level. Proceeding IRE 34(5), 335–348 (1946)
3. Balanis, C.A.: Antenna theory Analysis and Design, 3rd edn. John Wiley, New York (2005)
4. Mailloux, R.J.: Phased Array Antenna Handbook, 2nd edn. Artech House Inc. (2005)
5. Yan, K.K., Lu, Y.: Sidelobe reduction in arraypattern synthesis using genetic algorithm. IEEE Transactions on Antenna and Propagation 45(7), 1117–1121 (1997)
6. Khodier, M.M., Christodoulou, C.G.: Linear array geometry synthesis with minimum sidelobe level and null control using particle swarm optimization. IEEE Transactions on Antenna and Propagation 53(8), 2674–2679 (2005)
7. Chatterjee, S., Chatterjee, S., Poddar, D.R.: Side Lobe Level Reduction of a Linear Array using Tschebyscheff Polynomial and Particle Swarm Optimization. In: International Conference on Communications, Circuits and Systems, KIIT University (2012)
8. Kennedy, J., Eberhart, R.: Particle swarm optimization. In: Proceedings of IEEE International Conference Neural Networks, vol. IV, pp. 1942–1948 (1995)
9. Ratnaweera, A., Halgamuge, S.K., Watson, H.C.: Self-Organizing Hierarchical Particle Swarm Optimizer with Time-Varying AccelerationCoefficients. IEEE Transactions on Evolutionary Computation 8(3), 240–255 (2004)
10. Venter, G., Sobieszczanski-Sobieski, J.: Particle Swarm Optimization. AIAA Journal 41(8), 1583–1589 (2003)

Pervasive Diary in Music Rhythm Education: A Context-Aware Learning Tool Using Genetic Algorithm

Sudipta Chakrabarty[1], Samarjit Roy[2], and Debashis De[2]

[1] Department of Master of Computer Application, Techno India,
Kolkata – 7000091, West Bengal, India
[2] Department of Computer Science & Engineering, West Bengal University of Technology,
BF-142, Sector-I, Salt Lake City, Kolkata – 700064, West Bengal, India
chakrabarty.sudipta@gmail.com, samarjit.tech89@gmail.com,
dr.debashis.de@gmail.com

Abstract. Rhythm is the combination of beats which is most essential ingredient of music that contains the length of each note in a music composition. Knowledge of Rhythm structures and their application for new Music generation is very difficult task for the students of Musicology and the ratio between music teachers and music students is very low. In this paper a mechanism is introduced that efficiently selects the parent rhythms for creating offspring rhythm using Genetic Algorithm Optimization in Pervasive Education. Advancement of sensor technology and the wide use of social network services, music learning is now very easy comparing to earlier days. In this contribution m-learning is also selected to refer specifically to learning facilitated by mobile devices such as Personal Digital assistant (PDA) and mobile phones. The primary aim of m-learning is to provide the users with a learning environment which is not restricted to a specific location or time. Compared to a traditional classroom setting, m-learning increases the mobility of a learner, allowing him/her to learn. The ultimate goal of this study is to create a context awareness intelligent system tool for music rhythm creation for the application in Pervasive Teaching-Learning process. A software tool has also been implemented PERVASIVE DIARY to establish the processing algorithm.

1 Introduction

Music is meaningless without rhythm. Music is the meaningful combination of tempo along with the rhythm. A pervasive education learning environment is the framework by which pupils can befall entirely engrossed in the learning course of action. In a usual classroom, the instructor is the central source of information, and pupils are required to be in the same place at a time, whereas Pervasive education alters the traditional classroom perception of teaching and learning process.

Context awareness is defined complementary to but location awareness. Context awareness originated as a term from pervasive computing. This research uses the context-awareness knowledge structures to represent the learning environment; to

M.K. Kundu et al. (eds.), *Advanced Computing, Networking and Informatics - Volume 1*, 669
Smart Innovation, Systems and Technologies 27,
DOI: 10.1007/978-3-319-07353-8_77, © Springer International Publishing Switzerland 2014

identify the learning objectives that the user is really interested in; to propose learning activities to the user; and, to lead the user around the learning environment consist of various computing devices like PDA, wireless sensors, and servers.

Genetic Algorithm can be considered as multi-objective optimization techniques that implements optimization by simulating the biological natural law of evolution. A population has been initialized of arbitrarily generated solution. Each of the solution is evaluated to resolve the fitness value. A terminating condition has been test, if the solutions are good then terminate otherwise the solutions have to be optimized again. Hence, the best solutions are picked from the initial set of chromosomes. Then, the chromosomes having higher fitness values, exchange their information, to acquire the better solutions. This may then randomly be mutated some small percentage of the solutions thus obtained after crossover. Then again, each of the solution has been estimated, and checked the termination condition. Music Rhythmology as a context awareness learning process and applicable in the pervasive education as a huge number of students are interested to gather music knowledge.

The idea of the tool introduces XML schema as sensor data through sensor devices. A context rhythm editor is also designed by employing schema creating versatile rhythms using genetic Algorithm for supporting intelligent context mapping between sensor devices and context model.

2 Related Works

There are some previous research works that inspired to perform to these experiments. These works contributes a lot in automated musical research. Some works have been done on the use of pervasive computing in teaching and learning environment. In Paper [1], they present two approaches for the scalable tracking of mobile object trajectories and the efficient processing of continuous spatial range queries. At ASU, a Smart classroom [2-3] is built that used pervasive computing technology to enhance collaborative learning. Pervasive computing devices enable the children to utilize a pen reader to store text from a book into their PDA's Users like to employ the same device in diverse fields. In paper [4] judges numerous perceptual concerns in machine recognition perspective of musical patterns. This paper also proposed several measures of rhythm complexity and a novel technique to resolve a restricted tonal context. Further some other papers [5], [9] deal with the implementation of Musical pattern recognition by mathematical expressions. Again consider some other papers that are based on the creativity of music using Genetic Algorithm technique. The contributions describe the study of usefulness of Genetic Algorithm [6-7] for music composition. The paper [8] produces a new rhythm from a pre-defined rhythm applied to initial population using modified Genetic Algorithm operator. Another paper introduces a new concept of automated generating of realistic drum-set rhythm using Genetic Algorithm [10], [14]. Some researchers introduce a system for recognition of the patterns of Indian music by using key sequences with Median Filter and an effort suggested that three measurement techniques of rhythm complexity in a system for machine recognition of music pattern and also find pitch and rhythm error [13].

Additionally in several efforts by the authors have been described the musical pattern recognition and rhythmic features exposures by object-oriented manner [11-12]. A lot of discussions are available [15-16] about the features classification using Petri Nets and the way of implementations of musical complete composition of vocal and rhythmic cycles for the percussion-based tempo.

3 Fundamentals of Music Rhythm

Rhythm plays very important role in performing Music. Music is the systematic combination of tempo along with the rhythm. For generation of offspring rhythm about some common terminologies and their meanings in rhythm have to be known. They are given in Table 1.

Table 1. Rhythm terminologies and their meaning

Rhythm Terminology	Descriptions
Beat	A beat is a fragment of notes.
Note	Notes can have more or less than one beat.
Meter	It is the length of rhythm.
Tempo	This is the speed of the music.
Time Signature	This is a number that appears at the beginning of the music.
Combining Note Values	In 4/4 time, different note values can be combined in each measure as long as they equal four beats.
Even Meters	It is played in a meter of two that makes it an even meter.
Odd Meters	It is played in a meter of three that makes it an odd meter.

Rhythm content is based on a concept of rhythm which is the combination of meter, tempo, time signature, rhythm patterns superimposed on meter, etc. The basic three elements of rhythm are macro beats, micro beats, and melodic rhythm.

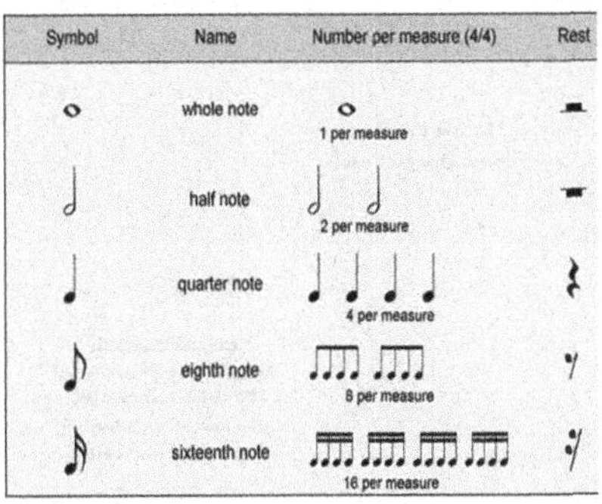

Fig. 1. Different rhythm symbols and their meanings

They are combined in perception, performance, or simultaneously in deriving rhythm syntax. Different Rhythm symbols, number per measure, rest symbol are depicted in Fig. 1.

4 Experimental Details

4.1 Overview of Pervasive Diary in Music Rhythm Education

Different rhythmic structures have been collected using various pervasive music sensors. Sensors in pervasive environments are not limited to physical or hardware sensors but also include virtual sensors. Then here in processing algorithm has been applied to produce high quality of offspring rhythms from that set of initial rhythm structures and build a recommended music knowledge base that apply on context awareness pervasive education. Some prime criteria of this effort are described below:

- Timely fast assessment of rhythm
- Variation in assessment on the rhythm elements
- Ease of creating input rhythm structures and Individualistic assessment
- Ease in creating output and editing rhythms
- Automatic comparing among different rhythms
- Both Automatic feedback and manual feedback

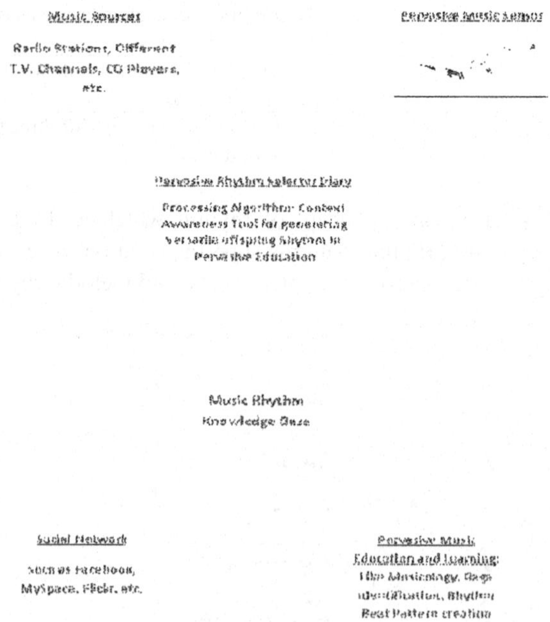

Fig. 2. Flow chart of pervasive diary

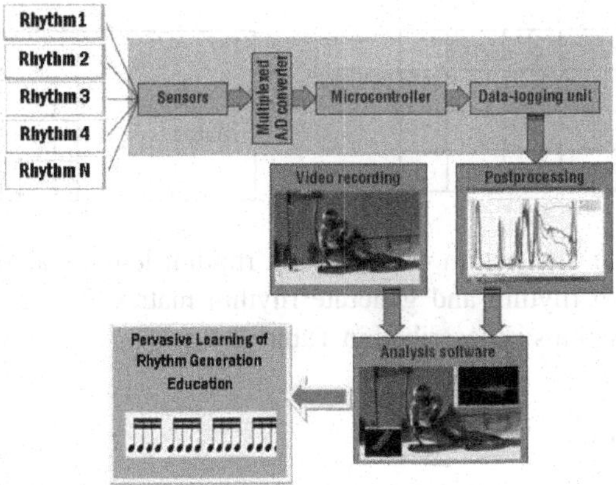

Fig. 3. Overall architecture of proposed work

Some applications of Music Rhythm Selector Diary in the field of Pervasive Music Education and Learning are: Raga Identification, Pattern Identification of music rhythm, versatile rhythm beat creation and Musicology

This proposed algorithm is applicable in the social network sites. The flowchart of proposed work is depicted in the Fig. 2 and the overall architecture of this anticipated concept is demonstrated in the Fig. 3.

4.2 Encoding Mechanism of Music Rhythm

Two types of notes have been considered in all the initial rhythm structures – single beat notes and double beat notes. The single beat note has been encoded as 0 and for double beat note = 1. For example,
Rhythm = Double Beat, Single Beat, Single Beat, Double Beat
After Encoding, Rhythm = 1001.

4.3 Generation of Rhythm Matrix

Music Rhythm Selector Diary is a context awareness tool that takes inputs of various rhythm structures of each length. If all the rhythm lengths are not equal, from the first beat has been gathered to the lowest number of music rhythm length of the initial rhythm structures of all individuals and generate the initial music rhythm matrix in Table 2.

Table 2. Basic rhythm beats

Rhythm 1	1	0	1	0	0	
Rhythm 2	1	1	0	0	0	1
Rhythm 3	1	0	1	0	1	
Rhythm 4	1	0	1	0		

Table 3. Rhythm matrix

Rhythm Matrix	1	0	1	0
	1	1	0	0
	1	0	1	0
	1	0	1	0

In the above Table 2, Rhythm 4 has lowest rhythm length, so first 4 beats have been taken from each rhythm and generate rhythm matrix. The creation of rhythm matrix from initial rhythms is described in Table 3.

4.4 Fitness Function

Fitness Function is derived from the objective function and it quantifies the feasibility of a solution. The following Fitness Function is being used for this proposed work:

Fitness Function = Decimal equivalent of each binary string

4.5 Parent Music Rhythm Selection Strategy Using Linear Rank Method

Selection is the first most important operator of Genetic Algorithm that applied on the initial population or chromosomes. Chromosomes are selected in the population to be parents to crossover and produce offspring. Linear Rank Selection mechanism is being utilized for selection process.

The Proposed Processing Algorithm

Step 1: Take binary encoded rhythm values from rhythm matrix.
Step 2: Convert binary to decimal of each rhythm, this is fitness value.
Step 3: Calculate rank values (Rank of worst fitness value = 1 and Rank of highest fitness value = 6 (n)). Then sort strings depend on rank values in ascending order.
Step 4: Fix two Selective Pressures (SP), where 2>= SP >=1. [SP = 1.9 and 1.1 considered]
Step 5: Calculate Scaled Rank using the following formula:

$$\text{Scaled Rank} = \text{sp} - 2 \times \frac{(\text{Rank} - 1)\,(\text{SP} - 1)}{(\text{N} - 1)} \tag{1}$$

Step 6: Choose two strings that have lowest Scaled Rank values in both SP = 1.9 and SP = 1.1.

5 Result Set Analysis and User Interface Design

Six initial rhythm structures has been taken, each of which meter length is sixteen (16) to establish proposed algorithm. Using the simple Binary Encoding mechanism the following six individual chromosomes are created as initial population. After getting all the initial chromosomes, Linear Rank Selection Mechanism has been applied for parent selection. The six chromosomes are:

String 1 = 1111111110000111
String 2 = 1000100000000000
String 3 = 1001100100000001
String 4 = 1001001111001100
String 5 = 1100110000001100
String 6 = 1001001100001100

Table 4. The experimental result of proposed algorithm

String No.	Population	Fitness (F_i)	Rank	Scaled Rank with SP =1.9	Scaled Rank with SP =1.1
2	1000100000000000	34816	1	1.9	1.1
6	1001001100001100	37644	2	1.54	1.06
4	1001001111001100	37836	3	1.18	1.02
3	1001100100000001	39169	4	0.82	0.98
5	1100110000001100	52236	5	0.46	0.94
1	1111111110000111	65415	6	0.1	0.9

From Table 4, the two fittest chromosomes for crossover operation have been found out that contains first and second lowest Scaled Rank. In the above Table 4, the two parent rhythms are chromosome1 and Chromosome5. Therefore String 1 and String 5 have been chosen as Parent 1 and Parent 2 for produce better Offspring Rhythm as they have lowest Scaled Rank Value in both Selective Pressure factors.
From design of Pervasive Diary is depicted in Fig. 4. It consists of three buttons for processing this work. Fig. 5 depicts the taken inputs from user in the Pervasive Diary Software. Fig. 6 represents the parent rhythm selection for rhythm generation using Linear Rank selection Mechanism.

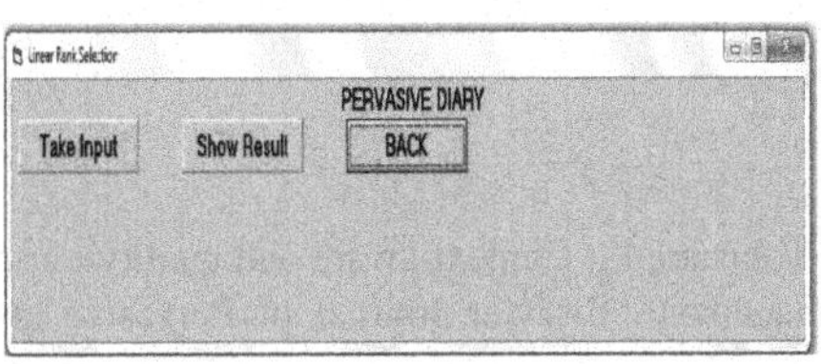

Fig. 4. Designing interface of pervasive diary

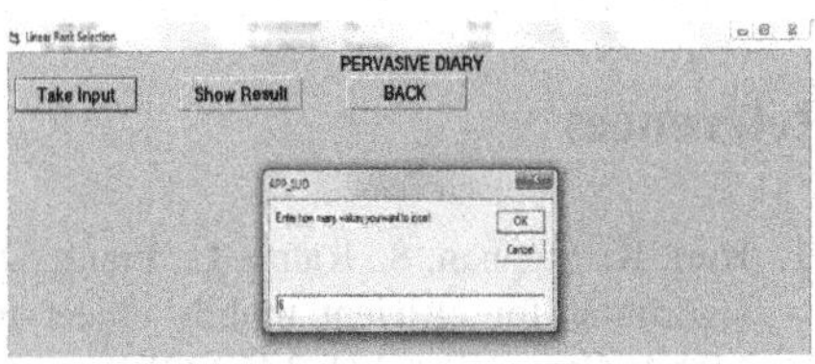

Fig. 5. User Interface of taking inputs

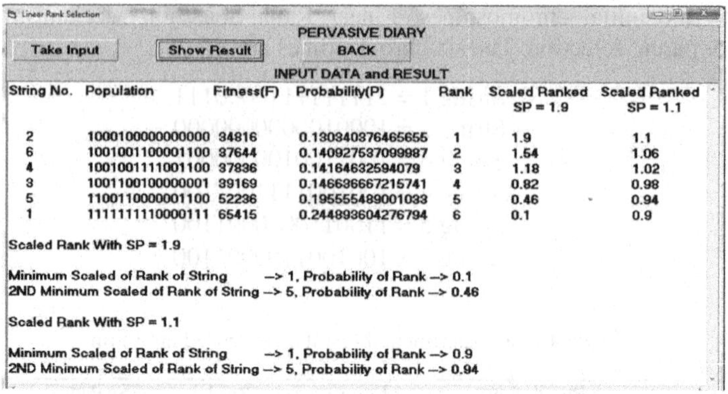

Fig. 5. User interface of parent rhythm selection using rank selection algorithm

6 Conclusion

In this effort, an intelligent prototype has been introduced of Web-based music rhythm learning system. In the experiments using a set of hardware and software tools the implemented system, all music rhythms have been evaluated exist at hand with the same meter and then evaluate which one is the appropriate for any song. On the technical side, the developments were found to be robust, and allowed for rapid prototyping of versatile music rhythm creation for pervasive music rhythm learning process.

The main objective of this contribution is to select the parent rhythms from a set of initial rhythm to produce offspring rhythms for practical implementation in World Music in context awareness pervasive music rhythm learning education. From the results of experiment on six initial rhythm structures, the study of parent rhythm selection has also been concluded using Linear Rank Selection Strategy and its importance. This contribution is implemented as a step towards developing tools to assist and evaluated each procedure and quality of each fundamental rhythmic structures and from that set of rhythms, the basic offspring rhythm has been searched which has more versatile features which contain more tempo.

References

1. Kurt, R., Stephan, S., Ralph, L., Frank, D., Tobiasm, F.: Context-aware and quality-aware algorithms for efficient mobile object management. Elsevier Journal of Pervasive and Mobile Computing, 131–146 (2012)
2. Bonwell, C., Eison, J.: Active learning: Creating excitement in the classroom. ASHE-ERIC Higher Education Report No. 1, George Washington University, Washington, DC (1991)

3. Yau, S.S., Gupta, S.K.S., Karim, F., Ahamed, S.I., Wang, Y., Wang, B.: Smart Classroom: Enhancing Collaborative Learning Using Pervasive Computing Technology. In: Proceedings of Annual Conference of American Society of Engineering Education (2003)
4. Shmulevich, I., Yli-Harja, O., Coyle, E.J., Povel, D.: Lemstrm K.: Perceptual Issues in Music Pattern Recognition Complexity of Rhythm and Key Finding. In: Computers and the Humanities, Kluwer Academic Publishers, pp. 23–35 (2001)
5. Chakraborty, S., De, D.: Pattern Classification of Indian Classical Ragas based on Object Oriented Concepts. International Journal of Advanced Computer engineering & Architecture 2, 285–294 (2012)
6. Gartland-Jones, A., Copley, P.: The Suitability of Genetic Algorithms for Musical Composition. Contemporary Music Review 22(3), 43–55 (2003)
7. Matic, D.: A Genetic Algorithm for composing Music. Proceedings of the Yugoslav Journal of Operations Research 20(1), 157–177 (2010)
8. Dostal, M.: Genetic Algorithms as a model of musical creativity – on generating of a human-like rhythimic accompainment. Computing and Informatics 22, 321–340 (2005)
9. Chakraborty, S., De, D.: Object Oriented Classification and Pattern Recognition of Indian Classical Ragas. In: Proceedings of the 1st International Conference on Recent Advances in Information Technology (RAIT), pp. 505–510. IEEE (2012)
10. Alfonceca, M., Cebrian, M., Ortega, A.: A Fitness Function for Computer-Generated Music using Genetic Algorithms. WSEAS Trans. on Information Science & Applications 3(3), 518–525 (2006)
11. De, D., Roy, S.: Polymorphism in Indian Classical Music: A pattern recognition approach. In: Proceedings of International Conference on communications, Devices and Intelligence System (CODIS), pp. 632–635. IEEE (2012)
12. De, D., Roy, S.: Inheritance in Indian Classical Music: An object-oriented analysis and pattern recognition approach. In: Proceedings of International Conference on RADAR, Communications and Computing (ICRCC), pp. 296–301. IEEE (2012)
13. Bhattacharyya, M., De, D.: An approach to identify thhat of Indian Classical Music. In: Proceedings of International Conference of Communications, Devices and Intelligence System (CODIS), pp. 592–595. IEEE (2012)
14. Chakrabarty, S., De, D.: Quality Measure Model of Music Rhythm Using genetic Algorithm. In: Proceedings of the International Conference on RADAR, Communications and Computing (ICRCC), pp. 203–208. IEEE (2012)
15. Roy, S., Chakrabarty, S., Bhakta, P., De, D.: Modelling High Performing Music Computing using Petri Nets. In: Proceedings of International Conference on Control, Instrumentation, Energy and Communication, pp. 757–761. IEEE (2013)
16. Roy, S., Chakrabarty, S., De, D.: A Framework of Musical Pattern Recognition using Petri Nets. Accepted In: Emerging Trends in Computing and Communication 2014, Springer-Link Digital Library, Submission No- 60 (2013)

An Elitist Binary PSO Algorithm for Selecting Features in High Dimensional Data

Suresh Dara and Haider Banka

Department of Computer Science and Engineering
Indian School of Mines, Dhanbad -826004, India
darasuresh@live.in, banka.h.cse@ismdhanbad.ac.in

Abstract. An elitist model of Binary Particle Swarm Optimization (BP SO) algorithm is proposed for feature selection from high dimensional data. Since the data are highly redundant, a fast pre-processing algorithm is employed to reduce features from high dimensions in a crude manner. The reduced feature subsets being still high dimensional, a further reduction is achieved by the proposed algorithm. The non-dominated sorting PSO algorithm is performed on the combined solutions of each two successive generations that also help to preserve the best solutions in a generation. The fitness functions are suitably formulated in multi objective framework for the conflicting objectives, i.e., to reduce the cardinality of the feature subsets and to increase the accuracy. The performance of the proposed algorithm is demonstrated on three high dimensional benchmark datasets, i.e., colon cancer, lymphoma and leukemia data.

1 Introduction

Many high dimensional datasets are like microarray gene expressions, text documents, digital images, clinical data, etc having thousands of features, many of these features are irrelevant or redundant. Unnecessary features increase the size of the search space and make generalization more difficult. This curse of dimensionality is a major obstacle in machine learning and data mining. When data is high dimension, the process of learning is very hard [1]. Hence, the feature reduction techniques are one of the important tool to reduce and select useful feature subsets that maximizes the prediction or classification accuracy.

There are two main reasons to keep the dimensionality reduction (i.e., the number of features) as small as possible: measurement cost and classification accuracy. A limited yet salient feature set simplifies both the pattern representation and the classifiers that are built on the selected representation. Consequently, the resulting classifier will be faster and will use less memory. Moreover, as stated earlier, a small number of features can alleviate the curse of dimensionality when the number of training samples is limited. On the other hand, a reduction in the number of features may lead to a loss in the discrimination power and thereby lower the accuracy of the resulting recognition system.

There are two main dimension reduction methods: i. Feature Extraction, and ii. Feature Selection. Feature extraction methods determine an appropriate subspace of dimensionality n (either in a linear or a nonlinear way) in the original

M.K. Kundu et al. (eds.), *Advanced Computing, Networking and Informatics - Volume 1*, Smart Innovation, Systems and Technologies 27,
DOI: 10.1007/978-3-319-07353-8_78,

feature space of dimensionality d ($d \leq n$). Feature selection is the problem of selecting a subset of d features from a set of n features based on some optimization techniques. FS can serve as a pre-processing tool of great importance before solving the classification problems. The selected feature subsets should be sufficient to describe the target concepts. The primary purpose of feature selection is to design a more compact classifier with little or no performance degradation possibly. FS has got considerable importance from last few decades in many areas including pattern classification, data analysis, multimedia information retrieval, biometrics, remote sensing, computer vision, medical data processing, machine learning, and data mining applications [2].

Feature selection methods can be categorised as Filter methods. Wrapper methods, Embedded/Hybrid methods, Ensemble methods [1]. Filter techniques [3], selects feature subsets independently of any learning algorithm, estimate a relevance score and a threshold scheme used to select the best features. The wrapper model [4] uses predictive accuracy of predetermined learning algorithms. The embedded techniques [5] allow interaction of different class of learning algorithms. More recently, the ensemble model [6] based on different sub sampling strategies, the learning algorithms run on a number of sub samples and the acquired features are united into a stable subset. However the feature selection techniques can be also categorized based on search strategies used, most commonly used strategies are forward selection, backward elimination, forward stepwise selection, backward stepwise selection and random mutation [7].

In this paper, we presented a meta-heuristic evolutionary based BPSO algorithm for select feature subset selection from microarray gene expression profiles. We used non-dominated sorting incorporated with BPSO which we called as elitist BPSO. For this purpose, we made preprocessing on microarray data, which is help us to reduce the dimensionality and removes the irrelevant, noise and redundant data. First stage, the microarray data is preprocessed and discretized to binary distinction table, reduce the dimensionality of gene samples are decrease the computational burden. In the second stage, elitist BPSO is used to select the significant feature subsets. External validation of selected feature subsets is done in terms of classification accuracy (%) using k-NN classifier.

The rest of this paper is organized as follows. Section 2 describes the preliminaries like microarray gene expression data, PSO algorithm, the non-dominated sorting algorithms and k-NN classifier. The preprocessing of gene expression data, fitness functions and the proposed PSO algorithm for feature selection is presented in Section 3. The experimental results on colon, lymphoma and leukemia data, along with the external validation using k-NN classifiers, are presented in Section 4. Finally, Section 5 concludes this paper.

2 Preliminaries

This section formally describes the basics of binary particle swarm optimization, the dominance criteria and non-dominated sorting algorithm along with the K-NN classifier as continuation and understanding of the present work.

2.1 Binary Particle Swarm Optimization (BPSO)

Particle Swarm Optimization (PSO) is a intelligence based multi-agent heuristic optimization technique is introduced by Kennedy et.al in 1995. The PSO initialisation of random population(swarm) of individuals(particles) in the n-dimensional search space. Each particle keeps two values (*present best* and *global best*) in its memory, its own best experience, one is best fitness values whose position P_i and it value P_{best}, and second one is best experience of whole swarm which position denotes P_g and its value *gbest*. The ith particle denote as $X_i = (X_{i1}, X_{i2}, X_{i3}, \ldots, X_{id}, \ldots, X_{in})$, also denotes the velocity $V_i = (V_{i1}, V_{i2}, \ldots, V_{id}, \ldots, V_{in})$. The position and velocity of each particles update in each iteration according to the (1) and (2)

$$V_{id}(t+1) = w * V_{id}(t) + c_1\rho_1(P_{id}(t) - X_{id}(t)) + c_2\rho_2(P_g(t) - X_{id}(t)) \quad (1)$$

$$X_{id}(t+1) = X_{id} + V_{id}(t+1) \quad (2)$$

where $d = 1, 2, 3, \ldots, n$ (i.e., dimension of each particle), w is inertia weight, it provides a balance between global and local exploration, and results in fewer iterations on average to find a sufficiently optimal solution, c_1 and c_2 are the same positive constants used in flock's simulations and are respectively called the *cognitive* and *social acceleration* coefficient, represent the weighting of the stochastic acceleration terms that pull each particle toward *pbest* and *gbest* positions. ρ_1 and ρ_2 are two random numbers in the range of [0, 1]. The update of pbest and gbest is compared to present particle, if it is less than then update, this process is continue for all particles.

PSO has been called as BPSO where each particle contains a combination of 0's and 1's [8]. Here, velocity V_{id} mapped into interval [0,1] via sigmoid function as $S(V_{id}) = \frac{1}{1+e^{-V_{id}}}$ and the velocity updates based on eq(1), and position updates based on eq.(3), where ρ is 0.5

$$X_{id} = 1, if \rho < S(V_{id}), otherwise X_{id} = 0 \quad (3)$$

The above update process implemented for all dimensions, and for all particles.

2.2 Non-dominated Sorting Algorithm

The concept of optimality, behind the multi objective optimization [9], deals with a set of solutions. The conditions for a solution to be dominated with respect to the other solutions are given as follows. If there are M objective functions, a solution s_1 is said to dominate another solution s_2, if both conditions 1 and 2 are true. 1). The solution s_1 is no worse than s_2 in all the M objective functions. 2). The solution s_1 is strictly better than s_2 in at least one of the M objective functions. Otherwise, the two solutions are non-dominating to each other. When a solution i dominates a solution j, then rank $r_i < r_j$. The major steps for finding the non-dominated set in a population P of size $|P|$ are outlined as follows in Algorithm 1.

Algorithm 1. Nondominated sorting Algorithm

Step:1	Set solution counter $i = 1$ and create an empty non-dominated set P'
Step:2	For a solution $j \in P(j \neq i)$, check if the solution j dominates the solution i. If yes then go to Step 4.
Step:3	If more solutions are left in P. increment j by one and go to Step 2. Else set $P' = P \cup \{i\}$.
Step:4	Increment i by one. If $i \leq \|P\|$ then go to Step 2 . Else declare P' as the non-dominated set

2.3 k-NN Classifier

K-nearest neighbor (k-NN): one of the simple instance based non parametric classifier, compares a new instance with its K neighbors (1, 3, 5, and 7, basically an odd number to avoid ties) in the training dataset and assigns it to the majority class among them. These nearest neighbors can be determined according Euclidean distance measure.

3 Proposed Approach

3.1 Preprocessing Data

Preprocessing directs to eliminating of ambiguously expressed genes as well as the constantly expressed genes across the tissue classes. The normalization is performed on each of the attributes so that it falls between 0.0 and 1.0. This helps us to give equal priority to each of the attributes as there is no way of knowing important or unimportant attributes/features. The gene data set, i.e., continuous attribute value table is normalized between range (0,1). Then we choose thresholds Th_i and Th_f, based on the idea of quartiles [10]. Then convert features values to binary (0 or 1 or *) as follows:

If $a' \leq Th_i$, **put** '0', **If** $a' \geq Th_f$ then **put** '1', **Else** put '*' don't care
We find the average of '*' occurrences over the feature table. Choose this as threshold Th_d. Those attributes are remove from the table which has the number of '$*$'s $\geq Th_d$. The distinction table is prepared accordingly as done in [10].

The distinction table consist of N columns, and rows corresponding to only those object pairs discern, where '1' signifies the present of gene and '0' as absence of gene. If a dataset contain two classes then the number of rows in the distinction table becomes $(C_1 * C_2) < \frac{C*(C-1)}{2}$, where $C_1 + C_2 = C$. A sample distinction table in shown in Table 1.

Here, assume that there are five conditional features $\{f_1, f_2, f_3, f_4, f_5\}$, the length of vector is $N = 5$. In a vector $\boldsymbol{v}$, the binary data '1' represents, if the corresponding feature is present, and a '0' represents its absence. The two classes are $C_1(C_{11}, C_{12})$ and $C_2(C_{21}, C_{22}, C_{23})$. The rows represent the object pairs and columns represent the features or attributes. The objective is to choose minimal number of column (features) from Table 1 that covers all the rows (i.e., object pairs in the table) as defined by the fitness functions.

Table 1. Distinction table model

	f_1	f_2	f_3	f_4	f_5
(C_{11}, C_{21})	1	0	1	0	1
(C_{11}, C_{22})	0	0	0	1	0
(C_{11}, C_{23})	1	0	1	0	0
(C_{12}, C_{21})	0	0	1	0	1
(C_{12}, C_{22})	0	1	0	1	0
(C_{12}, C_{23})	1	0	0	0	1

3.2 Fitness Function

We proposed two fitness functions F_1 and F_2. Where F_1 is used to finds number of features(i.e number of 1's), and F_2 decides the extent to which the feature can recognise among the objects pairs. The proposed fitness functions are as follows:

$$F_1(v) = \frac{N - O_v}{N} \quad (4), \qquad F_2(v) = \frac{R_v}{C_1 * C_2} \quad (5)$$

Here, v is the chosen feature subsets, O_v represents the number of 1's in v, C_1 and C_2 are the number of objects in the two classes, and R_v is the number of object pairs (i.e., rows in the distinction table) v can discern between. The fitness function F_1 gives the candidate credit for containing less number of features or attributes in v, and F_2 determines the extent to which the candidates can discern among objects pairs in the distinction table.

3.3 Elitist BPSO Method

We proposed elitist BPSO algorithm for feature subset selection. Here, the best non-dominated solutions of combined solutions are preserved at each generation. We consider the best 50% populations are as parents again, so that it generate a set of non-dominated solutions . The proposed approach is described in Algorithm 2.

Algorithm 2. Elitist BPSO algorithm for feature selection

Step:1 Initialize P no. of solutions with random velocity and positions
Step:2 Calculate fitness on P using equation (4) and (5)
Step:3 Update pbest
Update gbest
Step:4 Update velocities and coordinates of P using equation (1) – (3) to generate P'
Step:5 Calculate fitness values for P' using equation (4) and (5)
Step:6 Combine both P and P' as P''
Step:7 Perform non-dominated sorting on P'' using algorithm [1]
Step:8 Choose 50% best ranked solutions from P'' as P
Step:9 Repeat Step (3–8) for finite number of generations.

4 Results and Discussions

4.1 Gene Expression Data Sets

Micro array data provides access to expression levels of thousands of genes at once in order to identify co-expressed genes, relationships between genes, patterns of gene activity. The high dimensional gene expression datasets are used in our experiment, the details of number of sample sizes and test, train sample sizes described in Table 2.

Table 2. Details of the two-class microarray data

Datasets	Total Features	Reduced Features#	Classes	Samples
Colon[1]	2000	1102	Colon cancer	40
			Normal	22
Lymphoma[2]	4026	1867	Other type	54
			B-cell	42
Leukemia[3]	7129	3783	ALL	47
			AML	25

After Preprocessing

4.2 Results

We have implemented the elitist BPSO Algorithm to find minimal feature subsets on high dimensional cancer datasets; i.e., colon, lymphoma, and leukemia. We set two accelerator coefficients parameters (c_1 & c_2) to 2, and the minimum, and maximum of velocities were set to -4 and 4, respectively. The inertia weight (w) is one of the most important parameter in BPSO which can improve performance by properly balancing its local and global search [11]. The inertia weight(w) was set between 0.4 to 0.9. The varied population size was taken, set maximum runs as 50, also tested different population sizes like 10, 20, 30 and 50.

To check efficiency of the proposed algorithm, k–NN classifier is used as a validation tool. The results are taken in form of correct classification accuracy. The experimental results are carried out on three bench mark datasets as summarized in Table 3. Note that k is chosen to be an odd number to avoid the ties. The correct classification are reported to be $93.54\%, 95.85\%$ and 94.74% for those three datasets with varies swarm size and k values.

Table 4 depicts the comparative performance studies with simple GA and NSGA-II [10] for the same bench mark datasets i.e, colon, Leukemis, and Lymphoma data. Using GA with 15 gene set produces a classification of 77.42%, and for *lymphoma* data, 18 gene subset produces 93.76%, and for leukemia data using 19 gene subset, the classification score is 73.53%. The NSGA-II based feature selection method, using $k - NN$ classifier reported 90.3% on colon data with 9 gene set, for lymphoma 95.8% for 2 gene subset, and for leukemia 91.2% on 3

Table 3. Comparative performance on three data sets using k-NN classifier

Dataset	Population size	Selected feature subset	$k=1$ Cr	$k=1$ ICr	$k=3$ Cr	$k=3$ ICr	$k=5$ Cr	$k=5$ ICr	$k=7$ Cr	$k=7$ ICr
Colon:	10	10	100	0	83.87	16.13	83.87	16.13	80.65	19.35
# Genes 2000	20	9	100	0	83.87	16.13	83.87	16.13	83.87	16.13
Reduce to 1102	30	9	100	0	93.54	6.46	80.65	19.35	83.87	16.13
	50	9	100	0	90.32	9.68	80.65	19.35	87.09	12.91
Lymphoma:	10	20	100	0	93.75	6.25	93.75	6.25	89.75	10.25
# Genes 4026	20	22	100	0	95.85	4.15	91.66	8.34	91.66	8.34
Reduce to 1867	30	21	100	0	95.85	4.15	95.85	4.15	91.66	8.34
	50	15	100	0	93.75	6.25	93.75	6.25	91.66	8.34
Leukemia:	10	14	100	0	89.49	10.51	89.49	10.51	89.49	10.51
# Genes 7129	20	15	100	0	92.10	7.9	89.49	10.51	92.10	7.9
Reduce to 3783	30	14	100	0	94.75	5.25	89.49	10.51	89.49	10.51
	50	14	100	0	94.75	5.25	86.85	13.15	89.49	10.49

k-nearest neighbors classification (%) on test set.

Cr and **ICr** are Average correct and Incorrect classification score of two classes

gene subset. We got 100% classification score for all gene sets using one neighbor, for three data sets, 93.54%, 95.85% and 94.74% using 9, 20, and 14 gene subsets with respective data sets using $k = 3, 5, 7$.

Table 4. Comparative performance between proposed algorithm, NSGA-II and GA on three datasets using k-NN classifier

Dataset	feature subset size	Used Method	$k=1$ Cr	$k=1$ ICr	$k=3$ Cr	$k=3$ ICr	$k=5$ Cr	$k=5$ ICr	$k=7$ Cr	$k=7$ ICr
Colon	≤ 9	Proposed	**100**	**0**	**93.54**	**6.45**	**83.87**	**16.13**	**87.09**	**12.91**
	≤ 10	NSGA-II [10]	90.3	9.07	90.3	9.07	87.1	12.9	80.6	19.4
	≤ 15	GA [10]	71.0	29.00	58.10	41.90	48.40	51.60	61.30	38.70
Lymphoma	≤ 21	Proposed	**100**	**0**	**95.85**	**4.15**	**95.8**	**4.2**	**91.66**	**8.34**
	≤ 2	NSGA-II	93.8	16.2	95.8	4.2	95.8	4.2	95.8	4.2
	≤ 18	GA	89.59	10.41	89.59	10.41	93.76	6.24	93.76	6.24
Leukemia	≤ 14	Proposed	**100**	**0**	**94.74**	**5.26**	**89.49**	**10.51**	**92.10**	**7.9**
	≤ 5	NSGA-II	94.1	5.9	91.2	8.8	91.2	8.8	88.2	11.8
	≤ 19	GA	73.50	26.50	73.53	26.47	60.77	38.23	67.65	32.35

k-nearest neighbors classification (%) on test set.

5 Conclusion

In this paper, we proposed a elitist model BPSO to find future subset selection in gene expression microarray data. Non-dominating sorting helps to preserve Pareto-font solutions. Our preprocessing aids faster convergence along the search space and successfully employed to eliminate redundant, and irrelevant features.

The proposed approach is experimentally investigated of different parameters and population sizes. The main goal of the feature selection is selecting minimal number of features and get higher classification accuracy. Here we have achieved the goal through the implementation of two sub fitness functions. For three cancer data sets results reported in this paper demonstrating the feasibility and effectiveness of the proposed feature selection method. The performance of the proposed method and existed methods are compared using k-NN classifier and reported better classification accuracy.

Acknowledgment. This work was partially supported by Council of Scientific and Industrial Research (CSIR), New Delhi, India, under the Grant No. 22(0586)/12/EMR-II.

References

1. Lazar, C., et al.: A survey on filter techniques for feature selection in gene expression microarray analysis. IEEE/ACM Transactions on Computational Biology and Bioinformatics 9(4), 1106–1119 (2012)
2. Inza, I., Saeys, P.L.Y.: A review of feature selection techniques in Bioinformatics. International Journal of Computer Science (IAENG) 23(19), 2507–2517 (2007)
3. ElAlami, M.E.: A filter model for feature subset selection based on genetic algorithm. Knowledge-Based Systems 22(5), 356–362 (2009)
4. Sainin, M.S., Alfred, R.: A genetic based wrapper feature selection approach using nearest neighbour distance matrix. In: 2011 3rd Conference on Data Mining and Optimization (DMO), pp. 237–242 (2011)
5. Wahid, C.M.M., Ali, A.B.M.S., Tickle, K.S.: A novel hybrid approach of feature selection through feature clustering using microarray gene expression data. In: 2011 11th International Conference on Hybrid Intelligent Systems (HIS), pp. 121–126 (2011)
6. Nagi, S., Bhattacharyya, D.: Classification of microarray cancer data using ensemble approach. Network Modeling Analysis in Health Informatics and Bioinformatics, 1–15 (2013)
7. Mladenič, D.: Feature selection for dimensionality reduction. In: Saunders, C., Grobelnik, M., Gunn, S., Shawe-Taylor, J. (eds.) SLSFS 2005. LNCS, vol. 3940, pp. 84–102. Springer, Heidelberg (2006)
8. Kennedy, J., Eberhart, R.C.: A discrete binary version of the particle swarm algorithm. In: IEEE International Conference on Systems, Man, and Cybernetics, 1997 Computational Cybernetics and Simulation, vol. 5, pp. 4104–4108 (1997)
9. Deb, K.: Multi-objective optimization. Multi-objective Optimization Using Evolutionary Algorithms, 13–46 (2001)
10. Banerjee, M., Mitra, S., Banka, H.: Evolutionary rough feature selection in gene expression data. IEEE Transactions on Systems, Man, and Cybernetics, Part C: Applications and Reviews 37(4), 622–632 (2007)
11. Shi, Y., Eberhart, R.: Empirical study of particle swarm optimization. In: Proc. IEEE Congress, vol. 3, pp. 1945–1950 (1999)

An Immune System Inspired Algorithm for Protein Function Prediction

Archana Chowdhury[1], Amit Konar[1],
Pratyusha Rakshit[1], and Janarthanan Ramadoss[2]

[1] Electronics and Telecommunication Engineering Department
Jadavpur University, Kolkata, India
[2] Department of Compuiter Science Engineering, TJS College of Engineering
Chennai, India
{chowdhuryarchana,pratyusharl}@gmail.com,
konaramit@yahoo.co.in, srmjana_73@yahoo.com

Abstract. An important problem in the field of bioinformatics research is assigning functions to proteins that have not been annotated. The extent, to which protein function is predicted accurately, depends largely on the Protein-Protein interaction network. It has been observed that bioinformatics applications are benefited by comparing proteins on the basis of biological role. Similarity based on Gene Ontology is a good way of exploring the above mentioned fact. In this paper we propose a novel approach for protein function prediction by utilizing the fact that most of the proteins which are connected in Protein-Protein Interaction network, tend to have similar functions. Our approach, an immune system-inspired meta-heuristic algorithm, known as Clonal Selection Algorithm (CSA), randomly associates functions to unannotated proteins and then optimizes the score function which incorporates the extent of similarity between the set of functions of unannotated protein and annotated protein. Experimental results reflect that our proposed method outperforms other state of the art algorithms in terms of precession, recall and F-value, when utilized to predict the protein function of Saccharomyces Cerevisiae.

Keywords: protein function prediction, protein–protein interaction network, gene ontology, annotated protein, clonal selection algorithm.

1 Introduction

The main problem in molecular biology is to understand the function of a protein, as the function of most of the proteins is unknown. It has been observed that even the most studied species, Saccharomyces cerevisiae, is reported to have more than 26 percent of its proteins with unknown molecular functions [1]. Huge amount of data continue to accumulate due to the application of high throughput technologies in various genome projects. Protein-Protein Interaction (PPI) is an important source of information among these databases. The introduction of high-throughput techniques have resulted in an amazing number of new proteins been identified. However, the function of a large number of these proteins still remains unknown.

M.K. Kundu et al. (eds.), *Advanced Computing, Networking and Informatics - Volume 1*, Smart Innovation, Systems and Technologies 27,
DOI: 10.1007/978-3-319-07353-8_79, © Springer International Publishing Switzerland 2014

Several algorithms have been developed to predict protein functions, on the basic assumption that proteins with similar functions are more likely to interact. Among them, Deng proposed the Markov random field (MRF) model, which predicts protein functions based on the annotated proteins and the structure of the PPI network [2]andSchwikowski proposed neighbor counting approach [3]. In recent years, more and more research turned to predict protein functions semantically by combining the inter-relationships of function annotation terms with the topological structure information in the PPI network. To predict protein functions semantically, various methods were proposed to calculate functional similarities between annotation terms. Lord *et al.*[4] were the first to apply a measure of semantic similarity to GO annotations. Resink [5] used the concept of information content to calculate the semantic similarity between two GO terms.

In this paper, we aim to predict the function of an unannotated protein by using the topographical information of PPI network and function of annotated protein. The similarity of GO terms used to annotate proteins is measured using the information content of the respective terms as well as the terms which are common in the path from root to the GO terms. For this task we have employed the use of an immune system-inspired meta-heuristic algorithm, known as Clonal Selection Algorithm (CSA).The proposed method used a hypermutation strategy which provides the exploration capability to individual clone within the search space.

The rest of this paper is organized as follows: Section 2 give a brief idea about the definition and formulation of the problem as well as the scheme for solution representation. Section 3 provides an overview of the proposed. Experiments and Results are provided in Section 4.Section 5 concludes the paper.

2 Background of the Problem

2.1 Problem Definition

PPI network with N proteins is considered. The PPI network can be represented in the form of a binary data matrix $\mathbf{K}_{N\times N}$ where $k_{ij}=k_{ji}=1$and$k_{ij}=k_{ji}=0$ denotes the presence and absence of interaction between the proteins p_i and p_j respectively. The set of all functions for each protein p, is denoted as $F(p)$ thus, the set of all possible functions in the network is defined as $F= F(p_1)\text{U } F(p_2)\text{U}\ldots \text{ U } F(p_N)$ with number of all possible functions in the network $|F|=D$.

Given the PPI data matrix $\mathbf{K}_{N\times N}$, a protein function prediction algorithm tries to find a set of possible functions $F(p)$ of an unannotated protein p based on the functions $F(p')$ of all annotated proteins p' in the PPI network. Since the functions can be assigned to the unannotated protein p in a number of ways, a fitness function (measuring the accuracy of the function prediction) must be defined. The problem now turns out to be an optimization problem of finding a set of functions $F(p)$ of optimal adequacy as compared to all other feasible sets of functions for unannotated protein p.

2.2 Formulation of the Problem

The effectiveness of protein function prediction can be improved by taking the composite benefit of the topological configuration of the PPI network and the functional categories of annotated proteins through Gene Ontology (GO) [8], [9]. The protein functions are annotated using GO terms. GO is basically represented as a directed acyclic hierarchical structure in which a GO term may have multiple parents/ancestor GO terms. The probability, *prob*(t), for each GO term t in the GO tree is frequency of occurrence of the term and its children divided by the maximum number of terms in the GO tree. Thus the probability for each node/GO term will increase as we move up towards the root. The information content of a GO term in the GO tree is based on the *prob*(t) value and is given by

$$I(t) = -\log_{10} prob(t) \tag{1}$$

Here (1) shows that lesser is the probability of the GO term, more will be the information content associated with it. The similarity between two GO terms will be high if they share more information. The similarity between two terms, which is captured by the set of common ancestors, is the ratio of probability of the common terms between them to the probability of the individual terms.It can be represented as follows:

$$S(t_1,t_2) = \max_{t \in C(t_1,t_2)} \left(\frac{2 \log prob(t)}{\log prob(t_1) + \log prob(t_2)} \right) \tag{2}$$

Where $C(t_1, t_2)$ denotes the set of common ancestors of terms t_1 and t_2, $S(t_1,t_2)$ measures the similarity with respect to information content, in terms of the common ancestors of the terms t_1 and t_2.The value of the above similarity measure ranges between 0 and 1. With this representation scheme of protein functions, the similarity between a predicted function $f \in F(p)$ of unannotated protein p and a real function $f' \in F$ of the PPI network can be computed as follows.

$$sim(f, f') = S(t_1, t_2) \tag{3}$$

Where function $f$is annotated by t_1 and f' is annotated by t_2. Next, the score of the unannotated protein being annotated by the predicted function f is evaluated by (4).

$$score(p, f) = \sum_{\substack{j=1, \\ f' \in F(p_j)}}^{N} \frac{1}{dist(p, p_j)} \times sim(f, f') \tag{4}$$

Here $dist(p, p_j)$ represents the minimum number of edges between proteins p and p_j. Two important facts are included in (4), First is that, it assigns function f to protein p based on the similarity between f and all other protein functions available in the given network which conforms to the fact that theproteins with similar functions interact more frequently to construct the PPI network, secondly the term $1/dist(p, p_j)$ is included because proteins far away from p contribute less functional information than those having direct interaction with p. This is accomplished by assigning less weight to the proteins far away from p than its close neighbors. From (4), it is apparent that a high value of *score(p, f)* will indicate a higher adequacy in predicting *f* as a function of protein *p*.

2.3 Solution Representation and Cost Function Evaluation

In the proposed method a solution $\vec{X}_i$ is a vector of dimension D as Ddenotes maximum number of functions in the PPI network. The values of $\vec{X}_i$ belong to {0, 1} and the *j*-th parameter of $\vec{X}_i$ is interpreted as follows:

If $x_{i,j}$=1, then the *j*-th function f_j is predicted as a function of protein *p*. (5.a)

If $x_{i,j}$=0, then f_j is not predicted as a function of protein *p*. (5.b)

Let there be$D$=8 functions available in the network among which, the second, third, fifth and seventh have been predicted as assigned functions of unannotated protein *p*then the solution encoding scheme will be as shown in Fig.1.

0	1	1	0	1	0	1	0

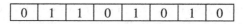

Fig. 1. Solution encoding scheme in the proposed method

In order to judge the quality of such a solution $\vec{X}_i$, for function prediction, the contribution by the entire set of predicted functions (denoted by set $F(p)$= {f_j| $x_{i,j}$=1 for j=[1, D]}) to annotate protein *p* is used for fitness function evaluation. Symbolically,

$$fit(\vec{X}_i) = \sum_{\forall f \in F(p)} score(p, f) \tag{6}$$

3 An Overview of Clonal Selection Algorithm (CSA)

CSA [6] is a population-based stochastic algorithm, which is inspired by the antigen driven affinity maturation process of B-cells in immune system. Overviews of the main steps of CSA are as follows:

A. Initialization

CSA starts with a population of *NP*,*D*-dimensional antibodies, $\vec{X}_i(t)=\{x_{i,1}(t), x_{i,2}(t), x_{i,3}(t),\ldots,x_{i,D}(t)\}$ for i= [1, *NP*] representing the candidate solutions, at the current generation t = 0 by randomly initializing in the range $[\vec{X}^{\min}, \vec{X}^{\max}]$.Thus the *j*-th component of the *i*-th antibody at t=0 is given by

$$x_{i,j}(0) = x_j^{\min} + rand_{i,j}(0,1) \times (x_j^{\max} - x_j^{\min}) \tag{7}$$

Where $rand_{i,j}(0, 1)$ is a uniformly distributed random number lying between 0 and 1. The affinity or the fitness $fit(\vec{X}_i(0))$ of the antibody $\vec{X}_i(0)$ is evaluated for i=[1, *NP*].

B. Selection of Antibodies for Cloning

The antibodies are sorted in descending order of fitness $fit(\vec{X}_i(t))$ for i= [1, NP] and the first n individuals of the population are selected for subsequent cloning phase.

C. Cloning

Each member $\vec{X}_k(t)$ of the selected subpopulation of n antibodies is allowed to produce clones for k= [1, n]. The number of clones c_k for the k-th individual is proportional to its affinity $fit(\vec{X}_k(t))$ and is calculated using (8). Here $fit_{\min}= fit(\vec{X}_n(t))$ and $fit_{\max}= fit(\vec{X}_1(t))$ represent the lowest and highest affinity of the sorted individuals in the subpopulation of n antibodies respectively. Similarly, $c_{\min}$ and $c_{\max}$ denote maximum and minimum number of clones.

$$c_k = \left\lfloor \frac{fit(\vec{X}_k(t)) - fit_{\min}}{fit_{\max} - fit_{\min}} \times (c_{\max} - c_{\min}) \right\rfloor + c_{\min} \tag{8}$$

D. Hypermutation

Each clone of $\vec{X}_k(t)$, denoted as $\vec{X}_k^l(t)$ for j= [1, D], k= [1, n] and l= [1, c_k] undergoes through the static hypermutation process using (9).

$$x_{k,j}^l(t+1) = x_{k,j}^l(t) + \alpha \times x_{k,j}^l(t) \times (x_j^{\max} - x_j^{\min}) \times G(0,\sigma) \tag{9}$$

Here α is a constant, however small and $G(0, \sigma)$ is a random Gaussian variable with zero mean and σ as the standard deviation. Usually, $\sigma$is taken as 1 [7]. The fitness $fit(\vec{X}_k^l(t))$ is evaluated for l= [1, c_k].

E. Clonal Selection

Let the set of matured clones (after hypermutation) corresponding to the k-th antibody, including itself, is denoted as S_k= $\{\vec{X}_k(t), \vec{X}_k^1(t+1), \vec{X}_k^2(t+1), ..., \vec{X}_k^{c_j}(t+1)\}$ for k= [1, n]. The best antibody in S_k with highest fitness is allowed to pass to the next generation. Symbolically,

$$\vec{X}_k(t+1) \leftarrow \arg\left(\max_{\forall \vec{X} \in S_k} (fit(\vec{X})) \right) \tag{10}$$

F. Replacement

The NP–n antibodies not selected for cloning operation are randomly re-initialized as in (7).

After each evolution, we repeat from step B until one of the following conditions for convergence is satisfied. The conditions include restraining the number of iterations, maintaining error limits, or the both, whichever occurs earlier.

4 Experiments and Results

The GO terms [8] and GO annotation dataset [9] used in the experiments were downloaded from Saccharomyces Genome Database (SGD). We filtered out all regulatory relationships, and maintain only the relationships resulting in 15 main functional categories for Saccharomyces cerevisiae as given in [10]. Protein-Protein interaction data of Saccharomyces cerevisiae were obtained from BIOGRID [11] database (http://thebiogrid.org/). To reduce the effect of noise, the duplicated interactions and self-interactions were removed. The final dataset consists of 69,331 interaction protein pairs involving 5386 annotated proteins. Let $\{f_{r1}, f_{r2}, \ldots, f_{rn}\}$ be the set of n real functions of protein p and $\{f_{p1}, f_{p2}, \ldots, f_{pm}\}$ denotes the set of m functions predicted by protein function assignment scheme. It is obvious that $1 \leq m, n \leq D$. The three performance metrics used to evaluate the effectiveness of our proposed method are:

$$Precision = \frac{\sum_{j=1}^{m} \max_{i=1}^{n}(sim(f_{r_i}, f_{p_j}))}{m} \tag{11}$$

$$Recall = \frac{\sum_{i=1}^{n} \max_{j=1}^{m}(sim(f_{r_i}, f_{p_j}))}{n} \tag{12}$$

$$F - value = \frac{2 \times Precision \times Recall}{Precision + Recall} \tag{13}$$

An algorithm having higher values for the above metrics supersedes others. The evaluation of these metrics were conducted on test datasets by varying the number of proteins in the network N= [10, 200] for a particular unannotated protein. We have used only biological process for our experiment as assigning biological process to unannotated protein includes biological experiments which are very costly.In our study, we have compared the relative performance of the proposed scheme with Firefly Algorithm (FA) [12], Particle Swarm Optimization (PSO) [13], and also with the existing methods like Indirect Neighbor Method (INM) [14] and Neighbor Counting (NC) [3] in Table 1 and Fig. 2, Fig. 3, Fig. 4, for predicting functions of protein YBL068W.We report here results for only the above mentioned protein in order to save space.The omitted results for different proteins follow a similar trend as those stated above. The proposed approach was applied on annotated proteins of the network as the real functions for the same will be known to us. For CSA, c_{min} and c_{max} are set to 2 and 10 respectively.For all the evolutionary/swarm algorithm-based prediction schemes, the population size is kept at 50 and the maximum function evaluations (FEs) is set as 300000 and also best parametric set-up for already existing method is used. It is evident from Table1 and Fig. 2-4 that our algorithm outperforms others with respect to the aforementioned performance metrics irrespective of number of proteins in the network.

Table 1. Comparative analysis for predicting functions of YBL068W with N=80

Protein	Real Function	Real Functions Predicted by Different Algorithms				
		CSA	FA	PSO	INM	NC
YBL068W	GO:0009116	GO:0009116	GO:0009116	GO:0009116	x	x
	GO:0006015	GO:0006015	X	x	x	x
	GO:0009165	x	GO:0009165	GO:0009165	GO:0009165	x
	GO:0044249	GO:0044249	X	GO:0044249	GO:0044249	GO:0044249
	GO:0031505	GO:0031505	GO:0031505	x	GO:0031505	x
	GO:0009156	x	X	x	x	GO:0009156
	GO:0016310	GO:0016310	GO:0016310	GO:0016310	x	x

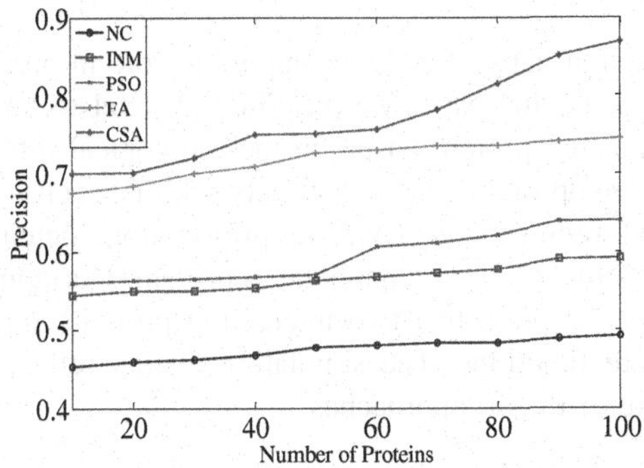

Fig. 2. Comparative analysis of precision plot

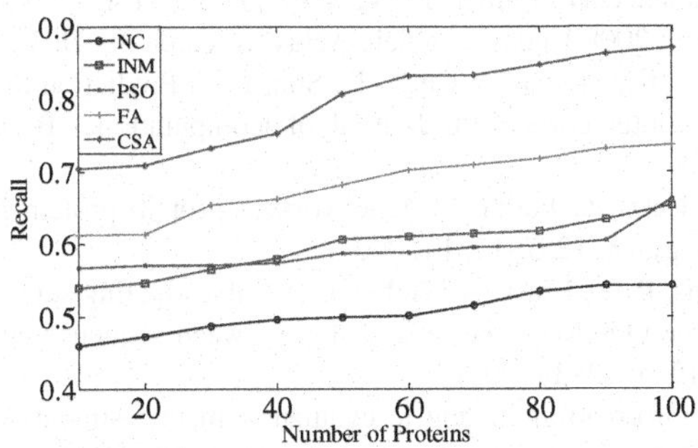

Fig. 3. Comparative analysis of recall plot

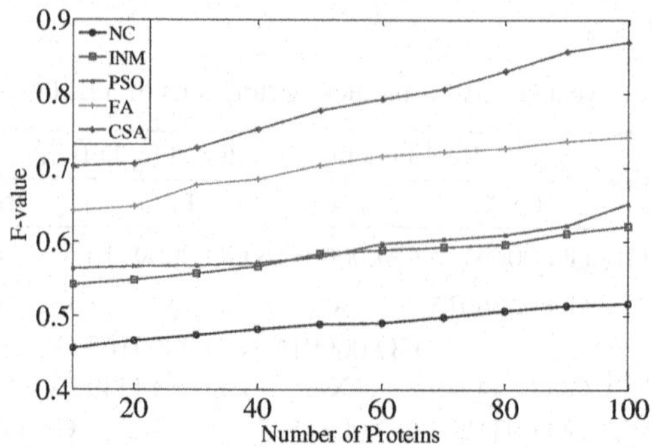

Fig. 4. Comparative analysis of F-value plot

5 Conclusion

Protein-Protein Interaction network play an important role in predicting the function of unannotated protein. In this paper we proposed a novel technique to predict the function of the unannotated protein based on the topological information as well as the functions of annotatedproteins of the PPI network of Saccharomyces Cerevisiae. Semantic similarity between proteins based on information content of the Go term is utilized to associate a function to an unannotated protein. Our approach does not entirely depend on the assumption that two interacting proteins are likely to have the same function or share functions. The simulation results reflect that our approach outperforms other state-of-the-art approaches.

References

1. Breitkreutz, B.J., Stark, C., Reguly, T., Boucher, L., Breitkreutz, A., Livstone, M., Oughtred, R., Lackner, D.H., Bähler, J., Wood, V., Dolinski, K., Tyers, M.: The BioGRID Interaction Database: 2008 Update. Nucleic Acids Research 36, D637– D640(2008)
2. Deng, M.H., Zhang, K., Mehta, S., Chen, T., Sun, F.Z.: Prediction of protein function using protein-protein interaction data. Journal of Computational Biology 10(6), 947–960 (2003)
3. Schwikowski, B., Uetz, P., Field, S.: A network of protein protein interactions in yeast. Nature Biotechnology 18, 1257–1261 (2000)
4. Lord, P.W., Stevens, R.D., Brass, A., Goble, C.A.: Investigating semantic similarity measures across the Gene Ontology: the relationship between sequence and annotation. Bioinformatics 19(10), 1275–1283 (2003)
5. Resnik, P.: Using information content to evaluate semantic similarity in a taxonomy. In: Proceedings of International Joint Conference for Artificial Intelligence, pp. 448–453 (1995)
6. Castro, D., Nunes, L., Zuben, F.J.V.: The clonal selection algorithm with engineering appli-cations. In: Proceedings of GECCO, pp. 36–39 (2000)

7. Felipe, C., Guimarães, F.G., Igarashi, H., Ramírez, J.A.: A clonal selection algorithm for optimization in electromagnetics. IEEE Transactions on Magnetics 41(5), 1736–1739 (2005)
8. Ashburner, M., Ball, C., Blake, J., Botstein, D., Butler, H., Cherry, J., Davis, A., Dolinski, K., Dwight, S., Eppig, J.: Gene ontology: tool for the unification of biology. Nature Genetics 25, 25–29 (2000)
9. Dwight, S., Harris, M., Dolinski, K., Ball, C., Binkley, G., Christie, K., Fisk, D., Issel Tarv-er, L., Schroeder, M., Sherlock, G.: Saccharomyces Genome Database (SGD) provides sec-ondary gene annotation using the Gene Ontology (GO). Nucleic Acids Research 30, 69–72 (2012)
10. Chowdhury, A., Konar, A., Rakshit, P., Janarthanan, R.: Protein Function Prediction Using Adaptive Swarm Based Algorithm. SEMCCO 2, 55–68 (2013)
11. Stark, C., Breitkreutz, B.J., Reguly, T., Boucher, L., Breitkreutz, A., Tyers, M.: BioGRID: a general repository for interaction datasets. Nucleic Acids Res. 34, D535–D539 (2006)
12. Yang, X.-S.: Firefly algorithms for multimodal optimization. In: Watanabe, O., Zeugmann, T. (eds.) SAGA 2009. LNCS, vol. 5792, pp. 169–178. Springer, Heidelberg (2009)
13. Kennedy, J.: Particle swarm optimization. In: Encyclopedia of Machine Learning, pp. 760–766 (2010)
14. Chua, H.N., Sung, W.-K., Wong, L.: Exploiting indirect neighbors antopological weight to predict protein function from protein protein interactions. Bioinformatics 22(13), 1623–1630 (2006)

Fast Approximate Eyelid Detection for Periocular Localization

Saharriyar Zia Nasim Hazarika[1], Neeraj Prakash[1], Sambit Bakshi[2], and Rahul Raman[2]

[1] Department of Computer Science and Engineering
Sikkim Manipal Institute of Technology, Sikkim - 737132, India
[2] Department of Computer Science and Engineering
National Institute of Technology, Rourkela, Odisha - 769008, India
{shazarika1991,sambitbaksi,rahulraman2}@gmail.com,
neeprkash@outlook.com

Abstract. Iris is considered to be one of the most reliable traits and is widely used in the present state-of-the-art biometric systems. However, iris recognition fails for unconstrained image acquisition. More precisely, the system cannot properly localize the iris from low quality noisy unconstrained image, and hence, the successive modules of biometric system fails. To achieve recognition from unconstrained iris images, the periocular region is considered. The periocular (periphery of ocular) region is proven to be a trait in itself and can serve as a biometric to recognize human, though with a lower accuracy compared to iris. In this paper, we propose a novel technique to localize periocular region on the basis of eyelid information extracted from eye image. The proposed method will perform periocular localization successfully even when iris detection fails. Our method detects the horizontal edges as eyelids and the rough map of eyelids gives the radius of iris, which is used to anthropometrically derive the periocular region. The proposed method has been validated on standard publicly available databases : UBIRISv1 and UBIRISv2, and is found to be satisfactory.

Keywords: Periocular recognition, Eyelid detection, Personal identification.

1 Introduction

Biometrics, as the name suggests, refers to certain characteristics of the human body. These characteristics are unique to every individual. The primary use of these traits or characteristics is in the identification and authentication of individuals [3]. In the field of computer science, biometrics is used as a means to maintain the identification process and allow access control. Biometrics is also an important tool for identifying individuals that are under surveillance. The biometric identifiers are often categorized into physiological and behavioral features. Physiological features are the features that are involved with the shape

M.K. Kundu et al. (eds.), *Advanced Computing, Networking and Informatics - Volume 1*, Smart Innovation, Systems and Technologies 27,

DOI: 10.1007/978-3-319-07353-8_80,

and size of the human body. Some examples include face, fingerprint, DNA, iris recognition etc. The behavioral features, on the other hand, are related to the pattern of behavior of a person, which include the person's speech, voice texture, gait etc. A good biometric trait is characterized by the use of the features that are highly unique, stable, easy to capture and prevents circumvention.

Out of all the biometric features available, only the iris is considered to be the most reliable. The reason is that reliability is particularly dependent on the ability to detect unique features easily that can be extracted and which remains invariant over a period of time. However, capturing of image of the iris takes place in a constrained environment. The various constraints include looking into the scanner of the iris camera, capturing of images from a distance, non-cooperative subjects etc. Also, one of the major problems is that the image of the iris is captured in the near infrared region (NIR). This limits the merits of the iris recognition method in the sense that the NIR wavelength does not allow any image to be captured in an outdoor environment. This leads to a possibility that a very good quality image of the iris might not be captured. This is due to the occurrence of occlusion i.e. some parts of the required features might be hidden due to the presence of spectacles or caps, or the presence of any other noisy factors. In such situations it might be possible to capture and recognize an individual through the image of the region around the eye.

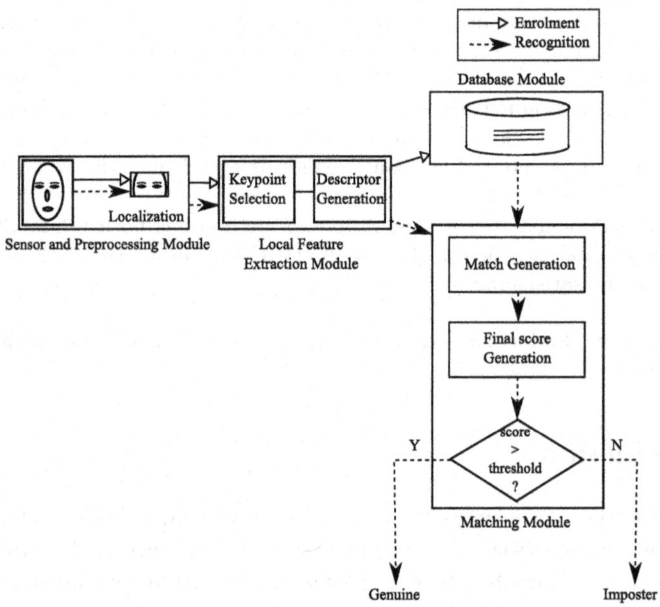

Fig. 1. Working model of biometric authentication

2 Related Works

To counter the challenges posed for unconstrained iris detection for identification of an individual, the region around the iris (periocular region) has been considered as an alternative trait. Features from regions outside the iris like blood vessels, eyelids, eyelashes, skin textures thus contribute as an integral part of periocular identification process. Various researchers have proposed several methods for the purpose of separating the eyelids from the rest of the image so that it can be used as a biometric feature. An overview of the related works by other researchers have been tabulated in Table 1. Masek [8] has proposed a method for the eyelid segmentation in which the iris and the eyelids have been separated through Hough transformation. In [14], Xu *et al.* have segmented out the upper and lower eyelid candidate region into 8 sub-blocks. Out of these 8 sub-blocks, the proper eyelid-eyelash model has been chosen based on maximum deviation from each block. This approach however bears a disadvantage. If the occluded region is wider than the region defined as eyelid-eyelash region in each block, then this approach produces wrong position of the features leading to mislocalization. In contrast to [8], Mahlouji and Noruzi [7] have used Hough transform to localize all the boundaries between the upper and lower eyelid regions. This method in [7] has a relatively lower mislocalization rate than [8], while the processing time of the former is also of the same order as the latter. He *et al.* in their work [5], have proposed a novel method for the eyelash, eyelid, and shadow localization. In [12], Thalji and Alsmadi have proposed an eyelid segmentation method wherein the pupil is detected first, and the limbic boundary is subsequently localized. After the localization of the limbic boundary, this algorithm successfully extracts the iris region excluding the eyelid and eyelashes and thus avoids occlusion. Adam *et al.* in [1] have presented two cumulative distribution functions to estimate the performance of the upper and lower eyelid segmentations, which yields $< 10\%$ mislocalization for 90% of the upper eyelids and 98% of the lower eyelids. Cui *et al.* [4] have proposed two algorithms for the detection of upper and lower eyelids and tested on CASIA dataset. This technique has been successful in detecting the edge of upper and lower eyelids after the eyelash and eyelids were segmented. Ling *et al.* proposed an algorithm capable of segmenting the iris from images of very low quality [6] through eyelid detection. The algorithm described in [6] being an unsupervised one, does not require a training phase. Radman *et al.* have used the live-wire technique to localize the eyelid boundaries based on the intersection points between the eyelids and the outer iris boundary [11].

3 Proposed Fast Eyelid Detection

The working model of a biometric system is shown in Fig. 1. Our proposed work deals with the first module shown in the system: sensor and preprocessing module. Our objective is to localize periocular region from the input eye image properly. For the said purpose we aim to detect the upper and lower eyelid

Table 1. Related works on eyelid segmentation

Year	Author(s)	Eyelid Segmentation Approach
2003	Masek [8]	Separated iris and eyelid region by horizontal Hough transform.
2006	Xu *et al.* [14]	Divided the upper and lower eyelid candidate region into 8 sub-blocks and chose the proper eyelid/eyelash model based on maximum deviation from each block.
2012	Mahmoud Mahlouji and Ali Noruzi [7]	The Hough transform is used to localize all the boundaries between upper and lower eyelids. The proposed method not only has relatively higher precision and but also compares with popular available methods in terms of processing time. (database used- CASIA)
2008	Zhaofeng He *et al.* [5]	An accurate and fast method for eyelash, eyelid and shadow localization is implemented which provides a novel prediction model for determining proper threshold for eyelash and shadow detection.
2013	Zyad Thalji and Mutasem Alsmadi [12]	The proposed iris image segmentation algorithm consists of separate modules for pupil detection, limbic bounary localization and eyelid/eyelash detection. The proposed algorithm successfully extracted the iris print from the images and also avoided the eyelash and eyelid which makes the noise.
2008	Mathieu Adam *et al.* [1]	Two cumulative distribution functions are presented to estimate the performances of upper and lower eyelid segmentation.
2004	Jiali Cui *et al.* [4]	The two algorithms are proposed for: Upper eyelid localization and lower eyelid detection. Edge of the upper and lower eyelids are detected after the eyelids and the eyelashes are segmented. (Database used : CASIA Iris database (version 1.0))
2010	Lee Luang Ling and Daniel Felix de Brito [6]	The algorithm is capable of segmenting the iris images of a very low quality. The major methods involved in this iris segmentation approach were: Pupillary detection, limbic boundary detection and eyelid/eyelash detection. Interesting feature of this algorithm is no training feature is required.
2013	Abduljalil Radman *et al.* [11]	The live-wire technique has been utilized to localize eyelid boundaries based on the intersection points between the eyelid and the outer iris boundary.

separately. In the first step, the eye image is smoothed to suppress the low-gradient edges. Wiener low-pass filter applied along the 3×3 neighborhood of every pixel is employed to smooth the image. The window size is chosen empirically as it suppresses negligible edges and retains strong edges as required by the proposed system. The smoothed image is subjected to an edge detection technique to find edge map existing in the image. Sobel, Prewitt, Roberts, Laplacian of Gaussian, zero cross, Canny or any other existing suitable edge detection technique can be used that efficiently detects edges. In implementation, we have used Canny edge detector with standard deviation of underlying Gaussian Filter as 1. Subsequently for every detected edge, a *horizontality factor* is calculated. This *horizontality factor* (denoted by hf) denotes how much horizontal an edge is through a fractional value lying within [0,1]. hf for i-th edge (denoted by hf_i) calculated through Equation 1.

$$hf_i = \frac{\text{Maximum } x \text{ coordinate of } i \text{ th edge} - \text{Minimum } x \text{ coordinate of } i \text{ th edge}}{\text{Number of unique pixels belonging to } i \text{ th edge}} \quad (1)$$

The horizontal factor of all edges that are mostly horizontal will be high (close to 1) and the same will be low for the edges that are vertical (close to 0). A threshold value is calculated by use of this horizontality function by Equation 2.

$$thresh = \min(hf) + 0.1 \times (\max(hf) - \min(hf)) \quad (2)$$

During this process maximum edges do have similar horizontality factor. Edges with most frequent horizontality factor is retained while other edges are removed. This assures removal of non-horizontal edges even when the input image is tilted. This process of eyelid detection is elaborated in Algorithm 1.

Successively, the horizontal edges thus generated are subjected to 2-means clustering method and the upper and lower eyelids are separated. Two cluster centers represent the approximate distance between two eyelids. Further, this distance is the least erroneous measure of the diameter of iris in this scenario. Based on this parameter, we choose a properly oriented rectangle around the eye which represents the periocular region [2]. A comprehensive flow of the whole proposed method has been illustrated in Fig. 2.

4 Experimental Results

Upon implementing the algorithm on UBIRISv1 and UBIRISv2 databases (Table 2), we observe that the entire eyelid has not been detected. Our algorithm results in a partial success in detecting the eyelids. The upper eyelid shows a partial localization of 68.13% and 65.86% for UBIRISv1 and UBIRISv2 respectively, while the lower eyelid shows a partial localization of 55.22% and 50.28% for the mentioned databases respectively. The experimental data have been tabulated in the Table 3. The significance of the results is that even with partial success in eyelid detection, we have been able to localize the periocular region.

Algorithm 1. Eyelid detection

```
Require: im
Ensure: cc2
A ← smooth(im)
B ← edge(A)
cc1 ← labelled_edge(B)
no_of_edges ← maximum(cc1)
for i = 1 to no_of_edges do
    hor_spr ← horizontal spread of the edge
    nop ← number of pixel in i^th edge
    horizontality(i) ← hor_spr / nop
end for
cc2 ← cc1
for i = 1 to max(cc1) do
    if horizontality(i) < 90% of horizontality of other edges then
        cc2 ← cc1 − i^th edge
    end if
end for
return cc2
```

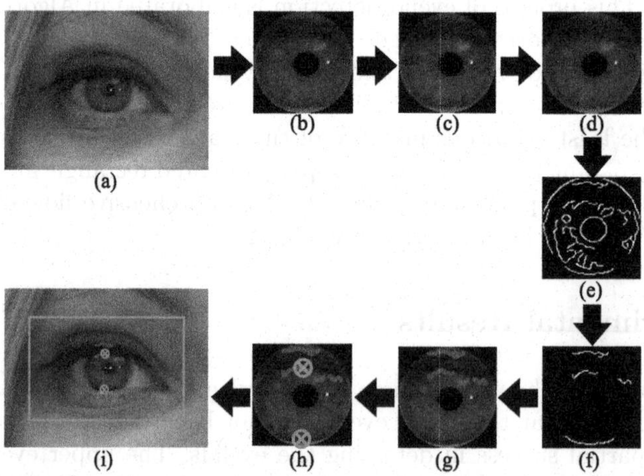

Fig. 2. Proposed periocular localization algorithm : (a) Coloured input image; (b) Noisy image of the iris; (c) Grayscale image of the iris; (d) Smoothed image of the iris; (e) Detection of edges of the image; (f) Horizontal edges retained; (g) Upper and lower eyelids detected through checking hf; (h) 2-means clustering performed on detected eyelid points; (i) Periocular region localized

Table 2. Detail of database for evaluation of proposed approach

Database	Developer	Version	Images	Subjects	Resolution	Color Model	Spectrum
UBIRIS	Soft Computing and Image Analysis (SOCIA) Group, Department of Computer Science, University of Beira Interior, Portugal	v1 [9]	1,877	241	800 × 600	RGB	VS
		v2 [10]	11,102	261	400 × 300	sRGB	

Table 3. Localization results on UBIRISv1 and UBIRISv2 databases

Accuracy →	Upper Eyelid		Lower Eyelid	
Databases ↓	Mislocalization (%)	Partial Localization (%)	Mislocalization (%)	Partial Localization (%)
UBIRISv1	22.48	68.13	35.79	55.22
UBIRISv2	30.07	65.86	40.18	50.28

5 Conclusion

Iris is considered to be one of the most reliable biometric traits, making it one of the most widely used in the present biometric scenario. However, in cases where unconstrained images are acquired, iris recognition fails to deliver desired accuracy. Hence, to achieve recognition in unconstrained images, the periocular region is considered as an alternative. Upon implementing the proposed eyelid detection algorithm on publicly available visual spectrum (VS) image database UBIRISv1 and UBIRISv2, it is observed that both the upper and lower eyelids show partial localization for most of the images. Even with the partial success in detecting the eyelids, we are able to approximately localize the periocular region.

References

1. Adam, M., Rossant, F., Amiel, F., Mikovikova, B., Ea, T.: Eyelid Localization for Iris Identification. Radioengineering 17(4), 82–85 (2008)
2. Bakshi, S., Sa, P.K., Majhi, B.: Optimised periocular template selection for human recognition. BioMed Research International 2013, 1–14 (2013)
3. Bakshi, S., Tuglular, T.: Security through human-factors and biometrics. In: 6th International Conference on Security of Information and Networks, pp. 463–463 (2013)
4. Cui, J., Wang, Y., Tan, T., Ma, L., Sun, Z.: A Fast and Robust Iris Localization Method Based on Texture Segmentation. In: Biometric Authentication and Testing, National Laboratory of Pattern Recognition, Chinese Academy of Sciences (2004)
5. He, Z., Tan, T., Sun, Z., Qiu, X.: Robust eyelid eyelash and shadow localization for iris recognition. In: 15th IEEE International Conference on Image Processing, pp. 265–268 (2008)
6. Ling, L.L., de Brito, D.F.: Fast and efficient iris image segmentation. Journal of Medical and Biological Engineering 30(6), 381–392 (2010)

7. Mahlouji, M., Noruzi, A.: Human Iris Segmentation for Iris Recognition in Unconstrained Environments. IJCSI International Journal of Computer Science Issues 9(1), 3, 149–155 (2012)
8. Masek, L.: Recognition of Human Iris Patterns for Biometric Identification. In: Bachelor of Engineering Thesis at The University of Western Australia (2003)
9. Proença, H., Alexandre, L.A.: UBIRIS: A noisy iris image database. In: Roli, F., Vitulano, S. (eds.) ICIAP 2005. LNCS, vol. 3617, pp. 970–977. Springer, Heidelberg (2005)
10. Proena, H., Filipe, S., Santos, R., Oliveira, J., Alexandre, L.: The UBIRISv2: A database of visible wavelength iris images captured on-the-move and at-a-distance. IEEE Transactions on Pattern Analysis and Machine Intelligence 32(8), 1529–1535 (2010)
11. Radman, A., Zainal, N., Ismail, M.: Efficient Iris Segmentation based on eyelid detection. Journal of Engineering Science and Technology 8(4), 399–405 (2013)
12. Thalji, Z., Alsmadi, M.: Iris Recognition Using Robust Algorithm for Eyelid, Eyelash and Shadow Avoiding. World Applied Sciences Journal 25(6), 858–865 (2013)
13. Valentina, C., Hartono, R.N., Tjahja, T.V., Nugroho, A.S.: Iris Localization using Circular Hough Transform and Horizontal Projection Folding. In: Proceedings of International Conference on Information Technology and Applied Mathematics (2012)
14. Xu, G., Zhang, Z., Ma, Y.: Improving the Performance of Iris Recognition System Using Eyelids and Eyelashes Detection and Iris Image Enhancement. In: 5th IEEE conference on Cognitive Informatics, pp. 871–876 (2006)

Prediction of an Optimum Parametric Combination for Minimum Thrust Force in Bone Drilling: A Simulated Annealing Approach

Rupesh Kumar Pandey and Sudhansu Sekhar Panda*

Department of Mechanical Engineering, Indian Institute of Technology Patna, India
rupeshiitp@gmail.com, sspanda@iitp.ac.in

Abstract. Minimally invasive drilling of bone has a great demand in orthopaedic surgical process as it helps in better fixation and quick healing of the damaged bones. The aim of the present study is to find out the optimal setting of the bone drilling parameters (spindle speed and feed rate) for minimum thrust force during bone drilling using simulated annealing (SA). The bone drilling experiments were carried out by central composite design scheme and based on the results obtained, a response surface model for thrust force as a function of drilling parameters is developed. This model is used as an objective function in the SA approach. The results of the confirmation experiments showed that the SA can effectively predict the optimal settings of spindle speed and feed rate for minimum thrust force during bone drilling. The suggested methodology can be very useful for orthopaedic surgeons to minimize the drilling induced bone tissue injury.

Keywords: Bone drilling, Thrust force, Response surface methodology, Simulated annealing.

1 Introduction

Drilling of bone is commonly employed during orthopaedic surgery to produce hole for the insertion of screws and wires to fix the damaged bone or for the installation of prosthetic device [1]. The thrust force induced to the bone during drilling is one of the major concerns as the exposure of the bone to their higher magnitudes can damage the bone cells or can even result in their death (osteonecrosis) leading to the loosening of the fixation and increased healing time [1-4]. Moreover, micro cracks and the drill bit breakage can also occur as an additional disadvantage of the higher mechanical forces. The micro cracks can lead to the further fracture of the damaged bone which can initiate large number of cracks resulting in the misalignment of the fixation or permanent failure [2]. The broken drill bit can obstruct the placement of other fixating devices and can cause adverse histological effects if it undergoes corrosion thus, demands for additional procedures to remove the broken drill bit [5-9] thereby increasing the operative time. The rate of heat generation is also high with higher drilling forces [10] which can result in thermal osteonecrosis. Therefore, minimization of the

* Corresponding author.

M.K. Kundu et al. (eds.), *Advanced Computing, Networking and Informatics - Volume 1*, Smart Innovation, Systems and Technologies 27,
DOI: 10.1007/978-3-319-07353-8_81, © Springer International Publishing Switzerland 2014

thrust force generated during bone drilling will result in better fixation of the broken bones and their quick recovery postoperatively.

Previously, many researchers have investigated the bone drilling process to study the effect of the various drilling parameters on the thrust force produced [1-4]. The early researches in this area were reported in late 1950s [3]. Spindle speed and feed rate were the main drilling parameters analyzed in most of the studies [3-4, 10-14]. There is a consensus among the researchers that the thrust force decreases with an increase in spindle speed [3-4, 10-14]. But, the drilling of bone with higher spindle speeds were reported with increased trauma [3, 14]. The analysis on the effect of the feed rate showed that the increase in feed rate increases the thrust force induced in bone drilling [1-2, 4, 13]. Despite of the above mentioned studies, there is a lack of a clear suggestion on the optimal settings of the feed rate and spindle speed for minimum thrust force generation during bone drilling.

In this work, a statistical model for bone drilling process to predict the thrust force as a function of feed rate (mm/min) and spindle speed (rpm) is developed using response surface methodology (RSM). Next, the model is used as a fitness function in SA algorithm to determine the optimal setting of feed rate and spindle speed for minimum thrust force during bone drilling. The adopted approach is then validated through the confirmation experiment.

RSM is a collection of mathematical and statistical tools which are easy, quick and effective for modeling the process in which several variables influence the response of interest [15-16]. In most of the real problems the relationship between the response and the independent variable is unknown. In RSM the relationship between the response and the independent process variables is represented as (1)

$$Y = f(A,B,C) + \varepsilon \tag{1}$$

Where Y is the desired response, f is the response function and ε represents the error observed in the response. A second order model is generally employed if the response function is nonlinear or not known, shown in (2) [15-16]

$$Y = \beta_0 + \sum_{i=1}^{k} \beta_i x_i + \sum_{i=1}^{k} \beta_{ii} x_i^2 + \sum_i \sum_j \beta_{ij} x_i x_j + \varepsilon \tag{2}$$

Where β_0 is the coefficient for constant term β_i, β_{ii} and β_{ij} are the coefficients for linear, square and interaction terms respectively.

Simulated annealing algorithm mimics the process of annealing which involves heating of a metal to a temperature beyond its melting point followed by a gradual cooling. In molten state, the particles are in random motion and when it is cooled gradually the particles rearrange themselves to attain minimum energy state. As the cooling is done gradually, lower and lower energy states are obtained until the lowest energy state is reached [17]. In the process of annealing the probability $P_r(E)$ of being at energy state is given by the Boltzmann distribution as (3):

$$P_r(E) = \frac{1}{Z(T)\exp(-E/KT)} \tag{3}$$

Where $Z(T)$ = normalized factor depending upon the temperature T and K is the Boltzmann constant. The probability $P_r(E)$ approaches one when the temperature T is high for all energy states. The probability of the higher energy states decreases as compared to the lower energy states as the temperature decreases.

Metropolis et al. [18] used the above criteria and proposed that if a random perturbation is applied to the present energy state of a solid with temperature T for the generation of the perturbed energy states and the difference of the energy ΔE between the two states is negative then the particles rearrange themselves such that they attain the low energy state. The probability of the acceptance of perturbed energy state as the new energy state is given as (4)

$$P_r(E) = \exp(-\Delta E / KT) \tag{4}$$

This criterion of acceptance for the new state is known as the Metropolis acceptance criteria. Kirkpatrick et al. [19] used the sequence of Metropolis algorithm evaluated by the sequence of reducing temperatures as simulated annealing minimization of an objective function. The objective function corresponds to the energy function used in Metropolis acceptance criteria. Recently, SA has been used successfully for the optimization of the various engineering problems [20-21]. It uses a number of points (N) to test the thermal equilibrium at a particular temperature before acquiring a reduced temperature state. The algorithm is stopped when the desirable minimum change in the value of the objective function is obtained or the temperature obtained is sufficiently small. The initial starting temperature T and the number of iterations N to be performed at each temperature are the two user defined parameters that governs the effective working of SA algorithm.

The step by step procedure of SA algorithm is discussed below [17].

- Initialize the starting temperature T and the termination temperature $T_{\min}$. Randomly generate initial starting point X . Also define the number of iterations N to be performed at each Temperature. Set the iteration counter as $i = 0$
- Calculate the value of the objective function $E = f(X)$.
- Update $E^{Best} = E$ and $X^{Best} = X$
- Generate the random neighborhood point X^* and calculate the objective function $E^* = f(X^*)$
- Evaluate the change in energy $\Delta E = E^* - E$
- If the change in energy $\Delta E < 0$, then update the point $X = X^*$ and if $E < E^{best}$,then $E^{best} = E$ and $X^{Best} = X$ .Update the iteration as $i = i + 1$ and go to next step. Else go to step number 3 and repeat the process.
- If $i > N$ go to the next step.
- Reduce the temperature by a factor α and update $T = \alpha T$.
- If $T \le T_{\min}$ then terminate the process and print X^{best} and E^{best} else, move to step 3.

2 Experimental Procedure

2.1 Experimental Design Based on Response Surface Methodology

The bone drilling parameters considered are feed rate (mm/min) and spindle speed (rpm) (shown in Table 1) and the response taken is thrust force (N). The central composite design (CCD) of RSM was employed to design the plan of experiments for studying the relationship between the response and the bone drilling parameters. Full factorial design for factors at two levels i.e. high (+1) and low (-1) corresponding to a face centered design with thirteen runs (four factorial points, four axial points and five central points) was used as shown in Table 2. The bone drilling parameters and their levels are considered based on the wide range of experiments reported in the literature [1-4, 10-14].

Table 1. Factor and levels considered for bone drilling.

	Control factor	*Low level (-1)*	*High level (+1)*
A	Feed rate (mm/min)	30	150
B	Spindle speed (rpm)	500	2500

2.2 Experimental Details

The work material used for conducting the bone drilling experiments was bovine femur, as the human bones are not easily available and it closely resembles the human bone, allowing the results to be extrapolated in the real surgical situations [19-20]. The bovine femur was obtained from a local slaughter house immediately after the slaughter and the experiments were done within few hours to maintain minimum loss in thermo-physical properties of the fresh bone [22-23]. No animal was scarified specifically for the purpose of this research.

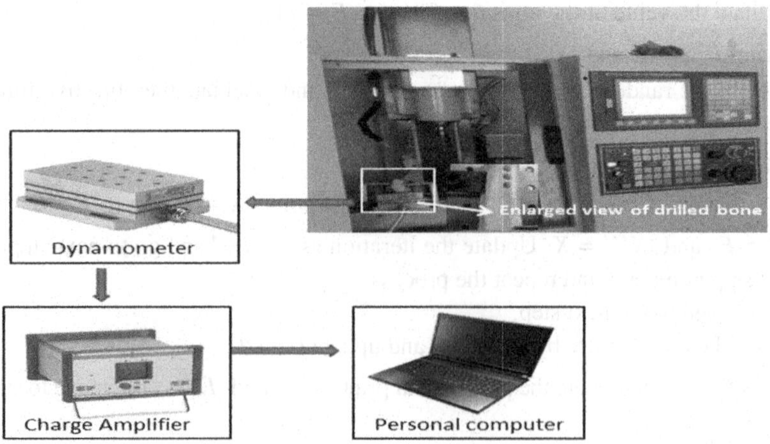

Fig. 1. Experimental set up

The experiments were carried out on 3 axis MTAB flex mill using 4.5mm HSS (high speed steel) drill bit. The drilling depth was 6mm. The drilling thrust force signals were measured using Kistler 9257B piezo electric dynamometer. The signals were acquired using 5070 multichannel charge amplifier and Dynoware software. The thrust force obtained for each experimental run is listed in the last column of Table 2.The experimental set up is shown in the Fig. 1.

Table 2. Experimental condition and result

Exp No.	*Feed rate (mm/min)*	*Spindle speed (rpm)*	*Thrust Force (N)*
1	90	1500	16.93
2	150	1500	24.55
3	90	500	29.25
4	150	500	45.43
5	30	2500	4.72
6	90	2500	14.31
7	30	1500	6.155
8	90	1500	16.79
9	90	1500	17.41
10	90	1500	17.29
11	150	2500	20.33
12	90	1500	17.01
13	30	500	11.32

3 Development of Mathematical Model

A mathematical model correlating the thrust force and drilling process parameters is developed based on (2) using design expert software version 8.0.1 [24]. A quadratic model is selected based on low standard deviation and high R squared value as mentioned in Table 3 [24].

Table 3. Model summary statistics.

Source	*Standard deviation*	*R-Squared*	*Adjusted R-Squared*	*Predicted R-Squared*	
Linear	4.08	0.8721	0.8466	0.6912	
2FI	3.00	0.9378	0.9171	0.7966	
Quadratic	1.44	0.9888	0.9808	0.8880	Suggested

The model is given by (5) as:

$$\begin{aligned} Force = {} & 8.796 + 0.3817 \times Feed\ rate - 0.0155 \times spindle\ speed \\ & - 7.7083 \times E - 5 \times Feed\ rate \times spindle\ speed \\ & - 4.2713 \times E - 4 \times Feed\ rate^2 \\ & + 4.8898 \times E - 6 \times spindle\ speed^2 \end{aligned} \quad (5)$$

Analysis of variance (ANOVA) carried out to find the significance of the developed model and individual model coefficients at 95% confidence interval is shown in Table 4.

Table 4. ANOVA table for the proposed model

Source	*DOF*	*SS*	*MS*	*P value*
Model	5	1287.77	257.55	<.0001
A-Feed rate	1	773.28	773.28	<.0001
B-Spindle speed	1	362.55	362.55	<.0001
AB	1	85.56	85.56	.0004
A2	1	6.53	6.53	.1200
B2	1	66.04	66.04	.0008
Residual	7	14.58	2.08	
Total	12	1302.35		

Where
DF= Degree of freedom
SS= Sum of squares
MS= Mean square
From the ANOVA table it can be seen that the model is significant as its p value is less than 0.0500. In this case A, B, AB, B2 are significant model terms. Values greater than 0.1000 indicate the model terms are not significant [24]. The comparison of the predicted thrust force values with the actual values is shown in the Fig. 2.

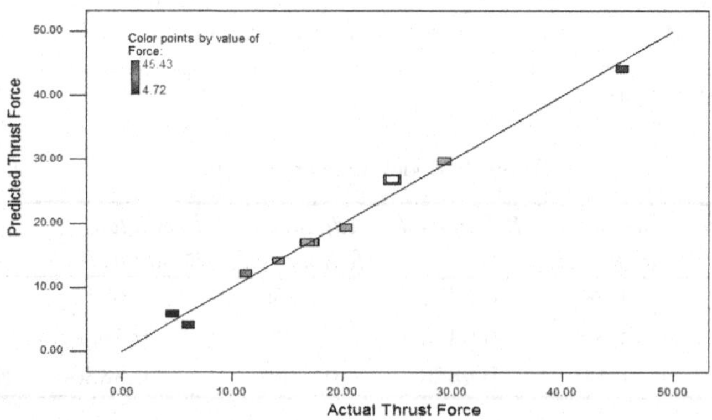

Fig. 2. Comparison between the predicted and the actual thrust force

4 Optimization of the Thrust Force with SA

The optimal setting of the spindle speed and feed rate for minimum thrust force during bone drilling is determined using SA. The developed response surface model of thrust force is taken as an objective function to be minimized. The SA algorithm is implemented using the following parameters:

Initial temperature $T = 100$
Termination criteria $T_{min} = 10^{-6}$
Boltzmann constant $K = 1$
Cooling rate $\alpha = 0.95$
Number of cycles at each temperature $N = 50$

The problem formulation is subjected to the boundaries (limitations) of the drilling parameters and is stated as follows:

$30 \leq$ Feed rate ≤ 150
$500 \leq$ Spindle speed ≤ 2500

The result obtained by SA is listed in Table 5. Fig. 3 shows the variation of the function value with number of iterations. From Table 5, it can be observed that the minimum value of thrust force is 3.6364 N with the feed rate of 30 (mm/min) and spindle speed of 1820.845 (rpm).

Table 5. Results obtained from SA analysis.

Parameters	*Value*
Minimum fitness function thrust force	3.6364 (N)
Optimal cutting conditions	
Feed rate	30
Spindle speed	1820.845

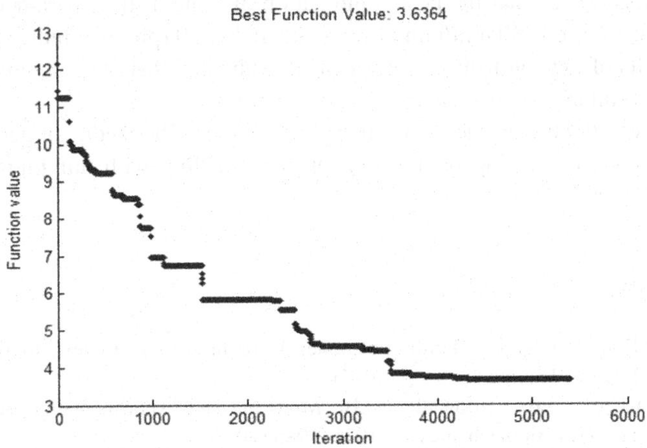

Fig. 3. Plot of the function value against iteration

To validate the result obtained from the above analysis, confirmation experiments were carried out. The result of the confirmation experiments are shown in Table 6.

Table 6. Confirmation experiments and results

No.	*Feed rate (mm/min)*	*Spindle speed (rpm)*	*Thrust force Exp.*	*Thrust force predicted*	*%Error*
1	30	1000	7.10	6.94	2.25
2	30	1820	3.87	3.63	6.20
3	60	1000	15.68	14.92	4.84
4	120	500	38.63	37.30	3.44

Four experiments were carried out within the range of the parameters studied for the confirmation of the obtained results. Three settings of feed rate and spindle speed were selected randomly whereas one optimal setting predicted by SA was analyzed. From Table 6 it is clear that the predicted values and those obtained from experiment are very close hence, RSM model can effectively predict the thrust force values whereas SA can be very useful to minimize the thrust force during bone drilling.

5 Conclusions

In the present investigation, an approach involving the integration of RSM with SA is used for the optimization of thrust force in bone drilling process. From the above analysis following conclusions are drawn:

- The developed response surface model can effectively predict the thrust force in bone drilling within the range of the parameters studied.
- SA results showed that the best combination of bone drilling parameters for minimum thrust force is 30 mm/min of feed rate and 1820 rpm of spindle speed.
- The results of the confirmation experiments validated that the combination of RSM and SA is suitable for optimizing the bone drilling process.
- The use of above approach can greatly assist the orthopaedic surgeons to decide the best level of drilling parameters for bone drilling with minimum mechanical damage.

References

1. Pandey, R.K., Panda, S.S.: Drilling of bone: A comprehensive review. Journal of Clinical Orthopedics and Trauma 4, 15–30 (2013)
2. Lee, J., Gozen, A.B., Ozdoganlar, O.B.: Modeling and experimentation of bone drilling forces. Journal of Biomechanics 45, 1076–1083 (2012)
3. Thompson, H.C.: Effect of drilling into bone. Journal of Oral Surgery 16, 22–30 (1958)
4. Wiggins, K.L., Malkin, S.: Drilling of bone. Journal of Biomechanics 9, 553–559 (1976)

5. Brett, P.N., Baker, D.A., Taylor, R., Griffiths, M.V.: Controlling the penetration of flexible bone tissue using the stapedotomy micro drill. Proceedings of the Institution of Mechanical Engineers, Part I: Journal of Systems and Control Engineering 218, 343–351 (2004)
6. Kendoff, D., Citak, M., Gardner, M.J., Stubig, T., Krettek, C., Hufner, T.: Improved accuracy of navigated drilling using a drill alignment device. Journal of Orthopaedic Research 25, 951–957 (2007)
7. Ong, F.R., Bouazza-Marouf, K.: The detection of drill-bit break-through for the enhancement of safety in mechatronic assisted orthopaedic drilling. Mechatronics 9, 565–588 (1999)
8. Price, M., Molloy, S., Solan, M., Sutton, A., Ricketts, D.M.: The rate of instrument breakage during orthopaedic procedures. International Orthopedics 26, 185–187 (2002)
9. Bassi, J.L., Pankaj, M., Navdeep, S.: A technique for removal of broken cannulated drill-bit: Bassi's method. Journal of Orthopaedic Trauma 22, 56–58 (2008)
10. Augustin, G., Davila, S., Mihoci, K., Udiljak, T., Vedrina, D.S., Antabak, A.: Thermal osteonecrosis and bone drilling parameters revisited. Archives of Orthopaedic and Trauma Surgery 128, 71–77 (2008)
11. Abouzgia, M.B., James, D.F.: Temperature rise during drilling through bone. International Journal of Oral and Maxillofacial Implants 12, 342–3531 (1997)
12. Hobkirk, J.A., Rusiniak, K.: Investigation of variable factors in drilling bone. Journal of Oral and Maxillofacial Surgery 35, 968–973 (1977)
13. Jacobs, C.H., Berry, J.T., Pope, M.H., Hoaglund, F.T.: A study of the bone machining process-drilling. Journal of Biomechanics 9, 343–349 (1976)
14. Albrektsson, T.: Measurements of shaft speed while drilling through bone. Journal of Oral and Maxillofacial Surgery 53, 1315–1316 (1995)
15. Myers, R.H., Montgomery, D.C.: Response surface methodology, 2nd edn. Wiley, New York (2002) ISBN 0-471-41255-4
16. Box, G.E.P., Hunter, J.S., Hunter, W.G.: Statistics for experimenters, 2nd edn. Wiley, New York (2005) ISBN 13978-0471-71813-0
17. Somashekhar, K.P., Mathew, J., Ramachandran, N.: A feasibility approach by simulated annealing on optimization of micro-wire electric discharge machining parameters. Int. J. Adv. Manuf. Technol. 61, 1209–1213 (2012)
18. Metropolis, N., Rosenbluth, A., Rosenbluth, N., Teller, A., Teller, E.: Equation of state calculation by fast computing machines. J. Chem. Phys. 21, 1087–1092 (1953)
19. Kirkpatrick, S., Gelatt, C.D., Vecchi, M.P.: Optimization by simulated annealing. Science 220, 671–680 (1983)
20. Glover, F., Gary, A.K.: Hand book of metaheuristics. Kluwer, London (2003)
21. Van, P.J., Laarhoven, E.H.A.: Simulated annealing: theory and applications. Kluwer, London (1987)
22. Karaca, F., Aksakal, B., Kom, M.: Influence of orthopaedic drilling parameters on temperature and histopathology of bovine tibia: An in vitro study. Medical Engineering & Physics 33(10), 1221–1227 (2011)
23. Lee, J., Ozdoganlar, O.B., Rabin, Y.: An experimental investigation on thermal exposure during bone drilling. Medical Engineering & Physics 34(10), 1510–1520 (2012)
24. Design Expert, `http://www.statease.com/dx8descr.html`

Author Index

Printed in the United States
By Bookmasters